Fields Institute Communications

Volume 83

Fields Institute Editorial Board:

Ian Hambleton, *Director of the Institute*

Huaxiong Huang, *Deputy Director of the Institute*

James G. Arthur, *University of Toronto*

Kenneth R. Davidson, *University of Waterloo*

Lisa C. Jeffrey, *University of Toronto*

Barbara Lee Keyfitz, *The Ohio State University*

Thomas S. Salisbury, *York University*

Juris Steprans, *York University*

Noriko Yui, *Queen's University*

The Communications series features conference proceedings, surveys, and lecture notes generated from the activities at the Fields Institute for Research in the Mathematical Sciences. The publications evolve from each year's main program and conferences. Many volumes are interdisciplinary in nature, covering applications of mathematics in science, engineering, medicine, industry, and finance.

More information about this series at http://www.springer.com/series/10503

Peter D. Miller • Peter A. Perry • Jean-Claude Saut
Catherine Sulem
Editors

Nonlinear Dispersive Partial Differential Equations and Inverse Scattering

 Springer

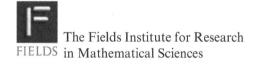 The Fields Institute for Research in Mathematical Sciences

Editors
Peter D. Miller
Department of Mathematics
University of Michigan–Ann Arbor
Ann Arbor, MI, USA

Peter A. Perry
Department of Mathematics
University of Kentucky
Lexington, KY, USA

Jean-Claude Saut
Département de Mathématiques
University of Paris-Sud
Orsay, France

Catherine Sulem
Department of Mathematics
University of Toronto
Toronto, ON, Canada

ISSN 1069-5265 ISSN 2194-1564 (electronic)
Fields Institute Communications
ISBN 978-1-4939-9805-0 ISBN 978-1-4939-9806-7 (eBook)
https://doi.org/10.1007/978-1-4939-9806-7

Mathematics Subject Classification: 35P25, 35Q53, 35Q55, 35Q15, 35Q35, 37K15, 37K40

© Springer Science+Business Media, LLC, part of Springer Nature 2019
This work is subject to copyright. All rights are reserved by the Publisher, whether the whole or part of the material is concerned, specifically the rights of translation, reprinting, reuse of illustrations, recitation, broadcasting, reproduction on microfilms or in any other physical way, and transmission or information storage and retrieval, electronic adaptation, computer software, or by similar or dissimilar methodology now known or hereafter developed.
The use of general descriptive names, registered names, trademarks, service marks, etc. in this publication does not imply, even in the absence of a specific statement, that such names are exempt from the relevant protective laws and regulations and therefore free for general use.
The publisher, the authors, and the editors are safe to assume that the advice and information in this book are believed to be true and accurate at the date of publication. Neither the publisher nor the authors or the editors give a warranty, expressed or implied, with respect to the material contained herein or for any errors or omissions that may have been made. The publisher remains neutral with regard to jurisdictional claims in published maps and institutional affiliations.

Cover illustration: Drawing of J.C. Fields by Keith Yeomans

This Springer imprint is published by the registered company Springer Science+Business Media, LLC, part of Springer Nature.
The registered company address is: 233 Spring Street, New York, NY 10013, U.S.A.

Preface

This volume contains lectures and invited papers from the Focus Program on "Nonlinear Dispersive Partial Differential Equations and Inverse Scattering" held at the Fields Institute from July 31 to August 18, 2017.

Our conference coincided with the fiftieth anniversary of the discovery by Gardner, Greene, Kruskal, and Miura[1] that the Korteweg-de Vries (KdV) equation could be integrated by exploiting a remarkable connection between KdV and the spectral theory of Schrödinger's equation in one space dimension. This led to the discovery of a number of completely integrable models of dispersive wave propagation including the one-dimensional cubic nonlinear Schrödinger (NLS) equation, the derivative NLS equation and, in two dimensions, the Davey-Stewartson, Kadomtsev-Petviashvili, and the Novikov-Veselov equations. These models have been extensively studied and, in some cases, the inverse scattering theory has been put on rigorous footing and used as a powerful analytical tool to study global well-posedness and elucidate long-time asymptotic behavior of the solutions, including dispersion, soliton solutions, and semiclassical limits.

Pioneering works in this literature are the papers of Deift and Zhou which establish the "nonlinear steepest descent" method for solving the Riemann-Hilbert problems at the heart of inverse scattering. More recently, rigorous treatments of inverse scattering have led to advances in the understanding of dispersive partial differential equations (PDEs), in particular addressing questions concerning asymptotic stability of solitons and the soliton resolution conjecture. Motivated by completely integrable models as well as by considerations stemming from the physical origin of the equations, the existence and stability properties of special solutions such as traveling waves and solitary waves have been thoroughly investigated by pure variational and PDE techniques. A common theme in many PDE results is that equations which are very far from the integrable cases nonetheless exhibit similar qualitative and asymptotic behavior.

---

[1] Clifford S. Gardner, John M. Greene, Martin D. Kruskal, and Robert M. Miura, *Phys. Rev. Lett.* **19**, 1095 (1967).

This conference brought together researchers in completely integrable systems and PDE with the goal of advancing the understanding of qualitative and long-time behavior in dispersive nonlinear equations.

Percy Deift's Coxeter Lectures, "Fifty Years of KdV: An integrable system," provided an introduction and overview of the completely integrable method and its applications in dynamical systems, probability, statistical mechanics, and many other areas of applied mathematics. The first week of the focus program consisted of expository lectures by Walter Craig, Patrick Gérard, Peter D. Miller, Peter A. Perry, and Jean-Claude Saut. Walter Craig presented a series of lectures on Hamiltonian PDEs. The notion of phase space, flows for PDEs and conserved integrals of motion were introduced as well as normal forms transformations and the concept of Nekhoroshev stability. Patrick Gérard's lectures on "Wave Turbulence and Complete Integrability" discussed a completely integrable model, the cubic Szegő equation, for which the growth of Sobolev norms, thought to characterize turbulent behavior, can be studied explicitly. Jean-Claude Saut provided a comprehensive survey of results for the Benjamin-Ono and intermediate long-wave equations, both completely integrable models for one-dimensional wave propagation, by inverse scattering and PDE methods. Peter A. Perry's lectures gave a rigorous treatment of the inverse scattering method for the cubic defocusing nonlinear Schrödinger equation in one dimension (based on the foundational work of Deift and Zhou) and for the defocusing Davey-Stewartson equation in two space dimensions (based on Perry's work and more recent work of Nachman, Regev, and Tataru). Peter D. Miller described some theory of Riemann-Hilbert problems, culminating in a description of the Deift-Zhou steepest descent method. The paper of Dieng, McLaughlin, and Miller in this volume provides a detailed exposition of a useful generalization, namely the $\bar{\partial}$-steepest descent method for Riemann-Hilbert problems, pioneered by Dieng and McLaughlin. This method has become an effective tool for attacking soliton resolution for completely integrable, dispersive PDEs.

The mini-school was followed by two workshops on various aspects of dispersive PDEs. The research papers collected here include new results on the focusing NLS equation, the massive Thirring model, and the BBM equation as dispersive PDE in one space dimension, as well as the KP-II equation, the Zakharov-Kuznetsov equation, and the Gross-Pitaevskii equation as dispersive PDE in two space dimensions.

The editors of this volume would like to thank the Fields Institute for Research in the Mathematical Sciences and its Director, Dr. Ian Hambleton, for their generous support. We are grateful to Esther Berzunza and Dr. Huaxiong Huang for their assistance with the organization of the conference as well as to Tyler Wilson and the Springer team for their assistance with the publication of this special volume. We are also grateful to the participants of the conference and to the authors for their contributions to this volume as well as to the referees for their invaluable help during the review process.

Finally, we dedicate this volume to our friend and colleague Walter Craig, who sadly passed away on January 18, 2019. Walter was a world renowned scholar for his work on nonlinear partial differential equations, infinite dimensional Hamiltonian systems, and their applications, in particular, to fluid dynamics. A constant source of inspiration, Walter was a generous mentor and a wonderful collaborator. He will be greatly missed by all who had the privilege of knowing him as a mathematician, colleague, and dear friend.

Ann Arbor, MI, USA Peter D. Miller
Lexington, KY, USA Peter A. Perry
Orsay, France Jean-Claude Saut
Toronto, ON, Canada Catherine Sulem

Contents

Part I Lectures

Three Lectures on "Fifty Years of KdV: An Integrable System" 3
Percy A. Deift

Wave Turbulence and Complete Integrability 39
Patrick Gérard

Benjamin-Ono and Intermediate Long Wave Equations: Modeling,
IST and PDE 95
Jean-Claude Saut

Inverse Scattering and Global Well-Posedness in One and Two
Space Dimensions 161
Peter A. Perry

Dispersive Asymptotics for Linear and Integrable Equations
by the $\bar{\partial}$ Steepest Descent Method 253
Momar Dieng, Kenneth D. T.-R. McLaughlin, and Peter D. Miller

Part II Research Papers

Instability of Solitons in the 2d Cubic Zakharov-Kuznetsov Equation 295
Luiz Gustavo Farah, Justin Holmer, and Svetlana Roudenko

On the Nonexistence of Local, Gauge-Invariant Birkhoff
Coordinates for the Focusing NLS Equation 373
Thomas Kappeler and Peter Topalov

Extended Decay Properties for Generalized BBM Equation 397
Chulkwang Kwak and Claudio Muñoz

Ground State Solutions of the Complex Gross Pitaevskii Equation 413
Slim Ibrahim

The Phase Shift of Line Solitons for the KP-II Equation 433
Tetsu Mizumachi

Inverse Scattering for the Massive Thirring Model 497
Dmitry E. Pelinovsky and Aaron Saalmann

Part I
Lectures

Three Lectures on "Fifty Years of KdV: An Integrable System"

Percy A. Deift

The goal in the first two Coxeter lectures was to give an answer to the question

"What is an integrable system?"

and to describe some of the tools that are available to identify and integrate such systems. The goal of the third lecture was to describe the role of integrable systems in certain numerical computations, particularly the computation of the eigenvalues of a random matrix. This paper closely follows these three Coxeter lectures, and is written in an informal style with an abbreviated list of references. Detailed and more extensive references are readily available on the web. The list of authors mentioned is not meant in any way to be a detailed historical account of the development of the field and I ask the reader for his'r indulgence on this score.

The notion of an integrable system originates in the attempts in the seventeenth and eighteenth centuries to integrate certain specific dynamical systems in some explicit way. Implicit in the notion is that the integration reveals the long-time behavior of the system at hand. The seminal event in these developments was Newton's solution of the two-body problem, which verified Kepler's laws, and by the end of the nineteenth century many dynamical systems of great interest had been integrated, including classical spinning tops, geodesic flow on an ellipsoid, the Neumann problem for constrained harmonic oscillators, and perhaps most spectacularly, Kowalewski's spinning top. In the nineteenth century, the general and very useful notion of *Liouville integrability* for Hamiltonian systems, was introduced: If a Hamiltonian system with Hamiltonian H and n degrees of freedom

---

P. A. Deift (✉)
Courant Institute of Mathematical Sciences, New York University, New York, NY, USA
e-mail: deift@cims.nyu.edu

has n independent, Poisson commuting integrals, $I_1, \ldots, I_n$, then the flow $t \mapsto z(t)$ generated by H can be integrated explicitly by quadrature, or symbolically,

$$\begin{cases} I_k(z(t)) = \text{const}, \ 1 \leq k \leq n, \ \text{rank} \ (dI_1, \ldots, dI_n) = n, \ \{I_k, I_j\} = 0, \\ 1 \leq j, k \leq n \quad \Rightarrow \quad \text{explicit integration}. \end{cases} \quad (1)$$

Around the same time the Hamilton-Jacobi equation was introduced, which proved to be equally useful in integrating systems.

The modern theory of integrable systems began in 1967 with the discovery by Gardner et al. [19] of a method to solve the Korteweg de Vries (KdV) equation

$$q_t + 6qq_x - q_{xxx} = 0 \quad (2)$$

$$q(x,t)_{t=0} = q_0(x) \to 0 \quad \text{as} \quad |x| \to \infty.$$

The method was very remarkable and highly original and expressed the solution of KdV in terms of the spectral and scattering theory of the Schrödinger operator $L(t) = -\partial_x^2 + q(x,t)$, acting in $L^2(-\infty < x < \infty)$ for each t. In 1968 Peter Lax [26] reformulated [19] in the following way. For $L(t) = -\partial_x^2 + q(x,t)$ and $B(t) \equiv 4\partial_x^3 - 6q\,\partial_x - 3q_x$.

$$\text{KdV} \equiv \partial_t L = [B, L] = BL - LB \quad (3)$$

$$\equiv \text{isospectral deformation of } L(t)$$

$$\Rightarrow \text{spec}\,(L(t)) = \text{spec}\,(L(0)) \Rightarrow \text{integrals of the motion for KdV}.$$

L, B are called *Lax pairs*: By the 1970s, Lax pairs for the Nonlinear Schrödinger Equation (NLS), the Sine-Gordon equation, the Toda lattice, ..., had been found, and these systems had been integrated as in the case of KdV in terms of the spectral and scattering theory of their associated "L" operators.

Over the years there have been many ideas and much discussion of what it means for a system to be integrable, i.e. explicitly solvable. Is a Hamiltonian system with n degrees of freedom integrable if and only if the system is Liouville integrable, i.e. the system has n independent, commuting integrals? Certainly as explained above, Liouville integrability implies explicit solvability. But is the converse true? If we can solve the system in some explicit fashion, is it necessarily Liouville integrable? We will say more about this matter further on. Is a system integrable if and only if it has a Lax pair representation as in (3)? There is, however, a problem with the Lax-pair approach from the get-go. For example, if we are investigating a flow on $n \times n$ matrices, then a Lax-pair would guarantee at most n integrals, viz., the eigenvalues, whereas an $n \times n$ system has $O(n^2)$ degrees of freedom—too little, a priori, for Liouville integrability. The situation is in fact even more complicated. Indeed, suppose we are investigating a flow on real skew-symmetric $n \times n$ matrices A—i.e. a flow for a generalized top. Such matrices constitute the dual Lie algebra of the orthogonal group O_n, and so carry a natural Lie-Poisson structure. The symplectic leaves of this structure are the co-adjoint orbits of O_n

$$\mathcal{A} = \mathcal{A}_A = \left\{ O\ A\ O^T : O \in O_n \right\} \tag{4}$$

Thus **any** Hamiltonian flow $t \to A(t)$ on $\mathcal{A}$, $A(t=0) = A$, must have the form

$$A(t) = O(t)\ A\ O(t)^T \tag{5}$$

for some $O(t) \in \mathcal{A}$ and hence has Lax-pair form

$$\frac{dA}{dt} = \dot{O}\ A\ O^T + O\ A\ \dot{O}^T = [B, A] \tag{6}$$

where

$$B = \dot{O}\ O^T = -B^T \tag{7}$$

The Lax-pair form guarantees that the eigenvalues $\{\lambda_i\}$ of A are constants of the motion. But we see from (4) that the co-adjoint orbit through A is simply specified by the eigenvalues of A. In other words the eigenvalues of A are just parameters for the symplectic leaves under considerations: They are of no help in integrating the system: Indeed $d\lambda_i|_{\mathcal{A}_A} = 0$ for all i. So for a Lax-pair formulation to be useful, we need

$$\text{Lax pair} + \text{``something''} \tag{8}$$

So, what is the "something"? A Lax-pair is a proclamation, a marker, as it were, on a treasure map that says "Look here!" The real challenge in each case is to turn the Lax-pair, if possible, into an effective tool to solve the equation. In other words, the real task is to find the "something" to dig up the treasure! Perhaps the best description of Lax-pairs is a restatement of Yogi Berra's famous dictum "If you come to a fork in the road, take it". So if you come upon a Lax-pair, take it!

Over the years, with ideas going back and forth, Liouville integrability, Lax-pairs, "algebraic integrability", "monodromy", the discussion of what is an integrable system has been at times, shall we say, quite lively. There is, for example, the story of Henry McKean and Herman Flashka discussing integrability, when one of them, and I'm not sure which one, said to the other: "So you want to know what is an integrable system? I'll tell you! You didn't think I could solve it. But I can!"

In this "wild west" spirit, many developments were taking place in integrable systems. What was not at all clear at the time, however, was that these developments would provide tools to analyze mathematical and physical problems in areas **far removed from their original dynamical origin**. These tools constitute what may now be viewed as an **integrable method (IM)**.

There is a picture that I like that illustrates, very schematically, the intersection of IM with different areas of mathematics. Imagine some high dimensional space, the "space of problems". The space contains a large number of "parallel" planes, stacked one on top of the other and separated. The planes are labeled as follows: dynamical systems, probability theory and statistical mechanics, geometry, combinatorics, statistical mechanics, classical analysis, numerical analysis, representation theory, algebraic geometry, transportation theory, ... In addition, there is another plane in the space labeled "the integrable method (IM)": Any problem lying on IM can be solved/integrated by tools taken from the integrable method. Now the fact of the matter is that the IM-plane intersects all of the parallel planes described above: Problems lying on the intersection of any one of these planes with the IM-plane are thus solvable by the integrable method (Fig. 1).

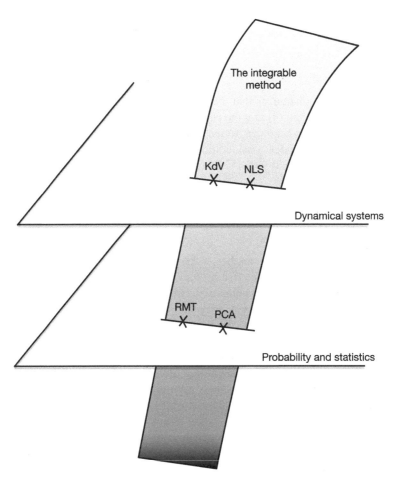

Fig. 1 Intersections of the integrable method

For each parallel plane we have, for example, the following intersection points:

- dynamical systems: Korteweg-de Vries (KdV), Nonlinear Schrödinger (NLS), Toda, Sine-Gordon, ...
- probability theory and statistics: Random matrix theory (RMT), Integrable probability theory, Principal component analysis (PCA), ...
- geometry: spaces of constant negative curvature R, general relativity in $1+1$ dimensions, ...
- combinatorics: Ulam's increasing subsequence problem, tiling problems, (Aztec diamond, hexagon tiling, ...), random particle systems (TASEP, ...), ...
- statistical mechanics: Ising model, XXZ spin chain, six vertex model, ...
- classical analysis: Riemann-Hilbert problems, orthogonal polynomials, (modern) special function theory (Painlevé equations), ...
- numerical analysis: QR, Toda eigenvalue algorithm, Singular value decomposition, ...
- representation theory: representation theory of large groups (S_∞, U_∞ ...), symmetric function theory, ...
- algebraic geometry: Schottky problem, infinite genus Riemann surfaces, ...
- transportation theory: Bus problem in Cuernavaca, Mexico, airline boarding, ...

The list of such intersections is long and constantly growing.

The singular significance of KdV is just that the **first intersection** that was observed and understood as such, was the junction of IM with dynamical systems, and that was at the point of KdV.

How do we come to such a picture? First we will give a precise definition of what we mean by an integrable system. Consider a simple harmonic oscillator:

$$\dot{x} = y \quad , \quad \dot{y} = -\omega^2 x \tag{9}$$

$$x(t)|_{t=0} = x_0 \quad , \quad y(t)|_{t=0} = y_0$$

The solution of (9) has the following form:

$$\begin{cases} x(t; x_0, y_0) = \frac{1}{\omega}\sqrt{w^2 x_0^2 + y_0^2} \sin\left(wt + \sin^{-1}\left(\frac{\omega x_0}{\sqrt{\omega^2 x_0^2 + y_0^2}}\right)\right) \\ y(t; x_0, y_0) = \sqrt{\omega^2 x_0^2 + y_0^2} \cos\left(wt + \sin^{-1}\left(\frac{\omega x_0}{\sqrt{\omega^2 x_0^2 + y_0^2}}\right)\right) \end{cases} \tag{10}$$

Note the following features of (10): Let $\varphi : \mathbb{R}^2 \to \mathbb{R}_+ \times (\mathbb{R}/2\pi\mathbb{Z})$

$$(\alpha, \beta) \longmapsto A = \frac{1}{\omega}\sqrt{\omega^2 \alpha^2 + \beta^2}, \ \theta = \sin^{-1}\left(\frac{\omega\alpha}{\sqrt{\omega^2\alpha^2 + \beta^2}}\right).$$

Then

$$\varphi^{-1} : \mathbb{R}_+ \times (\mathbb{R}/2\pi\mathbb{Z}) \to \mathbb{R}^2$$

has the form

$$\varphi^{-1}(A, \theta) = (A \sin \theta, \, \omega A \cos \theta)$$

Thus (10) implies

$$\begin{cases} \eta(t; \eta_0) = \varphi^{-1}\left(\varphi(\eta_0) + \omega \, t\right) \\ \text{where} \quad \eta(t) = (x(t), y(t)), \quad \eta_0 = (x_0, y_0), \quad \omega = (0, \omega) \end{cases} \tag{11}$$

In other words:

There exists a bijective change of variables $\eta \longmapsto \varphi(\eta)$ such that (12a)

$\eta(t, \eta_0)$ evolves according to (9) $\Rightarrow$ (12b)

$$\varphi(\eta(t); \eta_0) = \varphi(\eta_0) + t \, \omega$$

i.e., in the variables $(A, \theta) = \varphi(\alpha, \beta)$, solutions of (9) move linearly.

$$\begin{cases} \eta(t, \eta_0) \text{ is recovered from formula (11) via a map} \\ \varphi^{-1}(A, \theta) = (A \sin \theta, \, \omega A \cos \theta) \\ \\ \text{in which the behavior of } \sin \theta, \cos \theta \text{ is very well understood.} \\ \text{The same is true for } \varphi. \text{ What we learn, in particular, based on this} \\ \text{knowledge of } \varphi \text{ and } \varphi^{-1}, \quad \text{is that} \\ \\ \eta(t; \eta_0) \quad \text{evolves periodically in time with period } 2\pi/\omega \end{cases} \tag{12c}$$

We are led to the following:
We say that a dynamical system $t \mapsto \eta(t)$ is **integrable** if

$$\begin{cases} \text{There exists a bijective map } \varphi : \eta \mapsto \varphi(\eta) \equiv \zeta \\ \text{such that } \varphi \text{ linearizes the system} \\ \qquad \varphi\left(\eta(t)\right) = \varphi\left(\eta(t=0)\right) + \omega \, t \\ \text{and so} \\ \qquad \eta\left(t; \eta(t=0)\right) = \varphi^{-1}\left(\varphi\left(\eta(t=0)\right) + \omega \, t\right) \end{cases} \tag{13a}$$

AND

$$\begin{cases} \text{The behavior of } \varphi, \, \varphi^{-1} \text{ are well enough understood so that} \\ \text{the behavior of } \eta\left(t; \eta(t=0)\right) \text{ as } t \to \infty \text{ is clearly revealed.} \end{cases} \tag{13b}$$

More generally, we say a system η which depends on some parameters $\eta = \eta(a, b, \ldots)$ is **integrable** if

$$\begin{cases} \text{There exists a bijective change of variables } \eta \to \zeta = \varphi(\eta) \text{ such} \\ \text{that the dependence of } \zeta \text{ on } a, b, \ldots. \\ \qquad \zeta(a, b, \ldots) = \varphi(\eta(a, b, \ldots)) \\ \text{is simple/well-understood} \end{cases} \quad (14a)$$

and

$$\begin{cases} \text{The behavior of the function theory} \\ \qquad \eta \mapsto \zeta \equiv \varphi(\eta) \quad , \quad \zeta \mapsto \eta = \varphi^{-1}(\zeta) \\ \text{is well-enough understood so that the behavior of} \\ \qquad \eta(a, b \ldots) = \varphi^{-1}(\zeta(a, b \ldots,)) \end{cases} \quad (14b)$$

is revealed in an explicit form as $a, b, \ldots$ vary, becoming, in particular, large or small.

Notice that in this definition of an integrable system, various sufficient conditions for integrability such as commuting integrals, Lax-pairs, ..., are conspicuously absent. A system is integrable, if you can solve it, but subject to certain strict guidelines. This is a return to McKean and Flaschka, an institutionalization, as it were, of the "Wild West".

According to this definition, progress in the theory of integrable systems is made

EITHER

by discovering how to linearize a new system

$$\eta \to \zeta = \varphi(\eta)$$

using a **known** function theory φ. For example: Newton's problem of two gravitating bodies, is solved in terms of trigonometric functions/ellipses/parabolas— mathematical objects already well-known to the Greeks. In the nineteenth century, Jacobi solved geodesic flow on an ellipsoid using newly minted hyperelliptic function theory, and so on, ...

OR

by discovering/inventing a new function theory which linearizes the given problem at hand. For example: To facilitate numerical calculations in spherical geometry, Napier, in the early 1700s, realized that what he needed to do was to linearize multiplication

$$\eta \tilde{\eta} \longrightarrow \varphi(\eta \tilde{\eta}) = \varphi(\eta) + \varphi(\tilde{\eta})$$

which introduced a new function theory—the logarithm. Historically, **no** integrable system has had greater impact on mathematics and science, than multiplication!

There is a similar story for all the classical special functions, Bessel, Airy, ..., each of which was introduced to address a particular problem.

The following aspect of the above evolving integrability process is crucial and gets to the heart of the Integrable Method (IM): Once a new function theory has been discovered and developed, it enters the **toolkit** of IM, finding application in problems far removed from the original discovery.

Certain philosophical points are in order here.

(1) There is **no difference** in spirit, philosophically, between our definition of an integrable system and what we do in ordinary life. We try to address problems by rephrasing them (read "change of variables") so we can recognize them as something we know. After all, what else is a "precedent" in a law case? We introduce new words—a new "function theory"—to capture new developments and so extend and deepen our understanding. Recall that Adam's first cognitive act in Genesis was to give the animals names. The only difference between this progression in ordinary life versus mathematics, is one of degree and precision.

(2) This definition presents "integrability" **not as a predetermined** property of a system frozen in time. Rather, in this view the status of a system as integrable depends on the technology/function theory available **at the time**. If an appropriate new function theory is developed, the status of the system may change to integrable.

How does one determine if a system is integrable and how do you integrate it? Let me say at the outset, and categorically, that I believe there is no systematic answer to this question. Showing a system is integrable is **always** a matter of luck and intuition.

We do, however, have a **toolkit** which one can bring to a problem at hand.

At this point in time, the toolkit contains, amongst others, the following components:

(a) a broad and powerful **set of functions/transforms/constructions**

$$\eta \to \zeta = \varphi(\eta)$$

that can be used to convert a broad class of problems of interest in mathematics/physics, into "known" problems: In the simplest cases $\eta \to \varphi(\eta)$ linearizes the problem.

(b) **powerful techniques** to analyze φ, φ^{-1} such that the asymptotic behavior of the original η-system can be inferred explicitly from the known asymptotic behavior of the ζ-system, as relevant parameters, e.g. time, become large.

(c) a **particular, versatile** class of functions, the Painlevé functions, which play the same role in modern (nonlinear) theoretical physics that classical special functions played in (linear) nineteenth century physics. Painlevé functions form the **core of modern special function**, and their behavior is known with the same precision as in the case of the classical special functions. We note that the Painlevé equations **are themselves integrable** in the sense of Definition (14a).

(d) a class of **"integrable" stochastic models**—random matrix theory (RMT). Instead of modeling a stochastic system by the roll of a die, say, we now have the possibility to model a whole new class of systems by the **eigenvalues of a random matrix**. Thus RMT plays the role of a **stochastic special function theory**. RMT is "integrable" in the sense that key statistics such as the gap probability, or edge statistics, for example, are given by functions, e.g. Painlevé functions, that describe (deterministic) integrable problems as above. We will say more about this later.

We will now show how all this works in concrete situations. Note, however, by no means all known integrable systems can be solved using tools from the IM-toolkit. For example the beautiful system that Patrick Gérard et al. have been investigating recently (see e.g. [21]), seems to be something completely different. We will consider various examples. The first example is taken from dynamics, viz., the NLS equation.

To show that NLS is integrable, we **first** extract a particular tool from the toolkit—the Riemann-Hilbert Problem (RHP): Let $\Sigma \subset \mathbb{C}$ be an oriented contour and let $v : \Sigma \to G\ell(n, \mathbb{C})$ be a map (the "jump matrix") from Σ to the invertible $n \times n$ matrices, $v, v^{-1} \in L^\infty(\Sigma)$. By convention, at a point $z \in \Sigma$, the $(+)$ side (respectively $(-)$ side) lies to the left (respectively right) as one traverse Σ in the direction of the orientation, as indicated in Fig. 2. Then the (normalized) RHP (Σ, v) consists in finding an $n \times n$ matrix-valued function $m = m(z)$ such that

- $m(z)$ is analytic in $\mathbb{C}/\Sigma$
- $m_+(z) = m_-(z)\, v(z)$, $z \in \Sigma$

  where $m_\pm(z) = \lim\limits_{z' \to z_\pm} m(z')$
- $m(z) \to I_n$ as $z \to \infty$

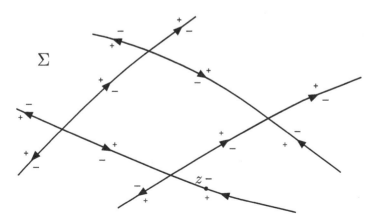

Fig. 2 Oriented contour Σ

Here "$z' \to z_\pm$" denotes the limit as $z' \in \mathbb{C}/\Sigma$ approaches $z \in \Sigma$ from the ($\pm$)-side, respectively. The particular contour Σ and the jump matrix v are tailored to the problem at hand.

There are many technicalities involved here: Does such an $m(z)$ exist? In what sense do the limits $m_\pm$ exist? And so on .... Here we leave such issues aside. RHP's play an analogous role in modern physics that integral representations play for classical special functions, such as the Airy function $Ai(z)$, Bessel function $J_n(z)$, etc. For example, $Ai(z) = \frac{1}{2\pi i} \int_{\mathscr{C}} \exp\left(\frac{t^3}{3} - zt\right) dt$ for some appropriate contour $\mathscr{C} \subset \mathbb{C}$, which makes it possible to analyze the behavior of $Ai(z)$ as $z \to \infty$, using the classical steepest descent method.

Now consider the defocusing NLS equation on $\mathbb{R}$

$$\begin{cases} i u_t + u_{xx} - 2|u|^2 u = 0 \\ u(x,t)\big|_{t=0} = u_0(x) \to 0 \quad \text{as} \quad |x| \to \infty. \end{cases} \quad (15)$$

In 1972, Zakharov and Shabat [35] showed that NLS has a Lax-pair formulation, as follows: Let

$$L(t) = (i\sigma)^{-1} (\partial_x - Q(t))$$

where

$$\sigma = \frac{1}{2}\begin{pmatrix} 1 & 0 \\ 0 & -1 \end{pmatrix}, \quad Q(t) = \begin{pmatrix} 0 & u(x,t) \\ -u(x,t) & 0 \end{pmatrix}.$$

For each t, $L(t)$ is a self-adjoint operator acting on vector valued function in $\left(L^2(\mathbb{R})\right)^2$. Then for some explicit $B(t)$, constructed from $u(x,t)$ and $u_x(x,t)$,

$$u(x,t) \text{ solves NLS} \iff \frac{dL(t)}{dt} = [B(t), L(t)]. \quad (16)$$

This the **second** tool we extract from our toolkit. So the Lax operator $L(t)$ marks a point, as it were, on our treasure map. How can one use $L(t)$ to solve the system?

One proceeds as follows: This crucial step was first taken by Shabat [30] in the mid-1970s in the case of KdV and developed into a general scheme for ordinary differential operators by Beals and Coifman [4] in the early 1980s.

The **map** φ in (13a) above for NLS is the scattering map constructed as follows: Suppose $u = u(x)$ is given, $u(x) \to 0$ sufficiently rapidly as $|x| \to \infty$. For *fixed* $z \in \mathbb{C}/\mathbb{R}$, there exists a *unique* 2×2 solution of the scattering problem

$$L\psi = z\psi \quad (17)$$

where
$$m = m(x, z; u) \equiv \psi(x, z) e^{-iz, x\sigma} \tag{18}$$

is bounded on $\mathbb{R}$ and
$$m(x, z; u) \to I \quad \text{as } x \to -\infty.$$

For **fixed** $x \in \mathbb{R}$, such so-called *Beals-Coifman solutions* also have the following properties:

$$\begin{aligned} m(x, z; u) \text{ is analytic in } z \text{ for } z \in \mathbb{C}/\mathbb{R} \\ \text{and continuous in } \overline{\mathbb{C}}_+ \text{ and in } \overline{\mathbb{C}}_- \end{aligned} \tag{19}$$

$$m(x, z; u) \to I \quad \text{as} \quad z \to \infty \quad \text{in} \quad \mathbb{C}/\mathbb{R}. \tag{20}$$

Now both $\psi_\pm(x, z; u) = \lim_{z' \to z_\pm} \psi(x, z; u)$, $z \in \mathbb{R}$ clearly solve $L\psi_\pm = z\psi_\pm$ which implies that there exists $v = v(z) = v(z; u)$ independent of x, such that for all $x \in \mathbb{R}$

$$\psi_+(x, z) = \psi_-(x, z) v(z) \quad , \quad z \in \mathbb{R} \tag{21}$$

or in terms of
$$m_\pm = \psi_\pm(x, z) e^{-ixz\sigma} \tag{22}$$

we have
$$m_+(x, z) = m_-(x, z) v_x(z) \quad , \quad z \in \mathbb{R} \tag{23}$$

where
$$v_x(z) = e^{ixz\sigma} v(z) e^{-ixz\sigma} \tag{24}$$

Said differently, for each $x \in \mathbb{R}$, $m(x, z)$ solves the normalized RHP (Σ, v_x) where $\Sigma = \mathbb{R}$, oriented from $-\infty$ to $+\infty$, and v is as above. In this way, a RHP enters naturally into the picture introduced by the Lax operator L.

It turns out that v has a special form

$$\begin{cases} v(z) = \begin{pmatrix} 1 - |r(z)|^2 & r(z) \\ -\overline{r(z)} & 1 \end{pmatrix} \\ v_x(z) = \begin{pmatrix} 1 - |r|^2 & r\, e^{ixz} \\ -\overline{r}\, e^{-ixz} & 1 \end{pmatrix} \end{cases} \tag{25}$$

where $r(z)$, the **reflection coefficient**, satisfies $\|r\|_\infty < 1$. We define the map φ for NLS as follows:

$$u \mapsto \varphi(u) \equiv r \qquad (26)$$

Suppose r is given and x fixed. To construct $\varphi^{-1}(r)$ we must solve the RHP $(\mathbb{R}, v_x)$ with v_x as in (25). If $m = m(x, z)$ is the solution of the RHP, then expanding at $z = \infty$, we have

$$m(x, z) = I + \frac{(m_1(x))}{z} + O\left(\frac{1}{z^2}\right), \qquad z \to \infty.$$

A simple calculation then shows that

$$u(x) = \varphi^{-1}(r) = -i(m_1(x))_{12}. \qquad (27)$$

Thus

$$\varphi \leftrightarrow \text{ scattering map } ; \quad \varphi^{-1} \leftrightarrow \text{RHP}.$$

Now the key fact of the matter is that

$$\varphi \text{ linearizes NLS}. \qquad (28)$$

Indeed if $u(t) = u(x, t)$ solves NLS with $u(x, t)\big|_{t=0} = u_0(x)$, then

$$r(t) = \varphi(u(t)) = r(z; u_0)\, e^{-itz^2} = \varphi(u_0)(z)\, e^{-itz^2} \qquad (29)$$

$$\text{or} \qquad \log r(t) = \log r(z, u_0) - i t z^2$$

which is linear motion!

This leads to the celebrated solution procedure

$$u(t) = \varphi^{-1}\left(\varphi(u_0)(\cdot)\, e^{-i t (\cdot)^2}\right). \qquad (30)$$

Thus condition (13a) for the integrability of NLS is established.

But condition (13b) is also satisfied. Indeed the analysis of the scattering map $u \to r = \varphi(u)$ is classical and well-understood. The inverse scattering map is **also** well-understood because of the nonlinear steepest descent method for RHP's introduced by Deift and Zhou in 1993 [13].[1] This is the **third tool** we extract from our toolkit. One finds, for example, that as $t \to \infty$

---

[1] This paper also contains some history of earlier approaches to analyze the behavior of solutions of integrable systems asymptotically.

$$u(x,t) = \frac{1}{\sqrt{t}} \alpha(z_0) \, e^{i x^2/4t - i\nu(z_0)\log 2t} \tag{31}$$

$$+ O\left(\frac{\ell n \, t}{t}\right)$$

where

$$z_0 = x/2t \, , \quad \nu(z) = -\frac{1}{2\pi} \log\left(1 - |r(z)|^2\right) \tag{32}$$

and

$\alpha(z)$ is an explicit function of r.

We see, in particular, that the long-time behavior of $u(x,t)$ is given with the same accuracy and detail as the solution of the linear Schröödinger equation $i u_t^0 + u_{xx}^0 = 0$ which can be obtained by applying the classical steepest descent method to the Fourier representation of $u^0(x,t)$

$$u^0(x,t) = \frac{1}{\sqrt{2\pi}} \int_\mathbb{R} \widehat{u^0}(z) \, e^{i(xz - tz^2)} \, dz$$

where $\widehat{u^0}$ is the Fourier transform of $u^0(x, t=0)$. As $t \to \infty$, one finds

$$u^0(x,t) = \frac{\widehat{u^0}(z_0)}{\sqrt{2it}} \, e^{ix^2/4t} + o\left(\frac{1}{\sqrt{t}}\right). \tag{33}$$

We see that NLS is an integrable system in the sense advertised in (13a,b). It is interesting and important to note that NLS is also integrable in the sense of Liouville. In 1974, Zakharov and Manakov in a celebrated paper [33], not only showed that NLS has a complete set of commuting integrals (read "actions"), but also computed the "angle" variables canonically conjugate to these integrals, thereby displaying NLS explicitly in so-called "action-angle" form. This effectively integrates the system by "quadrature" (see page 2). The first construction of action-angle variables for an integrable PDE is due to Zakharov and Faddeev in their landmark paper [32] on the Korteweg-de Vries equation in 1971.

We note that the asymptotic formula (31) and (32) for NLS was first obtained by Zakharov and Manakov in 1976 [34] using inverse scattering techniques, also taken from the IM toolbox, but without the rigor of the nonlinear steepest descent method.

The next example, taken from Statistical Mechanics, utilizes another tool from the toolkit, viz. the theory of integrable operators, IO's.

IO's were first singled out as a distinguished class of operators by Sakhnovich in the 1960s, 1970s and the theory of such operators was then fully developed by Its et al. [23] in the 1990s. Let Σ be an oriented contour in $\mathbb{C}$. We say an operator K acting on measurable functions h on Σ is *integrable* if it has a kernel of the form

$$K(x, y) = \frac{\sum_{i=1}^{n} f_i(x) g_i(y)}{x - y}, \quad n < \infty, \quad x, y \in \Sigma, \tag{34}$$

where

$$f_i, g_i \in L^\infty(\Sigma), \quad \text{and} \tag{35}$$

$$Kh(x) = \int_\Sigma K(x, y) h(y) \, dy.$$

If Σ is a "good" contour (i.e. Σ is a **Carleson curve**), K is bounded in $L^p(\Sigma)$ for $1 < p < \infty$.

Integral operators have many remarkable properties. In particular the integrable operators form an algebra and $(I + K)^{-1}$, if it exists, is also integrable if K is integrable. But most remarkably, $(I + K)^{-1}$ can be computed in terms of a naturally associated RHP on Σ. It works like this. If $K(x, y) = \sum_{i=1}^{n} f_i(x) g_i(y)/x - y$, then

$$(I + K)^{-1} = I + R \tag{36}$$

where $$R(x, y) = \sum_{i=1}^{n} F_i(x) G_i(y) \Big/ x - y$$

for suitable F_i, G_i. Now assume for simplicity that $\sum_{i=1}^{n} f_i(x) g_i(x) = 0$ and let

$$v(z) = I - 2\pi f(z) g(z)^T, \quad z \in \Sigma, \tag{37}$$

where $f = (f_1, \ldots, f_n)^T$, $g = (g_1, \ldots, g_n)^T$ and suppose $m(z)$ solves the normalized RHP (Σ, v). Then

$$F(z) = m_+(z) f(z) = m_-(z) f(z) \tag{38}$$

and

$$G(z) = \left(m_+^{-1}\right)^T g(z) = \left(m_-^{-1}\right)^T g(z) \tag{39}$$

Here is an example how integrable operators arise. Consider the spin-$\frac{1}{2}$ XY model in a magnetic field with Hamiltonian

$$H = -\frac{1}{2} \sum_{\ell \in \mathbb{Z}} \left(\sigma_\ell^x \sigma_{\ell+1}^x + \sigma_\ell^z \right) \tag{40}$$

where σ_ℓ^x, σ_ℓ^z are the standard Pauli matrices at the ℓth site of a 1-d lattice.

As shown by McCoy et al. [28] in 1983, the auto-correlation function $X(t)$

$$X(t) = \langle \sigma_0^x(t) \sigma_0^x \rangle_T = \frac{tr \left(e^{-\beta H} \left(e^{-iHt} \sigma_0^x e^{iHt} \right) \sigma_0^x \right)}{tr \, e^{-\beta H}}$$

where $\beta = \frac{1}{T}$, can be expressed as follows:

$$X(t) = e^{-t^2/2} \det(1 - K_t)$$

Here K_t is the operator on $L^2(-1, 1)$ with kernel

$$K_t(z, z') = \varphi(z) \frac{\sin it(z - z')}{\pi (z - z')} \quad , \quad -1 \le z, z' \le 1, \tag{41}$$

and

$$\varphi(z) = \tanh\left(\beta \sqrt{1 - z^2} \right) \quad , \quad -1 < z < 1. \tag{42}$$

Observe that

$$K_t(z, z') = \frac{\sum_{i=1}^{2} f_i(z) g_i(z')}{z - z'} \tag{43}$$

where

$$f = (f_1, f_2)^T = \left(\frac{-e^{tz} \varphi(z)}{2\pi i}, \frac{-e^{-tz} \varphi(z)}{2\pi i} \right)^T$$

$$g = (g_1, g_2)^T = \left(e^{-tz}, -e^{tz} \right)^T$$

so that K_t is an integrable operator. We have

$$v = v_t = I - 2\pi i \, f \, g^T = \begin{pmatrix} 1 + \varphi(z) & -\varphi(z) e^{2zt} \\ \varphi(z) e^{-2zt} & 1 - \varphi(z) \end{pmatrix}, \quad z \in (-1, 1). \tag{44}$$

As

$$\frac{d}{dt} \log \det(1 - K_t) = \frac{d}{dt} tr \log(1 - K_t)$$

$$= -tr\left(\frac{1}{1 - K_t} \dot{K}_t\right)$$

we see that $\frac{d}{dt} \log \det(1 - K_t)$, and ultimately $X(t)$, can be expressed via (36), (38), and (39) in terms of the solution m_t of the RHP $(\Sigma = (-1, 1), v_t)$

Applying the nonlinear steepest descent method to this RHP as $t \to \infty$, one finds (Deift and Zhou [14]) that

$$X(t) = \exp\left(\frac{t}{\pi} \int_{-1}^{1} \log|\tanh \beta s| \, ds + o(t)\right) \tag{45}$$

This shows that H in (40) is integrable in the sense that key statistics for H such as the autocorrelation function $X(t)$ for the spin σ_0^x is integrable in the sense of (14a,b)

$$X(t) \stackrel{\varphi}{\mapsto} K_t \in \text{integrable operators}$$

and φ^{-1} is computed with any desired precision using RH-steepest descent methods to obtain (45). Note that the appearance of the terms $\varphi(z) e^{\pm 2zt}$ in the jump matrix v_t for $K_t = \varphi(X(t))$, makes explicit the linearizing property of the map φ.

Another famous integrable operator appears in the bulk scaling limit for the gap probability for invariant Hermitian ensembles in random matrix theory. More precisely, consider the ensemble of $N \times N$ Hermitian matrix $\{M\}$ with invariant distribution

$$P_N(M) \, dM = \frac{e^{-N \, tr \, V(M)} \, dM}{\int e^{-N \, tr \, V(M)} \, dM},$$

where $V(x) \to +\infty$ as $|x| \to \infty$ and dM is Lebesgue measure on the algebraically independent entries of M.

Set $P_N([\alpha, \beta]) = $ gap probability $\equiv$ Prob $\{M$ has no eigenvalues in $[\alpha, \beta]\}$, $\quad \alpha < \beta$

We are interested in the scaling limit of $P_N([\alpha, \beta])$ i.e.

$$P(\alpha, \beta) = \lim_{N \to \infty} P_N\left(\left[\frac{\alpha}{\rho_N}, \frac{\beta}{\rho_N}\right]\right)$$

for some appropriate scaling $\rho_N \sim N$. One finds (and here RH techniques play a key role) that

$$P(\alpha, \beta) = \det(1 - K_s), \quad s = \beta - \alpha \tag{46}$$

where K_s has a kernel

$$K_s(x, y) = \frac{\sin(x - y)}{\pi(x - y)} \quad \text{acting on} \quad L^2(0, 2s).$$

Clearly $K_s(x, y) = \dfrac{e^{ix} e^{-iy} - e^{-ix} e^{iy}}{2\pi i(x - y)}$ is an integrable operator. The asymptotics of $P(\alpha, \beta)$ can then be evaluated asymptotically with great precision as $s \to \infty$, by applying the nonlinear steepest descent method for RHP's to the RHP associated with the integrable operator K_s, as in the case for K_t in (44) et seq.

Thus RMT is integrable in the sense that a key statistic, the gap probability in the bulk scaling limit, is an integrable system in the sense of (14a,b):

Scaled gap probability $P_{(\alpha,\beta)} \xrightarrow{\varphi} K_s(x, y) \in$ Integrable operators

φ^{-1} is then evaluated via the formula $\det(1 - K_s)$

which can be controlled precisely as $s \to \infty$.

The situation is similar for many other key statistics in RMT. It turns out that $P_{(\alpha,\beta)}$ solves the Painlevé V equation as a function of $s = \beta - \alpha$ (this is a famous, result of Jimbo et al. [25]). But the Painlevé V equation is a classically integrable Hamiltonian system which is also integrable in the sense of (14a,b). Indeed it is a consequence of the seminal work of the Japanese School of Sato et al. that all the Painlevé equations can be solved via associated RHP's (the RHP for Painlevé II in particular was also found independently by Flaschka and Newell), and hence are integrable in the sense of (14a,b) and amenable to nonlinear steepest descent asymptotic analysis, as described, for example, in the text, Painlevé Transcendents by Fokas et al. [18].

There is another perspective one can take on RMT as an integrable system. The above point of view is that RMT is integrable because key statistics are described by functions which are solutions of classically integrable Hamiltonian systems. But this point of view is unsatisfactory in that it attributes integrability in one area (RMT) to integrability in another (Hamiltonian systems). Is there a notion of integrability for stochastic systems that is intrinsic? In dynamics the simplest integrable system is free motion

$$\dot{x} = y, \quad \dot{y} = 0 \quad \Longrightarrow \quad x(t) = x_0 + y_0 t \quad , \quad y(t) = y_0. \tag{47}$$

Perhaps the simplest stochastic system is a collection of coins flipped independently. Now, we suggest, just as an integrable Hamiltonian system becomes (47) in new variables, the analogous property for a stochastic system should be that, in the appropriate variables, it is integrable if it just a product of independent spin flips.

Consider the scaled gap probability,

$$P_{(\alpha,\beta)} = \text{Prob } \{ \text{no eigenvalues in } (\alpha, \beta)\} = \det(1 - K_s) \tag{48}$$

But as the operator K_s is trace-class and $0 \leq K_s < 1$, it follows that

$$P_{\alpha,\beta} = \prod_{i=1}^{\infty} (1 - \lambda_i) \tag{49}$$

where $0 \leq \lambda_i < 1$ are the eigenvalues of K_s. Now imagine we have a collection of boxes, $B_1, B_2, \ldots$. With each box we have a coin: With probability λ_i a ball is placed in box B_i, or equivalently, with probability $1 - \lambda_i$ there is no ball placed in B_i. The coins are independent. Thus we see that the probability that there are no eigenvalue in (α, β), is the same as the probability of no balls being placed in all the boxes!

This is an intrinsic probabilistic view of RMT integrability. It applies to many other stochastic systems. For example, consider Ulam's longest increasing subsequence problem:

Let $\pi = \pi(1)\,\pi(2), \ldots \pi(N)$ be a permutation in the symmetric group S_N. If

$$i_1 < i_2 < \cdots < i_k \quad \text{and} \quad \pi(i_1) < \cdots < \pi(i_k) \tag{50}$$

we say that

$$\pi(i_1)\,\pi(i_2)\,\ldots,\pi(i_k) \tag{51}$$

is an increasing subsequence for π of length k. Let $\ell_N(\pi)$ denote the greatest length of any increasing subsequence for π, e.g. for $N = 6$, $\pi = 31,5624 \in S_6$ has $\ell_6(\pi) = 3$ and 356, 254 and 156 are all longest increasing subsequences for π. Equip S_N with uniform measure. Thus for $n \leq N$.

$$q_{n,N} \equiv \text{Prob } (\ell_N \leq n) \tag{52}$$
$$= \frac{\#\{\pi : \ell_N(\pi) \leq n\}}{N!}$$

Question How does $q_{n,N}$ behave as $n, N \to \infty$?

Theorem 1 (Baik et al. [1]) *Let $t \in \mathbb{R}$ be given. Then*

$$F(t) \equiv \lim_{N \to \infty} \text{Prob }\left(\ell_N \leq 2\sqrt{N} + t\,N^{1/6}\right) \tag{53}$$

exists and is given by $e^{-\int_t^\infty (x-t)\,u^2(x)\,dx}$ *where $u(x)$ is the (unique) Hastings-McLeod solution of the Painlevé II equation*

$$u'' = 2u^3 + xu \tag{54}$$

normalized such that

$$u(x) \sim Ai(x) = Airy\ function, \quad as\ x \to +\infty$$

(The original proof of this Theorem used RHP/steepest descent methods. The proof was later simplified by Borodin, Olshanski and Okounkov using the so-called Borodin-Okounkov-Case-Geronimo formula.)

Some observations:

(i) As Painlevé II is classically integrable, we see that the map

$$q_{n,N} \xrightarrow{\varphi} u^2(t) = -\frac{d^2}{dx^2} \log F(x)$$

transforms Ulam's longest increasing subsequence problem into an integrable system whose behavior is known with precision. There are many other classical integrable systems associated with $q_{n,N}$ but that is another story (see Baik et al. [2]).

(ii) The distribution $F(t) = e^{-\int_t^\infty (x,t)u^2(x)\,dx}$ is the famous Tracy-Widom distribution for the largest eigenvalue $\lambda_{\max}$ of a random Hermitian matrix in the edge-scaling limit. In other words, the length of the longest increasing subsequence behaves like the largest eigenvalue of a random Hermitian matrix. More broadly, what we are seeing here is an example of how RMT plays the role of a stochastic special function theory describing a stochastic problem from some other a priori unrelated area. This is no different, in principle, from the way the trigonometric functions describe the behavior of the simple harmonic oscillator. RMT is a very versatile tool in our IM toolbox—tiling problems, random particle systems, random growth models, the Riemann zeta function, ..., all the way back to Wigner, who introduced RMT as a model for the scattering resonances of neutrons off a large nucleus, are all problems whose solution can be expressed in terms of RMT.

(iii) $F(t)$ can also be written as

$$F(t) = \det(1 - A_t) \qquad (55)$$

where A_t is a particular trace class integrable operator, the Airy operator, with $0 \leq A_t < 1$. Thus $F(t) = \prod_{i=1}^{\infty} \left(1 - \tilde{\lambda}_i(t)\right)$ where $\{\tilde{\lambda}_i(t)\}$ are the eigenvalues of A_t. We conclude that $F(t)$, the (limiting) distribution for the length ℓ_N of the longest increasing subsequence, corresponds to an integrable system in the above intrinsic probabilistic sense.

(iv) It is of considerable interest to note that in recent work Gavrylenko and Lisovyy [20] have shown that the isomonodromic tau function for general Fuchsian systems can be expressed, up to an explicit elementary function, as a Fredholm determinant of the form $\det(1 - K)$ for some suitable trace class operator K. Expanding the determinant as a product of eigenvalues, we see that the general Fuchsian system, too, is integrable in the above intrinsic stochastic sense.

Another tool in our toolbox concerns the notion of a scattering system. Consider the Toda lattice in $\left(\mathbb{R}^{2n}, \omega = \sum_{i=1}^{n} dx_i \wedge dy_i\right)$ with Hamiltonian

$$H_T(x, y) = \frac{1}{2} \sum_{i=1}^{n} y_i^2 + \sum_{i=1}^{n-1} e^{(x_i - x_{i+1})} \tag{56}$$

giving rise to Hamilton's equations

$$\dot{x} = (H_T)_y \quad , \quad \dot{y} = -(H_T)_x. \tag{57}$$

The *scattering map* for a dynamical system maps the behavior of the system in the distant past onto the behavior of the system in the distant future. In my Phd, I worked on abstract scattering theory in Hilbert space addressing questions of asymptotic completeness for quantum systems and classical wave systems. When I came to Courant, I started to study the Toda system and I was amazed to learn that for this multi-particle system the scattering map could be computed explicitly. When I expressed my astonishment to Jürgen Moser, he said to me, "But every scattering system is integrable!" It took me some time to understand what he meant. It goes like this:

Suppose that you have a Hamiltonian system in $\left(\mathbb{R}^{2n}, \omega = \sum_{i=1}^{n} dx_i \wedge dy_i\right)$ with Hamiltonian H, and suppose that the solution

$$z(t) = (x(t), y(t)) \quad , \quad z(0) = (x(0), y(0)) = (x_0, y_0)$$

of the flow generated by H behaves asymptotically like the solutions $\hat{z}(t)$ of free motion with Hamiltonian

$$\widehat{H}(x, y) = \frac{1}{2} y^2$$

for which

$$\dot{x} = y, \quad \dot{y} = 0 \quad \text{with} \quad \hat{z}(0) = (\hat{x}_0, \hat{y}_0),$$

yielding

$$\hat{z}(t) = (\hat{x}_0 + \hat{y}_0 t, \hat{y}_0).$$

As $z(t) \sim \hat{z}(t)$ by assumption, we have as $t \to \infty$,

$$x(t) = t y^{\#} + x^{\#} + o(1) \tag{58a}$$

$$y(t) = y^\# + o(1) \tag{58b}$$

for some $x^\#$, $y^\#$.

Write

$$z(t) = U_t(z(0)), \quad \hat{z}(t) = \widehat{U}_t(\hat{z}(0)).$$

Then, provided $o(1) = o\left(\frac{1}{t}\right)$ in (58b),

$$\begin{aligned} W_t(z_0) &\equiv \widehat{U}_{-t} \circ U_t(z_0) \\ &= \widehat{U}_{-t}\left(ty^\# + x^\# + o(1), \quad y^\# + o\left(\frac{1}{t}\right)\right) \\ &= \left(ty^\# + x^\# + o(1) - t\left(y^\# + o\left(\frac{1}{t}\right)\right), \quad y^\# + o\left(\frac{1}{t}\right)\right) \\ &= \left(x^\# + o(1), \quad y^\# + o(1)\right). \end{aligned}$$

Thus

$$W_\infty(z_0) = \lim_{t \to \infty} W_t(z_0)$$

exists. Now

$$\begin{aligned} W_t \circ U_s &= \widehat{U}_{-t} \circ U_{t+s} \\ &= \widehat{U}_s \circ W_{t+s}, \end{aligned}$$

and letting $t \to \infty$, we obtain

$$W_\infty \circ U_s = \widehat{U}_s \circ W_\infty \tag{59}$$

so that W_∞ is an intertwining operator between U_s and $\widehat{U}_s$.

But clearly W_t is the composition of symplectic maps, and so is symplectic, and hence W_∞ is a symplectic map and hence W_∞^{-1} is symplectic. Thus from (59) we see that

$$U_s = W_\infty^{-1} \circ \widehat{U}_s \circ W_\infty \tag{60}$$

is symplectically equivalent to free motion, and hence is integrable. In particular if $\{\hat{\lambda}_k\}$ are the Poisson commuting integrals for $\widehat{H}$, then $\{\lambda_k = \hat{\lambda}_k \circ W_\infty\}$ are the (Poisson commuting) integrals for H.

What this computation is telling us is that if a system is scattering, or more generally, if the solution of one system *looks* asymptotically like some other system, then it is in fact (equivalent to) that system. Remember the famous story of Roy Cohn during the McCarthy hearings, when he was trying to convince the panel that a particular person was a Communist? He said: "If it looks like a duck, walks like a duck, and quacks like a duck, then it's a duck!"

Now direct computations, due originally to Moser, show that the Toda lattice is *scattering* in the sense of (58a,b). And so what Moser was saying is that the system is **necessarily** integrable. The Toda lattice is a rich and wonderful system and I spent much of the 1980s analyzing the lattice and its various generalizations together with Carlos Tomei, Luen-Chau Li and Tara Nanda. I will say much more about this system below. It was a great discovery of Flaschka [17] (and later independently, Manakov [27]) that the Toda system indeed had a Lax pair formulation (see (74) below).

The idea of a scattering system can be applied to PDE's. Some 15–20 years ago Xin Zhou and I [15] began to consider perturbations of the defocusing NLS equation on the line,

$$i\, u_t + u_{xx} - 2|u|^2 u - \epsilon |u|^\ell u = 0, \qquad \ell > 2 \qquad (61)$$

with

$$u(x, t=0) = u_0(x) \to 0 \quad \text{as} \quad |x| \to \infty.$$

In the spatially periodic case, $u(x,t) = u(x+1,t)$, solutions of NLS (the integrable case: $\epsilon = 0$) move linearly on a (generically infinite dimensional) torus. In the perturbed case ($\epsilon \neq 0$), KAM methods can be (extended and) applied (with great technical virtuosity) to show (here Kuksin, Pöschel, Kappeler have played the key role) that, as in the familiar finite dimensional case, some finite dimensional tori persist for (61) under perturbation. However, on the whole line with $u_0(x) \to 0$ as $|x| \to \infty$, the situation, as we now describe, is very different.

In the spirit of it "walks like a duck", what is the "duck" for solutions of (61)? The "duck" here is a solution $u^\#(x,t)$ of the NLS equation.

$$\begin{aligned} i\, u^\#_t + u^\#_{xx} - 2|u^\#|^2 u^\# &= 0 \\ u^\#(x,0) = u^\#_0(x) &\to 0 \quad \text{as} \quad |x| \to \infty. \end{aligned} \qquad (62)$$

Recall the following calculations from classical KAM theory in R^{2n}, say: Suppose that the flow with Hamiltonian H_0 is integrable and $H_\epsilon = H_0 + \epsilon \widehat{H}$ is a perturbation of H_0. Hamilton's equation for H_ϵ has the form

$$z_t = J \nabla H_\epsilon = J \nabla H_0 + \epsilon\, J \nabla \widehat{H}, \quad z(0) = z_0 \qquad (63)$$

with $J = \begin{pmatrix} 0 & I_n \\ -I_n & 0 \end{pmatrix}$. If $J \nabla H_0$ is linear in z, say $J \nabla H_0 = Az$, then we can solve (63) by D'Alembert's principle to obtain

$$z(t) = e^{At} z_0 + \epsilon \int_0^t e^{A(t-s)} J\nabla \widehat{H}(s) \, ds \tag{64}$$

to which an iteration procedure can be applied. If $J\nabla H_0$ is **not linear**, however, **no such D'Alembert formula exist**, and this is the reason that the starting point for any KAM investigation is to first write (63) in action-angle variables $z \mapsto \zeta$ for H_0: Then $J\nabla H_0$ is linear and (64) applies.

With this in mind, we used the linearizing map for NLS described in (26)

$$u(x,t) = u(t) \mapsto \varphi(u(t)) = r(t) = r(t;z)$$

as a change of variables for the perturbed equation (61). And although the map φ no longer linearizes the equation, it does transform the equation into the form

$$\frac{\partial r}{\partial t}(t,z) = -i z^2 r(t,z) + \epsilon F(z,t;r(t)) \tag{65}$$

to which D'Alembert's principle can be applied

$$\hat{r}(t,z) = r_0(z) + \epsilon \int_0^t F\left(z,s; \hat{r} e^{-is<\cdot>^2}\right) ds \tag{66}$$

where $r_0(z) = \varphi(z)$ and $\hat{r}(t,z) = r(t,s) e^{itz^2}$. The functional F depends on φ and φ^{-1}, and so, in particular, involves the RHP ($\Sigma = \mathbb{R}$, v_t). Fortunately this RHP can be evaluated with sufficient accuracy using steepest descent methods in order to obtain the asymptotics of $\hat{r}(t,z)$ as $t \to \infty$, and hence of $u(x,t) = \varphi^{-1}\left(\hat{r}(t) e^{-it<\cdot>^2}\right)$.

Let $U_t^\epsilon(u_0)$ be the solution of (61) and $U_t^{NLS}(u_0^\#)$ be the solution of NLS (62) with $u_0, u_0^\#$ in $H^{1,1} = \{f \in L^2(\mathbb{R}) : f' \in L^2(\mathbb{R}), \; x f \in L^2(\mathbb{R})\}$, respectively. Then the upshot of this analysis is, in particular, that

$$W^\pm(u_0) = \lim_{t \to \pm\infty} U_{-t}^{NLS} \circ U_t^\epsilon(u_0) \tag{67}$$

exist strongly which shows that as $t \to \pm\infty$,

$$U_t^\epsilon(u_0) \sim U_t^{NLS}\left(W^\pm(u_0)\right)$$

and much more. In particular, there are commuting integrals for (61), ...,

Three observations:

(a) As opposed to KAM where integrability is preserved on sets of high measure, here integrability is preserved on open subsets of full measure.
(b) As a tool in our IM toolbox, integrability makes it possible to analyze perturbations of integrable systems, via a D'Alembert principle.

(c) There is a Catch 22 in the whole story. Suppose you say, "my goal is describe the evolution of solutions of the perturbed equation (61) as $t \to \infty$". To do this one must have in mind what the solutions should look like as $t \to \infty$: Do they look like solutions of NLS, or perhaps like solutions of the free Schrödinger equation $i u_t + u_{xx} = 0$? Now suppose you disregard any thoughts on integrability and utilize any method you can think of, dynamical systems ideas, etc., to analyze the system and you find in the end that the solution indeed behaves like NLS. But here's the catch; if it looks like NLS, then the wave operators $W_\pm$ in (67) exist, and hence the system is integrable! It looks like a duck, walks like a duck and quacks like a duck, and so it's a duck! In other words, whatever methods you used, they would not have succeeded unless the system was integrable in the first place!

Finally, I would like to discuss briefly an extremely useful algebraic **tool** in the IM toolbox, viz., Darboux transforms/Backlund transformations. These are explicit transforms that convert solutions of one (nonlinear) equation into solutions of another equation, or into different solutions of the same equation. For example, the famous Miura transform, a particular Darboux/Backlund transform,

$$v(x,t) \to u(x,t) = v_x(x,t) + v^2(x,t)$$

converts solutions $v(x,t)$ of the modified KdV equation

$$v_t + 6v^2 v_x + v_{xxx} = 0$$

into solutions of the KdV equation

$$u_t + 6uu_x + u_{xxx} = 0.$$

Darboux transforms can be used to turn a solution of KdV without solitons into one with solitons, etc. Darboux/Backlund transforms also turn certain spectral problems into other spectral problems with (essentially) the same spectrum, for example,

$$H = -\frac{d^2}{dx^2} + q(x) \longrightarrow \tilde{H} = -\frac{d^2}{dx^2} + \tilde{q}(x)$$

where $\tilde{q} = q - 2\frac{d^2}{dx^2} \log \varphi$, and φ is any solution of $H\varphi = 0$,

constructs $\tilde{H}$ with (essentially) the same spectrum as H. Thus a Darboux/Backlund transform is a basic isospectral action. The literature on Darboux transforms is vast, and I just want to discuss one application to PDE's which is perhaps not too well known.

Consider the Gross-Pitaevskii equation in one-dimension,

$$i u_t + \frac{1}{2} u_{xx} + V(x) u + |u|^2 u = 0 \tag{68}$$

$$u(x, 0) = u_0(x).$$

For general V this equation is very hard to analyze. A case of particular interest is where

$$V(x) = q\,\delta(x), \qquad q \in \mathbb{R} \quad \text{and } \delta \text{ is the delta function.} \tag{69}$$

For such V, (68) has a particular solution

$$u_\lambda(x, t) = \lambda\, e^{i\lambda^2 t/2} \operatorname{sech}\left(\lambda|x| + \tanh^{-1}\left(\frac{q}{\lambda}\right)\right) \tag{70}$$

for any $\lambda > |q|$. This solution is called the Bose-Einstein condensate for the system.

Question Is u_λ asymptotically stable? In particular, if

$$u(x, t = 0) = u_\lambda(x, t = 0) + \epsilon\, w(x), \quad \epsilon \text{ small}, \tag{71}$$

does

$$u(x, t) \;\to\; u_\lambda(x, t) \quad \text{as} \quad t \to \infty?$$

In the case where $w(x)$ is even, one easily sees that the initial value problem (IVP) (68) with initial value given by (71) is equivalent to the initial boundary value problem (IBVP)

$$\begin{cases} i u_t + \tfrac{1}{2} u_{xx} + |u|^2 u = 0, & x > 0, \quad t > 0 \\ u(x, t = 0) = (71) \text{ for } x > 0 \\ \text{subject to the Robin boundary condition at } x = 0 \\ u_x(0, t) + q\, u(0, t) = 0. \end{cases} \tag{72}$$

Now NLS on $\mathbb{R}$ is integrable, but is NLS on $\{x > 0\}$ with boundary conditions as in (72) integrable? Remember that the origin of the boundary condition is the physical potential (read "force"!) $V(x)$. So we are looking at a dynamical system, which is integrable on $\mathbb{R}$, interacting with a new "force" V. It is not at all clear, a priori, that the **combined system** is integrable in the sense of (13a,b).

The stability question for u_λ was first consider by Holmer and Zworski [22], and using dynamical systems methods, they showed asymptotic stability of u_λ for times of order $|q|^{-2/7}$. But what about times larger than $|q|^{-2/7}$? Following on the work of Holmer and Zworski, Jungwoon Park and I [10] begin in 2009 to consider this question. Central to our approach was to try to show that the IBVP for NLS as in (72) was integrable, and then use RH/steepest-descent methods. In the linear case, a standard approach is to use the method of images: for Dirichlet and Neumann boundary conditions, one just reflects, $u(x) = -u(-x)$ or $u(x) = +u(-x)$ for $x < 0$, respectively.

For the Robin boundary condition in the linear case, the reflection is a little more complicated, but still standard. In this way one then gets an IVP on the line that can be solved by familiar methods. In the non-linear case, similar methods work for the Dirichlet and Neumann boundary conditions, but for the Robin boundary condition case, $q \neq 0$, how should one reflect across $x = 0$? It turns out that there is a beautiful method due to Bikbaev and Tarasov where they construct a particular Darboux transform version $b(x)$ of the initial data $u(x, t = 0)$, $x > 0$, and then define

$$\begin{cases} v(x) = b(-x) & x < 0 \\ = u(x, t = 0) & x > 0. \end{cases} \tag{73}$$

If $v(x, t)$ is the solution of (the integral equation) of NLS on $\mathbb{R}$ with initial conditions (73), then $v(x,t)\big|_{x>0}$ is a solution of the IBVP (72) for $t \geq 0$. In other words, the Darboux transform can function as a tool in our toolkits to show that a system is integrable.

Applying RH/steepest descent methods to $v(x, t)$, one finds that u_λ is asymptotically stable if $q > 0$, but for $q < 0$, generically, u_λ is not asymptotically stable: In particular, for times $t >> |q|^{-2}$, as $t \to \infty$, a second "soliton" emerges and one has a "two soliton" condensate.

We note that (72) can also be analyzed using Fokas' unified integration method instead of the Bikbaev-Tarasov transform, as in Its and Shepelsky [24].

Algorithms
As discussed above, the Toda lattice is generated by the Hamiltonian

$$H_T(x, y) = \frac{1}{2} \sum_{i=1}^{n} y_i^2 + \sum_{i=1}^{n-1} e^{(x_i - x_{i+1})}.$$

The key step in analyzing the Toda lattice was the discovery by Flaschka [17], and later independently by Manakov [27], that the Toda equations have a Lax-pair formulation

$$\begin{cases} \dfrac{dx}{dt} = H_{T,y}, \quad \dfrac{dy}{dt} = -H_{T,x} \\ \qquad\qquad \equiv \\ \qquad \dfrac{dM}{dt} = [M, B(M)] \end{cases} \tag{74}$$

where

$$M = \begin{pmatrix} a_1 & b_1 & & & \\ b_1 & \ddots & \ddots & & 0 \\ & \ddots & \ddots & \ddots & \\ & 0 & \ddots & & b_{n-1} \\ & & & b_{n-1} & a_n \end{pmatrix}, \qquad B(M) = \begin{pmatrix} 0 & -b_1 & & & \\ b_1 & 0 & -b_2 & 0 & \\ & b_2 & \ddots & \ddots & \\ & 0 & \ddots & & -b_{n-1} \\ & & & b_{n-1} & 0 \end{pmatrix}$$
$$= M_- - M_-^T$$

and

$$\begin{aligned} a_k &= -y_k/2 \,, & 1 \le k \le n \\ b_k &= \frac{1}{2} e^{\frac{1}{2}(x_k - x_{k+1})} \,, & 1 \le k \le n-1. \end{aligned} \tag{75}$$

In particular, the eigenvalues $\{\lambda_n\}$ of M are constants of the motion for Toda, $\{\lambda_n(t) = \lambda_n, t \ge 0\}$. Direct calculation shows that they are independent and Poisson commute, so that Toda is Liouville integrable. Now as $t \to \infty$, one can show, following Moser [29], that the off diagonal entries $b_k(t) \to 0$ as $t \to \infty$. As noted by Deift et al. [9], what this means is that Toda gives rise to an eigenvalue algorithm:

Let M_0 be given and let $M(t)$ solve the Toda equations (74) with $M(0) = M_0$. Then

- $t \mapsto M(t)$ is isospectral, spec $(M(t)) =$ spec (M_0).
- $M(t) \to \text{diag } (\lambda_1, \ldots, \lambda_n) \quad \text{as} \quad t \to \infty$. $\tag{76}$

Hence $\lambda_1, \ldots, \lambda_n$ must be the eigenvalues of M_0.

Note that $H_T(M) = \frac{1}{2} \text{tr } M^2$.

Now the default algorithm for eigenvalue computation is the QR algorithm. The algorithm without "shifts" works in the following way. Let $M_0 = M_0^T$, $\det M_0 \ne 0$, be given, where M_0 is $n \times n$.

Then M_0 has a unique QR-factorization

$$M_0 = Q_0 R_0 \,, \qquad Q_0 \text{ orthog, } R_0 \text{ upper triangular} \\ \text{with } (R_0)_{ii} > 0, \quad i = 1, \ldots, n. \tag{77}$$

Set

$$\begin{aligned} M_1 &\equiv R_0 Q_0 \\ &= Q_0^T M_0 Q_0 \end{aligned}$$

from which we see that

$$\text{spec}(M_1) = \text{spec } M_0.$$

Now M_1 has its own QR-factorization

$$M_1 = Q_1 R_1$$

Set

$$M_2 = R_1 Q_1$$
$$= Q_1^T M_1 Q_1$$

so that again spec M_2 = spec M_1 = spec M_0.

Continuing, we obtain a sequence of isospectral matrices

$$\text{spec } M_k = \text{spec } M_0 \quad , \quad k > 0,$$

and as $k \to \infty$, generically,

$$M_k \to \text{diag}(\lambda_1, \ldots, \lambda_n)$$

and again $\lambda_1, \ldots, \lambda_n$ must be the eigenvalues of M_0. If M_0 is tridiagonal, one verifies that M_k is tridiagonal for all k.

There is the following *Stroboscope Theorem* for the QR algorithm (Deift et al. [9]), which is motivated by earlier work of Symes [31]:

Theorem (QR: tridiagonal)

(78)

Let $M_0 = M_0^T$ be tri-diagonal. Then there exists a Hamiltonian flow $t \mapsto M_{QR}(t)$, $M_{QR}(0) = M_0$ with Hamiltonian

$$H_{QR}(M) = tr\ (M \log M - M) \tag{79}$$

with the properties

(i) *the flow is completely integrable*
(ii) *(Stroboscope property)* $M_{QR}(k) = M_k$, $k \geq 0$, where M_k are the QR iterates starting at M_0, $\det M_0 \neq 0$
(iii) $M_{QR}(t)$ *commutes with the Toda flow*
(iv) $\frac{dM}{dt} = [B(\log M), M]$, $B(\log M) = (\log M)_- - (\log M)_-^T$.

More generally, for any $G : \mathbb{R} \to \mathbb{R}$

$$H_G(M) = tr\ G(M)$$
$$\to \dot{M}_G = [M, B(g(M))] \quad , \quad g(M) = G'(M)$$

generates an eigenvalue algorithm, so in a concrete sense, we can say, at least in the tri-diagonal case, that eigenvalue computation is an integrable process.

Now the Lax equation (74) for Toda clearly generates a global flow $t \mapsto M(t)$ for all full symmetric matrices $M_0 = M_0^T$.

Question

(i) Is the Toda flow for general symmetric matrices M_0 Hamiltonian?
(ii) Is it integrable?
(iii) Does it constitute an eigenvalue algorithm i.e. spec $(M(t))$ = spec (M_0), $M(t) \to$ diagonal as $t \to \infty$?
(iv) Is there a stroboscope theorem for general M_0?

As shown in [5], the answer to all these questions is in the affirmative. Property (ii) is particularly novel. The Lax-pair for Toda only gives n integrals, viz. the eigenvalues of $M(t)$, but the dimension of the symplectic space for the full Toda is generically of dimension $2\left[\frac{n^2}{4}\right]$, so one needs of order $\frac{n^2}{4} \gg n$ Poisson commuting integrals. These are obtained in the following way: consider, for example, the case $n = 4$. Then $\left[\frac{n^2}{4}\right] = 4$

- $\det(M - z) = 0$ has 4 roots $\lambda_{01}, \lambda_{02}, \lambda_{03}, \lambda_{04}$
- $\det(M - z)_1 = 0$ has 2 roots $\lambda_{11}, \lambda_{12}$

where $(M - z)_1$ is obtained by chopping off the first row and last column of $M - z$

$$\begin{pmatrix} x-z & x & x & x \\ x & x-z & x & x \\ x & x & x-z & x \\ x & x & x & x-z \end{pmatrix}$$

Now $\lambda_{-1} + \lambda_{02} + \lambda_{03} + \lambda_{04} =$ trace M

and $\lambda_{11} + \lambda_{12} =$ "trace" of M_1

are the co-adjoint invariants that specify the $8 = 10 - 2 = 2\left[\frac{n^2}{4}\right]$ dimensional symplectic leaf $\mathcal{L}_{c_1, c_2} = \{M : tr\ M = c_1,\ tr\ M_1 = c_2\}$ on which the Toda flow is generically defined. The four independent integrals needed for integrability are then $\lambda_{01}, \lambda_{02}, \lambda_{03}, \lambda_{11}$. For general n, we keep chopping: $(M - z)_2$ is obtained by chopping off the first two rows and last two columns, etc. The existence of these "chopped" integrals, and their Poisson commutativity follows from the invariance properties of M under the actions of a tower of groups, $G_1 \subset G_2 \subset \ldots$. This shows that group theory is also a tool in the IM toolbox. This is spectacularly true in the work of Borodin and Okshanski on "big" groups like S_∞ and U_∞, and related matters.

Thus we conclude that eigenvalue computation in the full symmetric case is again an integrable process.

Remark The answer to Questions (i) ... (iv) is again in the affirmative for general, not necessarily symmetric matrices $M \in M(n, \mathbb{R})$. Here we need $\sim \frac{n^2}{2}$ integrals ..., but this is a whole other story (Deift et al. [6]).

The question that will occupy us in the remainder of this paper is the following: We have discussed two notions of integrability naturally associated with matrices: Eigenvalue algorithms and random matrix theory. What happens if we try to combine these two notions? In particular,

"What happens if we try to compute the eigenvalues of a random matrix?"
(80)

Let Σ_N denote the set of real $N \times N$ symmetric matrices. Associated with each algorithm $\mathcal{A}$, there is, in the discrete case, such as QR, a map $\varphi = \varphi_\mathcal{A} : \Sigma_N \to \Sigma_N$ with the properties

- isospectrality: spec $(\varphi_\mathcal{A}(H)) =$ spec (H)
- convergence: the iterates $X_{k+1} = \varphi_\mathcal{A}(X_k)$, $k \geq 0$, $X_0 = H$ given, converge to a diagonal matrix X_∞, $X_k \to X_\infty$ as $k \to \infty$

and in the continuum case, such as Toda, there exists a flow $t \mapsto X(t) \in \Sigma_N$ with the properties

- isospectrality :   spec $(X(t)) =$ spec $(X(0))$
- convergence : $X(t)$ converges to a diag. matrix X_∞ as $t \to \infty$.

In both cases, necessarily, the diagonal entries of X_∞ are the eigenvalues of H.

Given $\epsilon > 0$, it follows, in the discrete case, that for some m the off-diagonal entries of X_m are $O(\epsilon)$ and hence the diagonal entries of X_m give the eigenvalues of H to $O(\epsilon)$. The situation is similar for continuous flows $t \mapsto X(t)$. Rather than running the algorithm until all the off-diagonal entries are $O(\epsilon)$, it is customary to run the algorithm with *deflations* as follows: For an $N \times N$ matrix Y in block form

$$Y = \begin{pmatrix} Y_{11} & Y_{12} \\ Y_{21} & Y_{22} \end{pmatrix}$$

with Y_{11} of size $k \times k$ and Y_{22} of size $(N-k) \times (N-k)$ for some $k \in \{1, 2, \ldots, N-1\}$, the process of projecting

$$Y \mapsto \operatorname{diag}(Y_{11}, Y_{22})$$

is called *deflation*. For a

$$\text{given } \epsilon > 0, \text{ algorithm } \mathcal{A}, \text{ and matrix } H \in \Sigma_N \tag{81}$$

define the *k-deflation* time.

$$T^{(k)}(H) = T^{(k)}_{\epsilon, \mathcal{A}}(H), \quad 1 \leq k \leq N-1, \tag{82}$$

to be the *smallest* value of m such that X_m, the mth iterate of $\mathcal{A}$ with $X_0 = H$, has block form

$$X_m = \begin{pmatrix} X_{11}^{(k)} & X_{12}^{(k)} \\ X_{21}^{(k)} & X_{22}^{(k)} \end{pmatrix}$$

$X_{11}^{(k)}$ is $k \times k$, $X_{22}^{(k)}$ is $(N-k) \times (N-k)$ with

$$\|X_{12}^{(k)}\| = \|X_{22}^{(k)}\| < \epsilon. \tag{83}$$

The *deflation time* $T(H)$ is then defined as

$$T(H) = T_{\epsilon \mathcal{A}}(H) = \min_{1 \leq k \leq N-1} T^{(k)}_{\epsilon \mathcal{A}}(H) \tag{84}$$

If $\hat{k} \in \{1, 2, \ldots, N-1\}$ is such that

$$T(H) = T^{(\hat{k})}_{\epsilon \mathcal{A}}(H)$$

it follows that the eigenvalues of H are given by the eigenvalues of the block diagonal matrix diag $\left(X_{11}^{(\hat{k})}, X_{22}^{(\hat{k})}\right)$ to $O(\epsilon)$. After, running the algorithm to time $T(H)$, the algorithm restarts by applying the basic algorithm (in parallel) to the smaller matrices $X_{11}^{(\hat{k})}$ and $X_{22}^{(\hat{k})}$ until the next deflation time, and so on.

In 2009, Deift et al. [8] considered the deflation time $T = T_{\epsilon \mathcal{A}}$ for $N \times N$ matrices chosen from an ensemble $\mathcal{E}$. For a given algorithm $\mathcal{A}$ and ensemble $\mathcal{E}$ the authors computed $T(H)$ for 5000–15,000 samples of matrices H chosen from $\mathcal{E}$ and recorded the *normalized deflation time*

$$\tilde{T}(H) \equiv \frac{T(H) - <T>}{\sigma} \tag{85}$$

where $<T>$ is the sample average and σ^2 is the sample variance for $T(H)$ for the 5000–15,000 above samples. Surprisingly, the authors found that

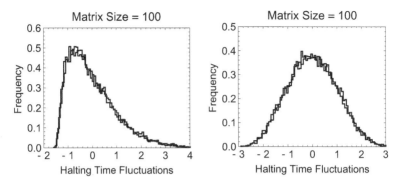

Fig. 3 Universality for $\tilde{T}$ when (**a**) $\mathcal{A}$ is the QR eigenvalue algorithm and when (**b**) $\mathcal{A}$ is the Toda algorithm. Panel (**a**) displays the overlay of two histograms for $\tilde{T}$ in the case of QR, one for each of the two ensembles $\mathcal{E} = $ BE, consisting of iid mean-zero Bernoulli random variables and $\mathcal{E} = $ GOE, consisting of iid mean-zero normal random variables. Here $\epsilon = 10^{-10}$ and $N = 100$. Panel (**b**) displays the overlay of two histograms for $\tilde{T}$ in the case of the Toda algorithm, and again $\mathcal{E} = $ BE or GOE. And here $\epsilon = 10^{-8}$ and $N = 100$

$$\begin{cases} \text{for a given } \epsilon \text{ and } N, \text{ in a suitable scaling regime } (\epsilon \text{ small}, N \text{ large}), \\ \text{the histogram of } \tilde{T} \text{ was } \textit{universal}, \\ \textit{independent of the ensemble } \mathcal{E}. \end{cases} \quad (86)$$

In other words the fluctuations in the deflation time T, suitably scaled, were universal independent of $\mathcal{E}$.

Here are some typical results of their calculations (displayed in a form slightly different from [8]) (Fig. 3).

Subsequently in 2014, Deift et al. [7] raised the question of whether the universality results of [8] were limited to eigenvalue algorithms for real symmetric matrices or whether they were present **more generally** in numerical computation. And indeed, the authors in [7], found similar universality results for a wide variety of numerical algorithms, including

- other eigenvalue algorithms such as QR with shifts, the Jacobi eigenvalue algorithm, and also algorithms applied to complex Hermitian ensembles
- conjugate gradient (CG) and GMRES (Generalized minimal residual algorithm) algorithms to solve linear $N \times N$ systems $Hx = b$ where $b = (b_1, \ldots, b_N)$ is i.i.d., and

$$H = XX^T, \quad X \text{ is } N \times m \text{ and random for CG}$$

and

$$H = I + X, \quad X \text{ is } N \times N \text{ is random for GMRES}$$

- an iterative algorithm to solve the Dirichlet problem $\Delta u = 0$ on a random star shaped region $\Omega \subset \mathbb{R}^2$ with random boundary data f on $\partial \Omega$. (Here the solution is constructed via the double layer potential method.)
- a genetic algorithm to compute the equilibrium measure for orthogonal polynomials on the line
- a decision process investigated by Bakhtin and Correll [3] in experiments using live participants.

All of the above results were numerical/experimental. In order to establish universality in numerical computation as a bona fide phenomenon, and not just an artifact suggested, however strongly, by certain computations as above, it was necessary to prove universality rigorously for an algorithm of interest. In 2016 Deift and Trogdon [11] considered the 1-deflation time $T(1)$ for the Toda algorithm. Thus one runs Toda $t \mapsto X(t), X(0) = H$, until time $t = T^{(1)}$ for which

$$E\left(T^{(1)}\right) = \sum_{j=2}^{N} \left|X_{1j}\left(T^{(1)}\right)\right|^2 < \epsilon^2.$$

Then $X_{11}\left(T^{(1)}\right)$ is an eigenvalue of H to $O(\epsilon)$. As Toda is a sorting algorithm, almost surely

$$\left|X_{11}\left(T^{(1)}\right) - \lambda_{\max}\right| < \epsilon \qquad (87)$$

where $\lambda_{\max}$ is the largest eigenvalue of H. Thus the Toda algorithm with stopping time given by the 1-deflation time is an algorithm to compute the largest eigenvalue of a real symmetric (or Hermitian) matrix.

Here is the result in [11] for $\beta = 1$ (real symmetric case) and $\beta = 2$ (Hermitian case). Order the eigenvalues of a real symmetric or Hermitian matrix by $\lambda_1 \leq \lambda_2 \leq \ldots, \lambda_N$. Then

$$F_\beta^{\text{gap}}(t) \equiv \lim_{N \to \infty} \text{Prob}\left(\frac{1}{C_V \, 2^{-2/3} \, N^{2/3} \, (\lambda_N - \lambda_{N-1})} \leq t\right), \quad t \geq 0 \qquad (88)$$

exists and is universal for a wide range of invariant and Wigner ensembles. $F_\beta^{\text{gap}}(t)$ is clearly the distribution function of the inverse of the top gap $\lambda_N - \lambda_{N-1}$ in the eigenvalues. Here C_V is an ensemble dependent constant.

Theorem 2 (Universality for $T^{(1)}$) *Let $\sigma > 0$ be fixed and let (ϵ, N) be in the scaling region*

$$\mathcal{L} \equiv \frac{\log \epsilon^{-1}}{\log N} \geq \frac{5}{3} + \frac{\sigma}{2}$$

Then if H is distributed according to any real ($\beta = 1$) or complex ($\beta = 2$) invariant or Wigner ensemble, we have

$$\lim_{N \to \infty} \text{Prob}\left(\frac{T^{(1)}}{C_V^{2/3} \, 2^{-2/3} \, N^{2/3} \left(\log \epsilon^{-1} - 2/3 \log N\right)} \leq t\right) = F_\beta^{gap}(t).$$

Thus $T^{(1)}$ behaves statistically like the inverse gap $(\lambda_N - \lambda_{N-1})^{-1}$ of a random matrix.

Now for (ϵ, N) in the scaling region,

$$N^{2/3}\left(\log \epsilon^{-1} - \frac{2}{3} \log N\right) = N^{2/3} \log N \, (\alpha - 2/3),$$

and it follows that $\text{Exp}\left(T^{(1)}\right) \sim N^{2/3} \log N$. This is the first such precise estimate for the stopping time for an eigenvalue algorithm: Mostly estimates are in the form of upper bounds, which are often too big because the bounds must take worst case scenarios into account.

Notes
- The proof of this theorem uses the most recent results on the eigenvalues and eigenvalues of invariant and Wigner ensembles by Yau, Erdös, Schlein, Bourgade ..., and others (see e.g. [16]).
- Similar universality results have now been proved (Deift and Trogdon [12]) for QR acting on positive definite matrices, the power method and the inverse power method.
- The theorem is relevant in that the theorem describes what is happening for "real life" values of ϵ and N. For example, for $\epsilon = 10^{-16}$ and $N \leq 10^9$, we have $\frac{\log \epsilon^{-1}}{\log N} \geq \frac{16}{9} > \frac{5}{3}$.
- Once again RMT provides a stochastic function theory to describe an integrable stochastic process, viz., 1-deflation. But the reverse is also true. Numerical algorithms with random data, raise new problems and challenges within RMT!

Acknowledgements The work of the author was supported in part by NSF Grant DMS–1300965.

References

1. J. Baik, P. Deift and K. Johansson, *On the distribution of the length of the longest increasing subsequence of random permutations*, J. Amer. Math. Soc. **12**, 1999, No. 4, 1119–1178.
2. J. Baik, P. Deift and T. Suidan, *Combinatorics and random matrix theory*, Graduate Studies in Mathematics **172**, American Mathematical Society, Providence, 2016.
3. Y. Bakhtin and J. Correll, *A neural computation model for decision-making times*, J. Math. Psychol. **56**(5), 2012, 333–340.

4. R. Beals and R. Coifman, *Scattering and inverse scattering for first order systems*, Comm. Pure Appl. Math., **37**, 1984, 39–90.
5. P. Deift, L-C. Li, T. Nanda and C. Tomei, *The Toda flow on a generic orbit is integrable*, Comm. Pure Appl. Math. **39**, 1986, 183–232.
6. P. Deift, L-C. Li and C. Tomei, *Matrix factorizations and integrable systems*, Comm. Pure Appl. Math. **42**, 1989, 443–521.
7. P. Deift, G. Menon, S. Olver and T. Trogdon, *Universality in numerical computations with random data*, Proc. Natl. Acad. Sci. USA 111, 2014, no. 42, 14973–14978.
8. P. Deift, G. Menon and C. Pfrang, *How long does it take to compute the eigenvalues of a random symmetric matrix?*, Random matrix theory, interacting particle systems, and integrable systems, 411–442, Math. Sci. Res. Inst. Publ. **65**, Cambridge Univ. Press, New York, 2014.
9. P. Deift, T. Nanda and C. Tomei, *Ordinary differential equations and the symmetric eigenvalue problem*, SIAM J. Numer. Anal. **20**, No. 1, 1983, 1–22.
10. P. Deift and J. Park, *Long-time asymptotics for solutions of the NLS equation with a delta potential and even initial data*, Intl. Math. Res. Not. IMRN 2011, no. 24, 5505–5624.
11. P. Deift and T. Trogdon, *Universality for the Toda algorithm to compute the largest eigenvalue of a random matrix*, Comm. Pure Appl. Math. **71**, 2018, no. 3, 505–536.
12. P. Deift and T. Trogdon, *Universality for eigenvalue algorithms on sample covariance matrices*, SIAM J. Numer. Anal. **55**, No. 6, 2017, 2835–2862.
13. P. Deift and X. Zhou, *A steepest descent method for oscillatory Riemann-Hilbert problems. Asymptotics for the MKdV equation*, Ann. of Math. Second Series **137**, No. 2 1993, 295–368.
14. P. Deift and X. Zhou, *Long-time asymptotics for the autocorrelation function of the transverse Ising chain at the critical magnetic field*, Singular limits of dispersive waves (Lyon, 1991), 183–201, NATO Adv. Sci. Inst. (Series B:Physics), **320**, Plenum, New York, 1994.
15. P. Deift and X. Zhou, *Perturbation theory for infinite-dimensional integrable systems on the line. A case study*, Acta Math. **188**, No. 2, 2002, 163–262.
16. L. Erdős and H. T. Yau, *A dynamical approach to random matrix theory*, Courant Lecture Notes **28**, American Math. Soc., Providence, 2017.
17. H. Flaschka, *The Toda lattice, I*, Phys. Rev. B **9**, 1974, 1924–1925.
18. A. Fokas, A. Its, A. Kitaev and V. Novokshenov, *Painlevé transcendents: The Riemann-Hilbert approach*, Math. Surveys and Monographs, Vol. 128, Amer. Math. Soc. Providence, 2006
19. C. Garner, J. Greene, M. Kruskal and R. Miura, *Method for solving the Korteweg-deVries equation*, Phys. Rev. Lett. **19**, 1967, 1095–1097.
20. P. Gavrylenko and O. Lisovyy, *Fredholm determinant and Nekrasov sum representations of isomonodromic tau functions*, Comm. Math. Phys. **363**, No. 1, 2018, 1–58.
21. P. Gérard and E. Lenzman, *A Lax pair structure for the half-wave maps equation*, Lett. Math. Phys. **108**, No. 7, 2018, 1635–1648.
22. J. Holmer and M. Zworski, *Breathing patterns in nonlinear relaxation*, Nonlinearity, **22**(6), 2009, 1259–1301.
23. A. Its, A. Izergin, V. Korepin and N. Slavnov, *Differential equations for quantum correlation functions*, Int. J. Mod. Physics, B4, 1990, 1003–1037.
24. A. Its and D. Shepelsky, *Initial boundary value problem for the focusing nonlinear Schrödinger equation with Robin boundary condition: half-line approach*, Proc. R. Soc. Lond. Ser A, Math. Phys. Eng. Sci. **469** 2013, 2149, 20120199, 14 pages.
25. M. Jimbo, T. Miwa, Y. Môri and M. Sato, *Density matrix of an impenetrable Bose gas and the fifth Painlevé transcendent*, Physica D, Vol. 1, No. 1, 1980, 80–158.
26. P. Lax, *Integrals of nonlinear equations of evolution and solitary waves*, Comm. Pure Appl. Math., **21**(5), 1968, 467–490.
27. S. Manakov, *Complete integrability and stochastization of discrete dynamical systems*, Sov. Phys. JETP **40**, 1974, 269–274.
28. B. McCoy, J. Perk and R. Schrock, *Time-dependent correlation functions of the transverse Ising chain at the critical magnetic field*, Nucl. Phys. **B220**, 1983, 35–47.

29. J. Moser, *Finitely many mass points on the line under the influence of an exponential potential—An integrable system*, Dynamical Systems Theory and Applications, J. Moser, ed., Springer-Verlag, New York, Berlin, Heidelberg, 1975, 467–497.
30. A.B. Shabat, *One-dimensional perturbations of a differential operator and the inverse scattering problem*, in Problems in Mechanics and Mathematical Physics, Nauka, Moscow, 1976.
31. W. Symes, *The QR algorithm and scattering for the finite nonperiodic Toda lattice*, Physica D, Vol. 4, No. 2, 1982, 275–280.
32. V. Zakharov and L. Faddeev, *Korteweg-de Vries equation: A completely integrable Hamiltonian system*, Funktsional. Anal. i Prilozhen, **5**:4, 1971, 18–27.
33. V. Zakharov and S. Manakov, *On the complete integrability of a nonlinear Schrödinger equation*, Journal of Theoretical and Mathematical Physics, 19(3), 1974, 551–559.
34. V. Zakharov and S. Manakov, *Asymptotic behavior of nonlinear wave systems integrated by the inverse method*, Zh. Eksp. Teor. Fiz., **71**, 1976, 203–215; Soviet Physics JETP **44**, No. 1, 1976, 106–112.
35. V. Zakharov and A. Shabat, *Exact theory of two-dimensional self-focusing and one-dimensional self-modulation of waves in nonlinear media*, Zh. Eksp. Teor. Fiz. **61**, 1971, 118–134; Soviet Physics JETP **34**, 1972, 62–69.

Wave Turbulence and Complete Integrability

Patrick Gérard

1 Introduction

Since the 1980s, the study of Hamiltonian evolution partial differential equations, such as nonlinear Schrödinger equations or nonlinear wave equations, has become a topic of growing importance in mathematical analysis. A fairly general theory of initial value problems is now available, and the main effort of the experts is therefore mainly directed towards the long term description of solutions, starting with the case of small initial data. Though this program is well advanced when the dispersive effects are maximal and generally lead to scattering for small data solutions, it is still widely unexplored when the dispersive effects are weaker, for instance if the physical space is a bounded domain or a closed manifold. In particular, a specific important question in the latter case is the possibility of appearance of small scales or strong oscillations of the solution as time becomes large, a crucial feature of what is called wave turbulence in Physics. Mathematically, this phenomenon may be detected by the long term growth of Sobolev norms of high regularity. Unfortunately, though this phenomenon is expected to appear generically, at this time the corresponding mathematical fact can be established in only very few cases. Natural candidates for studying this phenomenon are of course completely integrable models, since they allow very precise calculations of solutions. However, the most famous completely integrable PDEs, the Korteweg–de Vries equation and the one dimensional cubic nonlinear Schrödinger equation, do not display any growth of Sobolev norms, because the rich collection of their conservation laws is known to control all high Sobolev norms. Another integrable PDE, which has been

---

P. Gérard (✉)
Laboratoire de Mathématiques d'Orsay, Univ. Paris-Sud, CNRS, Université Paris-Saclay, Orsay, France
e-mail: patrick.gerard@math.u-psud.fr

introduced more recently as a normal form of some nonlinear wave equation, turns out to allow dramatic growth of Sobolev norms. This equation is posed on the Hardy space of holomorphic function on the unit disc and involves a very simple nonlinear term which combines a cubic expression with the action of the Szegő projector. The goal of these notes is to provide an elementary introduction to this equation and to its special integrability structure, which involves operators from classical analysis.

The content of these notes is based on a series of four lectures given at the Fields Institute during the Summer School on Nonlinear Dispersive PDEs and Inverse Scattering in August 2017. I wish to thank this institution, the organizers Peter D. Miller, Peter A. Perry, Jean-Claude Saut and Catherine Sulem, my collaborator Sandrine Grellier, with whom this theory has been developed, and my student Joseph Thirouin, who wrote a preliminary version of these notes dedicated to a mini-school for graduate students in the university of North Carolina in February 2016. The remaining typos are of my own.

2 Lecture 1: Growth of Sobolev Norms, Cubic Half-Wave and Szegő Equations

2.1 Set Up

Throughout these lectures, we shall deal with functions on the one dimensional torus

$$\mathbb{T} = \mathbb{R}/2\pi\mathbb{Z}$$

or equivalently 2π-periodic functions on the real line. The space $L^2(\mathbb{T})$ is endowed with the Hilbert inner product

$$(f|g) = \frac{1}{2\pi} \int_0^{2\pi} f(x)\overline{g(x)}\, dx$$

and L^p spaces are endowed with L^p norms corresponding to the same normalized Lebesque measure. The family

$$e_k(x) = e^{ikx}, \ k \in \mathbb{Z},$$

is an orthonormal basis of L^2, and, for $f \in L^2(\mathbb{T})$, the Fourier coefficient of f is defined as

$$\hat{f}(k) = (f|e_k),$$

so that

$$f = \sum_{k \in \mathbb{Z}} \hat{f}(k) e_k, \ (f|g) = \sum_{k \in \mathbb{Z}} \hat{f}(k) \overline{\hat{g}(k)}.$$

The former expansion can be extended to every distribution $f \in \mathcal{D}'(\mathbb{T})$, where the Fourier coefficients $\hat{f}(k)$ are growing at most polynomially as k tends to infinity.

Given a function $a : \mathbb{Z} \to \mathbb{C}$ with polynomial growth, we define the Fourier multiplier

$$a(D) : \mathcal{D}'(\mathbb{T}) \to \mathcal{D}'(\mathbb{T})$$

by

$$\widehat{a(D)f} := a(k)\hat{f}(k), \ k \in \mathbb{Z}.$$

Finally, for $s \in \mathbb{R}$, the Sobolev space $H^s(\mathbb{T})$ is made of distributions $f \in \mathcal{D}'(\mathbb{T})$ such that

$$\|f\|_{H^s}^2 := \sum_{k \in \mathbb{Z}} (1+k^2)^s |\hat{f}(k)|^2 < +\infty.$$

2.2 The Majda–Mac Laughlin–Tabak Model

In order to introduce the question of wave turbulence, let us start with a one-dimensional model introduced in 1997 by Majda et al. [12], which is a fractional nonlinear Schrödinger equation with a defocusing cubic nonlinearity,

$$\begin{cases} i\dfrac{\partial u}{\partial t} = |D|^\alpha u + |u|^2 u \\ u|_{t=0} = u_0 \in H^s(\mathbb{T}), \end{cases} \quad (1.1)$$

Notice that this model admits two fundamental conservation laws, namely the energy,

$$E(u) = (|D|^\alpha u | u) + \frac{1}{2}\|u\|_{L^4}^4$$

and the L^2 norm, which clearly control the $H^{\alpha/2}$ norm,

$$E(u) + \|u\|_{L^2}^2 \geq C\|u\|_{H^{\alpha/2}}^2.$$

As a consequence, it is easy to prove that, if $\alpha > 1$, the above initial value problem is globally well-posed on $H^s(\mathbb{T})$ for every $s \geq \alpha/2$, and that, for every such solution u,

$$\sup_{t \in \mathbb{R}} \|u(t)\|_{H^{\alpha/2}} < \infty.$$

In fact, this result can be extended to $\alpha \leq 1$, see [3] and [10] for $\alpha = 1$, and [16] for $\alpha > \frac{2}{3}$. The question we would like to address—following a similar question asked by Bourgain in [1] about the nonlinear Schrödinger equation on the multidimensional torus—is

Question Given $s > \alpha/2$, is Sobolev norm H^s of $u(t)$ bounded as t tends to infinity?

Suppose for instance that there exists $s > 1$ such that we have

$$\limsup_{t \to \infty} \|u(t, \cdot)\|_{H^s} = +\infty.$$

It means that for some sequence of times $t_n \to \infty$, we have at the same time

$$\sum_{k \in \mathbb{Z}^d} (1 + k^2)^s |\hat{u}(t_n, k)|^2 > n$$

and $\sum_{k \in \mathbb{Z}^d} (1 + k^2)^{\alpha/2} |\hat{u}(t_n, k)|^2 \leq C.$

In other words, at $t = t_n$, $|\hat{u}(t_n, k)|^2$ becomes big for big k's, so that the first series is big whereas the second one remains bounded, which reflects the fact that function $u(t_n, \cdot)$ oscillates. This is why such a phenomenon is called *transition to high frequencies*, or sometimes *wave turbulence*—in fact, wave turbulence is a much richer notion, which we will not pretend to describe here, but it is fair to say that transition to high frequencies is an important manifestation of it. What Physicists expect is that this phenomenon occurs quite frequently, as suggested by numerical simulations of [12] concerning solutions with random data. However, at this time we know very few mathematical facts establishing that such a phenomenon actually occurs.

First of all, let us notice that this phenomenon is typically nonlinear. If one restricts to the linear equation

$$i \frac{\partial u}{\partial t} = |D|^\alpha u$$

elementary Fourier analysis provides the following formula for the Fourier coefficient of the solution at time t,

$$\hat{u}(k, t) = e^{-it|k|^\alpha} \hat{u}_0(k) , k \in \mathbb{Z}.$$

Consequently, all the Sobolev norms are conserved.

Then, in order to study this nonlinear problem, it is natural to first focus on equations for which one can calculate explicitly the solutions, namely integrable equations. In the family of fractional cubic NLS above, the case $\alpha = 2$ has this property, due to the result of Zakharov and Shabat [20]. The equation

$$i \frac{\partial u}{\partial t} + \frac{\partial^2 u}{\partial x^2} = |u|^2 u, \quad x \in \mathbb{T},$$

admits plenty of conservation laws, which are of the form

$$\int_{\mathbb{T}} \left[|u^{(p)}(x)|^2 + F_p(u(x), \bar{u}(x), \ldots, u^{(p-1)}(x), \bar{u}^{(p-1)}(x)) \right] dx, \quad \forall p \in \mathbb{N},$$

where F_p is some polynomial, quadratic in the highest derivative. It is then easy to show that the trajectories of smooth solutions remain bounded in every Sobolev space.

What about the other values of α? This is a widely open problem. However, there seems to be a most favourable value of α for allowing this phenomenon, namely $\alpha = 1$. Indeed, this is the only value for which the equation fails to be dispersive. The next section is devoted to this case.

2.3 The Cubic Half-Wave Equation

In this section, we focus on the half-wave equation,

$$\begin{cases} i\dfrac{\partial u}{\partial t} = |D|u + |u|^2 u \\ u|_{t=0} = u_0 \in H^s(\mathbb{T}), \end{cases} \quad (1.2)$$

Equation (1.2) is globally well-posed on $H^s(\mathbb{T})$ for every $s \geq \frac{1}{2}$. This is of course a consequence of the already quoted conservation laws, which in this case control the $H^{1/2}$ norm. If $s > \frac{1}{2}$, it is then possible to apply a Brezis–Gallouet type estimate to conclude to global existence through a nonlinear Gronwall argument. If $s = \frac{1}{2}$, existence of solutions is obtained by a compactness argument, while uniqueness is a consequence of Trudinger–Sobolev type inequalities.

There is a simple way to reformulate (1.2) as a coupled system of two transport equations by introducing the Szegő projectors $\Pi_\pm$ defined as

$$\Pi_+ \left(\sum_{k \in \mathbb{Z}} c_k e^{ikx} \right) = \sum_{k=0}^{+\infty} c_k e^{ikx}, \quad \Pi_- \left(\sum_{k \in \mathbb{Z}} c_k e^{ikx} \right) = \sum_{k=-\infty}^{-1} c_k e^{ikx}. \quad (1.3)$$

Then, writing $u_\pm := \Pi_\pm u$, (1.2) reads

$$\begin{cases} i(\partial_t + \partial_x) u_+ = \Pi_+(|u|^2 u), \\ i(\partial_t - \partial_x) u_- = \Pi_-(|u|^2 u), \\ u = u_+ + u_- \end{cases}$$

The remarkable fact is that this systems decouples for small data on a nonlinear time interval.

Theorem 1 ([3]) *Assume $s > 1$, $u_0 = \Pi_+(u_0)$ and $\|u_0\|_{H^s} = \varepsilon$ where $\varepsilon > 0$ small enough. Denote by v the solution of the Cauchy problem*

$$i(\partial_t + \partial_x)v = \Pi_+(|v|^2 v), \quad v(0, x) = u_0(x). \tag{1.4}$$

Then, for some positive constant c_s,

$$\|u(t) - v(t)\|_{H^s} = O(\varepsilon^2), \quad |t| \leq \frac{c_s}{\varepsilon^2} \log\left(\frac{1}{\varepsilon}\right).$$

Of course, for $|t| \ll \frac{1}{\varepsilon^2}$, the nonlinear coupling is negligible and $u(t)$ is approximated by $u_0(x - t)$. The important feature in the above theorem is that Eq. (1.4) provides the first nonlinear correction on a bigger time interval. Here we shall not discuss the proof of Theorem 1, which relies on a Birkhoff normal form combined with the special structure of the so-called resonant triples associated to (1.2). See [3].

From Theorem 1, we learn that the first possible mechanism of growth of high Sobolev norms for solutions of (1.2) is to be sought in Eq. (1.4). After an elementary change of variable, we are reduced to the equation

$$i\partial_t w = \Pi(|w|^2 w),$$

where $w(t) \in H^s(\mathbb{T})$ and $\Pi_+ w(t) = w(t)$. This is precisely the cubic Szegő equation, which is the main topic of these lectures.

2.4 The Cubic Szegő Equation: Setting, Wave Turbulence Results and Strategy

2.4.1 Setting

If $E \subset \mathcal{D}'(\mathbb{T})$ is a subspace of distributions on the circle, we shall denote by E_+ the subspace

$$E_+ = \{u \in E : u = \Pi_+ u\} = \{u \in E : \forall k < 0, \hat{u}(k) = 0\}.$$

In the special case $E = L^2$, $L_+^2 = \Pi_+(L^2)$, and Π_+ is the orthogonal projector of L^2 onto L_+^2. Furthermore, elements of L_+^2 are those which can be extended, through the formula

$$\underline{u}(z) = \sum_{k=0}^{\infty} \hat{u}(k) z^k,$$

to holomorphic functions $\underline{u}$ on the disc satisfying

$$\sup_{r<1} \int_0^{2\pi} |\underline{u}(re^{ix})|^2 \, dx < \infty .$$

This space is the so-called Hardy space on the disc, and sometimes denoted by $H^2(\mathbb{D})$, but to avoid any confusion with Sobolev spaces, we keep the notation L_+^2. Similarly, if $s \geq 0$,

$$H_+^s = H^s \cap L_+^2 = \Pi_+(H^s) .$$

From now on, we shall set $\Pi := \Pi_+$.

The cubic Szegő is the following Hamiltonian evolution equation, introduced in [2].

$$\begin{cases} i\dfrac{\partial u}{\partial t} = \Pi(|u|^2 u), \\ u(0) = u_0 \in H_+^s. \end{cases} \qquad (1.5)$$

It admits the following conservation laws.

$$Q(u) = \|u\|_{L^2}^2 , \quad E(u) = \|u\|_{L^4}^4 , \quad M(u) = (Du|u) = \sum_{k=0}^{\infty} k|\hat{u}(k)|^2 .$$

Notice that, because of the spectral localization, the momentum $M(u)$ provides a valuable estimate of $H^{1/2}$ norm,

$$Q(u) + M(u) = \|u\|_{H^{1/2}}^2 .$$

Using this observation, it is easy to check that (1.5) is globally well-posed on H_+^s for every $s \geq \frac{1}{2}$.

2.4.2 Wave Turbulence Results

As a preliminary definition, recall that, if X is a complete metric space, a G_δ-subset of X is a countable intersection G of open subsets of X. If all these open subsets are dense in X, then Baire's theorem claims that G is dense: in this case, G is called a dense G_δ subset of X. Notice that, in a complete metric space, countable intersections of dense G_δ subsets are dense G_δ subsets. Our main result is the following.

Theorem 2 ([6]) *Let $s > \frac{1}{2}$. There exists a dense G_δ subset G_s of H_+^s such that, for every $u_0 \in G_s$, the solution of (1.5) satisfies*

$$\begin{cases} \limsup_{t \to \infty} \|u(t)\|_{H^s} = +\infty . \\ \liminf_{t \to \infty} \|u(t)\|_{H^s} < +\infty . \end{cases}$$

In other words, generic solutions of (1.5) have not only unbounded H^s norms, but display infinitely many transitions between low frequencies and high frequencies, sometimes called forward and backward cascades in the vocabulary of wave turbulence. The above theorem can be strengthened in different directions.

- Endowing the space $C_+^\infty = \cap_s H_+^s$ of its natural Fréchet space structure, it is possible to construct a dense G_δ-subset G of C_+^∞ such that initial data in G generate solutions of (1.5) satisfying the above properties for every $s > \frac{1}{2}$.
- Furthermore, the growth of Sobolev norms can be made superpolynomial,

$$\forall N > 0 \,, \ \limsup_{t \to \infty} \frac{\|u(t)\|_{H^s}}{|t|^N} = +\infty \,.$$

- The relative length of the time intervals where Sobolev norms are large is large enough. For instance, we are going to prove that

$$\limsup_{T \to +\infty} \frac{1}{T} \int_0^T \|u(t)\|_{H^1} \, dt = +\infty \,.$$

Notice that the above long time behaviour is very different from the one of solutions of the ODE

$$i\partial_t u = |u|^2 u \,,$$

from which (1.5) can be obtained by filtering nonnegative Fourier modes. Specifically, the explicit solution of the above ODE is

$$u(t, x) = u_0(x) \, e^{-it|u_0(x)|^2} \,,$$

so that, for generic initial data, for every $s > 0$,

$$\|u(t)\|_{H^s} \simeq |t|^s \,,$$

which is a much milder growth, with only one transition from low to high frequencies. This shows that filtering nonnegative Fourier modes has dramatic consequences on long time dynamics.

2.4.3 Strategy

The proof of Theorem 2 strongly relies on the complete integrability of Eq. (1.5), which will be established in the next lecture through a Lax pair structure. Then we will see how this structure leads to explicit formulas for the solution of (1.5), through an inverse spectral theorem for a special class of operators on the Hilbert space L_+^2. Finally, these explicit formulae will allow us to create growth of H^s norms for $s > \frac{1}{2}$ for carefully chosen data.

3 Lecture 2: The Lax Pair for the Cubic Szegő Equation on the Circle

3.1 Hankel Operators

The principle of *Lax pairs*, introduced by Peter Lax in his seminal paper on the KdV equation [11], is to translate a PDE in terms of the evolution of a family of operators. That is why, before stating the Lax pair theorem for the cubic Szegő equation (1.5), we must define a class of operators, called the *Hankel operators*.

3.1.1 Hankel Operators on Sequences

We denote by $\ell^2(\mathbb{N})$ the set of square-summable sequences of complex numbers indexed by $\mathbb{N} := \{0, 1, 2, \ldots\}$. On this space, there is a scalar product given by

$$(x|y) = \sum_{n=0}^{\infty} x_n \overline{y_n}.$$

Definition 2.1.1 Let $c = \{c_n\} \in \ell^2(\mathbb{N})$. The Hankel operator of symbol c is

$$\Gamma_c : \ell^2(\mathbb{N}) \longrightarrow \ell^2(\mathbb{N}), \ x = \{x_n\} \longmapsto y = \{y_n\},$$

where $\{y_n\}$ is defined by an "anti-convolution" product:

$$y_n = \sum_{p=0}^{\infty} c_{n+p} x_p, \quad \forall n \in \mathbb{N}.$$

As Γ_c is an operator given by a kernel $k(n, p) := c_{n+p}$, it is easy to compute its Hilbert-Schmidt norm: it is simply the L^2 norm of the kernel in the product space. More precisely,

$$\|\Gamma_c\|_{HS}^2 = \|k(n, p)\|_{\ell^2(\mathbb{N}\times\mathbb{N})}^2 = \sum_{n,p\geq 0} |c_{n+p}|^2 = \sum_{l=0}^{\infty}(1+l)|c_l|^2.$$

In the sequel, we make the assumption that $\|\Gamma_c\|_{HS} < \infty$.

Definition 2.1.2 The *shift operator* on $\ell^2(\mathbb{N})$ is the following (not onto) isometry

$$S : (x_0, x_1, x_2, \ldots) \longmapsto (0, x_0, x_1, x_2, \ldots).$$

The *anti-shift* is its adjoint with respect to the scalar product $(\cdot|\cdot)$:

$$S^* : (x_0, x_1, x_2, \dots) \longmapsto (x_1, x_2, x_3, \dots).$$

The following identities are direct consequences of the definitions:

$$S^*S = I,$$
$$SS^* = I - (\cdot|e_0)e_0, \quad \text{where } e_0 = (1, 0, 0, \dots),$$
$$S^*\Gamma_c = \Gamma_c S = \Gamma_{S^*c}.$$

The last identity even characterizes the Hankel operators among continuous operators on $\ell^2(\mathbb{N})$. We call $S^*\Gamma_c$ the *shifted Hankel operator* associated to Γ_c. We immediately infer a link between a Hankel operator and its associated shifted operator.

Lemma 2.1.3 $\Gamma_{S^*c}\Gamma_{S^*c}^* = \Gamma_c\Gamma_c^* - (\cdot|c)c.$

Proof The left hand side equals $\Gamma_c SS^*\Gamma_c^*$, and we apply the above formula on SS^*.
□

3.1.2 Hankel Operators on the Hardy Space

Using the isometric isomorphism

$$L^2_+ \xrightarrow{\sim} \ell^2(\mathbb{N})$$
$$u \longmapsto \{\hat{u}(n)\}_{n \geq 0},$$

we are going to define corresponding Hankel operators on L^2_+. For $u \in H^{1/2}_+$, we define $H_u : L^2_+ \to L^2_+, h \mapsto \Pi(u\bar{h})$. Observe that

$$\widehat{H_u(h)} = \Gamma_{\hat{u}}(\hat{h}),$$

and since $\Gamma_c^* = \Gamma_{\bar{c}}$,

$$\widehat{H_u^2(h)} = \Gamma_{\hat{u}}\Gamma_{\hat{u}}^*(\hat{h}).$$

Notice that H_u is $\mathbb{C}$-antilinear, and satisfies

$$(H_u(h_1)|h_2) = (u|h_1 h_2) = (H_u(h_2)|h_1),$$

whereas H_u^2 is $\mathbb{C}$-linear and self-adjoint positive. We similarly identify the conjugate of S, $u \mapsto e^{ix}u$, and the conjugate of its adjoint S^*, $u \mapsto \Pi(e^{-ix}u)$ (also denoted by S and S^* respectively). In this framework, the shifted Hankel operator is

$$K_u := S^* H_u = H_u S = H_{S^*u}.$$

Lemma 2.1.3 can be restated as

$$K_u^2 = H_u^2 - (\cdot|u)u. \tag{2.1}$$

We also need another class of operators, called *Toeplitz operators*.

Definition 2.1.4 Let $b \in L^\infty(\mathbb{T})$. For $h \in L_+^2$, we define $T_b(h) := \Pi(bh)$.

The $\mathbb{C}$-linear operator $T_b : L_+^2 \to L_+^2$ is called the Toeplitz operator of symbol b. It is bounded, and $T_b^* = T_{\bar{b}}$.

3.2 The Lax Pair Structure

The following theorem is the backbone of our analysis.

Theorem 3 ([2]) *If u is a H^s solution to (1.5) with $s > \frac{1}{2}$, then*

$$\boxed{\frac{d}{dt}H_u = [B_u, H_u]} \tag{2.2}$$

where

$$B_u := -iT_{|u|^2} + \frac{i}{2}H_u^2.$$

Notice that B_u is anti-self-adjoint (i.e. $B_u^* = -B_u$).

Proof We start with a crucial algebraic lemma.

Lemma 2.2.1 *Let $a, b, c \in H_+^s$, $s > \frac{1}{2}$. We have*

$$H_{\Pi(a\bar{b}c)} = T_{a\bar{b}}H_c + H_a T_{\bar{b}c} - H_a H_b H_c. \tag{2.3}$$

Indeed, for $h \in L_+^2$,

$$H_{\Pi(a\bar{b}c)}(h) = \Pi(\overline{\Pi(a\bar{b}c)}h) = \Pi(\overline{a\bar{b}c}h - \overline{(I-\Pi)(a\bar{b}c)}h) = \Pi(\overline{a\bar{b}c}h),$$

because $\overline{(I-\Pi)(a\bar{b}c)}h$ has only negative frequencies. Hence

$$H_{\Pi(a\bar{b}c)}(h) = \Pi(\bar{a}b\bar{c}h) = \Pi(\bar{a}b\Pi(\bar{c}h)) + \Pi(\bar{a}b(I-\Pi)(\bar{c}h)) = T_{\bar{a}b}H_c(h) + H_a(f),$$

where $f := b\overline{(I - \Pi)(c\bar{h})}$. Since $(I - \Pi)(c\bar{h})$ has only negative frequencies, f has only nonnegative frequencies, hence

$$f = \Pi(f) = \Pi(b\bar{c}h) - \Pi(b\overline{\Pi(c\bar{h})}) = T_{b\bar{c}}(h) - H_b H_c(h),$$

and the proof of the lemma is complete.

Assuming $i\dot{u} = \Pi(|u|^2 u)$, and applying Lemma 2.3 with $a = b = c = u$, we get

$$\frac{d}{dt} H_u = H_{\dot{u}} = -i H_{\Pi(|u|^2 u)}$$

$$= -i(T_{|u|^2} H_u + H_u T_{|u|^2} - H_u^3)$$

$$= -i T_{|u|^2} H_u - H_u(-i T_{|u|^2}) + \tfrac{i}{2} H_u^2 H_u - H_u(\tfrac{i}{2} H_u^2)$$

$$= [-i T_{|u|^2}, H_u] + [\tfrac{i}{2} H_u^2, H_u],$$

where $[A, B]$ stands for the commutator $AB - BA$. Notice that we used the antilinearity of the operator H_u. Finally, setting $B_u := -i T_{|u|^2} + \tfrac{i}{2} H_u^2$, we end up with (2.2). $\square$

A remarkable fact in this theory is that the latter Lax pair generates another one, which turns out to give independent conservation laws.

Theorem 4 ([5])

$$\boxed{\frac{d}{dt} K_u = [C_u, K_u]} \tag{2.4}$$

where $C_u := -i T_{|u|^2} + \tfrac{i}{2} K_u^2$ is also anti-self-adjoint.

Proof

$$\frac{d}{dt} K_u = -i K_{\Pi(|u|^2 u)}$$

$$= -i H_{\Pi(|u|^2 u)} S$$

$$= -i(T_{|u|^2} H_u S + H_u T_{|u|^2} S - H_u^3 S) .$$

Moreover, notice that

$$T_b(Sh) = ST_b(h) + (bSh|1) .$$

In the case $b = |u|^2$, this gives

$$T_{|u|^2} Sh = ST_{|u|^2} h + (|u|^2 Sh|1) .$$

Moreover,
$$(|u|^2 Sh | 1) = (u | u\overline{Sh}) = (u | K_u(h)) .$$

Consequently,
$$H_u T_{|u|^2} Sh = K_u T_{|u|^2} h + (K_u(h)|u)u .$$

We obtain
$$\frac{d}{dt} K_u = -i T_{|u|^2} K_u - i K_u T_{|u|^2} + i(H_u^2 - (\cdot|u)u) K_u .$$

Using identity (2.1) and antilinearity, this leads to (2.4). □

Observe that B_u, C_u are linear and antiselfadjoint. Following a classical argument due to Lax [11], we obtain the following consequence.

Corollary 2.2.2 *Under the conditions of Theorem 2.2, define* $U = U(t)$, $V = V(t)$ *to be the solutions of the following linear ODEs on* $\mathcal{L}(L_+^2)$,

$$\frac{dU}{dt} = B_u U , \quad \frac{dV}{dt} = C_u V , \quad U(0) = V(0) = I .$$

Then $U(t), V(t)$ *are unitary operators and*

$$H_{u(t)} = U(t) H_{u(0)} U(t)^* , \quad K_{u(t)} = V(t) K_{u(0)} V(t)^* .$$

Proof Just compute the time derivatives of

$$U(t)^* U(t), U(t) U(t)^*, V(t)^* V(t), V(t) V(t)^*, U(t)^* H_{u(t)} U(t), V(t)^* K_{u(t)} V(t) .$$

□

Since we also saw that H_u^2 and K_u^2 are positive trace class operators on L_+^2, we know that they have pure point spectrum, consisting of a sequence of nonnegative eigenvalues tending to 0—in fact with a convergent sum. The above corollary implies that these eigenvalues are conservation laws of the cubic Szegő evolution. We shall return to the study of isospectral sets of data in the next chapter, proving in particular that these two sequences can be chosen almost arbitrarily, and independently from each other. For the rest of this lecture, we show how this structure allows to derive an explicit formula for the solution of the Cauchy problem for (1.5).

3.3 The General Explicit Formula

Theorem 5 [5]) Let $u_0 \in H_+^{\frac{1}{2}}(\mathbb{T})$, and $u \in C(\mathbb{R}, H_+^{\frac{1}{2}}(\mathbb{T}))$ be the solution of Eq. (1.5) such that $u(0) = u_0$. Then, for $|z| < 1$,

$$\underline{u}(t,z) = \int_0^{2\pi} q_{t,z}(y) \frac{dy}{2\pi},$$

$$(I - z e^{-itH_{u_0}^2} e^{itK_{u_0}^2} S^*) q_{t,z} = e^{-itH_{u_0}^2}(u_0).$$

Notice that, since operator $e^{-itH_{u_0}^2} e^{itK_{u_0}^2} S^*$ has norm at most 1, function $q_{t,z}$ is well defined for every z such that $|z| < 1$.

Proof Our starting point is the following identity, valid for every $v \in L_+^2$,

$$\underline{v}(z) = ((I - zS^*)^{-1} v | 1), \quad z \in D. \tag{2.5}$$

Indeed, the Taylor coefficient of order k of the right hand side at $z = 0$ is

$$((S^*)^k v | 1) = (v | S^k 1) = \hat{v}(k),$$

which coincides with the Taylor coefficient of order k of the left hand side. Let $u \in C^\infty(\mathbb{R}, H_+^s)$ be a solution of (1.5), $s > \frac{1}{2}$. Applying (2.5) to $v = u(t)$ and using the unitarity of $U(t)$, we get

$$\underline{u}(t, z) = ((I - zS^*)^{-1} u(t) | 1) = (U(t)^* (I - zS^*)^{-1} u(t) | U(t)^* 1),$$

which yields

$$\underline{u}(t, z) = ((I - zU(t)^* S^* U(t))^{-1} U(t)^* u(t) | U(t)^* 1). \tag{2.6}$$

We shall identify successively $U(t)^* 1$, $U(t)^* u(t)$, and the restriction of $U(t)^* S^* U(t)$ on the range of H_{u_0}. We begin with $U(t)^* 1$,

$$\frac{d}{dt} U(t)^* 1 = -U(t)^* B_u(1),$$

and

$$B_u(1) = \frac{i}{2} H_u^2(1) - i T_{|u|^2}(1) = -\frac{i}{2} H_u^2(1).$$

Hence

$$\frac{d}{dt} U(t)^* 1 = \frac{i}{2} U(t)^* H_u^2(1) = \frac{i}{2} H_{u_0}^2 U(t)^* 1,$$

where we have used Corollary 2.2.2. This yields

$$U(t)^*1 = e^{i\frac{t}{2}H_{u_0}^2}(1) . \qquad (2.7)$$

Consequently,

$$U(t)^*u(t) = U(t)^*H_{u(t)}(1) = H_{u_0}U(t)^*(1) = H_{u_0}e^{i\frac{t}{2}H_{u_0}^2}(1) ,$$

and therefore

$$U(t)^*u(t) = e^{-i\frac{t}{2}H_{u_0}^2}(u_0) . \qquad (2.8)$$

Finally,

$$U(t)^*S^*U(t)H_{u_0} = U(t)^*S^*H_{u(t)}U(t) = U(t)^*K_{u(t)}U(t) ,$$

and therefore

$$U(t)^*S^*U(t)H_{u_0} = U(t)^*V(t)K_{u_0}V(t)^*U(t) . \qquad (2.9)$$

On the other hand,

$$\frac{d}{dt}U(t)^*V(t) = -U(t)^*B_{u(t)}V(t) + U(t)^*C_{u(t)}V(t) = U(t)^*(C_{u(t)}-B_{u(t)})V(t)$$

$$= \frac{i}{2}U(t)^*(K_{u(t)}^2 - H_{u(t)}^2)V(t) = \frac{i}{2}(U(t)^*V(t)K_{u_0}^2 - H_{u_0}^2U(t)^*V(t)) .$$

We infer

$$U(t)^*V(t) = e^{-i\frac{t}{2}H_{u_0}^2}e^{i\frac{t}{2}K_{u_0}^2} .$$

Plugging this identity into (2.9), we obtain

$$U(t)^*S^*U(t)H_{u_0} = e^{-i\frac{t}{2}H_{u_0}^2}e^{i\frac{t}{2}K_{u_0}^2}K_{u_0}e^{-i\frac{t}{2}K_{u_0}^2}e^{i\frac{t}{2}H_{u_0}^2}$$

$$= e^{-i\frac{t}{2}H_{u_0}^2}e^{itK_{u_0}^2}K_{u_0}e^{i\frac{t}{2}H_{u_0}^2}$$

$$= e^{-i\frac{t}{2}H_{u_0}^2}e^{itK_{u_0}^2}S^*H_{u_0}e^{i\frac{t}{2}H_{u_0}^2}$$

$$= e^{-i\frac{t}{2}H_{u_0}^2}e^{itK_{u_0}^2}S^*e^{-i\frac{t}{2}H_{u_0}^2}H_{u_0} .$$

We conclude that, on the range of H_{u_0},

$$U(t)^*S^*U(t) = e^{-i\frac{t}{2}H_{u_0}^2}e^{itK_{u_0}^2}S^*e^{-i\frac{t}{2}H_{u_0}^2} . \qquad (2.10)$$

It remains to plug identities (2.7), (2.8), (2.10) into (2.6). We finally obtain

$$\underline{u}(t,z) = ((I - ze^{-i\frac{t}{2}H_{u_0}^2}e^{itK_{u_0}^2}S^*e^{-i\frac{t}{2}H_{u_0}^2})^{-1}e^{-i\frac{t}{2}H_{u_0}^2}(u_0)|e^{i\frac{t}{2}H_{u_0}^2}(1))$$

$$= ((I - ze^{-itH_{u_0}^2}e^{itK_{u_0}^2}S^*)^{-1}e^{-itH_{u_0}^2}(u_0)|1),$$

which is the claimed formula in the case of data $u_0 \in H_+^s, s > \frac{1}{2}$. The case $u_0 \in H_+^{\frac{1}{2}}$ follows by a simple approximation argument. Indeed, we know from the wellposedness theory that, for every $t \in \mathbb{R}$, the mapping $u_0 \mapsto u(t)$ is continuous on $H_+^{\frac{1}{2}}$. On the other hand, the maps $u_0 \mapsto H_{u_0}, K_{u_0}$ are continuous from $H_+^{\frac{1}{2}}$ into $\mathcal{L}(L_+^2)$. Since $H_{u_0}^2, K_{u_0}^2$ are selfadjoint, the operator

$$e^{-itH_{u_0}^2}e^{itK_{u_0}^2}S^*$$

has norm at most 1. Hence, for $z \in D$, the right hand side of the formula is continuous from $H_+^{\frac{1}{2}}$ into $\mathbb{C}$. □

The remarkable feature of Theorem 5 is of course to reduce the nonlinear equation (1.5) to linear equations, namely the construction of the unitary groups $e^{itH_{u_0}^2}, e^{itK_{u_0}^2}$, and inversion of the linear equation giving $q_{t,z}$. By expanding in powers of z, notice that an equivalent formulation is that the Fourier coefficient of the solution at time t is given by

$$\hat{u}(t,k) = ((e^{-itH_{u_0}^2}e^{itK_{u_0}^2}S^*)^k e^{-itH_{u_0}^2}u_0 \mid 1).$$

However, despite the explicit feature of this formula, tracking the growth of the Sobolev norm

$$\|u(t)\|_{H^s} = \left(\sum_{k=0}^{\infty}(1+k^2)^s|\hat{u}(k,t)|^2\right)^{\frac{1}{2}}$$

does not seem to be a simple task. In the next two chapters, we shall see how this task can be addressed through the study of the spectral mapping associated to the pair (H_u, K_u).

4 Lecture 3: The Inverse Spectral Theorem

In the previous lecture, we have seen that the eigenvalues of H_u^2 and K_u^2 are conservation laws of the cubic Szegő evolution, and that they play a crucial role in calculating the solution of the equation from its initial datum, through the general explicit formula. Notice that the square roots of these quantities are precisely the singular values of the Hankel matrices $\Gamma_{\hat{u}}$ and $\Gamma_{S^*\hat{u}}$. In this lecture, we investigate

Wave Turbulence and Complete Integrability

more precisely the possible values that can take these eigenvalues, and we define a new system of coordinates including these eigenvalues, on which the evolution can be trivially described.

We begin with establishing a simple lemma about the eigenvalues of H_u^2 and K_u^2.

4.1 The Interlacement Property

Lemma 3.1.1 *For $u \in H_+^{1/2}$, let $(s_j^2)_{j \geq 1}$ and $(s_k'^2)_{k \geq 1}$ be the decreasing sequence formed by the eigenvalues of H_u^2 and K_u^2 respectively, written with multiplicities. Then we have*

$$s_1 \geq s_1' \geq s_2 \geq s_2' \geq \cdots \geq 0.$$

Proof The proof relies on the formula (2.1) and on a use of the min-max formula for compact self-adjoint operators: if A is a compact positive self-adjoint operator on some Hilbert space $\mathcal{H}$, and if $(\rho_j)_{j \in \mathbb{N}^*}$ denotes the decreasing sequence of its eigenvalues, then

$$\rho_j = \min_{\substack{F \subset \mathcal{H} \\ \dim F \leq j-1}} \max_{\substack{h \in F^\perp \\ \|h\|=1}} \|A(h)\|,$$

where F denotes a vector subspace of $\mathcal{H}$.

We are going to apply this principle to H_u^2 and K_u^2 on $\mathcal{H} = L_+^2$. First of all, notice that

$$\max_{\substack{h \in F^\perp \\ \|h\|=1}} \|K_u^2(h)\| = \max_{\substack{h \in F^\perp \\ \|h\|=1}} (K_u^2(h)|h).$$

We compute $(K_u^2(h)|h) = (H_u^2(h)|h) - |(u|h)|^2 \leq (H_u^2(h)|h)$. Taking the maximum and then the minimum over subspaces F on both sides finally yields $s_j'^2 \leq s_j^2$.

Next, if F is a subspace of L_+^2 of dimension at most $j - 1$, we have

$$\max_{\substack{h \in F^\perp \\ \|h\|=1}} \|K_u^2(h)\| \geq \max_{\substack{h \in (F+\mathbb{C}u)^\perp \\ \|h\|=1}} \|K_u^2(h)\| = \max_{\substack{h \in (F+\mathbb{C}u)^\perp \\ \|h\|=1}} \|H_u^2(h)\|,$$

because of formula (2.1.3). Now, $F + \mathbb{C}u$ is of dimension at most j. Hence, taking the minimum over F on both sides yields

$$s_j'^2 \geq \min_{\substack{F \subset L_+^2 \\ \dim F \leq j-1}} \max_{\substack{h \in (F+\mathbb{C}u)^\perp \\ \|h\|=1}} \|H_u^2(h)\| \geq \min_{\substack{\tilde{F} \subset L_+^2 \\ \dim \tilde{F} \leq j}} \max_{\substack{h \in \tilde{F}^\perp \\ \|h\|=1}} \|H_u^2(h)\| \geq s_{j+1}^2.$$

□

4.2 The Inverse Spectral Transform for Generic Finite Rank Hankel Operators

Now that we know, because of Corollary 2.2.2, that eigenvalues of H_u^2 and K_u^2 are conserved, there is a natural question to ask: given a set of positive interlaced real numbers, is it possible to find $u \in H_+^{1/2}$ such that the corresponding eigenvalues of H_u^2 and K_u^2 are precisely these numbers? In this section, we provide a positive answer in the finite rank case, when eigenvalues are distinct from one another.

4.2.1 The Kronecker Theorem

First of all, let us characterise the symbols of Hankel operators with finite rank.

Theorem 6 (Kronecker, 1877, See e.g. [13]) *The Hankel operator H_u has finite rank if and only if $\underline{u}(z)$ is a rational function of z with no pole in the closed unit disk.*

Rather than giving a complete proof of this fact, let us focus on the following facts.

- The rank 1 case. In that case, the proof is particularly easy, since $S^* H_u = H_u S$, we obtain $S^* u = \lambda u$, which, on the Fourier coefficients, means $\hat{u}(n+1) = \lambda \hat{u}(n)$, or

$$\hat{u}(n) = a\lambda^n .$$

Consequently, $|\lambda| < 1$ and

$$\underline{u}(z) = \frac{a}{1 - \lambda z} .$$

In this special case, the general explicit formula of Theorem 5—or a direct calculation—shows that u generates a traveling wave solution,

$$u(t, x) = e^{-i\omega t} u_0(x - ct) , \quad \omega = \frac{|a|^2}{(1 - |\lambda|^2)^2} , \quad c = \frac{|a|^2}{1 - |\lambda|^2} ,$$

which is in fact the ground state of the Gagliardo–Nirenberg type inequality

$$E(u) \leq Q(u)^2 + 2Q(u)M(u) .$$

- The general case is an adaptation of the above proof, which takes advantage of the fact that the sequence of Fourier coefficients satisfies a linear recurrent equation. Therefore a solution u such that H_u has finite rank can be interpreted as an exact multi-soliton.

- It is in fact possible to characterise completely the elements u such that

$$rk(H_u) + rk(K_u) = N.$$

This set is a Kähler complex submanifold $\mathcal{V}(N)$ of dimension N, on which the cubic Szegő evolution defines a Liouville integrable Hamiltonian system. Furthermore, one can prove [5, 6], that the corresponding trajectories on these submanifolds are quasiperiodic.
- In any $\mathcal{V}(N)$, one can prove [2] that the subset of functions u such that the non zero eigenvalues of H_u^2 and K_u^2 are all simple and distinct is a dense open subset, which we denote by $\mathcal{V}(N)_{\text{gen}}$. We call these elements generic in $\mathcal{V}(N)$.

4.2.2 The Inverse Spectral Theorem

Let us first define the direct spectral map. Let u be a generic element of, say, $\mathcal{V}(2N)$, which means that rk $H_u = N$ and rk $K_u = N$. We write We denote by $s := s_1 > \cdots > s_N > 0$ the singular values of $\Gamma_{\hat{u}}$, and by $s' := s'_1 > \cdots > s'_N > 0$ the ones of $\Gamma_{S^*\hat{u}}$.

Since $K_u = H_u S$, we have Ran $K_u \subseteq$ Ran H_u, and in fact, both spaces coincide, for their dimension is N. We call this space W. Setting

$$E_u(s_j) := \ker(H_u^2 - s_j^2 I),$$

$$F_u(s'_k) := \ker(K_u^2 - s_k'^2 I),$$

we infer from the hypothesis that for all $1 \le j, k \le N$, $\dim E_u(s_j) = \dim F_u(s'_k) = 1$, and besides,

$$W = \bigoplus_{j=1}^{N} E_u(s_j) = \bigoplus_{k=1}^{N} F_u(s'_k), \tag{3.1}$$

where the direct sums are orthonormal sums.

Denote by u_j (resp. u'_k) the orthogonal projection of u on $E_u(s_j)$ (resp. $F_u(s'_k)$). Observe that we must have $u_j \ne 0$ for all $j \in \{1, \ldots, N\}$. Indeed, in view of (2.1.3), if u was orthogonal to some $E_u(s_j)$, we would have

$$K_u^2(h) = H_u^2(h) - (h|u)u = H_u^2(h) = s_j^2 h,$$

for all $h \in E_u(s_j)$. But this cannot happen, since s_j^2 is not an eigenvalue of K_u^2. For the same reason, $u'_k \ne 0$.

Consider now the action of H_u on the subspaces $E_u(s_j)$, and note that $H_u(u_j) \in E_u(s_j)$. Since it is a one-dimensional space, we know that there exists $\lambda_j \in \mathbb{C}$ such that $H_u(u_j) = \lambda_j u_j$. Applying H_u to the equality, we get $H_u^2(u_j) = |\lambda_j|^2 u_j =$

$s_j^2 u_j$, so that we can write $\lambda_j = s_j e^{i\psi_j}$ for some $\psi_j \in \mathbb{T}$. The same holds for $K_u(u'_k)$, hence for all $1 \le j, k \le N$, we have found angles $\psi_j, \psi'_k \in \mathbb{T}$, such that

$$\begin{cases} H_u(u_j) = s_j e^{i\psi_j} u_j, \\ K_u(u'_k) = s'_k e^{i\psi'_k} u'_k. \end{cases}$$

Finally, we can define a map

$$\Phi_{2N} : u \in \mathcal{V}(N)_{\text{gen}} \mapsto (s_1, s'_1, \ldots, s_N, s'_N; \psi_1, \psi'_1, \ldots, \psi_N, \psi'_N)$$

valued into $\Omega_{2N} \times \mathbb{T}^{2N}$, where Ω_p denotes the subset

$$\{\sigma_1 > \cdots > \sigma_p > 0\} \subset \mathbb{R}^p.$$

Theorem 7 (The Finite Rank Case, No Multiplicity) *The map Φ_{2N} is bijective, and its inverse is given as follows. Given $(s_1, s'_1, \ldots, s_N, s'_N, \psi_1, \psi'_N, \ldots, \psi_N, \psi'_N) \in \Omega_{2N} \times \mathbb{T}^{2N}$, consider the $N \times N$ matrix $\mathscr{C}(z)$ given by*

$$\mathscr{C}(z)_{j,k} = \frac{s_j e^{i\psi_j} - s'_k e^{i\psi'_k} z}{s_j^2 - s'^2_k}$$

for $1 \le j, k \le N$. Then $\mathscr{C}(z)$ is invertible for any z in $\overline{\mathbb{D}}$ (the closed unit disc of $\mathbb{C}$), and

$$u(z) = \langle \mathscr{C}(z)^{-1}(\mathbf{1}_N), \mathbf{1}_N \rangle_{\mathbb{C}^N}, \quad \forall |z| \le 1,$$

where $\mathbf{1}_N$ is the vector of $\mathbb{C}^N$ each component of which is 1, and where $\langle \cdot, \cdot \rangle$ denotes the standard scalar product on $\mathbb{C}^N$.

Remark 1 The mapping Φ_{2N} turns out to be even a diffeomorphism, transforming the symplectic form on L_+^2 into

$$\sigma = \frac{1}{2} \left(\sum_{j=1}^N d(s_j^2) \wedge d\psi_j - \sum_{k=1}^N d((s'_k)^2) \wedge d\psi'_k \right).$$

In other words, Theorem 7 gives an explicit set of coordinates for the $2N$-torus of functions in $H_+^{1/2}$ whose associated Hankel and shifted Hankel operators have prescribed singular values, with an additional non-degeneracy condition contained in the strict inequalities.

Proof of the First Part of Theorem 7 We show how to recover u from its spectral data. Let us compute, for $1 \le j, k \le N$:

$$s_j^2(u_j|u_k') = (H_u^2(u_j)|u_k') = (u_j|H_u^2(u_k')) = \left(u_j|K_u^2(u_k') + (u_k'|u)u\right)$$
$$= s_k'^2(u_j|u_k') + \|u_k'\|^2 \|u_j\|^2.$$

Consequently,
$$(u_j|u_k') = \frac{\|u_k'\|^2 \|u_j\|^2}{s_j^2 - s_k'^2}.$$

Finally using (3.1), we get
$$\begin{cases} u_j = \|u_j\|^2 \sum_{k=1}^{N} \frac{u_k'}{s_j^2 - s_k'^2}, \\ u_k' = \|u_k'\|^2 \sum_{j=1}^{N} \frac{u_j}{s_j^2 - s_k'^2}. \end{cases}$$

Fix $j \in \{1, \ldots, N\}$. Noting that $SK_u = SS^* H_u = H_u - (u|\cdot)\mathbf{1}$, we write
$$s_j e^{i\psi_j} u_j(z) = H_u(u_j)(z) = SK_u(u_j)(z) + (u|u_j) = zK_u(u_j)(z) + \|u_j\|^2$$
$$= z\|u_j\|^2 \sum_{k=1}^{N} \frac{K_u(u_k')}{s_j^2 - s_k'^2} + \|u_j\|^2 = z\|u_j\|^2 \sum_{k=1}^{N} \frac{s_k' e^{i\psi_k'} u_k'}{s_j^2 - s_k'^2} + \|u_j\|^2,$$

On the other hand, we know that
$$s_j e^{i\psi_j} u_j = \|u_j\|^2 \sum_{k=1}^{N} \frac{u_k'}{s_j^2 - s_k'^2},$$

so we simplify by $\|u_j\|^2$ and eventually,
$$\sum_{k=1}^{N} \frac{s_j e^{i\psi_j} - s_k' e^{i\psi_k'} z}{s_j^2 - s_k'^2} u_k'(z) = 1,$$

which holds in fact for any $j \in \{1, \ldots, N\}$. This implies[1] that $\forall z \in \mathbb{C}$ such that $|z| \le 1$,

---

[1] For the sake of brevity, we assume here without further explanation that $\mathscr{C}(z)$ is invertible. See Sect. 4.3.1 for a proof of this non-trivial fact.

$$\begin{pmatrix} u'_1(z) \\ \vdots \\ u'_N(z) \end{pmatrix} = \mathscr{C}(z)^{-1} \begin{pmatrix} 1 \\ \vdots \\ 1 \end{pmatrix} = \mathscr{C}(z)^{-1}(\mathbf{1}_N).$$

To conclude, it remains to observe that $u = H_u(1)$, so $u \in W$. Thus $u(z) = \sum_{k=1}^{N} u'_k(z)$, and the formula

$$u(z) = \langle \mathscr{C}(z)^{-1}(\mathbf{1}_N) | \mathbf{1}_N \rangle$$

is proved. □

The proof of the surjectivity is more delicate, see Sect. 4.3.2 below.

Remark 2 Since $H_u^2 - K_u^2$ is a rank-one operator, and because the inclusion Ran $K_u \subseteq$ Ran H_u always holds true, we see that rk $K_u \in \{$rk $H_u,$ rk $H_u - 1\}$ whenever H_u has finite rank. The above results only deal with the case of equality, i.e. Ran $K_u =$ Ran H_u, but in fact, when Ran $K_u =$ Ran $H_u - 1$, namely $u \in \mathcal{V}(2N-1)$, the inverse spectral formula remains valid, simply setting $s'_N = 0$ in the matrix $\mathscr{C}(z)$.

4.3 The Evolution in New Coordinates

In the new coordinates we have just defined, it turns out that solving the Szegő equation becomes trivial. In the literature, those (s, s', ψ, ψ')-coordinates are related to the so-called *action-angle variables*: the actions s, s' are constant in time, and the angles ψ, ψ' evolve linearly with frequencies functions of s, s'. Here these functions are particularly simple, because the Hamiltonian function can be expressed in a simple way,

$$\|u\|_{L^4}^4 = \mathrm{Tr}(H_u^4) - \mathrm{Tr}(K_u^4) = \sum_j s_j^4 - \sum_k (s'_k)^4 .$$

Proposition 3.3.1 *Suppose $u_0 \in H_+^{1/2}$ satisfies the hypothesis of Theorem 7 and corresponds to a set $(s_0, s'_0, \psi_0, \psi'_0) \in \mathbb{R}^{2N} \times \mathbb{T}^{2N}$. If u_0 is assumed to be the initial data for the evolution problem (1.5), then for all $t \in \mathbb{R}$, $u = u(t)$ corresponds to a set $(s(t), s'(t), \psi(t), \psi'(t))$ satisfying*

$$\frac{ds_j}{dt} = 0, \quad \frac{ds'_k}{dt} = 0,$$

$$\frac{d\psi_j}{dt} = s_j^2, \quad \frac{d\psi'_k}{dt} = s_k'^2.$$

Proof We already know that the s_j's and s'_k's are conserved. Let us turn to ψ_j, for some $j \in \{1,\ldots,N\}$. Using the notations of the previous proof, we have for all $t \in \mathbb{R}$

$$H_{u(t)}(u_j(t)) = s_j e^{i\psi_j(t)} u_j(t). \tag{3.2}$$

But recall the Lax pair (2.2): it even implies that for any Borel function $f : \mathbb{R}_+ \to \mathbb{R}$,

$$\frac{d}{dt} f(H_u^2) = [B_u, f(H_u^2)] = [-iT_{|u|^2}, f(H_u^2)],$$

because $B_u = -iT_{|u|^2} + \frac{i}{2}H_u^2$, and H_u^2 commutes to $f(H_u^2)$. We are going to apply this identity to u with $f = \mathbf{1}_{\{s_j\}}$. This will give us the evolution of u_j, since $u_j = \mathbf{1}_{\{s_j\}}(H_u^2)u$. We thus have

$$\frac{d}{dt} u_j = \frac{d}{dt} f(H_u^2)u = [-iT_{|u|^2}, f(H_u^2)]u + f(H_u^2)(-i\Pi(|u|^2 u)) = -iT_{|u|^2} u_j.$$

We are ready to differentiate (3.2). The left hand side gives

$$\frac{d}{dt}\text{l.h.s.} = [B_u, H_u](u_j) + H_u(-iT_{|u|^2} u_j)$$
$$= -iT_{|u|^2} H_u(u_j) + \frac{i}{2}H_u^3(u_j) - H_u(\frac{i}{2}H_u^2(u_j))$$
$$= -iT_{|u|^2} H_u(u_j) + i s_j^2 H_u(u_j),$$

whereas for the right hand side:

$$\frac{d}{dt}\text{r.h.s.} = i\dot{\psi}_j s_j e^{i\psi_j} u_j + s_j e^{i\psi_j}(-iT_{|u|^2} u_j) = i\dot{\psi}_j H_u(u_j) - iT_{|u|^2} H_u(u_j).$$

Since $H_u(u_j) \neq 0$, we have $\dot{\psi}_j = s_j^2$.
Starting from (2.4) similarly leads to the law of the evolution of the ψ'_k's. □

4.3.1 Complement 1: $\mathscr{C}(z)$ Is Invertible

We prove the following proposition, which was included in the statement of Theorem 7:

Proposition 3.3.2 *Under the hypothesis of Theorem 7, the matrix $\mathscr{C}(z)$ is invertible for any $z \in \overline{\mathbb{D}}$.*

Proof Set $Q(z) := \det \mathscr{C}(z)$. The polynomial Q is of degree N exactly. Indeed, its dominant coefficient is

$$(-1)^N s'_1 \ldots s'_N \, e^{i(\psi'_1 + \cdots + \psi'_N)} \det\left(\frac{1}{s_j^2 - (s'_k)^2}\right) \neq 0,$$

because of the formula for the Cauchy determinant:

$$\det\left(\frac{1}{a_j + b_k}\right) = \frac{\prod_{i<j}(a_j - a_i)\prod_{k<\ell}(b_\ell - b_k)}{\prod_{j,k}(a_j + b_k)}. \tag{3.3}$$

Therefore Q has N zeroes in $\mathbb{C}$, counted with multiplicities. Assume one of these zeroes, say z_0, belongs to $\overline{\mathbb{D}}$. Since, by the Cramer formulae for $\mathscr{C}(z)^{-1}$ whenever $Q(z) \neq 0$,

$$u(z) = \langle \mathscr{C}(z)^{-1}(\mathbf{1}_N) | \mathbf{1}_N \rangle_{\mathbb{C}^N} = \frac{P(z)}{Q(z)},$$

where P is a polynomial of degree at most $N - 1$, and since $u \in L^2_+$, it is necessary that $P(z_0) = 0$. Therefore, after a finite number simplifications,

$$u(z) = \frac{\tilde{P}(z)}{\tilde{Q}(z)},$$

where $\tilde{Q}$ is a polynomial of degree at most $N - 1$ with no zeroes in the closed unit disc, and $\tilde{P}$ is a polynomial of degree at most $N - 2$. We can therefore decompose

$$u(z) = \sum_j \frac{\alpha_j}{(1 - p_j z)^{m_j}},$$

with $0 < |p_j| < 1$ and $\sum_j m_j \leq N - 1$. Now we use the following elementary lemma.

Lemma 3.3.3 *If*

$$v(z) = \frac{1}{(1 - pz)^m}$$

with $m \geq 1$ and $0 < |p| < 1$, then

$$\text{Ran}(H_u) = \text{span}\left\{\frac{1}{(1 - pz)^k}, k = 1, \ldots, m\right\}.$$

Let us prove the lemma. Given $h \in L^2_+$, we can expand

$$z^{m-1}h(z) = \sum_{k=0}^{m-1} c_k (z - \overline{p})^k + (z - \overline{p})^m g(z),$$

with $g \in L^2_+$ and $|z| < 1$. Taking the L^2 trace on $\mathbb{S}^1$, conjugating and multiplying by v, we infer

$$\frac{\overline{h\left(e^{ix}\right)}}{(1-pe^{ix})^m} = \sum_{k=0}^{m-1} \overline{c}_k \frac{e^{i(m-1-k)x}}{(1-pe^{ix})^{m-k}} + e^{-ix}\overline{g\left(e^{ix}\right)}.$$

Consequently,

$$\Pi(v\overline{h}) = \sum_{k=0}^{m-1} \overline{c}_k \frac{z^{m-k-1}}{(1-pz)^{m-k}},$$

and it is clear that one can fit the value of each coefficient c_k by an appropriate choice of h.

Using the lemma, we conclude that

$$\dim \mathrm{Ran}\, H_u = \sum_j m_j \leq N - 1,$$

which contradicts the assumption that H_u has rank N. □

4.3.2 Complement 2: Surjectivity of the Spectral Transform

We conclude this section by giving a proof of the fact that the mapping $u \mapsto (s, s', \psi, \psi')$ is onto (under the hypothesis of Theorem 7). The proof of the surjectivity of the mapping

$$u \mapsto (s, s', \psi, \psi')$$

in the most general case is given in [4, 6], by showing that this map is open and closed for appropriate topologies. Here we give a different, purely algebraic proof, based on a recent work in collaboration with A. Pushnitski.

First of all, the above calculations imply that, if u exists, its orthogonal projections u_j and u'_k onto the eigenspaces of H_u^2 and K_u^2 satisfy

$$\sum_{j=1}^{N} \frac{\|u_j\|^2}{s_j^2 - (s'_k)^2} = 1, \quad \sum_{k=1}^{N} \frac{\|u'_k\|^2}{s_j^2 - (s'_k)^2} = 1$$

In view of the above formula (3.3) for the Cauchy determinants, this imply that $\|u_j\|^2$ and $\|u'_k\|^2$ can be expressed in terms of the numbers s_i, s'_ℓ as

$$\|u_j\|^2 = (s_j^2 - (s'_j)^2) \prod_{i \neq j} \frac{s_j^2 - (s'_i)^2}{s_j^2 - s_i^2}, \quad \|u'_k\|^2 = (s_k^2 - (s'_k)^2) \prod_{\ell \neq k} \frac{s_\ell^2 - (s'_k)^2}{(s'_\ell)^2 - (s'_k)^2}.$$

Now, we fix once and for all $s_1 > s'_1 > s_2 > s'_2 > \cdots > s_N > s'_N > 0$, and we set similarly

$$\tau_j^2 := (s_j^2 - (s'_j)^2) \prod_{i \neq j} \frac{s_j^2 - (s'_i)^2}{s_j^2 - s_i^2}, \quad \kappa_k^2 := (s_k^2 - (s'_k)^2) \prod_{\ell \neq k} \frac{s_\ell^2 - (s'_k)^2}{(s'_\ell)^2 - (s'_k)^2},$$

so that we also have

$$\sum_{j=1}^{N} \frac{\tau_j^2}{s_j^2 - (s'_k)^2} = 1, \quad \sum_{k=1}^{N} \frac{\kappa_k^2}{s_j^2 - (s'_k)^2} = 1 \tag{3.4}$$

and moreover

$$\sum_{j=1}^{N} \frac{\tau_j^2 \kappa_k^2}{(s_j^2 - (s'_k)^2)(s_j^2 - (s'_\ell)^2)} = \delta_{k\ell}, \quad 1 \leq k, \ell \leq N. \tag{3.5}$$

Then we consider the following two N-dimensional Hermitian spaces. We denote by E and by E' the space $\mathbb{C}^N$ equipped respectively with the inner products

$$(\tilde{z}|z)_E := \sum_{j=1}^{N} \tau_j^2 \tilde{z}_j \bar{z}_j, \quad (\tilde{z}|z)_{E'} := \sum_{k=1}^{N} \kappa_k^2 \tilde{z}_k \bar{z}_k.$$

Consider the linear operator

$$\Omega : E' \to E, \quad (\Omega z)_j := \sum_{k=1}^{N} \frac{z_k \kappa_k^2}{s_j^2 - (s'_k)^2}.$$

We claim that Ω is a unitary operator. Indeed, for $w \in E$, $z \in E'$,

$$(\Omega z | w)_E = \sum_{j,k=1}^{N} \frac{z_k \bar{w}_j \tau_j^2 \kappa_k^2}{s_j^2 - (s'_k)^2} = (z | \Omega^* w)_{E'}$$

with

$$(\Omega^* w)_k := \sum_{j=1}^{N} \frac{w_j \tau_j^2}{s_j^2 - (s'_k)^2},$$

and (3.5) precisely means that $\Omega^* \Omega = I$. Also notice that

$$\Omega(\mathbf{1}_N) = \mathbf{1}_N. \tag{3.6}$$

Indeed, this is equivalent to

$$\sum_{k=1}^{N} \frac{\kappa_k^2}{s_j^2 - (s_k')^2} = 1, \ j = 1, \ldots, N,$$

the second identity in (3.4).

Then, given $(\psi_1, \psi_1', \psi_2, \psi_2', \ldots, \psi_N, \psi_N') \in \mathbb{T}^{2N}$, we consider the antilinear operators

$$H : E \to E, \ K' : E' \to E',$$

defined by

$$(Hz)_j = s_j e^{i\psi_j} \bar{z}_j, \ (K'z)_k = s_k' e^{i\psi_k'} \bar{z}_k.$$

Notice that H, K' satisfy $(H\tilde{z}|z)_E = (Hz|\tilde{z})_E$, and $(K'\tilde{z}|z)_{E'} = (K'z|\tilde{z})_{E'}$. In particular they are $\mathbb{R}$-linear operators which are symmetric with respect to the real scalar products defined by

$$(\tilde{z}, z)_E = \mathrm{Re}(\tilde{z}|z)_E, \ (\tilde{z}, z)_{E'} = \mathrm{Re}(\tilde{z}|z)_{E'}.$$

Moreover, $H^2, (K')^2$ are positive selfadjoint $\mathbb{C}$-linear operators on E, E' respectively. We set

$$K := \Omega K' \Omega^* : E \to E.$$

The following lemma establishes a crucial identity between the operators K^2 and H^2 on E.

Lemma 3.3.4 $K^2 = H^2 - (\cdot | \mathbf{1}_N)_E \mathbf{1}_N$.

Proof We compute

$$(\Omega(K')^2 z)_j = \sum_{k=1}^{N} \frac{\kappa_k^2 (s_k')^2 z_k}{s_j^2 - (s_k')^2}$$

$$= \sum_{k=1}^{N} \frac{\kappa_k^2 s_j^2 z_k}{s_j^2 - (s_k')^2} - \sum_{k=1}^{N} \kappa_k^2 z_k$$

$$= (H^2 \Omega z)_j - (z|\mathbf{1}_N)_{E'} \mathbf{1}_N.$$

Because of formula (3.6), this can be written as

$$\Omega(K')^2 = (H^2 - (\cdot | \mathbf{1}_N)_E \mathbf{1}_N) \Omega,$$

or

$$K^2 = H^2 - (\,.\,|\mathbf{1}_N)_E \mathbf{1}_N,$$

by setting $K := \Omega K' \Omega^* : E \to E$ as above. □

We now set

$$\Sigma := KH^{-1} : E \to E.$$

Since

$$\|Kz\|_E^2 = (K^2 z|z)_E = (H^2 z|z)_E - |(z|\mathbf{1}_N)_E|^2 \leq (H^2 z|z)_E = \|Hz\|_E^2,$$

we have

$$\forall z \in E, \ \|\Sigma z\|_E \leq \|z\|_E.$$

In fact, this contraction map Σ enjoys the following asymptotic stability property.

Lemma 3.3.5

$$\forall z \in E, \ \|\Sigma^n z\|_E \xrightarrow[n \to +\infty]{} 0.$$

Proof Since E is finite dimensional and since Σ is a contraction, the theory of the Jordan decomposition of matrices shows that it is enough to prove that Σ has no eigenvalue on the unit circle. Let $\omega \in \mathbb{S}^1$, and consider $F := \ker(\Sigma - \omega I)$. We claim that

$$F = \ker(\Sigma^* - \bar{\omega} I).$$

Indeed, since Σ is a contraction, the Hermitian form $B(z, \tilde{z}) := ((I - \Sigma^* \Sigma) z | \tilde{z})$ is non negative, hence

$$\|\Sigma z\|_E = \|z\|_E \iff \Sigma^* \Sigma z = z,$$

as a consequence of the Cauchy–Schwarz inequality for B. This implies $F \subset \ker(\Sigma^* - \bar{\omega} I)$. The reverse inclusion follows from a similar argument applied to the contraction Σ^*.

In order to study the space F, we recall that by definition $K = \Sigma H$. Hence, by symmetry of H and K for the real scalar product on E,

$$K = H \Sigma^*.$$

Therefore Lemma 3.3.4 can be reformulated as

$$\Sigma H^2 \Sigma^* = H^2 - (\,.\,|\mathbf{1})_E \mathbf{1}. \tag{3.7}$$

Given $z \in F$, we infer

$$\|Hz\|_E^2 = \|Hz\|_E^2 - |(z|\mathbf{1})_E|^2,$$

hence F is orthogonal to $\mathbf{1}_N$. Coming back to (3.7), we conclude that, if $z \in F$,

$$\bar{\omega}\Sigma H^2 z = H^2 z,$$

hence $H^2 z \in F$. Therefore F is a stable subspace for H^2, which is the diagonal matrix of s_j^2, $j = 1, \ldots, N$. Hence F has to be the direct sum of one dimensional eigenspaces of H^2. On the other hand, none of these lines is orthogonal to $\mathbf{1}_N$. The only possibility is therefore $F = \{0\}$. □

At this stage we are in position to construct $u \in H_+^{\frac{1}{2}}$ and a linear isometry $U : E \to L_+^2$ such that

$$UHU^* = H_u, \quad U\Omega K'\Omega^* U^* = K_u. \tag{3.8}$$

Notice that property (3.8) implies that (s, s', ψ, ψ') corresponds to u via Theorem 7.

First we give another reformulation of Lemma 3.3.4. Define $q \in E$ by

$$q_j = \frac{e^{i\psi_j}}{s_j}, \quad j = 1, \ldots, N,$$

so that $Hq = \mathbf{1}_N$. Plugging $K = H\Sigma^* = \Sigma H$ in $K^2 = H^2 - (\cdot|\mathbf{1}_N)_E \mathbf{1}_N$, we obtain

$$H\Sigma^*\Sigma H = H^2 - (\cdot|Hq)Hq,$$

hence, recalling that H is invertible,

$$\Sigma^*\Sigma = I - (\cdot|q)q. \tag{3.9}$$

Identity (3.9) combined with Lemma 3.3.5 has an important consequence. Indeed, iterating (3.9) yields, for every $z \in E$,

$$(\Sigma^*)^n \Sigma^n z = z - \sum_{k=0}^{n-1}(z|(\Sigma^*)^k q)(\Sigma^*)^k q.$$

Taking the scalar product with z in E, we obtain

$$\|\Sigma^n z\|_E^2 = \|z\|_E^2 - \sum_{k=0}^{n-1}|(z|(\Sigma^*)^k q)_E|^2.$$

Passing to the limit as $n \to \infty$, and using Lemma 3.3.5, we finally conclude

$$\forall z \in E, \ \|z\|_E^2 = \sum_{k=0}^{\infty} |(z|(\Sigma^*)^k q)_E|^2. \tag{3.10}$$

It allows us to define $U : E \to L_+^2$ by

$$Uz = \sum_{k=0}^{\infty} (z|(\Sigma^*)^k q)_E \, e^{ikx}.$$

In view of (3.10), the operator U is an isometry, and, for every $h \in L_+^2$,

$$U^* h = \sum_{k=0}^{\infty} \hat{h}(k)(\Sigma^*)^k q \ .$$

Consider

$$u := U(\mathbf{1}_N) = \sum_{k=0}^{\infty} (\mathbf{1}_N | (\Sigma^*)^k q)_E \, e^{ikx}.$$

Since Σ^* has eigenvalues only in the open unit disc—see the proof of Lemma 3.3.4, the Jordan decomposition implies that

$$|(z|(\Sigma^*)^k q)_E| \le C\beta^k$$

for some $\beta < 1$. Hence $u \in H_+^{1/2}$, so[2] we may study H_u and $K_u = S^* H_u = H_u S$. Given $z \in E$, consider

$$\widehat{UHz}(k) = (Hz|(\Sigma^*)^k q)_E = (\Sigma^k Hz|q)_E$$
$$= (H(\Sigma^*)^k z|q)_E = (Hq|(\Sigma^*)^k z)_E$$
$$= (\mathbf{1}_N|(\Sigma^*)^k z)_E.$$

Consequently, for every $h \in L_+^2$,

$$\widehat{UHU^*}h(k) = \left(\mathbf{1}_N \Big| \sum_{\ell=0}^{\infty} \hat{h}(\ell)(\Sigma^*)^{k+\ell} q\right)_E = \sum_{\ell=0}^{\infty} \hat{u}(k+\ell)\overline{\hat{h}(\ell)} \ .$$

We conclude

$$UHU^* = H_u \ .$$

---

[2] In fact u is even an analytic function on $\mathbb{T}$.

Furthermore, since

$$\widehat{U\Sigma z}(k) = \left(\Sigma z | (\Sigma^*)^k q\right)_E = (z|(\Sigma^*)^{k+1} q)_E = \widehat{S^* U z}(k),$$

we also have

$$UKU^* = U\Sigma H U^* = S^* U H U^* = S^* H_u = K_u,$$

or equivalently

$$U\Omega K'\Omega^* U^* = K_u.$$

4.4 Various Extensions

The above inverse spectral theorem admits several extensions. First of all, it is possible to extend to generic elements of $H_+^{1/2}$, defined as having simple singular values for both Hankel matrices. This set turns to be a dense G_δ subset of $H_+^{1/2}$, and the spectral map Φ is then a homeomorphism on $\Omega \times \mathbb{T}^\infty$, where Ω is the subset of ℓ^2 formed by strictly decreasing sequences of positive numbers. Furthermore, the inverse formula extends, by making the size of the matrix $\mathscr{C}(z)$ tend to infinity.

A more delicate extension takes into account the multiplicity of the singular values. A complete description can be found in the monograph [6]. As an application, it is proved that every trajectory in $H_+^{1/2}$ is almost periodic.

5 Lecture 4: Long Time Transition to High Frequencies

In our search of wave turbulence for the Szegő equation, the formula given by Theorem 7 does not help directly. On the contrary, it shows, together with Proposition 3.3.1, that initial data belonging to the finite-dimensional manifold $\mathcal{V}(N)$ give rise to a motion which is quasi-periodic in time: there exists $D \geq 1$, a smooth function $F : \mathbb{T}^D \to \mathcal{C}_+^\infty(\mathbb{T})$,[3] and a $\omega = (\omega_1, \ldots, \omega_D) \in \mathbb{R}^D$ such that

$$u(t) = F(\omega_1 t, \omega_2 t, \ldots, \omega_D t), \quad \forall t \in \mathbb{R}.$$

Such an orbit $t \mapsto u(t)$ remains bounded in every H^s.

However, this formula will allow us to infer the existence of transition to high frequencies for most solutions of the cubic Szegő equation, as we will now explain.

---

[3]Where $\mathcal{C}_+^\infty(\mathbb{T}) := \mathcal{C}^\infty(\mathbb{T}) \cap L_+^2$.

5.1 A Crucial Example: The Daisy Effect

Given $\varepsilon \in \mathbb{R}_+$, we define

$$u_0^\varepsilon(x) = e^{ix} + \varepsilon.$$

It is easy to check that $u_0^\varepsilon \in \mathcal{V}(3)$, hence the corresponding solution u^ε of (1.5) is valued in $\mathcal{V}(3)$, and consequently reads

$$u^\varepsilon(t,x) = \frac{a^\varepsilon(t)e^{ix} + b^\varepsilon(t)}{1 - p^\varepsilon(t)e^{ix}},$$

with $a^\varepsilon(t) \in \mathbb{C}^*, b^\varepsilon(t) \in \mathbb{C}, p^\varepsilon(t) \in D, a^\varepsilon(t) + b^\varepsilon(t)p^\varepsilon(t) \neq 0$. We are going to calculate these functions explicitly. We start with the special case $\varepsilon = 0$. In this case, $|u_0^0| = 1$, hence

$$u^0(t,x) = e^{-it}u_0^0(x)$$

so

$$a^0(t) = e^{-it}, \quad b^0(t) = 0, \quad p^0(t) = 0.$$

We come to $\varepsilon > 0$. The operators $H_{u_0}^2, K_{u_0}^2, S^*$ act on the range of $H_{u_0^\varepsilon}$, which is the two dimensional vector space spanned by $1, e^{ix}$. In this basis, the matrices of these three operators are respectively

$$\mathcal{M}(H_{u_0}^2) = \begin{pmatrix} 1+\varepsilon^2 & \varepsilon \\ \varepsilon & 1 \end{pmatrix}, \quad \mathcal{M}(K_{u_0}^2) = \begin{pmatrix} 1 & 0 \\ 0 & 0 \end{pmatrix}, \quad \mathcal{M}(S^*) = \begin{pmatrix} 0 & 1 \\ 0 & 0 \end{pmatrix}.$$

The eigenvalues of $H_{u_0}^2$ are

$$\rho_\pm^2 = 1 + \frac{\varepsilon^2}{2} \pm \varepsilon\sqrt{1 + \frac{\varepsilon^2}{4}},$$

hence the matrix of the exponential is given by

$$\mathcal{M}\left(e^{-itH_{u_0}^2}\right) = \frac{e^{-it\rho_+^2} - e^{-it\rho_-^2}}{\rho_+^2 - \rho_-^2}\mathcal{M}(H_{u_0}^2) + \frac{\rho_-^2 e^{-it\rho_+^2} - \rho_+^2 e^{-it\rho_-^2}}{\rho_-^2 - \rho_+^2}I$$

$$= \frac{e^{-i\Omega t}}{2\omega}\left(-2i\sin(\omega t)\mathcal{M}(H_{u_0}^2) + (2\omega\cos(\omega t) + 2i\Omega\sin(\omega t))I\right)$$

where $\omega := \varepsilon\sqrt{1 + \frac{\varepsilon^2}{4}}$, $\Omega := 1 + \frac{\varepsilon^2}{2}$.

We obtain

$$e^{-itH_{u_0}^2}(u_0) = \frac{e^{-i\Omega t}}{2\omega}\Big(-2i\varepsilon\Omega\sin(\omega t) + 2\varepsilon\omega\cos(\omega t) + (2\omega\cos(\omega t)$$
$$-i\varepsilon^2\sin(\omega t))e^{ix}\Big),$$

$$\mathcal{M}\left(e^{-itH_{u_0}^2}e^{itK_{u_0}^2}S^*\right) = \frac{e^{-it\frac{\varepsilon^2}{2}}}{2\omega}\begin{pmatrix} 0 & 2\omega\cos(\omega t) - i\varepsilon^2\sin(\omega t) \\ 0 & -2i\varepsilon\sin(\omega t) \end{pmatrix},$$

and finally

$$a^\varepsilon(t) = e^{-it(1+\varepsilon^2)},\ b^\varepsilon(t) = e^{-it(1+\varepsilon^2/2)}\left(\varepsilon\cos(\omega t) - i\frac{2+\varepsilon^2}{\sqrt{4+\varepsilon^2}}\sin(\omega t)\right)$$

$$p^\varepsilon(t) = -\frac{2i}{\sqrt{4+\varepsilon^2}}\sin(\omega t)\,e^{-it\varepsilon^2/2},\ \omega := \frac{\varepsilon}{2}\sqrt{4+\varepsilon^2}\,.$$

The important feature of such dynamics concerns the regime $\varepsilon \to 0$. Though $p^0(t) \equiv 0$, $p^\varepsilon(t)$ may visit small neighborhoods of the unit circle at large times (Fig. 1). Specifically, at time $t^\varepsilon = \pi/(2\omega) \sim \pi/(2\varepsilon)$, we have $1 - |p^\varepsilon(t)|^2 \sim \varepsilon^2/4$. A consequence is that the momentum density,

$$\mu_n(t^\varepsilon) := n|\hat{u}^\varepsilon(t^\varepsilon, n)|^2 = n|a^\varepsilon(t^\varepsilon) + b^\varepsilon(t^\varepsilon)p^\varepsilon(t^\varepsilon)|^2|p^\varepsilon(t^\varepsilon)|^{2(n-1)}$$
$$= n\frac{\varepsilon^4}{(4+\varepsilon^2)^2}\left(1 - \frac{\varepsilon^2}{4+\varepsilon^2}\right)^{n-1},$$

Fig. 1 The trajectory of p^ε for small ε

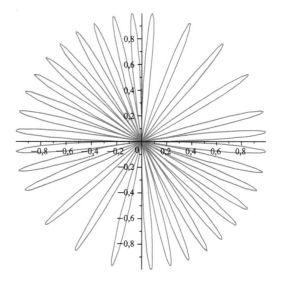

which satisfies

$$\sum_{n=1}^{\infty} \mu_n(t^\varepsilon) = Tr(K^2_{u^\varepsilon(t^\varepsilon)}) = Tr(K^2_{u^\varepsilon_0}) = 1 ,$$

becomes concentrated at high frequencies

$$n \simeq \frac{1}{\varepsilon^2}.$$

This induces the following instability of H^s norms

$$\|u^\varepsilon(t^\varepsilon)\|_{H^s} \simeq \frac{1}{(1 - |p^\varepsilon(t^\varepsilon)|^2)^{s-\frac{1}{2}}} \simeq \frac{1}{\varepsilon^{2s-1}} , \ s > \frac{1}{2}.$$

This proves in particular that conservation laws do not control H^s regularity for $s > \frac{1}{2}$. Notice that the family (u_0^ε) approaches u_0^0, which is a non generic element of $\mathcal{V}(3)$, since $H^2_{u_0}$ admits 1 as a double eigenvalue. Furthermore, on can also show that averages of the H^1 norm are growing,

$$\lim_{T \to +\infty} \frac{1}{T} \int_0^T \|u^\varepsilon(t)\|_{H^1} \, dt = \frac{1}{2\pi} \int_0^{2\pi} \frac{d\theta}{(\varepsilon^2 + \cos^2 \theta)^{\frac{1}{2}}} \simeq \log\left(\frac{1}{\varepsilon}\right).$$

Now let us try to explain this mechanism through the inverse spectral transform we have studied in the previous lecture. Fix a small $\varepsilon > 0$, and consider the following set of parameters,

$$s_1 = 1 + \varepsilon, \ s_1' = 1, \ s_2 = 1 - \varepsilon, \ s_2' = 0,$$
$$\psi_1 = 0, \ \psi_1' = 0, \ \psi_2 = \pi.$$

Through the formula of Lecture 3, this corresponds to a matrix

$$\mathscr{C}_\varepsilon(z) = \begin{pmatrix} \frac{1+\varepsilon-z}{(1+\varepsilon)^2 - 1} & \frac{1}{1+\varepsilon} \\ \frac{-(1-\varepsilon)-z}{(1-\varepsilon)^2 - 1} & -\frac{1}{1-\varepsilon} \end{pmatrix},$$

and by Theorem 7, the function

$$u_\varepsilon(z) := \langle \mathscr{C}_\varepsilon(z)^{-1}(1)|1\rangle_{\mathbb{C}^2} = \frac{2z(1-\varepsilon^2) + 3\varepsilon}{2 - \varepsilon z}$$

is such that rk $H_{u_\varepsilon} = 2$, rk $K_{u_\varepsilon} = 1$. Besides, $H_{u_\varepsilon}^2$ has s_1^2, s_2^2 as positive eigenvalues, and $K_{u_\varepsilon}^2$ has $s_1'^2$ only. We notice that the only pole of $u_\varepsilon(z)$ is $z = \frac{2}{\varepsilon}$, far away from the unit circle S^1.

Now modify the parameters in the following way:

$$s_1 = 1 + \varepsilon, \quad s_1' = 1, \quad s_2 = 1 - \varepsilon, \quad s_2' = 0,$$
$$\psi_1 = 0, \quad \psi_1' = 0, \quad \psi_2 = 0.$$

Only ψ_2 has changed, but now the matrix

$$\tilde{\mathscr{C}}_\varepsilon(z) = \begin{pmatrix} \frac{1+\varepsilon-z}{(1+\varepsilon)^2-1} & \frac{1}{1+\varepsilon} \\ \frac{(1-\varepsilon)-z}{(1-\varepsilon)^2-1} & \frac{1}{1-\varepsilon} \end{pmatrix},$$

and the corresponding function is

$$v_\varepsilon(z) := \langle \tilde{\mathscr{C}}_\varepsilon(z)^{-1}(\mathbf{1})|\mathbf{1}\rangle_{\mathbb{C}^2} = \frac{2 + \varepsilon^2 - 2z(1-\varepsilon^2)}{2 - (2-\varepsilon^2)z},$$

with a pole for $z = \frac{2}{2-\varepsilon^2}$, this time dramatically close to S^1. A simple computation shows that because of this pole,

$$\|v_\varepsilon\|_{H_+^s} \simeq \left(\frac{1}{\varepsilon}\right)^{2s-1} \gg 1.$$

In other words, if the singular values are close to each other, the Sobolev norms of u are very sensitive to the angles.

So if one considers the solution $t \mapsto u(t)$ of the cubic Szegő equation (1.5) with u_ε as an initial data, then by Proposition 3.3.1, we have $\psi_1(t) = t(1+\varepsilon)^2$ and $\psi_2(t) = \pi + t(1-\varepsilon)^2$. There is a time t_ε such that $\psi_1(t_\varepsilon) = \psi_2(t_\varepsilon)$:

$$t_\varepsilon = \frac{\pi}{4\varepsilon}.$$

Then the previous computation shows that $\|u(t_\varepsilon)\|_{H_+^s} \simeq (\frac{1}{\varepsilon})^{2s-1} \simeq (t_\varepsilon)^{2s-1}$ whenever $s > \frac{1}{2}$. Thus we found a solution with growing Sobolev norms, but the growth is polynomial in time, and the trajectories are bounded (even by great bounds) in any Sobolev space.

In the next section, we briefly explain how we can deduce for those growing but bounded solutions a weak turbulence theorem. In other words, we explain how to pass from families of solutions to the behaviour of one generic solution.

5.2 The General Instability Principle

The main step to transition to high frequencies is the following result.

Theorem 8 *For any $u_0 \in H_+^1$, and $M \geq 1$, there exists a sequence $\{u_0^n\}$ converging to u_0 in H_+^1, and sequences of times $T_n \to \infty$ and $\underline{t}_n \to \infty$ such that u^n, the solution to the cubic Szegő equation with $u^n(0) = u_0^n$, satisfies*

$$\frac{1}{T_n} \int_0^{T_n} \|u^n(t)\|_{H^1} \, dt \longrightarrow +\infty,$$

and

$$u^n(\underline{t}_n) \longrightarrow u_0 \quad \text{in } H_+^1,$$

as $n \to +\infty$.

Let us give a detailed proof of Theorem 8. The main idea is "to stick a small daisy to the dynamics". First of all, notice that it is enough to approximate initial data u_0 belonging to a dense subset of H_+^1, so we only consider u_0 such that for some $q \in \mathbb{N}$, $\operatorname{rk} H_{u_0} = \operatorname{rk} K_{u_0} = q$, with *simple* singular values which we denote by

$$s_1 > s_1' > s_2 > s_2' > \cdots > s_q > s_q'.$$

The idea is to approximate u_0 by adding new singular values, in the spirit of the examples of the previous section. Assume

$$u_0 := \Phi_{2q}^{-1}(s_1, s_1', \ldots, s_q, s_q'; \psi_1, \psi_1', \ldots, \psi_q, \psi_q').$$

We introduce the following class of rational functions.

We consider the family $u^{\delta,\varepsilon}$ for $\delta, \varepsilon \to 0$ with

$$u^{\delta,\varepsilon} = \Phi_{2q+3}^{-1}\left(s_1, s_1', \ldots, s_q, s_q', \delta(1+\varepsilon), \delta, \delta(1-\varepsilon); \psi_1, \psi_1', \ldots, \psi_q, \psi_q', 0, 0, \pi\right).$$

From the previous lecture, we can compute the corresponding solution to the cubic Szegő equation,

$$u^{\delta,\varepsilon}(z,t) = \left\langle \mathscr{C}_{\delta,\varepsilon}^{-1}(z,t) \begin{pmatrix} 1_q \\ 1_2 \end{pmatrix}, \begin{pmatrix} 1_q \\ 1_2 \end{pmatrix} \right\rangle \tag{4.1}$$

where

$$1_q = \begin{pmatrix} 1 \\ \vdots \\ 1 \end{pmatrix} \in \mathbb{R}^q.$$

Here, we let

$$T = \delta^2 t \tag{4.2}$$

and

$$\mathscr{C}_{\delta,\varepsilon}(z,t) = \begin{pmatrix} \mathscr{E}(z,t) & \mathscr{A}_{\delta,\varepsilon}(z,t) \\ \mathscr{B}_{\delta,\varepsilon}(z,t) & \frac{1}{\delta}\mathscr{C}_\varepsilon(z,T) \end{pmatrix}$$

with

$$\mathscr{E}(z,t) = \left(\frac{s_a e^{i(\psi_a + t s_a^2)} - s'_b z e^{i(\psi'_b + t(s'_b)^2)}}{s_a^2 - (s'_b)^2} \right)_{1 \leq a,b \leq q},$$

$$\mathscr{A}_{\delta,\varepsilon}(z,t) = \left(\frac{s_a e^{i(\psi_a + t s_a^2)} - \delta z e^{iT}}{s_a^2 - \delta^2}, \frac{e^{i(\psi_a + t s_a^2)}}{s_a} \right)_{1 \leq a \leq q},$$

$$\mathscr{B}_{\delta,\varepsilon}(z,t) = \begin{pmatrix} \frac{\delta(1+\varepsilon)e^{iT(1+\varepsilon)^2} - s'_b z e^{i(\psi'_b + t(s'_b)^2)}}{\delta^2(1+\varepsilon)^2 - (s'_b)^2} \\ \frac{-\delta(1-\varepsilon)e^{iT(1-\varepsilon)^2} - s'_b z e^{i(\psi'_b + t(s'_b)^2)}}{\delta^2(1-\varepsilon)^2 - (s'_b)^2} \end{pmatrix}_{1 \leq b \leq q},$$

$$\mathscr{C}_\varepsilon(z,T) = \begin{pmatrix} \frac{(1+\varepsilon)e^{i(1+\varepsilon)^2 T} - z e^{iT}}{(1+\varepsilon)^2 - 1} & \frac{e^{i(1+\varepsilon)^2 T}}{1+\varepsilon} \\ \frac{-(1-\varepsilon)e^{i(1-\varepsilon)^2 T} - z e^{iT}}{(1-\varepsilon)^2 - 1} & \frac{-e^{i(1-\varepsilon)^2 T}}{1-\varepsilon} \end{pmatrix}.$$

The crucial point is the following elementary behavior of the inverse of $\mathscr{C}_\varepsilon(z,T)$.

Lemma 4.2.1 *Uniformly for z in every compact disc of $\mathbb{C}$, we have*

$$(1 - p_\varepsilon(T)z)\mathscr{C}_\varepsilon(z,T)^{-1}$$
$$= e^{-iT(1+2\varepsilon^2)} \begin{pmatrix} \frac{\varepsilon \omega_{-\varepsilon}(T)}{\omega_{-\varepsilon}(T) + z - \varepsilon(\omega_{-\varepsilon}(T) - z)/2} & \frac{\varepsilon \omega_\varepsilon(T)}{-\omega_\varepsilon(T) + z - \varepsilon(\omega_\varepsilon(T) + z)/2} \end{pmatrix} + \mathcal{O}(\varepsilon^2),$$

where

$$p_\varepsilon(T) := e^{-i\varepsilon^2 T}\left[-i\left(1 - \frac{\varepsilon^2}{2} + \mathcal{O}(\varepsilon^4)\right)\sin(2\varepsilon T) + \left(\frac{\varepsilon}{2} + \mathcal{O}(\varepsilon^3)\right)\cos(2\varepsilon T) \right] \tag{4.3}$$

and
$$\omega_\varepsilon(T) := e^{iT(2\varepsilon+\varepsilon^2)}.$$

Furthermore, the following equalities hold uniformly for z in $\mathbb{D}$ independently of $t \in \mathbb{R}$

$$\mathscr{C}_\varepsilon(z, T)^{-1}(1_2) = \mathcal{O}(1)$$

$$^t\mathscr{C}_\varepsilon(z, T)^{-1}(1_2) = \mathcal{O}\left(\frac{1}{\varepsilon}\right).$$

Proof Let us write

$$\mathscr{C}_\varepsilon(z, T) = \frac{e^{iT}}{\varepsilon} \begin{pmatrix} \frac{(1+\varepsilon)e^{iT(2\varepsilon+\varepsilon^2)}-z}{2+\varepsilon} & \frac{\varepsilon e^{iT(2\varepsilon+\varepsilon^2)}}{1+\varepsilon} \\ \frac{(1-\varepsilon)e^{iT(-2\varepsilon+\varepsilon^2)}+z}{2-\varepsilon} & -\frac{\varepsilon e^{iT(-2\varepsilon+\varepsilon^2)}}{1-\varepsilon} \end{pmatrix}$$

so that

$$\det\left[\varepsilon e^{-iT}\mathscr{C}_\varepsilon(z, T)\right] = -\varepsilon e^{2iT\varepsilon^2}(1 + \mathcal{O}(\varepsilon^2))(1 - p_\varepsilon(T)z),$$

with

$$p_\varepsilon(T) = e^{-i\varepsilon^2 T}\left[-i\left(1 - \frac{\varepsilon^2}{2} + \mathcal{O}(\varepsilon^4)\right)\sin(2\varepsilon T) + \left(\frac{\varepsilon}{2} + \mathcal{O}(\varepsilon^3)\right)\cos(2\varepsilon T)\right]$$

Consequently, for $|z| \leq R$,

$$(1 - p_\varepsilon(T)z)\mathscr{C}_\varepsilon(z, T)^{-1}$$
$$= e^{-iT(1+2\varepsilon^2)}\begin{pmatrix} \varepsilon\omega_{-\varepsilon}(T) & \varepsilon\omega_\varepsilon(T) \\ \frac{\omega_{-\varepsilon}(T)+z-\varepsilon(\omega_{-\varepsilon}(T)-z)/2}{2} & \frac{-\omega_\varepsilon(T)+z-\varepsilon(\omega_\varepsilon(T)+z)/2}{2} \end{pmatrix} + \mathcal{O}(\varepsilon^2),$$

where

$$\omega_\varepsilon(T) := e^{iT(2\varepsilon+\varepsilon^2)}.$$

An important fact is the following estimate

$$1 - |p_\varepsilon(T)|^2 = \varepsilon^2 + \cos^2(2\varepsilon T)(1 + \mathcal{O}(\varepsilon^2)) + \mathcal{O}(\varepsilon^4).$$

In particular, on $\mathbb{D}$, $|1 - p_\varepsilon(T)z| \geq \frac{\varepsilon^2 + \cos^2(2\varepsilon T)}{4} \geq \frac{\varepsilon^2}{4}$. Hence, the coefficients of $\mathscr{C}_\varepsilon(z, T)^{-1}$ grow at most as $\mathcal{O}(\frac{1}{\varepsilon^2})$ in $\mathbb{D}$ uniformly in $t \in \mathbb{R}$.

For the other estimates, we compute

$$(1 - p_\varepsilon(T)z)\mathscr{C}_\varepsilon(z, T)^{-1}(1_2)$$

$$= e^{-iT(2\varepsilon+\varepsilon^2)} \begin{pmatrix} 2\varepsilon e^{iT\varepsilon^2}\cos(2\varepsilon T) \\ -ie^{iT\varepsilon^2}\sin(2\varepsilon T) + z - \frac{\varepsilon}{2}e^{iT\varepsilon^2}\cos(2\varepsilon T) \end{pmatrix} + \mathcal{O}(\varepsilon^2)$$

$$= e^{-iT(2\varepsilon+\varepsilon^2)} \begin{pmatrix} 2\varepsilon e^{iT\varepsilon^2}\cos(2\varepsilon T) \\ z - p_\varepsilon(T) \end{pmatrix} + \mathcal{O}(\varepsilon^2).$$

As

$$\frac{|\varepsilon \cos(2\varepsilon T)|}{|1 - p_\varepsilon(T)z|} \leq C \frac{|\varepsilon \cos(2\varepsilon T)|}{\varepsilon^2 + \cos^2(2\varepsilon T)} = \mathcal{O}(1)$$

we obtain $\mathscr{C}_\varepsilon(z, T)^{-1}(1_2) = \mathcal{O}(1)$.

Eventually, we compute

$$(1 - p_\varepsilon(T)z)^t \mathscr{C}_\varepsilon(z, T)^{-1}(1_2)$$

$$= e^{-iT(2\varepsilon+\varepsilon^2)} \begin{pmatrix} \varepsilon \omega_{-\varepsilon}(T) + \frac{1}{2}[\omega_{-\varepsilon}(T) + z - \frac{\varepsilon}{2}(\omega_{-\varepsilon}(T) - z)] \\ \varepsilon \omega_\varepsilon(T) + \frac{1}{2}[-\omega_\varepsilon(T) + z - \frac{\varepsilon}{2}(\omega_\varepsilon(T) + z)] \end{pmatrix} + \mathcal{O}(\varepsilon^2)$$

$$= \frac{e^{-iT(2\varepsilon+\varepsilon^2)}}{2} \begin{pmatrix} z - \overline{p_\varepsilon(T)} \\ z - p_\varepsilon(T) \end{pmatrix} + \mathcal{O}(\varepsilon + |\cos(2\varepsilon T)|)$$

and this leads to

$$^t\mathscr{C}_\varepsilon(z, T)^{-1}(1_2) = \mathcal{O}\left(\frac{1}{\varepsilon}\right).$$

$\square$

Observing that

$$\mathscr{A}_{\delta,\varepsilon}(z, t) = \left(\frac{e^{i(\psi_a + ts_a^2)}}{s_a} + \mathcal{O}(\delta)\right)_{1 \leq a \leq q} \otimes (1, 1),$$

$$\mathscr{B}_{\delta,\varepsilon}(z) = \begin{pmatrix} 1 \\ 1 \end{pmatrix} \otimes^t \left(z \frac{e^{i(\psi_b' + t(s_b')^2)}}{s_b'} + \mathcal{O}(\delta)\right)_{1 \leq b \leq q},$$

we can exploit estimates on $\mathscr{C}_\varepsilon(z, T)^{-1}$ in the following lemma.

Lemma 4.2.2 *The following matrix expansions hold in $L^\infty(\mathbb{D})$ as ε, δ tend to 0 uniformly in $t \in \mathbb{R}$.*

$$\mathscr{A}_{\delta,\varepsilon}(z,t)\mathscr{C}_\varepsilon(z,T)^{-1} = \left(\frac{e^{i(\psi_a+ts_a^2)}}{s_a} + \mathcal{O}(\delta)\right)_{1\leq a\leq q} \otimes^t \mathscr{C}_\varepsilon(z,T)^{-1}(1_2) = \mathcal{O}\left(\frac{1}{\varepsilon}\right)$$

$$\mathscr{C}_\varepsilon(z,T)^{-1}\mathscr{B}_{\delta,\varepsilon}(z,t) = \mathscr{C}_\varepsilon(z,T)^{-1}(1_2) \otimes^t \left(z\frac{e^{i(\psi_b'+t(s_b')^2)}}{s_b'} + \mathcal{O}(\delta)\right)_{1\leq b\leq q} = \mathcal{O}(1)$$

In particular, the vectors

$$\mathscr{A}_{\delta,\varepsilon}(z,t)\mathscr{C}_\varepsilon(z,T)^{-1}(1_2),$$

$$^t\mathscr{B}_{\delta,\varepsilon}(z,t)^t\mathscr{C}_\varepsilon(z,T)^{-1}(1_2)$$

and the matrix

$$\mathscr{A}_{\delta,\varepsilon}(z,t)\mathscr{C}_\varepsilon(z,T)^{-1}\mathscr{B}_{\delta,\varepsilon}(z,t)$$

have uniformly bounded coefficients for $(z,t) \in \mathbb{D} \times \mathbb{R}$, ε, δ *small.*

Let us introduce the following notation,

$$\mathscr{C}_{\delta,\varepsilon}(z,T)^{-1}\begin{pmatrix}1_q\\1_2\end{pmatrix} =: \begin{pmatrix}X_q^{\delta,\varepsilon}(z,t)\\Y^{\delta,\varepsilon}(z,t)\end{pmatrix}$$

and

$$^t\mathscr{C}_{\delta,\varepsilon}(z,T)^{-1}\begin{pmatrix}1_q\\1_2\end{pmatrix} =: \begin{pmatrix}\hat{X}_q^{\delta,\varepsilon}(z,t)\\\hat{Y}^{\delta,\varepsilon}(z,t)\end{pmatrix}.$$

We have

$$\begin{cases}\mathscr{E}(z,t)X_q^{\delta,\varepsilon}(z,t) + \mathscr{A}_{\delta,\varepsilon}(z,t)Y^{\delta,\varepsilon}(z,t) = 1_q\\ \mathscr{B}_{\delta,\varepsilon}(z,t)X_q^{\delta,\varepsilon}(z,t) + \frac{1}{\delta}\mathscr{C}_\varepsilon(z,T)Y^{\delta,\varepsilon}(z,t) = 1_2\end{cases}$$

$$\begin{cases}^t\mathscr{E}(z,t)\hat{X}_q^{\delta,\varepsilon}(z,t) + {}^t\mathscr{B}_{\delta,\varepsilon}(z,t)\hat{Y}^{\delta,\varepsilon}(z,t) = 1_q\\ {}^t\mathscr{A}_{\delta,\varepsilon}(z,t)\hat{X}_q^{\delta,\varepsilon}(z,t) + \frac{1}{\delta}{}^t\mathscr{C}_\varepsilon(z,T)\hat{Y}^{\delta,\varepsilon}(z,t) = 1_2\end{cases}$$

Hence, setting

$$\mathscr{J}_{\delta,\varepsilon}(z,t) := \mathscr{E}(z,t) - \delta\mathscr{A}_{\delta,\varepsilon}(z,t)\mathscr{C}_\varepsilon(z,T)^{-1}\mathscr{B}_{\delta,\varepsilon}(z,t),$$

we obtain

$$\mathscr{I}_{\delta,\varepsilon}(z,t) X_q^{\delta,\varepsilon}(z,t) = 1_q - \delta \mathscr{A}_{\delta,\varepsilon}(z,t) \mathscr{C}_\varepsilon(z,T)^{-1} 1_2 \tag{4.4}$$

$$^t\mathscr{I}_{\delta,\varepsilon}(z,t) \hat{X}_q^{\delta,\varepsilon}(z,t) = 1_q - \delta^t \mathscr{B}_{\delta,\varepsilon}(z,t)\,^t\mathscr{C}_\varepsilon(z,T)^{-1} 1_2 \tag{4.5}$$

$$Y^{\delta,\varepsilon}(z) = \delta \mathscr{C}_\varepsilon(z,T)^{-1}(1_2 - \mathscr{B}_{\delta,\varepsilon}(z,t) X_q^{\delta,\varepsilon}(z,t)) \tag{4.6}$$

$$\hat{Y}^{\delta,\varepsilon}(z,t) = \delta^t \mathscr{C}_\varepsilon(z,T)^{-1}(1_2 -^t \mathscr{A}_{\delta,\varepsilon}(z,t) \hat{X}_q^{\delta,\varepsilon}(z,t)). \tag{4.7}$$

In view of Lemma 4.2.2, and of formulae (4.4), (4.5), (4.6), (4.7), we observe that

$$\mathscr{I}_{\delta,\varepsilon}(z,t) = \mathscr{E}(z,t) - \delta \mathscr{A}_{\delta,\varepsilon}(z,t) \mathscr{C}_\varepsilon(z,T)^{-1} \mathscr{B}_{\delta,\varepsilon}(z,t) = \mathscr{E}(z,t) + \mathcal{O}(\delta)$$

is invertible for $z \in \mathbb{D}$ and δ and ε small enough, and we get the formulae

$$X_q^{\delta,\varepsilon}(z,t) = \mathscr{I}_{\delta,\varepsilon}(z,t)^{-1}(1_q - \delta \mathscr{A}_{\delta,\varepsilon}(z) \mathscr{C}_\varepsilon(z,T)^{-1} 1_2)$$
$$\hat{X}_q^{\delta,\varepsilon}(z,t) =\,^t\mathscr{I}_{\delta,\varepsilon}(z,t)^{-1}(1_q - \delta^t \mathscr{B}_{\delta,\varepsilon}(z,t)\,^t\mathscr{C}_\varepsilon(z,T)^{-1} 1_2)$$
$$Y^{\delta,\varepsilon}(z,t) = \delta \mathscr{C}_\varepsilon(z,T)^{-1}(1_2 - \mathscr{B}_{\delta,\varepsilon}(z,t) X_q^{\delta,\varepsilon}(z,t))$$
$$\hat{Y}^{\delta,\varepsilon}(z,t) = \delta^t \mathscr{C}_\varepsilon(z,T)^{-1}(1_2 -^t \mathscr{A}_{\delta,\varepsilon}(z,t) \hat{X}_q^{\delta,\varepsilon}(z,t)).$$

Summarizing, we have the following result.

Proposition 4.2.3 *For the norm $L^\infty(\mathbb{D})$, we have, uniformly in δ, ε such that $\varepsilon \ll \delta \ll 1$,*

$$X_q^{\delta,\varepsilon}(z,t) = \mathscr{E}(z,t)^{-1} 1_q + \mathcal{O}(\delta)$$
$$\hat{X}_q^{\delta,\varepsilon}(z,t) =\,^t\mathscr{E}(z,t)^{-1} 1_q + \mathcal{O}(\delta)$$
$$Y^{\delta,\varepsilon}(z,t) = \alpha(z,t)\delta \mathscr{C}_\varepsilon(z,T)^{-1}(1_2) + \mathcal{O}\left(\delta^2\right)$$
$$\hat{Y}^{\delta,\varepsilon}(z,t) = \beta(z,t)\delta\,^t\mathscr{C}_\varepsilon(z,T)^{-1}(1_2) + \mathcal{O}\left(\frac{\delta^2}{\varepsilon}\right).$$

where

$$\alpha(z,t) := 1 - \left\langle \mathscr{E}(z,t)^{-1}(1_q), \left(z \frac{e^{i(\psi'_b + t(s'_b)^2)}}{s'_b}\right)_{1 \le b \le q} \right\rangle,$$

$$\beta(z,t) := 1 - \left\langle^t\mathscr{E}(z,t)^{-1}(1_q), \left(\frac{e^{i(\psi_a + t s_a^2)}}{s_a}\right)_{1 \le a \le q} \right\rangle.$$

As a first consequence of these computations, we obtain the smoothness of $z \to u_{\varepsilon,\delta}(z, 0)$.

Corollary 4.2.4 *For ε, δ small, the following equality holds on some fixed neighborhood of the closed unit disc,*

$$u_{\varepsilon,\delta}(\cdot, 0) = u_0 + o(1).$$

In particular, the functions $z \to u_{\varepsilon,\delta}(z, 0)$ are uniformly analytic on $\overline{\mathbb{D}}$.

Proof This estimate is a direct consequence of the previous Proposition and of the fact that $\mathscr{C}_\varepsilon(z, 0)^{-1}$ is bounded uniformly in z for $|z| \leq 1 + \eta$ for some $\eta > 0$. Indeed, $p_\varepsilon(0) = \frac{\varepsilon}{2} + \mathcal{O}(\varepsilon^3)$ hence the only pole of $\mathscr{C}_\varepsilon(z, 0)^{-1}$ is $\frac{2}{\varepsilon}$. From Proposition 4.2.3 and Formula (4.1), we get

$$u_{\delta,\varepsilon}(z, 0) = \left\langle \begin{pmatrix} X_{q,\delta} \\ Y_{q,\delta} \end{pmatrix}, \begin{pmatrix} 1_q \\ 1_2 \end{pmatrix} \right\rangle$$

$$= \left\langle \begin{pmatrix} \mathscr{E}(z, 0)^{-1} 1_q + \mathcal{O}(\delta) \\ \alpha(z, 0)\delta \mathscr{C}_\varepsilon(z, 0)^{-1}(1_2) + \mathcal{O}(\delta^2) \end{pmatrix}, \begin{pmatrix} 1_q \\ 1_2 \end{pmatrix} \right\rangle$$

$$= u_0 + o(1) .$$

□

The second point is the existence of $\underline{t}_{\delta,\varepsilon} \to \infty$ such that $u_{\varepsilon,\delta}(\underline{t}_{\delta,\varepsilon}) \to u_0$ in H^1. This fact is elementary, because $t \mapsto u_{\delta,\varepsilon}(t) \in \mathcal{V}(2q + 3) \subset H^1_+$ is quasi-periodic, hence we can find $\underline{t}_{\delta,\varepsilon}$ as big as we wish such that

$$\|u_{\varepsilon,\delta}(\underline{t}_{\delta,\varepsilon}) - u_{\varepsilon,\delta}(0)\|_{H^1} < \varepsilon .$$

The claim then follows from the previous corollary.

We now turn to the main point, which is the behavior in mean time. We first establish the following lemma.

Lemma 4.2.5 *There exists a constant $c > 0$ such that, for any $z \in \mathbb{S}^1$ and any $t \in \mathbb{R}$,*

$$|\alpha(z, t)\beta(z, t)| \geq c .$$

Proof Using the compactness of the torus, it is enough to prove that the functions

$$\alpha(z, \psi, \psi') := 1 - \left\langle \mathscr{E}(z, \psi, \psi')^{-1}(1_q), \left(z \frac{e^{i\psi'_b}}{s'_b} \right)_{1 \leq b \leq q} \right\rangle,$$

$$\beta(z, \psi, \psi')) := 1 - \left\langle {}^t\mathscr{E}(z, \psi, \psi')^{-1}(1_q), \left(\frac{e^{i\psi_a}}{s_a} \right)_{1 \leq a \leq q} \right\rangle,$$

where

$$\mathscr{E}(z, \psi, \psi') = \left(\frac{s_a e^{i\psi_a} - s'_b z e^{i\psi'_b}}{s_a^2 - (s'_b)^2} \right)_{1 \leq a, b \leq q},$$

do not take the 0 value for $(z, \psi, \psi') \in \mathbb{S}^1 \times \mathbb{T}^{2q}$. Indeed, write $(Z_b(z, \psi, \psi'))_{1 \leq b \leq q}$ $:= \mathscr{E}(z, \psi, \psi')^{-1} 1_q$ so that

$$\sum_{b=1}^{q} \frac{s_a e^{i\psi_a} - s'_b z e^{i\psi'_b}}{s_a^2 - (s'_b)^2} Z_b(z, \psi, \psi') = 1, \ a = 1, \ldots, q. \tag{4.8}$$

Assume $\alpha(z, \psi, \psi') = 0$ for some $(z, \psi, \psi') \in \mathbb{S}^1 \times \mathbb{T}^{2q}$, namely

$$\sum_{b=1}^{q} Z_b(z, \psi, \psi') \frac{e^{i\psi'_b} z}{s'_b} = 1. \tag{4.9}$$

Then subtracting (4.8) from (4.9) leads to

$$s_a \sum_{b=1}^{q} \frac{s_a e^{-i\psi_a} - \bar{z} s'_b e^{-i\psi'_b}}{s_a^2 - (s'_b)^2} \frac{e^{i\psi'_b} z Z_b(z, \psi, \psi')}{s'_b} = 0, \ a = 1, \ldots, q.$$

This is a contradiction since, from our inverse spectral Theorem 7, the matrix

$$\left(\frac{s_a e^{-i\psi_a} - s'_b \zeta e^{-i\psi'_b}}{s_a^2 - (s'_b)^2} \right)_{1 \leq a, b \leq q}$$

is known to be invertible for any ζ of modulus one. A similar argument leads to $\beta \neq 0$. □

Finally, we are going to focus our analysis near the unit circle, where the singularity of $u^{\delta,\varepsilon}$ takes place. First observe that

$$\|u\|_{H^1}^2 \sim \int_0^{2\pi} |u'(e^{i\theta})|^2 d\theta.$$

Deriving with respect to z the formula (4.1) giving $u^{\delta,\varepsilon}$, we get

$$(u^{\delta,\varepsilon})'(z, t) = \left\langle \dot{\mathscr{C}}_{\delta,\varepsilon}(t) \mathscr{C}_{\delta,\varepsilon}(z, t)^{-1} \begin{pmatrix} 1_q \\ 1_2 \end{pmatrix}, {}^t \mathscr{C}_{\delta,\varepsilon}(z, t)^{-1} \begin{pmatrix} 1_q \\ 1_N \end{pmatrix} \right\rangle \tag{4.10}$$

where

$$\dot{\mathscr{C}}_{\delta,\varepsilon}(t) = \begin{pmatrix} \dot{\mathscr{E}}(t) & \dot{\mathscr{A}}_{\delta,\varepsilon}(t) \\ \dot{\mathscr{B}}_{\delta,\varepsilon}(t) & \frac{1}{\delta}\dot{\mathscr{C}}_{\varepsilon}(T) \end{pmatrix} \tag{4.11}$$

with

$$\dot{\mathscr{E}}(t) = \left(\frac{s'_b e^{i(\psi'_b + (s'_b)^2)t}}{s_a^2 - (s'_b)^2} \right)_{1 \le a,b \le q}, \tag{4.12}$$

$$\dot{\mathscr{A}}_{\delta,\varepsilon}(t) = \left(\frac{\delta e^{iT}}{s_a^2 - \delta^2}, 0 \right)_{1 \le a \le q} \tag{4.13}$$

$$\dot{\mathscr{B}}_{\delta,\varepsilon}(t) = \begin{pmatrix} \frac{s'_b e^{i(\psi'_b + (s'_b)^2)t}}{\delta^2(1+\varepsilon)^2 - (s'_b)^2} \\ \frac{s'_b e^{i(\psi'_b + (s'_b)^2)t}}{\delta^2(1-\varepsilon)^2 - (s'_b)^2} \end{pmatrix}_{\substack{1 \le j \le N \\ 1 \le b \le q}} \tag{4.14}$$

and

$$\dot{\mathscr{C}}_{\varepsilon}(T) = \frac{e^{iT}}{2\varepsilon} \begin{pmatrix} (1+\frac{\varepsilon}{2})^{-1} & 0 \\ -(1-\frac{\varepsilon}{2})^{-1} & 0 \end{pmatrix} \tag{4.15}$$

Hence, we get

$$(u^{\delta,\varepsilon})'(z) = \left\langle \dot{\mathscr{C}}_{\delta,\varepsilon}(t) \begin{pmatrix} X_q^{\delta,\varepsilon}(z) \\ Y^{\delta,\varepsilon}(z) \end{pmatrix}, \begin{pmatrix} \hat{X}_q^{\delta,\varepsilon}(z) \\ \hat{Y}^{\delta,\varepsilon}(z) \end{pmatrix} \right\rangle$$

$$= \left\langle \dot{\mathscr{C}}_{\delta,\varepsilon}(t) \begin{pmatrix} \mathscr{E}(z,t)^{-1} 1_q + \mathcal{O}(\delta) \\ \alpha\delta\mathscr{C}_{\varepsilon}(z,T)^{-1}(1_2) + \mathcal{O}(\delta^2) \end{pmatrix}, \begin{pmatrix} {}^t\mathscr{E}(z,t)^{-1} 1_q + \mathcal{O}(\delta) \\ \beta\delta {}^t\mathscr{C}_{\varepsilon}(z,T)^{-1}(1_2) + \mathcal{O}\left(\frac{\delta^2}{\varepsilon}\right) \end{pmatrix} \right\rangle.$$

Observing that

$$\dot{\mathscr{C}}_{\delta,\varepsilon}(t) = \begin{pmatrix} \mathcal{O}(1) & \mathcal{O}(\delta) \\ \mathcal{O}(1) & \frac{1}{\delta}\dot{\mathscr{C}}_{\varepsilon}(T) \end{pmatrix} = \begin{pmatrix} \mathcal{O}(1) & \mathcal{O}(\delta) \\ \mathcal{O}(1) & \mathcal{O}\left(\frac{1}{\delta\varepsilon}\right) \end{pmatrix},$$

we infer, if $\varepsilon \ll \delta$,

$$(u^{\delta,\varepsilon})'(z) =$$
$$= \left\langle \begin{pmatrix} \mathcal{O}(1) + \mathcal{O}(\delta^2) \\ \mathcal{O}(1) + \alpha \dot{\mathcal{C}}_\varepsilon(T)\mathcal{C}_\varepsilon(z,T)^{-1}(1_2) + \mathcal{O}(\frac{\delta}{\varepsilon}) \end{pmatrix}, \begin{pmatrix} \mathcal{O}(1) \\ \beta \delta^t \mathcal{C}_\varepsilon(z,T)^{-1}(1_2) + \mathcal{O}(\frac{\delta^2}{\varepsilon}) \end{pmatrix} \right\rangle$$
$$= \alpha(z,t)\beta(z,t)\delta \langle \dot{\mathcal{C}}_\varepsilon(T)\mathcal{C}_\varepsilon(z,T)^{-1}(1_2), {}^t\mathcal{C}_\varepsilon(z,T)^{-1}(1_2) \rangle + \mathcal{O}\left(\frac{\delta^2}{\varepsilon^2}\right).$$

Notice that the first term $\langle \dot{\mathcal{C}}_\varepsilon(T)\mathcal{C}_\varepsilon(z,T)^{-1}(1_2), {}^t\mathcal{C}_\varepsilon(z,T)^{-1}(1_2) \rangle$ is $U'_\varepsilon(z,T)$, where

$$U_\varepsilon(z,T) = \langle \mathcal{C}_\varepsilon(z,T)^{-1}(1_2), 1_2 \rangle$$

is precisely the daisy solution to the cubic Szegő equation. In view of the computations of the beginning of this chapter, we have

Lemma 4.2.6 *Let $\varepsilon > 0$ and consider U_ε the function corresponding to spectral data $((1+\varepsilon, 1, 1-\varepsilon), (1, 1, -1))$ and evolving under the cubic Szegő flow*

$$U_\varepsilon(z,T) = \langle \mathcal{C}_\varepsilon(z,T)^{-1}(1_2), 1_2 \rangle \tag{4.16}$$

then,

$$\frac{1}{\tau}\int_0^\tau \left(\int_0^{2\pi} |U'_\varepsilon(e^{i\theta}, T)|^2 d\theta\right)^{1/2} dT \to \infty.$$

as ε tends to 0 and $\varepsilon\tau$ tends to infinity.

Eventually, we use the decomposition

$$u'_{\delta,\varepsilon}(z,t) = \delta \alpha(z,t)\beta(z,t)U'_\varepsilon(z,T) + \mathcal{O}\left(\frac{\delta^2}{\varepsilon^2}\right)$$

and Lemma 4.2.6 to obtain

$$\frac{1}{\mathcal{T}}\int_0^{\mathcal{T}} \|u_{\delta,\varepsilon}(t)\|_{H^1} dt \geq \frac{1}{\mathcal{T}}\int_0^{\mathcal{T}} \left(\int_0^{2\pi} |u'_{\delta,\varepsilon}(e^{i\theta}, t)|^2 d\theta\right)^{1/2} dt$$
$$\gtrsim \delta \frac{1}{\delta^2 \mathcal{T}}\int_0^{\delta^2 \mathcal{T}} \left(\int_0^{2\pi} |U'_\varepsilon(e^{i\theta}, T)|^2 d\theta\right)^{1/2} dT.$$

Here we assumed $\delta \ll \varepsilon^2$ small enough to absorb the remaining term appearing in the expression of $u'_{\delta,\varepsilon}$. Then we get the result for $\delta^2 \varepsilon \mathcal{T}$ tending to infinity.

5.3 The Transition to High Frequencies

It is now easy to conclude the proof of Theorem 2, which we formulate here in terms of averages of the H^1 norm.

Theorem 9 (Transition to High Frequencies for the Cubic Szegő Equation) *There exists a dense G_δ-subset of H^1_+ denoted by G, such that for all u solution to the Szegő equation (1.5) with $u(0) \in G$, for all we have*

$$\limsup_{T \to \infty} \frac{1}{T} \int_0^T \|u(t)\|_{H^1}\, dt = \infty,$$

and on the other hand

$$\liminf_{t \to \infty} \|u(t)\|_{H^1} \leq \|u(0)\|_{H^1}.$$

Proof of Theorem 9 We adapt a strategy by Hani [8]. Let $p \geq 1$ be an integer. We define $\mathcal{O}_p$ to be the set of $u_0 \in H^1_+$ such that there exists $T > p$, $\underline{t} > p$ such that the solution u of the Szegő equation with $u(0) = u_0$ satisfies

$$\frac{1}{T} \int_0^T \|u(\bar{t})\|_{H^1}\, dt > p,$$

$$\|u(\underline{t}) - u_0\|_{H^1} < \frac{1}{p}.$$

It is clear that $\mathcal{O}_p$ is an open set, and by the Proposition 8, $\mathcal{O}_p$ is dense in H^1_+. By the Baire category theorem,

$$G := \bigcap_{p=1}^{\infty} \mathcal{O}_p$$

is a dense G_δ set, and elements of G satisfy the conclusions of Theorem 9. □

An Improvement In order to obtain superpolynomial growth, one introduces a larger daisy as follows. For $N \in \mathbb{N}$, pick real numbers $\xi_1 > \eta_1 > \xi_2 > \eta_2 > \cdots > \xi_{N-1} > \eta_{N-1} > \xi_N > 0$. If we take as singular values

$$1 + \varepsilon\xi_1 > 1 + \varepsilon\eta_1 > 1 + \varepsilon\xi_2 > 1 + \varepsilon\eta_2 > \ldots,$$

and 0 as angles, we find a matrix

$$\widetilde{\mathscr{C}}_\varepsilon(z) = \left[\frac{(1 + \varepsilon\xi_j) - (1 + \varepsilon\eta_k)z}{(1 + \varepsilon\xi_j)^2 - (1 + \varepsilon\eta_k)^2} \right]_{1 \leq j,k \leq N}.$$

One can prove that corresponding $\tilde{u}_\varepsilon \in \mathcal{V}(2N - 1)$ given by the inverse spectral Theorem 7 then satisfies, for a generic choice of (ξ, η),

$$\|\tilde{u}_\varepsilon\|_{H^s} \simeq \frac{1}{\varepsilon^{(N-1)(2s-1)}} \simeq (t_\varepsilon)^{(N-1)(2s-1)}.$$

As before, it is also possible to find, changing the angles along the Szegő trajectory, a matrix $\mathscr{C}_\varepsilon(z)$ such that the associated function u_ε is a rational fraction with no pole close to the unit disc. We refer to [6] for the complete proof.

5.4 Related Equations

We conclude this lecture by mentioning some related results for equations connected to the cubic Szegő equation.

5.4.1 The Cubic Szegő Equation on the Line

The cubic Szegő equation can be stated similarly on the real line, where Π denotes the orthogonal projector of $L^2(\mathbb{R})$ onto the subspace of functions with Fourier transform supported into $\mathbb{R}_+$. In this case, Hankel operators H_u can still be defined and they satisfy a similar Lax pair evolution. However, the shift operator S does not exist anymore, nor does the shifted operator K_u. There is nevertheless a second Lax pair identity for

$$L_u = H_u^2 - (.|u)u \, .$$

We mention here only some results of this theory which is entirely due to Pocovnicu [14, 15].

Theorem 10 (Pocovnicu [14, 15]) *The cubic Szegő equation on the real line*

$$i\partial_t u = \Pi(|u|^2 u),$$

satisfies the following properties.

- *The traveling wave solutions, namely the solutions $Q \in H_+^{\frac{1}{2}}(\mathbb{R})$ of*

$$cDQ + \omega Q = \Pi(|u|^2 u)$$

for some $(c, \omega) \in \mathbb{R}^2$, are exactly given by

$$Q(x) = \frac{a}{x + \lambda},$$

with $a \in \mathbb{C}$ and $\lambda \in \mathbb{C}$, $\mathrm{Im}\lambda > 0$.

- *If*

$$u_0(x) = \sum_{j=1}^{N} \frac{a_j}{x + \lambda_j},$$

then $u(t, x)$ conserves the same N-soliton form, where unknowns the coefficients a_j, λ_j satisfy a completely integrable ODE. If $N \geq 2$, some of these solutions satisfy, as $t \to +\infty$,

$$\forall s > \frac{1}{2}, \ \|u(t, .)\|_{H^s} \sim t^{2s-1}.$$

Notice that the transition to high frequencies is much moire explicit here than on the circle. It is also of quite a different nature.

5.4.2 Back to the Half-Wave Equation

It is tempting to combine Theorems 1.4 and 2 to obtain some wave turbulence result for the half-wave equation. Unfortunately, this strategy only leads to the following relatively weak result.

Theorem 11 *Given $0 < \eta \ll 1$, $K \gg 1$ and $s > \frac{1}{2}$, there $T > 0$ and a solution of*

$$i\partial_t u = |D|u + |u|^2 u$$

satisfying

$$\|u(0)\|_{H^s} \leq \eta, \ \|u(T)\|_{H^s} \geq K.$$

In the case of the line, however, it is possible to guarantee moreover the saturated estimate [7]

$$\forall t \geq T, \ \|u(t)\|_{H^s} \sim K.$$

The proof is based on a careful construction of a two-soliton solution whose profiles are perturbations of the traveling waves of the cubic Szegő equation.

5.4.3 A System in Two Space Dimensions

It is in fact possible to take a greater advantage of Theorems 1.4 and 2 by introducing an additional variable on the line, as did Hani et al. [9] in their remarkable study of the cubic nonlinear Schrödinger equation on the cylinder $\mathbb{R} \times \mathbb{T}^2$. Following the modified scattering method developed by these authors, Haiyan Xu proved the following theorem.

Theorem 12 (Xu [19]) *Consider the equation*

$$i\partial_t u = -\partial_x^2 u + |D_y|u + |u|^2 u, \quad (x, y) \in \mathbb{R} \times \mathbb{T} \tag{4.17}$$

with $u(t, x, y + \pi) = -u(t, x, y)$. For initial data which are sufficiently smooth, decaying and small, the long time behavior of the solutions is given by

$$e^{it(-\partial_x^2 + |D_y|)} u(t) - G(\pi \log(t)) \to 0, \quad t \to +\infty, \tag{4.18}$$

where G is a solution of the following decoupled system,

$$i\partial_t \hat{G}_{\pm} = \Pi_{\pm}(|\hat{G}_{\pm}|^2 \hat{G}_{\pm}), \tag{4.19}$$

where $G_{\pm}(t, x, y) := \Pi_{\pm} G(t, x, y)$ denote the Szegő projections in the y variable, and $\hat{F}(t, \xi, y)$ denotes the Fourier transform of $F(t, x, y)$ with the respect to the x variable.

Conversely, given any solution G of (4.19), with initial data sufficiently smooth, decaying and small, there exists a solution u of Eq. (4.17) such that (4.18) holds.

Notice that the Fourier variable $\xi \in \mathbb{R}$ is just a parameter in the above decoupled system of cubic Szegő equations. Applying the superpolynomial version of Theorem 2, one can prove

Theorem 13 (Xu [19]) *There exist solutions u of Eq. (4.17) such that*

$$\forall N \geq 1, \quad \limsup_{t \to +\infty} \frac{\|u(t)\|_{L_x^2(\mathbb{R}, H_y^1(\mathbb{T}))}}{(\log t)^N} dt = +\infty,$$

and on the other hand

$$\liminf_{t \to \infty} \|u(t)\|_{L_x^2(\mathbb{R}, H_y^1(\mathbb{T}))} < +\infty.$$

5.4.4 An Integrable Perturbation of the Cubic Szegő Equation

Another remarkable property of the cubic Szegő equation is that it admits a one parameter family of perturbations which is still integrable. Consider the following Hamiltonian equation on $H_+^s(\mathbb{T})$

$$i\partial_t u = \Pi(|u|^2 u) + \alpha \int_{\mathbb{T}} u \, dx. \tag{4.20}$$

Because the action of the antishift operator S^* kills the second term in the right hand side, it is not difficult to check that the Szegő Lax pair for K_u is still valid for this equation. On the other hand, the Lax pair for H_u does not hold. It turns out that this equation is still integrable, as proved by Xu in [18]. Moreover, let us mention interesting qualitative differences of the dynamics compared to with the cubic Szegő equation.

Theorem 14 (Xu [17, 18]) *If $\alpha > 0$, there exists rational solutions u of (4.20) such that*

$$\forall s > \frac{1}{2}, \ \|u(t)\|_{H^s} \sim e^{c_s t}, \ t \to +\infty$$

with $c_s > 0$. Furthermore, such a phenomenon does not happen for rational solutions if $\alpha < 0$.

Notice that, in view of Corollary A.2 in the appendix, the exponential growth is optimal.

Appendix A: The L^∞ Estimate and Its Consequences

In this section, we show how the lax pair structure leads to the following a priori estimate for solutions of the cubic Szegő equation.[4]

Theorem A.15 ([2]) *Assume $u_0 \in H^s_+$ for some $s > 1$. Then the corresponding solution u of the cubic Szegő equation satisfies*

$$\sup_{t \in \mathbb{R}} \|u(t)\|_{L^\infty} < +\infty.$$

The proof of this theorem relies on the following proposition.

Proposition A.1 *Given $u \in H^{\frac{1}{2}}_+$, we denote by $\{s_j(u)\}_{j \geq 1}$ the sequence of singular values of H_u, repeated according to multiplicity. The following double inequality holds:*

$$\frac{1}{2} \sum_{n=0}^{\infty} |\hat{u}(n)| \leq \sum_{j=1}^{\infty} s_j(u) \leq \sum_{n=0}^{\infty} \left(\sum_{\ell=0}^{\infty} |\hat{u}(n+\ell)|^2 \right)^{\frac{1}{2}}$$

Furthermore, the right hand side is controlled by

$$C_s \|u\|_{H^s}$$

for every $s > 1$.

---

[4] In view of the Lax pair for K_u, this estimate is also valid for Eq. (4.20).

Remark A.3 This proposition can be interpreted as a double inequality for the trace norm of the Hankel operator H_u. A more complete characterization of functions u such that the trace norm of H_u is finite is given in Peller [13]. Here we provide an elementary proof.

Assuming the proposition, let us show the theorem. From the second equality in the proposition, we have

$$\sum_{j=1}^{\infty} s_j(u_0) < +\infty.$$

Since each $s_j(u)$ is a conservation law, we have, for every $t \in \mathbb{R}$,

$$\sum_{j=1}^{\infty} s_j(u(t)) = \sum_{j=1}^{\infty} s_j(u_0).$$

Finally, by the first inequality in the proposition,

$$\sup_{t \in \mathbb{R}} \sum_{n \geq 0} |\hat{u}(t,n)| \leq 2 \sup_{t \in \mathbb{R}} \sum_{j=1}^{\infty} s_j(u(t)) = 2 \sum_{j=1}^{\infty} s_j(u_0).$$

The proof is completed by the elementary observation that

$$\|u\|_{L^\infty} \leq \sum_{n \geq 0} |\hat{u}(n)|.$$

We now pass to the proof of the proposition.

Proof of Proposition A.1 We denote by $\{e_j\}_{j \geq 1}$ an orthonormal basis of $\overline{\mathrm{Ran}(H_u)} = \overline{\mathrm{Ran}(H_u^2)}$ such that $H_u^2 e_j = s_j^2 u$. Such a basis exists because H_u^2 is a compact selfadjoint operator. Notice that

$$\|H_u(e_j)\|^2 = (H_u^2(e_j)|e_j) = s_j^2.$$

We set, for every $n \geq 0$, $\varepsilon_n(x) = e^{inx}$.

Let us prove the first inequality. Observe that

$$\hat{u}(2n) = (H_u(\varepsilon_n)|\varepsilon_n), \quad \hat{u}(2n+1) = K_u(\varepsilon_n|\varepsilon_n).$$

On the other hand,

$$(H_u(\varepsilon_n)|\varepsilon_n) = \sum_j (H_u(\varepsilon_n)|e_j)(e_j|\varepsilon_n),$$

and $(H_u(\varepsilon_n)|e_j) = (H_u(e_j)|\varepsilon_n)$, so that

$$\sum_{n=0}^{\infty} |\hat{u}(2n)| \le \sum_{n,j} |(H_u(e_j)|\varepsilon_n)(e_j|\varepsilon_n)|$$

$$\le \sum_j \left(\sum_{n=0}^{\infty} |(H_u(e_j)|\varepsilon_n)|^2\right)^{\frac{1}{2}} \left(\sum_{n=0}^{\infty} |(e_j|\varepsilon_n)|^2\right)^{\frac{1}{2}}$$

$$= \sum_j \|H_u(e_j)\| \|e_j\| = \sum_j s_j.$$

Arguing similarly with K_u, we obtain

$$\sum_{n=0}^{\infty} |\hat{u}(2n+1)| \le \sum_k s'_k.$$

Summing up, we have proved

$$\sum_{n=0}^{\infty} |\hat{u}(n)| \le \sum_j s_j + \sum_k s'_k \le 2 \sum_j s_j.$$

We now pass to the second inequality. Notice that

$$(H_u(e_j), H_u(e_{j'})) = (H_u^2(e_{j'}), e_j) = s_j^2 \delta_{jj'}.$$

In other words, the sequence $\{H_u(e_j)/s_j\}$ is orthonormal. We then define the following antilinear operator on L_+^2,

$$\Omega_u(h) = \sum_j (e_j, h) \frac{H_u(e_j)}{s_j}.$$

Notice that, due to the orthonormality of both systems $\{e_j\}$ and $\{H_u(e_j)/s_j\}$, $\|\Omega_u(h)\| \le \|h\|$. Similarly, we define

$$^t\Omega_u(h) = \sum_j \frac{(H_u(e_j), h)}{s_j} e_j,$$

so that

$$\forall h, h' \in L_+^2, \quad (\Omega_u(h)|h') = (^t\Omega_u(h')|h).$$

We next observe that

$$s_j = (H_u(e_j)|\Omega_u(e_j)) = \sum_{n=0}^{\infty}(H_u(e_j)|\varepsilon_n)(\varepsilon_n|\Omega_u(e_j))$$

$$= \sum_{n=0}^{\infty}\sum_{\ell=0}^{\infty} \hat{u}(n+\ell)(\varepsilon_\ell|e_j)(\varepsilon_n|\Omega_u(e_j))$$

But using the transpose of Ω_u, we get

$$\sum_j (\varepsilon_\ell|e_j)(\varepsilon_n|\Omega_u(e_j)) = \sum_j (\varepsilon_\ell|e_j)(e_j|{}^t\Omega_u(\varepsilon_n)) = (\varepsilon_\ell|{}^t\Omega_u(\varepsilon_n)) = (\varepsilon_n|\Omega_u(\varepsilon_\ell)).$$

Consequently,

$$\sum_j s_j = \sum_{n,\ell\geq 0} \hat{u}(n+\ell)(\varepsilon_n|\Omega_u(\varepsilon_\ell)).$$

Applying the Cauchy–Schwarz inequality to the sum on ℓ, we infer

$$\sum_j s_j \leq \sum_{n=0}^{\infty} \|\Omega_u(\varepsilon_\ell)\| \left(\sum_{\ell=0}^{\infty} |\hat{u}(k+\ell)|^2\right)^{\frac{1}{2}},$$

and the claim follows from the fact that $\|\Omega_u(\varepsilon_\ell)\| \leq \|\varepsilon_\ell\| = 1$.

We finally need to control the right hand side of this last inequality. By the Cauchy–Schwarz inequality in the n sum, we have, for every $s > 1$,

$$\sum_{n=0}^{\infty}\left(\sum_{\ell=0}^{\infty}|\hat{u}(n+\ell)|^2\right)^{\frac{1}{2}} \leq \left(\sum_{n=0}^{\infty}(1+n)^{1-2s}\right)^{\frac{1}{2}} \left(\sum_{n,\ell\geq 0}(1+n)^{2s-1}|\hat{u}(n+\ell)|^2\right)^{\frac{1}{2}}$$

$$\leq \left(\frac{s}{s-1}\right)^{\frac{1}{2}} \left(\sum_{n,\ell\geq 0}(1+n+\ell)^{2s-1}|\hat{u}(n+\ell)|^2\right)^{\frac{1}{2}}$$

$$\leq C_s \|u\|_{H^s},$$

and Proposition A.1 is proved. □

Corollary A.2 *For $s > 1$ and $u_0 \in H_+^s$, the corresponding solution $t \mapsto u(t)$ of the Szegő equation[5] satisfies*

$$\|u(t)\|_{H^s} \leq C_s e^{C_s'|t|}, \quad \forall t \in \mathbb{R},$$

where C_s, C_s' are positive constants which only depend on s and $\|u_0\|_{H^s}$.

Proof We compute $\frac{d}{dt}\|D^s u(t)\|_{L^2}^2$ and use the boundedness of the L^∞ norm to write

$$\left|\frac{d}{dt}\|D^s u(t)\|_{L^2}^2\right| \leq C\|u\|_{H^s}^2,$$

for an appropriate constant C depending on the norm of u_0 in H_+^s. A Gronwall inequality then completes the proof of the corollary. □

References

1. J. Bourgain. Problems in Hamiltonian PDE's. *Geom. Funct. Anal.*, (Special Volume, Part I):32–56, 2000. GAFA 2000 (Tel Aviv, 1999).
2. P. Gérard and S. Grellier. The cubic Szegő equation. *Ann. Sci. Éc. Norm. Supér. (4)*, 43(5):761–810, 2010.
3. P. Gérard and S. Grellier. Effective integrable dynamics for a certain nonlinear wave equation. *Anal. PDE*, 5(5):1139–1155, 2012.
4. P. Gérard and S. Grellier. Invariant tori for the cubic Szegö equation. *Inventiones mathematicae*, 187(3):707–754, 2012.
5. P. Gérard and S. Grellier. An explicit formula for the cubic Szegő equation. *Transactions of the American Mathematical Society*, 367(4):2979–2995, 2015.
6. P. Gérard and S. Grellier. *The cubic Szegő equation and Hankel operators*, volume 389 of *Astérisque*. Société mathématique de France, 2017.
7. P. Gérard, E. Lenzmann, O. Pocovnicu, and P. Raphaël. A two-soliton with transient turbulent regime for the cubic half-wave equation on the real line. *Ann. PDE* 4(1), Art. 7, 2018.
8. Z. Hani. Long-time instability and unbounded Sobolev orbits for some periodic nonlinear Schrödinger equations. *Arch. Ration. Mech. Anal.*, 211(3):929–964, 2014.
9. Z. Hani, B. Pausader, N. Tzvetkov, and N. Visciglia. Almost sure global well-posedness for fractional cubic Schrödinger equation on torus. *Forum Math. Pi3*, 2015.
10. J. Krieger, E. Lenzmann, and P. Raphaël. Nondispersive solutions to the l2-critical half- wave equation. *Arch. Ration. Mech. Anal.*, 209(1):61–129, 2013.
11. P. D. Lax. Integrals of nonlinear equations of evolution and solitary waves. *Comm. Pure Appl. Math.*, 21:467–490, 1968.
12. A Majda, D. McLaughlin, and E. Tabak. A one dimensional model for dispersive wave turbulence. *Nonlinear Science*, 6:9–44, 1997.
13. V. Peller. *Hankel operators and their applications*. Springer Science & Business Media, 2012.
14. O. Pocovnicu. Explicit formula for the solution of the szegő equation on the real line and applications. *Discrete Contin. Dyn. Syst. A*, 31(3):607–649, 2011.

---

[5]In view of the Lax pair for K_u, this estimate is also valid for Eq. (4.20).

15. O. Pocovnicu. Traveling waves for the cubic szegő equation on the real line. *Anal. PDE*, 4(3):379–404, 2011.
16. J. Thirouin. On the growth of Sobolev norms of solutions of the fractional defocusing NLS equation on the circle. *Annales de l'Institut Henri Poincaré (C), Nonlinear Analysis*, 34(509–531), 2016.
17. H. Xu. Large time blowup for a perturbation of the cubic Szegő equation. *Analysis & PDE*, 7(3):717–731, 2014.
18. H. Xu. The cubic Szegő equation with a linear perturbation. Preprint, arXiv:1508.01500, August 2015.
19. H. Xu. Unbounded sobolev trajectories and modified scattering theory for a wave guide nonlinear schrödinger equation. *Mathematische Zeitschrift*, 286(1–2):443–489, 2017.
20. V. E. Zakharov and A. B. Shabat. Exact theory of two-dimensional self-focusing and one-dimensional self-modulation of waves in nonlinear media. *Ž. Èksper. Teoret. Fiz.*, 61(1):118–134, 1971.

Benjamin-Ono and Intermediate Long Wave Equations: Modeling, IST and PDE

Jean-Claude Saut

1 Introduction

In order to illustrate the links and interactions between PDE and Inverse Scattering methods, we have chosen to focus on two one-dimensional examples that have a physical relevance (in the context of internal waves) and that lead to yet unsolved interesting issues.

The Benjamin-Ono (BO) and Intermediate Long Wave (ILW) equations are two classical examples of completely integrable one-dimensional equations, maybe not so well-known as the Korteweg de Vries or the cubic nonlinear Schrödinger equations though. A striking fact is that a complete rigorous resolution of the Cauchy problem by IST techniques is still incomplete for the BO equations for arbitrary large initial data while it is widely open in the ILW case, even for small data. On the other hand, both those problems can be solved by "PDE" techniques, for arbitrary initial data in relatively big spaces but no general result on the long time behavior of "large" solutions is known with the notable exception of stability issues of solitons and multisolitons. In particular the so-called *soliton resolution conjecture* although expected has not been proven yet.

Those two scalar equations are one-dimensional, one-way propagation asymptotic models for internal waves in an appropriate regime. We have thus three different viewpoints on BO and ILW equations and this article aims to review them and emphasize their possible links. Since we do not want to ignore the modeling aspects, we first recall the derivation of the equations in the context of internal waves in a two-layer system. The modeling of internal waves displays a variety

---

J.-C. Saut (✉)
Laboratoire de Mathématiques, UMR 8628, Université Paris-Saclay, Paris-Sud et CNRS, Orsay, France
e-mail: jean-claude.saut@u-psud.fr

of fascinating scientific problems. We refer for instance to the survey article [98] for the physical modeling aspects and to [38, 61, 219, 241] for the rigorous derivation of asymptotic models.

The Benjamin-Ono equation was first formally derived by Benjamin, [30] (and independently in [62] where one can find also numerical simulations and experimental comparisons), and later by Ono [221], to describe the propagation of long weakly nonlinear internal waves in a stratified fluid such that two layers of different densities are joined by a thin region where the density varies continuously (pycnocline), the lower layer being infinite.[1] Benjamin also wrote down the explicit algebraically decaying solitary wave solution and also the periodic traveling wave.

The Intermediate Long Wave equation was introduced by Kubota et al. [137] to describe the propagation of a long weakly nonlinear internal wave in a stratified medium of finite total depth. A formal derivation is also given in Joseph [111] who used the dispersion relation derived in [230] in the context of the Whitham non local equation [266]. Joseph derived furthermore the solitary wave solution.

We also refer to [29, 58, 136, 155, 162, 222, 233, 246] for the relevance of the ILW equation in various oceanic or atmospheric contexts.

The ILW equation reduces formally to the BO equation when the depth of the lower layer tends to infinity.

A rigorous derivation (in the sense of consistency) is given in [38, 61] using a two-layer system, that is a system of two layers of fluids of different densities, the density of the total fluid being discontinuous though (see below). Interesting comparisons with experiments can be found e.g. in [136].[2] Bidirectional versions are derived in [38, 56, 57, 61], see below for a quick description of a rigorous derivation in the sense of consistency.

Both equations belong to the general class of equations of the type (see [149])

$$u_t + uu_x - \mathcal{L}u_x = 0, \tag{1.1}$$

where $\mathcal{L}$ is defined after Fourier transform by $\widehat{\mathcal{L}f}(\xi) = p(\xi)\hat{f}(\xi)$ where p is a real symbol, with $p(\xi) = p_\delta(\xi) = \xi\coth(\delta\xi) - \frac{1}{\delta}, \delta > 0$ for the ILW equation and $p(\xi) = |\xi|$ for the BO equation.

They can alternatively be written respectively,

$$u_t + uu_x - Hu_{xx} = 0, \tag{1.2}$$

where H is the Hilbert transform, that is the convolution with $PV(\frac{1}{x})$ and

$$u_t + uu_x + \frac{1}{\delta}u_x + \mathcal{T}(u_{xx}) = 0, \tag{1.3}$$

---

[1] One can find interesting comparisons with experiments in [183].
[2] Recall however that BO and ILW equations are weakly nonlinear models and they do not fit well with the modeling of higher amplitude waves, see e.g. the experiments in [250].

where

$$T = PV \int_{-\infty}^{\infty} \coth\left(\frac{x-y}{\delta}\right) u(y) dy.$$

The BO equation has the following scaling and translation invariance: if u is a solution de BO, so is v defined by

$$v(x, t) = cu(c(x - x_0), c^2 t)), \quad \forall c > 0, x_0 \in \mathbb{R}.$$

As aforementioned, the ILW equation reduces formally to the BO equation when $\delta \to \infty$ and it was actually proven in [1] that the solution u_δ of (1.3) with initial data u_0 converges as $\delta \to +\infty$ to the solution of the Benjamin-Ono equation (1.2) with the same initial data in suitable Sobolev spaces.

Furthermore, if u_δ is a solution of (1.3) and setting

$$v_\delta(x, t) = \frac{3}{\delta} u_\delta(x, \frac{3}{\delta} t),$$

v_δ tends as $\delta \to 0$ to the solution u of the KdV equation

$$u_t + u u_x + u_{xxx} = 0. \tag{1.4}$$

Both the BO and ILW equations conserve formally the L^2 norm of the initial data. Moreover they have an Hamiltonian structure

$$u_t + \partial_x \mathcal{H} u = 0,$$

where

$$\mathcal{H} u = \frac{1}{2} \int_{\mathbb{R}} (\frac{u^3}{3} - (|D|^{1/2} u)^2) dx$$

for the BO equation and

$$\mathcal{H} u = \frac{1}{2} \int_{\mathbb{R}} (\frac{u^3}{3} - \frac{1}{2} |T_\delta^{1/2} u|^2) dx,$$

where

$$\widehat{T_\delta f}(\xi) = p_\delta(\xi) \hat{f}(\xi), \text{ where } p_\delta(\xi) = \xi \coth(\delta \xi) - \frac{1}{\delta},$$

for the ILW equation.

We now recall now the long process leading to ILW and BO equations following [38] that will lead to a rigorous justification of the equations in the sense of consistency.

The physical context is that of the two-layer system of two inviscid incompressible fluids of different densities $\rho_1 < \rho_2$, see picture below.

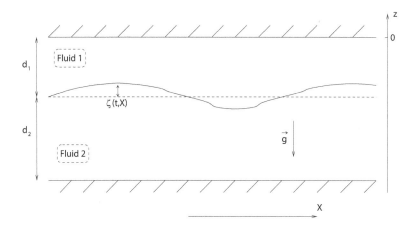

We refer to [38] for the derivation of the complete two-layer system (extending the classical water wave system, see [140]) describing the evolution of the interface ζ and of a suitable velocity variable and to [141] for a deep analysis of this system.

The asymptotic models are derived from the two-layer system[3] [38] (see also [241]) after introducing scaling parameters, namely

- **a** = typical amplitude of the deformation of the interface, λ = typical wavelength.
- Dimensionless independent variables

$$\widetilde{X} := \frac{X}{\lambda}, \quad \widetilde{z} := \frac{z}{d_1}, \quad \widetilde{t} := \frac{t}{\lambda/\sqrt{gd_1}},$$

are introduced. Likewise, we define the dimensionless unknowns

$$\widetilde{\zeta} := \frac{\zeta}{a}, \quad \widetilde{\psi}_1 := \frac{\psi_1}{a\lambda\sqrt{g/d_1}},$$

---

[3] Both the BO and ILW equations were first derived in the context of a continuously stratified fluid to model long internal waves on a pycnocline (boundary separating two liquid layers of different densities), see [30, 62, 137]. The rigorous justification in this context is studied in [70].

as well as the dimensionless parameters[4]

$$\gamma := \frac{\rho_1}{\rho_2}, \quad d := \frac{d_1}{d_2}, \quad \epsilon := \frac{a}{d_1}, \quad \mu := \frac{d_1^2}{\lambda^2};$$

Though they are redundant, it is also notationally convenient to introduce two other parameters ε_2 and μ_2 defined as

$$\varepsilon_2 = \frac{a}{d_2} = \varepsilon d, \quad \mu_2 = \frac{d_2^2}{\lambda^2} = \frac{\mu}{d^2}.$$

The range of validity of the various regimes is summarized in the following table.

	$\epsilon = O(1)$	$\epsilon \ll 1$
$\mu = O(1)$	Full equations	$d \sim 1$: FD/FD eq'ns
$\mu \ll 1$	$d \sim 1$: SW/SW eq'ns	$\mu \sim \epsilon$ and $d^2 \sim \epsilon$: B/FD eq'ns
	$d^2 \sim \mu \sim \epsilon_2^2$: SW/FD eq'ns	$\mu \sim \epsilon$ and $d \sim 1$: B/B eq'ns
		$d^2 \sim \mu \sim \epsilon^2$: ILW eq'ns
		$d = 0$ and $\mu \sim \epsilon^2$: BO eq'ns

The ILW regime is thus obtained when $\mu \sim \varepsilon^2 \ll 1$ and $\mu_2 \sim 1$ (and thus $d^2 \sim \mu \sim \varepsilon_2$); in this case, one gets the following expansion of the nonlocal *interface operator* (see [38] for details), where $|D| = \sqrt{-\Delta}$:

$$\mathbf{H}^{\mu,d}[\varepsilon\zeta]\psi_1 = -\sqrt{\mu}|D|\coth(\sqrt{\mu_2}|D|)\nabla\psi_1 + O(\mu). \tag{1.5}$$

In the BO regime one has $\mu \ll 1$ and $d = 0$ (and thus $\mu_2 = \infty$, $\varepsilon_2 = 0$), and one gets the approximation

$$\mathbf{H}^{\mu,d}[\varepsilon\zeta]\psi_1 \sim -\sqrt{\mu}|D|\nabla\psi_1. \tag{1.6}$$

The BO regime is thus the limit of the ILW one when the depth of the lower layer is infinite, $d \to 0$, or $\mu_2 \to \infty$.

Using those scalings one derives (see [38]) from the full two-layers system the ILW system, written below in horizontal spatial dimension $N = 1, 2$.

$$\begin{cases} [1 + \sqrt{\mu}\frac{\alpha}{\gamma}|D|\coth(\sqrt{\mu_2}|D|)]\partial_t\zeta + \frac{1}{\gamma}\nabla \cdot ((1 - \epsilon\zeta)\mathbf{v}) \\ \quad - (1 - \alpha)\frac{\sqrt{\mu}}{\gamma^2}|D|\coth(\sqrt{\mu_2}|D|)\nabla \cdot \mathbf{v} = 0, \\ \partial_t\mathbf{v} + (1 - \gamma)\nabla\zeta - \frac{\epsilon}{2\gamma}\nabla|\mathbf{v}|^2 = 0. \end{cases} \tag{1.7}$$

---

[4]Note that $d \sim \frac{1}{\delta}$, where δ is as in the previous notation.

Remark 1 The coefficient $\alpha \geq 0$ in (1.7) is a free modeling parameter stemming from the use of the Benjamin-Bona-Mahony (BBM) trick. In dimension $N = 1$ and with $\alpha = 0$, (1.7) corresponds to (5.47) of [61] which is obtained by expanding the Hamiltonian of the full system with respect to ϵ and μ. However this system is not linearly well-posed. It is straightforward to ascertain that the condition $\alpha \geq 1$ insures that (1.7) is linearly well-posed for either $N = 1$ or $N = 2$.

Remark 2 The ILW equation derived formally in [111, 137] is obtained as the unidirectional limit of the one dimensional ($d = 1$) version of (1.7) when $\alpha = 0$, see for instance §5.5 in [61].

Using the above approximation of the interface operator, leads to the BO system:

$$\begin{cases} [1 + \sqrt{\mu}\frac{\alpha}{\gamma}|D|]\partial_t \zeta + \frac{1}{\gamma}\nabla \cdot ((1 - \epsilon\zeta)\mathbf{v}) - (1-\alpha)\frac{\sqrt{\mu}}{\gamma^2}|D|\nabla \cdot \mathbf{v} = 0, \\ \partial_t \mathbf{v} + (1-\gamma)\nabla\zeta - \frac{\epsilon}{2\gamma}\nabla|\mathbf{v}|^2 = 0 \end{cases} \quad (1.8)$$

where α has the same significance as in the previous remark and which again reduces to the BO equation for one-dimensional, unidirectional waves.

The well-posedness of the Cauchy problem for (1.7), (1.8) (in space dimension one and two) on long time scales $O(1/\epsilon)$ has been established in [271] together with the limit of solutions of the ILW system to those of the BO system when $\mu_2 \to \infty$. We also refer to [25] for a study of solitary wave solutions to both the one-dimensional ILW and BO systems and to [37] for numerical simulations.

Remark 3 The above derivation was performed for purely gravity waves. Surface tension effects result in adding a third order dispersive term in the asymptotic models. One gets for instance the so-called Benjamin equation (see [31, 32]):

$$u_t + uu_x - Hu_{xx} - \delta u_{xxx} = 0, \quad (1.9)$$

where $\delta > 0$ measures the capillary effects.

This equation, which is in some sense *close* to the KdV equation, is not known to be integrable. Its solitary waves, the existence of which was proven in [31, 32] by the degree-theoretic approach, present oscillatory tails. We refer to [146] for the Cauchy problem in L^2 and to [12, 21] for further results on the existence and stability of solitary wave solutions and to [47] for numerical simulations.

In presence of surface tension, the ILW equation has to be modified in the same way. We are not aware of mathematical results on the resulting equation.

Remark 4 The BO equation was fully justified in [219] as a model of long internal waves in a two-fluid system by taking into account the influence of the surface tension at the interface. The existence time of the full two-fluid system is proportional to the surface tension coefficient (which is very small in real oceanographic systems), making the approximation valid only for very short time.

Remark 5 Natural generalizations of the ILW and BO equations arise when looking for weak transverse effects, aiming for instance to understand the transverse instability of the BO or ILW solitons and the oblique interactions of such solitary waves, see Sect. 6.8. In a weakly transverse regime this leads to Kadomtsev-Petviashvili (KP) versions of the BO and ILW equations (see [7, 91, 175] for the derivation, [93, 150, 152] for a mathematical study). We will go back in more details to this issue in Sect. 6.8.

Remark 6 In [179] Matsuno considers the two-fluid system when the upper layer is large with respect to the lower one and in presence of a non trivial topography of the fixed bottom. In this context he derives formally a forced ILW and a forced BO (fBO) equation and describes the effect on the soliton dynamics in the case of the fBO equation.

The paper will be organized as follows. The next section will recall the formal framework of Inverse Scattering for both the BO and ILW equations. Section 3 will be devoted on results obtained by pure PDE techniques while Sect. 4 will describe the rigorous results on the Cauchy problem obtained by IST methods for the BO equation.

Finally in the last section we will briefly comment on the periodic case and on related (non integrable) equations and systems, making conjectures, based on numerical simulations on the long time behavior of solutions to ILW, BO and related equations. We will also discuss various issues related to BO and ILW equations: zero dispersion limit, controllability, transverse stability of solitary waves, modified and higher order equations, interaction of solitary waves.

We provide an extensive bibliography since we aim to cover the relevant papers dealing, under various aspects, with the BO and ILW equations.

Notations We will denote $|\cdot|_p$ the norm in the Lebesgue space $L^p(\mathbb{R}^2)$, $1 \leq p \leq \infty$ and $\|\cdot\|_s$ the norm in the Sobolev space $H^s(\mathbb{R})$, $s \in \mathbb{R}$. We will use the weighted Sobolev space $H^{s,k}(\mathbb{R}) = \{f \in \mathcal{S}'(\mathbb{R}), |\langle x \rangle^k \langle i\partial_x \rangle^s f|_2 < \infty\}$, where $\langle x \rangle = (1+x^2)^{1/2}$. We will denote $\hat{f}$ or $\mathcal{F}(f)$ the Fourier transform of a tempered distribution f. For any $s \in \mathbb{R}$, we define $D^s f$ by its Fourier transform

$$\widehat{D^s f}(\xi) = |\xi|^s \hat{f}(\xi).$$

2 An Overview of the Inverse Scattering Framework for the ILW and BO Equations

It turns out that the ILW and BO equations are among the relatively few physically relevant equations possessing an inverse scattering formalism. We give in this Section a brief historical overview and refer to Sect. 4 for technical details and rigorous and more recent results.

The IST formalism for the BO equation has been given in [4, 6], see also [118, 270] for extensions to a larger class of potentials. However, before those pioneering works, some facts have been discovered showing the (unexpected) rich structure of the BO equation.

R.I. Joseph, K.M. Case, A. Nakamura and Y. Matsuno seem to have been the first authors in the late seventies to notice the specific properties of the BO equation that lead to conjecture its complete integrability. Joseph [112] derived the expression of the 2-soliton. Case [49] derived also the 2-soliton and in [48, 50] suggested the existence of an infinite number of conserved quantities I_n and computed the first non trivial ones, namely

$$I_4 = \int \{u^4/4 + \frac{3}{2}u^2 H(u_x) + 2(u_x)^2\}dx, \tag{2.1}$$

$$I_5 = \int \{u^5/5 + [\frac{4}{3}u^3 H(u_x) + u^2 H(uu_x)] + [2u(H(u_x))^2 + 6u(u_x)^2] \\ - 4u_{xx} H((u_x))\}dx, \tag{2.2}$$

$$I_6 = \int \{u^6/6 + [\frac{5}{4}u^4 H(u_x) + \frac{5}{3}u^3 H(uu_x)] \\ + \frac{5}{2}[5u^2(u_x)^2 + u^2(H(u_x)^2 + 2uH(u_x)H(uu_x)] \\ - 10[(u_x)^2 H(u_x) + 2uu_{xx} H(u_x)] + 8(u_{xx})^2\}dx. \tag{2.3}$$

This lead Case to conjecture the complete integrability of BO, in particular the existence of a Lax pair.

We refer to [193] for a nice review of the various methods used to derive the BO conservation laws.

Remark 7 The "standard" invariants I_2 and I_3 are respectively given by the L^2 norm and the hamiltonian, namely

$$I_2 = \frac{1}{2}\int u^2 dx,$$

$$I_3 = \int \left(\frac{1}{2}|D^{1/2}u|^2 - \frac{1}{6}u^3\right)dx.$$

Nakamura [211] proved the existence of a Bäcklund transform (see also [35]) and the existence of an infinite number of conserved quantities. Furthermore he gave an inverse scattering transform of the BO equation.

In [77] the Benjamin-Ono equation is shown to possess two non-local linear operators, which generate its infinitely many commuting symmetries and constants of the motion in involution. These symmetries define the hierarchy of the BO equation, each member of which is a hamiltonian system. The above operators are the nonlocal analogues of the Lenard operator and its adjoint for the Korteweg-de Vries equation, see the discussion in [234].

Further progress were made by Y. Matsuno. In [163] he followed Hirota's method [100] to transform the BO equation into a bilinear form and deduced the explicit expression of the N-soliton. Investigating the asymptotic behavior of the N-soliton as $t \to \infty$ he noticed that, contrary to the case of the KdV equation, no phase-shift appears as the result of collisions of solitons. Still using the bilinear transformation method he gave in [171] the exact formula for the N-soliton solution to the higher order BO equation

$$u_t = \partial_x(\text{grad } I_5(u)).$$

In [174] Matsuno proved, using a recurrence formula derived from the Bäcklund transformation of the BO equation that the functional derivative of a conserved quantity of the BO equation is a conserved quantity. The interaction of N-solitons is studied in more details in [164]. More specifically, the nature of the interaction of 2-solitons is characterized by the amplitudes of the two solitons. This paper contains also a precise analysis of the asymptotics as $t \to \infty$ of solutions to the linearized equation (analogous of the Airy equation for the KdV equation). In [181] Kaup and Matsuno consider the linearization of the BO equation about the N-soliton solution, proving the linear stability of the N-soliton against infinitesimal perturbations. They also obtain a formal large time asymptotics of solutions to dissipative perturbations of the BO equation. Finally in [172, 173] one finds the asymptotic behavior of the number density function of solitons in the small dispersion limit.

In the survey article [168] a detailed description is given to the interaction process of two algebraic solitons using the pole expansion of the solution, in particular to the effects of small perturbations on the overtaking collision of two BO solitons by employing a direct multisoliton perturbation theory. It is shown that the dynamics of interacting algebraic solitons reveal new aspects which have never been observed in the interaction process of usual solitons expressed in terms of exponential functions.

In addition to [168] we refer to the book [165] for a good description of many "algebraic" aspects of the BO and ILW equations: use of the Hirota bilinear transform method [100], Bäcklund transforms, multi-solitons, BO and ILW hierarchies...

The Lax pair of the BO equation was derived in [212], and [35] while the action-angle variables for the BO equation and their Poisson brackets are determined in [119]. The direct and inverse transforms for BO was formulated by Fokas, Anderson and Ablowitz [4, 6]. The direct problem is a differential Riemann-Hilbert problem while the inverse problem consists in solving linear Fredholm equations. Pure soliton solutions are obtained by solving a linear algebraic system whose coefficients depend linearly on x, t.

More precisely, the Lax pair for the BO equation writes, following [4, 6]

$$\begin{aligned} i\Phi_x^+ + \lambda(\Phi^+ - \Phi^-) &= -u\Phi^+, \\ i\Phi_t^\pm - 2i\lambda\Phi_x^\pm + \Phi_{xx}^\pm - i[u]_x^\pm\Phi^\pm &= -\nu\Phi^\pm, \end{aligned} \qquad (2.4)$$

where

$$[u]^{\pm} = \pm\frac{u}{2} + \frac{1}{2i}Hu$$

and where λ is a constant interpreted as a spectral parameter, ν is an arbitrary constant and $[u]^+$ and $[u]^-$ are the boundary values of functions analytic in the upper and lower half complex z-planes respectively.

The core of the analysis of the IST for BO is the linear spectral problem associated to the first equation in (2.4) which can be interpreted as a differential Riemann-Hilbert problem. This equation yields unique solutions for Φ^+ and Φ^- provided one imposes some boundary conditions as $z \to \infty$ say in the upper half plane. The choice made in [4] is that either $\Phi^+(z, t, \lambda) \to 0$ or 1 as $z \to \infty$, $\Im z > 0$.

One can only consider the (+) functions and thus drop the superscript $^+$. Let (Jost functions) $M, \overline{M}$ denote the "left" eigenfunctions while $N, \overline{N}$ denote the "right" ones. They are specified by the boundary conditions:

$$M \to 1, \overline{M} \to e^{i\lambda x} \text{ as } x \to -\infty; \quad \overline{N} \to 1, N \to e^{i\lambda x} \text{ as } x \to \infty.$$

One then can establish the "scattering equation":

$$M = \overline{N} + \beta(\lambda, t)\theta(\lambda)N, \quad \beta(\lambda, t) = i \int_{-\infty}^{\infty} u(y, t)M(y, t, \lambda)e^{-i\lambda y}dy, \quad (2.5)$$

where $\theta(y) = 1, \lambda > 0$ and $0, \lambda < 0$.

The evolution of the scattering data is given by

$$\rho(k; t) = ik^2 \exp(ik^2 t),$$

$$f(k, t) = f(k, 0) \exp(-ik^2 t).$$

The solution of the "inverse" problem consists essentially in solving (2.5). The evolution of the scattering data is simple and the solution of the inverse problem is characterized by a linear Fredholm equation.

The soliton solutions are obtained by taking

$$\rho(k; t) = 0, \quad N(x, t; k) = 1 - i\sum_{j=1}^{n} \frac{\phi_j}{k - k_j}.$$

The IST formalism for ILW has been given in [133, 134] (see also [235–237]) but, as in the case of the BO equation, facts suggesting the (formal) complete integrability of the equation have been discovered before, for instance the existence of multi-solitons (see [113, 215]).

The ILW equation possesses an infinite sequence of conserved quantities (see e.g. [145, 165]) which leads to a ILW hierarchy. The first non trivial one is

$$J_4(u) = \int_{-\infty}^{\infty} \left(\frac{1}{4}u^4 + \frac{3}{2}u^2 \mathcal{T}(u_x) + \frac{1}{2}u_x^2 + \frac{3}{2}[\mathcal{T}(u_x)]^2 \right. \\ \left. + \frac{1}{\delta}\left[\frac{3}{2}u^3 + \frac{9}{2}u\mathcal{T}(u_x)\right] + \frac{3}{2\delta^2} \right) dx. \tag{2.6}$$

The direct scattering problem is associated with a Riemann-Hilbert problem in a strip of the complex plane.

More precisely, the Lax pair for the ILW equation is given (see [133, 134] and Chapter 4 in [3]):

$$\begin{aligned} iv_x^+ + (u - \lambda)v^+ &= \mu v^-, \\ iv_t^+ + i(2\lambda + \delta^{-1})v_x^\pm + v_{xx}^\pm + (\pm iu_x - \mathcal{T}(u_x) + v)v^\pm &= 0, \end{aligned} \tag{2.7}$$

where λ, μ are parametrize as

$$\lambda(k) = -\frac{1}{2}k \coth(k\delta) \quad \text{and} \quad \mu(k) = \frac{1}{2}k \operatorname{cosech}(k\delta).$$

k is a constant which is interpreted as a spectral parameter and v is an arbitrary constant. Given u, the first equation in (2.7) defines a Riemann-Hilbert problem in the horizontal strip, more precisely, $v^\pm(x)$ represent the boundary values of analytic functions in the strip between $\Im(z) = 0$ and $\Im(z) = 2\delta$, $z = x + iy$ and periodically extended in the vertical direction.

Using the operator $\mathcal{T}$, $v^\pm$ may be written as

$$\begin{aligned} v^+(x) &= \lim_{y \to 0} v(x) = \frac{1}{2}(I - i\mathcal{T})\psi(x), \\ v^-(x) &= \lim_{y \to 0} v(x) = \frac{1}{2}(I + i\mathcal{T})\psi(x), \end{aligned} \tag{2.8}$$

where $\psi \in L^1(\mathbb{R})$, is Hölder of exponent α and for $\Im z \neq 0 \pmod{2\delta}$, $v(z)$ is given by

$$v(z) = \frac{1}{4i\delta} \int_{-\infty}^{\infty} \coth\{\frac{\pi}{2\delta}(y - z)\}\psi(y) dy.$$

Defining

$$W^\pm(x; k) = v^\pm(x; k) \exp\{\frac{1}{2}ik(x \mp i\delta)\},$$

(2.8) can be rewritten as

$$iW_x^+ + [\zeta_+ + 1/(2\delta)](W^+ - W^-) = -uW^+,$$
$$iW_t^\pm - 2i\zeta_+ W_x^\pm + W_{xx}^\pm + [\pm iu_x - \mathcal{T}(u_x)) + \rho]W^\pm = 0, \quad (2.9)$$

where

$$\zeta_\pm(k) = \frac{1}{2}k \pm \frac{1}{2}k\coth(k\delta) \mp 1/(2\delta),$$

and

$$\rho = -k\zeta_+ + \frac{1}{4}k^2 + \nu.$$

The solution of (2.9) is given by the integral equation

$$W^+(x;k) = W_0(x;k) + \int_{-\infty}^{\infty} G(x,y;k)u(y)W^+(y;k)dy \quad (2.10)$$

where $W_0(x;k)$ is the solution of (2.9) corresponding to $u = 0$ and G is the Green function satisfying

$$i\frac{\partial}{\partial x}\{G^+(x,y;k)\} + [\zeta_+ + 1/(2\delta)][G^+(x,y;k) - G^-(x,y;k)] = -\delta(x-y),$$
$$(2.11)$$

where $G^\pm(x,y;k) = G(x \mp i\delta, y; k)$.

We refer to [3] for an integral representation of $G_\pm$. As in the case of the BO equation one needs also the eigenfunction $W^+(x;k)$. We also denote $M, \overline{M}$ the "left" eigenfunctions and $N, \overline{N}$ the "right" eigenfunctions, that have the following asymptotic behavior:

$$M(x;k) \sim 1, \quad \overline{M}(x:k) \sim e^{ikx+k\delta}, \quad \text{as } x \to -\infty$$

$$N(x;k) \sim e^{ikx+k\delta}, \quad \overline{N}(x,k) \sim 1, \quad \text{as } x \to \infty.$$

The eigenfunctions $M, N, \overline{N}$ are related through the completeness relation

$$M(x;k) = a(k)\overline{N}(x;k) + b(k)N(x;k), \quad (2.12)$$

where $a(k)$ and $b(k)$ have the integral representation:

$$a(k) = 1 + \frac{1}{2i\zeta_+(k)}\int_{-\infty}^{\infty} u(y)M(y;k)dy,$$
$$b(k) = -\frac{1}{2i\zeta_+(k)}\theta(\zeta_+ + 1/(2\delta))\int_{-\infty}^{\infty} u(y)M(y;k)e^{-iky-k\delta}dy. \quad (2.13)$$

One can prove that the evolution of the scattering data is simple, namely

$$a(k, t) = a(k, 0), \quad b(k, t) = b(k, 0) \exp\{ik[k \coth(k\delta) - 1/\delta]t\},$$

so that

$$\rho(k, t) = \rho(k, 0) \exp\{ik[k \coth(k\delta) - 1/\delta]\}.$$

Similarly,

$$k_j(t) = k_j(0),$$

$$C_j(t) = C_j(0) \exp\{ik_j[k_j \coth(k_j\delta) - 1/\delta]t\},$$

for $j = 1, 2, \ldots, n$.

The inverse scattering problem is based on Eq. (2.12) and consist in reconstructing M, N, $\overline{N}$ from the knowledge of $a(k), b(k)$ together with appropriate information about the ground states. The formal solution of this Riemann-Hilbert problem is given in [3].

Remark 8 The pure soliton solutions are recovered by taking $\rho(k; t) = 0$.

All the results in this subsection are formal. In particular Fokas and Ablowitz [4] have to make some key spectral assumptions in their definition of the scattering data of the IST for the BO equation. Also, their IST framework does not behave well enough to be solved by iteration.

We refer to Sect. 4 for a review on rigorous results for the Cauchy problem and related issues using IST techniques.

3 Rigorous Results by PDE Methods

3.1 *The Linear Group*

We will recall here the various dispersive estimates satisfied by the linear BO and ILW equations. The linear part of both the BO and the ILW equation defines a unitary group $S_{BO}(t)$, resp. $S_{ILW}(t)$ in all Sobolev spaces $H^s(\mathbb{R})$, $s \geq 0$ which is by Plancherel unitarily equivalent in L^2 to the multiplication by, respectively, $e^{it\xi|\xi|}$ and $e^{it(\xi^2 \coth(\delta\xi) - \xi/\delta)}$.

The fundamental solutions:

$$G_{BO}(x,t) = \mathcal{F}^{-1}(e^{it\xi|\xi|})(x) \quad \text{and} \quad G_{ILW}(x,t) = \mathcal{F}^{-1}(e^{it(\xi^2 \coth(\delta\xi) - \xi/\delta)})(x),$$

play an important role in the dispersive properties of the BO and ILW equations. $G_{BO}(x, 1)$ is easily seen to be a bounded C^∞ function. Its asymptotic behavior is obtained in [247] (particular case of Theorem 3.1):

$$G_{BO}(x, 1) \sim \sqrt{\pi} \cos\left(\left(\frac{|x|}{2}\right)^2 + \frac{\pi}{4}\right), \quad \text{as } x \to -\infty,$$

and

$$G_{BO}(x, 1) \sim \sum_{k=0}^{\infty} (-1)^k \frac{[2(2k+1)]!}{(2k+1)! x^{2(2k+1)+1}} \quad \text{as } x \to +\infty.$$

On the other hand one proves that:

$$|G_{ILW}(x,t)| \lesssim \begin{cases} t^{-\frac{1}{3}} \left(t^{-1/3} x\right)^{-1/4} & \text{for } x \leq t \\ t^{-\frac{1}{2}} & \text{for } x \geq t. \end{cases} \quad (3.1)$$

In both cases one obtains the dispersive estimate

$$|S_{BO}(t)\phi|_\infty, \ |S_{ILW}(t)\phi|_\infty \lesssim \frac{1}{t^{1/2}} |\phi|_1 \quad (3.2)$$

yielding for instance the same Strichartz estimates as the linear one-dimensional Schrödinger group.

The BO and ILW groups display a Kato type local smoothing property [60, 124]. The optimal results for the group $S_{BO}(t)$ are gathered in the next theorem from [125]:

Theorem 1 *There exist constants c_0, c_1 such that*

$$\left(\int_{-\infty}^{\infty} |D^{1/2} S_{BO}(t) u_0(x) dt\right)^{1/2} = c_0 |u_0|_2. \quad (3.3)$$

for any $x \in \mathbb{R}$.

$$|D^{1/2} \int_{-\infty}^{\infty} S_{BO}(t) g(., t) dt|_2 \leq c_1 \int_{-\infty}^{\infty} \left(\int_{-\infty}^{\infty} |g(x,t)|^2 dt\right)^{1/2} dx, \quad (3.4)$$

and

$$\sup_x \left(\int_{-\infty}^{\infty} |\partial_x \left(\int_0^t S_{BO}(t-\tau) f(.,\tau) d\tau \right)|^2 dt \right)^{1/2}$$
$$\leq c_1 \int_{-\infty}^{\infty} \left(\int_{-\infty}^{\infty} |f(x,t)|^2 dt \right)^{1/2} dx. \quad (3.5)$$

As aforementioned the Strichartz estimates are consequences of the dispersive estimate (3.2). We state here the version in [125]

Theorem 2 *Let $p \in [2, \infty)$ and q be such that $2/q = 1/2 - 1/p$. Then*

$$\left(\int_{-\infty}^{\infty} |S_{BO}(t)u_0|_p^q dt \right)^{1/q} \leq c|u_0|_2 \quad (3.6)$$

and

$$\left(\int_{-\infty}^{\infty} \left| \int_0^t S_{BO}(t-\tau) f(.,\tau) d\tau \right|_p^q \right)^{1/q} \leq C \left(\int_{-\infty}^{\infty} |f(.,\tau)|_{p'}^{q'} dt \right)^{1/q'} \quad (3.7)$$

where $1/p + 1/p' = 1/q + 1/q' = 1$.

The last group of dispersive estimates are those on the maximal function $\sup_t S_{BO}(t)$. Then, see [125]:

Theorem 3

$$\left(\int_{-\infty}^{\infty} \sup_{-\infty < t < \infty} |S_{BO}(t)u_0(x)|^4 dx \right)^{1/4} \leq c|D^{1/4} u_0|_2, \quad (3.8)$$

and

$$\left(\int_{-\infty}^{\infty} \sup_{-\infty < t < \infty} |\int_0^t S_{BO}(t-\tau) f(.,\tau) d\tau|^4 dx \right)^{1/4}$$
$$\leq c \left(\int_{-\infty}^{\infty} \left(\int_{-\infty}^{\infty} |D^{1/2} f(x,t)| dt \right)^{4/3} dx \right)^{3/4}. \quad (3.9)$$

Moreover for $s > \frac{1}{2}$ and $\rho > \frac{3}{4}$

$$\left(\int_{-\infty}^{\infty} \sup_{0 \leq t \leq T} |S_{BO}(t)u_0(x)|^2 \right)^{1/2} \leq c(1+T)^\rho ||u_0||_s. \quad (3.10)$$

Remark 9 As will be discussed below, and contrary to the case of the KdV equation, the above estimates cannot be used to define a functional space on which one could implement an iterative scheme based on the Duhamel formulation

$$u(t) = S_{BO}(t)u_0 + \int_0^t S_{BO}(t-\tau)uu_x(\tau)d\tau.$$

This is due to the quasilinear nature of the BO equation.

3.2 An Easy Result

Being skew-adjoint perturbations of the inviscid Burgers equation, the ILW and BO equations are easily seen to be locally well-posed in $H^s(\mathbb{R})$, $s > \frac{3}{2}$, [104, 240]. It turns out that they are actually *globally* well posed in the same range of Sobolev spaces.

The argument goes back to [240] and was applied in this context in [1]. We describe it briefly for the BO equation. The following computations are formal but can be easily justified by smoothing the equation and/or the initial data. We start from the identity

$$\frac{1}{2}\frac{d}{dt}|D^s u|_0^2 + (D^s(uu_x), D^s u) = 0. \tag{3.11}$$

For $s > \frac{3}{2}$ one deduces after some manipulations using commutator estimates and the Sobolev imbedding

$$|u_x|_\infty \leq \frac{c}{\sqrt{\eta}}||u||_{3/2+\eta}, \quad \forall \eta > 0,$$

that

$$|(D^s(uu_x), D^s u)| \leq \frac{c}{\sqrt{\eta}}||u||_{3/2+\eta}||u||_s^2,$$

holding for any $\eta > 0$ such that $\frac{3}{2} + \eta < s$. For such an η we have the interpolation inequality

$$||u||_{3/2+\eta} \leq c||u||_s^{2\gamma\eta}||u||_{3/2}^{1-2\gamma\eta}$$

where $1 - \theta = \frac{2\eta}{2s-3} =: 2\gamma\eta$ and c is a constant independent of η.

On the other hand, using the conservation law I_5 one can establish the uniform bound

$$||u||_{L^\infty(\mathbb{R};H^{3/2}(\mathbb{R}))} \leq ca(||u_0||_{3/2}). \tag{3.12}$$

Combining those estimates yields

$$\frac{1}{2}\frac{d}{dt}\|u\|_s^2 \leq C\left(\frac{1}{\sqrt{\eta}}\|u\|_s^{2+2\gamma\eta}\right), \tag{3.13}$$

where the constant C is defined by $C = [ca(\|u_0\|_{3/2})]^{1-2\gamma\eta}$.
Integrating this last inequality in time leads to the estimate

$$\|u(\cdot,t)\|_s^2 \leq y(t), \tag{3.14}$$

where y is the solution of the differential equation

$$y'(t) = \frac{C}{\sqrt{\eta}} y(t)^{1+\gamma\eta}, \quad y(0) = \|u_0\|_s^2$$

on its maximal interval of existence $[0, T(\eta)]$. Here $\gamma = 1/(2s-3)$. This equation is easily integrated, finding that

$$y(t) = (\|u_0\|_s^{-2\gamma\eta} - \gamma\sqrt{\eta}Ct)^{-1/\gamma\eta},$$

whence

$$T(\eta) = \frac{1}{\gamma C\sqrt{\eta}}\|u_0\|_s^{-2\gamma\eta} \to \infty \quad \text{as } \eta \to 0.$$

For any fixed $T > 0$ we can choose $\eta > 0$ so small that $T < \frac{1}{2}T(\eta)$. Then it follows that for $0 \leq t \leq T$,

$$y(t) \leq c(T; \|u_0\|_{3/2})\|u_0\|_s^2.$$

This implies an a priori bound that is crucial to prove the existence of a unique global solution $u \in L^\infty_{loc}(\mathbb{R}, H^s(\mathbb{R}))$ emerging from an initial data $u_0 \in H^s(\mathbb{R})$, $s > 3/2$.

The strong continuity in time and the continuity of the flow map is established by using the Bona-Smith method [41].

Remark 10

1. A similar global result holds true for the ILW equation, see [1].
2. As aforementioned, it is proved in [1] that the H^s, $s > 3/2$ solutions of the ILW equation converge to that of the BO (resp. KdV) equation when $\delta \to +\infty$ (resp. $\delta \to 0$). Recall that in the KdV case, (and this is often missed in the literature...) one has to rescale the ILW solution u_δ as

$$v_\delta(x,t) = \frac{3}{\delta}u_\delta(x, \frac{3}{\delta}t),$$

and then let $\delta \to 0$, to obtain a KdV solution.

3.3 Global Weak Solutions

By deriving local smoothing properties *à la Kato* for the nonlinear equation and various delicate commutator estimates, Ginibre and Velo [86–89] proved various global existence results of weak solutions for a class of generalized BO equations. For the BO equation itself (and also for the ILW equation[5]), this implies the existence of a global weak solution in $L^\infty(\mathbb{R}; L^2(\mathbb{R})) \cap L^2_{\text{loc}}((\mathbb{R}; H^{1/2}_{\text{loc}}(\mathbb{R}))$ for any initial data in $L^2(\mathbb{R})$ and similar results for data with higher regularity, e.g. $H^1(\mathbb{R})$. Those results hold also true for the ILW equation (see [89, Section 6]).

For initial data in the energy space $H^{1/2}(\mathbb{R})$ the existence of global weak solutions in the space $L^\infty(\mathbb{R}; H^{1/2}(\mathbb{R}))$ has been established in [240] for both the BO and ILW equations.

3.4 Semilinear Versus Quasilinear

Deciding whether a nonlinear dispersive equation is semilinearly well-posed is somewhat subtle. The distinction plays a fundamental role in the choice of the method for solving the Cauchy problem.

To start with we consider the case of the Benjamin-Ono equation following [206]. Since they are relatively simple we will give complete proofs.

We thus consider the Cauchy problem

$$\begin{cases} u_t - Hu_{xx} + uu_x = 0, \ (t,x) \in \mathbb{R}^2, \\ u(0,x) = \phi(x). \end{cases} \quad (3.15)$$

Setting $S(t) = e^{tH\partial_x^2}$, we write (3.15) as an integral equation:

$$u(t) = S(t)\phi - \int_0^t S(t-t')(u_x(t')u(t'))dt'. \quad (3.16)$$

The main result is the following

Theorem 4 *Let $s \in \mathbb{R}$ and T be a positive real number. Then there does not exist a space X_T continuously embedded in $C([-T,T], H^s(\mathbb{R}))$ such that there exists $C > 0$ with*

$$\|S(t)\phi\|_{X_T} \le C\|\phi\|_{H^s(\mathbb{R})}, \quad \phi \in H^s(\mathbb{R}), \quad (3.17)$$

and

$$\left\|\int_0^t S(t-t')\left[u(t')u_x(t')\right]dt'\right\|_{X_T} \le C\|u\|_{X_T}^2, \quad u \in X_T. \quad (3.18)$$

---

[5]The results hold also for a rather general class of nonlocal dispersive equations, see [89].

Note that (3.17) and (3.18) would be needed to implement a Picard iterative scheme (actually the second iteration) on (3.16), in the space X_T. As a consequence of Theorem 4 we can obtain the following result.

Theorem 5 *Fix $s \in \mathbb{R}$. Then there does not exist a $T > 0$ such that (3.15) admits a unique local solution defined on the interval $[-T, T]$ and such that the flow-map data-solution*

$$\phi \longmapsto u(t), \quad t \in [-T, T],$$

for (3.15) is C^2 differentiable at zero from $H^s(\mathbb{R})$ to $H^s(\mathbb{R})$.

Remark 11 This result implies that the Benjamin-Ono equation is "quasilinear". It has been precised in [131] where it is shown that the flow map cannot even be locally Lipschitz in H^s for $s \geq 0$. This is of course in strong contrast with the KdV equation.

3.4.1 Proof of Theorem 4

Suppose that there exists a space X_T such that (3.17) and (3.18) hold. Take $u = S(t)\phi$ in (3.18). Then

$$\left\| \int_0^t S(t - t') \left[(S(t')\phi)(S(t')\phi_x) \right] dt' \right\|_{X_T} \leq C \| S(t)\phi \|_{X_T}^2.$$

Now using (3.17) and that X_T is continuously embedded in $C([-T, T], H^s(\mathbb{R}))$ we obtain for any $t \in [-T, T]$ that

$$\left\| \int_0^t S(t - t') \left[(S(t')\phi)(S(t')\phi_x) \right] dt' \right\|_{H^s(\mathbb{R})} \lesssim \| \phi \|_{H^s(\mathbb{R})}^2. \tag{3.19}$$

We show that (3.19) fails by choosing an appropriate ϕ.

Take ϕ defined by its Fourier transform as[6]

$$\widehat{\phi}(\xi) = \alpha^{-\frac{1}{2}} \mathbf{1}_{I_1}(\xi) + \alpha^{-\frac{1}{2}} N^{-s} \mathbf{1}_{I_2}(\xi), \quad N \gg 1, \; 0 < \alpha \ll 1,$$

where I_1, I_2 are the intervals

$$I_1 = [\alpha/2, \alpha], \quad I_2 = [N, N + \alpha].$$

Note that $\|\phi\|_{H^s} \sim 1$. We will use the next lemma, similar to Lemma 4 in [205]:

---

[6]The analysis below works as well for $\mathcal{R}e\, \phi$ instead of ϕ (some new harmless terms appear).

Lemma 1 *The following identity holds:*

$$\int_0^t S(t-t')\left[(S(t')\phi)(S(t')\phi_x)\right] dt' =$$

$$\int_{\mathbb{R}^2} e^{ix\xi+itp(\xi)} \xi \, \hat{\phi}(\xi_1)\hat{\phi}(\xi-\xi_1) \frac{e^{it(p(\xi_1)+p(\xi-\xi_1)-p(\xi))} - 1}{p(\xi_1) + p(\xi-\xi_1) - p(\xi)} d\xi d\xi_1, \quad (3.20)$$

where $p(\xi) = \xi|\xi|$.

According to the above lemma,

$$\int_0^t S(t-t')\left[(S(t')\phi)(S(t')\phi_x)\right] dt' = c(f_1(t,x) + f_2(t,x) + f_3(t,x)),$$

where, from the definition of ϕ, we have the following representations for f_1, f_2, f_3:

$$f_1(t,x) = \frac{c}{\alpha} \int_{\substack{\xi_1 \in I_1 \\ \xi-\xi_1 \in I_1}} \xi \, e^{ix\xi+it\xi|\xi|} \frac{e^{it(\xi_1|\xi_1|+(\xi-\xi_1)|\xi-\xi_1|-\xi|\xi|)} - 1}{\xi_1|\xi_1| + (\xi-\xi_1)|\xi-\xi_1| - \xi|\xi|} d\xi d\xi_1,$$

$$f_2(t,x) = \frac{c}{\alpha N^{2s}} \int_{\substack{\xi_1 \in I_2 \\ \xi-\xi_1 \in I_2}} \xi \, e^{ix\xi+it\xi|\xi|} \frac{e^{it(\xi_1|\xi_1|+(\xi-\xi_1)|\xi-\xi_1|-\xi|\xi|)} - 1}{\xi_1|\xi_1| + (\xi-\xi_1)|\xi-\xi_1| - \xi|\xi|} d\xi d\xi_1,$$

$$f_3(t,x) = \frac{c}{\alpha N^s} \int_{\substack{\xi_1 \in I_1 \\ \xi-\xi_1 \in I_2}} \xi \, e^{ix\xi+it\xi|\xi|} \frac{e^{it(\xi_1|\xi_1|+(\xi-\xi_1)|\xi-\xi_1|-\xi|\xi|)} - 1}{\xi_1|\xi_1| + (\xi-\xi_1)|\xi-\xi_1| - \xi|\xi|} d\xi d\xi_1$$

$$+ \frac{c}{\alpha N^s} \int_{\substack{\xi_1 \in I_2 \\ \xi-\xi_1 \in I_1}} \xi \, e^{ix\xi+it\xi|\xi|} \frac{e^{it(\xi_1|\xi_1|+(\xi-\xi_1)|\xi-\xi_1|-\xi|\xi|)} - 1}{\xi_1|\xi_1| + (\xi-\xi_1)|\xi-\xi_1| - \xi|\xi|} d\xi d\xi_1.$$

Set

$$\chi(\xi,\xi_1) := \xi_1|\xi_1| + (\xi-\xi_1)|\xi-\xi_1| - \xi|\xi|.$$

Then clearly

$$\mathcal{F}_{x\mapsto\xi}(f_1)(t,\xi) = \frac{c\xi e^{it\xi|\xi|}}{\alpha} \int_{\substack{\xi_1 \in I_1 \\ \xi-\xi_1 \in I_1}} \frac{e^{it\chi(\xi,\xi_1)} - 1}{\chi(\xi,\xi_1)} d\xi_1,$$

$$\mathcal{F}_{x\mapsto\xi}(f_2)(t,\xi) = \frac{c\xi e^{it\xi|\xi|}}{\alpha N^{2s}} \int_{\substack{\xi_1 \in I_2 \\ \xi-\xi_1 \in I_2}} \frac{e^{it\chi(\xi,\xi_1)} - 1}{\chi(\xi,\xi_1)} d\xi_1,$$

$$\mathcal{F}_{x\mapsto\xi}(f_3)(t,\xi)$$
$$= \frac{c\xi e^{it|\xi|}}{\alpha N^s}\left(\int_{\substack{\xi_1\in I_1 \\ \xi-\xi_1\in I_2}} \frac{e^{it\chi(\xi,\xi_1)}-1}{\chi(\xi,\xi_1)}d\xi_1 + \int_{\substack{\xi_1\in I_2 \\ \xi-\xi_1\in I_1}} \frac{e^{it\chi(\xi,\xi_1)}-1}{\chi(\xi,\xi_1)}d\xi_1\right).$$

Since the supports of $\mathcal{F}_{x\mapsto\xi}(f_j)(t,\xi)$, $j=1,2,3$, are disjoint, we have

$$\left\|\int_0^t S(t-t')[(S(t')\phi)(S(t')\phi_x)]dt'\right\|_{H^s(\mathbb{R})} \geq \|f_3(t,\cdot)\|_{H^s(\mathbb{R})}.$$

We now give a lower bound for $\|f_3(t,\cdot)\|_{H^s(\mathbb{R})}$. Note that for $(\xi_1,\xi-\xi_1)\in I_1\times I_2$ or $(\xi_1,\xi-\xi_1)\in I_2\times I_1$ one has $|\chi(\xi,\xi_1)| = 2|\xi_1(\xi-\xi_1)| \sim \alpha N$. Hence it is natural to choose α and N so that $\alpha N = N^{-\epsilon}$, $0<\epsilon\ll 1$. Then

$$\left|\frac{e^{it\chi(\xi,\xi_1)}-1}{\chi(\xi,\xi_1)}\right| = |t| + O(N^{-\epsilon})$$

for $\xi_1\in I_1, \xi-\xi_1\in I_2$ or $\xi_1\in I_2, \xi-\xi_1\in I_1$. Hence for $t\neq 0$,

$$\|f_3(t,\cdot)\|_{H^s(\mathbb{R})} \gtrsim \frac{N\,N^s\,\alpha\,\alpha^{\frac{1}{2}}}{\alpha N^s} = \alpha^{\frac{1}{2}}N.$$

Therefore we arrive at

$$1 \sim \|\phi\|^2_{H^s(\mathbb{R})} \geq \|f_3(t,\cdot)\|_{H^s(\mathbb{R})} \geq \alpha^{\frac{1}{2}}N \sim N^{\frac{1-\epsilon}{2}},$$

which is a contradiction for $N\gg 1$ and $\epsilon\ll 1$. This completes the proof of Theorem 4.

3.5 Proof of Theorem 5

Consider the Cauchy problem

$$\begin{cases} u_t - Hu_{xx} + uu_x = 0, \\ u(0,x) = \gamma\phi, \quad \gamma\ll 1, \quad \phi\in H^s(\mathbb{R}). \end{cases} \tag{3.21}$$

Suppose that $u(\gamma,t,x)$ is a local solution of (3.21) and that the flow map is C^2 at the origin from $H^s(\mathbb{R})$ to $H^s(\mathbb{R})$. We have successively

$$u(\gamma,t,x) = \gamma S(t)\phi + \int_0^t S(t-t')u(\gamma,t',x)u_x(\gamma,t',x)dt'$$

$$\frac{\partial u}{\partial \gamma}(0, t, x) = S(t)\phi(x) =: u_1(t, x)$$

$$\frac{\partial^2 u}{\partial \gamma^2}(0, t, x) = -2 \int_0^t S(t - t') \left[(S(t')\phi)(S(t')\phi_x)\right] dt'.$$

The assumption of C^2 regularity yields

$$\left\| \int_0^t S(t - t') \left[(S(t')\phi)(S(t')\phi_x)\right] dt' \right\|_{H^s(\mathbb{R})} \lesssim \|\phi\|_{H^s(\mathbb{R})}^2.$$

But the above estimate is (3.19), which has been shown to fail in Sect. 3.4.1.

Remark 12 The previous results are in fact valid in a more general context. We consider now the class of equations

$$u_t + uu_x - Lu_x = 0, \quad u(0, x) = \phi(x), \quad (t, x) \in \mathbb{R}^2, \tag{3.22}$$

where L is defined via the Fourier transform

$$\widehat{Lf}(\xi) = \omega(\xi)\hat{f}(\xi).$$

Here $\omega(\xi)$ is a continuous real-valued function. Set $p(\xi) = \xi \omega(\xi)$. We assume that $p(\xi)$ is differentiable and such that, for some $\gamma \in \mathbb{R}$,

$$|p'(\xi)| \lesssim |\xi|^\gamma, \quad \xi \in \mathbb{R}. \tag{3.23}$$

The next theorem shows that (3.22) shares the bad behavior of the Benjamin–Ono equation with respect to iterative methods.

Theorem 6 *Assume that (3.23) holds with $\gamma \in [0, 2[$. Then the conclusions of Theorems 4, 5 are valid for the Cauchy problem (3.22).*

The proof follows the considerations of the previous section. The main point in the analysis is that for $\xi_1 \in I_1, \xi - \xi_1 \in I_2$ one has

$$|p(\xi_1) + p(\xi - \xi_1) - p(\xi)| \lesssim \alpha N^\gamma, \quad \alpha \ll 1, \quad N \gg 1.$$

We choose α and N such that $\alpha N^\gamma = N^{-\epsilon}, 0 < \epsilon \ll 1$. We take the same ϕ as in the proof of Theorem 4 and arrive at the lower bound

$$1 \sim \|\phi\|_{H^s(\mathbb{R})}^2 \geq \alpha^{\frac{1}{2}} N = N^{1 - \frac{\gamma + \epsilon}{2}},$$

which fails for $0 < \epsilon \ll 1$, $\gamma \in [0, 2[$.

Here we give several examples where Theorem 6 applies.

- Pure power dispersion:

$$\omega(\xi) = |\xi|^\gamma, \quad 0 \leq \gamma < 2.$$

This dispersion corresponds to a class of models for vorticity waves in the coastal zone (see [248]). It is interesting to notice that the case $\gamma = 2$ corresponds to the KdV equation which can be solved by iterative methods, see e.g. [126]. Therefore Theorem 6 is sharp for a pure power dispersion. However, the Cauchy problem corresponding to $1 \leq \gamma < 2$ has been proven in [99] to be locally well-posed by a compactness method combined with sharp estimates on the linear group for initial data in $H^s(\mathbb{R})$, $s \geq (9 - 3\gamma)/4$.

- Perturbations of the Benjamin-Ono equation:

$\omega(\xi) = (|\xi|^2 + 1)^{\frac{1}{2}}$. This case corresponds to an equation introduced by Smith [248] for continental shelf waves.

$\omega(\xi) = \xi \coth(\xi)$. This corresponds to the Intermediate long wave equation.

Remark 13 As it is clear from the above proof, the "ill-posedness" of the Benjamin-Ono equation is due to the "bad" interactions of very small and very large frequencies. This phenomena does not occur in the periodic case, for initial data say of zero mean, see Sect. 5.2 below.

This is in contrast with similar "ill-posedness" results for the KP-I equation (see [205]) which are due to the large zero set of a resonant function and which persist in the periodic case.

Remark 14 The generalized BO equation

$$u_t + u^p u_x - H u_{xx} = 0, \quad p \geq 2, \tag{3.24}$$

is in fact *semilinear* since the Cauchy problem for small data in suitable Sobolev spaces was proven in [125] to be locally-well posed by a contraction method.

3.6 Global Well-Posedness in L^2

The global well-posedness result in H^s, $s > 3/2$ does not use any dispersive estimate (but it uses the existence of a non trivial invariant). The results of Ginibre and Velo use a Kato type (dispersive) local smoothing but they concern only weak solutions. The next step was to use in a more crucial way the dispersion properties to obtain the local well-posedness (LWP) of the Cauchy problem in larger Sobolev spaces, aiming to reach at least the energy space $H^{1/2}(\mathbb{R})$.

The first significant result in that direction was from Koch and Tzvetkov [132] who proved the LWP in $H^s(\mathbb{R})$, $s > 5/4$.[7] The main idea of the proof is to improve the dispersive estimates (Strichartz estimates) by localizing them in space frequency dependent time intervals together with classical energy estimates. This method was improved by Kenig and Koenig [121] who obtained LWP in the space $H^s(\mathbb{R})$, $s > 9/8$.

A breakthrough was made by Tao [255] who obtained the LWP (and thus the global well-posedness using the first non trivial invariant) in $H^1(\mathbb{R})$. The new ingredient is to apply a gauge transformation (a variant of the classical Cole-Hopf transform for the Burgers equation) in order to eliminate the terms involving the interaction of very low and very high frequencies, where the derivative falls on the very high frequencies. Note that those interactions are responsible of the lack of regularity of the flow map, see last paragraph and [131].

More precisely, to obtain the estimates at a H^1 level, Tao introduces the new unknown

$$w = \partial_x P_{+\text{hi}}(e^{-1/2F}),$$

where F is some spacial primitive of u and $P_{+\text{hi}}$ is the projection on high positive frequencies. Then w satisfies the equation

$$\partial_t w - i\partial_x^2 w = -\partial_x P_{+\text{hi}}(\partial_x^{-1} w P_- \partial_x u) + \text{negligible terms},$$

where P_- is the projection on negative frequencies.

Thanks to the frequency projections, the nonlinear term appearing in the right hand side does not involve any low-high frequency interaction terms. Finally, to invert this gauge transform, one gets an equation of the form

$$u = 2i e^{\frac{i}{2}F} w + \text{negligible terms}. \tag{3.25}$$

The gauge transformation is not related to the complete integrability of the BO equation since a variant of the gauge transform is used in [99] in order to establish the global well-posedness in L^2 of the (non integrable) fractional KdV equation

$$u_t + u u_x - D^\alpha u_x = 0 \tag{3.26}$$

for $\alpha \in (1, 2)$.

Further improvements were brought by Burq and Planchon [45] who proved LWP in $H^s(\mathbb{R})$, $s > 1/4$ and thus global-wellposedness in the energy space $H^{1/2}(\mathbb{R})$.

Finally Kenig and Ionescu [103] proved the local (and thus global) well-posedness in the space $H^s(\mathbb{R})$, $s \geq 0$. Uniqueness however is obtained in the

---

[7]Ponce [232] used dispersive properties but only reached $H^{3/2}(\mathbb{R})$.

class of limits of smooth functions. Both [103] and [45] use Tao's ideas in the context of Bourgain spaces. The main difficulty is that Bourgain's spaces do not enjoy an algebra property so that one looses regularity when estimating u in terms of w in (3.25). To overcome this difficulty Burq and Planchon first paralinearize the equation and then use a localized version of the gauge transform on the worst nonlinear term. On the other hand, Ionescu and Kenig decompose the solution in two parts: the first one is the smooth solution of BO evolving from the low frequency part of the initial data while the second one solves a dispersive system renormalized by a gauge transformation involving the first part. This system is solved via a fixed point argument in a dyadic version of Bourgain's space with a special structure in low frequencies.

In [200] Molinet and Pilod simplified the proof of Ionescu and Kenig and furthermore proved stronger uniqueness properties, in particular they obtain the unconditional uniqueness in the space $L^\infty(0, T; H^s(\mathbb{R}))$ for $s > 1/4$.

Methods using gauge transformations are very good to obtain low regularity results but behave badly with respect to perturbations of the BO equation. In particular it is not clear if they apply to the ILW equation.

Molinet and Vento [209] proposed a method which is less powerful to get low regularity results (say in $L^2(\mathbb{R})$) but allows to deal with perturbations of the BO equation, in particular to the ILW equation. Their approach combines classical energy estimates with Bourgain type estimates on a time interval that does not depend on the space frequency. It yields local well-posedness in $H^s(\mathbb{R}), s \geq 1/2$, the unconditional uniqueness holding only when $s > 1/2$. The method works as well in the periodic case and also for more general dispersion symbols, in particular similar results apply to the ILW equation. Since it does not rely on the gauge transform the method allows to prove strong convergence results in the energy space for solutions of viscous perturbations of the BO equation (see e.g. [223] for physical examples and [176] for a formal multi-soliton dissipative perturbation of BO).

Combining this technique with refined Strichartz estimates and modified energies, Molinet et al. [203] extended this result to the regularity $H^s(\mathbb{R}), s > \frac{1}{4}$. Moreover their method also applies to fractional KdV equation with low dispersion, that is to

$$u_t + u u_x - D^\alpha u_x = 0,$$

with $0 < \alpha \leq 1$ yielding a local well-posedness result in $H^s(\mathbb{R}), s > \frac{3}{2} - \frac{5\alpha}{4}$, and thus the global well-posedness in the energy space $H^{\alpha/2}(\mathbb{R})$ for $\alpha > \frac{6}{7}$. We refer to [127, 129] for numerical simulations.

Recently Ifrim and Tataru [101] revisited the global well-posedness of BO in $L^2(\mathbb{R})$ by a normal form approach. More precisely they split the quadratic part into two parts, a milder one and a paradifferential one. The normal form correction is then constructed in two steps, a direct quadratic correction for the milder part and a renormalization type correction for the paradifferential part. The second step use a paradifferential version of Tao's renormalization.

3.7 Long Time Dynamics

We have seen that the Cauchy problem is well understood in Sobolev spaces $H^s(\mathbb{R})$, $s \geq 0$. The long time dynamics is much less understood (see Sect. 5.4 below for conjectures on solutions emerging from initial data of arbitrary size). The case of small initial data is considered in [101] where the following result is proved

Theorem 7 *Assume that the initial data u_0 for the BO equation satisfies*

$$|u_0|_2 + |xu_0|_2 \leq \epsilon \ll 1.$$

Then the global associated solution u satisfies the dispersive decay bounds

$$|u(t,x)| + |Hu(x,t)| \lesssim \epsilon |t|^{-1/2} \langle x_- t^{-1/2} \rangle^{-1/2}, \text{ where } x_- = -\min(x, 0)$$

up to time

$$|t| \lesssim T_\epsilon := e^{\frac{c}{\epsilon}}, \quad c \ll 1.$$

Remark 15 We recall that the linear BO flow satisfies the decay property

$$||e^{-tH\partial_x^2}||_{L^1 \to L^\infty} \lesssim t^{-1/2}.$$

The better decay rate in the region $x < 0$ is due to the positivity of the phase velocity $|\xi|$ which send the propagating waves to the right and the dispersive waves to the left.

We refer to [102] for similar results on ILW.

3.8 Solitary Waves

Both the BO and ILW equations possess explicit solitary wave solutions. As aforementioned, Benjamin [30] found the profile of the BO one, namely

$$\phi(x) = \frac{4}{1+x^2}, \qquad (3.27)$$

leading to the family of solitary waves $\phi_c(x - ct)$ where c is a positive constant and

$$\phi_c(y) = \frac{4c}{c^2 y^2 + 1}.$$

Note that since the symbol of the dispersion operator is not smooth, Paley-Wiener type arguments imply that the solitary wave cannot be exponentially decreasing, see [39].

Note also that $|\phi_c|_1 \equiv 4\pi, \forall c > 0$ while $|\phi_c|_2, |\phi_c|_\infty \to 0$ as $c \to 0$, thus BO possesses arbitrary small solitary waves.

On the other hand the symbol in the ILW equation is smooth and the explicit solitary wave found in [111] decays exponentially, namely for arbitrary $C > 0$ and $\delta > 0$

$$\phi_{C,\delta}(x) = \frac{2a \sin(a\delta)}{\cosh(ax) + \cos(a\delta)}, \tag{3.28}$$

where a is the unique solution of the transcendental equation

$$a\delta \cot(a\delta) = (1 - C\delta), \quad a \in (0, \pi/\delta).$$

Before considering the stability properties of those solitary waves, we review important *uniqueness* results. Amick and Toland ([19], see also [8, 18]), using the maximum principle for linear elliptic equations, estimates on a Green function and the Cauchy-Riemann equation, proved the uniqueness of the BO solitary wave in the sense that the only solutions to the pseudo-differential equation

$$u(x)^2 - u(x) = \mathcal{G}(u)(x), \quad x \in \mathbb{R},$$

which satisfies the boundary condition

$$u(x) \to 0 \text{ as } |x| \to \infty,$$

where

$$\mathcal{G}(f)(x) = \frac{1}{2\pi} \int_{-\infty}^{\infty} |\xi| e^{-i\xi x} \left(\int_{-\infty}^{\infty} f(\eta) e^{i\xi \eta} d\eta \right) d\xi$$

are the functions

$$u(x) = 0 \quad \text{and} \quad u_a(x) = \frac{4}{1 + (x-a)^2}, \quad a \in \mathbb{R}.$$

A similar result holds true for the ILW equation. Let

$$(\mathcal{N}_\delta + \gamma)\phi = \phi^2 \tag{3.29}$$

where[8]

$$\mathcal{F}(\mathcal{N}_\delta u)(\xi) = (\xi \coth \xi\delta)\mathcal{F}(u)\xi$$

be the equation of a solitary wave $u(x, t) = \phi(x - Ct)$ to the ILW equation, where $\gamma = C - 1/\delta$.

---

[8] Note that $\mathcal{F}(\mathcal{N}_\infty u)(\xi) = |\xi|\hat{u}(\xi)$.

Then Albert and Toland ([16] and also [8]) proved that for $\delta > 0$ and $C > 0$ be given, if $\phi \in L^2(\mathbb{R})$ is a non trivial solution of (3.29), then there exists $b \in \mathbb{R}$ such that $\phi(x) = \phi_{C,\delta}(x+b)$.

The proof in [8] relies on two special properties of the operator $\mathcal{N}_\delta$ that degenerate into classical ones (linked to the Hilbert transform) when $\delta = \infty$, providing another proof of the Amick-Toland uniqueness result for the BO equation.

We turn now to stability issues. The orbital stability of the BO and ILW solitary waves can be obtained by the classical Cazenave-Lions method in [51] (minimization of the Hamiltonian with fixed L^2 norm) providing orbital stability in the energy space $H^{1/2}(\mathbb{R})$ but the first known proofs were by using the Souganidis-Strauss method, see [11, 33, 42, 43] for BO and [9–11] for ILW.

The H^1 orbital stability of the Benjamin-Ono 2-soliton is proven in [217]. The 2-soliton of velocities $c_1 > 0, c_2 > 0$ with $c_1 < c_2$ is explicitly given in [164] as

$$\phi_{c_1,c_2}(t,x) = 4\frac{c_2\theta_1^2 + c_1\theta_2^2 + (c_1+c_2)c_{12}}{(\theta_1\theta_2 - c_{12})^2 + (\theta_1 + \theta_2)^2},$$

where $\theta_n = c_n(x - c_n t), n = 1, 2$ and $c_{12} = \left(\frac{c_1+c_2}{c_1-c_2}\right)^2$.

The proof in [217] relies on the integrability of BO, and is reminiscent of the similar one for the stability of the KdV 2-soliton (see [158]), namely it uses the fact that the 2-soliton locally minimize the invariant I_4 subject to given values of I_3 and I_2.

An alternative characterization of the 2-soliton uses the self-adjoint operator M that appears in the Lax pair associated to the integrable equation (KdV or BO). Recall that the Lax pair for BO was constructed in [4]. The 2-solitons are the potentials for which the self-adjoint operator M has two eigenvalues.

The proof involves the spectral analysis of the one-parameter family of self-adjoint operators $L(t)$ which are the linearization of

$$I_4'(u) + \alpha I_3'(u) + \beta I_2'(u) = 0$$

at the double soliton. Contrary to the KdV case where $L(t)$ is a fourth order self-adjoint linear ordinary differential operator, for BO one obtains a nonlocal operator since the Hilbert transform appears in it. This makes the spectral analysis more complicated. The new approach in [217] consists in making a simplification in the spectral problem to reduce the spectral analysis of the one-parameter family $L(t)$ to the analysis of the spectra of two stationary operators L_1 and L_2. The proof is then reduced to proving the two facts:

1. L_1 has one negative eigenvalue and L_2 has no negative eigenvalue;
2. zero is a simple eigenvalue of L_1 and L_2.

The asymptotic stability of the BO solitons in the energy space $H^{1/2}(\mathbb{R})$ is proven in [92, 122].

We describe briefly the statement of [122], denoting by

$$Q(x) = \frac{4}{1+x^2}$$

the profile of the BO soliton and $Q_c(x) = cQ(cx)$.

Theorem 8 *There exist $C, \alpha_0 > 0$, such that if $u_0 \in H^{\frac{1}{2}}(\mathbb{R})$ satisfies $\|u_0 - Q_c\|_{\frac{1}{2}} = \alpha \leq \alpha_0$, then there exist $c^+ > 0$ with $|c^+ - 1| \leq C\alpha$ and a C^1 function $\rho(t)$ such that the solution of BO with $u(0) = u_0$ satisfies*

$$u(t, . + \rho'(t)) \rightharpoonup Q_{c^+} \quad \text{in } H^{\frac{1}{2}} \text{ weak}, \quad |u(t) - Q_{c^+}(. - \rho(t))|_{L^2(x > \frac{t}{10})} \to 0,$$

$$\rho(t) \to c^+ \quad \text{as } t \to +\infty.$$

Remark 16 The convergence of $u(t)$ to Q_{c^+} as $t \to \infty$ holds in fact strongly in L^2 in the region $x > \epsilon t$, for any $\epsilon > 0$ provided $\alpha_0 = \alpha_0(\epsilon)$ is small enough. This result is optimal in L^2 since $u(t)$ could contain other small (and then slow) solitons and since in general $u(t)$ does not go to 0 in L^2 for $x < 0$.

For instance, if $\|u(t) - Q_{c^+}(. - \rho(t))\|_{\frac{1}{2}} \to 0$ as $t \to \infty$, then $E(u) = E(Q_{c^+})$ and moreover $\int_\mathbb{R} u^2 dx = \int_\mathbb{R} Q_{c^+}^2 dx$ so that by the variational characterization of $Q(x)$, $u(t) = Q_{c^+}(x - x_0 - c^+ t)$ is a solution.

It is expected but not proved yet that the convergence in the same local sense $x > \epsilon t$ holds in $H^{\frac{1}{2}}(\mathbb{R})$.

The proof of Theorem 5 is based on the corresponding one ([159] and the references therein) for the generalized KdV equation where the stability is deduced from a Liouville type theorem. There are however two new difficulties.

1. The proof of the L^2 monotonicity property is more subtle because of the nonlocal character of the BO equation.
2. The proof of the linear Liouville theorem which requires the analysis of some linear operators related to Q.

Similar arguments and the strategy used for the asymptotic stability in the energy space of the sum of N solitons for the subcritical generalized KdV equation yield a similar result for N-solitons of the BO equation [159]:

Theorem 9 *Let $N \geq 1$ and $0 < c_1^0 < \ldots < c_N^0$. There exist $L_0 > 0, A_0 > 0$ and $\alpha_0 > 0$ so that if $u_0 \in H^{1/2}$ satisfies for some $0 \leq \alpha < \alpha_0, L \geq L_0$,*

$$\|u_0 - \sum_{j=1}^N Q_{c_j^0}(\cdot - y_j^0)\|_{H^{1/2}} \leq \alpha \quad \text{where} \quad \forall j \in (2, \ldots N), \; y_j^0 - y_{j-1}^0 \geq L,$$

and if $u(t)$ is the solution of BO corresponding to $u(0) = u_0$, then there exist $\rho_1(t), \ldots, \rho_N(t)$ such that the following hold

(a) *Stability of the sum of N decoupled solitons,*

$$\forall t \geq 0, \ \|u(t) - \sum_{j=1}^{N} Q_{c_j^0}(x - \rho_j(t))\|^{1/2} \leq A_0 \left(\alpha + \frac{1}{L}\right).$$

(b) *Asymptotic stability of the sum of N solitons. There exist $c_1^+, \ldots, c_N^+$, with $|c_j^+ - c_j^0| \leq A_0(\alpha + \frac{1}{L})$ such that*

$$\forall j, \quad u(t, . + \rho_j(t))) \rightharpoonup Q_{c_j^+} \quad \text{in } H^{1/2} \text{ weak as } t \to +\infty,$$

$$\|u(t) - \sum_{j=1}^{N} Q_{c_j^+}(. - \rho_j(t))\|_{L^2(x \geq \frac{1}{10 c_1^0(t)})} \to 0, \ \rho_j'(t) \to c_j^+ \text{ as } t \to +\infty.$$

Recall that the BO equation possesses explicit multi-soliton solutions. Let $U_N(x; c_j, y_j)$ denotes the explicit family of N-soliton profiles. Using the previous theorem and the continuous dependence of the solution in $H^{1/2}$ one obtains the following corollary [159].

Let $N \geq 1, 0 < c_1^0 < \ldots < c_N^0$ and set

$$d_N(u) = \inf\{\|u - U_N(.; c_j^0, y_j)\|_{1/2}, y_j \in \mathbb{R}\}.$$

Then

Corollary 1 (Stability in $H^{1/2}$ of Multi-Solitons) *For all $\delta > 0$, there exists $\alpha > 0$ such that if $d_N(u_0) \leq \alpha$ then for all $t \in \mathbb{R}$, $d_N(u(t)) \leq \delta$.*

As aforementioned, the proof of Theorem 8 is based on a rigidity theorem:

Theorem 10 *There exist $C, \alpha_0 > 0$, such that if $u_0 \in H^{\frac{1}{2}}$ satisfies $\|u_0 - Q_c\|_{H^{\frac{1}{2}}} = \alpha \leq \alpha_0$, and if the solution $u(t)$ of BO with $u(0) = u_0$ satisfies for some function $\rho(t)$*

$$\forall \epsilon > 0, \exists A_\epsilon > 0, \ \text{such that } \int_{|x| > A_\epsilon} u^2(t, x + \rho(t)) dx < \epsilon,$$

then there exist $c_1 > 0, x_1 \in \mathbb{R}$ such that

$$u(t, x) = Q_{c_1}(x - x_1 - c_1 t), \quad |c_1 - 1| + |x_1| \leq C\alpha.$$

The proof of the rigidity theorem (Liouville theorem) requires in particular the analysis of some linear operators related to Q that uses the fact that $Q(x)$ is explicit and some known results on the linearization of the BO equation around Q [33, 264].

Remark 17 Similar asymptotic stability results—though expected- do not seem to be known for the ILW equation.

3.9 A Result on Long Time Asymptotic

As already noticed, the complete description asymptotic behavior of the BO or ILW solutions with arbitrary large initial data is unknown (see Sect. 5.4 for the so-called soliton resolution conjecture).

A significant progress was made in [210] which is concerned with the long time behavior of solutions. It concerns global solutions u of the Benjamin-Ono equation satisfying

$$u \in C(\mathbb{R}; H^1(\mathbb{R})) \cap L^\infty_{\text{loc}}(\mathbb{R}; L^1(\mathbb{R})), \tag{3.30}$$

and moreover:

$$\exists a \in [0, 1/2), \, \exists c_0 > 0 \quad \text{such that}$$

$$\forall T > 0, \quad \sup_{t \in [0,T]} \int_{-\infty}^{\infty} |u(x,t)| dx \leq c_0(1+T^2)^{a/2}. \tag{3.31}$$

The main result in [210] is then

Theorem 11 *Under the above assumption,*

$$\int_{10}^{\infty} \frac{1}{\log t} \left(\int_{-\infty}^{\infty} \phi'\left(\frac{x}{\lambda(t)}\right) \left(u^2 + (D^{1/2}u)^2\right)(x,t) dx \right) dt < \infty. \tag{3.32}$$

Hence,

$$\liminf_{t \to +\infty} \int_{-\infty}^{\infty} \phi'\left(\frac{x}{\lambda(t)}\right) \left(u^2 + (D^{1/2}u)^2\right)(x,t) dx = 0, \tag{3.33}$$

with

$$\lambda(t) = \frac{ct^b}{\log t}, \quad a+b=1, \quad \text{and } \phi'(x) = \frac{1}{1+x^2}, \tag{3.34}$$

for any fixed $c > 0$.

Remark 18

1. This result discards the existence of non trivial time periodic solutions (in particular breathers) and of solutions moving with a speed slower that a soliton.
2. The theorem implies that there exists a sequence of times $\{t_n; n \in \mathbb{N}\}$ with $t_n \to +\infty$ as $n \to \infty$ such that

$$\lim_{n\infty} \int_{|x| \leq \frac{ct_n^b}{\log(t_n)}} (u^2 + (D^{1/2}u)^2)(x, t_n)dx = 0. \tag{3.35}$$

3. The solitons

$$u(x,t) = Q_c(x - ct), \quad Q_c(x) = \frac{4c}{1 + c^2x^2},$$

belongs to the class (3.30) and they also satisfy (3.33).

Remark 19 The above result does not use the integrability of the BO equation. It is likely to hold for the ILW equation.

The proof of Theorem 11 relies in particular on the estimates obtained in [122] and uses the fact that ϕ' is a multiple of the soliton.

4 Rigorous Results by IST Methods

4.1 The BO Equation

An attempt to solve the Cauchy problem of the BO equation by IST methods is due to Anderson and Taflin [20]. They obtained a formal series linearization of the BO equation that can be interpreted as a distorted Fourier transform associated to a singular perturbation of $\frac{1}{i}\frac{d}{dx}$, from which they deduce a Lax pair for BO and under suitable assumptions on the scattering data, a power series for the inverse transform. Unfortunately, as proved in [59], both the direct and inverse problems considered in [20] are not analytic for generic potential, leading to divergent series in general.

The first rigorous results on the Cauchy problem by IST methods were obtained by Coifman and Wickerhauser in [59], (see also [27, 28]). They used a more complicated regularized IST formalism and solved it by iteration, yielding the global well-posedness of the Cauchy problem for small initial data. More precisely they use constructive methods to investigate the spectral theory of the Benjamin-Ono equation. Since the linearization series used previously is singular (see above), they replace it with an improved series obtained by finite-rank renormalization. This introduces additional scattering data, which are shown to be dependent upon a single function, though not the usual one. They then prove the continuity of the direct and inverse scattering transforms defined by the improved series for small complex potentials.

Let $w(x) = (1 + |x|)$. The global well-posedness in [59] writes:

Theorem 12 *Let q be a real function such that $w^{n+1}q$ is a small function in $L^1(\mathbb{R})$ for some $n > 0$. Suppose also that $w^n q'$ and $w^n q''$ are also small in $L^1(\mathbb{R})$. Then there exists a unique solution of the Benjamin-Ono equation with initial data q.*

Some interesting issues arise also in [59] linked to the generation of soliton solutions. We comment them briefly in connection with [172, 188] and specially [226, 227] that we will follow closely. Let the initial data for the BO equation renormalized as $u_0 = U_0 U(x/L_0)$, where U_0 is a characteristic amplitude and L_0 is a characteristic length. Then, the number of solitons depends on the sole parameter $\sigma = U_0 L_0$ (Ursell number). In the limit $\sigma \gg 1$ the initial potential generates a large number N of solitons. An approximation for N was found in [172] and later confirmed by Miloh et al. [188] (see [194] for the ILW equation),

$$N = \frac{1}{2\pi} \int_{u_0(x) \geq 0} u_0(x) dx.$$

An important question is that of the existence of a threshold for the generation of solitons, that is whether a small initial perturbation of an algebraic profile $u(x) = \frac{a}{1+x^2}$ ie in the limit $\sigma \ll 1$, $(a \ll 1)$ can support propagation of at least one soliton.

Note that the mass $\mathcal{M}[u_c] = \int_{\mathbb{R}} u_c(x) dx$ of the BO soliton $u_c(x - ct) = \frac{4c}{1+c^2(x-ct)^2}$ is constant, $\mathcal{M}[u_c] = 4\pi$.

For the modified KdV equation, such a property is related to the existence of a threshold on the soliton generation, in this case perturbations with $\mathcal{M}[u] \leq \frac{1}{2}\mathcal{M}_{\text{sol}}$ do not support solitons.

For the BO equation, this issue depends on the possible non genericity of the potentials and on the very special structure of the Jost function. By definition generic potentials u are those for which $n_0 \neq 0$ where

$$n_0 = \frac{1}{2\pi} \int_{\mathbb{R}} u(x) n(x) dx.$$

Here $n(x)$ is the limiting Jost function that we describe now. For generic potentials, the Jost functions vanish as $k \to 0^+$ according to the approximation

$$N(x, k) \to \frac{n(x)}{1 + n_0(\gamma + \ln(ik))} + O\left(\frac{k}{\ln k}\right),$$

where γ is the Euler constant and $n(x)$ satisfies

$$i n_x = -P^+(u n_x)$$

where $P^+(v) = \frac{1}{2}(v - iHv)$.

The problem at the center of the analysis is the spectral problem

$$i\phi_x^+ + k(\phi^+ - \phi^-) = -u(x)\phi^+. \tag{4.1}$$

Coifman and Wickerhauser [59] proved that the scattering problem (4.1) has no bound states in a neighborhood of the origin, if $u(x) \to O(|x|^{-1-\mu})$, $\mu > 0$. In fact this result is valid for generic potentials in the previous sense.

For non generic potentials, *ie* satisfying $n_0 = 0$, the limiting Jost function is bounded in the limit $k \to 0^+$ and properties of the scattering problem are modified. The zero potential $u(x) = 0$ is non generic as well as the soliton solutions.

The main results in [226] concern the study of the perturbation of non generic potentials where the number of bound states may change depending on the type of the perturbation. More precisely the authors consider a potential in the form $u^\epsilon = u(x) + \epsilon \eta(x)$ where $\epsilon \ll 1$ and $u(x)$ satisfies the constraint $n_0 = 0$. In the particular case $u(x) = 0$, this reduces to the problem of soliton generation by a small initial data. They derive a criterion for a new eigenvalue to emerge from the edge of the continuum spectrum at $k = 0$. In particular, new eigenvalues may always appear due to perturbations of the zero background and the soliton solutions. They derive the leading order term for the new eigenvalue and for the associated bound state at short and long distances, as well as for the variation of the continuous spectrum. Similar issues seem unknown for the ILW equation.

Further progress of the IST theory for BO is due to Miller and Wetzel [191] who studied the direct scattering problem of the Fokas-Ablowitz IST theory when the potential is a rational function with simple poles and obtained the explicit formula for the scattering data.

A breakthrough on the IST for the BO equation appeared recently in the works by Wu [268, 269] who in particular solved completely the direct scattering problem for data of arbitrary size. Let us first introduce a few notations.

In what follows, $\mathbb{H}^{p,+}$ is the L^p Hardy space of the upper plane. more precisely $f \in \mathbb{H}^{p,+}$, $1 < p \leq \infty$ if it is the L^p (and a.e.) boundary value of an analytic function $F(x + iy)$ in the upper plane $y > 0$ such that $\sup_{y>0} |F(. + iy)|_p < \infty$. $\mathbb{H}^{2,+}$ is also denoted H^+.

In [268] Wu studies the operator L_u arising in the Lax pair for the BO equation and proved that its spectrum is discrete and simple. More precisely, we first rewrite the Fokas-Ablowitz formalism in a slightly different way, denoting first

$$C_\pm \phi = \frac{\phi \pm iH\phi}{2}$$

the Cauchy projections. In other words, $\widehat{C_\pm f} = \chi_{\mathbb{R}_\pm} \hat{f}$.

When they act on $L^2(\mathbb{R})$ the ranges are $H^\pm$, the Hardy spaces of L^2 functions whose Fourier transform are supported on the positive and negative half lines. The Lax pair writes in those notations, on H^+

$$L_u \phi = \frac{1}{i}\phi_x - C_+(uC_+\phi) = \lambda \phi$$

$$B_u\phi = 2\lambda\phi_x + i\phi_{xx} + 2(C_+u_x)(C_+\phi),$$

and in H^- :

$$L_u\phi = \frac{1}{i}\phi_x - C_-(uC_-\phi) = \lambda\phi$$

$$B_u\phi = 2\lambda\phi_x + i\phi_{xx} + 2(C_-u_x)(C_-\phi).$$

Since when u is real the equations on H^- are just the complex conjugate of the equations on H^+ one will assume that u is real and focus on the H^+ part of the Lax pair. The scattering data of the IST are closely related to the spectrum of the operator L_u.

The next two theorems in [268] provide useful informations on the spectral properties of L_u.

Theorem 13 *Suppose that $u \in L^2(\mathbb{R}) \cap L^\infty(\mathbb{R})$. Then L_u is a relatively compact perturbation of $\frac{1}{i}\partial_x$ and is self-adjoint on H^+ with domain $H^+ \cap H^1(\mathbb{R})$.*

Theorem 14 *Suppose that $u \in L^1(\mathbb{R}) \cap L^\infty(\mathbb{R})$ and $xu \in L^2(\mathbb{R})$. Then the operator L_u has only finitely many negative eigenvalues and the dimension of each eigenspace is 1.*

Remark 20 By Weyl's theorem and standard spectral theory, the essential spectrum of L_u is the same as that of $\frac{1}{i}\partial_x$, that is $\mathbb{R}^+ \cup \{0\}$. However it is not clear that the eigenvalues are simple and even if there are finitely many of them. On the other hand those spectral properties are crucial for the construction of scattering data in the Fokas-Ablowitz IST method in [4].

A crucial step in the proof of simplicity is the discovery of a new identity connecting the L^2 norm of the eigenvector to its inner product with the scattering potential.

More precisely, assuming that $u \in L^2(\mathbb{R}) \cap L^\infty(\mathbb{R})$, and $xu \in L^2(\mathbb{R})$, then if $\lambda < 0$ is an eigenvalue of L_u and ϕ is an eigenvector, then

$$\left|\int_\mathbb{R} \phi u\, dx\right|^2 = 2\pi\lambda \int_\mathbb{R} |\phi|^2 ds.$$

The proof for finiteness is an extension of ideas involved in the Bergman-Schwinger bound for Schrödinger operators.

As aforementioned, the direct scattering problem is completely justified in [269] where Yulin Wu examined the full spectrum of L_u establishing existence, uniqueness and asymptotic properties of the Jost solutions to the scattering problem, providing possible directions for the correct setup for the inverse problem.

Recalling that $w(x) = (1 + |x|)$, we will use the notations
$L^p_s(\mathbb{R}) = \{f,\, w^s f \in L^p(\mathbb{R})\}$ and

$$H^s_s(\mathbb{R}) = \{f;\, f \in L^2_s(\mathbb{R}) \text{ and } \hat{f} \in L^2_s(\mathbb{R})\},$$

Two Jost functions m_1 and m_e are considered in [269]. They are solutions of the following equations with suitable boundary conditions stated in the next lemma:

$$\frac{1}{i}\partial_x m_1 - C_+(um_1) = k(m_1 - 1),$$

and

$$\frac{1}{i}\partial_x m_e - C_+(um_e) = \lambda m_e.$$

Here $\lambda \pm 0i \in \mathbb{R}^+ \pm 0i$, and

$$k \in \rho(L_u) \cup (\mathbb{R}^+ \pm 0i) = (\mathbb{C} \setminus \{\lambda_1 \ldots \lambda_N\} \setminus [0, \infty)) \cup (\mathbb{R}^+ \pm 0i),$$

which is the resolvent set glued with two copies of the positive real axis. Following again [269] we provide here the translation of notation:

$$M(x, k) = m_1(x, \lambda + 0i), \quad \bar{M}(x, \lambda) = m_e(x, \lambda + 0i),$$

$$N(x, \lambda) = m_e(x, \lambda - 0i), \quad \bar{N}(x, \lambda) = m_1(x, \lambda - 0i).$$

The following lemma may be considered as the definition of m_1 and m_e. In what follows, one uses the integral operators

$$G_k(x) = \frac{1}{2\pi} \int_0^\infty \frac{e^{ix\xi}}{\xi - k} d\xi,$$

for $k \in \mathbb{C} \setminus [0, \infty)$, and

$$\tilde{G}_k(x) = \frac{1}{2\pi} \int_{-\infty}^0 \frac{e^{ix\xi}}{\xi - k} d\xi,$$

for $k \in \mathbb{C} \setminus (-\infty, 0]$.

For $\epsilon > 0$ one has

$$G_{\lambda \pm 0i}(x) = \frac{1}{2\pi} \int_{-\infty}^\infty \frac{e^{ix\xi}}{\xi - (\lambda \pm i\epsilon)} d\xi - \tilde{G}_{\lambda \pm i\epsilon}(x) = \pm i e^{\mp \epsilon x} e^{i\lambda x} \chi_{\mathbb{R}^\pm}(x) - \tilde{G}_{\lambda \pm i\epsilon}(x),$$

(4.2)

with

$$G_{\lambda \pm 0i}(x) = \lim_{\epsilon \to 0} G_{\lambda \pm i\epsilon}(x) = \pm i e^{i\lambda x} \chi_{\mathbb{R}^\pm}(x) - \tilde{G}_\lambda(x), \quad (4.3)$$

for $\lambda > 0$.

The limit in (4.3) holds in the following sense: the first term in (4.2) converges pointwise and the second term in (4.2) converges in $L^{p'}$ for every $p' \in [2, \infty)$. The latter is checked when observing that $\tilde{G}_{\lambda \pm i\epsilon}$ is the inverse Fourier transform of $\frac{\chi_{\mathbb{R}} - \xi}{\xi - (\lambda \pm i\epsilon)}$, which converges to $\frac{\chi_{\mathbb{R}} - \xi}{\xi - \lambda}$ in every L^p for $p \in (1, 2]$ assuming $\lambda > 0$.

Lemma 2 *Let $p > 1$ and $s > s_1 > 1 - \frac{1}{p}$ be given and let $u \in L_s^p(\mathbb{R})$. Assume that $m_1(x, k), m_e(x, \lambda \pm 0i) \in L^\infty_{-(s-s_1)}(\mathbb{R})$ for fixed $k \in (\mathbb{C} \setminus [0, \infty)) \cup (\mathbb{R}^+ \pm 0i)$ and $\lambda \in \mathbb{R}^+$, then the following are equivalent:*

(a) $m_1(x, k), m_e(x, \lambda \pm 0i)$ solve

$$\frac{1}{i} \partial_x m_1 - C_+(um_1) = k(m_1 - 1),$$

$$\frac{1}{i} \partial_x m_e - C_+(um_e) = \lambda m_e,$$

together with the asymptotic conditions

$$m_1(x, k) - 1 \to 0 \quad \begin{cases} \text{as } |k| \to \infty & \text{if } k \in \mathbb{C} \setminus [0, \infty) \\ \text{as } x \to \mp\infty & \text{if } k = \lambda \pm 0i \in \mathbb{R}^+ \pm 0i, \end{cases} \quad (4.4)$$

$$m_e(x, \lambda \pm 0i) - e^{i\lambda x} \to 0 \quad \text{as } x \to \mp\infty.$$

The above asymptotic conditions should be read with either the upper or the lower sign.

(b) $m_1(x, k), m_e(x, \lambda \pm 0i)$ solve the following integral equations:

$$m_1(x, k) = 1 + G_k \star (um_1(., k))(x), \quad (4.5)$$

$$m_e(x, \lambda \pm 0i) = e^{i\lambda x} + G_{\lambda \pm 0i} \star (um_e(., \lambda \pm 0i))(x). \quad (4.6)$$

Moreover, if either (a) or (b) holds, one has the stronger bounds

$$m_1(x, k) - 1 \in L^\infty(\mathbb{R}) \cap \mathbb{H}^{p,+}$$

for fixed $k \in \mathbb{C} \setminus [0, \infty)$ and

$$m_1(x, \lambda \pm 0i), m_e(x, \pm 0i) \in L^\infty(\mathbb{R})$$

for fixed $\lambda \in \mathbb{R}^+$.

Here is the theorem establishing the existence and uniqueness of Jost solutions, as stated in [269].

Theorem 15 *Let $s > s_1 > \frac{1}{2}$ and $u \in L^2_s(\mathbb{R})$. Let $\rho(L_u)$ be the resolvent set of $L_u = \frac{1}{i}\partial_x - C_+ u C_+$, regarded as an operator on $\mathbb{H}^+$. Then for every $k \in \rho(L_u) \cup (\mathbb{R}^+ \pm 0i)$, and every $\lambda > 0$, there exist unique $m_1(x,k)$ and $m_e(x, \lambda \pm 0i) \in L^\infty_{-(s-s_1)}(\mathbb{R})$ solving (4.5), (4.6) respectively, with improved bounds $m_1(x,k), m_e(x, \lambda \pm 0i) \in L^\infty(\mathbb{R})$. Furthermore, the mapping $k \to m_1(k)$ is analytic from $\rho(L_u)$ to $L^\infty_{-(s-s_1)}$, and $m_1(k) \in C^{0,\gamma}_{loc}((\rho(L_u) \cup \mathbb{R}^+ \pm 0i)), L^\infty_{-(s-s_1)}(\mathbb{R})$ while $m_e(\lambda \pm 0i) \in C^{0,\gamma}_{loc}(\mathbb{R})$.*
Here γ is some number between 0 and 1.

When one studies the asymptotic behavior of the Jost solutions and scattering coefficients as k approaches to 0 within the set $\rho(L_u) \cup (\mathbb{R}^+ \pm 0i)$ one notices that the convolution kernel G_k has a logarithmic singularity at $k = 0$ and so does the operator T_k. By subtracting a rank one operator from T_k the modified operator has a limit at $k = 0$. The asymptotic behavior of the Jost functions can be recovered from *modified* Jost functions, we refer to [269] for details. One recovers rigorously the asymptotics in [4, 118]. Those asymptotic formulas are useful to clarify the global behavior of the scattering coefficients. One also finds in [269] asymptotic formulas in the limit $k \to \infty$ and a formal derivation of the time evolution of the Jost functions and scattering coefficients provided u is a smooth and decaying solution of the BO equation.

Remark 21 A large class of *quasi-periodic* solutions of the BO equation has been found in [238] by the Hirota method and in [73] by the method of the theory of finite-zone integration. More precisely the quasi-periodic solution of the BO equation found in [73] writes

$$u(x,t) = c + \sum_{j=1}^{N}(a_j - b_j) - 2\,\text{Im}\,\frac{\partial}{\partial x} \ln \det M(x,t), \qquad (4.7)$$

where the $N \times N$ matrix $M(x,t)$ has the elements

$$M_{jm} = c_m \delta_{jm} e^{i(a_m - b_m) - i(a_m^2 - b_m^2)t} - \frac{1}{b_j - a_m}. \qquad (4.8)$$

The constants a_j, b_j, c satisfy the inequalities $c \leq a_1 \leq b_1 \leq a_2 \leq b_2 \leq \ldots \leq a_n \leq b_n$ and the constants c_i are defined by

$$|c_i|^2 = -(b_i - c_i) \prod_{j \neq i}^{N} \frac{(a_i - a_j)(b_i - b_j)}{(a_i - c)\prod_{j=1}^{N}(b_i - a_j)(a_i - b_j)}. \qquad (4.9)$$

This multiply periodic N-phase solution has been used in [110] to construct the modulation solution of BO corresponding to a step like function:

$$u(x,0) = \begin{cases} A & x < 0,\ A > 0 \\ 0 & x > 0, \end{cases} \qquad (4.10)$$

or a smoothly decreasing function (leading to a shock formation for the Burgers equation):

$$u(x,0) = \alpha\left(1 - \frac{2}{\pi}\arctan\beta x\right), \quad \alpha, \beta > 0, \tag{4.11}$$

or a modulated wavetrain which has a compressive wavenumber modulation in the wavenumber $g(x)$:

$$u(x,0) = -2g(x)\left(\frac{1 - \sqrt{2}\cos(sg(x))}{3 - 2\sqrt{2}\cos(xg(x))}\right) \tag{4.12}$$

with

$$g(x) = \alpha\left(1 - \frac{2}{\pi}\arctan\beta x\right) + b_0.$$

4.2 The ILW Equation

As recalled in the previous section, the formal framework of inverse scattering for the ILW equation was given in [133, 134]. We are not aware of rigorous results using IST to solve the Cauchy problem even for small initial data.

5 Related Results and Conjectures

5.1 The Modified Cubic BO and ILW Equations

The modified cubic BO and ILW equations write respectively[9]

$$u_t + u^2 u_x - H u_{xx} = 0 \tag{5.1}$$

and

$$u_t + u^2 u_x + \frac{1}{\delta} u_x + \mathcal{T}(u_{xx}) = 0. \tag{5.2}$$

Contrary to the modified KdV equation they do not seem to be completely integrable. We will not describe in details the local well-posedness results since most

---

[9]Note that equations (5.1) and (5.2) are different from the modified BO and ILW equations which are linked to the usual BO and ILW equations via a Miura transform, see Sect. 6.6.

of the used methods are close to that of the original one. We refer for instance to [128, 204] and the references therein for the local Cauchy problem in $H^s(\mathbb{R})$, $s \geq \frac{1}{2}$ and to [202] for the local Cauchy problem in $H^1(\mathbb{T})$. Scattering results of small solutions can be found in [95, 96].

Using the factorization technique of Hayashi and Naumkin, Naumkin and Sanchez-Suarez [216] study the large time asymptotics of small solutions of the modified ILW equation in suitable weighted Sobolev spaces. Difficulties arise from the non-homogeneity of the symbol.

More precisely, the main result in [216] states that for any initial data $u_0 \in H^3(\mathbb{R}) \cap H^{2,1}(\mathbb{R})$ with $\|u_0\|_{H^3 \cap H^{2,1}} \leq \epsilon$, and $\int_{\mathbb{R}} u_0(s)dx = 0$, there exists a global solution $u \in C([0, \infty); H^3(\mathbb{R}))$ of the modified ILW satisfying

$$|u(t)|_\infty \leq C\epsilon t^{-1/2}.$$

There exist moreover unique modified final states describing the asymptotics of $u(t)$ for large times.

Equations (5.1) and (5.2) are specially interesting from a mathematical point of view since they are L^2 critical, as is the quintic generalized KdV equation and one thus expects a finite type blow-up phenomena. This has been established for the modified BO equation in [161] as follows.

Let $Q \in H^{1/2}(\mathbb{R})$, $Q > 0$ be the unique ground state solution of the "elliptic" equation

$$D^1 Q + Q = Q^3, \quad D^1 = |\partial_x|$$

constructed by variational arguments (see [14, 264]) and whose uniqueness was established in [82]. Then it is proven in [161] that there exists a solution u of (5.1) satisfying $|u(t)|_2 = |Q|_2$ and

$$u(t) - \frac{1}{\lambda^{\frac{1}{2}}t} Q\left(\frac{\cdot - x(t)}{\lambda(t)}\right) \to 0 \quad \text{in } H^{1/2}(\mathbb{R}) \text{ as } t \to 0$$

where

$$\lambda(t) \sim t, \quad x(t) \sim -|\ln t| \quad \text{and} \quad \|u(t)\|_{\dot{H}^{1/2}} \sim t^{-\frac{1}{2}} \|Q\|_{\dot{H}^{\frac{1}{2}}} \quad \text{as } t \to 0.$$

This result obviously implies the orbital instability of the ground state. Also this blow-up behavior is unstable since any solution merging from an initial data with $|u_0|_2 < |Q|_2$ is global and bounded in L^2. The proof is inspired by a corresponding one for the L^2 critical generalized KdV (see [160]). There are however new difficulties to overcome. First, the slow spatial decay of the solitary wave Q creates serious difficulties to construct a blow-up profile. Next difficulties arise when considering localized versions of basic quantities such as mass or energy. Contrary to the case of the L^2 critical KdV equation, standard commutator estimates are not enough and one has to use suitable localization arguments as in [122, 123].

Remark 22 A similar result is not known but expected to be true for the modified ILW equation (5.2).

5.2 The Periodic Case

In order to maintain a reasonable size, we focussed in the present paper on the Cauchy problem in the real line. For the sake of completeness, we however briefly review here the state of the art for the Cauchy problem posed in the circle $\mathbb{T}$ which leads to many interesting issues, both on the IST and PDE sides.

The periodic BO equation writes

$$u_t + uu_x - \mathcal{H}u_{xx} = 0, \tag{5.3}$$

where

$$\mathcal{H}f(x) = \frac{1}{L} PV \int_{-L/2}^{L/2} \cot\left(\frac{\pi(x-y)}{L}\right) f(y) dy,$$

while the periodic ILW equation writes

$$u_t + 2uu_x + \delta^{-1} u_x - (\mathcal{T}_\delta u)_{xx} = 0, \quad \delta > 0, \tag{5.4}$$

where

$$\mathcal{T}_\delta u(x) = \frac{1}{L} PV \int_{-L/2}^{L/2} \Gamma_{\delta,L}(x-y) u(y) dy$$

with

$$\Gamma_{\delta,L}(\xi) = -i \sum_{n \neq 0} \coth\left(\frac{2\pi n \delta}{L}\right) e^{2in\pi\xi/L}.$$

We are not aware of rigorous results via inverse scattering methods for the periodic ILW equation. On the other hand a spectacular progress was recently achieved in [85] for the BO equation.

We recall that the formalism of IST for the periodic ILW can be found in [5].

On the other hand many results have been obtained by PDE methods.

The results in [1] still remain true since they do not rely on dispersive properties of the linear group, yielding global well-posedness of the Cauchy problem for the periodic BO or ILW equations in $H^s(\mathbb{T})$, $s > \frac{3}{2}$.

Concerning lower regularity results, Molinet [197] proved the global well-posedness in the energy space $H^{1/2}(\mathbb{T})$ by combining Tao's normal form and estimates in Bourgain spaces. He also proved that the flow map $u_0 \to u(.,t)$ is not uniformly continuous in $H^s(\mathbb{T})$, $s > 0$. The proof of this fact is inspired by a corresponding result in [132] for the Cauchy problem in $H^s(\mathbb{R})$, $s > 0$ but with a simpler proof, using the property that if u is a solution of the BO equation with initial data u_0, so is $u(\cdot + \omega t, t) + \omega, \forall \omega \in \mathbb{R}$ with initial data $u_0 + \omega$. Actually this lack of uniform continuity has nothing to do with dispersion and is linked to the Burgers equation (nonlinear transport equation)

$$u_t + u u_x = 0.$$

Actually the locally Lipschitz property of the flow map is recovered in the space $H_0^s(\mathbb{T})$ of $H^s(\mathbb{T})$ functions with zero mean or more generally on hyperplanes of functions with fixed mean value.

This result is improved in [198] where the global well-posedness in $L^2(\mathbb{T})$ is proven, the flow map being Lipschitz (in fact real analytic) on every bounded set of $H_0^s(\mathbb{T}), s \geq 0$. Note that uniqueness then holds in a space smaller than $C([0, T]; L^2(\mathbb{T}))$ but containing the limit in $C([0, T]; L^2(\mathbb{T}))$ of smooth solutions of BO.

Note that one also finds in [201] a simplified proof of the global well-posedness in $L^2(\mathbb{T})$ with unconditional uniqueness in $H^{1/2}(\mathbb{T})$.

As aforementioned, the results in [209] apply to the ILW equation, yielding global well-posedness in the energy space $H^{1/2}(\mathbb{T})$. We are not aware of well-posedness results in $L^2(\mathbb{T})$ for the ILW equation.

Both the BO and ILW equations possess periodic solitary waves. Explicit formulas can be found in [54, 113, 215] but, as noticed in [5] they appear to be incorrect. An explicit formulation is given in [187]. The L-periodic traveling wave of the BO equation is in the formulation of [24]:

$$\phi_c(x) = \frac{4\pi}{L} \frac{\sinh(\gamma)}{\cosh(\gamma) - \cos(2\pi x/L)}, \tag{5.5}$$

where $\gamma > 0$ satisfies $\tanh(\gamma) = \frac{2\pi}{cL}$, implying that $c > \frac{2\pi}{L}$. The profile of an even, zero mean, periodic traveling wave of the ILW equation is given by (see [23] and also [215]):

$$\phi_c(x) = \frac{2K(k)i}{L}\left(Z(\frac{2K(k)i}{L}(x - i\delta); k) - Z(\frac{2K(k)i}{L}(x + i\delta); k)\right), \tag{5.6}$$

where $K(k)$ denotes the complete elliptic integral of first kind, Z is the Jacobi zeta function and $k \in (0, 1)$. Note that for fixed L and δ, the wave speed c and the elliptic modulus k must satisfy specific constraints.

We also refer to [225] for another derivation and a clear and complete discussion of the periodic solutions of the ILW equation.

The orbital stability of the periodic traveling waves is proven in [24] for the BO equation and in [23] for the ILW equation.

A program (see [68, 69, 258–261]) has been devoted to the construction of an infinite sequence of invariant measures of Gaussian type associated to the conservation laws of the Benjamin-Ono equation.

Those invariant measures $\{\mu_n\}$ on (L^2, ϕ_t) where ϕ_t is the BO flow on $L^2(\mathbb{T})$ satisfy

$$\mu_n \text{ is concentrated on } H^s(\mathbb{T}) \text{ for } s < n - \frac{1}{2},$$

$$\mu_n(H^{n-1/2}(\mathbb{T})) = 0.$$

μ_n is formally defined as a renormalization of

$$d\mu_n(u) = e^{-E_n(u)} du = e^{R_n(u)} e^{\|u\|_n^2} du,$$

where we have denoted

$$E_n(u) = \|u_n\|^2 + R_n(u)$$

the sequence of conservation laws of the BO equation, $\|.\|_n$ denoting the homogeneous Sobolev norm of order n and R_n a lower order term.

These results do not apply to infinitely smooth solutions since $\mu_n(C^\infty(\mathbb{T})) = 0$ for all n. Sy [252] has constructed a measurable dynamical system for BO on the space $C^\infty(\mathbb{T})$, namely a probabilistic measure μ invariant by the BO flow, defined on $H^3(\mathbb{T}))$ and satisfying in particular $\mu(C^\infty(\mathbb{T})) = 1$.

Remark 23 No similar results seem to be available for the ILW equation.

5.3 Zero Dispersion Limit

Interesting issues arise when looking at the zero dispersion limit of the BO equation, that is the behavior of solutions to

$$u_t + uu_x - \epsilon H u_{xx} = 0, \ \epsilon > 0 \qquad (5.7)$$

when $\epsilon \to 0$.

The corresponding problem for the KdV equation has been first extensively studied by Lax and Levermore in the eighties [144], see also [63–67, 262] and the survey article [189]. For the BO equation, as for the KdV equation, at the advent of a shock in the dispersionless Burgers equation, the solution of (5.7) is regularized by a dispersive shock wave (DSW). In the DSW region, the solution of (5.7) may be formally approximated using Whitham modulation theory, see [265]. Unlike the case of the KdV equation the modulation equations for the BO equation are fully uncoupled (see [73]), consisting of several independent copies of the inviscid Burgers equation.

A formalism for matching the Whitham modulation approximation for the DSW onto inviscid Burgers equation in the domain exterior to the DSW was developed by Matsuno [177, 178] and Jorge et al. in [110].

Recent rigorous results have been obtained by P. D. Miller and co-authors. Miller and Xu [192] partially confirmed the above formal results by computing rigorously the weak limit (modulo an approximation of the scattering data) of the solution of the Cauchy problem for (5.7) for a class of positive initial data, using IST techniques, and by developing an analogue for the BO equation of the method of Lax-Levermore for the KdV equation.

Using the exact formulas for the scattering data of the BO equation valid for general rational potential with simple poles that they obtained in [191], Miller and Wetzel [190] analyzed rigorously the scattering data in the small dispersion limit, deducing in particular precise asymptotic formulas for the reflection coefficient, the location of the eigenvalues and their density, and the asymptotic dependence of the phase constant associated with each eigenvalue on the eigenvalue itself. Such an analysis seems to be unknown for more general potentials.

We refer to [193] for a study of the BO hierarchy with positive initial data in a suitable class in the zero-dispersion limit. We also mention [75] for a description of dispersive shock waves for a class of nonlinear wave equations with a nonlocal BO type dispersion which does not use the integrability of the equation.

We are not aware of similar small dispersion limit results for the ILW equation.

5.4 The Soliton Resolution Conjecture

Since IST methods provide so far rigorous results on the Cauchy problem only for small enough initial data, one cannot use them to prove the soliton resolution conjecture (see [254]) as in the KdV case (see e.g. [242] and the references therein). However we strongly believe that it holds true for the BO and ILW equations, as suggested for instance by the following numerical simulations in [130] (see also [188, 228] for other illuminating simulations and [229] for a theoretical analysis of the spectral method in [228]).[10]

---

[10]Note that the soliton resolution conjecture might also be valid for the Benjamin equation (1.9) which is not integrable.

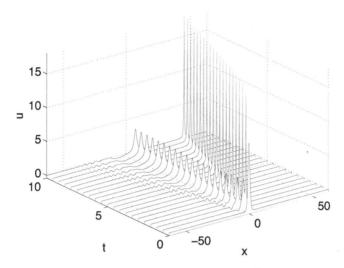

Fig. 1 Solution to the BO equation for the initial data $u_0 = 10\mathrm{sech}^2 x$

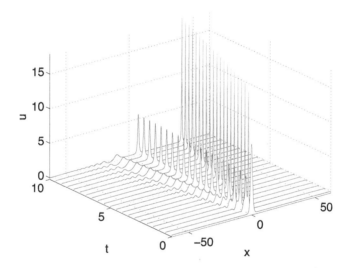

Fig. 2 Solution to the ILW equation with $\delta = 1$ for the initial data $u_0 = 10\mathrm{sech}^2 x$

We first show the formation of solitons from localized initial data for the BO equation in Fig. 1. Note a tail of dispersive oscillations propagating to the left.

A similar decomposition of localized initial data into solitons and radiation is shown for the ILW equation in Fig. 2. Note that this case is numerically easier to treat with Fourier methods since the soliton solutions are more rapidly decreasing (exponentially instead of algebraically) than for the fKdV, fBBM and BO equations. The different shape of the solitons is also noticeable in comparison to Fig. 1.

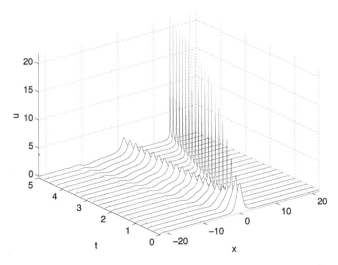

Fig. 3 Solution to the fKdV equation with $\alpha = 0.6$ for the initial data $u_0 = 5\text{sech}^2 x$

More generally, we conjecture that some form of soliton resolution holds for the class of fractional KdV equations

$$u_t + uu_x - |D|^\alpha u_x = 0, \tag{5.8}$$

when $\alpha > \frac{1}{2}$, the Cauchy problem being then globally well-posed in the energy space $H^{\alpha/2}(\mathbb{R})$.[11] A first step towards this conjecture is achieved in [203] where global well-posedness is established in the energy space when $\alpha > \frac{6}{7}$.

The simulation in Fig. 3 for the fKdV with $\alpha = 0.6$ suggests the decomposition into solitons plus radiation.

6 Varia

We describe here a few more qualitative results on the Benjamin-Ono and the Intermediate Long Wave equations and some of their natural extensions (higher order equations, two-dimensional versions...).

6.1 Damping of Solitary Waves

In the real physical world, various dissipative mechanisms affect the propagation of dispersive waves. The dissipative term can be local e.g. $-\partial_{xx}$ or non local, see for

---

[11] Recall that the ground state solution (that exists and is unique when $\alpha > \frac{1}{3}$) is orbitally stable if and only if $\alpha > \frac{1}{2}$ and spectrally unstable when $\frac{1}{3} < \alpha < \frac{1}{2}$, see [22, 151].

instance [74, 90, 115, 223, 224] for physical examples. The Cauchy problem for the dissipative BO or ILW equations is studied in [240] and more recently in [199, 209]. Of course the dissipative term destroys the Hamiltonian structure and the complete integrability. The solutions of the Cauchy problem then tends to zero as $t \to \infty$ (see [40, 71, 72] and the references therein for a description of the large time asymptotics of solutions of the Benjamin-Ono-Burgers and related equations).

However we are not aware of rigorous results on the quantitative influence of dissipation in the solitonic structure of the BO, or ILW equation.[12]

In [182] the Whitham method is applied to the Benjamin-Ono-Burgers equation

$$u_t + uu_x - Hu_{xx} = \epsilon u_{xx}, \epsilon \ll 1 \tag{6.1}$$

to give a formal description of the solution of the Cauchy problem corresponding to a steplike initial data as $\epsilon \to 0$.

6.2 Weighted Spaces

The Cauchy problem for the BO equation in weighted spaces is studied in [76–81, 97, 105] yielding smoothing and unique continuation properties [106]. A result on the propagation of regularity for solutions to the Cauchy problem associated to the Benjamin-Ono equation is proved in [107]. Essentially, if $u_0 \in H^{3/2}(\mathbb{R})$ is such that its restriction belongs to $H^m(b, \infty)$ for some $m \in \mathbb{Z}, m \geq 2$ and some $b \in \mathbb{R}$, then the restriction of the corresponding solution $u(., t)$ belongs to $H^m(\beta, \infty)$ for any $\beta \in \mathbb{R}$ and any $t > 0$. This shows that the regularity of the datum travels to the left with infinite speed.

Another interesting result is in [108] where the Cauchy problem for BO is solved with bore-like initial data.

6.3 Control

Control and stabilization issues were investigated by Linares and Rosier [153], in $H^s(\mathbb{T})$, $s > \frac{1}{2}$, and in [143] in $L^2(\mathbb{T})$ for the periodic BO equation.

A typical exact controllability (via a forcing term) result from [143] is as follows. In order to keep the mass conserved, the control input is chosen in the form

$$(\mathcal{G}h)(x, t) = a(x) \left(h(x, t) - \int_{\mathbb{T}} a(y) h(x, t) dy \right),$$

where a is a given smooth nonnegative function such that $\{x \in \mathbb{T}; a(x) > 0\} = \omega$ and with mass one.

---

[12] Actually this question seems to be open also for the KdV equation.

Theorem 16

(i) *(Small data) For any $T > 0$ there exists some $\delta > 0$ such that for any $u_0, u_1 \in L^2(\mathbb{T})$ with*

$$|u_0|_2 \leq \delta, \ |u_1|_2 \leq \delta \ and \ \int_{\mathbb{T}} u_0 = \int_{\mathbb{T}} u_1,$$

one can find a control input $h \in L^2(0, T; L^2(\mathbb{T}))$ such that the solution u of the system

$$\begin{cases} u_t - Hu_{xx} + uu_x = \mathcal{G}h \\ u(x, 0) = u_0(x), \end{cases} \quad (6.2)$$

satisfies $u(x, T) = u_1(x)$ on $\mathbb{T}$.

(ii) *(Large data) For any $R > 0$ there exists a positive $T = T(R)$ such that the above property holds for any $u_0, u_1 \in L^2(\mathbb{T})$ with*

$$|u_0|_2 \leq R, \ |u_1|_2 \leq R \ and \ \int_{\mathbb{T}} u_0 = \int_{\mathbb{T}} u_1$$

This result relies strongly on the bilinear estimates proved in [201].
We do not know of corresponding results for the ILW equation.

Remark 24

1. We refer to [143] for stabilizations results.
2. Related results for the *linear* problem were obtained in [147].

6.4 Initial-Boundary Value Problems

The global well-posedness of the initial- boundary value problem the BO and the ILW equation on the half-line with zero boundary condition at $x = 0$ is proven respectively in [94] and in [26], for small initial data in suitable weighted Sobolev space. Moreover the long time asymptotic is given, for instance, one gets for the ILW equation [17]

$$u(x, t) = \frac{1}{3\pi \delta t} \frac{x}{(\sigma t)^{1/3}} \operatorname{Ai}\left(\frac{x}{(\delta t)^{1/3}}\right) + \min\left(1, \frac{x}{\sqrt{t}}\right) O\left(t^{-1-a/2}\right),$$

where $a > 0$ and Ai is the Airy function.

6.5 Transverse Stability Issues

The Kadomtsev-Petviashvili (KP) equation was introduced heuristically in [114] to study the transverse stability of the KdV soliton with respect to long, weakly transverse perturbations. Since their formal analysis does not depend on the dispersive term in the KdV equation, it applies as well as to the BO and ILW equations, yielding the KP-II version of those one dimensional equations, that is, respectively

$$u_t + u_x + uu_x - Hu_{xx} + \partial_x^{-1}u_{yy} = 0, \tag{6.3}$$

and

$$u_t + uu_x + \frac{1}{\delta}u_x + \mathcal{T}(u_{xx}) + \partial_x^{-1}u_{yy} = 0 \tag{6.4}$$

Those equations (that are not known to be integrable) were in fact derived formally in the context of internal waves in [7, 56, 91, 175].

Note however that they suffer from two shortcomings of the usual KP-II equation. First, as was, noticed in [139], the (artificial) singularity at $\xi_1 = 0$ of the symbol $\xi_1^{-1}\xi_2^2$ of $\partial_x^{-1}\partial_{yy}$ induces a poor error estimate when comparing the KP-II equation with the full water wave system in the appropriate regime. Roughly speaking, one get (see [139] and [142]) an error estimate of the form, in suitable Sobolev norms:

$$\|u_{KP} - u_{WW}\| = o(1),$$

the "correct" error should be, as in the KdV case, $O(\epsilon^2 t)$ where ϵ is the small parameter measuring the weak nonlinear and long wave effects. It is very likely that such a poor precision also holds for the KP-BO and KP-ILW equations.

The second shortcoming concerns the zero mass in x constraint also arising from the singularity of the symbol (see [208]). This of course do not exclude that those KP type equations predict at least qualitatively the behavior of real waves, see e.g. [2, 245].

Coming back to rigorous results, the local Cauchy problem (for a larger class of equations including KP-BO and KP-ILW) is studied in [150] and the global existence and scattering of small solutions is proven in [93].

It was proven in [151] that (6.3) does not possess any non trivial (lump type) *localized* solitary wave. The proof extends as well to (6.4).

Recall that the KdV soliton is transversally L^2 stable with respect to weak transverse perturbations described by the KP II equation [195, 196]. This result is unconditional since it was established in [207] that the Cauchy problem for KP II is globally well-posed in $H^s(\mathbb{R} \times \mathbb{T}), s \geq 0$, or for all initial data of the form $u_0 + \psi_c$ where $u_0 \in H^s(\mathbb{R}^2), s \geq 0$ and $\psi_c(x - ct, y)$ is a solution of the KP-II equation such that for every $\sigma \geq 0$, $(1 - \partial_x^2 - \partial_y^2)^{\sigma/2}\psi_c$ is bounded and belongs to $L_x^2 L_y^\infty(\mathbb{R}^2)$. Note that such a condition is satisfied by the value at $t = 0$ of any soliton or N-soliton of the KdV equation.

A priori this condition is not satisfied by a function ψ that is not decaying along a line $\{(x, y); x - vy = x_0\}$ such as for instance a KP-II line soliton that writes $\psi(x - vy - ct)$. However, one checks that the change of variable

$$X = x + vy - v^2 t, \, Y = y - vt, \, T = t \tag{6.5}$$

leaves the KP-II equation invariant and thus the global well-posedness results hold for an initial data that is localized perturbation of the KP-II line soliton. Observe that the same transformation leaves also invariant the KP-BO and KP-ILW equations.

It is natural to conjecture that a similar stability result holds true for the BO and ILW soliton with respect to (6.3) and (6.4). We refer to [7] for formal arguments in favor of this conjecture.

Such a result would be however a conditional one since no global existence result is available for both (6.3) and (6.4).

It is worth noticing however that it was observed in [249], following the method of [138] that the *periodic* BO solitary waves could be spectrally transversally unstable for some ranges of frequencies. This deserves further investigations.

6.6 Modified BO and ILW Equations

It is well known that a real solution of the focusing modified KdV equation

$$v_t + 6v^2 v_x + v_{xxx} = 0$$

is sent by the Miura transform

$$u = v^2 + i v_x$$

to a complex solution of the KdV equation

$$u_t + 6u u_x + u_{xxx} = 0.$$

It turns out that a similar fact holds for the BO and ILW equations, see [165, 213, 214, 239] and specially [243] to which we refer for an explanation of the mathematical origin of the MILW equation.

The modified ILW and BO equations write respectively

$$v_t + \beta v_x (e^v - 1) + \frac{1}{\delta} v_x + v_x \mathcal{T}(v_x) + \mathcal{T}(v_{xx}) = 0, \tag{6.6}$$

$$q_t + \alpha q_x (e^q - 1) + q_x H(q_x) + H(q_{xx}) = 0, \tag{6.7}$$

where α, β are real constants.

Remark 25 We are not aware of results on the Cauchy problem for (6.7) or (6.6) by PDE methods.

Both Eqs. (6.7) and (6.6) are integrable soliton equations. In particular the MILW possesses an infinite number of conserved quantities, [239], a linear scattering theory [239], a Bäcklund transform [213, 239] and multi-soliton solutions, [213]. The formal IST for the MBO and MILW is studied in [243, 244] respectively.

One can check, see [243] that the Miura type transformation

$$u = \frac{1}{2}\{\mathcal{T}(v_x) + \beta(e^v - 1) + iv_x\}$$

maps real-valued solutions of (6.6) with $\alpha = -\beta$ into complex-valued solution of the ILW equation.

A similar transformation holds in the BO case, see [244].

6.7 Higher Order BO and ILW

Higher order BO and ILW equations (as higher order KdV) appear in two different contexts.

1. First when going to next orders in the expansion of the two-layer system in the ILW /BO regime, see Introduction. For instance, one gets in the notations of [61] (see also [175] for a similar equation) the next equation in the BO regime

$$u_t = \frac{\rho h_1^2}{2\rho_1^2} A^2 H(u_{xx}) + \frac{3\sqrt{2}}{4\rho_1} A u u_x$$
$$- \frac{\sqrt{2}}{2} \epsilon \frac{\rho h_1^2}{\rho_1^2} A[\partial_x(uH(u_x)) + H\partial_x(uu_x)] + \frac{\epsilon}{2}\left(\frac{\rho^2 h_1^2}{\rho_1^2} - \frac{h_1^2}{3}\right) A^2 u_{xxx},$$
(6.8)

where $A = \left(\frac{g\rho_1(\rho-\rho_1)}{h_1}\right)^{1/2}$, and where $\rho > \rho_1$ are the densities of the two fluid layers, h_1 is the height of the upper layer, g the constant of gravity and $\epsilon > 0$ is a small parameter.

This process could actually be continued to obtain an infinite sequence of higher order BO equations. It applies as well to the ILW equation, yielding the following equation, in the above notations of [61], where here $\mathcal{T}_h$ is the Fourier multiplier with symbol $-i\coth(\epsilon h|\xi|)$, h being the depth of the lower layer:

$$u_t = \frac{\rho h_1^2}{2\rho_1^2} A^2 \mathcal{T}_h(u_{xx}) + \frac{3\sqrt{2}}{4\rho_1} A u u_x -$$

$$\frac{\sqrt{2}}{2} \epsilon \frac{\rho h_1^2}{\rho_1^2} A[\partial_x(u\mathcal{T}_h(u_x)) + \mathcal{T}_h \partial_x(uu_x)] - \frac{\epsilon}{2}\left(\frac{\rho^2 h_1^2}{\rho_1^2}(\mathcal{T}_h)^2 + \frac{h_1^2}{3}\right) A^2 u_{xxx}.$$
(6.9)

An equation like (6.8) was also derived in [167] where a solitary-wave solution of the equation is obtained by means of a singular perturbation method. The characteristics of the solution are discussed in comparison with those for a higher-order BO equation of the Lax type, see next section.

2. Second, one gets a hierarchy of (integrable) higher order BO or ILW equations by considering the Hamiltonian flows of the successive conservation laws, see e.g. [165].

For instance, the next equations in the hierarchy are for the ILW and BO equation given respectively by

$$u_t = \partial_x \operatorname{grad} I_4(u)$$

and

$$u_t = \partial_x \operatorname{grad} J_4(u),$$

where I_4 and J_4 were defined in chapter "Wave Turbulence and Complete Integrability", that is

$$u_t - 3u^2 u_x + 4u_{xxx} + 3(uHu_x)_x + 3H(uu_x)_x = 0 \qquad (6.10)$$

and

$$u_t - 3u^2 u_x - 3(u\mathcal{T} u_x)_x - 3\mathcal{T}(u_{xx}) + u_{xxx} - 3\mathcal{T}^2(u_{xxx})$$
$$-\frac{1}{\delta}[18uu_x + 9\mathcal{T} u_{xx}] - \frac{3}{\delta^2} u_x = 0 \qquad (6.11)$$

We are not aware of rigorous results on (6.10), (6.11) by IST methods.

On the other hand, it was shown in [231] that (6.8) is quasilinear in the sense that, as for the BO equation, the flow map cannot be C^2 in any Sobolev space $H^s(\mathbb{R})$, $s \in \mathbb{R}$ (the proof applies as well to (6.10)). In [148] the Cauchy problem for (6.8) is proven to be locally well-posed in $H^s(\mathbb{R}), s \geq 2$ and in some weighted Sobolev spaces.

This result was significantly improved in [200] where the global well-posedness of (6.8) was established in $H^s(\mathbb{R}), s \geq 1$, and thus in particular in the energy space $H^1(\mathbb{R})$. The main difficulty is of course to obtain the local well-posedness in $H^1(\mathbb{R})$. This is achieved by introducing, as in [254], a gauge transformation that weakens the high-low frequency interactions in the nonlinear terms. Such a transformation was already used in [148] to obtain the H^2 well-posedness but it is combined here with a Besov version of Bourgain spaces, together with the full Kato smoothing effect for functions that are localized in space frequencies.

Another result in [200] concerns the limiting behavior of solutions u_ϵ when $\epsilon \to 0$. A direct standard compactness method (as used for instance in the BO-Burgers equation in the zero dissipation limit) does not seems to work since the two leading terms (of order one) in the Hamiltonian I_4 have opposite signs. However it is shown that the solution of (6.8) converges in $L^\infty(0, T; H^1(\mathbb{R}))$, $\forall T > 0$ to the corresponding solution of BO as $\epsilon \to 0$ provided the ratio of density $\frac{\rho_1}{\rho}$ equals $\frac{1}{\sqrt{3}}$.

This shortcoming of (1.2) might not occur when using the method of [38] to derive asymptotic models of internal waves since it is more flexible and leads to (equivalent in the sense of consistency) *families* of models. In particular at least one member of the family of higher order BO equations might contain one for which the limit to BO holds, for any ratio of densities.

As mentioned in [200], it is likely that the previous results hold true for the higher order ILW equation (6.9).

We are not aware of similar results for the next equations in the BO or ILW hierarchy (6.10) and (6.11) but it is likely that the method in [200] could be applied.

For any of those higher order equations ((6.10), (6.11), (6.8), (6.9)) questions about existence and stability of solitary wave solutions seem to be widely open except for the Lax hierarchy of the BO equation for which Matsuno [170] has established the Lyapunov stability of the N-soliton in the hierarchy, using in particular results from the IST. One should also mention [167] where by using a multisoliton perturbation theory it is proven analytically that the overtaking collision between two-solitary waves exhibits a phase shift, the amplitudes being not altered after interaction.

Remark 26 Concerning the periodic problem, Tanaka [253] has recently proven that the Cauchy problem for third order type Benjamin-Ono equations is locally well posed in $H^s(\mathbb{T})$, $s > \frac{5}{2}$.

6.8 Interaction of Solitary Waves

As noticed in [263] where one can find a nice overview of the subject and a fascinating analysis of observed (via imaging technique) oblique wave-wave interactions on internal waves in the strait of Georgia, "although nonlinear interactions that occur when two large internal waves collide at oblique angles are often observed in the natural world, quantitative and theoretical aspects of these interactions are only poorly understood".

This important topic has been first extensively studied for shallow water gravity waves, mainly in the context of the KP-II equation. Although KP-II has no fully localized solitary wave solutions [44], it has a large variety of two-dimensional exact solutions, those wave patterns being generated by nonlinear interactions among several obliquely propagating solitary waves. In particular the resonant interactions among those solitary waves play a fundamental role in multidimensional wave phenomenon, leading to very complicated patterns. We refer for instance to [34, 52, 245] and to the book [135] and the references therein for an extensive description of those waves. We also refer to [267] for a study of the direct scattering theory for KP II on the background of a line soliton.

It is worth noticing that some of those patterns look very much like real observed waves [2] and the classical picture below of interaction of line solitons on the Oregon coast.

According to [184, 185] the oblique interaction of solitary waves propagating at different directions $\mathbf{n}_1$ and $\mathbf{n}_2$ can be classified into two types. The first one occurs when the two solitary waves propagate in almost the same direction ($1 - \mathbf{n}_1.\mathbf{n}_2 \gg \epsilon$ where ϵ is the ratio between a typical amplitude of the waves and a typical depth of the fluid)) and interact for a relatively long time (strong interaction). The second type corresponds to the interaction of solitary waves propagating in almost opposite directions $(1 - \mathbf{n}_1.\mathbf{n}_2 \simeq O(\epsilon))$ and hence the interaction is relatively short (weak interaction).

Typically the two waves evolves according to their own one-dimensional equation (KdV for surface waves ILW or BO for internal waves) in the case of weak interactions and according to a KP-II like equation in the case of strong interactions.

Such exact multiline solitons are not known to exist for the BO and ILW versions of the KP equation since they are not integrable. Nevertheless it makes prefect sense to look for oblique interactions of internal line solitons. Most of the existing work consists in numerical simulations.

For internal waves the question of oblique interaction of solitary waves has been addressed in particular by Oikawa and Tsuji, in the context of BO (infinite depth) [220, 256] and ILW (finite depth [257]). One finds in [256] numerical simulations of the strong interaction of nonlinear long waves whose propagation directions are very close to each other. Two initial settings are considered, first a superposition of two BO solitons with the same amplitude and with different directions, and the second one is an oblique reflection of a BO soliton at a vertical wall. It is observed that the Mach reflection does occur for small incident angles and for some incident angles very large stem waves (see below) are observed.

The case of finite depth is considered in [257] under the assumption of a small but finite amplitude. When the angle θ between the wave normals of two solitons is not small, it is shown by a perturbation method that in the lowest order of approximation the solution is a superposition of two ILW solitons and in the next order of approximation the effect of the interaction appears as position phase shifts and as an increase in amplitude at the interaction center of two solitons. When θ is small, it is shown that the interaction is described approximately by (6.4). By solving it numerically for a V-shaped initial wave that is an appropriate initial value for the oblique reflection of a soliton due to a rigid wall, it is shown that for a relatively large angle of incidence θ the reflection is regular, but for a relatively small θ the reflection is not regular and a new wave called stem is generated. The results are also compared with those of the Kadomtsev-Petviashvili (KP) equation and of the two-dimensional Benjamin-Ono equation (6.3).

Remark 27 Recall that BO or ILW equations have the limitation of weak nonlinearity. Other internal waves models without the assumption of weak non-linearity have been derived in [57, 175]. The system derived in [57] has solitary wave solutions which are in good agreement with the experimental results in [136].

On the other hand, we recall that Matsuno in particular, [166, 175], has extended the (formal) analysis leading to the BO equation to next order in amplitude and derived a higher order equation similar to a higher order BO equation, namely, say for the wave elevation:

$$\zeta_t + \zeta_x + \frac{3}{2}\alpha\zeta\zeta_x + \frac{1}{2}\alpha\delta H\zeta_{xx} - \frac{3}{8}\alpha^2\zeta^2\zeta_x$$
$$+ \frac{1}{2}\sigma\alpha\delta\left[\frac{5}{4}\zeta H\zeta_{xx} + \frac{9}{4}H(\zeta\zeta_x)_x\right] - \frac{3}{8}\left(\sigma^2 - \frac{4}{9}\right)\delta^2\zeta_{xxx} = 0, \quad (6.12)$$

where here $\sigma = \frac{\rho_1}{\rho_2}$ is the ratio of densities, $\alpha = \frac{a}{h}$ the ratio of a typical amplitude of the wave over the depth of the upper layer and $\delta = \frac{h}{l}$ is the ratio of h over a typical wavelength.

The same issues of transverse stability of 1D- solitary waves or of interaction of line solitons arise for those higher order models but have not be addressed yet as far as we know.

The situation is different when considering the interaction of long internal gravity waves propagating on say two neighborhood pycnoclines in a stratified fluid, see [156, 157] where a system of coupled ILW equations is derived. We refer to [83, 84] for the shallow-water regime, leading to a system of coupled KdV or KP-II type equations.

For deep water systems, Grimshaw and Zhu [91] have shown that in the strong interaction case, each wave is governed by its own ILW equation (BO equation in the infinite depth case), the main effect being a phase shift of order $O(\epsilon)$, ϵ measuring the wave amplitude. In the case of weak interactions, and when Δ_1, Δ_2 are of order $O(\epsilon)$, the interaction is governed by a coupled two-dimensional KP-II type ILW or BO equations. Here $\Delta_1 = |c_m/c_n - \cos\delta|$, $\Delta_2 = |c_n/c_m - \cos\delta|$, where δ is the angle between the two directions of propagation and c_m, c_n are the linear, long wave speeds for wave with mode numbers m, n. We refer to [151] for a mathematical study of those systems.

In any case the rigorous mathematical analysis of oblique interactions of internal waves deserves further investigations.

Acknowledgements The Author thanks heartfully the co-organizers of the program, Peter D. Miller, Peter A. Perry and Catherine Sulem for their involvement in the success of the Program. He also thanks Jaime Angulo, Felipe Linares and Didier Pilod for very useful comments on a preliminary draft.

References

1. L. ABDELOUHAB, J. BONA, M. FELLAND AND J.-C.SAUT, *Non local models for nonlinear dispersive waves*, Physica D Nonlinear Phenomena, **40**, (1989), 360–392.
2. M.J. ABLOWITZ AND D.E. BALDWIN, *Photographs and videos at http://www.markablowitz. com/line-solitons and http://www.douglasbaldwin.com/nl-waves.html*.
3. M.J. ABLOWITZ AND P.A. CLARKSON, *Solitons, nonlinear evolution equations and inverse scattering*, London Mathematical Society Lecture Notes series **149**, Cambridge University Press, (1991).
4. M.J. ABLOWITZ AND A.S. FOKAS, *The inverse scattering transform for the Benjamin-Ono equation: a pivot to multidimensional problems*, Stud. App. Math. **68** (1983), 1–10.
5. M.J. ABLOWITZ, A.S. FOKAS, J. SATSUMA AND H. SEGUR, *On the periodic intermediate long wave equation*, J. Phys. A Math. Gen. **15** (1982), 781–786.
6. M.J. ABLOWITZ, A.S. FOKAS AND R. ANDERSON, *The direct linearizing transform and the Benjamin-Ono equation*, Physics Letters A **93** (8) (1983), 375–378.
7. M.J. ABLOWITZ AND H. SEGUR, *Long internal waves in fluids of great depth*, Studies Appl. Math. **62** (1980), 249–262.
8. J. ALBERT, *Positivity properties and uniqueness of solitary wave solutions of the intermediate long-wave equation*, in Evolution equations (Baton Rouge, LA, (1992)), 11–20, Lecture Notes in Pure and Appl. Math., 168, Dekker, New York, 1995.

9. J. ALBERT, *Positivity properties and stability of solitary-wave solutions of model equations for long waves*, Comm. Partial Differential Equations **17** (1–2) (1992), 1–22.
10. J. P. ALBERT AND J. L. BONA, *Total Positivity and the Stability of Internal Waves in Stratified Fluids of Finite Depth* IMA J. Appl. Math. **46** (1–2) (1991), 1–19.
11. J.P. ALBERT, J. L BONA AND D. HENRY, *Sufficient conditions for stability of solitary-wave solutions of model equations for long waves*, Physica D **24**, (1987), 343–366.
12. J.ALBERT, J.L. BONA AND J. M. RESTREPO, *Solitary-wave solution of the Benjamin equation*, SIAM J. Appl. Math. **59** (6) (1999), 2139–2161.
13. J.P. ALBERT, J.L. BONA AND N.V. NGUYEN, *On the stability of KdV multi-solitons*, Diff. Int. Equations **20** (8) (2007), 841–878.
14. J. ALBERT, J.L. BONA AND J.-C. SAUT, *Model equations for waves in stratified fluids*, Proc. Royal Soc. London A, **453** (1997), 1233–1260.
15. J. ALBERT AND F. LINARES, *Stability and symmetry of solitary-wave solutions to systems modeling interactions of long waves*, J. Math. Pures Appl. **79** (3) (2000), 195–226.
16. J.P.ALBERT AND J.F. TOLAND, *On the exact solutions of the intermediate long-wave equation*, Diff. Int. Eq. **7** (3–4) (1994), 601–612.
17. M.P.A. ALEJANDRE AND E.I KAIKINA, *Intermediate long wave on a half-line*, J. Evol. Equ. **11** (2011), 743–770.
18. C.J. AMICK AND J. TOLAND, *Uniqueness and related analytic properties for the Benjamin-Ono equation-a nonlinear Neumann problem in the plane*, Acta Math. **167** (1991), 107–126.
19. C.J. AMICK AND J. TOLAND, *Uniqueness of Benjamin's solitary wave solutions of the Benjamin-Ono equation*, IMA J. of Appl. Math. **46** (1–2) (1991), 21–28.
20. R.L. ANDERSON AND E. TAFLIN, *The Benjamin-Ono equation-Recursivity of linearization maps-Lax pairs*, Letters Math. Phys. **9** (1985), 299–311.
21. J. ANGULO PAVA, *Existence and stability of solitary wave solutions of the Benjamin equation*. J. Diff. Eq. **152**, (1999), 136–159.
22. J. ANGULO PAVA, *Stability properties of solitary waves for fractional KdV and BBM equations*, Nonlinearity **31** (3) (2018), 920–956.
23. J. ANGULO PAVA, E. CARDOSO JR. AND F. NATALI, *Stability properties of periodic traveling waves for the intermediate long wave equation*, Rev. Mat. Iberoam. **33** (2) (2017), 417–448.
24. J. ANGULO PAVA AND F.NATALI, *Positivity properties of the Fourier transform and the stability of periodic traveling-wave solutions*, SIAM J. Math. Anal. **40** (3) (2008), 1123–1151.
25. J. ANGULO PAVA AND J.-C. SAUT, *Existence of solitary waves solutions for internal waves in two-layers systems*, arXiv:1804.02328v1 6 Apr 2018 and Quarterly of Applied Math. (2019).
26. M.P. A. ALEJANDRE AND E. I. KAIKINA, INTERMEDIATE LONG-WAVE EQUATION ON A HALF-LINE, J. Evol. Equ. **11** (2011), 743–770.
27. R. BEALS AND R.R. COIFMAN, *Scattering, transformations spectrales et équations d'évolution non linéaires I,II*, Séminaire Goulaouic-Meyer-Schwartz 1980/81, exposé XXII, and 1981/1982, exposé XXI, Ecole Polytechnique, Palaiseau.
28. R. BEALS AND R.R. COIFMAN, *Scattering and inverse scattering for first-order systems*, Comm. Pure Appl. Math. **37** (1984), 39–90.
29. D.J. BENNEY AND D.R.S. KO *The propagation of long large amplitude intrnal waves*, Studies Appl. Math. **59** (1978), 187–199.
30. T. B. BENJAMIN, *Internal waves of permanent form in fluids of great depth*, J. Fluid Mech., **29** (1967), 559–592.
31. T. B. BENJAMIN, *A new kind of solitary wave*, J. Fluid Mech. **2455** (1992), 401–411.
32. T.B. BENJAMIN, *Solitary and periodic waves of a new kind*, Philos. Trans. Roy. London Ser. A **340** (1996), 1775–1806.
33. D.P. BENNETT, R.W. BROWN, S.E. STANSFIELD, J.D. STROUGHAIR AND J.L. BONA, *The stability of internal solitary waves*, Math. Proc. Camb. Phil. Soc. **94** (1983) 351–379.
34. G. BIONDINI, K.-I. MARUNO, M. OIKAWA AND H. TSUJI, *Soliton interaction of the Kadomtsev-Petviashvili equation and generation of large-amplitude water waves*, Studies Appl; math. **122** (2009), 377–394.

35. T.L. BOCK AND M.D. KRUSKAL, *A two-parameter Miura transformation of the Benjamin-Ono equation*, Phys. Lett. A **74** (1979), 173–176.
36. J. L. BONA, M. CHEN AND J.-C. SAUT, *Boussinesq equations and other systems for small-amplitude long waves in nonlinear dispersive media I: Derivation and the linear theory*, J. Nonlinear Sci., **12** (2002), 283–318.
37. J.L. BONA, A. DURAN AND D. MITSOTAKIS, *Solitary wave solutions of Benjamin-Ono and other systems for internal waves. I. Approximations*, preprint March 2018.
38. J. L. BONA, D. LANNES AND J.-C. SAUT, *Asymptotic models for internal waves*, J. Math. Pures. Appl. **89** (2008) 538–566.
39. J.L. BONA AND YI A. LI, *Decay and analyticity of solitary waves*, J. Mathématiques Pures et Appl. **76** (1997), 377–430.
40. J.L. BONA AND L. LUO, *Large time asymptotics of the generalized benjamin-Ono-Burgers equation*, DCDS **4** (1) (2011), 15–50.
41. J.L. BONA AND R. SMITH, *The initial value problem for the Korteweg-de Vries equation*, Philos. Trans. R. Soc. Lond., Ser. A, **278** (1975), 555–601.
42. J. L. BONA AND A. SOYEUR, *On the stability of solitary-wave solutions of model equations for long-waves*, J. Nonlinear Sci. **4** (1994), 449–470.
43. J. L. BONA, P. E. SOUGANIDIS AND W. A. STRAUSS, *Stability and instability of solitary waves of KdV type*, Proc. Roy. Soc. London A **411** (1987), 395–412.
44. A. DE BOUARD AND J.-C. SAUT, *Solitary waves of the generalized KP equations*, Ann. IHP Analyse Non Linéaire **14**, 2 (1997), 211–236.
45. N. BURQ AND F. PLANCHON, *The Benjamin-Ono equation in energy space*, Phase space analysis of partial differential equations, 55–62, Progr. Nonlinear Differential Equations Appl., **69**, Birkhäuser Boston, Boston, MA.
46. F. CALOGERO AND A. DEGASPERIS, *Spectral transforms and solitons*, North-Holland, Amsterdam, New-York (1982).
47. D.C. CALVO AND T.R. AKYLAS, *On interfacial gravity-capillary solitary waves of Benjamin type and their stability*, Physics of Fluids **15** (5) (2003), 1261–1270.
48. K.M. CASE, *Properties of the Benjamin-Ono equation*, J. Math. Phys. **20** (1979) 972–977.
49. K.M. CASE, *The N-soliton solution of the Benjamin-Ono equation*, Proc.Nat. Acad. Sci. **75** (1978), 3562–3563.
50. K.M. CASE, *Benjamin-Ono-related equations and their solutions*, Proc. Nat. Acad.Sci. USA **76** (1–3) (1979), 1–3.
51. T. CAZENAVE AND P.-L. LIONS, *Orbital stability of standing waves for some nonlinear Schrödinger equations*, Commun. Math. Phys. **85** (4) (1982), 549–561.
52. S. CHAKRAVARTY AND Y. SATSUMA, *Soliton solutions of the KP equation and application in shallow water waves*, Studies Appl. Math. **123** (2009), 83–151.
53. H.H. CHEN AND D.J. KAUP, *Conservation laws of the Benjamin-Ono equation*, J. Math. Phys. **21** (1) (1980), 19–20.
54. H.H. CHEN AND Y.C. LEE, *Internal wave solitons of fluid with finite depth*, Phys.Rev. Lett. **43** (1979), 264.
55. , H.H. CHEN, R. HIROTA AND Y.C. LEE, *Inverse scattering problem for internal waves with finite fluid depth*, Physics Lett. **75A** (1980) 254–256.
56. W. CHOI AND R. CAMASSA, *Weakly nonlinear internal waves in a two-fluid system*, J. Fluid. Mech. **313** (1996), 83–103.
57. W. CHOI AND R. CAMASSA, *Long internal waves of finite amplitude*, Physics Rev. Letters **77** (9) (1996), 1759–1996.
58. D.R. CHRISTIE, K. MUIRHEAD AND A. HALES, *On solitary waves in the atmosphere*, J. Atmos. Sc. **35** (1978), 805.
59. R.R. COIFMAN AND M.V. WICKERHAUSER, *The scattering transform for the Benjamin-Ono equation*, Inverse Problems **6** (5) (1990), 825–862.
60. P. CONSTANTIN AND J.-C. SAUT, *Local smoothing properties of dispersive equations*, J. Amer. Math. Soc. **1** (1988), 413–439.
61. W. CRAIG, P. GUYENNE ANF H. KALISCH, *Hamiltonian long-wave expansions for free surfaces and interfaces*, Comm. Pure. Appl. Math. **58** (2005)1587–1641.

62. R.E. DAVIS AND A. ACRIVOS, *Solitary internal waves in deep water*, J. Fluid Mech. **29** (1967), 593–607.
63. P.A. DEIFT, A.R. ITS AND X. ZHOU, *Long-time asymptotics for integrable nonlinear wave equations*, in *Important developments in soliton theory*, 181–204, Springer series in Nonlinear Dynamics, A.S. Fokas and V. E. Zakharov Eds., Springer, Berlin 1993.
64. P. A. DEIFT, S. VENAKIDES AND X. ZHOU, *New results in small dispersion KdV by an extension of the steepest descent method for Riemann-Hilbert problems*, Int. Math. Res. Notes **6** (1997), 286–299.
65. P. A. DEIFT, S. VENAKIDES AND X. ZHOU, *An extension of the steepest descent method for Riemann-Hilbert problems: the small dispersion limit of the Korteweg-de Vries equation*, Proc. Nat. Acad. Sci. USA **95** (2) (1998), 450–454.
66. P. A. DEIFT AND X. ZHOU, *A steepest descent method for oscillatory Riemann-Hilbert problems. Asymptotics for the MKdV equation*, Ann. of Math. **137** (2) (1993), 295–398.
67. P. A. DEIFT AND X. ZHOU, *Long-time asymptotics for integrable systems. Higher order theory*, Comm. Math. Phys. **165** (1) (1994), 175–191.
68. Y. DENG, *Invariance of the Gibbs measure for the Benjamin-Ono equation*, J. Eur. Math. Soc. **17** (5) (2015), 1107–1198.
69. Y. DENG, N. TZVETKOV AND N. VISCIGLIA, *Invariant measures and long time behavior for the Benjamin-Ono equation III*, Comm. Math. Phys. **339** (3) (2015), 815–857.
70. B. DESJARDINS, D. LANNES AND J.-C. SAUT, *in preparation*.
71. D.B. DIX, *Temporal decay of solutions to the Benjamin-Ono-Burgers equation*, J. Diff. Eq. **90** (1991), 238–287.
72. D.B. DIX, *The dissipation of nonlinear dispersive waves: the case of asymptotically weak nonlinearity*, Commun. In PDE **17** (9–10), (1992), 1665–1693.
73. S.YU DOBROKHOTOV AND I.M. KRICHEVER, *Multi-phase solutions of the Benjamin-Ono equation*, Math. Notes **49** (1991), 583–594.
74. P.M. EDWIN AND B. ROBERTS, *The Benjamin-Ono-Burgers equation: an application tin solar physics*, Wave motion **8** (2) (1986), 151–158.
75. G.A. EL, L.T.K. NGUYEN AND N.F. SMYTH, *Dispersive shock waves with nonlocal dispersion of Benjamin-Ono type*, Nonlinearity **31** (2018), 1392–1416.
76. C. FLORES, *On decay properties of solutions to the IVP for the Benjamin-Ono equation*, J. Dyn. Diff. Eq. **25** (2013), 907-.
77. A.S. FOKAS AND B. FUCHSSTEINER, *The hierarchy of the Benjamin-Ono equation*, Phys. Lett. **86 A** (1981), 341–345.
78. G. FONSECA AND F. LINARES, *The Benjamin-Ono equation with unbounded data*, J. Math. Anal. Appl. **247** (2) (2000), 426–447.
79. G. FONSECA, F. LINARES AND G.PONCE, *The IVP for the Benjamin-Ono equation in weighted spaces II*, J. Funct. Anal. **262** (2012), 2031–2049.
80. G. FONSECA, F. LINARES AND G.PONCE, *The IVP for the dispersion generalized Benjamin-Ono equation in weighted spaces*, Ann. Inst. H. PoincaréAnal. Non Linéaire **30** (5) (2013), 763–790.
81. G. FONSECA, AND G.PONCE, *The IVP for the Benjamin-Ono equation in weighted spaces*, J. Funct. Anal. **260** (2010), 436–459.
82. R. FRANK AND E. LENZMANN, *Uniqueness of nonlinear ground states for fractional Laplacians in $\mathbb{R}^n$*, Acta. Math. **210** (2013), 261–318.
83. J.A. GEAR, *Strong interactions between solitary waves belonging to different wave modes*, Studies in Appl. Math. **72** (1985), 95–124.
84. J.A. GEAR AND R. GRIMSHAW, *Weak and strong interactions between internal solitary waves*, Studies in Appl. Math. **70** (1984), 235–258.
85. P. GÉRARD AND T. KAPPELER, arXiv: 1905.01849, 6 May 2019.
86. J. GINIBRE AND G. VELO, *Propriétés de lissage et existence de solutions pour l'équation de Benjamin-Ono généralisée*, C. R. Acad. Sci. Paris **308**, Série I (1989), 309–314.
87. J. GINIBRE AND G. VELO, *Smoothing properties and retarded estimates for some dispersive evolution equations*, Commun. Math. Phys. **144** (1993), 163–188.

88. J. GINIBRE AND G. VELO, *Commutator expansions and smoothing properties of generalized Benjamin-Ono equations*, Ann. Inst. Henri Poincaré, Physique Théorique **51** (2) (1989), 221–229.
89. J. GINIBRE AND G. VELO, *Smoothing properties and existence of solutions for the generalized Benjamin-Ono equation*, J. Diff. Equations **93** (1991), 150–212.
90. R. H. GRIMSHAW, N.F. SMYTH AND Y.A STEPANYANNTS, *Decay of Benjamin-Ono solitons under the influence of dissipation*, Wave Motion **78** (2018), 98–115.
91. R. H. GRIMSHAW AND Y. ZHU, *Oblique interaction between internal solitary waves*, Studies Appl. Math. **92** (1994), 249–270.
92. S. GUSTAFSON, H. TAKAOKA AND T-P. TSAI, *Stability in $H^{1/2}$ of the sum of K solitons for the Benjamin-Ono equation*, J. Math. Phys. **50** (1) (2009), 013101–01315.
93. B. HARROP-GRIFFITHS AND J.L. MARZUOLA, *Small data global solutions of the Camassa-Chï equation*, Nonlinearity **31** (2018), 1868–1904.
94. N.HAYASHI AND E.I KAIKINA, *The Benjamin-Ono equation on a half-line*, Int. J. of Math. and Math. Sci. (2010), 1–38.
95. N. HAYASHI AND P. NAUMKIN, *Large time asymptotics of solutions to the generalized Benjamin-Ono equation*, Transactions AMS **351** (1) (1999), 109–130.
96. N. HAYASHI AND P. NAUMKIN, *On the reduction of the modified Benjamin-Ono equation to the cubic derivative nonlinear Schrödinger equation*, Disc. Cont. Dyn. Syst. **8** (1) (2002), 237–255.
97. N.HAYASHI, K. KATO AND T. OZAWA, *Dilation method and smoothing effects of solutions to the Benjamin-Ono equation*, Proc.Roy. Soc. Edinburgh **126 A** (1996), 273–285.
98. K.R. HELFRICH AND W.K. MELVILLE, *Long nonlinear internal waves*, Ann. Rev. Fluid Mech. **38** (2006), 395–425.
99. S. HERR, A.D. IONESCU, C.E. KENIG AND H. KOCH, *A para-differential renormalization technique for nonlinear dispersive equations*, Communications in PDE **35** (2010), 1827–1875.
100. R. HIROTA, *Exact solution of the Korteweg-de Vries equation for multiple collisions of solitons*, Phys. Rev. Lett. **27** (118) (1971), 1192.
101. M.IFRIM AND D. TATARU, *Well-posednes and dispersive decay of small data solutions for the Benjamin-Ono equation*, arXiv:1701.08476v2 [math.AP] 20 Feb 2017 and Annales Sci. ENS to appear.
102. M.IFRIM AND J.-C. SAUT, in preparation.
103. A. IONESCU AND C.E. KENIG, *Global well-posedness of the Benjamin-Ono equation in low-regularity spaces,* J. Amer. Math. Soc., **20** (2007), 753–798.
104. R.J. IÓRIO JR., *On the Cauchy problem for the Benjamin-Ono equation*. Comm. Partial Differential Equations **11** (10) (1986), 1031–1081.
105. R.J. IÓRIO JR., *The Benjamin-Ono equation in weighted Sobolev spaces*, J. Math. Anal and Appl. **157** (2) (1991), 577–590.
106. R.J. IÓRIO JR., *Unique continuation principle for the Benjamin-Ono equation*, Diff. Int. Equations **16** (2003), 1281–1291.
107. P. ISAZA, F. LINARES AND G. PONCE, *On the propagation of regularities in solutions of the Benjamin-Ono equation*, J. Funct. Anal. **270** (2018), 976–1000.
108. R. IORIO, F. LINARES AND M. SCALOM, *KdV and BO equations witthn bore-like data*, Diff. Int;
109. YUHAN JIA, *Numerical study of the KP solitons and higher order Miles theory of the Mach reflection in shallow water*, PhD thesis, Ohio Stae University, (2014).
110. M.C. JORGE, A.A. MINZONI AND N.F. SMYTH, *Modulation solutions for the Benjamin-Ono equation*, Phys. D **132** (1999), 1–18.
111. R.I. JOSEPH, *Solitary waves in a finite depth fluid*, J. Physics A Mathematics and General **10** (12), (1977), L225-L228.
112. R.I. JOSEPH, *Multisoliton-like solutions to the Benjamin-Ono equation*, J. Math. Phys. **18** (12) (1977), 2251–2258.

113. R.I. JOSEPH AND R. EGRI, *Multi-soliton solutions in a finite depth fluid*, J. Physics A Mathematics and General **11** (5) (1978), L97-L102.
114. B. KADOMTSEV AND V.I. PETVIASHVILI, *On the stability of solitary waves in weakly dispersing media*, Sov. Phys. Dokl. **15** (1970), 539–541.
115. T. KAKUTANI AND K. MATSUUCHI, *Effect of viscosity of long gravity waves*, J. Phys. Soc. Japan **39**(1) (1975), 237–246.
116. L.A KALYAKIN, *Long -wave asymptotics. Integrable equations as the asymptotic limit of nonlinear systems*, Russian Math; Surveys **44** (1) (1989), 3–42.
117. D.J. KAUP AND Y. MATSUNO, *Initial value problem of the linearized Benjamin-Ono equation and its applications*, J. Math. Phys. **38** (10) (1997), 5198–5224.
118. D.J. KAUP AND Y. MATSUNO, *The inverse scattering for the Benjamin-Ono equation*, Stud. Appl. Math. **101** (1998), 73–98.
119. D.J. KAUP, T.I. LAKOBA AND Y. MATSUNO, *Complete integrability of the Benjamin-Ono equation by means of action-angle variables*, Physics Letters A**238** (1998), 123–133.
120. D.J. KAUP, T.I. LAKOBA AND Y. MATSUNO, *Perturbation theory for the Benjamin-Ono equation*, Inverse Problems **15** (1999), 215–240.
121. C.E. KENIG AND K.D. KOENIG, *On the local well-posedness of the Benjamin-Ono and modified equations*, Math. Res. Lett. **10** (2003), 879–896.
122. C.E. KENIG AND Y. MARTEL, *Asymptotic stability of solitons for the Benjamin-Ono equation*, Revista Matematica Iberoamericana **25** (2009), 909–970.
123. C.E. KENIG, Y. MARTEL AND L. ROBBIANO, *Local well-posedness and blow-up in the energy space for a class of L^2 critical dispersion generalized Benjamin-Ono equations*, Ann. I. H. Poincaré, **142** (1) (1994), 853–887.
124. C.E. KENIG, G. PONCE AND L. VEGA, *Oscillatory integrals and regularity of dispersive equations,* Indiana Univ. Math. J., **40** (1991), 33–69.
125. C.E. KENIG, G. PONCE AND L. VEGA, *On the generalized Benjamin-Ono equation* Trans. Amer. Math. Soc., **342** (1) (1994), 155–172.
126. C.E. KENIG, G. PONCE AND L. VEGA, *Well-posedness and scattering results for the generalized Korteweg- de Vries equation via the contraction principle*, Comm. Pure Appl. Math. **46** (1993), 527–620.
127. C. KLEIN AND R. PETER, *Numerical study of blow-up in solutions to generalized Korteweg-de Vries equations*. Preprint available at `arXiv:1307.0603`
128. C.E. KENIG AND H. TAKAOKA, *Global well-posedness of the modified Benjamin-Ono equation with initial data in $H^{1/2}$.*, International Math. Res. (2006), 1–44.
129. C. KLEIN AND J.-C.SAUT, *A numerical approach to blow-up issues for dispersive perturbations of Burgers equation*, Physica D **295–296** (2015), 46–65.
130. C. KLEIN AND J.-C. SAUT, *IST versus PDE, a comparative study*, in *Hamiltonian Partial Differential Equations and Applications*, Fields Institute Communications, **75** (2015), 383–449.
131. H. KOCH AND N. TZVETKOV, *Nonlinear wave interactions for the Benjamin-Ono equation*. Int. Math. Res. Not. **30** (2005), 1449–1464.
132. H. KOCH AND N. TZVETKOV, *On the local well-posedness of the Benjamin-Ono equation in $H^s(\mathbb{R})$*, IMRN **26** (2003) 55–68.
133. Y. KODAMA, J. SATSUMA AND M.J. ABLOWITZ, *Nonlinear intermediate long-wave equation: analysis and method of solution*, Phys. Rev. Lett. **46** (1981), 687–690.
134. Y. KODAMA, M.J. ABLOWITZ AND J. SATSUMA, *Direct and inverse scattering problems of the nonlinear intermediate long wave equation*, J. Math. Physics **23** (1982), 564–576.
135. Y. KODAMA *Solitons in two-dimensional shallow-water*, book in preparation (2018).
136. C.G. KOOP AND G. BUTLER, *An investigation of internal solitary waves in a two-fluid system*, J. Fluid Mech. **119** (1981), 225–251.
137. T. KUBOTA, D.R.S KO AND L.D. DOBBS, *Weakly nonlinear, long internal gravity waves in stratified fluids of finite depth*, J. Hydronautics **12** (1978), 157–165.
138. E.A. KUZNETSOV, M.D. SPECTOR AND G.E. FAL'KOVICH, *On the stability of nonlinear waves in integrable models*, Physica D **10** (1984), 379–386.

139. D LANNES, *Consistency of the KP approximation*, Discrete Cont. Dyn. Syst. (2003) Suppl. 517–525.
140. D. LANNES, *Water waves: mathematical theory and asymptotics*, Mathematical Surveys and Monographs, vol 188 (2013), AMS, Providence.
141. D. LANNES, *A stability criterion for two-fluid interfaces and applications*, Arch. Rational Mech. Anal. **208** (2013) 481–567.
142. D. LANNES AND J.-C.SAUT, *Weakly transverse Boussinesq systems and the KP approximation*, Nonlinearity **19** (2006), 2853–2875.
143. C. LAURENT, F. LINARES AND L. ROSIER, *Control and stabilization of the Benjamin-Ono equation in $L^2(\mathbb{T})$*, Arch. Rational Mech. Anal. **218** (2015), 1531–1575.
144. P.D. LAX AND C.D. LEVERMORE, *The small dispersion limit of the Korteweg de Vries equation*, Comm. Pure Appl. Math. **36**, (1983), 253–290 (Part I), 571–593 (Part II), 809–929 (Part III).
145. D.R LEBEDEV AND A.O. RADUL, *Generalized internal long waves equations, construction, Hamiltonian structure, and conservation laws*, Commun. Math. Phys. **91** (1983) 543–555.
146. F. LINARES, *L^2 global well-posedness of the initial value problem associated to the Benjamin equation*, J. Diff. Eq. **152** (1999), 377–399.
147. F. LINARES AND J.H. ORTEGA, *On the controlability and stabilization of the linearized Benjamin-Ono equation*, ESAIM Control Optim. Calc. Var. **11** (2004), 204–218.
148. F. LINARES, D. PILOD AND G. PONCE, *Well-posedness for a higher order Benjamin-Ono equation*, J. Diff. Eq. **250** (2011), 450–475.
149. F. LINARES, D. PILOD AND J.-C. SAUT, *Dispersive perturbations of Burgers and hyperbolic equations I: local theory*, SIAM J. Math. Anal. **46** (2) (2014), 1505–1537.
150. F. LINARES, D. PILOD AND J.-C. SAUT, *The Cauchy problem for the fractional Kadomtsev-Petviashvili equations*, SIAM J. Math. Anal. **50** (3) (2018, 3172–3209.
151. F. LINARES, D. PILOD AND J.-C. SAUT, *Remarks on the orbital stability of ground state solutions of fKdV and related equations*, Advances Diff. Eq. **20** (9/10), (2015), 835–858.
152. F. LINARES, D. PILOD AND J.-C. SAUT, in preparation.
153. F. LINARES AND L. ROSIER, *Control and stabilization of the Benjamin-Ono equation on a periodic domain*, Trans. Amer. Math. Soc. **367** (7), (2015), 4595–4626.
154. A. K. LIU, *Interaction of solitary waves in stratified fluids*, in Advances in nonlinear waves, Vol. I, Res. Notes in Math., vol. 95, Pitman, Boston, MA, 1984.
155. A.K. LIU, J.R. HOLBROOK AND J.R. APEL, *Nonlinear internal wave evolution in the Sulu Sea*, J. Phys. Oceanography **15** (1985), 1613.
156. A.K. LIU, T. KUBOTA AND D. KO, *Resonant transfer of energy between nonlinear waves in neighbouring pycnoclines*, Stud. Appl. Math., **63**(1980), 25–45.
157. A. LIU, N. R. PEREIRA AND D. KO, *Weakly interacting internal solitary waves in neighbouring pycnoclines*, J. Fluid. Mech., **122**(1982), 187–194.
158. J. MADDOCKS AND R. SACHS, *On the stability of KdV multi-solitons*, Commun. Pure Appl. Math. **46**, (1993), 867–902.
159. Y. MARTEL AND F. MERLE, *Asymptotic stability of solitons of the subcritical generalized KdV equations revisited*, Nonlinearity **18** (2005), 55–80.
160. Y. MARTEL, F. MERLE AND P. RAPHAËL, *Blow-up for the critical generalized Korteweg-de Vries equation: minimal mass dynamics*, J. Eur. Math. Soc. **17** (2015), 1855–1925.
161. Y. MARTEL AND D. PILOD, *Construction of a minimal mass blow up solution of the modified Benjamin-Ono equation*. Math. Ann. **369** (1–2) (2017), 153–245.
162. S.A. MASLOWE AND L.G. REDEKOPP, *Long internal waves in stratified shear flows*, J. Fluid Mech. **101** (2) (1980), 321–348.
163. Y.MATSUNO, *Exact multi-solitons of the Benjamin-Ono equation*, J. Physics A Mathematics and General **12** (4) (1979), 619–62.
164. Y.MATSUNO, *Interaction of the Benjamin-Ono solitons*, J. Physics A Mathematics and General **13** (5) (1980), 1519–1536.
165. Y. MATSUNO, *Bilinear Transformation Method*, Academic Press, New York, 1984.

166. Y. MATSUNO, *Phase shift of interacting algebraic solitary waves in a two-layer system*, Phys. Rev. Lett. **73** (10) (1994), 1316–1319.
167. Y. MATSUNO, *Higher order nonlinear evolution equation for interfacial waves in a two-layer fluid system*, Phys. Rev. E **49** (1994), 2091-
168. Y. MATSUNO, *Dynamics of interacting algebraic solitons*, International J. of Modern Physics, **9** (17) (1995), 1985–2018.
169. Y. MATSUNO, *New integrable nonlinear integro-differential equations and related finite-dimensional dynamical systems*, J. Math. Phys. **29** (1) (1988), 49–56.
170. Y. MATSUNO, *The Lyapunov stability of the N-soliton solutions in the Lax hierarchy of the Benjamin-Ono equation*, J. Math. Phys. **47** (2006), 103505.
171. Y. MATSUNO, *Solutions of the higher order Benjamin-Ono equation*, J. Phys. Soc. Japan **46** (1980), 1024–1028.
172. Y. MATSUNO, *Asymptotic properties of the Benjamin-Ono equation*, J. Phys. Soc. Japan **51** (2) (1982), 667674.
173. Y. MATSUNO, *Number density function of Benjamin-Ono solitons*, Physics Letters **87A** (1–2) (1981), 15–17.
174. Y. MATSUNO, *Recurrence formula and conserved quantities of the Benjamin-Ono equation*, J. Phys. Soc. Japan **52** (2) (1983), 2955–2958.
175. Y. MATSUNO, *Oblique interaction of interfacial solitary waves in a two-layer deep fluid*, Proc. R. Soc. Lond. A **454** (1998), 835–856.
176. Y. MATSUNO, *Multisoliton perturbation theory for the Benjamin-Ono equation and its application to real physical systems*, Phys. Rev. E **51** (2) (1995), 1471–1483.
177. Y. MATSUNO, *The small dispersion limit of the Benjamin-Ono equation and the evolution of a step initial data*, J. Phys. Soc. Japan **67** (1998), 1814–1817.
178. Y. MATSUNO, *Nonlinear modulation of periodic waves in the small dispersion limit of the Benjamin-Ono equation*, Phys. Rev. E **58** (1998), 7934–7940.
179. Y. MATSUNO, *Forced Benjamin-Ono equation and its application to soliton dynamics*, Phys. Rev. E **52** (6) (1995), 6333–6343.
180. Y. MATSUNO AND D.J. KAUP, *Linear stability of multiple internal solitary waves in fluid of great depth*, Physics Letters A **228** (1997), 176–181.
181. Y. MATSUNO AND D.J. KAUP, *Initial value problem of the linearized Benjamin-Ono equation and its applications*, J. Math. Physics **38** (10) (1997), 5198–5224.
182. Y. MATSUNO, V.S. SHCHESNOVICH, A.M. KAMCHATNOV AND R.A. KRAENKEL, *Whitham method for the Benjamin-Ono-Burgers equation and dispersive shocks*, Phys. Rev. E **75** 2007), 016637.
183. T. MAXWORTHY, *On the formation of nonlinear internal waves from the gravitational collapse of mixed regions in two and three dimensions*, J. Fluid Mech. **96** (1) (1980), 47–64.
184. J.W. MILES, *Obliquely interacting solitary waves* J. Fluid Mech. **79** (1977), 157–169.
185. J.W. MILES, *Resonantly interacting solitary waves*, J. Fluid Mech. **79** (1977), 171–179.
186. J.W. MILES, *On internal solitary waves*, Tellus **31** (1979), 456–462.
187. T. MILOH, *On periodic and solitary wavelike solutions of the intermediate long wave equation*, J. Fluid Mech. **211** (1990), 617–627.
188. T. MILOH, M. PRESTIN, L. SHTILMAN AND M.P. TULIN, *A note on the numerical and N-soliton solutions of the Benjamin-Ono evolution equation*, Wave Motion **17** (1993), 1–10.
189. P.D. MILLER, *On the generation of dispersive shock waves*, Physica D **333** (2016), 66–83.
190. P.D. MILLER AND A.N. WETZEL, *The scattering transform for the Benjamin-Ono equation in the small dispersion limit*, Physica D **333** (2016), 185–199.
191. P.D. MILLER AND A.N. WETZEL, *Direct scattering for the Benjamin-Ono equation with rational initial data*, Stud. Appl. Math. **137** (1) (2016), 53–69
192. P.D. MILLER AND Z. XU, *On the zero-dispersion limit of the Benjamin-Ono problem for positive initial data*, Comm. Pure Appl. Math. **64** (2011), 205–270.
193. P.D. MILLER AND Z. XU, *The Benjamin-Ono hierarchy with asymptotically reflectionless initial data in the zero-dispersion limit*, Commun. Math. Sci. **10** (1) (2012), 117–130.

194. A.A. MINZONI AND T. MILOH, *On the number of solitons for the intermediate long wave equation*, Wave Motion **20** (1993), 131–142.
195. T. MIZUMACHI, *Stability of line solitons for the KP-II equation in $\mathbb{R}^2$*, Memoirs of the AMS, vol. 238, number 1125, (2015).
196. T. MIZUMACHI AND N. TZVETKOV, *Stability of the line soliton of the KP-II equation under periodic transverse perturbations*, Math. Ann. **352** (3) (2012), 659–690.
197. L. MOLINET, *Global well-posedness in the energy space for the Benjamin-Ono equation on the circle*, Math. Annalen **337** (2007), 353–383.
198. L. MOLINET, *Global well-posedness in L^2 for the periodic Benjamin-Ono equation*, Amer. J. Math. **130** (3) (2008), 635–683.
199. L. MOLINET, *A note on the inviscid limit of the Benjamin -Ono-Burgers equation in the energy space*, Proc.of the AMS **141** (8) (2013), 2793–2798.
200. L. MOLINET AND D. PILOD, *Global Well-Posedness and Limit Behavior for a Higher-Order Benjamin-Ono Equation*, Comm. in Partial Differential Equations, **37** (2012), 2050–2080.
201. L. MOLINET AND D. PILOD, *The Cauchy problem for the Benjamin-Ono equation in L^2 revisited*. Anal. and PDE **5** (2) (2012), 365–395.
202. L. MOLINET AND F. RIBAUD, *Well-posedness in H^1 for generalized Benjamin-Ono equations on the circle*, Disc;.and Cont. Dyn. Syst. **23** (4) (2009), 1291–1307.
203. L. MOLINET, D. PILOD AND S. VENTO, *On well-posedness for some dispersive perturbations of the Burgers equation*, Ann. I.H. Poincaré ANL, (2018), to appear.
204. L. MOLINET AND F. RIBAUD, *Well-posedness results for the generalized Benjamin-Ono equation with arbitrary large initial data*, IMRN **70** (2004), 3757–3795.
205. L. MOLINET, J. C. SAUT, AND N. TZVETKOV, *Well-posedness and ill-posedness results for the Kadomtsev-Petviashvili-I equation*, Duke Math. J. **115** no. 2 (2002), 353–384.
206. L. MOLINET, J.-C. SAUT AND N. TZVETKOV, *Ill-posedness issues for the Benjamin-Ono and related equations*, SIAM J. Math. Anal. **33**, 4, (2001), 982–988.
207. L. MOLINET, J.C. SAUT AND N. TZVETKOV, *Global well-posedness for the KP-II equation on the background of a non localized solution*, Annales IHP, Analyse Non Linéaire, **28**, 5 (2011), 653–676.
208. L. Molinet, J. C. Saut, and N. Tzvetkov, *Remarks on the mass constraint for KP type equations*, SIAM J. Math. Anal. **39**, no.2 (2007), 627–641.
209. L. MOLINET AND S. VENTO, *Improvement of the energy method for strongly non resonant dispersive equations and applications*, Anal. and PDE **8** (6) (2015), 1455–1495. (2015),
210. C. MÑUNOZ AND G. PONCE, *On the asymptotic behavior of solutions to the Benjamin-Ono equation*, arxiv:1810.02329v1 A Oct 2018.
211. A. NAKAMURA, *Bäcklund transform and conservation laws of the Benjamin-Ono equation*, J. Phys. Soc. Japan **47** (4) (1979), 1335–1340.
212. A. NAKAMURA, *A direct method of calculating periodic wave solutions to nonlinear evolution equations. I. Exact two-periodic wave solution*, J. Phys. Soc. Japan **47** (4) 1701–1705.
213. A. NAKAMURA, *Exact N-soliton solution of the modified finite depth fluid equation*, J. Phys. Soc. Japan **47** (1979), 2043–2044.
214. A. NAKAMURA, *N-periodic wave and N-soliton solutions of the modified Benjamin-Ono equation*, J. Phys. Soc. Japan **47** (1979), 2045–2046.
215. A. NAKAMURA AND Y. MATSUNO, *Exact one-and two periodic wave solutions of fluids of finite depth*, J. Phys. Soc. Japan **48** (2) (1980), 653–656.
216. P.I. NAUMKIN AND I. SANCHEZ-SUARES, *On the modified intermediate long wave equation*, Nonlinearity **31** (2018), 980–1008.
217. A. NEVES AND O. LOPES, *Orbital stability of double solitons for the Benjamin-Ono equation*, Commun. Math. Phys. **262** (2006), 757–791.
218. S. NOVIKOV, S. V. MANAKOV, L. P. PITAEVSKII, V. E. ZAKHAROV, *Theory of solitons. The inverse scattering method*, Contemporary Soviet Mathematics, Consultant Bureau, New York and London, 1984.
219. K. OHI AND T. IGUCHI, *A two-phase problem for capillary-gravity waves and the Benjamin-Ono equation*, Discr. Cont. Dyn. Syst. **23** (4) (2009), 1205–1240.

220. M. OIKAWA AND H. TSUJI, *Oblique interaction of weakly nonlinear long waves in dispersive systems*, Fluid Dynamics Research, **38** (12) (2006), 868–898.
221. H. ONO, *Algebraic solitary waves in stratified fluids*, J. Physical Soc. Japan, **39** (1975), 1082–1091.
222. A.R. OSBORNE AND T.L. BURCH, *Internal solitons in the Andaman sea*, Science **208** (1980), 451.
223. E. OTT AND R.N. SUDAN, *Damping of solitary waves*, Phys. Fluids **13** (3) (1970), 1431–1434.
224. E. OTT AND R.N. SUDAN, *Nonlinear theory of ion acoustic waves with Landau damping*, Phys. Fluids **12** (1969), 2388–2394.
225. A. PARKER, *Periodic solutions of the intermediate long-wave equation: a nonlinear superposition principle*, J. Phys. A: Math. Gen. **25** (1992), 2005–2032.
226. D. PELINOVSKY AND C. SULEM, *Bifurcations of new eigenvalues for the Benjamin-Ono equation*, J. Math Physics **39** (12) (1998), 6552–6572.
227. D. PELINOVSKY AND C. SULEM, *Eigenfunctions and eigenvalues for a scalar Riemann-Hilbert problem associated to inverse scattering*, Commun. Math. Phys. **208** (2000), 713–760.
228. B. PELLONI AND V.A. DOUGALIS, *Numerical solution of some nonlocal, nonlinear dispersive wave equations*, J. Nonlinear Sci. **10** (1) (2000), 1–22.
229. B. PELLONI AND V.A. DOUGALIS, *Error estimates for a fully discrete spectral scheme for a class of nonlinear, nonlocal dispersive wave equation*, Appl. Num. Math. **37** (2001), 95–107.
230. O.M. PHILLIPS, *The dynamics of the upper ocean*, Cambridge University Press (1966).
231. D. PILOD, *On the Cauchy problem for higher-order nonlinear dispersive equations*, J. Diff. Eq. **245** (2005), 2055–2077.
232. G. PONCE, *On the global well-posedness of the Benjamin-Ono equation*, Diff. Int. Eq., **4** (1991), 527–542.
233. N.N. ROMANOVA *Long nonlinear waves in layers of drastic wind velocity changes*, Bull. USSR Acad. Sci. Atmos. Oceanic Phys. **20** (1984), 6.
234. J.A. SANDERS AND JING PING WANG, *Integrability and their recursion operators*, Nonlinear Analysis TMA **47** (8) (2001), 5213–5240.
235. P.M. SANTINI, M.J. ABLOWITZ AND A.S. FOKAS, *On the limit from the intermediate long wave equation to the Benjamin-Ono equation*, J. Math. Phys. **25** (4) (1984), 892–899.
236. J. SATSUMA, M.J. ABLOWITZ AND Y. KODAMA, *On an internal wave equation describing a stratified fluid with finite depth*, Physics Letters **73A**, (4) (1979), 283–286.
237. J. SATSUMA AND M.J. ABLOWITZ, *Solutions of an internal wave equation describing a stratified fluid with finite depth*, in Nonlinear Partial Differential Equations in Engineering and Applied Science (eds) R.L. Sternberg, A.J. Kalinowski, and J.S. Popadakis, New York, Marcel Dekker (1980), 397–414.
238. J. SATSUMA AND A. ISHIMORI, *Periodic wave and rational solutions of the Benjamin-Ono equation*, J. Phys. Soc. Japan **46** (2) (1979),
239. J. SATSUMA, T.R. TAHA AND M.J. ABLOWITZ, *On a Bäcklund transformation and scattering problem for the modified intermediate long wave equation*, J. Math. Phys. **25** (4) (1994), 900904.
240. J.-C. SAUT, *Sur quelques généralisations de l'équation de Korteweg-de Vries*, J. Math. Pures Appl. **58** (1979), 21–61.
241. J.-C. SAUT, *Lectures on Asymptotic Models for Internal Waves*, in Lectures on the Analysis of Nonlinear Partial Differential Equations Vol.2 MLM2, Higher Education Press and International Press, Beijing-Boston (2011), 147–201.
242. P.C. SCHUUR, *Asymptotic analysis of soliton problems. An inverse scattering approach*, Lecture Notes in Mathematics 1232, Springer-Verlag 1986.
243. G. SCOUFIS AND C. M. COSGROVE, *An application of the inverse scattering transform to the modified long wave equation*, J. Math. Physics **46** (10) (2005), 103501.
244. G. SCOUFIS AND C. M. COSGROVE, *On the initial value problem for the modified Benjamin-Ono equation*, J. Math. Physics **36** (1995), 5753–5759.
245. H. SEGUR, *Who cares about integrability*, Physica D **51** (1991), 343–359.

246. H. SEGUR AND J.L. HAMMACK, *Soliton models of long internal waves*, J. Fluid Mech. **118** (1982), 285.
247. A. SIDI, C. SULEM AND P.-L. SULEM, *On the long time behavior of a generalized KdV equation*, Acta Applicandae Mathematicae **7**, (1986), 35–47.
248. R. SMITH, *Nonlinear Kelvin and continental-shelf waves*, J. Fluid Mech. **57** (1972), 379–391.
249. M.D. SPECTOR AND T. MILOH, *Stability of nonlinear periodic internal waves in a deep fluid*, SIAM J. Appl. Math. **54** (3) (1994), 688–707.
250. A. P. STAMP AND M. JACKAI, *Deep-water internal solitary waves*. J. Fluid Mech. **305** (1995), 347–371
251. S. TANAKA, *On the N-tuple wave solutions of the Korteweg-de Vries equation*, Publ.R.I.M.S. Kyoto Univ. **8** (1972), 419–427.
252. M. SY, *Invariant measure and long time behavior of regular solutions of the Benjamin-Ono equation*, Analysis and PDE **11** (8) (2018), 1841–1879.
253. T. TANAKA, *Local well-posedness for third order Benjamin-Ono type equations on the torus*, arXiv:1812.03477v1, 9 Dec 2018.
254. T. TAO, *Why are solitons stable?*, Bull. AMS **46** (1) (2009), 1–33.
255. T. TAO, *Global well-posedness of the Benjamin-Ono equation in H^1*, J. Hyperbolic Diff. Equations, **1** (2004) 27–49.
256. H. TSUJI AND M. OIKAWA, *Oblique interaction of internal solitary waves in a two-layer system of infinite depth*, Fluid Dynamics Research **29** (4) (2001), 251–267
257. H. TSUJI AND M. OIKAWA, *Two-dimensional interaction of solitons in a two-layer fluid of finite amplitude*, Fluid Dynamics Research **42** (6) (2010), 06506.
258. N. TZVETKOV, *construction of a Gibbs measure associated to the periodic Benjamin-Ono equation*, Probab. Theory Relat. Fields **146** (2010), 481–514.
259. N. TZVETKOV AND N. VISCIGLIA, *Gaussian measures associated to the higher order conservation laws of the Benjamin-Ono equation*, Ann. Sci. Éc. Norm. Supér. **46** (2) (2013), 249–299.
260. N. TZVETKOV AND N. VISCIGLIA, *Invariant measures and long-time behavior for the Benjamin-Ono equation*, Int. Math. Res. Not. **17** (2014), 4679–4714.
261. N. TZVETKOV AND N. VISCIGLIA, *Invariant measures and long time behaviour for the Benjamin-Ono equation, II*, J. Math. Pures Appl. **103** (1) (2015), 102–141.
262. S. VENAKIDES, *The Korteweg-de Vries equation with small dispersion: higher order Lax-Levermore theory*, Comm. Pure Appl. Math. **43** (3) (1990), 335–361.
263. C. WANG AND R. PAWLOWICZ, *Oblique interactions of nonlinear near-surface waves in the strait of Georgia*, J. of Geophysical Research **117** (2012), C06031.
264. M.I. WEINSTEIN, *Existence and dynamic stability of solitary wave solutions of equations arising in long wave propagation*, Comm. Part. Diff. Eq. **12** (1987), 133–1173.
265. G.B. WHITHAM, *Linear and nonlinear waves*, Wiley, New York 1974.
266. G. B. WHITHAM, *Variational methods and applications to water waves*, Proc.R. Soc. Lond. Ser. A., **299** (1967), 6–25.
267. DERCHYI WU, *The direct problem of perturbed Kadomtsev-Petviashvili II line solitons*, arXiv:1807.01420v, 4 Jul 2018.
268. YILUN WU, *Simplicity and finiteness of discrete spectrum of the Benjamin-Ono scattering operator*, SIAM J. Math. Anal. **48** (2) (2016), 1348–1367.
269. YILUN WU, *Jost solutions and the direct scattering problem for the Benjamin-Ono equation*, arXiv:1704.01692v1[math.AP] 6 Apr 2017.
270. Z. XU, *Analysis and numerical analysis of the Benjamin-Ono equation*, PhD thesis, University of Michigan, 2010.
271. LI XU, *Intermediate long waves systems for internal waves*, Nonlinearity, 25 (2012), 597–640.

Inverse Scattering and Global Well-Posedness in One and Two Space Dimensions

Peter A. Perry

1 Introduction

These notes are a considerably revised and expanded version of lectures given at the Fields Institute workshop on "Nonlinear Dispersive Partial Differential Equations and Inverse Scattering" in August 2017. These lectures, together with lectures of Walter Craig, Patrick Gerard, Peter D. Miller, and Jean-Claude Saut, constituted a week-long introduction to recent developments in inverse scattering and dispersive PDE intended for students and postdoctoral researchers working in these two areas.

The goal of my lectures was to give a complete and mathematically rigorous exposition of the inverse scattering method for two dispersive equations: the defocussing cubic nonlinear Schrödinger equation (NLS) in one space dimension, and the defocussing Davey-Stewartson II (DS II) equation in two space dimensions. Each is arguably the simplest example in its class since neither admits solitons; moreover, the long-time behavior of solutions to each equation has been rigorously deduced from inverse scattering [17, 36, 38]. Inverse scattering for the defocussing NLS provides an introduction to the Riemann-Hilbert method further discussed in the contribution of Dieng, McLaughlin and Miller in this volume [19]. Inverse scattering for the defocussing Davey-Stewartson II equation provides an introduction to the $\overline{\partial}$-methods used extensively in two-dimensional inverse scattering (see, for example, the surveys [10, 23, 27], the monograph [1], and references therein).

In Lecture 1 (Sect. 2), I give an overview of the inverse scattering method for these two equations, focusing on the formal (i.e., algebraic) aspects of the theory. I motivate the solution formulas given by inverse scattering, seen as a composition of nonlinear maps and a linear time-evolution of scattering data.

P. A. Perry (✉)
Department of Mathematics, University of Kentucky, Lexington, KY, USA
e-mail: peter.perry@uky.edu

In Lecture 2 (Sect. 3), I analyze inverse scattering for the defocussing cubic NLS in one dimension in depth, based on the seminal paper of Deift and Zhou [17]. We study the direct map via Volterra integral equations for the Jost solutions for the operator (2.6), and the inverse map via the Riemann-Hilbert Problem 2.3 using the approach of Beals and Coifman [7].

In Lecture 3 (Sect. 4), I discuss the inverse scattering method for the Davey-Stewartson II equation in depth. This lecture has been completely rewritten in light of the recent work of Nachman et al. [36] which introduced a number of new ideas and techniques from harmonic analysis and pseudodifferential operators to the study of scattering maps. Using these techniques, the authors proved that the defocussing DS II equation is globally well-posed in $L^2(\mathbb{R})$ and that all solutions scatter to solutions of the associated linear problem. Their results significantly improve earlier work of mine [38], which proved global well-posedness of the defocussing DS II equation in the weighted Sobolev space $H^{1,1}(\mathbb{R})$ and obtained large-time (dispersive) asymptotics of solutions in L^∞-norm. In the revised Sect. 4, I give a pedagogical proof of the results in [38] using some of the ideas of [36] to streamline proofs significantly.

At the end of each lecture, I've added exercises which supplement the text and develop key ideas.

I hope that these lectures will appeal to a wide audience interested in recent progress in this rapidly developing field.

2 Introduction to Inverse Scattering

Among dispersive PDE's that describe wave propagation are the *completely integrable* PDE's. These equations—which include the Korteweg-de Vries and cubic NLS equations in one space dimension, and the Davey-Stewartson and Kadomtsev-Petviashvilli equation in two space dimensions—are equivalent to simple linear flows by conjugation with an invertible, nonlinear map adapted to the PDE and very strongly dependent on its special structure. This nonlinear map is called a *scattering transform* and serves the same function for these equations that the Fourier transform does for the linear Schrödinger equation.

To understand what this means, let's consider the Cauchy problem for linear Schrödinger equation in one dimension.

$$\begin{cases} i\dfrac{\partial q}{\partial t} + \dfrac{\partial^2 q}{\partial x^2} = 0, \\ q(x,0) = q_0(x). \end{cases} \quad (2.1)$$

assuming for simplicity that $q_0 \in \mathscr{S}(\mathbb{R})$. The Fourier transform

$$(\mathcal{F}q)(\xi) = \widehat{q}(\xi) = \int e^{-ix\xi} f(x)\, dx$$

reduces the Cauchy problem (2.1) to the trivial flow

$$i\frac{\partial}{\partial t}\widehat{q}(\xi, t) = |\xi|^2 \widehat{q}(\xi, t)$$

leading to the solution formula

$$q(x,t) = \frac{1}{2\pi}\int_{\mathbb{R}} e^{it\theta}\widehat{q_0}(\xi)\, d\xi, \quad \theta_0(\xi; x, t) = \xi x/t - \xi^2. \qquad (2.2)$$

We can also write the solution as $q(t) = e^{it\Delta}q_0$ where $e^{it\Delta}$ is the solution operator

$$\left(e^{it\Delta} f\right)(x) = \mathcal{F}^{-1}\left(e^{-it(\cdot)^2}(\mathcal{F}f)\right)(x). \qquad (2.3)$$

Since the phase function θ_0 in (2.2) has a single, nondegenerate critical point at $\xi_0 = x/2t$, the solution has large-time asymptotics

$$q(x,t) \sim \frac{1}{\sqrt{4\pi i t}} e^{ix^2/(4t)} \widehat{q_0}\left(\frac{x}{2t}\right) + \mathcal{O}\left(t^{-3/4}\right).$$

The map $\mathcal{F}$ is linear, has a well-behaved and explicit inverse, and yields a well-behaved solution formula that extends to initial data in Sobolev spaces. The representation formula, combined with stationary phase methods, leads to a complete description of long-time asymptotic behavior.

Similar results may be obtained for integrable systems provided that the scattering transforms are well-controlled and have well-behaved inverses. In these lectures we will discuss two examples in depth: the defocussing, cubic nonlinear Schrödinger equation in one space dimension, and the defocussing Davey-Stewartson II equation in two space dimensions. Neither of these equations admits solitons, so that the dynamics are purely dispersive.

In each case, the scattering transform is the "next best thing to linear": it is a diffeomorphism when restricted to appropriate function spaces (and, in each case, its Fréchét derivative at zero is a Fourier-like transform—see Remarks 3.4 and 3.14 for the NLS scattering maps, and see (4.44) and the accompanying discussion for the DS II scattering map). It also has global Fourier-like properties which facilitate the analysis of large-time asymptotics of the solution.

2.1 The Defocussing Cubic Nonlinear Schrödinger Equation

Zakharov and Shabat [46] showed that the Cauchy problem for the cubic nonlinear Schrödinger equation

$$\begin{cases} i\dfrac{\partial q}{\partial t} + \dfrac{\partial^2 q}{\partial x^2} - 2|q|^2 q = 0 \\ q(x,0) = q_0(x) \end{cases} \quad (2.4)$$

is integrable by inverse scattering. To describe the direct and inverse scattering maps, we will follow the conventions of [19]. These conventions differ slightly but inessentially from those of Deift [16] and Deift-Zhou [17].[1]

Equation (2.4) is the consistency condition for the overdetermined system

$$\begin{cases} \Psi_x = -i\lambda\sigma_3\Psi + \mathbf{Q}_1\Psi, \\ \Psi_t = \left(-2i\lambda^2\sigma_3 + 2\lambda\mathbf{Q}_1 + \mathbf{Q}_2\right)\Psi \end{cases} \quad (2.5)$$

where

$$\mathbf{Q}_1 = \begin{pmatrix} 0 & q \\ \bar{q} & 0 \end{pmatrix}, \quad \mathbf{Q}_2 = \begin{pmatrix} -i|q|^2 & iq_x \\ -i\overline{q_x} & i|q|^2 \end{pmatrix}$$

and Ψ is an unknown 2×2 matrix-valued function of (x,t). Note that the first of equations (2.4) is an eigenvalue problem for the self-adjoint operator

$$\mathcal{L} := i\sigma_3 \dfrac{d}{dx} + \mathbf{Q}(x), \quad \sigma_3 := \begin{pmatrix} 1 & 0 \\ 0 & -1 \end{pmatrix}, \quad \mathbf{Q}(x) = \begin{pmatrix} 0 & -iq(x) \\ iq(x) & 0 \end{pmatrix} \quad (2.6)$$

acting on $L^2(\mathbb{R}, M_2(\mathbb{C}))$, the square-integrable, 2×2-matrix valued functions (see Exercise 2.7). This equation is sometimes called the ZS-AKNS equation after the fundamental papers of Ablowitz et al. [2] and Zakharaov-Shabat [46]. Note that, if $q = 0$, the operator $\mathcal{L}$ has continuous spectrum on the real line and bounded, matrix-valued eigenfunctions $\Psi_0(x,\lambda) = e^{-i\lambda x\sigma_3}$.

The scattering transform of $q \in L^1(\mathbb{R})$ is defined as follows. There exist unique 2×2 matrix-valued solutions $\Psi^\pm(x,\lambda)$ of $\mathcal{L}\psi = \lambda\psi$ with

$$\lim_{x\to\pm\infty} \Psi^\pm(x,\lambda)e^{i x\lambda\sigma_3} = \mathbb{I}, \quad (2.7)$$

---

[1] Deift and Zhou write the ZS-AKNS equation as $\psi_x = i\lambda\sigma\psi + \mathbf{Q}_1\psi$ where $\sigma = (1/2)\sigma_3$. This results in various sign changes and changes in factors of 2 throughout. The conventions of [19] also make the scattering maps linearize to antilinear Fourier-type transforms, whereas those of [17] linearize to the usual Fourier transform.

where $\mathbb{I}$ denotes the identity matrix. Matrix-valued solutions of $\mathcal{L}\psi = \lambda\psi$ have the properties that (1) $\det \psi(x)$ is independent of x and (2) any two nonsingular solutions ψ_1 and ψ_2 are related by $\psi_1 = \psi_2 A$ for a constant matrix A (see Exercises 3.10 and 3.11). For this reason, there is a matrix $T(\lambda)$ with

$$\Psi^+(x,\lambda) = \Psi^-(x,\lambda)T(\lambda). \tag{2.8}$$

Clearly $\det T(\lambda) = 1$ and, by a symmetry argument (see Exercises 3.13 and 3.14),

$$T(\lambda) = \begin{pmatrix} a(\lambda) & \overline{b(\lambda)} \\ b(\lambda) & \overline{a(\lambda)} \end{pmatrix}, \quad |a(\lambda)|^2 - |b(\lambda)|^2 = 1 \tag{2.9}$$

The *direct scattering map* is the map $\mathcal{R} : q \to r$ where

$$r(\lambda) = -b(\lambda)/\overline{a(\lambda)}. \tag{2.10}$$

The direct scattering map linearizes the flow (2.4) in the sense that, if $q(x,t)$ solves the initial value problem (2.4) with initial data q_0 and $r_0 = \mathcal{R}(q_0)$, then

$$\mathcal{R}(q(\cdot,t))(\lambda) = e^{4it\lambda^2} r_0(\lambda).$$

We give a heuristic proof of this law of evolution, based on the Lax representation (2.5), at the beginning of Sect. 3.4.

The inverse of $\mathcal{R}$ is determined as follows. Denote by $\psi(x,z)$ a solution to $\mathcal{L}\psi = z\psi$, written as

$$\frac{d}{dx}\psi = -iz\sigma_3\psi + \mathbf{Q}_1(x)\psi, \quad \mathbf{Q}_1(x) = \begin{pmatrix} 0 & q(x) \\ \overline{q(x)} & 0 \end{pmatrix},$$

where $z \in \mathbb{C}$, and factor $\psi(x,z) = \mathbf{M}(x,z)e^{-ixz\sigma_3}$. Then $\mathbf{M}$ obeys the differential equation

$$\frac{d}{dx}\mathbf{M}(x,z) = -iz \, \text{ad}\,\sigma_3\,(\mathbf{M}(x,z)) + \mathbf{Q}_1(x)\mathbf{M}(x,z) \tag{2.11}$$

where, for a 2×2 matrix A,

$$\text{ad}\,\sigma_3(A) = [\sigma_3, A].$$

One can show that (2.11) admits special solutions, the *Beals-Coifman solutions*, which are piecewise analytic in $\mathbb{C} \setminus \mathbb{R}$, have distinct boundary values $\mathbf{M}_\pm(x,\lambda)$ on $\mathbb{R}$, and for each z obey the asymptotic conditions

$$\mathbf{M}(x,z) \to \mathbb{I} \text{ as } x \to +\infty, \quad \mathbf{M}(x,z) \text{ is bounded as } x \to -\infty.$$

The Beals-Coifman solutions are unique and, moreover, $q(x)$ can be recovered from their asymptotic behavior:

$$q(x) = \lim_{z \to \infty} 2iz \mathbf{M}_{12}(x, z). \tag{2.12}$$

The boundary values satisfy a jump relation

$$\begin{cases} \mathbf{M}_+(x, \lambda) = \mathbf{M}_-(x, \lambda) \mathbf{V}(\lambda; x), \\ \mathbf{V}(\lambda; x) = \begin{pmatrix} 1 - |r(\lambda)|^2 & -\overline{r(\lambda)} e^{-2i\lambda x} \\ r(\lambda) e^{2i\lambda x} & 1 \end{pmatrix}. \end{cases} \tag{2.13}$$

The asymptotics of $\mathbf{M}(x, z)$ together with the jump relation (2.13) define a Riemann-Hilbert problem for $\mathbf{M}(x, z)$, which we write as $\mathbf{M}(z; x)$ to emphasize that x plays the role of a parameter in the RHP.

Riemann-Hilbert Problem 2.1 For given r, and $x \in \mathbb{R}$, find $\mathbf{M}(z; x)$ so that

(i) $\mathbf{M}(z; x)$ is analytic in $\mathbb{C} \setminus \mathbb{R}$ for each x,
(ii) $\lim_{z \to \infty} \mathbf{M}(z; x) = \mathbb{I}$,
(iii) $\mathbf{M}(z; x)$ has continuous boundary values $\mathbf{M}_\pm(\lambda; x)$ on $\mathbb{R}$
(iv) The jump relation

$$\mathbf{M}_+(\lambda; x) = \mathbf{M}_-(\lambda; x) V(\lambda; x), \quad \mathbf{V}(\lambda; x) = \begin{pmatrix} 1 - |r(\lambda)|^2 & -\overline{r(\lambda)} e^{-2i\lambda x} \\ r(\lambda) e^{2i\lambda x} & 1 \end{pmatrix}$$

holds.

Remark 2.2 In condition (ii) above, the limit is meant to be uniform in proper subsectors of the upper and lower half planes. That is, for any $\varepsilon > 0$,

$$\lim_{R \to \infty} \sup_{\substack{|z| \geq R \\ \arg(z) \in (\alpha, \beta)}} (\mathbf{M}(z; x) - \mathbb{I}) = 0$$

if (α, β) is a proper subinterval of $(0, \pi)$ or $(\pi, 2\pi)$.

The *inverse scattering map* $\mathcal{I} : r \to q$ determined by Riemann-Hilbert Problem 2.1 and the reconstruction formula (2.12).

Thus, to implement the solution formula

$$q(x, t) = \mathcal{I}\left(e^{4i(\cdot)^2 t} \mathcal{R}(q_0)(\cdot)\right)(x), \tag{2.14}$$

we compute the scattering transform $r_0 = \mathcal{R}(q_0)$ and solve the following Riemann-Hilbert problem (RHP).

Riemann-Hilbert Problem 2.3 For given r_0 and parameters x, t, find $\mathbf{M}(z; x, t)$ so that

(i) $\mathbf{M}(z; x, t)$ is analytic in $\mathbb{C} \setminus \mathbb{R}$ for each x, t,
(ii) $\lim_{z \to \infty} \mathbf{M}(z; x, t) = \mathbb{I}$,
(iii) $\mathbf{M}(z; x, t)$ has continuous boundary values $\mathbf{M}_\pm(\lambda; x, t)$ on $\mathbb{R}$
(iv) The jump relation

$$\mathbf{M}_+(\lambda; x, t) = \mathbf{M}_-(\lambda; x, t) \mathbf{V}(\lambda; x, t),$$

$$\mathbf{V}(\lambda; x, t) = \begin{pmatrix} 1 - |r_0(\lambda)|^2 & -\overline{r_0(\lambda)} e^{-2it\theta} \\ r_0(\lambda) e^{2it\theta} & 1 \end{pmatrix}$$

holds, where

$$\theta(\lambda; x, t) = 2\lambda^2 + x\lambda/t. \qquad (2.15)$$

Given the solution of RHP 2.3, we can then compute

$$q(x, t) = \lim_{z \to \infty} 2iz \mathbf{M}_{12}(z; x, t)$$

where the limit is meant in the sense of Remark 2.2.

Denote by $\mathscr{S}(\mathbb{R})$ the Schwartz class functions on $\mathbb{R}$ and let

$$\mathscr{S}_1(\mathbb{R}) = \{ r \in \mathscr{S}(\mathbb{R}) : \|r\|_\infty < 1 \}.$$

Beals and Coifman [7] proved:

Theorem 2.4 *Suppose that $q_0 \in \mathscr{S}(\mathbb{R})$. Then $r_0 \in \mathscr{S}_1(\mathbb{R})$, RHP 2.3 has a unique solution for each $x, t \in \mathbb{R}$, and (2.14) defines a classical solution of the Cauchy problem (2.4).*

We will give a complete proof of Theorem 2.4 in Sect. 3.4.

The solution formula (2.14) defines a continuous solution map provided that its component maps are continuous. To describe the mapping properties of $\mathcal{R}$ and $\mathcal{I}$, we define

$$H^{1,1}(\mathbb{R}) := \left\{ f \in L^2(\mathbb{R}) : f', xf \in L^2(\mathbb{R}) \right\} \qquad (2.16)$$

and

$$H^{1,1}_1(\mathbb{R}) := \left\{ f \in H^{1,1}(\mathbb{R}) : \|f\|_\infty < 1 \right\}.$$

Note that $H^{1,1}_1(\mathbb{R})$ is an open subset of $H^{1,1}(\mathbb{R})$ since $\|f\|_\infty \leqslant c\|f\|_{H^{1,1}}$ (see Exercise 3.2). The following result is proved by Deift and Zhou in [17, §3] and also follows from Zhou's analysis [47] of Sobolev mapping properties of the scattering transform.

Theorem 2.5 ([17]) *The maps $\mathcal{R} : H^{1,1}(\mathbb{R}) \to H^{1,1}_1(\mathbb{R})$ and $\mathcal{I} : H^{1,1}_1(\mathbb{R}) \to H^{1,1}(\mathbb{R})$ are Lipschitz continuous maps with $\mathcal{R} \circ \mathcal{I} = \mathcal{I} \circ \mathcal{R} = I$.*

A consequence of Theorems 2.4, 2.5, and local well-posedness theory for the NLS is:

Theorem 2.6 *The Cauchy problem 2.4 is globally well-posed in $H^{1,1}(\mathbb{R})$ with solution* (2.14).

Theorem 2.6 is of interest not because of the global well-posedness result: far superior results are available through PDE methods—see, for example, [34] and [42] and references therein—and most recently through a very different approach to complete integrability pioneered by Koch-Tataru [32], Killip-Visan-Zhang [30], and Killip-Visan [31] which give conserved quantities and well-posedness results in the presence of very rough initial data. Rather, Theorem 2.6 is of interest because the solution map so constructed can be used to study large-time asymptotics of solutions with initial data in $H^{1,1}(\mathbb{R})$. Deift and Zhou [17] gave a rigorous proof of long-time dispersive behavior for the solution of (2.4), motivated by formal results of Zakharov and Manakov [45]. Their proof is an application of the Deift-Zhou steepest descent method [15]. Dieng and McLaughlin [18] gave a different proof using the '$\overline{\partial}$-steepest descent method' and obtained a sharp remainder estimates. This result is discussed in a companion paper by Dieng, McLaughlin and Miller in this volume [19].

Theorem 2.7 ([19]) *The unique solution to* (2.4) *with initial data $q_0 \in H^{1,1}(\mathbb{R})$ has the asymptotic behavior*

$$q(x,t) \sim t^{-1/2}\alpha(z_0)e^{ix^2/(4t) - i\nu(z_0)\log(2t)} + \mathcal{O}\left(t^{-3/4}\right)$$

where $z_0 = -x/(4t)$, $\nu(z) = -\dfrac{1}{2\pi}\log\left(1 - |r(z)|^2\right)$, $|\alpha(z)|^2 = \nu(z)/2$, and

$$\arg\alpha(z) = \frac{1}{\pi}\int_{-\infty}^{z} \log(z-s)\,d\left(\log\left(1 - |r(s)|^2\right)\right) + \frac{\pi}{4} + \arg(i\nu(z)) + \arg r(z).$$

The remainder term is uniform in $x \in \mathbb{R}$.

2.2 The Defocussing Davey-Stewartson II Equation

The Cauchy problem for the defocussing Davey-Stewartson II (DSII) equation is

$$\begin{cases} iq_t + 2(\partial_z^2 + \partial_{\bar{z}}^2)q + (g + \bar{g})q = 0, \\ \qquad\qquad \partial_{\bar{z}} g = -4\partial_z\left(|q|^2\right), \\ \qquad\qquad q(z,0) = q_0(z). \end{cases} \qquad (2.17)$$

Here $z = x_1 + ix_2$ and

$$\partial_{\bar{z}} = \frac{1}{2}\left(\frac{\partial}{\partial x_1} + i\frac{\partial}{\partial x_2}\right), \quad \partial_z = \frac{1}{2}\left(\frac{\partial}{\partial x_1} - i\frac{\partial}{\partial x_2}\right).$$

Here and in what follows, the notation $f(z)$ for a function of $z = x_1 + ix_2$ does *not* imply that f is an analytic function of z.

Ablowitz and Segur [3, Chapter 2, §2.1.d] showed that the Davey-Stewartson II equation is completely integrable. The solution of DS II by inverse scattering was developed by Beals-Coifman [6, 8, 9] and Fokas-Ablowitz [21–23]. A rigorous analysis of the scattering maps, including the case $q_0 \in \mathscr{S}(\mathbb{R}^2)$ was carried out by Sung in a series of three papers [39–41].

The DSII flow is linearized by a zero-energy spectral problem for the operator

$$\mathcal{L} = \begin{pmatrix} \partial_{\bar{z}} & 0 \\ 0 & \partial_z \end{pmatrix} - \mathbf{Q}(z), \quad \mathbf{Q}(z) = \begin{pmatrix} 0 & q(z) \\ \overline{q(z)} & 0 \end{pmatrix}. \qquad (2.18)$$

To define the scattering transform for $q \in \mathscr{S}(\mathbb{R}^2)$, we look for solutions

$$\begin{pmatrix} \psi_1(z,k) \\ \psi_2(z,k) \end{pmatrix} = \begin{pmatrix} m^1(z,k)e^{ikz} \\ m^2(k,z)e^{ikz} \end{pmatrix}$$

of $\mathcal{L}\psi = \mathbf{0}$, where kz denotes complex multiplication of $k = k_1 + ik_2$ by $z = x_1 + ix_2$. We assume that $m^1(z,k) \to 1$ and $m^2(z,k) \to 0$ for each $k \in \mathbb{C}$ as $|z| \to \infty$. Such unbounded solutions ψ_1, ψ_2 are sometimes called *complex geometric optics* (CGO) solutions and were introduced in scattering theory by Faddeev [20]. An easy computation shows that

$$\begin{cases} \partial_{\bar{z}} m^1(z,k) = q(z) m^2(z,k), \\ (\partial_z + ik) m^2(z,k) = \overline{q(z)} m^1(z,k), \\ \lim_{|z|\to\infty} (m^1(z,k), m^2(z,k)) = (1,0). \end{cases} \qquad (2.19)$$

The system (2.19) is formally equivalent[2] to a system of integral equations:

---

[2] Convolution with $(\pi z)^{-1}$ (resp. $(\pi \bar{z})^{-1}$) is a formal inverse to $\partial_{\bar{z}}$ (resp. ∂_z). See Sect. 4.1, Eqs. (4.7) and (4.8) and the accompanying discussion and references.

$$\begin{cases} m^1(z,k) = 1 + \dfrac{1}{\pi} \displaystyle\int_{\mathbb{C}} \dfrac{1}{z-w} q(w) m^2(w,k)\, dw \\ m^2(z,k) = \dfrac{1}{\pi} \displaystyle\int_{\mathbb{C}} \dfrac{e_{-k}(z-w)}{\bar{z}-\bar{w}} \overline{q(w)} m^1(z,k)\, dw \end{cases} \quad (2.20)$$

where

$$e_k(z) = e^{i(kz + \bar{k}\bar{z})}. \quad (2.21)$$

For $q \in \mathscr{S}(\mathbb{R}^2)$, m^1 and m^2 admit large-z expansions of the form

$$m^1(z,k) \sim 1 + \sum_{j \geq 1} \frac{a_j(k)}{z^j},$$

$$m^2(z,k) \sim e_{-k}(z) \sum_{j \geq 1} \frac{b_j(k)}{\bar{z}^j}.$$

The scattering transform $\mathcal{S}q$ of $q \in \mathscr{S}(\mathbb{R}^2)$ is $-ib_1(k)$. From the integral equations (2.20) we can see that

$$(\mathcal{S}q)(k) = -\frac{i}{\pi} \int_{\mathbb{C}} e_k(z) \overline{q(z)} m^1(z,k)\, dz. \quad (2.22)$$

This map is a perturbation of the antilinear 'Fourier transform'

$$(\mathcal{F}_a q) = -\frac{i}{\pi} \int_{\mathbb{C}} e_k(z) \overline{q(z)}\, dz$$

which satisfies $\mathcal{F}_a \circ \mathcal{F}_a = I$. We will see that, remarkably, the same holds for $\mathcal{S}$. The scattering transform $\mathcal{S}$ linearizes the DSII equation (2.17) in the following sense: if $q(z,t)$ solves (2.17) and $q(\,\cdot\,,t) \in S(\mathbb{R}^2)$ for each t, then

$$\mathcal{S}(q(\,\cdot\,,t))(k) = e^{2i(k^2 + \bar{k}^2)t} \mathcal{S}(q_0))(k).$$

Thus, a putative solution by inverse scattering is given by

$$q(z,t) = \mathcal{S}^{-1}\left(e^{2i((\,\cdot\,)^2 + (\overline{\,\cdot\,})^2)} (\mathcal{S}q_0)(k) \right)(z). \quad (2.23)$$

To implement the solution formula (2.23), we compute the scattering transform $\mathbf{s}(k) = \mathcal{S}(q_0)(k)$ and, for each t, solve the system

$$\begin{cases} \partial_{\overline{k}} n^1(z,k,t) = \mathbf{s}(k) e^{2i(k^2+\overline{k}^2)t} n^2(z,k,t) \\ (\partial_k + iz) n^2(z,k,t) = \overline{\mathbf{s}(k)} e^{-2i(k^2+\overline{k}^2)t} n^1(z,k,t). \end{cases} \qquad (2.24)$$

The solution $q(z,t)$ is given by

$$q(z,t) = -\frac{i}{\pi} \int_{\mathbb{C}} e^{it\varphi} \overline{\mathbf{s}(k)} n^1(z,k,t)\, dk. \qquad (2.25)$$

where

$$\varphi(k;z,t) = 2\left(k^2 + \overline{k}^2\right) - \frac{kz + \overline{k}\overline{z}}{t}.$$

The results of Beals-Coifman, Fokas-Ablowitz, and Sung imply the following analogue of Theorem 2.4.

Theorem 2.8 *Suppose that $q_0 \in \mathcal{S}(\mathbb{R}^2)$. Then $\mathcal{S}q_0 \in \mathcal{S}(\mathbb{R}^2)$. Moreover, the system (2.24) has a unique solution for each (z,t), and (2.25) defines a classical solution of the Cauchy problem 2.17.*

Nachman et al. [36] proved the following remarkable result on the scattering transform $\mathcal{S}$. Recall that the Hardy-Littlewood maximal function of $f \in L^p(\mathbb{R}^n)$, $1 \leq p \leq \infty$, is given by

$$(\mathcal{M}f)(x) = \sup_{r>0} \frac{1}{|B(x,r)|} \int_{B(x,r)} |f(y)|\, dy,$$

where $B(x,r)$ denotes the ball of radius r about $x \in \mathbb{R}^n$ and $|A|$ denotes the Lebesgue measure of the measurable set $A \subset \mathbb{R}^n$. For $q \in L^2(\mathbb{R}^2)$, set

$$\widehat{q}(k) = \frac{1}{\pi} \int e_k(z) q(z)\, dz.$$

Theorem 2.9 ([36]) *The scattering transform $\mathcal{S}$ extends to a diffeomorphism from $L^2(\mathbb{R}^2)$ onto itself with $\mathcal{S} \circ \mathcal{S} = I$. Moreover, the following estimates hold:*

(i) $\|\mathcal{S}q\|_{L^2(\mathbb{R}^2)} = \|q\|_{L^2(\mathbb{R}^2)}$
(ii) $|(\mathcal{S}(q))(k)| \leq C\left(\|q\|_{L^2}\right) |\mathcal{M}\widehat{q}(k)|$

Theorem 2.9 considerably extends earlier work of Brown [11] and Perry [38], who considered the scattering map respectively for small data in $L^2(\mathbb{R}^2)$ and data in a weighted space $H^{1,1}(\mathbb{R}^2)$ analogous to the space $H^{1,1}(\mathbb{R})$ for the NLS. It also illuminates other work of Astala et al. [4] and Brown et al. [12] on the Fourier-like mapping properties of $\mathcal{S}$. The maximal function estimate is particularly important for the analysis of scattering since it implies that the solution of DSII by inverse scattering is bounded *pointwise* by a maximal function for the solution of the linear problem. This means, for example, that Strichartz-type estimates for the linear problem imply Strichartz-type estimates for the nonlinear problem.

As a consequence of Theorems 2.8 and 2.9, Nachman, Regev, and Tataru obtain a complete characterization of the dynamics for DSII. Denote by $U(t)$ the (nonlinear) solution operator for (2.17), and by $V(t)$ the solution operator for the linearization of (2.17) at $q = 0$, i.e.,

$$\begin{cases} v_t + 2\left(\partial_z^2 + \partial_{\bar{z}}^2\right)v = 0, \\ v(z, 0) = v_0(z). \end{cases} \quad (2.26)$$

Theorem 2.10 ([36]) *The Cauchy problem for (2.17) is globally well-posed in $L^2(\mathbb{R}^2)$ with*

$$\|q(t)\|_{L^2(\mathbb{R}^2)} = \|q(0)\|_{L^2(\mathbb{R}^2)}$$

for all t. Moreover, all solutions scatter in the sense that, for any $q_0 \in L^2(\mathbb{R}^2)$, there is a function $v_0 \in L^2(\mathbb{R}^2)$ so that

$$\lim_{t \to \pm\infty} \|U(t)q_0 - V(t)v_0\|_{L^2} = 0.$$

The function v_0 is given by

$$v_0 = \mathcal{F}_a \mathcal{S} q_0.$$

Note that the scattering is *trivial* because the $t \to -\infty$ (past) and $t \to +\infty$ (future) asymptotes are the same.

Perry [38] obtained pointwise asymptotics under somewhat more restrictive conditions on the initial data. Let

$$H^{1,1}(\mathbb{R}^2) = \left\{ q \in L^2(\mathbb{R}^2) : \nabla q, |\cdot| q(\cdot) \in L^2(\mathbb{R}^2) \right\}. \quad (2.27)$$

Theorem 2.11 *Suppose that $q_0 \in L^1(\mathbb{R}^2) \cap H^{1,1}(\mathbb{R}^2)$ and that $(\mathcal{S}q_0)(0) \neq 0$. Then*

$$q(x, t) \sim v(z, t) + o\left(t^{-1}\right)$$

where $v(z, t)$ solves the linearized equation (2.26) with initial data $v_0 = \mathcal{F}_a \mathcal{S} q_0$.

Exercises for Sect. 2 In the following exercises, the Fourier transforms $\mathcal{F}$ and $\mathcal{F}^{-1}$ are defined by

$$(\mathcal{F}f)(\xi) = \int e^{-ix\xi} f(x)\, dx \quad (2.28)$$

Inverse Scattering and Global Well-Posedness

$$\left(\mathcal{F}^{-1}g\right)(x) = \frac{1}{2\pi} \int e^{ix\xi} g(\xi)\, d\xi. \tag{2.29}$$

Exercise 2.1 Show that, with the conventions (2.28) and (2.29),

$$\mathcal{F}(f * g)(\xi) = (\mathcal{F}f)(\xi)\,(\mathcal{F}g)(\xi).$$

Exercise 2.2 Suppose that $f(x) = e^{-zx^2}$ for some z with $\operatorname{Re} z > 0$. Show that

$$(\mathcal{F}f)(\xi) = \sqrt{\frac{\pi}{z}} e^{-\xi^2/4z}.$$

Use the formula

$$\int_{-\infty}^{\infty} e^{-zx^2}\, dx = \sqrt{\frac{\pi}{z}}$$

and made a contour shift in the integration.

Exercise 2.3 The distribution inverse Fourier transform of $e^{\pm it\xi^2}$ may be computed as

$$\mathcal{F}^{-1}\left(e^{\pm it\xi^2}\right) = \lim_{\varepsilon \to 0^+} \mathcal{F}^{-1}\left(e^{\pm it\xi^2} e^{-\varepsilon \xi^2}\right).$$

Using the result of Exercise 2.2, show that

$$\mathcal{F}^{-1}\left(e^{\pm it\xi^2}\right) = \frac{1}{\sqrt{\mp 4\pi i t}} e^{\mp ix^2/(4t)}$$

where we take the principal branch of the square root function.

Exercise 2.4 Use the result of Exercise 2.3 and the convolution theorem from Exercise 2.1 to show from the solution formula (2.2) that

$$q(x,t) = \frac{1}{\sqrt{4\pi i t}} \int_{-\infty}^{\infty} e^{i(x-y)^2/(4t)} q_0(y)\, dy$$

for $q_0 \in \mathscr{S}(\mathbb{R})$.

Exercise 2.5 Suppose that ψ is a twice continuously differentiable, $N \times N$ matrix-valued solution to the system

$$\psi_x = A(x,t)\psi$$
$$\psi_t = B(x,t)\psi$$

where $A(x,t)$ and $B(x,t)$ are continuously differentiable $N \times N$ matrix-valued functions of x, t. Suppose further that $\det \psi(x, t) \neq 0$ for all (x, t). Show that

$$A_t - B_x + [A, B] = 0.$$

Hint: cross-differentiate the equations and use the equality $\psi_{tx} = \psi_{tx}$ (Clairaut's Theorem).

Exercise 2.6 A *fundamental* solution of (2.5) is a twice-differentiable 2×2 matrix-valued solution $\psi(x, t)$ with $\det \psi(x, t) \neq 0$ for all (x, t). Using the result of Exercise 2.5, show that if (2.5) admits a fundamental solution for a given smooth function $q(x, t)$, then $q(x, t)$ solves (2.4).

Exercise 2.7 Let $\mathcal{L}$ be the operator (2.6). Show that, for any smooth, compactly supported, 2×2 matrix-valued functions $\psi(x)$ and $\varphi(x)$, the identity $(\psi, \mathcal{L}\varphi) = (\mathcal{L}\psi, \varphi)$ holds, where the inner product is defined by

$$(\psi, \phi) = \int_{\mathbb{R}} \mathrm{Tr}\left(\psi^*(x)\phi(x)\right) dx.$$

Exercise 2.8 Consider the alternative Lax representation (from the original paper of Zakharov and Shabat [46])

$$L = \begin{pmatrix} i\partial_x & q \\ \bar{q} & -i\partial_x \end{pmatrix}$$

$$B = \begin{pmatrix} 2i\partial_x^2 - i|q|^2 & q_x + 2q\partial_x \\ \bar{q}_x + 2\bar{q}\partial_x & -2i\partial_x^2 + i|q|^2 \end{pmatrix}$$

Show that (2.4) is equivalent to the operator identity

$$\dot{L} = [B, L].$$

Remark The operator L is formally self-adjoint and B is formally skew-adjoint. This structure corresponds to the Lax representation for KdV.

Exercise 2.9 Suppose given a family of smooth solutions $\psi_1(z, k, t), \psi_2(z, k, t)$ of (4.3)–(4.4), indexed by $k \in \mathbb{C}$, so that[3]

(i) $\lim_{k \to \infty} e^{-ikz + ik^2 t} \psi_1(z, t, k) = 1$,
(ii) $\lim_{k \to \infty} e^{-ikz + ik^2 t} \psi_2(z, t, k) = 0$,
(iii) for each (t, z), $\psi_2(z, k, t) \neq 0$ for at least one k.

---

[3]These conditions are motivated by what one can actually prove about the solutions $m^1(z, k, t) = e^{-ikz + ik^2 t} \psi_1(z, k, t)$ and $m^2(z, k, t) = e^{-ikz + ik^2 t} m^2(z, k, t)$!

Cross-differentiate the first equations of (4.3) and (4.4) and equate mixed partials to show that

$$-2i\varepsilon (q\partial_z \overline{q})\psi_1 + (\dot{q} - i\overline{g}q)\psi_2 =$$
$$2i\left(2\varepsilon\overline{q}\partial_z q + i\varepsilon q\partial_z\overline{q} + \frac{1}{2}\partial_{\overline{z}}g\right)\psi_1 + 2i\left(\partial_z^2 q + \partial_{\overline{z}}^2 q + \frac{1}{2}gq\right)\psi_2 \quad (2.30)$$

Conclude that the compatibility condition (2.17) holds. To be really thorough (!), you should check that cross-differentiating the second equations of (4.3)–(4.4) gives the same relation.

Hint: Use *both* equations (4.3) to eliminate $\overline{z}$-derivatives of ψ_1 and z-derivatives of ψ_2. Expressions involving 'irreducible' derivatives such as $\partial_z \psi_1$ and $\partial_{\overline{z}}\psi_2$ should cancel, leading to (2.30). Then use the asymptotic conditions to argue that the coefficients of ψ_1 and ψ_2 must both be zero.

3 The Defocussing Cubic Nonlinear Schrödinger Equation

This lecture largely follows the analysis of Deift-Zhou [17, esp. §3] with a few inessential changes. We will analyze the direct and inverse scattering maps for NLS and, for completeness, give a proof of Beals-Coifman's result that the solution formula via inverse scattering generates a classical solution of the defocusing NLS equation (2.4) if $q_0 \in \mathscr{S}(\mathbb{R})$.

We will solve the NLS equation in the sense that we find a solution of the integral equation

$$q(t) = e^{it\Delta}q_0 - i\int_0^t e^{i(t-s)\Delta}\left(2|q(s)|^2 q(s)\right)ds \quad (3.1)$$

on $H^1(\mathbb{R})$, where $e^{it\Delta}$ is the solution operator (2.3) for the linear Schrödinger equation. Here

$$H^1(\mathbb{R}) = \left\{u \in L^2(\mathbb{R}) : u' \in L^2(\mathbb{R})\right\}. \quad (3.2)$$

Although (3.1) can be solved in much weaker spaces (see, for example [34] or [42, Chapter 3]), the space $H^1(\mathbb{R})$ will serve our purpose of showing that the inverse scattering method produces a continuous solution map on $H^{1,1}(\mathbb{R})$. The following lemma shows that, to show that (2.14) solves (3.1), it suffices to show that (2.14) produces a classical solution of (2.4) for initial data in $\mathscr{S}(\mathbb{R})$.

Lemma 3.1 *Let $q_0 \in H^1(\mathbb{R})$ and suppose that $\{q_n\}$ is a sequence from $\mathscr{S}(\mathbb{R})$ with $\|q_n - q_0\|_{H^1} \to 0$ as $n \to \infty$. Suppose that $q_n(z, t)$ solves (3.1) with initial data q_n*

and that $q_n(z,t) \to q(z,t)$ in the sense that $\sup_{t \in (0,T)} \|q_n(\cdot,t) - q(\cdot,t)\|_{H^1} \to 0$ as $n \to \infty$. Then $q(z,t)$ solves (3.1) with initial data q_0.

We leave the proof as Exercise 3.6.

3.1 The Direct Scattering Map

In this subsection we'll construct the direct scattering map by studying solutions $\Psi^\pm$ of the problem $\mathcal{L}\psi = \lambda\psi$. Here $\mathcal{L}$ is the ZS-AKNS operator (2.6), ψ is 2×2 matrix-valued, $\lambda \in \mathbb{R}$, and $\Psi^\pm$ satisfy the asymptotic conditions

$$\lim_{x \to \pm\infty} \Psi^\pm(x,\lambda) e^{i\lambda x \sigma_3} = \mathbb{I}.$$

It is well-known that the Jost solutions exist and are unique for $q \in L^1(\mathbb{R})$, and that $\det \Psi^\pm(x,\lambda) = 1$.

We begin with some reductions. A straightforward computation shows that for any $z \in \mathbb{C}$, the solution space of $\mathcal{L}\psi = z\psi$ is invariant under the mapping

$$\psi(x,z) \mapsto \sigma_1 \overline{\psi(x,\bar{z})} \sigma_1^{-1}, \quad \sigma_1 = \begin{pmatrix} 0 & 1 \\ 1 & 0 \end{pmatrix} \tag{3.3}$$

(Exercise 3.13). From this symmetry and the uniqueness of Jost solutions, it follows that the matrix-valued Jost solutions take the form

$$\Psi(x,\lambda) = \begin{pmatrix} \Psi_{11}(x,\lambda) & \overline{\Psi_{21}(x,\lambda)} \\ \Psi_{21}(x,\lambda) & \overline{\Psi_{11}(x,\lambda)} \end{pmatrix} \tag{3.4}$$

and that the matrix $T(\lambda)$ defined in (2.8) takes the form (2.9). From the relation $|a(\lambda)|^2 - |b(\lambda)|^2 = 1$ it follows that $|a(\lambda)| \geq 1$, and that

$$r(\lambda) = -b(\lambda)/\overline{a(\lambda)}$$

is a well-defined function with $|r(\lambda)| < 1$. We will prove:

Theorem 3.2 *The map $q \mapsto r$ is locally Lipschitz continuous from $H^{1,1}(\mathbb{R})$ to $H^{1,1}_1(\mathbb{R})$.*

The approach we'll take here is inspired by the analysis of the scattering transform by Muscalu et al. in [35], which also contains an interesting discussion of the Fourier-like mapping properties of the scattering transform. In order to obtain effective formulas for the scattering data $a(\lambda)$ and $b(\lambda)$, we make the change of variables

$$\Psi^+(x,\lambda) = e^{-i\lambda x \sigma_3} \mathbf{N}(x,\lambda). \tag{3.5}$$

It follows from the equation $\mathcal{L}\Psi^+ = \lambda \Psi^+$ that

$$\begin{cases} \dfrac{d}{dx}\mathbf{N}(x,\lambda) = \begin{pmatrix} 0 & e^{2i\lambda x}q(x) \\ e^{-2i\lambda x}\overline{q(x)} & 0 \end{pmatrix} \mathbf{N}(x,\lambda) \\ \lim_{x\to+\infty} \mathbf{N}(x,\lambda) = \mathbb{I} \end{cases} \quad (3.6)$$

while, by (2.8),

$$\lim_{x\to-\infty} \mathbf{N}(x,\lambda) = T(\lambda). \quad (3.7)$$

By the symmetry (3.4), we have

$$\mathbf{N}(x,\lambda) = \begin{pmatrix} N_{11}(x,\lambda) & \overline{N_{21}(x,\lambda)} \\ N_{21}(x,\lambda) & \overline{N_{11}(x,\lambda)} \end{pmatrix}$$

so it suffices to construct N_{11} and N_{21}. Equation (3.6) is equivalent to the integral equation

$$\mathbf{N}(x,\lambda) = \mathbb{I} - \int_x^\infty \begin{pmatrix} 0 & e^{2i\lambda y}q(y) \\ e^{-2i\lambda y}\overline{q(y)} & 0 \end{pmatrix} \mathbf{N}(y,\lambda)\, dy$$

which has a convergent Volterra series solution for $q \in L^1(\mathbb{R})$. Indeed, setting

$$N_{11}(x,\lambda) = a(x,\lambda), \quad N_{21}(x,\lambda) = b(x,\lambda),$$

we have

$$a(x,\lambda) = 1 + \sum_{n=1}^\infty A_{2n}(x,\lambda), \quad (3.8)$$

$$b(x,\lambda) = -\sum_{n=0}^\infty A_{2n+1}(x,\lambda). \quad (3.9)$$

Here

$$A_n(x,\lambda) = \int_{x<y_1<y_2<\ldots<y_n} Q_n(y_1,\ldots,y_n) e^{2i\lambda \phi_n(y)}\, dy_n \ldots dy_1$$

where

$$Q_n(y_1,\ldots,y_n) = \begin{cases} \prod_{j=1}^m q(y_{2j-1})\overline{q(y_{2j})}, & n=2m, \\ \overline{q(y_1)} \prod_{j=1}^m q(y_{2j})\overline{q(y_{2j+1})}, & n=2m+1 \end{cases}$$

and (with the convention that $\phi_0 = 0$)

$$\phi_{2m}(y_1,\ldots,y_{2m}) = \sum_{j=1}^m (y_{2j-1} - y_{2j}),$$

$$\phi_{2m+1}(y_1,\ldots,y_{2m+1}) = -y_1 + \phi_{2m}(y_2,\ldots,y_{2m+1})$$
$$= -\phi_{2m}(y_1,\ldots,y_{2m}) - y_{2m+1}$$

The bound

$$\int_{y_1<y_2\ldots<y_n} |Q_n(y)|\, dy_n\, dy_{n-1}\ldots dy_1 \leq \frac{\|q\|_1^n}{n!}$$

shows that the Volterra series converge uniformly in $x \in \mathbb{R}$ and q in bounded subsets of $L^1(\mathbb{R})$. By (3.7) and dominated convergence, we obtain the following representations of the maps $q \mapsto a$ and $q \mapsto b$:

$$a(\lambda) = 1 + \sum_{n=1}^\infty A_{2n}(\lambda) \tag{3.10}$$

$$b(\lambda) = -\sum_{n=0}^\infty A_{2n+1}(\lambda) \tag{3.11}$$

where

$$A_n(\lambda) = \int_{y_1<y_2<\ldots<y_n} Q_n(y_1,\ldots,y_n) e^{2i\lambda\phi_n(y)}\, dy_n\ldots dy_1. \tag{3.12}$$

From this representation we obtain an $L^1 \to L^\infty$ mapping property of the scattering transform.

Proposition 3.3 *The map $q \mapsto r$ is locally Lipschitz continuous from $L^1(\mathbb{R})$ to $L^\infty(\mathbb{R})$.*

Proof It suffices to show that $q \mapsto a$ and $q \mapsto b$ are locally Lipschitz continuous. If so, this continuity and the lower bound $|a(\lambda)| \geq 1$ imply local Lipschitz continuity of $q \mapsto r$. If $M_n : (L^1(\mathbb{R}))^n \to L^\infty(\mathbb{R})$ is a multilinear map and

$$F_n(q) = M_n(q,\ldots q, \overline{q},\ldots \overline{q})$$

with m entries of q and $n-m$ entries of $\overline{q}$, then

$$F_n(q_1) - F_n(q_2) = \sum_{j=1}^{m} M_n(q_2,\ldots,q_2,\underbrace{q_1-q_2}_{j\text{th entry}},,q_1,\ldots q_1,\overline{q_1},\ldots\overline{q_1})$$

$$+ \sum_{j=m+1}^{n} M_n(q_2,\ldots,q_2,\overline{q_2},\ldots,\overline{q_2},\underbrace{\overline{q_1-q_2}}_{j\text{th entry}},\overline{q_1},\ldots,\overline{q_1})$$

so that, setting

$$\gamma = \max\left(\|q_1\|_{L^1}, \|q_2\|_{L^1}\right),$$

we have

$$\|F_n(q_1) - F_n(q_2)\|_{L^\infty} \leqslant \|M_n\|_{(L^1)^n \to L^\infty}\, n\gamma^{n-1}\, \|q_1 - q_2\|_{L^1}$$

Thus, referring to (3.12), we have

$$\|A_n(\lambda; q_1) - A_n(\lambda; q_2)\|_{L^\infty} \leqslant \frac{1}{(n-1)!}\gamma^{n-1}\|q_1 - q_2\|_{L^1}.$$

We conclude that

$$\|a(\,\cdot\,; q_1) - a(\,\cdot\,, q_2)\|_{L^\infty} \leqslant e^\gamma \|q_1 - q_2\|_{L^1}$$
$$\|b(\,\cdot\,; q_1) - b(\,\cdot\,, q_2)\|_{L^\infty} \leqslant e^\gamma \|q_1 - q_2\|_{L^1}.$$

$\square$

Remark 3.4 From the above analysis, it is easy to see that the Fréchét derivative of $\mathcal{R}$ at $q=0$ is the "antilinear Fourier transform"

$$(\mathcal{F}_a q)(\lambda) = -\int e^{-2ix\lambda}\overline{q(x)}\,dx.$$

With a bit more work, we can prove:

Proposition 3.5 *The map $q \mapsto r$ is locally Lipschitz continuous from $L^1(\mathbb{R}) \cap L^2(\mathbb{R})$ into $L^2(\mathbb{R})$.*

Proof In what follows we use the fact that

$$\|f\|_{L^2(\mathbb{R})} = \sup_{\varphi \in C_0^\infty(\mathbb{R})} \left|\int \varphi(\lambda) f(\lambda)\, d\lambda\right|.$$

It follows from (3.12) and the trivial inequalities

$$\int |\widehat{\varphi}(\phi_{2n-1}(y))q(y_{2n-1})| \, dy_{2n-1} \leqslant \|\varphi\|_2 \|q\|_2,$$

$$\int |\widehat{\varphi}(\phi_{2n}(y))q(y_{2n})| \, dy_{2n} \leqslant \|\varphi\|_2 \|q\|_2,$$

that

$$\|A_n\|_{L^2(\mathbb{R})} \leqslant \frac{\|q\|_{L^1}^{n-1}}{(n-1)!} \|q\|_{L^2}.$$

Thus the power series representations for $a - 1$ and b converge for $X = L^1(\mathbb{R}) \cap L^2(\mathbb{R})$, $Y = L^2(\mathbb{R})$, showing that $q \mapsto a$ and $q \mapsto b$ are locally Lipschitz continuous as maps from $L^1(\mathbb{R}) \cap L^2(\mathbb{R})$ into $L^2(\mathbb{R})$. It now follows that $q \mapsto r$ has the same continuity. □

As in the theory of the Fourier transform, additional smoothness of q implies additional decay of r.

Proposition 3.6 *The map $q \mapsto r$ is locally Lipschitz continuous from $H^{1,1}(\mathbb{R})$ to $H^{0,1}(\mathbb{R})$.*

Proof It suffices to exhibit an L^2-convergent power series for $\lambda b(\lambda)$. We will assume for the moment that $q \in \mathcal{S}(\mathbb{R})$ and begin with the formula

$$\lambda b(\lambda) = \sum_{n=1}^{\infty} \lambda A_{2n-1}(\lambda).$$

Using the integration by parts identity

$$\int_{y_{2n-2}}^{\infty} \overline{q(z)}(2i\lambda) e^{2i\lambda(-\phi_{2n-2}(y)+z)} \, dz =$$

$$\left[\overline{q(z)} e^{2i\lambda(-\phi_{2n-2}(y)+z)} \right]\Big|_{y_{2n-2}}^{\infty} - \int_{y_{2n-2}}^{\infty} \overline{q'(z)} e^{2i\lambda(-\phi_{2n-2}(y)+z)} \, dz$$

we conclude that

$$\int \lambda b(\lambda) \varphi(\lambda) \, d\lambda = \int \varphi(\lambda) \widehat{\overline{q}}'(\lambda) \, d\lambda + \sum_{n=2}^{\infty} \left(I_{1,2n-1} + I_{2,2n} \right),$$

where, for $n \geqslant 2$,

$$|I_{1,2n-1}| \leq$$

$$\int_{y_1<\ldots<y_{2n-2}} \left(\prod_{k=1}^{2n-2} |q(y_j)|\right) |\widehat{\varphi}(\phi_{2n-3}(y))| |q(y_{2n-2})| \, dy_{2n-2}\ldots dy_1$$

and

$$|I_{2,2n-1}| \leq$$

$$\int_{y_1<\ldots<y_{2n-1}} \left(\prod_{k=1}^{2n-2} |q(y_j)|\right) |\widehat{\varphi}(\phi_{2n-1}(y))| |q'(y_{2n-1})| \, dy_{2n-1}\ldots dy_1.$$

It follows that

$$|I_{1,2n-1}| \leq \frac{1}{(2n-3)!} \|q\|_{L^1}^{2n-3} \|q\|_{L^\infty} \|q\|_{L^2} \|\varphi\|_{L^2}$$

$$|I_{2,2n-1}| \leq \frac{1}{(2n-2)!} \|q\|_{L^1}^{2n-2} \|q'\|_{L^2} \|\varphi\|_{L^2}$$

Taking suprema over φ with $\|\varphi\|_{L^2} = 1$ and recalling that $\|q\|_{L^\infty} \lesssim \|q\|_{H^1}$ (Exercise 3.2), we recover the estimate

$$\|(\cdot)b(\cdot)\|_{L^2} \leq \|q\|_{H^{1,0}} + \sum_{n=2}^\infty \frac{1}{(2n-3)!} \|q\|_{L^1}^{2n-3} \left(\|q\|_{L^1} + \|q\|_{L^2}\right) \|q\|_{H^{1,0}}$$

which is finite because $\|q\|_{L^1} \lesssim \|q\|_{H^{1,1}}$ and the series

$$\sum_{n=2}^\infty \frac{x^{2n-3}}{(2n-3)!} = \sinh(x)$$

converges for all x. The local Lipschitz continuity is proven by continuity of multilinear functionals as in Proposition 3.3. □

Finally:

Proposition 3.7 *The map $q \mapsto r'$ is locally Lipschitz continuous from $H^{0,1}(\mathbb{R})$ to $L^2(\mathbb{R})$.*

Proof By the quotient rule and the lower bound on $|a|$, it suffices to show that the map $q \mapsto (a', b')$ is locally Lipschitz continuous from $H^{0,1}(\mathbb{R})$ to $L^2(\mathbb{R}) \times L^2(\mathbb{R})$ From (3.10)–(3.11) we have

$$-i\frac{\partial a}{\partial \lambda}(\lambda) = \sum_{n=1}^{\infty} \int_{y_1 < \ldots < y_{2n}} 2Q_{2n}(y) \phi_{2n}(y) e^{2i\lambda \phi_{2n}(y)} \, dy$$

$$-i\frac{\partial b}{\partial \lambda}(\lambda) = \sum_{n=1}^{\infty} \int_{y_1 < \ldots < y_{2n-1}} 2Q_{2n-1}(y) \phi_{2n-1}(y) e^{2i\lambda \phi_{2n-1}(y)} \, dy$$

so integrating against $\varphi \in C_0^{\infty}(\mathbb{R})$ we obtain

$$\left| \int \varphi \frac{\partial a}{\partial \lambda} d\lambda \right| \leq \sum_{n=1}^{\infty} \int_{y_1 < \ldots < y_{2n}} |Q_{2n}(y)| \, |\phi_{2n}(y)| \, |\widehat{\varphi}(\phi_{2n}(y))| \, dy \qquad (3.13)$$

$$\leq \sum_{n=1}^{\infty} \int_{y_1 < \ldots < y_{2n}} \sum_{j=1}^{2n} |y_j| \, |Q_{2n}(y)| \, |\widehat{\varphi}(\phi_{2n}(y))| \, dy$$

$$\left| \int \varphi \frac{\partial b}{\partial \lambda} d\lambda \right| \leq \sum_{n=1}^{\infty} \int_{y_1 < \ldots < y_{2n-1}} |Q_{2n-1}(y)| \, |\phi_{2n-1}(y)| \, |\widehat{\varphi}(\phi_{2n-1}(y))| \, dy$$

$$(3.14)$$

$$\leq \sum_{n=1}^{\infty} \int_{y_1 < \ldots < y_{2n-1}} \sum_{j=1}^{2n-1} |y_j| \, |Q_{2n-1}(y)| \, |\widehat{\varphi}(\phi_{2n-1}(y))| \, dy$$

As before, we will bound the left-hand integrals in (3.13) and (3.14) by norms of q times $\|\varphi\|_{L^2}$ and take the supremum over φ with $\|\varphi\|_{L^2} = 1$.

To bound the right-hand side of (3.13), first note that the integrand is symmetric under interchange of $(n-1)$ pairs of indices (the pair containing y_j is excluded), so we may write

$$\left| \int \varphi \frac{\partial a}{\partial \lambda} d\lambda \right| \leq$$

$$\sum_{n=1}^{\infty} \frac{1}{(n-1)!} \sum_{j=1}^{2n} \int_{\mathbb{R}^{2n}} \left(\prod_{k=1}^{j-1} |q(y_k)| \right) (|y_j q(y_j)|) \left(\prod_{k=j+1}^{2n} |q(y_k)| \right) |\widehat{\varphi}(\phi_{2n}(y))| \, dy$$

Using Young's inequality $\|f * g\|_2 \leq \|f\|_1 \|g\|_2$ repeatedly beginning with the y_{2n} integration, we have

$$\left\| \int_{\mathbb{R}^{2n-j}} |\widehat{\varphi}(\phi_{2n}(y))| \left(\prod_{k=j+1}^{2n} |q(y_k)| \right) dy_{2n} \ldots dy_{j+1} \right\|_{L^2(dy_j)} \leq \|q\|_{L^1}^{2n-j} \|\varphi\|_{L^2}$$

so that

$$\left\|\frac{\partial a}{\partial \lambda}\right\|_{L^2} \leq \sum_{j=1}^{\infty} \frac{2n}{(n-1)!} \|q\|_{H^{0,1}} \|q\|_{L^1}^{2n-1}$$

which is uniformly bounded for q in bounded subsets of $H^{0,1}$ since

$$\sum_{j=1}^{\infty} \frac{2n}{(n-1)!} x^{2n-1}$$

converges for all x.

To bound the right hand side of (3.14), we first note that the $n = 1$ term is trivially bounded by $\|q\|_{H^{0,1}} \|\varphi\|_{L^2}$. For $n \geq 2$, the integrand is symmetric under $(n-2)!$ interchanges of pairs so we may estimate the remaining terms on the right-hand side of (3.14) by

$$\sum_{n=2}^{\infty} \frac{1}{(n-2)!} \sum_{j=1}^{2n-1} \int_{\mathbb{R}^{2n-1}} |y_j| |Q_{2n-1}(y)| |\widehat{\varphi}(\phi_{2n-1}(y))| \, dy$$

Writing

$$\int_{\mathbb{R}^{2n-1}} |y_j| |Q_{2n-1}(y)| |\widehat{\varphi}(\phi_{2n-1}(y))| \, dy =$$

$$\int_{\mathbb{R}^{2n-1}} \left(\prod_{k=1}^{j-1} |q(y_k)|\right) (|y_j||q(y_j)|) \left(\prod_{k=j+1}^{2n-1} |q(y_k)|\right) |\widehat{\varphi}(\phi_{2n-1}(y))| \, dy$$

we may use the estimate

$$\left\|\int_{\mathbb{R}^{2n-j-1}} \left(\prod_{k=j+1}^{2n-1} |q(y_k)|\right) |\widehat{\varphi}(\phi_{2n-1}(y))| \, dy\right\|_{L^2(dy_j)} \leq \|q\|_{L^1}^{2n-j-1} \|\varphi\|_{L^2}$$

(which again follows by repeated applications of Young's inequality) to conclude that

$$\left\|\frac{\partial b}{\partial \lambda}\right\|_{L^2} \leq \|q\|_{H^{1,0}} + \sum_{n=2}^{\infty} \frac{2n-1}{(n-2)!} \|q\|_{L^1}^{2n-2} \|q\|_{H^{0,1}}.$$

The right-hand side is again bounded uniformly for q in a bounded subset of $H^{0,1}(\mathbb{R})$ since the series

$$\sum_{n=2}^{\infty} \frac{2n-1}{(n-2)!} x^{2n-2}$$

converges for all x.

We have shown that $\partial a/\partial \lambda$ and $\partial b/\partial \lambda$ have convergent series representations. We can use multilinearity of the terms as in the proof of Proposition 3.3 to obtain local Lipschitz continuity. □

3.2 Beals-Coifman Solutions

Beals and Coifman [7] identified solutions of (2.11) which have piecewise analytic continuations to $\mathbb{C} \setminus \mathbb{R}$ and solve a Riemann-Hilbert problem determined completely by the scattering data. It follows from the definition of $\mathbf{M}$ that two nonsingular solutions $\mathbf{M}_1$ and $\mathbf{M}_2$ of (2.11) are related by

$$\mathbf{M}_1(x, \lambda) = \mathbf{M}_2(x, \lambda) e^{-ix\lambda \, \text{ad} \, \sigma_3} A \tag{3.15}$$

where A is a constant matrix and

$$e^{t \, \text{ad} \, \sigma_3} \begin{pmatrix} a & b \\ c & d \end{pmatrix} = \begin{pmatrix} a & e^{2t} b \\ e^{-2t} c & d \end{pmatrix}.$$

(see Exercises 3.7 and 3.12). We will use this fact repeatedly in what follows.

We now construct the Beals-Coifman solutions from solutions $\mathbf{M}^\pm$ of (2.11) corresponding to the Jost solutions $\Psi^\pm$. By the factorization $\psi(x, \lambda) = \mathbf{M}(x, \lambda) e^{-i\lambda x \sigma_3}$ and the symmetry (3.4), solutions of (2.11) normalized by either of the two conditions $\lim_{x \to \pm\infty} \mathbf{M}(x, z) = \mathbb{I}$ take the form

$$\mathbf{M}(x, \lambda) = \begin{pmatrix} m_{11}(x, \lambda) & \overline{m_{21}(x, \lambda)} \\ m_{21}(x, \lambda) & \overline{m_{11}(x, \lambda)} \end{pmatrix}$$

so we need only study $m_1 := m_{11}$ and $m_2 := m_{21}$. Moreover, it is clear from (2.11) and (3.5) that

$$\mathbf{M}(x, \lambda) = e^{-i\lambda x \sigma_3} \mathbf{N}(x, \lambda) e^{i\lambda x \sigma_3}. \tag{3.16}$$

Let $\Psi^+(x, \lambda) = \mathbf{M}^+(x, \lambda) e^{-i\lambda x \sigma_3}$. It follows from (3.16) that

$$m_1^+(x, \lambda) := m_{11}^+(x, \lambda) = a(x, \lambda)$$
$$m_2^+(x, \lambda) := m_{21}^+(x, \lambda) = e^{2i\lambda x} \overline{b(x, \lambda)}$$

It follows from (3.8)–(3.9) that

$$m_1^+(x, \lambda) = 1 + \sum_{n=1}^{\infty} \int_{x<y_1<...<y_{2n}} Q_{2n}(y) e^{2i\lambda \phi_{2n}(y_1,...,y_{2n})} \, dy \qquad (3.17)$$

$$m_2^+(x, \lambda) = -\sum_{n=1}^{\infty} \int_{x<y_1<...<y_{2n-1}} \overline{Q_{2n+1}(y)} e^{2i\lambda(\phi_{2n}(x,y_1,...,y_{2n-1})} \, dy \qquad (3.18)$$

Since the phase functions

$$\phi_{2n}(y) := \phi_{2n}(y_1, \ldots, y_{2n})$$

and

$$\phi_{2n}(x, y) := \phi_{2n}(x, y_1, \ldots, y_{2n-1})$$

are nonpositive over their respective domains of integration, m_1^+ and m_2^+ continue to analytic functions $m_1^+(x, z)$ and $m_2^+(y, z)$ for Im $z < 0$ obeying the bounds

$$|m_1^+(x, z) - 1| + |m_2^+(x, z)| \leqslant \gamma_+(x) e^{\gamma_+(x)}$$

where

$$\gamma_+(x) = \int_x^\infty |q(y)| \, dy.$$

It follows that $a(\lambda)$ also has a bounded analytic continuation to the lower half-plane which we denote by $a(z)$. Using (3.17) and (3.18) with λ replaced by z we can deduce the large-x asymptotics

$$\lim_{x \to -\infty} \begin{pmatrix} m_1^+(x, z) \\ m_2^+(x, z) \end{pmatrix} = \begin{pmatrix} a(z) \\ 0 \end{pmatrix}, \quad \lim_{x \to \infty} \begin{pmatrix} m_1^+(x, z) \\ m_2^+(x, z) \end{pmatrix} = \begin{pmatrix} 1 \\ 0 \end{pmatrix}$$

for each fixed z with Im $z < 0$.

We can also use the Volterra series to analyze the large-z behavior of the extensions $m_1^+(x, z)$ and $m_2^+(x, z)$ for fixed x. Observe that

$$\left| e^{2iz\phi_{2n}(y)} \right| \leqslant e^{-2\operatorname{Im}(z)\phi_{2n}(y)}$$

and

$$\left| e^{2iz(\phi_{2n}(x,y))} \right| \leqslant e^{-2\operatorname{Im}(z)\phi_{2n}(x,y)}.$$

An argument using the dominated convergence theorem together with the absolute and uniform convergence of the Volterra series for m_1 and m_2 shows that

$$\lim_{z \to \infty} \begin{pmatrix} m_1^+(x, z) \\ m_2^+(x, z) \end{pmatrix} = \begin{pmatrix} 1 \\ 0 \end{pmatrix} \tag{3.19}$$

and

$$\lim_{z \to \infty} a(z) = 1$$

where the limit is taken as $|z| \to \infty$ in any proper subsector of the lower half-plane. In a similar way, if $\Psi^-(x, \lambda) = \mathbf{M}^-(x, \lambda) e^{-i\lambda x \sigma_3}$, we can use the Volterra series

$$m_1^-(x, \lambda) = 1 + \sum_{n=1}^{\infty} \int_{y_{2n} < \ldots < y_1 < x} Q_{2n}(y) e^{2i\lambda \phi_{2n}(y)} \, dy$$

$$m_2^-(x, \lambda) = -\sum_{n=1}^{\infty} \int_{y_{2n-1} < \ldots < y_1 < x} \overline{Q_{2n+1}(y)} e^{2i\lambda \phi_{2n}(x,y)} \, dy$$

and the fact that $\phi_{2n}(y)$ and $\phi_{2n}(x, y)$ are nonnegative on their respective domains of integration to show that $m_1^-(x, \lambda)$ and $m_2^-(x, \lambda)$ continue to analytic functions $m_1^-(x, z)$ and $m_2^-(x, z)$ for $\operatorname{Im} z > 0$ with

$$|m_1^-(x, z) - 1| + |m_2^-(x, z)| \leq \gamma_-(x) e^{\gamma_-(x)}$$

where

$$\gamma_-(x) = \int_{-\infty}^{x} |q(y)| \, dy.$$

We can deduce the asymptotics

$$\lim_{x \to +\infty} \begin{pmatrix} m_1^-(x, z) \\ m_2^-(x, z) \end{pmatrix} = \begin{pmatrix} \overline{a(\overline{z})} \\ 0 \end{pmatrix}, \quad \lim_{x \to -\infty} \begin{pmatrix} m_1^-(x, z) \\ m_2^-(x, z) \end{pmatrix} = \begin{pmatrix} 1 \\ 0 \end{pmatrix}. \tag{3.20}$$

It will be important to know that $a(z)$ has no zeros in $\operatorname{Im} z < 0$. It follows from (2.8) that

$$a(\lambda) = \begin{vmatrix} \Psi_{11}^+(x, \lambda) & \Psi_{12}^-(x, \lambda) \\ \Psi_{21}^+(x, \lambda) & \Psi_{22}^-(x, \lambda) \end{vmatrix}$$

so the same holds true for λ replaced by z with $\operatorname{Im} z < 0$ by analytic continuation. Thus $a(z) = 0$ if and only if the columns are linearly dependent. Since $(\Psi_{11}^+, \Psi_{21}^+)$ decay exponentially as $x \to +\infty$ and $(\Psi_{12}^-, \Psi_{22}^-)$ decay exponentially as $x \to -\infty$, it is easy to show that this condition leads to a square-integrable solution of $\mathcal{L}\psi = z\psi$ with imaginary eigenvalue z, which is forbidden by the self-adjointness of the operator $\mathcal{L}$ in (2.6) (see Exercise 2.7). Hence $a(z)$ has no zeros in $\operatorname{Im} z < 0$.

We can now construct piecewise analytic solutions of (2.11), normalized so that $\mathbf{M}^r(x, z) \to \mathbb{I}$ as $x \to +\infty$ (the "r" is for "right-normalized"), by the formulas

$$\mathbf{M}^r(x, z) = \begin{pmatrix} m_1^-(x, z) & \overline{m_2^+(x, \bar{z})} \\ m_2^-(x, z) & \overline{m_1^+(x, \bar{z})} \end{pmatrix} \begin{pmatrix} 1/\overline{a(\bar{z})} & 0 \\ 0 & 1 \end{pmatrix}, \qquad \operatorname{Im} z > 0 \qquad (3.21)$$

(recall the first asymptotic relation of (3.20)) and

$$\mathbf{M}^r(x, z) = \sigma_1 \overline{\mathbf{M}^r(x, \bar{z})} \sigma_1, \qquad \operatorname{Im} z < 0 \qquad (3.22)$$

(recall (3.3)). The piecewise analytic function $\mathbf{M}^r(x, z)$ admits boundary values $\mathbf{M}_\pm^r(x, \lambda)$ as $\pm \operatorname{Im} z \to 0$ and $\operatorname{Re} z \to \lambda$.

Since $\mathbf{M}_\pm^r(x, \lambda)$ solve (3.6), it follows from (3.15) that there is a jump matrix $\mathbf{V}_r(\lambda)$ with

$$\mathbf{M}_+^r(x, \lambda) = \mathbf{M}_-^r(x, \lambda) e^{-i\lambda x \, \mathrm{ad}\, \sigma_3} \mathbf{V}_r(\lambda).$$

To compute the jump matrix, first note that, from (2.8) and the definition of $\mathbf{M}^\pm$,

$$\mathbf{M}^+(x, \lambda) = \mathbf{M}^-(x, \lambda) e^{-i x \lambda \, \mathrm{ad}\, \sigma_3} \begin{pmatrix} a(\lambda) & \overline{b(\lambda)} \\ b(\lambda) & \overline{a(\lambda)} \end{pmatrix} \qquad (3.23)$$

Write

$$f(x) \underset{x \to \pm\infty}{\sim} g(x)$$

if

$$\lim_{x \to \pm\infty} |f(x) - g(x)| = 0.$$

Since $\mathbf{M}^\pm(x, \lambda) \to \mathbb{I}$ as $x \to \pm\infty$, it follows from (3.23) that

$$\begin{pmatrix} m_1^+(x, \lambda) \\ m_2^+(x, \lambda) \end{pmatrix} \underset{x \to -\infty}{\sim} \begin{pmatrix} a(\lambda) \\ e^{2ix\lambda} b(\lambda) \end{pmatrix}, \qquad \begin{pmatrix} m_1^+(x, \lambda) \\ m_2^+(x, \lambda) \end{pmatrix} \underset{x \to +\infty}{\sim} \begin{pmatrix} 1 \\ 0 \end{pmatrix},$$

$$\begin{pmatrix} m_1^-(x,\lambda) \\ m_2^-(x,\lambda) \end{pmatrix} \underset{x \to -\infty}{\sim} \begin{pmatrix} 1 \\ 0 \end{pmatrix}, \qquad \begin{pmatrix} m_1^-(x,\lambda) \\ m_2^-(x,\lambda) \end{pmatrix} \underset{x \to +\infty}{\sim} \begin{pmatrix} \overline{a(\lambda)} \\ -e^{2ix\lambda} b(\lambda) \end{pmatrix}.$$

From these asymptotic relations, (3.21) (for $\mathbf{M}_-^r$), and (3.22) (for $\mathbf{M}_+^r$), we conclude that

$$\mathbf{M}_-^r(x,\lambda) \underset{x \to \infty}{\sim} e^{-i\lambda x \,\mathrm{ad}\,\sigma_3} \begin{pmatrix} 1 & -\overline{b(\lambda)}/a(\lambda) \\ 0 & 1 \end{pmatrix}$$

$$\mathbf{M}_+^r(x,\lambda) \underset{x \to -\infty}{\sim} e^{-i\lambda x \,\mathrm{ad}\,\sigma_3} \begin{pmatrix} 1 & 0 \\ -b(\lambda)/\overline{a(\lambda)} & 1 \end{pmatrix},$$

so that

$$\mathbf{V}^r(\lambda) = \begin{pmatrix} 1 - |r(\lambda)|^2 & -\overline{r(\lambda)} \\ r(\lambda) & 1 \end{pmatrix}$$

where

$$r(\lambda) = -b(\lambda)/\overline{a(\lambda)}.$$

The Beals-Coifman solutions $\mathbf{M}^r(x, z)$ play a fundamental role in the inverse problem. To describe their large-z asymptotic behavior, recall (cf. Remark 2.2) that $\lim_{z \to \infty} F(z) = A$ uniformly in proper subsectors of $\mathbb{C} \setminus \mathbb{R}$ if

$$\lim_{R \to \infty} \sup_{\substack{|z| \geq R \\ \arg(z) \in (\alpha, \beta)}} |F(z) - A| = 0$$

for any proper subinterval (α, β) of $(0, \pi)$ or $(\pi, 2\pi)$.

Theorem 3.8 (Right-Normalized Beals-Coifman Solutions) *Suppose that* $q \in L^1(\mathbb{R})$. *For each* $z \in \mathbb{C} \setminus \mathbb{R}$, *there exists a unique solution to the problem*

$$\begin{cases} \dfrac{d}{dx}\mathbf{M} = -iz\,\mathrm{ad}\,\sigma_3(\mathbf{M}) + \mathbf{Q}_1(x)\mathbf{M}, \\ \qquad\quad \lim_{x \to +\infty} \mathbf{M}(x,z) = \mathbb{I}, \\ \qquad\quad \mathbf{M}(x,z) \text{ is bounded as } x \to -\infty. \end{cases} \qquad (3.24)$$

The unique solution $\mathbf{M}^r(x, z)$ *has the asymptotic behavior*

$$\lim_{z \to \infty} \mathbf{M}^r(x, z) = \mathbb{I}$$

as $z \to \infty$ *in any proper subsector of the upper or lower half-planes. Moreover,* $\mathbf{M}^r(x, z)$ *has continuous boundary values* $\mathbf{M}^r_\pm$ *as* $\pm \operatorname{Im} z \downarrow 0$ *and* $\operatorname{Re} z \to \lambda \in \mathbb{R}$ *that satisfy the jump relation*

$$\mathbf{M}^r_+(x, \lambda) = \mathbf{M}^r_-(x, \lambda) e^{-i\lambda x \operatorname{ad} \sigma_3} \mathbf{V}^r(\lambda) \tag{3.25}$$

where

$$\mathbf{V}^r(\lambda) = \begin{pmatrix} 1 - |r(\lambda)|^2 & -\overline{r(\lambda)} \\ r(\lambda) & 1 \end{pmatrix} \tag{3.26}$$

and $r(\lambda) = -b(\lambda)/\overline{a(\lambda)}$. *If* $q \in L^1(\mathbb{R}) \cap L^2(\mathbb{R})$, *then, for each* x,

$$\mathbf{M}^r_\pm(x, \lambda) - \mathbb{I} \in L^2(\mathbb{R}). \tag{3.27}$$

Finally, if $q \in L^1(\mathbb{R}) \cap C(\mathbb{R})$,

$$q(x) = \lim_{z \to \infty} 2iz \left(\mathbf{M}^r\right)_{12}(x, z) \tag{3.28}$$

where the limit is taken as $|z| \to \infty$ *in any proper subsector of the upper or lower half-plane.*

Proof We have already computed the jump relation; the claimed large-z asymptotic behavior follows from (3.19) and the analogous statement for m_1^- and m_2^-. It remains to show that the Beals-Coifman solutions are unique, to show that (3.27) holds, and to prove the reconstruction formula (3.28).

To prove uniqueness, suppose that, for given z, $\mathbf{M}$ and $\mathbf{M}^\sharp$ solve (3.24). We'll assume that $\operatorname{Im} z < 0$ since the proof for $\operatorname{Im} z > 0$ is similar.

Since $\mathbf{M}$ and $\mathbf{M}^\sharp$ both solve (2.11), there is a constant matrix

$$A = \begin{pmatrix} a_{11} & a_{12} \\ a_{21} & a_{22} \end{pmatrix}$$

so that

$$\mathbf{M}(x, z) = \mathbf{M}^\sharp(x, z) \begin{pmatrix} a_{11} & e^{-2ixz} a_{12} \\ e^{2ixz} a_{21} & a_{22} \end{pmatrix}.$$

Since $\left|e^{-2ixz}\right| \to \infty$ as $x \to +\infty$ while both $\mathbf{M}, \mathbf{M}^\sharp \to \mathbb{I}$ as $x \to +\infty$, it follows that $a_{12} = 0$. Since $\left|e^{2ixz}\right| \to \infty$ as $x \to -\infty$ while both $\mathbf{M}$ and $\mathbf{M}^\sharp$ are bounded, we conclude that $a_{21} = 0$. Using the normalization at $+\infty$ again we conclude that $a_{11} = a_{22} = 1$ and $\mathbf{M} = \mathbf{M}^\sharp$.

The property (3.27) follows from the series representations for $m_1^\pm$ and $m_2^\pm$ and an argument similar to the proof of Proposition 3.7.

Finally, we consider the reconstruction formula (3.32). We give the proof for $\operatorname{Im} z > 0$ since the proof for $\operatorname{Im} z < 0$ is similar. First, note that

$$\left(\mathbf{M}^r\right)_{12}(x, z) = \overline{m_2^+(x, \bar{z})},$$

it suffices to show that

$$q(x) = \lim_{z \to \infty} 2iz\overline{m_2^+(x, \bar{z})}.$$

To see this we use the absolutely and uniformly convergent Volterra series representation (3.18), the fact that

$$\lim_{z \to \infty} 2iz \int_x^\infty q(y) e^{2i\bar{z}(x-y)} \, dy = q(x),$$

and the fact that, for any $n \geq 2$,

$$\lim_{z \to \infty} 2iz \int_{x < y_1 < \ldots < y_{2n-1}} Q_{2n-1}(y) e^{2i\bar{z}(x - \phi_{2n-1}(y))} \, dy = 0$$

by dominated convergence. $\square$

We can also construct "left" Beals-Coifman solutions normalized at $-\infty$ as follows:

$$\mathbf{M}^\ell(x, z) = \begin{pmatrix} m_1^+(x, z) & \overline{m_2^-(x, \bar{z})} \\ m_2^+(x, z) & \overline{m_1^-(x, \bar{z})} \end{pmatrix} \begin{pmatrix} 1/a(z) & 0 \\ 0 & 1 \end{pmatrix}, \qquad \operatorname{Im} z < 0$$

and

$$\mathbf{M}^\ell(x, z) = \sigma_1 \overline{\mathbf{M}^\ell(x, \bar{z})} \sigma_1, \qquad \operatorname{Im} z > 0.$$

Theorem 3.9 (Left-Normalized Beals-Coifman Solutions) *Suppose that $q \in L^1(\mathbb{R})$. For each $z \in \mathbb{C} \setminus \mathbb{R}$, there exists a unique solution to the problem*

$$\begin{cases} \dfrac{d}{dx}\mathbf{M} = -iz\,\mathrm{ad}\,\sigma_3(\mathbf{M}) + \mathbf{Q}_1(x)\mathbf{M}, \\ \lim_{x\to-\infty}\mathbf{M}(x,z) = \mathbb{I}, \\ \mathbf{M}(x,z) \text{ is bounded as } x\to+\infty. \end{cases}$$

The unique solution $\mathbf{M}^\ell(x,z)$ has the asymptotic behavior

$$\lim_{z\to\infty} \mathbf{M}^\ell(x,z) = \mathbb{I}$$

as $z\to\infty$ in any proper subsector of the upper or lower half-plane. Moreover, $\mathbf{M}^\ell(x,z)$ has continuous boundary values $\mathbf{M}^\ell_\pm$ on $\mathbb{R}$ as $\pm\,\mathrm{Im}\,z\downarrow 0$ and $\mathrm{Re}\,z\to\lambda\in\mathbb{R}$ that satisfy the jump relation

$$\mathbf{M}^\ell_+(x,\lambda) = \mathbf{M}^\ell_-(x,\lambda) e^{-i\lambda x\,\mathrm{ad}\,\sigma_3}\mathbf{V}^\ell(z) \tag{3.29}$$

where

$$\mathbf{V}^\ell(z) = \begin{pmatrix} 1 & -\overline{\check{r}(\lambda)} \\ \check{r}(\lambda) & 1 - |\check{r}(\lambda)|^2 \end{pmatrix} \tag{3.30}$$

and $\check{r}(\lambda) = -b(\lambda)/a(\lambda)$. If $q \in L^1(\mathbb{R}) \cap L^2(\mathbb{R})$, then, for each x,

$$\mathbf{M}^\ell_\pm(x,\lambda) - \mathbb{I} \in L^2(\mathbb{R}). \tag{3.31}$$

Finally, if $q \in L^1(\mathbb{R}) \cap C(\mathbb{R})$,

$$q(x) = \lim_{z\to\infty} 2iz\left(\mathbf{M}^\ell\right)_{12}(x,z) \tag{3.32}$$

where the limit is taken as $|z|\to\infty$ in any proper subsector of the upper or lower half-plane.

We omit the proof.

The large-z asymptotics of $\mathbf{M}^r(x,z)$ (resp. $\mathbf{M}^\ell(x,z)$) together with the jump relation (3.25) (resp. the jump relation (3.29)) define a *Riemann-Hilbert problem*. We will see that, properly formulated, these Riemann-Hilbert problems have unique solutions given the data r and $\check{r}$, offering a means of recovering q from r and $\check{r}$.

In fact, r uniquely determines $\check{r}$, a, and b, so that the Riemann-Hilbert problem for $\mathbf{M}^r$ can be conjugated to the Riemann-Hilbert problem for $\mathbf{M}^\ell$. To see this, consider the analytic function F on $\mathbb{C}\setminus\mathbb{R}$ defined by

$$F(z) = \begin{cases} 1/\overline{a(\overline{z})} & \text{Im}(z) > 0, \\ a(z) & \text{Im}(z) < 0. \end{cases} \quad (3.33)$$

From what has already been proved, $F(z)$ is piecewise analytic in $\mathbb{C} \setminus \mathbb{R}$, $F(z) \to 1$ as $|z| \to \infty$ in any proper subsector of the upper or lower half-plane, and $F(z)$ has continuous boundary values $F_\pm$ on the real axis with

$$F_+(\lambda) = F_-(\lambda)(1 - |r(\lambda)|^2)$$

as follows from the definitions of F and r together with the relation (2.9). Taking logarithms we see that

$$\log F_+(\lambda) - \log F_-(\lambda) = \log\left(1 - |r(\lambda)|^2\right).$$

Recall that, if $f \in H^1(\mathbb{R})$, the function

$$W(z) = \frac{1}{2\pi i} \int_{-\infty}^{\infty} \frac{f(\zeta)}{\zeta - z} d\zeta \quad (3.34)$$

is the unique function on $\mathbb{C} \setminus \mathbb{R}$ with $W(z) \to 0$ as $z \to \infty$ and $W_+ - W_- = f$, where $W_\pm$ are the boundary Motivated by this fact, we set

$$G(z) = \exp\left(\frac{1}{2\pi i} \int_{-\infty}^{\infty} \frac{1}{s - z} \log\left(1 - |r(s)|^2\right) ds\right).$$

Note that, for $r \in L^\infty \cap L^2$ with $\|r\|_\infty < 1$, we have

$$G(z) = 1 + \mathcal{O}\left(\frac{1}{z}\right)$$

as $z \to \infty$. The function G satisfies the jump and boundary conditions and is analytic in $\mathbb{C} \setminus \mathbb{R}$. Moreover, the function $H(z) = F(z)/G(z)$ is analytic in the same region, continuous across the real axis, and $\lim_{z \to \infty} H(z) = 1$. It follows from Liouville's theorem that $F(z) = G(z)$, which shows that a is uniquely determined by r. Since $\check{r}(\lambda) = \overline{r(\lambda)a(\lambda)}/a(\lambda)$ we see that r determines $\check{r}$.

In what follows, it will be important to note that the boundary values $F_\pm(\lambda)$ satisfy the identity

$$F_+(\lambda) F_-(\lambda) = a(\lambda)/\overline{a(\lambda)}$$

as follows easily from the definition.

One can also conjugate the Riemann-Hilbert problem for $\mathbf{M}^r$ to that for $\mathbf{M}^\ell$ as follows. Given a function $\mathbf{M}^r(x, z)$ solving the "right" Riemann-Hilbert problem (i.e., the jump condition (3.25) and the normalization condition (3.27)), the function

$$\mathbf{M}(x, z) = \mathbf{M}^r(x, z) \begin{pmatrix} F(z)^{-1} & 0 \\ 0 & F(z) \end{pmatrix}$$

is easily seen to solve the Riemann-Hilbert problem for $\mathbf{M}^\ell$ (i.e., the jump condition (3.29) and the normalization condition (3.31)) since the additional factor doesn't change the large-z asymptotic behavior of the solution, while the jump matrices $\mathbf{V}^r$ and $\mathbf{V}^\ell$ are related by the identity $\mathbf{V}^r = F_-^{-\sigma_3} \mathbf{V}^\ell F_+^{\sigma_3}$.

3.3 The Inverse Scattering Map

To reconstruct q we will solve the following Riemann-Hilbert problem (compare RHP 2.1).

Riemann-Hilbert Problem 3.10 Given $r \in H_1^{1,1}(\mathbb{R})$ and $x \in \mathbb{R}$, find a function $\mathbf{M}(x, z) : \mathbb{C} \setminus \mathbb{R} \to SL(2, \mathbb{C})$ so that:

(i) $\mathbf{M}(x, z)$ is analytic in $\mathbb{C} \setminus \mathbb{R}$ for each x,
(ii) $\mathbf{M}(x, z)$ has continuous boundary values $\mathbf{M}_\pm(x, \lambda)$ on $\mathbb{R}$,
(iii) $\mathbf{M}^\pm(x, \lambda) - \mathbb{I}$ in $L^2(\mathbb{R})$, and
(iv) The jump relation

$$\mathbf{M}_+(x, \lambda) = \mathbf{M}_-(x, \lambda) e^{-i\lambda x \, \text{ad}\,\sigma_3} \mathbf{V}(\lambda)$$

holds, where

$$\mathbf{V}(\lambda) = \begin{pmatrix} 1 - |r(\lambda)|^2 & -\overline{r(\lambda)} \\ r(\lambda) & 1 \end{pmatrix}.$$

We will recover $q(x)$ from

$$q(x) = \lim_{z \to \infty} 2iz\, (\mathbf{M})_{12}(x, z).$$

Riemann-Hilbert problem 3.10 may usefully be thought of as an elliptic boundary value problem (the analyticity condition means that $\overline{\partial}\mathbf{M} = 0$ on $\mathbb{C} \setminus \mathbb{R}$). For this reason one should be able to reformulate RHP 3.10 as a boundary integral equation, much as the Dirichlet problem on bounded domain may be reduced to a boundary integral equation. We now describe such a formulation, due to Beals and Coifman [7].

First, observe that the jump matrix $\mathbf{V}(\lambda)$ admits a factorization of the form

$$\mathbf{V}(\lambda) = \left(I - w^-(\lambda)\right)^{-1} \left(I + w^+(\lambda)\right)$$

where

$$w^+(\lambda) = \begin{pmatrix} 0 & 0 \\ r(\lambda) & 0 \end{pmatrix}, \quad w^-(\lambda) = \begin{pmatrix} 0 & -\overline{r(\lambda)} \\ 0 & 0 \end{pmatrix}.$$

so that

$$e^{-i\lambda x \, \mathrm{ad}\, \sigma_3} \mathbf{V}(\lambda) = \left(I - w_x^-(\lambda)\right)^{-1} \left(I + w_x^+(\lambda)\right)$$

where

$$w_x^+(\lambda) = \begin{pmatrix} 0 & 0 \\ e^{2i\lambda x} r(\lambda) & 0 \end{pmatrix}, \quad w_x^-(\lambda) = \begin{pmatrix} 0 & -e^{-2i\lambda x}\overline{r(\lambda)} \\ 0 & 0 \end{pmatrix}.$$

Note that, if $r \in H_1^{1,1}(\mathbb{R})$, then $w_\pm \in L^\infty(\mathbb{R}) \cap L^2(\mathbb{R})$ with $\|w^\pm\|_\infty < 1$. Next, introduce the unknown matrix-valued function

$$\mu(x, \lambda) = \mathbf{M}_+(x, \lambda) \left(I + w_x^+(\lambda)\right)^{-1} = \mathbf{M}_-(x, \lambda) \left(I - w_x^-(\lambda)\right)^{-1}$$

and observe that

$$\mathbf{M}_+(x, \lambda) - \mathbf{M}_-(x, \lambda) = \mu(x, \lambda) \left(w_x^+(\lambda) + w_x^-(\lambda)\right).$$

Recalling (3.34) and the asymptotic condition on $\mathbf{M}(x, z)$, we conclude that

$$\mathbf{M}(x, z) = \mathbb{I} + \frac{1}{2\pi i} \int \frac{\mu(x, s) \left(w_x^+(\lambda) + w_x^-(\lambda)\right)}{s - z} ds. \tag{3.35}$$

Using this representation, we can derive a boundary integral equation for the unknown function $\mu(x, \lambda)$ which, if solvable, uniquely determines $\mathbf{M}(x, z)$ from the Cauchy integral formula. For $f \in H^1(\mathbb{R})$, define the Cauchy projectors $C_\pm$ by

$$(C_\pm f)(\lambda) = \lim_{\varepsilon \downarrow 0} \frac{1}{2\pi i} \int \frac{f(s)}{s - (\lambda \pm i\varepsilon)} ds. \tag{3.36}$$

The projectors $C_\pm$ extend to isometries of $L^2(\mathbb{R})$ with $C_+ - C_- = I$, the identity operator on $L^2(\mathbb{R})$. Moreover, $C_\pm$ act in Fourier representation as multiplication by the respective functions $\chi_{(0,\infty)}$ and $-\chi_{(-\infty,0)}$, where χ_A denotes the characteristic function of the set A. Taking limits in (3.35) we recover

$$\mathbf{M}_+(x, z) = \mathbb{I} + C_+ \left(\mu w_x^+ + \mu w_x^-\right).$$

Since
$$\mathbf{M}_+ = \mu\left(I + w_x^+\right)$$

and
$$C_+(\mu w_x^+) = \mu w_x^+ + C_-\left(\mu w_x^+\right)$$

we conclude that
$$\mu = \mathbb{I} + \mathcal{C}_w(\mu) \tag{3.37}$$

where for a matrix-valued function h and $w = (w_+, w_-)$
$$\mathcal{C}_w(h) = C_+(hw_x^-) + C_-(hw_x^+). \tag{3.38}$$

The integral operator $\mathcal{C}_w$ is called the *Beals-Coifman integral operator*, and Eq. (3.37) is called the *Beals-Coifman integral equation*. For $r \in L^\infty(\mathbb{R}) \cap L^2(\mathbb{R})$, the operator $\mathcal{C}_w$ is a bounded operator on matrix-valued $L^2(\mathbb{R})$ functions; moreover, since the Beals-Coifman solutions are expected to have boundary values $\mathbf{M}_\pm$ with $\mathbf{M}_\pm(x, \cdot) - \mathbb{I}$ belonging to $L^2(\mathbb{R})$ (see Theorem 3.9), it is reasonable to impose the condition $\mu(x, \cdot) - \mathbb{I} \in L^2(\mathbb{R})$.

Proposition 3.11 *Suppose that $x \in \mathbb{R}$ and $r \in H_1^{1,1}(\mathbb{R})$. There exists a unique solution μ of the Beals-Coifman integral equation (3.37) with $\mu(x, \cdot) - \mathbb{I} \in L^2(\mathbb{R})$. Moreover, $\mu(x, \cdot) - \mathbb{I} \in H^1(\mathbb{R})$ with*

$$\left\| \frac{\partial \mu}{\partial \lambda}(x, \cdot) \right\|_{L^2} \lesssim \frac{\|r\|_{H^{1,0}}}{1 - \|r\|_{L^\infty}} \|\mu - 1\|_{L^2}. \tag{3.39}$$

where the implied constant depends only on x.

Proof Norm the matrix-valued functions on $L^2(\mathbb{R})$ by

$$\|F\|_{L^2}^2 = \int \left(|F_{11}(\lambda)|^2 + |F_{12}(\lambda)|^2 + |F_{21}(\lambda)|^2 + |F_{22}(\lambda)|^2\right) d\lambda$$

i.e., $\|F\|_{L^2}^2 = \int |F(\lambda)|^2 d\lambda$ where $|A|$ is the Frobenius norm on 2×2 matrices. Since $\|C_\pm\|_{L^2 \to L^2} = 1$ and $\|w_x^\pm\|_{L^\infty} = \|r\|_{L^\infty}$, it follows that

$$\|\mathcal{C}_w\|_{L^2 \to L^2} = \|r\|_{L^\infty} < 1.$$

Hence $(I - \mathcal{C}_w)^{-1}$ exists as a bounded operator on $L^2(\mathbb{R})$. Setting $\mu^\sharp = \mu - \mathbb{I}$, (3.37) becomes

$$\mu^\sharp = \mathcal{C}_w(\mathbb{I}) + \mathcal{C}_w(\mu^\sharp) \tag{3.40}$$

where $\mathcal{C}_w(\mathbb{I}) \in L^2$ since $r \in L^2$. Hence

$$\mu^\sharp = (I - \mathcal{C}_w)^{-1} \mathcal{C}_w \mathbb{I} \tag{3.41}$$

and $\mu = \mathbb{I} + \mu^\sharp$. Any two solutions μ_1 and μ_2 with $\mu_1 - \mathbb{I}, \mu_2 - \mathbb{I} \in L^2$ satisfy $(\mu_1 - \mu_2) = \mathcal{C}_w(\mu_1 - \mu_2)$ so that $\mu_1 = \mu_2$.

Next, we show that, for each x, $\mu(x, \cdot) - \mathbb{I} \in H^1(\mathbb{R})$, following closely the argument in [17, §3]. First suppose that $r \in C_0^\infty(\mathbb{R})$. In (3.40), the first right-hand term is actually a smooth function since $C_\pm$ preserve Sobolev spaces. An argument with difference quotients shows that the derivative of $\mu^\sharp$ with respect to λ exists as a vector in L^2 and

$$\frac{\partial \mu^\sharp}{\partial \lambda} = \mathcal{C}_{dw/d\lambda} \mathbb{I} + \mathcal{C}_{dw/d\lambda} \mu^\sharp + \mathcal{C}_w\left(\frac{\partial \mu^\sharp}{\partial \lambda}\right) \tag{3.42}$$

where

$$\mathcal{C}_{dw/d\lambda} h = C_+\left(h \frac{\partial w_x^+}{\partial \lambda}\right) + C_-\left(h \frac{\partial w_x^-}{\partial \lambda}\right).$$

To obtain an effective bound on $\partial \mu^\sharp / \partial \lambda$ we first note that

$$\left\|\frac{\partial w_x^\pm}{\partial \lambda}\right\|_{L^2} \leq c \|r\|_{H^{1,0}}$$

where c depends linearly on x. Next, we recall that for any $\varepsilon > 0$,

$$\|f\|_{L^\infty} \lesssim \varepsilon \|f'\|_{L^2} + \varepsilon^{-1} \|f\|_{L^2}.$$

It now follows from (3.42) that

$$\|(\mu^\sharp)'\|_{L^2} \leq c \|\mu^\sharp\|_{L^\infty} \|r\|_{H^{1,0}} + \|r\|_{L^\infty} \left\|\frac{\partial \mu^\sharp}{\partial \lambda}\right\|_{L^2}$$

$$\leq c\varepsilon \|(\mu^\sharp)'\|_{L^2} \|r\|_{H^{1,0}} + c\varepsilon^{-1} \|\mu^\sharp\|_{L^2} \|r\|_{H^{1,0}} + \|r\|_{L^\infty} \left\|\frac{\partial \mu^\sharp}{\partial \lambda}\right\|_{L^2}$$

For ε with $c\varepsilon + \|r\|_{L^\infty} < 1$ we conclude that (3.39) holds if $r \in C_0^\infty(\mathbb{R})$.

To complete the argument, suppose $r \in H^{1,0}(\mathbb{R})$ and $\{r_n\}$ is a sequence from $C_0^\infty(\mathbb{R})$ with $r_n \to r$ in $H^{1,0}(\mathbb{R})$. Let $\mu_n^\sharp$ correspond to r_n. Using the second resolvent formula, it is easy to see that $(I - \mathcal{C}_{w_n})^{-1} \to (I - \mathcal{C}_w)^{-1}$ as operators on L^2 so that $\mu_n^\sharp \to \mu^\sharp$ as $n \to \infty$. It is now easy to see that $\mu^\sharp$ has a bounded weak L^2 derivative obeying (3.39). □

For subsequent use, we note a simple but very important consequence of the proof of Proposition 3.11.

Proposition 3.12 (Vanishing Theorem for RHP 3.10) *Suppose that $r \in H^{1,0}(\mathbb{R})$ with $\|r\|_{L^\infty} < 1$ and $x \in \mathbb{R}$. Suppose that $\mathbf{n}(x,z) : \mathbb{C} \setminus \mathbb{R} \to SL(2,\mathbb{C})$ so that*

(i) $\mathbf{n}(x,z)$ *is analytic in* $\mathbb{C} \setminus \mathbb{R}$,
(ii) $\mathbf{n}(x,z)$ *has continuous boundary values* $\mathbf{n}_\pm(x,\lambda)$ *on* $\mathbb{R}$,
(iii) $\mathbf{n}_\pm(x,\cdot) \in L^2(\mathbb{R})$,
(iv) *The jump relation*

$$\mathbf{n}_+(x,\lambda) = \mathbf{n}_-(x,\lambda) e^{-i\lambda x \, \mathrm{ad}\, \sigma_3} \mathbf{V}(\lambda)$$

holds, with $\mathbf{V}(\lambda)$ *as in RHP 3.10.*

Then $\mathbf{n}(x,z) \equiv \mathbf{0}$.

Proof Given such a function $\mathbf{n}(x,z)$, let

$$\nu(x,\lambda) = \mathbf{n}_+(x,\lambda) \left(I + w_x^+(\lambda)\right)^{-1} = \mathbf{n}_-(x,\lambda) \left(I - w_x^-(\lambda)\right)^{-1}.$$

Mimicking the arguments that lead to the Beals-Coifman integral equation we conclude that

$$\nu = \mathcal{C}_w \nu$$

which shows that $\nu \equiv 0$ since $\|\mathcal{C}_w\|_{L^2 \to L^2} < 1$. It now follows from (3.35) with μ replaced by ν that $\mathbf{n}(x,z) \equiv 0$. □

A piecewise analytic function $\mathbf{n}(x,z)$ satisfying (i)–(iv) above is called a *null vector* for RHP 3.10. Proposition 3.12 asserts that RHP 3.10 has no nontrivial null vectors.

Since the solution of RHP 3.10 is unique, any transformation of $\mathbf{M}$ that leaves the solution space invariant is a symmetry of the solution. Since

$$\sigma_1 \overline{\mathbf{V}(\lambda)} \sigma_1 = \mathbf{V}(\lambda)^{-1},$$

it follows that the map

$$\mathbf{M}(x,z) \mapsto \sigma_1 \overline{\mathbf{M}(x,\overline{z})} \sigma_1$$

preserves the solution space. This symmetry implies that

$$\mu(x,\lambda) = \sigma_1 \overline{\mu(x,\lambda)} \sigma_1. \tag{3.43}$$

We now define

$$\mathbf{M}_+(x,\lambda) = \mu(x,\lambda)\left(I + w_x^+(\lambda)\right), \quad \mathbf{M}_-(x,\lambda) = \mu(x,\lambda)\left(I - w_x^-(\lambda)\right).$$

The next proposition shows that $\mathbf{M}(x, z)$ are Beals-Coifman solutions for a potential q determined by the asymptotics of $\mathbf{M}(x, z)$.

Proposition 3.13 *Suppose that $r \in H_1^{1,1}(\mathbb{R})$, denote by $\mathbf{M}(x, z)$ the unique solution of RHP 3.10, and by $\mathbf{M}_\pm(x, \lambda)$ the boundary values of $\mathbf{M}(x, z)$. Then*

$$\frac{d}{dx}\mathbf{M}(x,z) = -iz\,\mathrm{ad}\,\sigma_3\,(\mathbf{M}) + \mathbf{Q}_1(x)\mathbf{M}(x,z)$$

where

$$\mathbf{Q}_1(x) = \frac{1}{2\pi}\,\mathrm{ad}\,\sigma_3\left(\int \mu(x,s)\left(w_x^+(s) + w_x^-(s)\right)\right)ds \tag{3.44}$$

takes the form

$$\mathbf{Q}_1(x) = \begin{pmatrix} 0 & q(x) \\ \overline{q(x)} & 0 \end{pmatrix}.$$

Proof First, by differentiating the solution formula $\mu - \mathbb{I} = (I - \mathcal{C}_w)^{-1}\mu$ and using the fact that $r \in H^{0,1}(\mathbb{R})$, it is easy to see that $(d\mu/dx)(x, \cdot) \in L^2(\mathbb{R})$. It follows from the representation

$$\mathbf{M}_\pm(x,\lambda) - \mathbb{I} = C_\pm\left(\mu(x,\cdot)\left(w_x^+(\cdot) + w_x^-(\cdot)\right)\right)(\lambda)$$

the same is true for $\mathbf{M}^\pm(x,\lambda) - \mathbb{I}$. We will differentiate the jump relation for $\mathbf{M}$ and use Proposition 3.12. From the jump relation for $\mathbf{M}_\pm$, we have

$$\left(\frac{d\mathbf{M}_+}{dx} + i\lambda\,\mathrm{ad}\,\sigma_3(\mathbf{M}_+)\right) = \left(\frac{d\mathbf{M}_-}{dx} + i\lambda\,\mathrm{ad}\,\sigma_3(\mathbf{M}_-)\right)e^{-ix\lambda\,\mathrm{ad}\,\sigma_3}\mathbf{V}(\lambda) \tag{3.45}$$

where we used $\mathbf{V} = \mathbf{M}_-^{-1}\mathbf{M}^+$ and the Leibniz rule for the derivation $A \mapsto \mathrm{ad}\,\sigma_3(A)$. Using the identity

$$i\lambda(C_\pm f)(\lambda) = -\frac{1}{2\pi}\int f(s)\,ds + C_\pm\left((\cdot)f(\cdot)\right)(\lambda) \tag{3.46}$$

and the fact that $r \in H^{1,1}(\mathbb{R})$, we see that

$$i\lambda\,\mathrm{ad}\,\sigma_3(\mathbf{M}_\pm) + \mathbf{Q}_1(x) \in L^2(\mathbb{R})$$

where $\mathbf{Q}_1$ is the bounded continuous function of x given by (3.44). We conclude that

$$\mathbf{n}(x, z) := \frac{d}{dx}\mathbf{M}(x, z) + i\lambda \operatorname{ad} \sigma_3(\mathbf{M}(x, z)) - \mathbf{Q}_1(x)\mathbf{M}(x, z)$$

is a null vector for RHP (3.10) for each x, hence identically zero by Proposition 3.12. Finally, the diagonal components of $\mathbf{Q}_1$ are zero since $\mathbf{Q}_1$ lies in the range of $\operatorname{ad} \sigma_3(\cdot)$, and $(\mathbf{Q}_1)_{21} = \overline{(\mathbf{Q}_1)_{12}}$ owing to the symmetry (3.43). □

Tracing through the definitions we obtain the reconstruction formula

$$q(x) = -\frac{1}{\pi}\int \overline{r(s)}e^{-2ixs}\mu_{11}(x, s)\, ds \qquad (3.47)$$

which together with RHP 3.10 defines the inverse scattering map $\mathcal{I} : r \to q$.

Remark 3.14 The Fréchet derivative of the map $\mathcal{I}$ at $r = 0$ is clearly the map

$$q(x) = -\frac{1}{\pi}\int e^{-2ix\lambda}\overline{r(\lambda)}\, d\lambda.$$

This map is the inverse map for the Fréchet derivative of $\mathcal{R}$ (see Remark 3.4).

In the sequel, it will be important to know that

$$\frac{d}{dx}\mu = -i\lambda \operatorname{ad} \sigma_3(\mu) + \mathbf{Q}_1(x)\mu. \qquad (3.48)$$

This is a simple consequence of Proposition 3.13 and the Leibniz rule for $\operatorname{ad} \sigma_3$ (see Exercise 3.14).

We'll first show that $q \in H^{1,1}(\mathbb{R})$, and then show that the map $r \mapsto q$ is a locally Lipschitz continuous map from $H_1^{1,1}(\mathbb{R})$ to $H^{1,1}(\mathbb{R})$. To aid the analysis, note that (3.37) has 11 and 12 components

$$\mu_{11}(x, \lambda) = 1 + C_-\left(\mu_{12}(x, \cdot)e^{2i(\cdot)x}r(\cdot)\right)(\lambda) \qquad (3.49)$$

$$\mu_{12}(x, \lambda) = -C_+\left(\mu_{11}(x, \cdot)e^{-2i(\cdot)x}\overline{r(\cdot)}\right) \qquad (3.50)$$

so that $\mu_{11} - 1 \in \operatorname{Ran} C_-$. The following lemma [17, Lemma 3.4] will play a critical role.

Lemma 3.15 *Suppose that $r \in H^{1,0}(\mathbb{R})$. For any $x > 0$, the estimates*

$$\left\|C_+ e^{-2ix(\cdot)}\overline{r(\cdot)}\right\|_{L^2} \lesssim \|r\|_{H^{1,0}} (1 + x^2)^{-1/2},$$
$$\left\|C_- e^{2ix(\cdot)}r(\cdot)\right\|_{L^2} \lesssim \|r\|_{H^{1,0}} (1 + x^2)^{-1/2} \qquad (3.51)$$

hold.

Proof By Plancherel's theorem and the fact that C_+ acts in Fourier transform representation as multiplication by $\chi_+(\xi) := \chi_{(0,\infty)}(\xi)$, we may estimate

$$\left\| C_+ e^{-2ix(\cdot)} \overline{r(\cdot)} \right\|_{L^2} = \left\| \chi_+ \widehat{\overline{r}}(\cdot + x) \right\|_{L^2}$$

$$\lesssim (1 + |x|^2)^{-1/2} \|\widehat{\overline{r}}\|_{H^{0,1}}$$

where in the last step we used $(1 + |x|^2)^{1/2}(1 + |\xi + x|^2)^{-1/2} \leq 1$ for $x > 0$ and $\xi > 0$. The other proof is similar. $\square$

Using Lemma 3.15, we can obtain "one-sided" control over the inverse scattering map.

Proposition 3.16 *Suppose that* $r \in H_1^{1,1}(\mathbb{R})$. *Then* q *as defined by (3.47) belongs to* $H^{1,1}(\mathbb{R}^+)$, *and the map* $r \mapsto q$ *is locally Lipschitz continuous from* $H_1^{1,1}(\mathbb{R})$ *to* $H^{1,1}(\mathbb{R}^+)$.

Proof We write (3.47) as $q(x) = q_0(x) + q_1(x)$ where

$$q_0(x) = -\frac{1}{\pi} \int \overline{r(s)} e^{-2isx} \, ds$$

and

$$q_1(x) = -\frac{1}{\pi} \int \overline{r(s)} e^{-2ixs} (\mu_{11}(x, s) - 1) \, ds.$$

Clearly, $q_0 \in H^{1,1}(\mathbb{R})$ with the correct continuity so it suffices to study $q_1(x)$. From (3.49) we may write

$$q_1(x) = -\frac{1}{\pi} \int C_+ \left(\overline{r(\cdot)} e^{-2ix(\cdot)} \right) (s) (\mu_{11}(x, s) - 1) \, ds$$

for $x < 0$, where we used the facts that $\mu_{11} - 1 \in \text{Ran } C_-$, that

$$\int (C_- f)(s)(C_- f(s)) \, ds = 0$$

and that $C_+ - C_- = I$. From the solution formula (3.41) we have the estimate

$$\|m_{11}(x, \cdot) - 1\|_{L^2} \leq \frac{\|\mathcal{C}_w \mathbb{I}\|_{L^2}}{1 - \|r\|_\infty} \leq (1 + x^2)^{-1/2} \frac{\|r\|_{H^{1,0}}}{1 - \|r\|_\infty}$$

where in the last step we used (3.51). By this estimate, Lemma 3.15, and the Schwartz inequality, we conclude that for $x > 0$,

$$|q_1(x)| \lesssim \frac{1}{(1+x^2)} \frac{\|r\|_{H^{1,0}}}{1-\|r\|_\infty}$$

so that in particular $q_1 \in H^{0,1}(\mathbb{R}^+)$.

To show that $q_1' \in L^2$, we differentiate and use (3.48) to conclude that

$$q_1'(x) = -q(x)\left(\frac{1}{\pi}\int \overline{r(s)}e^{-2isx}\mu_{21}(x,s)\,ds\right). \tag{3.52}$$

Since $r \in L^2$, $\mu_{21}(x,\lambda) = \overline{\mu_{12}(x,\lambda)}$, and $\mu_{12}(x,\cdot) \in L^2$ with bounds uniform in x, we can bound the integral uniformly in x by the Schwartz inequality and conclude that $q_1' \in L^2(\mathbb{R}^+)$ as required.

To obtain the local Lipschitz continuity, first note that $r \mapsto q_0$ has the required mapping properties, so it suffices to consider the map $r \mapsto q_1$. To show that $r \mapsto q_1$ is locally Lipschitz continuous into $H^{0,1}(\mathbb{R}^+)$, it suffices, by estimates already given, to show that $r \mapsto (1+|x|^2)^{1/2}(\mu_{11}(x,\cdot)-1)$ is locally Lipschitz continuous. It follows from (3.49)–(3.50) that

$$\mu_{11} = 1 - A_r \mu_{11}$$

where

$$(A_r h)(\lambda) = C_-\left(C_+\left(h(\diamond)e^{2i(\diamond)x}\overline{r(\diamond)}\right)(\cdot)e^{-2i(\cdot)x}r(\cdot)\right)(\lambda)$$

Since $r \in H_1^{1,1}(\mathbb{R})$, $A_r 1 \in L^2$ and the operator A_r is bounded from L^2 to itself with norm $\|r\|_\infty^2 < 1$ so that μ_{11} is given by the L^2-convergent Neumann series

$$\mu_{11} - 1 = \sum_{n=1}^\infty A_r^n(1)$$

The map $r \mapsto A_r^n(1)$ takes the form $F_n(r,\ldots,r,\bar{r},\ldots,\bar{r})$ where $F_n : \left(H_1^{1,1}(\mathbb{R})\right)^{2n} \to L^2(\mathbb{R})$ is a multilinear function obeying the bound

$$\|F_n(r_1,\ldots,r_{2n})\|_{L^2(\mathbb{R})} \leq (1+|x|^2)^{-1/2}\left(\prod_{i=1}^{2n-1}\|r_i\|_{L^\infty}\right)\|r_{2n}\|_{H^{1,0}}.$$

The required local Lipschitz continuity for $\mu_{11} - 1$ now follows as in the proof of Proposition 3.3.

To show that $r \mapsto q_1$ is locally Lipschitz from $H_1^{1,1}(\mathbb{R})$ to $H^{0,1}(\mathbb{R}^+)$, it suffices by (3.52) to show that $r \mapsto \mu_{21}(x,\cdot)$ is locally Lipschitz from $H_1^{1,1}(\mathbb{R})$ to $L^2(\mathbb{R})$ with bounds uniform in $x \in \mathbb{R}^+$. Since $\mu_{21} = \overline{\mu_{12}}$, we can use the continuity result for μ_{11} and (3.50) to obtain the necessary result. □

The results obtained so far show that the map $r \mapsto q$ is locally Lipschitz from $H_1^{1,1}(\mathbb{R})$ to $H^{1,1}(\mathbb{R}^+)$ and so gives "half" of the desired result. To obtain the full local Lipschitz continuity result, first note that, by trivial modifications of the proofs, we can show that $r \mapsto q$ is locally Lipschitz continuity from $H_1^{1,1}(\mathbb{R})$ to $H^{1,1}(c, \infty)$ for any $c \in \mathbb{R}$. To finish the analysis, we consider the Riemann-Hilbert problem satisfied by the "left" Beals-Coifman solutions from Theorem 3.9.

Riemann-Hilbert Problem 3.17 Given $r \in H_1^{1,1}(\mathbb{R})$ and $x \in \mathbb{R}$, find a function $\mathbf{M}(x, z) : \mathbb{C} \setminus \mathbb{R} \to SL(2, \mathbb{C})$ so that:

(i) $\mathbf{M}(x, z)$ is analytic in $\mathbb{C} \setminus \mathbb{R}$ for each x,
(ii) $\mathbf{M}(x, z)$ has continuous boundary values $\mathbf{M}_\pm(x, \lambda)$ on $\mathbb{R}$,
(iii) $\mathbf{M}^\pm(x, \lambda) - \mathbb{I}$ in $L^2(\mathbb{R})$, and
(iv) The jump relation

$$\mathbf{M}_+(x, \lambda) = \mathbf{M}_-(x, \lambda) e^{-i\lambda x \, \mathrm{ad}\,\sigma_3} \mathbf{V}(\lambda)$$

holds, where

$$\mathbf{V}(\lambda) = \begin{pmatrix} 1 & -\overline{\check{r}(\lambda)} \\ \check{r}(\lambda) & 1 - |\check{r}(\lambda)|^2 \end{pmatrix}.$$

The associated reconstruction formula is:

$$\check{q}(x) = \lim_{z \to \infty} 2iz \, (\mathbf{M})_{12}(x, z). \tag{3.53}$$

We can analyze RHP 3.17 in much the same way as RHP 3.10 and prove:

Proposition 3.18 *Suppose that $r \in H_1^{1,1}(\mathbb{R})$. Then the map $\check{r} \mapsto \check{q}$ is locally Lipschitz continuous from $H_1^{1,1}(\mathbb{R})$ to $H^{1,1}(\mathbb{R}^-)$.*

Indeed, the same result holds true of $H^{1,1}(\mathbb{R}^-)$ is replaced by $H^{1,1}((-\infty, c))$. Since the map $r \mapsto \check{r}$ is locally Lipschitz continuous, it remains only to prove that $q = \check{q}$. To do so we recall that the respective solutions $\mathbf{M}^r(x, z)$ and $\mathbf{M}^\ell(x, z)$ of RHP's 3.10 and 3.17 are related by

$$\mathbf{M}^\ell(x, z) = \mathbf{M}^r(x, z) \begin{pmatrix} F(z) & 0 \\ 0 & F(z)^{-1} \end{pmatrix}$$

where $F(z)$ was defined in (3.33) and show to satisfy $F(z) = 1 + \mathcal{O}(1/z)$ as $z \to \infty$. It follows that

$$\lim_{z \to \infty} 2iz \left(\mathbf{M}^\ell\right)_{12}(x, z) = \lim_{z \to \infty} 2iz \left(\mathbf{M}^r\right)_{12}(x, z)$$

so that

$$q(x) = \breve{q}(x).$$

Propositions 3.16, and 3.18, and these observations prove:

Proposition 3.19 *The map $r \mapsto q$ defined by RHP 3.17 and the reconstruction formula (3.47) defines a locally Lipschitz continuous map from $H^{1,1}_1(\mathbb{R})$ to $H^{1,1}(\mathbb{R})$.*

To finish the proof of Theorem 2.5, it remains to show that the maps $\mathcal{R}$ and $\mathcal{I}$ are one-to-one and mutual inverses.

Let $r \in H^{1,1}(\mathbb{R})$. By solving RHP 3.10 we construct the unique Beals-Coifman solutions for the potential $q = \mathcal{I}(r)$. From the Riemann-Hilbert problem satisfied by the solutions, we read off that q has scattering transform $\mathcal{R}(q) = r$, showing that $\mathcal{R} \circ \mathcal{I}$ is the identity map on $H^{1,1}_1(\mathbb{R})$.

Next, we claim that $\mathcal{R}$ is one-to-one. Suppose that $q_1, q_2 \in H^{1,1}(\mathbb{R})$ and $\mathcal{R}(q_1) = \mathcal{R}(q_2) = r$. If $\mathbf{M}^{(1)}(x, z)$ and $\mathbf{M}^{(2)}(x, z)$ are the respective Beals-Coifman solutions for q_1 and q_2, each satisfies RHP 3.10 and so the difference satisfies a homogeneous RHP as in Proposition 3.12. It now follows from Proposition 3.12 that $\mathbf{M}^{(1)}(x, z) = \mathbf{M}^{(2)}(x, z)$. Since q can be recovered from large-z asymptotics of $\mathbf{M}(x, z)$, it now follows that $q_1 = q_2$.

3.4 Solving NLS for Schwartz Class Initial Data

In this subsection we prove Theorem 2.4. We will use the complete integrability of NLS in the following form: a smooth function $q(x, t)$ solves NLS if and only the overdetermined system (2.5) admits a 2×2 matrix-valued fundamental solution $\Psi(x, t, \lambda)$. Recall that a joint solution $\Psi(x, t, \lambda)$ is a fundamental solution if $\det \Psi(x, t, \lambda) > 0$ for all (x, t). Given such a fundamental solution, one can cross-differentiate the system (2.5) and equate coefficients of Ψ_{xt} and Ψ_{tx} to obtain (2.4).

We can also give a heuristic derivation of the evolution equations for the scattering data a and b from (2.5), assuming that $q(x, t) \in \mathscr{S}(\mathbb{R})$ as a function of x. Let $\Psi^+(x, t, \lambda)$ denote the Jost solution for $q(x, t)$. For each t,

$$\Psi^+(x, t, \lambda) \underset{x \to \infty}{\sim} e^{-i\lambda x \sigma_3}$$

and $\Psi_t(x, t, \lambda) \to 0$ as $x \to +\infty$. On the other hand,

$$\Psi^+(x, t, \lambda) \underset{x \to -\infty}{\sim} e^{-i\lambda x \sigma_3} T(\lambda, t),$$

where $T(\lambda)$ is given by (2.9) with $a = a(\lambda, t)$ and $b = b(\lambda, t)$. A joint solution of (2.5) must take the form $\Psi(x, t, \lambda) = \Psi^+(x, t, \lambda) C(t)$ for a matrix-valued function $C(t)$. From the second equation of (2.5) we obtain

$$(\Psi^+)_t C(t) + \Psi^+ C'(t) = -2i\lambda^2 \sigma_3 \Psi^+ C + o(1) \tag{3.54}$$

where $o(1)$ denotes terms that vanish as $x \to \pm\infty$ for each fixed t owing to the decay of q and its derivatives. Taking $x \to +\infty$ in (3.54), we obtain $C'(t) = -2i\lambda^2 \sigma_3 C(t)$ so that, normalizing to $C(0) = \mathbb{I}$, we have $C(t) = e^{-2i\lambda^2 \sigma_3 t}$. Taking $x \to -\infty$ in (3.54), we obtain

$$T'(\lambda)e^{-i\lambda x \sigma_3}C(t) + T(\lambda)e^{-i\lambda x \sigma_3}(-2i\lambda^2 \sigma_3)C(t) = -2i\lambda^2 \sigma_3 e^{i\lambda x \sigma_3} T(\lambda)C(t)$$

or

$$T'(\lambda) = -2i\lambda^2 \operatorname{ad} \sigma_3 T(\lambda)$$

which implies that

$$\dot{a}(\lambda, t) = 0, \quad \dot{b}(\lambda, t) = 4i\lambda^2 b(\lambda).$$

We consider the solution $\mathbf{M}(z; x, t)$ of RHP 2.3 and the recovered potential

$$q(x,t) = -\frac{1}{\pi}\int \overline{r_0(s)} e^{-2it\theta} \mu_{11}(s; x, t)\, ds \tag{3.55}$$

where r_0 is the scattering transform of the initial data q_0 and θ is the phase function (2.15). We denote by $\mathbf{M}_\pm(z; x, t)$ the boundary values of the solution to RHP 2.3. Note that, by construction, $\det \mathbf{M}_\pm(\lambda; x, t) = 1$ for all (λ, x, t). To prove that (3.55) solves the NLS equation, we will show that the functions

$$\Psi_\pm(\lambda; x, t) = \mathbf{M}_\pm(\lambda; x, t) e^{-i(\lambda x + 2\lambda^2 t)\sigma_3},$$

which again have determinant one, solve the overdetermined system (2.5).

To do this, it suffices to show that $\mathbf{M}_\pm$ solve

$$\begin{cases} \mathbf{M}_x = (-i\lambda \operatorname{ad} \sigma_3 + \mathbf{Q}_1)\mathbf{M} \\ \mathbf{M}_t = \left(-2i\lambda^2 \operatorname{ad} \sigma_3 + 2\lambda \mathbf{Q}_1 + \mathbf{Q}_2\right)\mathbf{M} \end{cases} \tag{3.56}$$

We will prove:

Theorem 3.20 *Suppose that $q_0 \in \mathscr{S}(\mathbb{R})$, let $r = \mathcal{R}(q)$, let $\mathbf{M}_\pm(\lambda; x, t)$ be the boundary values of the solution to RHP 2.3, and let q be given by (3.55). Then q is a classical solution of the defocussing NLS equation (2.4) with $q(x, 0) = q_0(x)$.*

Proof We have already shown that $\mathbf{M}_\pm$ solves the first of equations (3.56) in Proposition 3.13 by differentiating RHP 3.17 with respect to the parameter x and using Proposition 3.12, the vanishing theorem for RHP 3.17. We will show that the second equation in (3.56) holds by differentiating the time-dependent RHP 2.3 with respect to t and using an analogous vanishing theorem.

The jump matrix in RHP 2.3 may be written

$$\mathbf{V}(\lambda; x, t) = e^{-it\theta \,\mathrm{ad}\, \sigma_3} \mathbf{V}(\lambda), \quad \mathbf{V}(\lambda) = \begin{pmatrix} 1 - |r_0(\lambda)|^2 & -\overline{r_0(\lambda)} \\ r_0(\lambda) & 1 \end{pmatrix}$$

Differentiating the jump relation for $\mathbf{M}_\pm$ and using the Leibniz rule for $\mathrm{ad}\,\sigma_3(\,\cdot\,)$, we obtain

$$\left(\frac{\partial}{\partial t} + 2i\lambda^2 \,\mathrm{ad}\,\sigma_3\right) \mathbf{M}_+(\lambda; x, t) = \left(\frac{\partial}{\partial t} + 2i\lambda^2 \,\mathrm{ad}\,\sigma_3\right) \mathbf{M}_-(\lambda; x, t) \mathbf{V}(\lambda; x, t)$$

We will show that $\partial \mathbf{M}_\pm/\partial t$ and $2i\lambda^2 \,\mathrm{ad}\,\sigma_3(\mathbf{M}_\pm) - 2\lambda \mathbf{Q}_1 \mathbf{M}_\pm - \mathbf{Q}_2 \mathbf{M}_\pm$ are L^2 boundary values of functions analytic in $\mathbb{C} \setminus \mathbb{R}$, so that

$$\mathbf{n}_\pm(\lambda; x, t) := \left(\frac{\partial}{\partial t} + 2i\lambda^2 \,\mathrm{ad}\,\sigma_3 - 2\lambda \mathbf{Q}_1 - \mathbf{Q}_2\right) \mathbf{M}_\pm$$

satisfy the hypothesis of Proposition 3.12 for each t. It will then follow that the functions $\mathbf{M}_\pm(\lambda; x, t)$ satisfy the second of equations (3.56), showing that $q(x, t)$ is a classical solution of NLS. It follows from Theorem 2.5 that $q(x, 0) = q_0(x)$, so that $q(x, t)$ satisfies the initial value problem.

It remains to show that $\partial \mathbf{M}_\pm/\partial t$ and $2i\lambda^2 \,\mathrm{ad}\,\sigma_3(\mathbf{M}_\pm) - 2\lambda \mathbf{Q}_1 \mathbf{M}_\pm - \mathbf{Q}_2 \mathbf{M}_\pm$ have the required properties. This is accomplished in Lemmas 3.21 and 3.22 below. □

In what follows we will write $h_\pm \in \partial C(L^2)$ for a pair of L^2 functions (h_-, h_+) if $h_\pm$ are the boundary values of a function h analytic in $\mathbb{C} \setminus \mathbb{R}$. In this language, conditions (i)–(iii) of Proposition 3.12 state that $\mathbf{n}_\pm \in \partial C(L^2)$.

Lemma 3.21 *Suppose that $r \in \mathscr{S}_1(\mathbb{R})$ and let $\mathbf{M}_\pm(x, t, \lambda)$ be boundary values of the unique solution of the RHP 2.3. Then $\partial \mathbf{M}_\pm/\partial t \in \partial C(L^2)$.*

Proof First we study $\partial \mu/\partial t$ where μ solves the Beals-Coifman integral equation

$$\mu = \mathbb{I} + \mathcal{C}_{w_{x,t}} \mu \tag{3.57}$$

where

$$w_{x,t}^\pm(\lambda) = e^{-it\theta \,\mathrm{ad}\,\sigma_3} w^\pm(\lambda)$$

and

$$\mathcal{C}_{w_{x,t}} h = C_-\left(h w_{x,t}^+\right) + C_+\left(h w_{x,t}^-\right).$$

Differentiating (3.57) we see that

$$\frac{\partial \mu}{\partial t} = \mathcal{C}_{\partial w_{x,t}/\partial t}(\mu) + \left(\mathcal{C}_{w_{x,t}} \frac{\partial \mu}{\partial t}\right),$$

Since $(I - \mathcal{C}_{w_{x,t}})$ is invertible, this equation can be solved to show that $\partial \mu/\partial t \in L^2(\mathbb{R})$ provided the inhomogeneous term

$$\mathcal{C}_{\partial w_{x,t}/\partial t}\mu = C_+\left(\mu \frac{\partial w_{x,t}^-}{\partial t}\right) + C_-\left(\mu \frac{\partial w_{x,t}^+}{\partial t}\right)$$

belongs to L^2 as a function of λ. Since $\mu - I \in L^2$ it suffices to show that $\partial w_{x,t}^\pm/\partial t \in L^\infty \cap L^2$. Since $\partial w_{x,t}^+/\partial t = i\theta r e^{it\theta}$ and $\partial w_{x,t}^-/\partial t = -i\theta \bar{r} e^{-it\theta}$ and θ is a quadratic polynomial in λ, this follows from the fact that $r \in \mathscr{S}(\mathbb{R})$.

Since $\mathbf{M}_\pm - I = C_\pm\left(\mu(w_{x,t}^+ + w_{x,t}^-)\right)$ we have

$$\frac{\partial \mathbf{M}_\pm}{\partial t} = C_\pm\left[\frac{\partial \mu}{\partial t}\left(w_{x,t}^- + w_{x,t}^+\right) + \mu\left(\frac{\partial w_{x,t}^-}{\partial t} + \frac{\partial w_{x,t}^+}{\partial t}\right)\right]$$

It follows from the facts that $\partial \mu/\partial t \in L^2$ and $r \in \mathscr{S}_1(\mathbb{R})$ that the expression in square brackets is an L^2 function. This shows that $\partial \mathbf{M}_\pm/\partial t \in \partial C(L^2)$.

$\square$

In the proof of the next lemma, we will make use of the following large-z asymptotic expansion for the right-normalized Beals-Coifman solution for $r \in \mathscr{S}_1(\mathbb{R})$. Since the Beals-Coifman solution solves the Riemann-Hilbert problem, we have (compare (3.35))

$$\mathbf{M}(z; x) = \mathbb{I} + \frac{1}{2\pi i}\int \frac{1}{s-z} f(s; x)\, ds \tag{3.58}$$

where

$$f(s; x) = \mu(s; x)\left(w_x^-(s) + w_x^+(s)\right). \tag{3.59}$$

If $r \in \mathscr{S}_1(\mathbb{R})$ then, since $\mu(\cdot\,; x) - \mathbb{I} \in L^2(\mathbb{R})$ for each x, the asymptotic expansion

$$\mathbf{M}(z; x, t) \sim \mathbb{I} + \sum_{j \geq 0} \frac{m_j(x)}{z^{j+1}} \tag{3.60}$$

holds. Substituting (3.60) into the differential equation (3.24) we obtain the relations

$$i\,\text{ad}\,\sigma_3(m_0(x)) = \mathbf{Q}_1(x)$$
$$m_j'(x) = -i\,\text{ad}\,\sigma_3(m_{j+1}) + \mathbf{Q}m_j(x), \qquad j \geq 0$$

One can compute the coefficients $m_i(x)$ by deriving using these relations together with the boundary condition

$$\lim_{x \to +\infty} m_i(x) = 0.$$

Given all coefficients up to m_{j-1}, one first computes $\operatorname{ad}\sigma_3(m_j)$ and then uses $\operatorname{ad}\sigma_3(m_j)$ to find the diagonal of m_j. We will only need the following identities:

$$m_0(x) = \begin{pmatrix} \dfrac{i}{2}\int_{+\infty}^{x} |q(s)|^2\, ds & -\dfrac{i}{2}q(x) \\ \dfrac{i}{2}\overline{q(x)} & -\dfrac{i}{2}\int_{+\infty}^{x} |q(s)|^2\, ds \end{pmatrix}, \qquad (3.61)$$

$$-i\operatorname{ad}\sigma_3(m_1(x)) = \begin{pmatrix} 0 & \dfrac{i}{2}q(x)\int_{+\infty}^{x} |q(s)|^2\, ds \\ -\dfrac{i}{2}\overline{q(x)}\int_{+\infty}^{x} |q(s)|^2\, ds & 0 \end{pmatrix} \qquad (3.62)$$

$$+ \begin{pmatrix} 0 & -\dfrac{i}{2}q_x(x) \\ \dfrac{i}{2}\overline{q_x(x)} & 0 \end{pmatrix}.$$

We can identify

$$m_j(x) = -\frac{1}{2\pi i}\int s^{j-1} f(s;x)\, ds,$$

where f is given by (3.59), using Eq. (3.58). In the application, μ, $w_\pm$, f, m_j and f_j also depend parametrically on t.

Lemma 3.22 *Fix $r \in \mathscr{S}_1(\mathbb{R})$, and let $\mathbf{M}_\pm(x, t, \lambda)$ be boundary values of the unique solution to RHP 2.3. Then*

$$2i\lambda^2 \operatorname{ad}\sigma_3(\mathbf{M}_\pm) - 2\lambda \mathbf{Q}_1 \mathbf{M}_\pm - \mathbf{Q}_2 \mathbf{M}_\pm \in \partial C(L^2).$$

Proof In what follows we write $f_\pm \doteq g_\pm$ if $f_\pm - g_\pm \in \partial C(L^2)$. In this notation, we seek to prove that

$$2i\lambda^2 \operatorname{ad}\sigma_3(\mathbf{M}_\pm) \doteq 2\lambda \mathbf{Q}_1 \mathbf{M}_\pm + \mathbf{Q}_2 \mathbf{M}_\pm.$$

We compute

$$2i\lambda^2 \operatorname{ad} \sigma_3(\mathbf{M}_\pm) = 2i\lambda^2 \operatorname{ad} \sigma_3(\mathbf{M}_\pm - \mathbb{I})$$
$$= \operatorname{ad} \sigma_3 \left(2i\lambda^2 C_\pm f\right)$$

where f is given by (3.59) (but now μ and $w_x^\pm$ also depend on t). Using the identity

$$\lambda^2 (C_\pm h)(\lambda) = C_\pm \left((\cdot)^2 h(\cdot)\right)(\lambda) - \frac{\lambda}{2\pi i} \int h(s)\, ds - \frac{1}{2\pi i} \int s h(s)\, ds \quad (3.63)$$

and identifying $m_j(x)$ with the moments of f, we conclude that

$$\lambda^2 C_\pm f \doteq \lambda m_0(x,t) + m_1(x,t)$$

so that

$$2i\lambda^2 \operatorname{ad} \sigma_3(\mathbf{M}_\pm) \doteq 2\lambda Q_1 + 2i \operatorname{ad} \sigma_3(m^{(1)}) \quad (3.64)$$
$$\doteq 2\lambda Q_1 \mathbf{M}_\pm + 2\lambda Q_1(\mathbb{I} - \mathbf{M}_\pm) + 2i \operatorname{ad} \sigma_3(m_1)\mathbf{M}_\pm$$

where we used the facts that $\mathbb{I} - \mathbf{M}_\pm \doteq 0$ and that Q_1 is a bounded function of x. We compute the second right-hand term in (3.64):

$$2\lambda Q_1(\mathbb{I} - \mathbf{M}_\pm) \doteq -2Q_1 \lambda C_\pm f \quad (3.65)$$
$$\doteq -2Q_1 m_0$$
$$= \begin{pmatrix} -i|q|^2 & iq \int_\infty^x |q|^2 \\ -i\bar{q} \int_\infty^x |q|^2 & i|q|^2 \end{pmatrix}$$
$$\doteq \begin{pmatrix} -i|q|^2 & iq \int_\infty^x |q|^2 \\ -i\bar{q} \int_\infty^x |q|^2 & i|q|^2 \end{pmatrix} \mathbf{M}_\pm$$

Combining (3.62), (3.64), and (3.65), we conclude that

$$2i\lambda^2 \operatorname{ad} \sigma_3(\mathbf{M}_\pm) \doteq 2\lambda Q_1 \mathbf{M}_\pm + Q_2 \mathbf{M}_\pm$$

as claimed. □

Exercises for Sect. 3

Exercise 3.1 Show that if $f \in H^{0,1}(\mathbb{R})$, then $f \in L^p(\mathbb{R})$ for $1 \leq p \leq 2$ with

$$\|f\|_p \leq \left(\int (1+x^2)^{-p/(2-p)} dx\right)^{(2-p)/2p} \|f\|_{H^{0,1}}.$$

Exercise 3.2 Recall the space $H^1(\mathbb{R})$ defined in (3.2). Show that, if $f \in H^1(\mathbb{R})$, then f is bounded and Hölder continuous with $\|f\|_\infty \leq c\|f\|_{H^1}$ and $|f(x) - f(y)| \leq \|f\|_{H^1} |x-y|^{1/2}$. Show also that $H^1(\mathbb{R})$ is an algebra, i.e., if $f, g \in H^1(\mathbb{R})$, then $fg \in H^1(\mathbb{R})$.

Exercise 3.3 Prove the identities (3.46) and (3.63). You can either use the definition of $C_\pm$ as a limit of Cauchy integrals or use their definition as Fourier multipliers.

Exercises 3.4–3.5 outline a proof of local well-posedness for NLS viewed as the integral equation (3.1).

Exercise 3.4 Let $X = C((0, T); H^1(\mathbb{R}))$, the space of continuous $H^1(\mathbb{R})$-valued functions on $(0, T)$. Fix $q_0 \in H^1(\mathbb{R})$ and define a mapping $\Phi : X \to X$ by

$$\Phi(q) = e^{it\Delta} q_0 - i \int_0^t e^{i(t-s)\Delta} \left(2|q(s)|^2 q(s)\right) ds.$$

Using the result of Exercise 3.2, show that the estimates

$$\|\Phi(q)\|_X \leq \|q_0\|_{H^1} + 2c^2 T \|q\|_X^3$$

$$\|\Phi(q_1) - \Phi(q_2)\|_X \leq 2c^2 T \left(\|q_1\|_X + \|q_2\|_X\right)^2 \|q_1 - q_2\|_X$$

hold, where c is the constant in the inequality of Exercise 3.2.

Exercise 3.5 The solution of (3.1) is a fixed point for the map $\Phi(q)$. For $\alpha > 0$, denote by B_α the ball of radius α in X.

(i) Show that for $\|q_0\|_{H^1} < \alpha/2$ and $2c^2 T < 1/(8\alpha^2)$ (i.e., T sufficiently small depending on $\|q_0\|_{H^1}$), Φ maps B_α into itself.
(ii) Show that, under the same conditions, Φ is a contraction on B_α.

Conclude that, for T sufficiently small, Φ is a contraction on the ball of radius α and so has a unique fixed point.

Exercise 3.6 Prove Lemma 3.1. *Hints*: Note that $e^{it\Delta}$ is an isometry of $H^1(\mathbb{R})$. Use the fact that $H^1(\mathbb{R})$ is an algebra (see Exercise 3.2) to conclude that $|q_n(s)|^2 q_n(s) \to |q(s)|^2 q(s)$ in H^1 uniformly in $s \in [0, T]$, and take limits in (3.1).

Exercise 3.7 Show that

$$\operatorname{ad}\sigma_3 \begin{pmatrix} a & b \\ c & d \end{pmatrix} = \begin{pmatrix} 0 & 2b \\ -2c & 0 \end{pmatrix}$$

and conclude that $A \mapsto \operatorname{ad}\sigma_3(A)$ is a linear map on 2×2 matrices with eigenvalues 2, 0, and -2. Find the eigenvectors and show that

$$\exp(t \operatorname{ad}\sigma_3) \begin{pmatrix} a & b \\ c & d \end{pmatrix} = \begin{pmatrix} a & e^{2t}b \\ e^{-2t}c & d \end{pmatrix}.$$

Check that

$$\exp(t \operatorname{ad}\sigma_3)(A) = e^{t\sigma_3} A e^{-t\sigma_3}. \tag{3.66}$$

Exercise 3.8 Show that $\operatorname{ad}\sigma_3(\cdot)$ obeys the Leibniz rule

$$\operatorname{ad}\sigma_3(AB) = \operatorname{ad}\sigma_3(A)B + A \operatorname{ad}\sigma_3(B)$$

and use this to verify (3.45).

Exercise 3.9 Prove *Jacobi's formula for differentiation of determinants*:

$$\frac{d}{dx} \det \Phi(x) = \sum_{i=1}^{n} \det \begin{pmatrix} \Phi_{1,1}(x) & \Phi_{1,2}(x) & \cdots & \Phi_{1,n}(x) \\ \Phi_{2,1}(x) & \Phi_{2,2}(x) & \cdots & \Phi_{2,n}(x) \\ \vdots & \vdots & & \vdots \\ \Phi'_{i,1}(x) & \Phi'_{i,2}(x) & \cdots & \Phi'_{i,n}(x) \\ \vdots & \vdots & & \vdots \\ \Phi_{n,1}(x) & \Phi_{n,2}(x) & \cdots & \Phi_{n,n}(x) \end{pmatrix}.$$

Show that if we define the *adjugate matrix* of a nonsingular matrix A by

$$A(\operatorname{adj} A) = \det(A) I$$

(where I is the $n \times n$ identity matrix), then Jacobi's formula may be written

$$\frac{d}{dx} \det \Phi(x) = \operatorname{tr}\left(\operatorname{adj}(\Phi(x)) \frac{d\Phi}{dx}(x) \right).$$

Exercise 3.10 Using Jacobi's formula, show that if $\Psi(t)$ is a differentiable, $N \times N$ matrix-valued function and $\Psi'(t) = B(t)\Psi(t)$ for a traceless matrix $B(t)$, then $\det \Psi(t)$ is independent of t. *Hint*: recall that $\operatorname{Tr}(AB) = \operatorname{Tr}(BA)$ for any $n \times n$ matrices A, B.

Exercise 3.11 Show that, if Ψ_1 and Ψ_2 are 2×2 nonsingular matrix-valued solutions of $\mathcal{L}\psi = z\psi$, then $\Psi_2^{-1}\Psi_1$ is independent of x.

Exercise 3.12 Using the result of Exercise 3.11, show that (3.15) holds for any two nonsingular solutions $\mathbf{M}_1$ and $\mathbf{M}_2$ of (2.11) (see (3.66)).

Exercise 3.13 Show that the map $\Psi \mapsto \sigma_1 \overline{\Psi(x, \bar{z})} \sigma_1^{-1}$ preserves the solution space of $\mathcal{L}\psi = z\psi$.

Exercise 3.14 Using the fact that $M_\pm$ satisfy (2.11) for $z = \lambda$, show that the same is true of μ. You will need to use the Leibniz rule from Exercise 3.8 together with the fact that $(d/dx)w_x^+ = -i\lambda \operatorname{ad} \sigma_3(w_x^+)$.

4 The Defocussing DS II Equation

In this lecture we will solve the defocussing Davey-Stewartson equation by inverse scattering method. The original lecture in August 2017 was based on Perry's [38] earlier work, which solved the DS II equation for initial data in $H^{1,1}(\mathbb{R}^2)$. Subsequently, Nachman et al. [36] used the inverse scattering method to prove global well-posedness in $L^2(\mathbb{R})$. In this lecture we will "compromise" by solving DS II in the space $H^{1,1}(\mathbb{R}^2)$ but use some of the tools introduced in [36] to simplify the proof. In particular, we will avoid entirely the resolvent expansions and multilinear estimates which make the proof in [38] somewhat complicated.

The DS II equation is the nonlinear dispersive equation[4]

$$\begin{cases} i\partial_t q + 2\left(\partial_z^2 + \partial_{\bar{z}}^2\right)q + (g + \bar{g})q = 0, \\ \partial_{\bar{z}} g + 4\varepsilon \partial_z \left(|q|^2\right) = 0, \\ q(z, 0) = q_0(z) \end{cases} \quad (4.1)$$

where $\varepsilon = +1$ for the defocussing equation, and $\varepsilon = -1$ for the focusing equation, and

$$\partial_{\bar{z}} = \frac{1}{2}\left(\partial_{x_1} + i\partial_{x_2}\right) \quad \partial_z = \frac{1}{2}\left(\partial_{x_1} - i\partial_{x_2}\right). \quad (4.2)$$

We will describe the formal inverse scattering theory for either sign of ε, but only solve the defocussing case ($\varepsilon = +1$) for initial data in $H^{1,1}(\mathbb{R}^2)$. The DSII equation is the compatibility condition for the following system of equations:

---

[4]We have rescaled q to agree with the conventions of [36].

$$\begin{cases} \partial_{\bar{z}}\psi_1 = q\psi_2 \\ \partial_z \psi_2 = \varepsilon \bar{q}\psi_1 \end{cases} \quad (4.3)$$

$$\begin{cases} \partial_t \psi_1 = 2i\partial_z^2 \psi_1 + 2i(\partial_{\bar{z}}q) - 2iq\partial_{\bar{z}}\psi_2 + ig\psi_1 \\ \partial_t \psi_2 = -2i\partial_{\bar{z}}^2 \psi_2 - 2i\varepsilon(\partial_z \bar{q})\psi_1 + 2i\varepsilon \bar{q}\partial_z \psi_1 - i\bar{g}\psi_2 \end{cases} \quad (4.4)$$

Motivated by the Lax representation (4.3)–(4.4) for the defocussing ($\varepsilon = 1$) DS II equation and the formal inverse scattering theory of Sect. 4.3, we will establish the existence of a scattering transform $\mathcal{S} : H^{1,1}(\mathbb{R}^2) \to H^{1,1}(\mathbb{R}^2)$ associated to the linear system (4.3) which linearizes the defocussing DS II equation. Using (4.4), we will see that if $\mathbf{s}(t) = \mathcal{S}q(t)$ for a solution $q(t)$ of the defocussing DS II equation, then $\mathbf{s}(t)$ obeys the linear evolution equation

$$\dot{\mathbf{s}}(k, t) = 2i(k^2 + \overline{k}^2)\mathbf{s}(k, t).$$

We will show that the scattering transform $\mathcal{S}$ satisfies $\mathcal{S}^{-1} = \mathcal{S}$ so that a putative solution to the defocussing DS II equation is given by

$$q_{\text{inv}}(z, t) = \mathcal{S}\left(e^{\left(2it\left((\cdot)^2 + \overline{(\cdot)}^2\right)\right)} (\mathcal{S}q_0)(\cdot)\right)(z) \quad (4.5)$$

The mapping properties of $\mathcal{S}$ established in Sect. 4.4 imply that $q_{\text{inv}}(z, 0) = q_0$ and that $(t, q_0) \mapsto q_{\text{inv}}(\cdot, t; q_0)$ is a continuous map from $(-T, T) \times H^{1,1}(\mathbb{R}^2)$ to $H^{1,1}(\mathbb{R}^2)$ for any $T > 0$, Lipschitz continuous in q_0. We will then show that q_{inv} solves the DS II equation for initial data $q_0 \in \mathscr{S}(\mathbb{R}^2)$ by constructing solutions of the system (4.3)–(4.4), where $q = q_{\text{inv}}$, with prescribed asymptotic behavior. It will follow from Exercise 2.9 that q_{inv} solves the DS II equation for $q_0 \in \mathscr{S}(\mathbb{R}^2)$. The Lipschitz continuity of $\mathcal{S}$ and local well-posedness theory for the DS II equation then imply that q_{inv} solves the integral equation form (4.24) of DS II for initial data $q_0 \in H^{1,1}(\mathbb{R}^2)$.

To keep the exposition of reasonable length, we will take as given the results of Beals-Coifman [8–10] and Sung [39–41] that the scattering transform $\mathcal{S}$ maps $\mathscr{S}(\mathbb{R}^2)$ into itself. Our emphasis is on the estimates that extend the map $\mathcal{S}$ to $H^{1,1}(\mathbb{R}^2)$ which enable us to apply the formula (4.5) to initial data in this space. One can use the techniques developed in these lectures to give a simpler proof Sung's results, but we will not carry this out here.

4.1 Preliminaries

As already outlined in the first lecture, both the direct and inverse scattering transforms are defined via a system of $\bar{\partial}$ equations. In this subsection we collect some useful estimates on the solid Cauchy transform (see (4.7)), the Beurling transform (see (4.19)), and other useful integral operators.

The *Hardy-Littlewood-Sobolev inequality* plays a fundamental role in the analysis of $\bar{\partial}$ problems and also in the proof of dispersive estimates in the local well-posedness theory for the DS II equation. For a proof, see for example [34, Section 2.2]. A sharp constant for the Hardy-Littlewood-Sobolev inequality together with an explicit maximizer is given in [33]; see [24] for a simplified proof of the optimal inequality.

Theorem 4.1 (Hardy-Littlewood-Sobolev Inequality) *Suppose that $0 < \alpha < n$, $1 < p < q < \infty$, and*

$$\frac{1}{q} = \frac{1}{p} - \frac{\alpha}{n}.$$

If $f \in L^p(\mathbb{R}^n)$, the integral

$$(I_\alpha f)(x) = \int \frac{f(y)}{|x-y|^{n-\alpha}} \, dy$$

converges absolutely for a.e. x, and the estimate

$$\|I_\alpha(f)\|_{L^q} \lesssim_{n,p,\alpha} \|f\|_{L^p} \tag{4.6}$$

holds.

The *solid Cauchy transform* is the integral operator

$$\left(\partial_{\bar{z}}^{-1} f\right)(z) = \frac{1}{\pi} \int \frac{1}{z-w} f(w) \, dw \tag{4.7}$$

initially defined on $C_0^\infty(\mathbb{R}^2)$ and extended by density to $L^p(\mathbb{R}^2)$ for $p \in (1, 2)$ by (4.9). Proofs of the following fundamental estimates may be found, for instance, in [44, Chapter I.6] or [5, section 4.3]. Some are exercises at the end of this section. We leave the formulation of similar results for the conjugate solid Cauchy transform

$$\left(\partial_z^{-1} f\right)(z) = \frac{1}{\pi} \int \frac{1}{\bar{z}-\bar{w}} f(w) \, dw \tag{4.8}$$

to the reader.

1. Fractional integration and L^∞ estimates. Let $p \in (1, 2)$ and let p^* be the Sobolev conjugate exponent $(p^*)^{-1} = p^{-1} - 1/2$ for $n = 2$. Then, as a consequence of the Hardy-Littlewood-Sobolev inequality (4.6),

$$\left\|\partial_{\bar{z}}^{-1} f\right\|_{L^{p^*}} \lesssim_p \|f\|_{L^p}. \tag{4.9}$$

On the other hand, an easy argument with Hölder's inequality (Exercise 4.3) shows that for $1 < p < 2 < r < \infty$,

$$\left\| \partial_{\bar{z}}^{-1} f \right\|_{L^\infty} \lesssim_{q,r} \| f \|_{L^p(\mathbb{R}^2) \cap L^r(\mathbb{R}^2)}. \tag{4.10}$$

2. Hölder continuity and asymptotic behavior. If $p \in (2, \infty)$, if p' is the Hölder conjugate of p and if $f \in L^p \cap L^{p'}$, then $\partial_{\bar{z}}^{-1} f$ is continuous and

$$\left| \left(\partial_{\bar{z}}^{-1} f \right)(z) - \left(\partial_{\bar{z}}^{-1} f \right)(z') \right| \lesssim_p \| f \|_{L^p} |z - z'|^{1 - 2/p}. \tag{4.11}$$

Again assuming $f \in L^p \cap L^{p'}$,

$$\lim_{|z| \to \infty} \left(\partial_{\bar{z}}^{-1} f \right)(z) = 0.$$

Next, we consider the model operator

$$S : f \to \partial_{\bar{z}}^{-1}(qf) \tag{4.12}$$

which occurs in the analysis of the scattering transform. An important consequence of (4.9) is that for any $q \in L^2$ and any $p > 2$, the operator S is a bounded operator from L^p to itself with operator bound

$$\| S \|_{L^p \to L^p} \lesssim_p \| q \|_{L^2}, \tag{4.13}$$

so that $\ker_{L^p}(I - S)$ is trivial for $\| q \|_{L^2}$ sufficiently small.

The operator S is also a compact operator. Recall that a subset V of a metric space is called *precompact* if the closure of V is compact, and that a bounded operator A on a Banach space X is *compact* if A maps bounded subsets of X into precompact subsets of X. To prove that S is compact, we first discuss the Kolmogorov-Riesz theorem that characterizes compact subsets of $L^p(\mathbb{R}^n)$. Our discussion draws on [28, 29] which provides a very readable exposition of the history and proof of this theorem.

Recall that a metric space (X, d) is said to be *totally bounded* if, for any $\varepsilon > 0$, X admits a finite cover by ε-balls. A metric space is compact if and only if it is complete and totally bounded, and a subset of a metric space is precompact if and only if it is totally bounded.

Theorem 4.2 (Kolmogorov-Riesz) *A subset F of $L^p(\mathbb{R}^n)$ is totally bounded if, and only if:*

(i) *F is bounded,*
(ii) *(uniform decay) For every $\varepsilon > 0$ there is an $R > 0$ so that*

$$\int_{|x|\geq R} |f(x)|^p\, dx < \varepsilon^p \text{ for all } f \in F,$$

and

(iii) *(L^p-equicontinuity) For every $\varepsilon > 0$ there is a $\delta > 0$ so that for every $f \in F$ and every $h \in \mathbb{R}^n$ with $|h| < \delta$,*

$$\int_{\mathbb{R}^n} |f(x+h) - f(x)|^p\, dx < \varepsilon^p.$$

Lemma 4.3 *The operator $S : L^p(\mathbb{R}^2) \to L^p(\mathbb{R}^2)$ is compact for any $q \in L^2(\mathbb{R}^2)$ and any $p > 2$.*

Proof We need to show that, for any $p > 2$ and any bounded subset B of $L^p(\mathbb{R}^2)$, the set

$$\{Sf : f \in B\}$$

is totally bounded. Since S is a bounded operator by (4.13), (i) of Theorem 4.2 is obvious. To prove (ii), let χ_R denote the characteristic function of the set $\{x : |x| \leq R\}$. Then

$$(1 - \chi_{4R})(x)(Sf)(x) =$$
$$(1 - \chi_{4R})\frac{1}{\pi}\int \frac{1}{x-y}\left(\chi_R(y) + (1 - \chi_R(y))\right) q(y) f(y)\, dy \quad (4.14)$$

The first right-hand term of (4.14) is bounded by a constant times $R^{2/p-1} \|f\|_{L^p}$ where we used Hölder's inequality and the estimate

$$(1 - \chi_{4R})|x - y|^{-1}\chi_R(y) \lesssim R^{-1}.$$

The second right-hand term is bounded by a constant times $\|(1 - \chi_R)q\|_{L^2} \|f\|_{L^p}$. This shows that (ii) holds.

Finally, to show (iii), let $\varepsilon > 0$ be given. By Exercise 4.12 we may write $q = q_n + q_s$ where q_n is a smooth function of compact support and $\|q_s\|_{L^2} < \varepsilon$. We may write

$$(Sf)(x) = \frac{1}{\pi}\int \frac{1}{x-y}(q_n(y) + q_s(y))\, f(y)\, dy = (S_n f)(x) + (S_s f)(x)$$

and estimate $\|S_s f\|_{L^p} \lesssim_p \varepsilon \|f\|_{L^p}$ by (4.13). On the other hand, we may compute

$$(S_n f)(x+h) - (S_n f)(x) = h\int \frac{1}{x+h-y}\frac{1}{x-y} q_n(y) f(y)\, dy.$$

It follows from Young's inequality that

$$\|(S_n f)(\cdot + h) - (S_n f)(\cdot)\|_{L^p} \leq |h| \left\|(x+h)^{-1} x^{-1}\right\|_{L^{p'}} \|q_n\|_{L^p} \|f\|_{L^p}$$

$$\lesssim |h|^{1-2/p} \|q_n\|_{L^p} \|f\|_{L^p}.$$

Hence

$$\|(Sf)(\cdot + h) - (Sf)(\cdot)\|_{L^p} \lesssim_p \left(2\varepsilon + |h|^{1-2/p} \|q_n\|_{L^p}\right) \|f\|_{L^p}$$

which implies the required bound. □

Next, we will discuss an estimate on fractional integrals due to Nachman, Regev, and Tataru. Our Theorem 4.4 is a special case of [36, Theorem 2.3]; as we will see, this estimate plays a critical role in the analysis of the scattering transform. We will give a simple direct proof of Theorem 4.4 suggested by Adrian Nachman; in Exercise 4.4, we outline a complete proof of [36, Theorem 2.3] by the same method.

To state the estimate and introduce some key ingredients of Nachman's proof, we first recall that the Hardy-Littlewood maximal function for a locally integrable function f on $\mathbb{R}^n$ is given by

$$\mathcal{M}f(x) = \sup_{r>0} \left(\frac{1}{|B(x,r)|} \int_{B(x,r)} |f(y)|\, dy\right)$$

where $B(x, r)$ is the ball of radius r about $x \in \mathbb{R}^n$, and $|\cdot|$ denotes Lebesgue measure. The maximal function is a bounded sublinear operator from $L^p(\mathbb{R}^n)$ to itself for $p \in (1, \infty]$ so that

$$\|\mathcal{M}f\|_{L^p} \lesssim_p \|f\|_{L^p}, \quad p \in (1, \infty]. \tag{4.15}$$

In particular, if $f \in L^p$ for $p \in (1, \infty]$, then $(\mathcal{M}f)(x)$ is finite for almost every x.

Next, recall that an *approximate identity* is a family of nonnegative functions $K_t \in L^1(\mathbb{R}^n)$, indexed by $t \in (0, \infty)$, with

(i) $\int K_t(x)\, dx = 1$,
(ii) $|K_t(x)| \leq t^{-n}$, and
(iii) $|K_t(x)| \leq A t |x|^{-(n+1)}$.

It is not difficult to see that the estimate

$$|(K_t * f)(x)| \lesssim \mathcal{M}f(x)$$

holds, where the implied constant is independent of t. One example of an approximate identity is the Poisson kernel

$$P_t(x) = c_n \frac{t}{\left(|x|^2 + t^2\right)^{(n+1)/2}}$$

where c_n is chosen to normalize the integral of K_t to 1.

The action of the Poisson kernel by convolution may be viewed as the action of a Fourier multiplier with symbol $e^{-t|\xi|}$. That is, denoting by $\mathcal{F}$ the transform

$$(\mathcal{F}f)(\xi) = \int e^{-ix\cdot\xi} f(x)\, dx,$$

we have

$$\mathcal{F}[(P_t * f)](\xi) = e^{-t|\xi|} \widehat{f}(\xi). \tag{4.16}$$

Denote by $|D|^{-1}$ the Fourier multiplier with symbol $|\xi|^{-1}$. By the identity

$$|\xi|^{-1} = \int_0^\infty e^{-t|\xi|}\, dt$$

it follows that

$$|D|^{-1} f = \int_0^\infty (P_t * f)(x)\, dt.$$

We can now state and prove:

Theorem 4.4 ([36]) *Suppose that $p \in (1, 2]$ and $f \in L^p(\mathbb{R}^2)$. The estimate*

$$\left|\left(\partial_{\bar{z}}^{-1} f\right)(x)\right| \lesssim (\mathcal{M}f(x))^{1/2} \left(\mathcal{M}\widehat{f}(0)\right)^{1/2} \tag{4.17}$$

holds, where $\widehat{f}$ denotes the Fourier transform of f.

Proof (Suggested by Adrian Nachman) The Poisson kernel is an approximate identity so by standard theory

$$|(P_t * f)(x)| \lesssim \mathcal{M}f(x)$$

with the implied constant independent of $t > 0$. We now write

$$|D|^{-1} f(x) = \int_0^R (P_t * f)(x)\, dt + \frac{1}{(2\pi)} \int_R^\infty e^{i\xi\cdot x} e^{-t|\xi|} \widehat{f}(\xi)\, d\xi$$
$$= I_1(x) + I_2(x)$$

where in the second term we used (4.16). We may estimate

$$|I_1(x)| \lesssim R \mathcal{M} f(x)$$

$$|I_2(x)| \lesssim \int \frac{1}{|\xi|} e^{-R|\xi|} |\widehat{f}(\xi)| \, d\xi$$

$$\lesssim \sum_{j=-\infty}^{\infty} 2^j e^{-R2^j} 2^{-2j} \int_{2^{j-1} < |\xi| < 2^j} |\widehat{f}(\xi)| \, d\xi$$

$$\lesssim \left(\sum_{j=-\infty}^{\infty} 2^j e^{-R2^j} \right) \mathcal{M} \widehat{f}(0)$$

$$\lesssim R^{-1} \mathcal{M} \widehat{f}(0).$$

where we introduced a dyadic decomposition in the ξ variable. Thus

$$\left| |D|^{-1} f(x) \right| \lesssim R \mathcal{M} f(x) + R^{-1} \mathcal{M} \widehat{f}(0).$$

Optimizing in R, we obtain the desired bound (4.17). $\square$

We will usually use this estimate in the form

$$\left| \left(\partial_{\bar{z}}^{-1} e_k f \right)(x) \right| \lesssim (\mathcal{M} f(x))^{1/2} \left(\mathcal{M} \widehat{f}(k) \right)^{1/2} \qquad (4.18)$$

where

$$\widehat{f}(k) = \frac{1}{\pi} \int e_k(z) f(z) \, dz$$

is the natural Fourier transform in this setting.

This estimate is of particular importance because it captures, in a quantitatively precise way, the effect of the oscillatory factor e_k on the behavior of the fractional integral (In this context, see in particular Lemma 4.13 and the subsequent analysis of the scattering transform in Sect. 4.4; in [36], see particularly section 4). It replaces less precise estimates, based on integration by parts, that were used in [38] to capture the behavior of solutions as a function of k.

The *Beurling operator* is defined on $C_0^\infty(\mathbb{R}^2)$ as the principal value integral

$$(\mathbf{S} f)(z) = -\frac{1}{\pi} \lim_{\varepsilon \downarrow 0} \int_{|z-w| > \varepsilon} \frac{f(w)}{(z-w)^2} \, dw \qquad (4.19)$$

and extends to bounded operator on $L^p(\mathbb{R}^2)$ for all $p \in (1, \infty)$. It is an isometry on L^2. We define

$$\overline{\mathbf{S}} f = \overline{\mathbf{S} \overline{f}}. \qquad (4.20)$$

The Beurling operator has the property that

$$\mathbf{S}(\partial_{\bar{z}} f) = \partial_z f \qquad (4.21)$$

for functions $f \in C_0^\infty(\mathbb{R}^2)$. By density this extends property to functions $f \in W^{1,p}(\mathbb{R}^2)$ for $p \in (1, \infty)$. For a full discussion, see for example, [5, Chapter 4].

4.2 Local Well-Posedness

Next, we review the local well-posedness theory for the DS II equation due to Ghidaglia and Saut [26]. The results in this subsection hold for either sign of ε. We first recast (4.1) as an integral equation using the solution operator $V(t)$ for the linear problem

$$i\partial_t v + 2\left(\partial_z^2 + \partial_{\bar{z}}^2\right) v = 0 \qquad (4.22)$$

which is a linear dispersive equation. To formulate the integral equation, observe that (4.1) may be reformulated as a nonlinear Schrödinger-type equation with nonlocal nonlinearity:

$$\begin{cases} i\partial_t q + 2\left(\partial_z^2 + \partial_{\bar{z}}^2\right) q + 4\varepsilon \left(\mathbf{S}\left(|q|^2\right) + \bar{\mathbf{S}}\left(|q|^2\right)\right) q = 0, \\ q(z,0) = q_0(z) \end{cases} \qquad (4.23)$$

where $\mathbf{S}$ is the *Beurling operator* (4.19) and $\bar{\mathbf{S}}$ is the conjugate Beurling operator (4.20).

We will say that a function $q \in C([0, T], L_z^2(\mathbb{R}^2)) \cap L^4(\mathbb{R}_z^2 \times [0, T])$ solves the Cauchy problem (4.1) if q solves the integral equation

$$q(t) = V(t) q_0 + 4i\varepsilon \int_0^t V(t-s) \left[q(s) \left(\mathbf{S}(|q|^2)(s) + \bar{\mathbf{S}}(|q|^2)(s) \right) \right] ds \qquad (4.24)$$

as an integral equation in the space

$$X = C((0, T), L^2(\mathbb{R}^2)) \cap L^4(\mathbb{R}^2 \times (0, T)). \qquad (4.25)$$

This integral equation is motivated by Duhamel's formula (see Exercise 4.5) and makes sense in this space because of the Strichartz estimates discussed below. Ghidaglia and Saut [26, Theorem 2.1] prove:

Theorem 4.5 *For any $q_0 \in L^2(\mathbb{R}^2)$, there is a $T^* > 0$ and a unique solution $q(t)$ to (4.24) belonging to $C((0, T^*), L^2(\mathbb{R}^2)) \cap L^4(\mathbb{R}^2 \times (0, T^*))$ with $q(t) = q(0)$ and $\|q(t)\|_{L^2} = \|q_0\|_{L^2}$.*

Note that the proof of Theorem 4.5 is insensitive to the sign of ε, but does not guarantee global existence. This is to be expected since there are solutions of the focussing ($\varepsilon = -1$) DS II equation whose L^2-mass concentrates to a point in finite time [37].

The idea of the proof is to show that the mapping

$$\Phi(u) = V(t)q_0 + 4i\varepsilon \int_0^t V(t-s)\left[q(s)\left(\mathbf{S}(|q|^2)(s) + \overline{\mathbf{S}}(|q|^2)(s)\right)\right] ds \quad (4.26)$$

is a contraction on the space X for some $T > 0$ depending on the initial data q_0. One can reconstruct a complete proof by tracing through standard arguments used to show that the L^2-critical nonlinear Schrödinger equation

$$iu_t + \Delta u - |u|^2 u = 0$$

in two space dimensions is locally well-posed (see for example the text of Ponce and Linares [34, Section 5.1] or the original paper of Cazenave and Weissler [14]); dispersive estimates for $V(t)$ are essentially the same as those for the unitary group $\exp(it\Delta)$, while the nonlinear term in (4.23) is "morally cubic" owing to the fact that $\overline{\mathbf{S}}$ preserves $L^p(\mathbb{R}^2)$ for any $p \in (1, \infty)$. We will give an outline based on [34, Section 5.1].

To carry out the proof of Theorem 4.5, we will need the following Strichartz estimates on $V(t)$.

Proposition 4.6 *Let $V(t)$ be the solution operator for the linear equation (4.22). The following estimates hold.*

$$\|V(t)f\|_{L^4_{z,t}} \lesssim \|f\|_{L^2} \quad (4.27)$$

$$\left\|\int_{-\infty}^{\infty} V(t-s)g(s)\, ds\right\|_{L^4_{z,t}} \lesssim \|g\|_{L^{4/3}_{z,t}} \quad (4.28)$$

$$\left\|\int_{-\infty}^{\infty} V(t)g(t)\, dt\right\|_{L^2_z} \lesssim \|g\|_{L^{4/3}_{z,t}} \quad (4.29)$$

These estimates are consequences of the basic dispersive estimate

$$\|V(t)f\|_{L^\infty} \lesssim t^{-1} \|f\|_{L^1} \quad (4.30)$$

which follows from the representation of $V(t)f$ as a Fourier integral (Exercise 4.6). One can prove (4.27)–(4.29) for $V(t)$ by mimicking the proof of the analogous estimates for $V(t)$ replaced by $e^{it\Delta}$, the solution semigroup for the Schrödinger equation in two space dimensions, given in [34, Section 4.2]. The proofs are essentially identical since $\exp(it\Delta)$ and $V(t)$ both obey the basic dispersive estimate (4.30). The reader is asked to prove (4.28) in Exercise 4.7.

The first step in the proof of Theorem 4.5 is to show that the mapping (4.26) preserves a ball in X. Suppose that $\|q\|_X < \alpha$. Using the Strichartz estimate (4.29) on the second right-hand term of (4.26) and the fact that $V(t)$ is unitary on L^2 on the first right-hand term, we may estimate

$$\sup_{t\in(0,T)} \|\Phi(q(t))\|_{L^2} \lesssim \|q_0\|_{L^2} + T^{1/4} \|q\|_{L^4(\mathbb{R}^2\times(0,T))}^3$$

$$\lesssim \|q_0\|_{L^2} + T^{1/4}\alpha^3$$

where in the first step we used (4.29) and then used Hölder's inequality in the integration over t. Similarly, using (4.27) and (4.28) respectively on the first and second right-hand terms of (4.26), we obtain an estimate of the same form for $\|\Phi(q)\|_{L^4(\mathbb{R}^2\times(0,T))}$. Hence, for any q with $\|q\|_X < \alpha$,

$$\|\Phi(q)\|_X \lesssim \|q_0\|_{L^2} + T^{1/4}\alpha^3.$$

Choosing $\alpha \gtrsim \|q_0\|_{L^2}$ and $T^{1/4} \lesssim \alpha^{-2}$, we obtain that $\|\Phi(q)\|_X \leqslant \alpha/2$. Note that the 'guaranteed' time of existence decreases with the L^2 norm of the initial data.

The next step is to show that Φ is a contraction in the sense that

$$\|\Phi(q_1) - \Phi(q_2)\|_X \leqslant \frac{1}{2} \|q_1 - q_2\|_X$$

for T sufficiently small. Using the Strichartz estimates and the multilinearity of the map

$$(q_1, q_2, q_3) \mapsto q_1 \overline{\mathbf{S}}(q_2 q_3)$$

we have

$$\|\Phi(q_1) - \Phi(q_2)\|_X \lesssim T^{1/4}\alpha^2 \|q_1 - q_2\|_X$$

for any q_1, q_2 with $\|q_1\|_X, \|q_2\|_X < \alpha$. By shrinking T if necessary we can assure that Φ is a contraction, and hence (4.24) has a unique solution.

4.3 Complete Integrability

In this subsection we will sketch the formal inverse scattering theory for the DS II equations, tacitly assuming that $q(\,\cdot\,, t) \in \mathscr{S}(\mathbb{R}^2)$ and that the scattering transform $\mathbf{s} \in \mathscr{S}(\mathbb{R}^2)$, so that various asymptotic expansions make sense. Sung [39–41] proved rigorously that the scattering transform $q \mapsto \mathbf{s}$ maps $\mathscr{S}(\mathbb{R}^2)$ to itself. It follows from these mapping properties that the putative solution q_{inv} defined by (4.5) belongs to $C((-T, T); \mathscr{S}(\mathbb{R}^2))$ for any $T > 0$. These facts imply that the functions

$m^1(z,t,k)$ and $m^2(z,t,k)$ which we will construct below are bounded smooth functions with asymptotic expansions separately in z for fixed k or in k for fixed z.

Equation (4.1) is the compatibility condition for the following system of equations for unknowns $\psi_1(z,t,k)$ and $\psi_2(z,t,k)$:

$$\partial_{\bar{z}}\psi_1 = q\psi_2 \qquad (4.31\text{a})$$

$$\partial_z \psi_2 = \varepsilon \bar{q} \psi_1 \qquad (4.31\text{b})$$

$$\partial_t \psi_1 = 2i\partial_z^2 \psi_1 + 2i(\partial_{\bar{z}} q)\psi_2 - 2iq\partial_{\bar{z}}\psi_2 + ig\psi_1 \qquad (4.32\text{a})$$

$$\partial_t \psi_2 = -2i\partial_{\bar{z}}^2 \psi_2 - 2i\varepsilon(\partial_z \bar{q})\psi_1 + 2i\varepsilon \bar{q}\partial_z \psi_1 - i\bar{g}\psi_2 \qquad (4.32\text{b})$$

Cross-differentiating (4.31a) and (4.32a), assuming (ψ_1, ψ_2) is a joint solution and that ψ_1 and ψ_2 are linearly independent, one finds that the DSII equation

$$\begin{cases} iq_t + 2\left(\partial_z^2 + \partial_{\bar{z}}^2\right)q + (g + \bar{g})q = 0 \\ \partial_{\bar{z}} g + 4\varepsilon \partial_z \left(|q|^2\right) = 0 \end{cases} \qquad (4.33)$$

emerges as a compatibility condition (see Exercise 2.9).

To define and implement the scattering transform, we'll consider solutions of (4.31a)–(4.31b) with asymptotics specified by a complex parameter k: we seek solutions of the form[5]

$$\psi_1 = C_1(k,t)e^{ikz}m^1, \quad \psi_2 = C_2(k,t)e^{ikz}m^2$$

where for each fixed t and k,

$$\left(m^1(z,k,t), m^2(z,k,t)\right) \to (1,0) \text{ as } |z| \to \infty. \qquad (4.34)$$

A calculation similar to the one carried out in Sect. 3.4 shows that

$$C_1(k,t) = C_2(k,t) = e^{-2ik^2 t}.$$

We outline the computation in Exercise 4.8. In the new variables, we find

$$\partial_{\bar{z}} m^1 = qm^2 \qquad (4.35\text{a})$$

$$(\partial_z + ik) m^2 = \varepsilon \bar{q} m^1 \qquad (4.35\text{b})$$

---

[5] We follow the conventions of Nachman et al. [36] and denote the renormalized forms of ψ_1 and ψ_2 respectively by m^1 and m^2; the superscripts are not exponents!

$$\partial_t m^1 = 2i\left(\partial_z^2 + 2ik\partial_z\right)m^1 + 2i\left(\partial_{\bar{z}}q\right)m^2 - 2iq\partial_{\bar{z}}m^2 + igm^1 \tag{4.35c}$$

$$\partial_t m^2 = -2i\partial_{\bar{z}}^2 m^2 + 2ik^2 m^2 - 2i\varepsilon\left(\partial_z\bar{q}\right)m^1 + 2i\varepsilon\bar{q}(\partial_z + ik)m^1 \tag{4.35d}$$
$$- i\bar{g}m^2$$

As we will show (see Lemma 4.19), for each fixed time t and position z, the solutions of (4.35a)–(4.35b) obeying the asymptotic condition (4.34) also obey the dual equations

$$\begin{cases} \partial_{\bar{k}} m^1 = e_{-k}\mathbf{s}\overline{m^2} \\ \partial_{\bar{k}} m^2 = e_{-k}\mathbf{s}\overline{m^1} \\ \left(m^1(z,k,t), m^2(z,k,t)\right) \to (1,0) \text{ as } |k| \to \infty. \end{cases} \tag{4.36}$$

where

$$e_k(z) = e^{i(kz + \bar{k}\bar{z})}$$

and the scattering transform $\mathbf{s}(k,t)$ of $q(z,t)$ is defined by

$$m^2(z,k,t) = e_{-k}\frac{i}{z}\mathbf{s}(k,t) + \mathcal{O}\left(|z|^{-2}\right). \tag{4.37}$$

Assuming that $\mathbf{s}(\cdot, t) \in \mathscr{S}(\mathbb{R}^2)$ and that m^1 and m^2 are bounded, it follows from (4.36) that m^1 and m^2 have large-k asymptotic expansions of the form (see Exercise 4.2)

$$\begin{cases} m^1(z,k,t) \sim 1 + \sum_{j \geq 1} \dfrac{\alpha_j(z,t)}{k^j} \\ m^2(z,k,t) \sim \sum_{j \geq 1} \dfrac{\beta_j(z,t)}{k^j} \end{cases}$$

for each fixed z, t. Substituting these expansions into (4.35a)–(4.35b) shows that

$$q(z,t) = -i\varepsilon\overline{\beta_1(z,t)} \tag{4.38}$$

(see Exercise 4.9).

Thus, to recover $q(z,t)$, we need (i) an equation of motion for the scattering transform $\mathbf{s}(k,t)$ and (ii) a way of reconstructing $m^1(z,k,t)$ and $m^2(z,k,t)$ from the scattering transform $\mathbf{s}(k,t)$.

We can derive an equation of motion for **s** formally as follows. If we assume that $q(\cdot, t) \in \mathcal{S}(\mathbb{R}^2)$, we expect m^1 and m^2 to have large-z (differentiable) asymptotic expansions of the form

$$\begin{cases} m^1(z, k, t) \sim 1 + \dfrac{a(k, t)}{z} + \mathcal{O}\left(|z|^{-2}\right) \\ m^2(z, k, t) \sim e_{-k}(z) \dfrac{b(k, t)}{\bar{z}} + \mathcal{O}\left(|z|^{-2}\right) \end{cases} \quad (4.39)$$

Note that, comparing the second equation of (4.39) with (4.37), we have $b(k, t) = i\mathbf{s}(k, t)$. Substituting these expansions into (4.35c)–(4.35d) and taking $|z| \to \infty$, we see that

$$\dot{a}(k, t) = 0, \quad \dot{b}(k, t) = 2i \left(k^2 + \bar{k}^2\right) b(k, t).$$

Thus, formally, the map $q \mapsto (a, b)$ gives action-angle variables for the flow (4.33). In particular, if $q(z, t)$ solves the DSII equation, then the scattering data obeys the linear evolution

$$\mathbf{s}(k, t) = e^{2it\left(k^2 + \bar{k}^2\right)} \mathbf{s}(k, 0).$$

It remains to show how $q(z, t)$, the solution of the DSII equation, may be recovered from $\mathbf{s}(k, t)$. Here we use the fact that m^1 and m^2, now regarded also as functions of time, obey the equations

$$\begin{cases} \partial_{\bar{k}} m^1 = e^{it\varphi} \mathbf{s} \overline{m^2} \\ \partial_{\bar{k}} m^2 = e^{it\varphi} \mathbf{s} \overline{m^1} \\ m^1(z, k) - 1, m^2(z, k) \to 0 \text{ as } |k| \to \infty \end{cases} \quad (4.40)$$

where $\mathbf{s}(k)$ is the scattering transform of the initial data $q(z, 0)$, and

$$\varphi(z, k, t) = 2\left(k^2 + \bar{k}^2\right) - \frac{kz + \bar{k}\bar{z}}{t} \quad (4.41)$$

is a phase function formed from e_{-k} and the evolution for **s**. We can then reconstruct $q(z, t)$ from the asymptotics of $m^2(z, k, t)$ using (4.38).

The proof that $q(z, t)$ so defined in fact solves (4.1) uses the Lax representation (4.35a)–(4.35d). In the case $\varepsilon = 1$, we will show that $m^1(z, k, t)$ and $m^2(z, k, t)$ defined by (4.40) generate a solution of the Lax equations (4.31a)–(4.32b) where $q(\cdot, t)$ is the scattering transform of $\mathbf{s}(\cdot, t)$. It will then follow that $q(z, t)$, defined as $\mathcal{S}(\mathbf{s}(\cdot, t))$, solving the DSII equation.

In what follows, we will study the scattering transform $\mathcal{S}$ in depth to obtain the Lipschitz mapping property (Sect. 4.4). In order to prove the solution formula (2.23), it suffices to check initial data $q_0 \in \mathscr{S}(\mathbb{R}^2)$. In Sect. 4.5, we use the Lax representation (4.3)–(4.4) to show that (2.23) does indeed generate a solution to (2.17).

4.4 The Scattering Map

We now define the scattering transform $\mathcal{S} : q \to \mathbf{s}$ more precisely. Given $q \in H^{1,1}(\mathbb{R}^2)$ and $k \in \mathbb{C}$, one first solves the linear system

$$\begin{cases} \partial_{\bar{z}} m^1(z,k) = q(z) m^2(z,k) \\ (\partial_z + ik) m^2(z,k) = \overline{q(z)} m^1(z,k) \\ m^1(\,\cdot\,,k) - 1, \ m^2(\,\cdot\,,k) \in L^4(\mathbb{R}^2). \end{cases} \quad (4.42)$$

One then computes the scattering transform from the integral representation

$$\mathbf{s}(k) = (\mathcal{S}q)(k) := -\frac{i}{\pi} \int_{\mathbb{R}^2} e_k(z) \overline{q(z)} m^1(z,k) \, dz. \quad (4.43)$$

This definition accords with the definition (4.37) given by asymptotic expansion of $m_2(z,k,t)$ if $q \in \mathscr{S}(\mathbb{R}^2)$ because one can compute the first term in the large-z asymptotic expansion for $m^2(z,k,t)$ explicitly (Exercise 4.10; in keeping with the emphasis of this section, t-dependence is suppressed). Note that, in this normalization, $\mathcal{S}$ is an antilinear map. Its linearization at $q = 0$ is an "antilinear Fourier transform"

$$\widehat{f}(k) = (\mathcal{F}_a f)(k) := -\frac{i}{\pi} \int_{\mathbb{R}^2} e_k(z) \overline{f(z)} \, dz. \quad (4.44)$$

It is easy to check, using standard Fourier theory, that $\mathcal{F}_a = \mathcal{F}_a^{-1}$ defines an isometry from $L^2(\mathbb{R}^2)$ onto itself and a Lipschitz continuous map from $H^{1,1}(\mathbb{R}^2)$ onto itself (Exercise 4.11). Thus,

$$(\mathcal{S}q)(k) = (\mathcal{F}_a q)(k) - \frac{i}{\pi} \int e_k(z) \overline{q(z)} \left(m^1(z,k) - 1 \right) dz. \quad (4.45)$$

Equation (4.45) provides a useful way to understand the scattering transform: it is a perturbation of the linear Fourier transform in which the integral transform also depends on q. In [36], the authors exploit the fact that the second term may be viewed as a pseudodifferential operator whose mapping properties can be controlled by estimates on the 'symbol' $a(x, \xi) = m^1(\xi, x) - 1$ (the reversal of arguments (x, ξ) in m^1 is deliberate!).

We will prove:

Theorem 4.7 *The scattering transform S is a locally Lipschitz continuous map from $H^{1,1}(\mathbb{R}^2)$ onto itself. Moreover, $S = S^{-1}$*

Proof We begin with a reduction. Suppose we can prove that $\|Sq_1 - Sq_2\|_{H^{1,1}} \lesssim \|q_1 - q_2\|_{H^{1,1}}$ for $q_1, q_2 \in \mathscr{S}(\mathbb{R}^2)$ with constants uniform in $q_1, q_2 \in \mathscr{S}(\mathbb{R}^2)$ having $H^{1,1}(\mathbb{R}^2)$ norm bounded by a fixed constant, We can then extend the map S by density to a nonlinear mapping on $H^{1,1}(\mathbb{R}^2)$ with the same continuity properties. Similarly, if $S = S^{-1}$ on $\mathscr{S}(\mathbb{R}^2)$, this identity extends by density to $H^{1,1}(\mathbb{R}^2)$.

The claimed mapping properties of S for $q_1, q_2 \in \mathscr{S}(\mathbb{R}^2)$ are proved in Propositions 4.14, 4.17, and 4.18 of what follows. The property $S = S^{-1}$ on $\mathscr{S}(\mathbb{R}^2)$ is proved in Proposition 4.20. □

The proofs of Propositions 4.14, 4.17, 4.18, and 4.20 rest on a careful analysis of the solutions to (4.42). Some of the results along the way are proved for $q \in L^2(\mathbb{R}^2)$ or $q \in H^{1,1}(\mathbb{R}^2)$. Although we follow the outline of [38], we use ideas of [36] at a number of points to simplify the proofs.

4.4.1 Existence and Uniqueness of Solutions

First, we will show that (4.42) has a unique solution for each $q \in H^{1,1}(\mathbb{R}^2)$ and $k \in \mathbb{C}$. The following "vanishing theorem" for $\bar{\partial}$-problems is originally due to Vekua [44], was used by Beals-Coifman [8], and was improved to the form stated here by Brown and Uhlmann [13]. The short and elegant proof we give here is taken from the paper of Nachman et al. [36, proof of Lemma 3.2].

Theorem 4.8 *Suppose that $a \in L^2(\mathbb{R}^2)$, $u \in L^p(\mathbb{R}^2)$ for some $p > 2$, and $\partial_{\bar{z}} u = a\bar{u}$ in distribution sense. Then $u = 0$.*

Proof ([36]) Define

$$a_n(x) = \begin{cases} a(x), & |x| < n \text{ and } |a(x)| < n \\ 0, & \text{otherwise} \end{cases}$$

and $a_s = a - a_n$. We then have $a = a_n + a_s$ where $a_n \in L^{p_1} \cap L^{p_2}$ for some $1 < p_1 < 2 < p_2$ and $\|a_s\|_{L^2}$ is small for n large. Let

$$v(z) = \begin{cases} \exp\left(-\partial_{\bar{z}}^{-1} a_n \dfrac{\bar{u}}{u}\right)(z), & u(z) \neq 0 \\ 1, & u(z) = 0 \end{cases}$$

The function uv obeys

$$\partial_{\bar{z}}(uv) = a_s(uv)$$

so that, choosing n large enough, we may conclude that $uv = 0$, by the remarks following (4.12). On the other hand, v and v^{-1} belong to L^∞ by (4.10). Hence, $u = 0$. □

A short computation using the operator identity

$$(\partial_z + ik) = e_{-k}\partial_z e_k \tag{4.46}$$

shows that the functions

$$m_\pm(z, k) = m^1(z, k) \pm e_{-k}\overline{m^2(z, k)} \tag{4.47}$$

solve the system

$$\begin{cases} \partial_{\bar{z}} m^\pm(z, k) = \pm e_{-k} q(z) \overline{m^\pm(z, k)}, \\ m_\pm - 1 \in L^4(\mathbb{R}^2) \end{cases} \tag{4.48}$$

Proposition 4.9 *There exists a unique solution of* (4.42) *for any* $k \in \mathbb{C}$ *and* $q \in L^2(\mathbb{R}^n)$.

Proof We prove uniqueness first. Suppose that (m^1, m^2) and (n^1, n^2) solve (4.42) for $q \in L^2$. We claim that $m^1 = n^1$ and $m^2 = n^2$. Setting

$$w^1 = m^1 - n^1, \quad w^2 = m^2 - n^2,$$

we obtain a solution (w^1, w^2) of (4.42) with $w^1(\,\cdot\,, k)$ and $w^2(\,\cdot\,, k)$ in $L^4(\mathbb{R}^2)$, so that the same is true of $w_\pm$ under the change of variable (4.47). By Theorem 4.8, $w_\pm = 0$, so $(w^1, w^2) = 0$.

In order to prove existence of solutions to (4.42), it suffices to solve (4.48). To this end, consider the equation

$$\partial_{\bar{z}} w + e_{-k} u \overline{w} = -e_{-k} u \tag{4.49}$$

where one should think of w as $m^\pm - 1$ and u as $\mp q$. This equation is equivalent to the integral equation

$$w - Tw = \partial_{\bar{z}}^{-1}(e_{-k} u)$$

where

$$Tf = \partial_{\bar{z}}^{-1}\left(e_{-k} u \overline{f}\right).$$

The operator T is the composition of the operator S with complex conjugation (the factor e_{-k} can be absorbed into the definition of q in (4.12)). Hence, by Lemma 4.3, T is a compact operator. Since T is compact, $(I - T)$ is a Fredholm operator.

We claim that, by Theorem 4.8, $\ker_{L^4}(I - T)$ is trivial. If so, it follows from the Fredholm alternative that $(I - T)^{-1}$ is a bounded operator on L^4 and that

$$w = (I - T)^{-1} \left(\partial_{\bar{z}}^{-1}(e_{-k}u) \right) \tag{4.50}$$

solves (4.49). Suppose then that $f \in \ker_{L^4}(I - T)$, i.e., $Tf = f$. Then f is a weak solution of the equation $\partial_{\bar{z}} f = e_{-k} q \overline{f}$ and hence, by Theorem 4.8, $f = 0$. This finishes the proof. □

We end this subsection with a resolvent estimate on $(I - T)^{-1}$. This is one of the key points where we use the smoothness of $q \in H^{1,1}(\mathbb{R}^2)$. Very different techniques are used in [36, §3] to control the resolvent assuming only that $q \in L^2(\mathbb{R}^2)$.

We will exploit the integration by parts formula

$$\frac{1}{\pi} \int \frac{1}{z - w} e_{-k} f(w) \, dw = -\frac{e_{-k}(z) f(z)}{i\bar{k}} + \frac{1}{i\bar{k}} \int \frac{1}{z - w} e_{-k}(w) (\partial_{\bar{z}} f)(w) \, dw \tag{4.51}$$

(see Exercise 4.13).

From this identity it follows that

$$(Tf)(z) = -\frac{1}{i\bar{k}} e_{-k}(z) q(z) \overline{f(z)} + \frac{1}{i\bar{k}} \int \frac{1}{z - w} e_{-k}(w) \partial_{\bar{z}}(q\overline{f})(w) \, dw. \tag{4.52}$$

Using the estimate (4.9) and (4.52), we see that

$$\|Tf\|_{L^4} \lesssim \frac{1}{|k|} \left(\|q\overline{f}\|_{L^4} + \|\partial_{\bar{z}}(q\overline{f})\|_{L^{4/3}} \right)$$

so that

$$\left\| T^2 f \right\|_{L^4} \lesssim \frac{1}{|k|} \left(\|qTf\|_{L^4} + \|\partial_{\bar{z}}(qTf)\|_{L^{4/3}} \right)$$

$$\lesssim \frac{1}{|k|} \left(\|q\|_{L^8} \|Tf\|_{L^8} + \|\partial_{\bar{z}}q\|_{L^2} \|Tf\|_{L^4} + \|q\|_{L^4}^2 \|f\|_{L^4} \right)$$

$$\lesssim \frac{1}{|k|} \left(\|q\|_{L^8} \|q\|_{L^{8/3}} + \|\partial_{\bar{z}}q\|_{L^2} \|q\|_{L^2} + \|q\|_{L^4}^2 \right) \|f\|_{L^4}$$

Since $H^1(\mathbb{R}^2)$ is continuously embedded in $L^p(\mathbb{R}^2)$ for all $p \geq 2$ (see Exercise 4.15), it follows that

$$\left\| T^2 \right\|_{L^4 \to L^2} \lesssim \frac{1}{|k|} \|q\|_{H^1}^2 \tag{4.53}$$

From (4.53) and the identity $(I - T) = (I - T^2)^{-1}(I + T)$, we immediately obtain the following large-k resolvent estimate.

Lemma 4.10 *Fix $R > 0$. There is an $N = N(R)$ so that for all $k \in \mathbb{C}$ with $|k| \geqslant N$ and all $q \in H^1(\mathbb{R}^2)$ with $\|q\|_{H^1} \leqslant R$, the estimate $\left\|(I - T)^{-1}\right\|_{L^4 \to L^4} \leqslant 2$ holds.*

To obtain uniform resolvent estimates (i.e., estimates valid for all $k \in \mathbb{C}$ and q in a bounded subset of $H^{1,1}(\mathbb{R}^2)$), we now follow the ideas of [38]. Using a different approach, Nachman, Regev, and Tataru obtain similarly uniform estimates for q in a bounded subset of L^2 (see [36, Section 3]).

In our case, Lemma 4.10 gives uniform control for q in a bounded subset of $H^{1,1}(\mathbb{R}^2)$ and sufficiently large $|k|$. It remains to control the resolvent for (k, q) with k in a bounded subset of $\mathbb{C}$ and q in a bounded subset of $H^{1,1}(\mathbb{R}^2)$.

Lemma 4.11 *Let B be a bounded subset of $H^{1,1}(\mathbb{R}^2) \times \mathbb{C}$. Then*

$$\sup_{(q,k)\in B} \left\|(I - T)^{-1}\right\|_{L^4 \to L^4} < \infty.$$

Proof Write T as $T(q, k)$ to show the dependence of the operator on $q \in L^2(\mathbb{R}^2)$ and $k \in \mathbb{C}$. We prove the required estimate in two steps. First, we show that the mapping

$$L^2(\mathbb{R}^2) \times \mathbb{C} \ni (q, k) \mapsto (I - T(q, k))^{-1} \in \mathcal{B}(L^4) \tag{4.54}$$

is continuous. Second, we show that if B is a bounded subset of $H^{1,1}(\mathbb{R}^2) \times \mathbb{C}$, then B is a pre-compact subset in $L^2(\mathbb{R}^2) \times \mathbb{C}$. Thus the resolvents $\{(I - T)^{-1} : (q, k) \in B\}$, as the image of a pre-compact set under a continuous map, form a bounded subset of $\mathcal{B}(L^4)$.

First we consider continuity of the map (4.54). By the second resolvent formula, it suffices to show that the map $(q, k) \mapsto T(k, q)$ is continuous from $L^2(\mathbb{R}^2) \times \mathbb{C}$ to $\mathcal{B}(L^4)$. But

$$\left\|T(k, q) - T(k', q')\right\|_{L^4 \to L^4} \leqslant \left\|T(k, q) - T(k', q)\right\|_{L^4 \to L^4} \tag{4.55}$$
$$+ \left\|T(k', q) - T(k', q')\right\|_{L^4 \to L^4}$$
$$\lesssim \left\|(e_k - e_{k'})q\right\|_{L^2} + \left\|q - q'\right\|_{L^2}$$

where in the second step we used (4.13) (where q now includes the factor e_k) and the linearity of S in q. The continuity is immediate.

Pre-compactness of B as a subset of $L^2(\mathbb{R}^2) \times \mathbb{C}$ follows from the Kolmogorov-Riesz Theorem and is left as Exercise 4.16. □

We can also prove Lipschitz continuity of the resolvent as a function of q.

Lemma 4.12 *Fix $R > 0$ and $q_1, q_2 \in H^{1,1}(\mathbb{R}^2)$ with $\|q_i\|_{H^{1,1}} \leqslant R$, $i = 1, 2$. Then*

$$\sup_{k \in \mathbb{C}} \left\| (I - T(q_1, k))^{-1} - (I - T(q_2, k))^{-1} \right\|_{L^4 \to L^4} \lesssim_R \|q_1 - q_2\|_{L^2}.$$

Proof This is a consequence of Lemma 4.11, the estimate (4.55), and the second resolvent formula. □

4.4.2 Estimates on the Scattering Transform

In order to analyze the scattering transform, we need estimates on the regularity in k and large-k behavior of the scattering solutions $m^1(z, k)$ and $m^2(z, k)$. In essence, this entails understanding the joint (z, k) behavior of solutions to the model equation (4.49).

In order to do this, we need (1) estimates on the resolvent $(I - T)^{-1}$ uniform in k and (2) estimates on the joint (z, k) behavior of the function $\partial_{\bar{z}}^{-1}(e_{-k}q)$. In [38] both of these steps were accomplished using the smoothness and decay of q (i.e., using $q \in H^{1,1}(\mathbb{R}^2)$). In [36], the authors need only assume that $q \in L^2$: they use ideas of concentration compactness [25] to obtain the required control of the resolvent, and use the fractional integral estimates from Theorem 4.4 to control $\partial_{\bar{z}}^{-1}(e_{-k}q)$.

In these notes, we will take an intermediate route and borrow insights from [36] to provide a cleaner and more concise proof of the main results in [38]. In particular, by exploiting Theorem 4.4, we will avoid the multilinear estimates and resolvent expansions used in [38]. A number of calculations below also exploit the ideas behind [36, Theorem 2.3], a sharp L^2 boundedness theorem for non-smooth pseudodifferential operators.

We begin with a mixed-L^p estimate which actually holds for $q \in L^2$ (see [36, Lemma 4.1]). The technique of proof is borrowed from [36, Lemma 4.1], with our weaker resolvent estimate from Lemma 4.11 used instead of their stronger L^2 estimate [36, Theorem 1.1].

Lemma 4.13 *Suppose that $q \in H^{1,1}(\mathbb{R}^2)$ and that $(\mathcal{M}\widehat{q})(k)$ is finite. Let m^1 and m^2 be the unique solutions of (4.42). Then*

$$\left\| m^1(\cdot, k) - 1 \right\|_{L^4} + \left\| m^2(\cdot, k) \right\|_{L^4} \leqslant C \left(\|q\|_{H^1} \right) (\mathcal{M}\widehat{q})(k)^{1/2} \quad (4.56)$$

Moreover, the maps $q \mapsto m^1$ and $q \mapsto m^2$ are locally Lipschitz continuous as maps from $H^{1,1}(\mathbb{R}^2)$ to $L^4(\mathbb{R}_z^2 \times \mathbb{R}_k^2)$.

Proof By the definition (4.47) of $m_\pm$ and Eq. (4.48) obeyed by $m_\pm$, it suffices to prove the estimate

$$\|w\|_{L^4} \leqslant C \left(\|q\|_{H^{1,1}} \right) (\mathcal{M}\widehat{q})(k)^{1/2}$$

for solutions w of the model equation (4.49).

From the solution formula (4.50), we estimate

$$\|w\|_{L^4} \lesssim \left\|(I-T)^{-1}\right\|_{L^4 \to L^4} \|\partial_{\bar{z}}(e_{-k}u)\|_{L^4}$$

$$\lesssim C\left(\|q\|_{H^{1,1}}\right) \|u\|_{L^2}^{1/2} (\mathcal{M}\widehat{u})(k)^{1/2}$$

where we used Lemma 4.11 and the fractional integral estimate (4.18). This estimate now implies (4.56).

An immediate consequence of (4.56) is the estimate

$$\left\|m^1 - 1\right\|_{L^4(\mathbb{R}_z^2 \times \mathbb{R}_k^2)} + \left\|m^2\right\|_{L^4(\mathbb{R}_z^2 \times \mathbb{R}_k^2)} \leqslant C\left(\|q\|_{H^{1,1}}\right) \|q\|_{L^2}^{1/2}. \quad (4.57)$$

The Lipschitz continuity follows from Lemma 4.12, the solution formula (4.50), and (4.15). □

We can now prove:

Proposition 4.14 *The scattering transform $\mathcal{S}$ is bounded and Lipschitz continuous from $H^{1,1}(\mathbb{R}^2)$ to $L^2(\mathbb{R}^2)$.*

Proof As already discussed, it suffices to prove the Lipschitz continuity estimates for $q \in \mathscr{S}(\mathbb{R}^2)$. We use the fact that $q \in \mathscr{S}(\mathbb{R}^2)$ in the computations leading to (4.59).

By Eq. (4.45), it suffices to show that the integral

$$I(k) = -\frac{i}{\pi} \int e_k(z) \overline{q(z)} \left(m^1(z,k) - 1\right) dz \quad (4.58)$$

defines an L^2 function of k, locally Lipschitz as a function of q. From (4.42), we may write $m^1(z,k) - 1 = \partial_{\bar{z}}^{-1}\left(q(\cdot) m^2(\cdot, k)\right)$ and change orders of integration to obtain

$$I(k) = -\frac{i}{\pi} \int \left[\partial_{\bar{z}}^{-1}(e_k \bar{q})(z)\right] q(z) m^2(z,k) \, dz \quad (4.59)$$

and conclude from the estimate (4.17) and Lemma 4.13 that

$$|I(k)| \lesssim C\left(\|q\|_{H^1}\right) (\mathcal{M}\widehat{\bar{q}}(k))^{1/2} \int (\mathcal{M}\bar{q}(z))^{1/2} |q(z)| |m^2(z,k)| \, dz$$

$$\leqslant C\left(\|q\|_{H^1}\right) (\mathcal{M}\widehat{\bar{q}}(k))^{1/2} \|q\|_{L^2}^{3/2} \left\|m^2(\cdot, k)\right\|_{L^4}$$

where in the second line we used (4.15). Using (4.57) and Hölder's inequality, we conclude that $I \in L^2$ with

$$\|I\|_{L^2} \lesssim C\left(\|q\|_{H^1}\right) \|q\|_{L^2}^2.$$

To show Lipschitz continuity, we note that, by (4.59),

$$I(k; q_1) - I(k; q_2) = \qquad (4.60)$$

$$\frac{i}{\pi} \int \left[q_1(z) \partial_{\bar{z}}^{-1}(e_k \overline{q_1})(z) - q_2(z) \partial_{\bar{z}}^{-1}(e_k \overline{q_2})(z) \right] m^2(z, k; q_1) \, dz$$

$$+ \frac{i}{\pi} \int \left[q_2(z) \partial_{\bar{z}}^{-1}(e_k \overline{q_2})(z) \right] \left(m^2(z, k; q^1) - m^2(z, k; q_2) \right) dz$$

The map $q \mapsto q \partial_{\bar{z}}^{-1}(e_k \overline{q})$ is Lipschitz continuous from $L^2(\mathbb{R}^2)$ into $L^4(\mathbb{R}_k^2; L^{4/3}(\mathbb{R}_z^2)))$ by multilinearity, (4.15), and (4.18) (see Exercise 4.17). The map $q \mapsto m^2(z, k; q)$ is Lipschitz continuous from $H^{1,1}(\mathbb{R}^2)$ into $L^4(\mathbb{R}_z^2 \times \mathbb{R}_k^2)$ by Lemma 4.13. □

Remark 4.15 The "integration by parts" that transforms (4.58)–(4.59) is also one of the key ideas behind the proof of the L^2 boundedness theorem for pseudodifferential operators with non-smooth symbols, Theorem 2.3, in [36]. Tracing through the argument used to estimate $I(k)$, it is easy to see that the same argument proves that

$$I(f, k) = -\frac{i}{\pi} \int e_k(z) f(z) \left(m^1(z, k) - 1 \right) dz$$

satisfies the estimate

$$|I(k, f)| \lesssim C \left(\|q\|_{H^1} \right) \left((\mathcal{M} \widehat{f}(k) \right)^{1/2} \|f\|_{L^2} \left\| m^2(\cdot, k) \right\|_{L^4}$$

so that

$$\|I(\cdot, f)\|_{L^2} \lesssim C \left(\|q\|_{H^1} \right) \|q\|_{L^2} \|f\|_{L^2}. \qquad (4.61)$$

To prove Theorem 4.7, it remains to show that, for $q \in H^{1,1}(\mathbb{R}^2)$, $Sq \in H^1(\mathbb{R}^2)$ and $Sq \in L^{2,1}(\mathbb{R}^2)$, and that the corresponding maps are locally Lipschitz continuous. As a first step, we show that, if $q \in H^{1,1}(\mathbb{R}^2)$, then $Sq \in L^p(\mathbb{R}^2)$ for all $p \in [2, \infty)$.

Proposition 4.16 *For any $p \in [2, \infty)$, the scattering transform S is locally Lipschitz continuous from $H^{1,1}(\mathbb{R}^2)$ to $L^p(\mathbb{R}^2)$.*

Proof The Fourier transform has this mapping property by the Hausdorff-Young inequality and the fact that $H^{1,1} \hookrightarrow L^q$ for $q \in (1, 2]$ (see Exercise 4.14). Hence, owing to (4.45), it suffices to prove that the map $q \mapsto I(k)$ defined by (4.58) has the required continuity.

Using (4.59), the fractional integral estimate (4.18), and the a priori estimate (4.56) on m^2, we may estimate

$$|I(k)| \lesssim \int \left|\partial_{\bar{z}}^{-1}(e_k \bar{q})(z)\right| |q(z)| \left|m^2(z,k)\right| dz$$

$$\lesssim C\left(\|q\|_{H^{1,1}}\right) \left[(\mathcal{M}\widehat{q})(k)^{1/2}(Mq)(z)^{1/2}\right]^2 |q(z)| dz$$

$$\lesssim C\left(\|q\|_{H^{1,1}}\right) \|q\|_{L^2}^2 (\mathcal{M}\widehat{q})(k)$$

which shows that $I \in L^p(\mathbb{R}^2)$ for any $p > 2$ by the Hausdorff-Young inequality again. The proof of Lipschitz continuity uses (4.60) and analogous estimates. □

Proposition 4.16 allows us to prove:

Proposition 4.17 *The map S is locally Lipschitz continuous from $H^{1,1}(\mathbb{R}^2)$ to $H^1(\mathbb{R}^2)$.*

Proof It suffices to prove the Lipschitz estimate for $q \in \mathscr{S}(\mathbb{R}^2)$. In view of Proposition 4.14, property (4.21) of the Beurling transform, and the boundedness of the Beurling transform on L^p, it suffices to show that the map $q \mapsto \partial_{\bar{z}} I$ (where the differentiation is with respect to k) is locally Lipschitz continuous from $H^{1,1}(\mathbb{R}^2)$ into $L^2(\mathbb{R}^2)$. In Lemma 4.19, we will show that, for $q \in \mathscr{S}(\mathbb{R}^2)$, m^1 and m^2 also solve the $\bar{\partial}_k$-problem (4.62). Thus, for $q \in \mathscr{S}(\mathbb{R}^2)$ we may compute

$$\partial_{\bar{k}} \mathbf{s}(k) = \frac{1}{\pi} \int e_k(z) \overline{zq(z)} m^1(x,k) \, dz - \frac{i}{\pi} \mathbf{s}(k) \int \overline{q(z)} m^2(z,k) \, dz$$

$$= I_1 + I_2$$

Tracing through the proof of Proposition 4.14 with $\bar{q}$ replaced by $\overline{zq}$, we conclude that I_1 defines an L^2 function of k, Lipschitz continuous in q. It remains to estimate I_2.

By Proposition 4.16, $\mathbf{s} \in L^4(\mathbb{R}^2)$, so it suffices to show that the integral defines a Lipschitz map from $q \in H^{1,1}(\mathbb{R}^2)$ to $L^4(\mathbb{R}_k^2)$. Since $m^2 \in L^4(\mathbb{R}_z^2 \times \mathbb{R}_k^2)$ and $q \in H^{1,1}(\mathbb{R}^2) \subset L^{4/3}(\mathbb{R}_z^2)$, this is a consequence of Hölder's inequality and Lemma 4.13. □

To complete the proof of Theorem 4.7, we show that $Sq \in L^{2,1}(\mathbb{R}^2)$ with appropriate Lipschitz continuity.

Proposition 4.18 *The map S is locally Lipschitz continuous from $H^{1,1}(\mathbb{R}^2)$ to $L^{2,1}(\mathbb{R}^2)$.*

Proof We need only show that $q \mapsto I(k)$ has the above property, where $I(k)$ is defined by (4.58). We begin with a computation for $q \in \mathscr{S}(\mathbb{R}^2)$, using the trivial identity $\partial_{\bar{z}} e_k = i\bar{k} e_k$ and integration by parts:

$$\bar{k} I(k) = -\frac{1}{\pi} \int e_k(z) \partial_{\bar{z}} \left(\overline{q(z)} \left(m^1(z,k) - 1 \right) \right) dz$$

$$= I_1 + I_2$$

where

$$I_1 = -\frac{1}{\pi} \int e_k(z) \, (\partial_{\bar{z}} \overline{q})(z) \, m^1(z,k) \, dz,$$

$$I_2 = -\frac{1}{\pi} \int e_k(z) |q(z)|^2 m^2(z,k) \, dz,$$

and we used the first equation in (4.42) to simplify I_2.

The integral I_1 defines an L^2 function of k since $\partial_{\bar{z}} \overline{q} \in L^2(\mathbb{R}^2)$ by an argument similar to the proof of Proposition 4.14. To analyze I_2, we use the second equation in (4.42) to write

$$I_2 = \int |q(z)|^2 \partial_z^{-1} \left(e_k(\cdot) q(\cdot) m^1(\cdot, k) \right)(z) \, dz$$

$$= I_{21} + I_{22}$$

where

$$I_{21} = \int |q(z)|^2 \partial_z^{-1} (e_k q)(z) \, dz,$$

$$I_{22} = \int |q(z)|^2 \partial_z^{-1} \left(e_k q(\cdot) \left(m^1(\cdot, k) - 1 \right) \right)(z) \, dz.$$

By "integration by parts" we have

$$I_{21} = -\int e_k(z) q(z) \partial_z^{-1} \left(|q|^2 \right)(z) \, dz$$

which exhibits I_{21} as the Fourier transform of an L^2 function since $|q|^2$ in $L^{4/3}(\mathbb{R}^2)$. On the other hand

$$I_{22} = -\int e_k(z) q(z) \left[\partial_z^{-1} \left(|q(\cdot)|^2 \right) \right](z) \left(m^1(z,k) - 1 \right) dz$$

which exhibits I_{22} in the form $I(k, f)$ (see Remark 4.15) where f is the L^2 function $q \partial^{-1} |q|^2$. The needed L^2 bound is a direct consequence of (4.61).

As usual, the proof of Lipschitz continuity rests on the multilinearity of explicit expressions involving q and the Lipschitz continuity of m^1 and m^2 viewed as functions of q. To prove that I_1 is locally Lipschitz continuous, one mimics the proof that I is Lipschitz beginning with (4.60) in the proof of Proposition 4.14. To show that I_2 is Lipschitz continuous, one notes that I_{21} is an explicit multilinear function of q, while I_{22} can be controlled by the same method used to prove Lipschitz continuity of I on the proof of Proposition 4.14. □

It now follows that $\mathcal{S}$, initially defined on $\mathscr{S}(\mathbb{R}^2)$, extends to a locally Lipschitz continuous map from $H^{1,1}(\mathbb{R}^2)$ to itself. It remains to prove that $\mathcal{S}^{-1} = \mathcal{S}$.

By the Lipschitz continuity of $\mathcal{S}$ on $H^{1,1}(\mathbb{R}^2)$, it suffices to prove that $\mathcal{S} = \mathcal{S}^{-1}$ on the dense subset $\mathscr{S}(\mathbb{R}^2)$. The idea of the proof is to use uniqueness of solutions to the system (4.42) together with the fact that, for $q \in \mathscr{S}(\mathbb{R}^2)$, the functions (m^1, m^2) satisfy *both* the system (4.42) and the following system of $\overline{\partial}_k$-equations.

Lemma 4.19 *Suppose that $q \in \mathscr{S}(\mathbb{R}^2)$ and let (m^1, m^2) be the unique solutions to (4.42). Then, for each $z \in \mathbb{C}$,*

$$\begin{cases} \partial_{\overline{k}} m^1(z,k) = e_{-k} s(k) \overline{m^2(z,k)} \\ \partial_{\overline{k}} m^2(z,k) = e_{-k} s(k) \overline{m^1(z,k)} \\ m^1(z,k) - 1, \; m^2(z,k) = \mathcal{O}\left(|k|^{-1}\right) \end{cases} \quad (4.62)$$

where $s(k)$ is given by (4.43)

Proof For $q \in \mathscr{S}(\mathbb{R}^2)$, the solutions (m^1, m^2) of (4.42) have the large-z asymptotic (differentiable) expansions

$$m^1(z,k) = 1 + \mathcal{O}\left(|z|^{-1}\right)$$

$$m^2(z,k) = e_{-k}(z) \frac{is(k)}{\overline{z}} + \mathcal{O}\left(|z|^{-2}\right)$$

where s is given by (4.43) (see Exercise 4.10 and the comments after (4.43)). If $v^1 = \partial_{\overline{k}} m^1$ and $v^2 = \partial_{\overline{k}} m^2$ then, differentiating (4.42) with respect to $\overline{k}$ we recover

$$\partial_{\overline{z}} v^1 = q v^2$$

$$(\partial_z + ik) v^2 = \overline{q} v^1$$

It follows from the asymptotic expansions for m^1 and m^2 above that $v^1 = \mathcal{O}\left(|z|^{-1}\right)$ but $v^2 = e_{-k} s(k) + \mathcal{O}\left(|z|^{-1}\right)$. Hence, in order to use the uniqueness theorem for solutions of (4.42), we need to make a subtraction to remove the constant term in v^2. Setting

$$w^1(z,k) = \partial_{\overline{k}} m^1 - e_{-k} s \overline{m^2}, \quad w^2 = \partial_{\overline{k}} m^2 - e_{-k} s \overline{m^1},$$

and using (4.42), we conclude that

$$\partial_{\overline{z}} w^1 = q w^2$$

$$(\partial_z + ik) w^2 = \overline{q} w^1$$

where w_1 and w_2 are $\mathcal{O}(|z|^{-1})$ as $|z| \to \infty$. Hence by Proposition 4.9, $w^1 = w^2 = 0$. Since w^1 and w^2 are smooth functions of z and k, it follows that the first two of equations (4.62) hold for each z.

It remains to show that, for each fixed z, $m_1(z, k) - 1$ and $m^2(z, k)$ are $\mathcal{O}(|k|^{-1})$ as $|k| \to \infty$. For $q \in \mathscr{S}(\mathbb{R}^2)$, the functions m^1 and m^2 are smooth functions of z and k with bounded derivatives (see Sung [39, section 2]). From the integral formulas

$$m^1(z, k) = 1 + \frac{1}{\pi} \int \frac{1}{z - \zeta} q(\zeta) m^2(\zeta, k) \, d\zeta \qquad (4.63)$$

$$e_k(z) m^2(z, k) = \frac{1}{\pi} \int \frac{1}{\bar{z} - \bar{\zeta}} e_k(\zeta) \overline{q(\zeta)} m^1(\zeta, k) \, d\zeta \qquad (4.64)$$

we first note that it is enough to prove that $m^2(z, k) = \mathcal{O}(|k|^{-1})$ uniformly in z since it will then follow from (4.63) that $m^1(z, k) - 1 = \mathcal{O}(|k|^{-1})$. We can integrate by parts in (4.64) to see that

$$e_k(z) m^2(z, k) = \frac{1}{ik} e_k(z) \overline{q(z)} m^1(z, k) - \frac{1}{\pi i k} \int \frac{1}{\bar{z} - \bar{\zeta}} e_k(\zeta) \partial_\zeta \left(\overline{q(\zeta)} m^1(\zeta, k) \right) d\zeta$$

which shows that $m^2(z, k) = \mathcal{O}(|k|^{-1})$. □

Given $\mathbf{s}(k) \in \mathscr{S}(\mathbb{R}^2)$, the inverse scattering transform $\mathcal{S}\mathbf{s}$ is computed by solving the system

$$\begin{cases} \partial_{\bar{k}} n^1(k, z) = \mathbf{s}(k) n^2(k, z) \\ (\partial_k + iz) n^2(k, z) = \overline{\mathbf{s}(k)} n^1(k, z) \\ n^1(k, z) - 1, n^2(k, z) = \mathcal{O}(|k|^{-1}) \end{cases} \qquad (4.65)$$

and extracting $\mathcal{S}\mathbf{s}$ from the asymptotic expansion

$$n^2(k, z) = e_{-z}(k) \frac{i}{\bar{k}} (\mathcal{S}\mathbf{s})(z) + \mathcal{O}(|k|^{-2}).$$

The system (4.65) is uniquely solvable by Proposition 4.9. On the other hand, if m^1 and m^2 solve (4.42) for given $q \in \mathscr{S}(\mathbb{R}^2)$ and $\mathbf{s} = \mathcal{S}q$, these functions also solve (4.36). A short computation shows that $n^1(k, z) = m^1(z, k)$, $n^2(k, z) = e_{-z}(k) \overline{m^2(z, k)}$ solve the system (4.65). Since this solution is unique, we may compute $\mathcal{S}\mathbf{s}$ using the large-k expansion of $m^2(z, k)$ (see Exercise 4.9):

$$n^2(k, z) = e_{-z}(k) \overline{m^2(z, k)}$$

$$= e_{-z}(k) \left(\frac{iq(z)}{\bar{k}} + \mathcal{O}(|k|^{-2}) \right)$$

$$= e_{-z}(k)\frac{iq(z)}{\overline{k}} + \mathcal{O}\left(|k|^{-2}\right)$$

to conclude that $\mathcal{S}\mathbf{s} = q$. We have proved:

Proposition 4.20 *Suppose that $q \in \mathscr{S}(\mathbb{R}^2)$. Then $\mathcal{S}(\mathcal{S}(q)) = q$.*

4.5 Solving the DSII Equation

In this subsection we use the scattering transform to solve (4.33) with initial data $q_0 \in H^{1,1}(\mathbb{R}^2)$. The putative solution q_{inv} is given by (4.5); note that $q_{\text{inv}}(z, 0; q_0) = q_0$ by Proposition 4.20. We will prove:

Theorem 4.21 *The function (4.5) is the unique global solution of (4.1) for any $q_0 \in H^{1,1}(\mathbb{R}^2)$.*

We begin with an important reduction.

Proposition 4.22 *Suppose that, for each $q_0 \in \mathscr{S}(\mathbb{R}^2)$, $q_{\text{inv}}(z, t; q_0)$ solves the integral equation (4.24). Then $q_{\text{inv}}(z, t; q_0)$ solves (4.24) for any $q_0 \in H^{1,1}(\mathbb{R}^2)$.*

Proof Observe that the map $q_0 \mapsto q_{\text{inv}}(\,\cdot\,,\,\cdot\,; q_0)$ is a continuous map from $H^{1,1}(\mathbb{R}^2)$ to $C((0, T); H^{1,1}(\mathbb{R}^2))$ for any $T > 0$, and recall from Exercise 4.15 that $H^{1,1}(\mathbb{R}^2)$ is continuously embedded in $L^4(\mathbb{R}^2)$. It follows that $r \mapsto q_{\text{inv}}(\,\cdot\,, t; r)$ is a continuous map from $H^{1,1}(\mathbb{R}^2)$ into the space X (see (4.25)) for any $T > 0$.

Let $q_0 \in H^{1,1}(\mathbb{R}^2)$ and let $\{q_{0,n}\}$ be a sequence from $\mathscr{S}(\mathbb{R}^2)$ with $q_{0,n} \to q_0 \in H^{1,1}(\mathbb{R}^2)$. Then $q_{\text{inv}}(\,\cdot\,,\,\cdot\,; q_n) \to q_{\text{inv}}(\,\cdot\,,\,\cdot\,; q)$ in X as $n \to \infty$. The result now follows from the fact that (4.24) takes the form $q = \Phi(q)$ where Φ is continuous on X. □

Given this reduction, it suffices to prove that $q_{\text{inv}}(z, t; q_0)$ solves (4.24) for any $q_0 \in \mathscr{S}(\mathbb{R}^2)$. Recall that, by Sung's work [39–41], the map $\mathcal{S}$ restricts to a continuous map from $\mathscr{S}(\mathbb{R}^2)$ to itself, $\mathbf{s} = \mathcal{S}q_0$ is also a Schwartz class function, and the function $t \mapsto q_{\text{inv}}(z, t; q_0)$ is continuously differentiable as a map from $\mathbb{R}$ to $\mathscr{S}(\mathbb{R}^2)$. It then suffices to show that $q_{\text{inv}}(z, t; q_0)$ is a classical solution to (4.33). In the remainder of this section, we will use the complete integrability of (4.33) to prove this fact by showing that the solution $(m^1(z, t, q_0), m^2(z, t, q_0)$ of the $\overline{\partial}_k$-problem (4.40) generates a joint classical solution of Eqs. (4.35a)–(4.35d). We will then show that, as a consequence, q is a classical solution of (4.33).

Consider the $\overline{\partial}$-problem

$$\begin{cases} \left(\partial_{\overline{k}} m^1\right)(z, k, t) = e_{-k}(z)\mathbf{s}(k, t)\overline{m^2(z, k, t)}, \\ \left(\partial_{\overline{k}} m^2\right)(z, k, t) = e_{-k}(z)\mathbf{s}(k, t)\overline{m^1(z, k, t)}, \\ m^1(z, \,\cdot\,, t) - 1, \; m^2(z, \,\cdot\,, t) = \mathcal{O}\left(|k|^{-1}\right) \end{cases} \quad (4.66)$$

where

$$\mathbf{s}(k,t) = e^{2i(k^2+\bar{k}^2)t}\mathbf{s}(k). \quad (4.67)$$

Note that the L^4 condition on (m^1, m^2) is replaced by an asymptotic condition since, for $\mathbf{s} \in \mathscr{S}(\mathbb{R}^2)$, the solutions are bounded smooth functions and have complete asymptotic expansions in k (see Exercise 4.1).

We will show that (m^1, m^2) is a joint solution of Eqs. (4.35a)–(4.35d) where $q(z,t) = q_{\text{inv}}(z,t;q_0)$ and that, moreover,

$$\lim_{|k|\to\infty} m^1(z,k,t) = 1$$

for all (z,t) and $m^2(z,k,t) \neq 0$ for all (z,t) and some $k \in \mathbb{C}$. These facts, together with the identity (2.30) from Exercise 2.9 can then be used to show that $q(z,t)$ so defined solves the DS II equation.

In analogy to the Riemann-Hilbert problem for defocussing NLS, we will base our proof that the solutions of (4.66) furnish solutions of the Lax equations (4.35a)–(4.35d) on a vanishing lemma, this time for the $\bar{\partial}$-system. We state it in greater generality than is needed here.

Lemma 4.23 *Suppose that w_1, w_2 are solutions of the system*

$$\begin{cases} (\partial_{\bar{k}} w_1)(z,k) = e_{-k}\mathbf{s}(k)\overline{w_2(z,k)}, \\ (\partial_{\bar{k}} w_2)(z,k) = e_{-k}\mathbf{s}(k)\overline{w_1(z,k)} \end{cases}$$

for $\mathbf{s} \in L^2$ and $w_1, w_2 \in L_k^4(\mathbb{R}^2)$. Then $w_1 = w_2 = 0$.

This lemma is an easy consequence of Theorem 4.8 if one considers the functions $w_\pm = w_1 \pm w_2$.

First, we'll show that a solution of (4.66) also solves (4.35a)–(4.35b) with

$$m^1(\cdot,k) - 1, \; m^2(\cdot,k) \in L_z^4(\mathbb{R}^2)$$

for each k. For notational convenience we suppress dependence on t.

Proposition 4.24 *Suppose that $m^1(z,k), m^2(z,k)$ solve (4.36) for each $z \in \mathbb{C}$. Then $m^1(\cdot,k)-1, \; m^2(\cdot,k) \in L_z^4(\mathbb{R}^2)$ for each $k \in \mathbb{C}$, and (m_1, m_2) solve (4.35a)–(4.35b) for each z, where $q(z)$ is defined by*

$$q(z) = -\frac{i}{\pi}\int e_k(z)\overline{\mathbf{s}(k)}m^1(z,k)\,dk. \quad (4.68)$$

Proof Differentiating (4.36) we compute (see Exercise 4.18)

$$\begin{cases} \partial_{\bar{k}}\left(\partial_{\bar{z}}m^1\right) = e_{-k}\mathbf{s}\,\overline{(\partial_z + ik)m^2} \\ \partial_{\bar{k}}\left((\partial_z + ik)m^2\right) = e_{-k}\mathbf{s}\,\overline{\partial_{\bar{z}}m^1} \end{cases} \qquad (4.69)$$

(the pointwise differentiation makes sense because, for $\mathbf{s} \in \mathscr{S}(\mathbb{R}^2)$, the functions m^1 and m^2 are smooth functions of both variables). From the large-k asymptotics of m^1, we see that $\partial_{\bar{z}}m^1 = \mathcal{O}\left(|k|^{-1}\right)$, so that $\partial_{\bar{z}}m^1 \in L^4(\mathbb{R}^2)$. On the other hand,

$$(\partial_z + ik)\,m^2 - c(z) \in L^4_k(\mathbb{R}^2)$$

where

$$c(z) = \lim_{|k| \to \infty} ikm^2(z, k) = \frac{i}{\pi} \int e_{-k}(z)\mathbf{s}(k)m^1(z, k)\,dk$$

Making a subtraction in (4.69) we have

$$\begin{cases} \partial_{\bar{k}}w_1 = e_{-k}\mathbf{s}(k)\overline{w_2} \\ \partial_{\bar{k}}w_2 = e_{-k}\mathbf{s}(k)\overline{w_1} \end{cases} \qquad (4.70)$$

where

$$w_1 = \partial_{\bar{z}}m^1 - \overline{c(z)}m^2, \quad w_2 = (\partial_z + ik)\,m^2 - c(z)m^1$$

(see Exercise 4.18). We can now apply Lemma 4.23 to conclude that m^1 and m^2 satisfy (4.35a)–(4.35b) with q as defined in (4.68). □

Remark 4.25 Since $q \in \mathscr{S}(\mathbb{R}^2)$, it follows that m^1 and m^2 have complete large-z asymptotic expansions for each fixed k.

Next, we show that m^1 and m^2 satisfy (4.35c)–(4.35d) by a similar technique, now tracking the dependence of m^1 and m^2 on time.

Proposition 4.26 *Suppose that $m^1(z, k, t)$ and $m^2(z, k, t)$ solve (4.66). Then m^1 and m^2 solve (4.35c)–(4.35d) where q is defined by (4.68) and g is given by $g = -4\partial_{\bar{z}}^{-1}\left(\partial_z|q|^2\right)$.*

Proof At top order the Lax equations (4.35c)–(4.35d) (taking $\varepsilon = +1$ here and in what follows) imply that

$$v_1 := \left(\partial_t - 2i\partial_z^2 + 4k\partial_z\right)m_1 \sim 0, \quad v_2 := \left(\partial_t + 2i\partial_{\bar{z}}^2 - 2ik^2\right)m^2 \sim 0$$

where the corrections vanish as $|z| \to \infty$. Motivated by this observation, we differentiate (4.66) and compute

$$\partial_{\bar{k}} v_1 = e^{it\varphi} s \overline{v_2} \qquad (4.71)$$

$$\partial_{\bar{k}} v_2 = e^{it\varphi} s \overline{v_1} \qquad (4.72)$$

(see Exercise 4.19) where φ is given by (4.41). If v_1 and v_2 were decreasing at infinity as functions of k, Lemma 4.23 would allow us to conclude that $v_1 = v_2 = 0$. This is not the case since $k^2 m^2$ is of order k as $k \to \infty$ and $4k\partial_z m_1$ is of order 1 as $k \to \infty$. For this reason we must make a subtraction using the asymptotic expansions of m^1 and m^2 which will lead to the remaining, lower-order terms in (4.35c)–(4.35d). From Exercise 4.9, we have

$$m^1(z, k) = 1 - \frac{i \partial_{\bar{z}}^{-1} (|q|^2)}{k} + \mathcal{O}(k^{-1}) \qquad (4.73)$$

$$m^2(z, k) = \frac{-i\bar{q}}{k} + \frac{-\bar{q} \partial_{\bar{z}}^{-1} (|q|^2) + \partial_z \bar{q}}{k^2} + \mathcal{O}(k^{-3}) \qquad (4.74)$$

so that

$$v_1 = -4i \partial_z \partial_{\bar{z}}^{-1} (|q|^2) + \mathcal{O}(k^{-1}), \qquad (4.75)$$

$$v_2 = -2\bar{q} k - 2i \left(-q \partial_{\bar{z}}^{-1} (|q|^2) + \partial_z \bar{q} \right) + \mathcal{O}(k^{-1}). \qquad (4.76)$$

Thus if

$$w_1 = v_1 - 2i(\partial_{\bar{z}} q) m^2 + 2iq \partial_{\bar{z}} m^2 - igm^1, \qquad (4.77)$$

$$w_2 = v_2 + 2i(\partial_z \bar{q}) m^1 - 2i\bar{q}(\partial_z + ik) m^1 + i\bar{g} m_2, \qquad (4.78)$$

it follows from the asymptotic expansions (4.73)–(4.74) and (4.75)–(4.76) that $w_1 = \mathcal{O}(k^{-1})$ and $w_2 = \mathcal{O}(k^{-1})$ (see Exercise 4.20), while a straightforward computation (Exercise 4.21) shows that

$$\begin{cases} \partial_{\bar{k}} w_1 = e^{it\varphi} s \overline{w_2} \\ \partial_{\bar{k}} w_2 = e^{it\varphi} s \overline{w_1} \end{cases} \qquad (4.79)$$

We can now use Lemma 4.23 to conclude that $w_1 = w_2 = 0$ and (4.35c)–(4.35d) hold. □

Proof of Theorem 4.21 It follows from Proposition 4.20 that $q_{\text{inv}}(z, 0; q_0) = q_0(z)$, so it suffices to show that q_{inv} is a classical solution of (4.1). By Proposition 4.22 it suffices to prove that this is the case for $q_0 \in \mathscr{S}(\mathbb{R}^2)$. By Propositions 4.24–4.26, the functions m^1 and m^2 solve (4.35a)–(4.35d). If we now set $\psi_1 = e^{i(kz - k^2 t)} m^1$ and $\psi_2 = e^{i(kz - k^2 t)} m^2$, it will follow from the computations in Exercise 2.9 that (2.30) holds with $q = q_{\text{inv}}$ provided we can show that ψ_1 and ψ_2 satisfy conditions (i)–(iii) given there.

Conditions (i) and (ii) are proved in Proposition 4.24. Condition (iii) is equivalent to the statement that, for each (t, z), $m^2(z, k, t) \neq 0$ for at least one k. If not, it follows from Lemma 4.19 that $\partial_{\bar{k}} m^1(z, k) \equiv 0$ so $m^1(z, k, t) \equiv 1$ and $m^2(z, k, t) \equiv 0$ for this fixed (z, t) and all k. It then follows from the ∂_k equation for m^2 that $\mathbf{s}(k, t) \equiv 0$ for this fixed t. But then, since $\mathbf{s}$ evolves linearly, we get $\mathbf{s}(k, t) \equiv 0$, hence $q(z, t) \equiv 0$ by the invertibility of $\mathcal{S}$. Hence, $m^2(z, k, t) \neq 0$ for some k, and conditions (i)–(iii) hold. □

4.6 L^2 Global Well-Posedness and Scattering: The Work of Nachman, Regev, and Tataru

In this section we discuss briefly the results of Nachman et al. [36]. Their first result is a remarkable strengthening of Theorem 4.7.

Theorem 4.27 ([36, Theorem 1.2]) *The scattering transform $\mathcal{S}$ is a diffeomorphism from $L^2(\mathbb{R}^2)$ onto itself with $\mathcal{S}^{-1} = \mathcal{S}$. Moreover, $\|\mathcal{S}q\|_{L^2} = \|q\|_{L^2}$, and the pointwise bound*

$$|(\mathcal{S}q)(k)| \leqslant C\left(\|q\|_{L^2}\right) \mathcal{M}\widehat{q}(k) \tag{4.80}$$

holds.

Theorem 4.27 rests on the following resolvent estimate which is proven using concentration compactness methods. Denote by $\dot{H}^{1/2}(\mathbb{R}^2)$ the homogeneous Sobolev space of order $1/2$, which embeds continuously into $L^4(\mathbb{R}^2)$, and denote by $\dot{H}^{-1/2}(\mathbb{R}^2)$ its topological dual. The authors consider the model equation

$$L_q u = f, \quad L_q u = \partial_{\bar{z}} u + q\bar{u}$$

(compare (4.49)) for $u \in \dot{H}^{1/2}(\mathbb{R}^2)$, where $f \in \dot{H}^{-1/2}(\mathbb{R}^2)$.

Theorem 4.28 ([36, Theorem 1.1]) *The estimate*

$$\left\|L_q^{-1}\right\|_{\dot{H}^{-1/2}(\mathbb{R}^2) \to \dot{H}^{1/2}(\mathbb{R}^2)} \leqslant C\left(\|q\|_{L^2}\right)$$

where $t \to C(t)$ is an increasing, locally bounded function on $[0, \infty)$.

As an immediate consequence of Theorem 4.27, we have:

Theorem 4.29 ([36, Theorem 1.4]) *For any Cauchy data $q_0 \in L^2(\mathbb{R}^2)$, the defocussing DS II equation has a unique global solution in $C(\mathbb{R}, L^2(\mathbb{R}^2) \cap L^4(\mathbb{R}^2 \times \mathbb{R}))$.*

The authors' Theorem 2.4 also includes stability estimates, pointwise bounds, and a global bound on the L^4 norm of the solution in space and time.

The pointwise bound (4.80) plays a crucial role in the authors' analysis of scattering for the DS II equation. Applied to the solution $q_{\text{inv}}(z, t)$ given by (2.23) it implies that

$$|q_{\text{inv}}(z, t)| \lesssim C \left(\|Sq_0\|_{L^2}\right) \mathcal{M}q_{\text{lin}}(z, t)$$

where

$$q_{\text{lin}}(z, t) = \mathcal{F}_a \left(e^{it((\cdot)^2 + (\overline{\cdot})^2)} (Sq_0) \right)(z).$$

which is exactly the solution to the linear problem (4.22) with initial data

$$v_0 = (\mathcal{F}_a \circ S)(q_0).$$

Since the maximal function is bounded between L^p-spaces for $p \in (1, \infty)$, this implies immediately that L^p estimates in space or mixed $L^p - L^q$ estimates in space and time which hold for the linear problem, automatically hold for the nonlinear problem for q_0 on bounded subsets of $L^2(\mathbb{R}^2)$. In particular, it follows that $q \in L^4(\mathbb{R}^2 \times \mathbb{R})$.

Using these estimates, Nachman, Regev and Tataru show that all solutions scatter in $L^2(\mathbb{R}^2)$ and that, indeed the scattering is trivial in the sense that past and future asymptotics are equal. Denote by $U(t)$ the nonlinear evolution

$$U(t)f(z) = \mathcal{S}^{-1} \left(e^{it((\cdot)^2 + (\overline{\cdot})^2)} (Sf) \right)(z).$$

and by $V(t)$ the linear evolution

$$V(t)f(z) = \mathcal{F}_a^{-1} \left(e^{it((\cdot)^2 + (\overline{\cdot})^2)} (\mathcal{F}_a f) \right)(z).$$

In scattering theory we seek initial data $v_\pm$ for the linear equation so that

$$\lim_{t \to \pm\infty} \|U(t)q_0 - V(t)v_\pm\|_{L^2} = 0$$

as vectors in $L^2(\mathbb{R}^2)$. Formally we have

$$v_\pm(t) = \lim_{t \to \pm\infty} V(-t)U(t)q_0$$

if the limit exists. The limiting maps, if they exist, are the nonlinear wave operators $W^\pm$.

To show the convergence, it suffices to show that

$$\frac{d}{dt} V(-t)U(t)q_0 = V(-t)N(q(t))$$

is integrable as an L^2-valued function of t, where $N(q) = q(g \mid \bar{g})$ is the nonlinear term in the DS II equation. Since $q(z,t) \in L^4(\mathbb{R}^2 \times \mathbb{R}^2)$, it follows that $N(q(t)) \in L^{4/3}(\mathbb{R}^2 \times \mathbb{R}^2)$, since the Beurling operator $\bar{\partial}^{-1}\partial$ is bounded from L^p to itself for any $p \in (1, \infty)$. From the Strichartz estimate (4.29), it follows that $V(-t)N(q(t))$ is integrable as an L^2-valued function of t, so that the asymptotes $v_\pm$ exist. Hence:

Theorem 4.30 ([36, Lemma 5.5]) *The nonlinear wave operators $W^\pm$ exist and are Lipschitz continuous maps on $L^2(\mathbb{R}^2)$.*

It can be shown that $v_+ = v_-$ so that the scattering is trivial.

Exercises for Sect. 4

Exercise 4.1 Using integration by parts, show that for any $f \in C_0^\infty(\mathbb{R}^2)$,

$$\frac{1}{\pi} \int \frac{1}{z-w} (\partial_{\bar{w}} f)(w)\, dw = f(z).$$

Hint: Develop a Green's formula for the $\bar{\partial}$ operator analogous to the corresponding formula for the Laplacian, and use the fact that

$$\int \frac{1}{z-w} f(w)\, dw = \lim_{\varepsilon \downarrow 0} \int_{\mathbb{R}^2 \setminus B(z,\varepsilon)} \frac{1}{z-w} f(w)\, dw.$$

Exercise 4.2 Suppose that f is a measurable function with $\int |x|^N |f(x)|\, dx < \infty$ for all nonnegative integers N. Show that $u(z) = \left(\partial_{\bar{z}}^{-1} f\right)(z)$ has a large-z asymptotic expansion of the form

$$u(z) \sim \sum_{j \geq 1} \frac{a_j}{z^j}$$

and give an explicit remainder estimate for $u - \sum_{j=1}^N a_j z^{-j}$.

Remark The equation $\partial_{\bar{z}} u = f \in \mathscr{S}(\mathbb{R}^2)$ implies that u is 'almost' analytic near infinity; the expansion above shows that f 'almost' has a Taylor series near the point at infinity.

Exercise 4.3 Prove (4.10) by writing

$$\int \frac{1}{z-w} f(z) \, dz = \left(\int_{|z-w| \leqslant 1} + \int_{|z-w|>1} \right) \frac{1}{z-w} f(w) \, dw$$

and using Hölder's inequality.

Exercise 4.4 (Proof Suggested by Adrian Nachman) The purpose of this exercise is to prove Theorem 2.3 of [36], which asserts the following. For $0 < \alpha < n$ and $f \in L^p(\mathbb{R}^n)$, $1 < p \leqslant 2$, the estimate

$$\left| |D|^{-\alpha} f(x) \right| \lesssim \left(\lambda^{n-\alpha} M\widehat{f}(0) + \lambda^\alpha Mf(0) \right). \tag{4.81}$$

holds for any $\lambda > 0$, where $|D|^{-\alpha}$ is the Fourier multiplier with symbol $|\xi|^{-\alpha}$.

(a) Prove that

$$\left(|D|^{-\alpha} f \right)(x) = c_\alpha \int_0^\infty t^{\alpha-1} (P_t * f)(x) \, dx.$$

(b) Prove estimate (4.81) by splitting $\int_0^\infty = \left(\int_0^\lambda + \int_\lambda^\infty \right)$ and mimicking the proof of Theorem 4.4.

(c) Optimize in λ to show that

$$\left| \left(|D|^{-\alpha} f \right)(x) \right| \lesssim \left(M\widehat{f}(0) \right)^{\alpha/n} \left(Mf(x) \right)^{1-\alpha/n}.$$

Exercise 4.5 (Duhamel's Formula) Suppose that $f \in \mathscr{S}(\mathbb{R}^2)$ and $F \in C(\mathbb{R}; \mathscr{S}(\mathbb{R}^2))$. Show that the initial value problem

$$iv_t + 2(\partial_z^2 + \partial_{\bar{z}}^2)v = F, \quad v(0) = f$$

is solved by $v \in C(\mathbb{R}; \mathscr{S}(\mathbb{R}^2))$ given by

$$v(t) = V(t)f - i \int_0^t V(t-s) F(s) \, ds.$$

Exercise 4.6 The purpose of this exercise is to prove the basic dispersive estimate (4.30). In this exercise we define

$$\widehat{f}(\xi_1, \xi_2) = \int f(x_1, x_2) e^{-i(\xi_1 x_1 + \xi_2 x_2)} \, dx_1 \, dx_2$$

so that the inverse Fourier transform is

$$\check{g}(x_1, x_2) = \frac{1}{(2\pi)^2} \int g(\xi_1, \xi_2) e^{i(\xi_1 x_1 + \xi_2 x_2)} \, d\xi_1 \, d\xi_2.$$

(a) Let $h(\xi) = \frac{1}{2}(\xi_2^2 - \xi_1^2)$, and let

$$\widehat{f}(\xi_1, \xi_2) = \int_{\mathbb{R}^2} e^{-ix \cdot \xi} f(x) \, dx$$

be the usual (unitary) Fourier transform on $L^2(\mathbb{R}^2)$. Show that, if $f \in \mathscr{S}(\mathbb{R}^2)$, then the unique solution to (4.22) with $v(x, 0) = f(x)$ is given by

$$v(x, t) = \frac{1}{(2\pi)^2} \int e^{ix \cdot \xi} e^{ith(\xi)} \widehat{f}(\xi) \, d\xi$$

(b) Compute the distribution Fourier transform of $e^{ith(\xi)}$ using the result of Exercise 2.3 and separation of variables.
(c) Conclude that

$$v(x, t) = \int K(t, x - y) f(y) \, dy$$

where

$$|K(t, x)| \lesssim t^{-1}.$$

Exercise 4.7 The purpose of this exercise is to prove the Strichartz estimate (4.28).

(a) Using the dispersive estimate $\|V(t)f\|_{L^\infty} \lesssim t^{-1} \|f\|_{L^1}$, the trivial estimate $\|V(t)(f)\|_{L^2}$, and real interpolation, prove that for any $p > 2$,

$$\|V(t)f\|_{L^p} \lesssim_p t^{2/p - 1} \|f\|_{L^{p'}}.$$

(b) Regarding $\int_{-\infty}^{\infty} V(t - s) g(s) \, ds$ as a joint convolution in s, z, use part (a) with $p = 4$ and the Hardy-Littlewood-Sobolev inequality (4.6) with $n = 1$ and $\alpha = 1/2$ to prove (4.28).

Exercise 4.8 The purpose of this exercise is to find the correct normalization for time-dependent joint solutions of (4.31a)–(4.32b). Suppose that

$$\Psi_1(z, k, t) = m_1(z, k, t) e^{ikz}, \quad \Psi_2(z, k, t) = m_2(z, k, t) e^{ikz}$$

solves (4.31a)–(4.31b) with

$$\left(m^1(z,k,t), m^2(z,k,t)\right) \to (1,0) \text{ as } |z| \to \infty.$$

Let

$$\psi_1(z,k,t) = C_1(k,t)\Psi_1(z,k,t), \quad \psi_2(z,k,t) = C_2(k,t)\Psi_2(z,k,t)$$

be a joint solution of (4.31a)–(4.32b). Assuming that

$$\partial m_1(z,k,t)/\partial t, \ \partial m_2(z,k,t)/\partial t \to 0 \text{ as } |z| \to \infty$$

and that $q(z,t)$, $g(z,t) \to 0$ as $|z| \to \infty$, show that $C_1(k,t) = C_2(k,t) = e^{-2ik^2 t}$.

Exercise 4.9 Suppose that the solutions m^1 and m^2 of (4.35a)–(4.35b) admit (differentiable) asymptotic expansions of the form

$$m^1(z,k,t) \sim 1 + \sum_{j \geq 1} \frac{\alpha_j(z,t)}{k^j},$$

$$m^2(z,k,t) \sim \sum_{j \geq 1} \frac{\beta_j(z,t)}{k^j}$$

for each (z,t). Use (4.35a)–(4.35b) to show that

$$i\beta_1(z,t) = \overline{\varepsilon q(z,t)},$$

$$\left(\partial_{\bar{z}} \alpha_j\right)(z,t) = q(z,t) \beta_j(z,t),$$

$$\left(\partial_z \beta_j\right)(z,t) + i\beta_{j+1}(z,t) = \overline{\varepsilon q(z,t)} \alpha_j(z,t)$$

and compute β_1, α_1, and β_2.

Exercise 4.10 The *conjugate Solid Cauchy transform* is the integral operator

$$\left(\partial_{\bar{z}}^{-1} f\right)(z) = \frac{1}{\pi} \int \frac{1}{\bar{z} - \bar{\zeta}} f(\zeta) \, d\zeta$$

and is a solution operator for the equation $\partial u = f$. Suppose that $q \in \mathscr{S}(\mathbb{R}^2)$ and that Eq. (4.42) admit bounded solutions m^1, m^2. Show that, in the large-z asymptotic expansion (4.37), the function $\mathbf{s}(k)$ is given by (4.43). *Hint*: Remember the identity (4.46).

Exercise 4.11 The unitary normalization for the Fourier transform on $\mathbb{R}^2$ is given by

$$(\mathcal{F}_0 f)(\xi) = \frac{1}{2\pi} \int_{\mathbb{R}^2} e^{-ix \cdot \xi} f(x) \, dx$$

so that

$$\left(\mathcal{F}_0^{-1} g\right)(x) = \frac{1}{2\pi} \int_{\mathbb{R}^2} e^{ix\cdot\xi} g(\xi)\, d\xi.$$

With this normalization, $\mathcal{F}_0$ is a unitary map from $L^2(\mathbb{R}^2)$ to itself. Show that if $k = k_1 + ik_2$,

$$(\mathcal{F}_a f)(k) = 2\left(\mathcal{F}_0 \overline{f}\right)(2k_1, -2k_2)$$

so that $\mathcal{F}_a$ is also an isometry of $L^2(\mathbb{R}^2)$. Show also that $\mathcal{F}_a^{-1} = \mathcal{F}_a$.

Exercise 4.12 Show that, given $q \in L^2(\mathbb{R}^2)$ and any $\varepsilon > 0$, we may write

$$q = q_n + q_s$$

where q_n is a smooth function of compact support, and $\|q_s\|_{L^2} < \varepsilon$. *Hint*: mollify q and truncate to a large ball using a smooth cutoff function.

Exercise 4.13 Prove (4.51) for $f \in C_0^\infty(\mathbb{R}^2)$ using the result of Exercise 4.1 and the identity

$$(i\overline{k}) \partial_{\overline{z}}(e_k)(z) = e_k(z).$$

By a density argument and (4.9), show that the same identity holds true as functions in $L^p(\mathbb{R}^2)$ provided $f \in L^p(\mathbb{R}^2) \cap L^{2p/(p+2)}(\mathbb{R}^2)$ and $\partial_{\overline{z}} f \in L^{2p/(2p+2)}(\mathbb{R}^2)$.

Exercise 4.14 Denote by $L^{2,1}(\mathbb{R}^2)$ the space of measurable complex-valued functions with

$$\|f\|_{L^{2,1}} := \left(\int (1+|x|^2)|f(x)|^2\, dx\right)^{1/2} < \infty.$$

Show that for any $f \in L^{2,1}(\mathbb{R}^2)$ and any $p \in (1, 2]$, $\|f\|_{L^p} \lesssim \|f\|_{L^{2,1}}$.

Exercise 4.15 Show that for any $f \in H^1(\mathbb{R}^2)$ and any $p \in [2, \infty)$, $\|f\|_{L^p} \lesssim_p \|f\|_{H^1}$. *Hint*: By the Hausdorff-Young inequality, it suffices to show that $\|f\|_{L^p} \lesssim_p \|f\|_{L^{2,1}}$ for $p \in (1, 2]$ (why?). Then, use the result of Exercise 4.14.

Exercise 4.16 Using Theorem 4.2, show that $H^{1,1}(\mathbb{R}^2)$ is compactly embedded in $L^2(\mathbb{R}^2)$. That is, show that bounded subsets of $H^{1,1}(\mathbb{R}^2)$ are identified with subsets of $L^2(\mathbb{R}^2)$ having compact closure.

Exercise 4.17 Show that the map $q \mapsto q\, \partial_{\overline{z}}^{-1}(e_k \overline{q})$ is Lipschitz continuous from $L^2(\mathbb{R}^2)$ into $L^4(\mathbb{R}^2_k, L^{4/3}(\mathbb{R}^2_z))$. *Hint*: Use (4.18), bilinearity in q, and Hölder's inequality.

Exercise 4.18 Using the commutation relation

$$\partial_{\bar{z}} e_{-k} = e_{-k} \left(\partial_{\bar{z}} - i\bar{k} \right)$$

show that (4.69) and (4.70) hold.

Exercise 4.19 Recall that

$$\varphi(z, k, t) = 2 \left(k^2 + \bar{k}^2 \right) t + \frac{kz + \bar{k}\bar{z}}{t}.$$

Using the commutation relations

$$\partial_z e^{it\varphi} = e^{it\varphi} \left(\partial_z - ik \right), \quad \partial_{\bar{z}} e^{it\varphi} = e^{it\varphi} \left(\partial_{\bar{z}} - i\bar{k} \right),$$

$$\partial_t e^{it\varphi} = e^{it\varphi} \left(\partial_t + 2i(k^2 + \bar{k}^2) \right),$$

derive (4.71)–(4.72) from the $\partial_{\bar{k}}$ equations (4.66) and the time evolution (4.67).

Exercise 4.20 Use the asymptotic expansions (4.73)–(4.74) to show that w_1 and w_2 as defined in (4.77)–(4.78) are of order k^{-1} as $k \to \infty$.

Exercise 4.21 Show that (4.79) holds using (4.71)–(4.72) and the fact that (4.66) holds. Note that $e_{-k}\mathbf{s}(k,t) = e^{it\varphi}\mathbf{s}$.

Acknowledgements I am grateful to the University of Kentucky for sabbatical support during part of the time these lectures were prepared, and to Adrian Nachman and Idan Regev for many helpful discussions about their work. I thank the participants in a Fall 2017 working seminar on [36]— Russell Brown, Joel Klipfel, George Lytle, and Mihai Tohaneanu—for helping me understand this paper in greater depth. I have benefited from conversations with Deniz Bilman about Lax representations and Riemann-Hilbert problems. I have also benefited from course notes for Percy Deift's 2008 course at the Courant Institute on the defocusing NLS equation,[6] his 2015 course there on integrable systems,[7] and Peter Miller's Winter 2018 course at the University of Michigan on integrable systems and Riemann-Hilbert problems.[8] For supplementary material, I have drawn on several excellent sources for material on dispersive equations, Riemann-Hilbert problems, and $\bar{\partial}$-problems, namely the monograph of Astala et al. [5], the textbook of Ponce and Linares [34], and the monograph of Trogdon and Olver [43]. This work was supported by a grant from the Simons Foundation/SFARI (359431, PAP).

---

[6] I am grateful to Percy Deift for kindly sharing his handwritten notes for this course.
[7] Handwritten notes may be found at https://www.math.nyu.edu/faculty/deift/RHP/.
[8] Please see the course website http://www.math.lsa.umich.edu/~millerpd/CurrentCourses/651_Winter18.html.

References

1. Mark J. Ablowitz, *Nonlinear dispersive waves*, Cambridge Texts in Applied Mathematics, Cambridge University Press, New York, 2011, Asymptotic analysis and solitons. MR 2848561
2. Mark J. Ablowitz, David J. Kaup, Alan C. Newell, and Harvey Segur, *The inverse scattering transform-Fourier analysis for nonlinear problems*, Studies in Appl. Math. **53** (1974), no. 4, 249–315. MR 0450815
3. Mark J. Ablowitz and Harvey Segur, *Solitons and the inverse scattering transform*, SIAM Studies in Applied Mathematics, vol. 4, Society for Industrial and Applied Mathematics (SIAM), Philadelphia, Pa., 1981. MR 642018
4. Kari Astala, Daniel Faraco, and Keith M. Rogers, *On Plancherel's identity for a two-dimensional scattering transform*, Nonlinearity **28** (2015), no. 8, 2721–2729. MR 3382582
5. Kari Astala, Tadeusz Iwaniec, and Gaven Martin, *Elliptic partial differential equations and quasiconformal mappings in the plane*, Princeton Mathematical Series, vol. 48, Princeton University Press, Princeton, NJ, 2009. MR 2472875
6. R. Beals and R. R. Coifman, *Scattering and inverse scattering for first order systems [MR0728266 (85f:34020)]*, Surveys in differential geometry: integral systems [integrable systems], Surv. Differ. Geom., vol. 4, Int. Press, Boston, MA, 1998, pp. 467–519. MR 1726933
7. Richard Beals and Ronald R. Coifman, *Scattering and inverse scattering for first order systems*, Comm. Pure Appl. Math. **37** (1984), no. 1, 39–90. MR 728266
8. _____, *Multidimensional inverse scatterings and nonlinear partial differential equations*, Pseudodifferential operators and applications (Notre Dame, Ind., 1984), Proc. Sympos. Pure Math., vol. 43, Amer. Math. Soc., Providence, RI, 1985, pp. 45–70. MR 812283
9. _____, *The D-bar approach to inverse scattering and nonlinear evolutions*, Phys. D **18** (1986), no. 1–3, 242–249, Solitons and coherent structures (Santa Barbara, Calif., 1985). MR 838329
10. _____, *Linear spectral problems, nonlinear equations and the $\bar{\partial}$-method*, Inverse Problems **5** (1989), no. 2, 87–130. MR 991913
11. R. M. Brown, *Estimates for the scattering map associated with a two-dimensional first-order system*, J. Nonlinear Sci. **11** (2001), no. 6, 459–471. MR 1871279
12. R. M. Brown, K. A. Ott, and P. A. Perry, *Action of a scattering map on weighted Sobolev spaces in the plane*, J. Funct. Anal. **271** (2016), no. 1, 85–106. MR 3494243
13. Russell M. Brown and Gunther A. Uhlmann, *Uniqueness in the inverse conductivity problem for nonsmooth conductivities in two dimensions*, Comm. Partial Differential Equations **22** (1997), no. 5–6, 1009–1027. MR 1452176
14. Thierry Cazenave and Fred B. Weissler, *Some remarks on the nonlinear Schrödinger equation in the critical case*, Nonlinear semigroups, partial differential equations and attractors (Washington, DC, 1987), Lecture Notes in Math., vol. 1394, Springer, Berlin, 1989, pp. 18–29. MR 1021011
15. P. Deift and X. Zhou, *A steepest descent method for oscillatory Riemann-Hilbert problems. Asymptotics for the MKdV equation*, Ann. of Math. (2) **137** (1993), no. 2, 295–368. MR 1207209
16. Percy Deift, *Fifty years of kdv: An integrable system (coxeter lectures, august 2017)*, 2018.
17. Percy Deift and Xin Zhou, *Long-time asymptotics for solutions of the NLS equation with initial data in a weighted Sobolev space*, Comm. Pure Appl. Math. **56** (2003), no. 8, 1029–1077, Dedicated to the memory of Jürgen K. Moser. MR 1989226
18. M. Dieng and K. D. T. . McLaughlin, *Long-time Asymptotics for the NLS equation via dbar methods*, ArXiv e-prints (2008).
19. M. Dieng, P. Miller, and K. D. T. McLaughlin, *Dispersive asymptotics for linear and integrable equations by the $\bar{\partial}$ method of steepest descent*, Nonlinear Dispersive Partial Differential Equations and Inverse Scattering, Fields Institute Communications, **83**, Springer, New York, 2019, pp. 253–291. https://doi.org/10.1007/978-1-4939-9806-7_5

20. L. D. Faddeev, *Increasing solutions of the Schrödinger equation*, Sov. Phys. Doklady **10** (1965), 1033–1035.
21. A. S. Fokas and M. J. Ablowitz, *Method of solution for a class of multidimensional nonlinear evolution equations*, Phys. Rev. Lett. **51** (1983), no. 1, 7–10. MR 711737
22. _____, *On the inverse scattering transform of multidimensional nonlinear equations related to first-order systems in the plane*, J. Math. Phys. **25** (1984), no. 8, 2494–2505. MR 751539
23. Athanassios S. Fokas and Mark J. Ablowitz, *The inverse scattering transform for multidimensional (2 + 1) problems*, Nonlinear phenomena (Oaxtepec, 1982), Lecture Notes in Phys., vol. 189, Springer, Berlin, 1983, pp. 137–183. MR 727861
24. Rupert L. Frank and Elliott H. Lieb, *A new, rearrangement-free proof of the sharp Hardy-Littlewood-Sobolev inequality*, Spectral theory, function spaces and inequalities, Oper. Theory Adv. Appl., vol. 219, Birkhäuser/Springer Basel AG, Basel, 2012, pp. 55–67. MR 2848628
25. Patrick Gérard, *Description du défaut de compacité de l'injection de Sobolev*, ESAIM Control Optim. Calc. Var. **3** (1998), 213–233. MR 1632171
26. Jean-Michel Ghidaglia and Jean-Claude Saut, *On the initial value problem for the Davey-Stewartson systems*, Nonlinearity **3** (1990), no. 2, 475–506. MR 1054584
27. P. G. Grinevich, *The scattering transform for the two-dimensional Schrödinger operator with a potential that decreases at infinity at fixed nonzero energy*, Uspekhi Mat. Nauk **55** (2000), no. 6(336), 3–70. MR 1840357
28. Harald Hanche-Olsen and Helge Holden, *The Kolmogorov-Riesz compactness theorem*, Expo. Math. **28** (2010), no. 4, 385–394. MR 2734454
29. _____, *Addendum to "The Kolmogorov-Riesz compactness theorem" [Expo. Math. 28 (2010) 385–394] [MR2734454]*, Expo. Math. **34** (2016), no. 2, 243–245. MR 3494284
30. Rowan Killip, Monica Vişan, and Xiaoyi Zhang, *Low regularity conservation laws for integrable PDE*, Geom. Funct. Anal. **28** (2018), no. 4, 1062–1090. MR 3820439
31. Rowan Killip and Monica Visan, *KdV is wellposed in H^{-1}*, arXiv e-prints (2018), arXiv:1802.04851.
32. Herbert Koch and Daniel Tataru, *Conserved energies for the cubic nonlinear Schrödinger equation in one dimension*, Duke Math. J. **167** (2018), no. 17, 3207–3313. MR 3874652
33. Elliott H. Lieb, *Sharp constants in the Hardy-Littlewood-Sobolev and related inequalities*, Ann. of Math. (2) **118** (1983), no. 2, 349–374. MR 717827
34. Felipe Linares and Gustavo Ponce, *Introduction to nonlinear dispersive equations*, second ed., Universitext, Springer, New York, 2015. MR 3308874
35. Camil Muscalu, Terence Tao, and Christoph Thiele, *A Carleson theorem for a Cantor group model of the scattering transform*, Nonlinearity **16** (2003), no. 1, 219–246. MR 1950785
36. A. I. Nachman, I. Regev, and D. I. Tataru, *A Nonlinear Plancherel Theorem with Applications to Global Well-Posedness for the Defocusing Davey-Stewartson Equation and to the Inverse Boundary Value Problem of Calderon*, ArXiv e-prints (2017).
37. Tohru Ozawa, *Exact blow-up solutions to the Cauchy problem for the Davey-Stewartson systems*, Proc. Roy. Soc. London Ser. A **436** (1992), no. 1897, 345–349. MR 1177134
38. Peter A. Perry, *Global well-posedness and long-time asymptotics for the defocussing Davey-Stewartson II equation in $H^{1,1}(\mathbb{C})$*, J. Spectr. Theory **6** (2016), no. 3, 429–481, With an appendix by Michael Christ. MR 3551174
39. Li-Yeng Sung, *An inverse scattering transform for the Davey-Stewartson II equations. I*, J. Math. Anal. Appl. **183** (1994), no. 1, 121–154. MR 1273437
40. _____, *An inverse scattering transform for the Davey-Stewartson II equations. II*, J. Math. Anal. Appl. **183** (1994), no. 2, 289–325. MR 1274142
41. _____, *An inverse scattering transform for the Davey-Stewartson II equations. III*, J. Math. Anal. Appl. **183** (1994), no. 3, 477–494. MR 1274849
42. Terence Tao, *Nonlinear dispersive equations*, CBMS Regional Conference Series in Mathematics, vol. 106, Published for the Conference Board of the Mathematical Sciences, Washington, DC; by the American Mathematical Society, Providence, RI, 2006, Local and global analysis. MR 2233925

43. Thomas Trogdon and Sheehan Olver, *Riemann-Hilbert problems, their numerical solution, and the computation of nonlinear special functions*, Society for Industrial and Applied Mathematics (SIAM), Philadelphia, PA, 2016. MR 3450072
44. I. N. Vekua, *Generalized analytic functions*, Pergamon Press, London-Paris-Frankfurt; Addison-Wesley Publishing Co., Inc., Reading, Mass., 1962. MR 0150320
45. V. E. Zakharov and S. V. Manakov, *Asymptotic behavior of non-linear wave systems integrated by the inverse scattering method*, Z. Èksper. Teoret. Fiz. **71** (1976), no. 1, 203–215. MR 0673411
46. V. E. Zakharov and A. B. Shabat, *Exact theory of two-dimensional self-focusing and one-dimensional self-modulation of waves in nonlinear media*, Ž. Èksper. Teoret. Fiz. **61** (1971), no. 1, 118–134. MR 0406174
47. Xin Zhou, L^2-*Sobolev space bijectivity of the scattering and inverse scattering transforms*, Comm. Pure Appl. Math. **51** (1998), no. 7, 697–731. MR 1617249

Notation Index

ad σ_3	Operator $A \mapsto [\sigma_3, A]$ on 2×2 matrices
$e_k(z)$	Unimodular phase function (see (2.21), p. 170) where $k, z \in \mathbb{C}$
$e^{it\Delta}$	Solution operator for linear Schrödinger equation (2.1)
m^1, m^2	Solutions to (4.35a)–(4.35b) obeying the asymptotic condition (4.34)
$\mathbf{n}(x, z)$	Null vector for a Riemann-Hilbert problem (see Proposition 3.12)
p^*	Sobolev conjugate exponent $(2 - p)/2p$
q_{inv}	Solution of DS II by inverse scattering (see (4.5), p. 212)
q_{lin}	Solution of linearized DS II equation (see (4.22), p. 219)
r	"Right" reflection coefficient for NLS $r(\lambda) = -b(\lambda)/\overline{a(\lambda)}$ (see (3.29), p. 191)
$\check{r}$	"Left" reflection coefficient for NLS $\check{r}(\lambda) = -b(\lambda)/a(\lambda)$ (see (3.25), p. 189)
$\mathbf{s}$	Scattering transform $\mathbf{s} = \mathcal{S}q$ for DSII equation (see (4.43), p. 225)
$\mathcal{B}(Y)$	The Banach algebra of bounded operators on a Banach space Y
$C_\pm$	Cauchy projectors for $L^2(\mathbb{R})$ (see (3.36), p. 194)
$\mathcal{C}_w$	Beals-Coifman integral operator (see (3.38), p. 195) with weights $w = (w^+, w^-)$
$\mathcal{F}$	Fourier transform (2.28) (lectures 1 and 2)
$\mathcal{F}_a$	Antilinear Fourier transform (see (4.44), p. 225)
$H^1(\mathbb{R})$	Sobolev space (see (3.2), p. 175)
$H^{1,1}(\mathbb{R})$	Weighted Sobolev space (see (2.16), p. 167)
$H^{1,1}_1(\mathbb{R})$	Open subset of $H^{1,1}(\mathbb{R})$ consisting of those $r \in H^{1,1}(\mathbb{R})$ with $\|r\|_\infty < 1$.
$H^{1,1}(\mathbb{R}^2)$	Weighted Sobolev space (see (2.27), p. 172)
$\mathbb{I}$	The 2×2 identity matrix
$\mathcal{I}$	Inverse scattering map for NLS (see RHP 2.1 and (2.12))
$\mathcal{L}$	Linear operator for NLS spectral problem (see (2.6), p. 164) or DS II spectral problem (see (2.18), p. 169)

M	Normalized solution $\mathbf{M}(x,z) = \psi(x,z)e^{-ixz\sigma_3}$ (see (2.11), p. 165)
$\mathbf{M}^\pm$	Normalized Jost solution defined by $\Psi^\pm = \mathbf{M}^\pm e^{-ix\lambda\sigma_3}$
$\mathbf{M}^\ell, \mathbf{M}^r$	Left and right Beals-Coifman solutions (see Theorems 3.9 and 3.8)
$\mathbf{M}_\pm$	Boundary values of a Beals-Coifman solution
$\mathcal{M}f$	Hardy-Littlewood maximal function of $f \in L^p(\mathbb{R}^n)$
N	Normalized Jost solution (see (3.5), p. 176)
Q	Potential matrix in $\mathcal{L}$ (see (2.6), p. 164) (NLS) or ((2.18), p. 169) (DS II)
$\mathbf{Q}_1, \mathbf{Q}_2$	Matrices in Lax representation for NLS (see (2.5), p. 164)
$\mathcal{R}$	Direct scattering map for NLS (see (2.10), p. 165)
$\mathscr{S}(\mathbb{R}), \mathscr{S}(\mathbb{R}^2)$	Schwartz class of C^∞ functions of rapid decrease on $\mathbb{R}, \mathbb{R}^2$
$\mathscr{S}_1(\mathbb{R})$	Functions $r \in \mathscr{S}(\mathbb{R})$ with $\|r\|_\infty < 1$
S	Model compact operator (see (4.12), p. 214)
$\mathcal{S}$	Scattering map for DS II equation (see (2.22), p. 170)
$\mathbf{S}$	Beurling transform (see (4.19), p. 218)
$\overline{\mathbf{S}}$	Conjugate Beurling transform (see (4.20), p. 218)
$T(\lambda)$	Transition matrix (see (2.9), p. 165)
$V(t)$	Solution operator for linearized DS II equation (4.22)
$\mathbf{V}^\ell, \mathbf{V}^r$	Jump matrices for Riemann-Hilbert problems satisfied by left and right Beals-Coifman solutions (see respectively (3.30) and (3.26))
θ	Phase function for the RHP that solves NLS (see (2.15), p. 167)
φ	Phase function for the $\bar{\partial}$-problem that solves DS II (see (4.41), p. 224)
$\Psi^\pm$	Jost solutions to $\mathcal{L}\psi = \lambda\psi$ (see (2.7), p. 164)
ψ	Generic solution to $\mathcal{L}\psi = z\psi$
μ	Solution to Beals-Coifman integral equation (see (3.37), p. 195)
$\mu^\sharp$	$\mu - \mathbb{I}$
$\sigma_1, \sigma_2, \sigma_3$	Pauli matrices $\begin{pmatrix} 0 & 1 \\ 1 & 0 \end{pmatrix}, \begin{pmatrix} 0 & -i \\ i & 0 \end{pmatrix}, \begin{pmatrix} 1 & 0 \\ 0 & -1 \end{pmatrix}$
$\partial_{\bar{z}}^{-1}$	Solid Cauchy transform (see (4.7), p. 213)
∂_z^{-1}	Conjugate solid Cauchy transform (see (4.8), p. 213)
$\partial C(L^2)$	The linear space of pairs (h_+, h_-) with $h_\pm = C_\pm h$ for some $h \in L^2(\mathbb{R})$
$\partial_z, \partial_{\bar{z}}$	Wirtinger ($\bar{z}$ and z) derivatives (see (4.2), p. 211)

Dispersive Asymptotics for Linear and Integrable Equations by the $\bar{\partial}$ Steepest Descent Method

Momar Dieng, Kenneth D. T.-R. McLaughlin, and Peter D. Miller

1 Introduction

The long time behavior of solutions $q(x, t)$ of the Cauchy initial-value problem for the defocusing nonlinear Schrödinger (NLS) equation

$$\mathrm{i}\frac{\partial q}{\partial t} + \frac{\partial^2 q}{\partial x^2} - 2|q|^2 q = 0, \tag{1}$$

with initial data decaying for large x:

$$q(x, 0) = q_0(x) \to 0, \quad |x| \to \infty, \tag{2}$$

has been studied extensively, under various assumptions on the smoothness and decay properties of the initial data q_0 [3, 5, 6, 8, 10, 19, 20]. The asymptotic behavior takes the following form: as $t \to +\infty$, one has

$$q(x, t) = t^{-1/2} \alpha(z_0) \mathrm{e}^{\mathrm{i}x^2/(4t) - \mathrm{i}\nu(z_0) \ln(8t)} + \mathcal{E}(x, t), \tag{3}$$

M. Dieng
Department of Mathematics, University of Arizona, Tucson, AZ, USA

K. D. T.-R. McLaughlin
Department of Mathematics, University of Arizona, Tucson, AZ, USA

Department of Mathematics, Colorado State University, Fort Collins, CO, USA
e-mail: kenmcl@rams.colostate.edu

P. D. Miller (✉)
Department of Mathematics, University of Michigan–Ann Arbor, Ann Arbor, MI, USA
e-mail: millerpd@umich.edu

© Springer Science+Business Media, LLC, part of Springer Nature 2019
P. D. Miller et al. (eds.), *Nonlinear Dispersive Partial Differential Equations and Inverse Scattering*, Fields Institute Communications 83,
https://doi.org/10.1007/978-1-4939-9806-7_5

where $\mathcal{E}(x,t)$ is an error term and for $z \in \mathbb{R}$, $\nu(z)$ and $\alpha(z)$ are defined by

$$\nu(z) := -\frac{1}{2\pi}\ln(1-|r(z)|^2), \quad |\alpha(z)|^2 = \frac{1}{2}\nu(z), \tag{4}$$

and

$$\arg(\alpha(z)) = \frac{1}{\pi}\int_{-\infty}^{z}\ln(z-s)\,d\ln(1-|r(s)|^2) + \frac{\pi}{4} + \arg(\Gamma(i\nu(z))) - \arg(r(z)). \tag{5}$$

Here $z_0 = -x/(4t)$, Γ is the gamma function, and $r(z)$ is the so-called reflection coefficient associated to the initial data q_0. The connection between the initial data $q_0(x)$ and the reflection coefficient $r(z)$ is achieved through the spectral theory of the associated self-adjoint Zakharov-Shabat differential operator

$$\mathcal{L} := i\sigma_3\frac{d}{dx} + \mathbf{Q}(x), \quad \sigma_3 := \begin{pmatrix} 1 & 0 \\ 0 & -1 \end{pmatrix}, \quad \mathbf{Q}(x) := \begin{pmatrix} 0 & -iq_0(x) \\ iq_0(x) & 0 \end{pmatrix},$$

acting in $L^2(\mathbb{R};\mathbb{C}^2)$ as described, for example, in [6]. See also the contribution of Perry in this volume: [17, Section 2].

The modulus $|\alpha(z_0)|$ of the complex amplitude $\alpha(z_0)$ as written in (4) was first obtained by Segur and Ablowitz [19] from trace formulæ under the assumption that $q(x,t)$ has the form (3) where $\mathcal{E}(x,t)$ is small for large t. Zakharov and Manakov [20] took the form (3) as an ansatz to motivate a kind of WKB analysis of the reflection coefficient $r(z)$ and as a consequence were able to also calculate the phase of $\alpha(z_0)$, obtaining for the first time the phase as written in (5). Its [10] was the first to observe the key role played in the large-time behavior of $q(x,t)$ by an "isomonodromy" problem for parabolic cylinder functions; this problem has been an essential ingredient in all subsequent studies of the large-t limit and as we shall see it is a non-commutative analogue of the Gaussian integral that produces the familiar factors of $\sqrt{2\pi}$ in the stationary phase approximation of integrals. The first time that the form (3) itself was rigorously deduced from first principles (rather than assumed) and proven to be accurate for large t (incidentally reproducing the formulæ (4)–(5) in an ansatz-free fashion) was in the work of Deift and Zhou [3] (see [6] for a pedagogic description) who brought the recently introduced nonlinear steepest descent method [4] to bear on this problem. Indeed, under the assumption of high orders of smoothness and decay on the initial data q_0, the authors of [3] proved that $\mathcal{E}(x,t)$ satisfies

$$\sup_{x \in \mathbb{R}}|\mathcal{E}(x,t)| = \mathcal{O}\left(\frac{\ln(t)}{t}\right), \quad t \to +\infty. \tag{6}$$

It is reasonable to expect that any estimate of the error term $\mathcal{E}(x,t)$ would depend on the smoothness and decay assumptions made on q_0, and so it is natural to ask what happens to the estimate (6) if the assumptions on q_0 are weakened. Early in this millennium, Deift and Zhou developed some new tools for the analysis of Riemann-Hilbert problems, originally aimed at studying the long time behavior of perturbations of the NLS equation [7]. Their methods allowed them to establish long time asymptotics for the Cauchy problem (1)–(2) with essentially minimal assumptions on the initial data [8]. Indeed, they assumed the initial data q_0 to lie in the weighted Sobolev space

$$H^{1,1}(\mathbb{R}) := \left\{ f \in L^2(\mathbb{R}) : xf, f' \in L^2(\mathbb{R}) \right\}. \tag{7}$$

It is well known that if $q_0 \in H^{1,1}(\mathbb{R})$, then the associated reflection coefficient[1] satisfies $r \in H_1^{1,1}(\mathbb{R})$, where

$$H_1^{1,1}(\mathbb{R}) := \left\{ f \in H^{1,1}(\mathbb{R}) : \sup_{z \in \mathbb{R}} |f(z)| < 1 \right\}, \tag{8}$$

and more generally the spectral transform $\mathcal{R}$ associated with the Zakharov-Shabat operator $\mathcal{L}$ (6) is a map $\mathcal{R} : H^{1,1}(\mathbb{R}) \to H_1^{1,1}(\mathbb{R})$, $q_0 \mapsto r = \mathcal{R}q_0$ that is a bi-Lipschitz bijection [21]. The result of [8] is then that the Cauchy problem (1)–(2) for $q_0 \in H^{1,1}(\mathbb{R})$ has a unique weak solution for which (3) holds with an error term $\mathcal{E}(x, t)$ that satisfies, for any fixed κ in the indicated range,

$$\sup_{x \in \mathbb{R}} |\mathcal{E}(x, t)| = \mathcal{O}\left(t^{-\left(\frac{1}{2}+\kappa\right)} \right), \quad t \to +\infty, \quad 0 < \kappa < \frac{1}{4}. \tag{9}$$

Subsequently, McLaughlin and Miller [13, 14] developed a method for the asymptotic analysis of Riemann-Hilbert problems in which jumps across contours are "smeared out" over a two-dimensional region in the complex plane, resulting in an equivalent $\bar{\partial}$ problem that is more easily analyzed. In this paper we adapt and extend this method to the Riemann-Hilbert problem of inverse-scattering associated to the Cauchy problem (1)–(2). The main point of our work is this: by using the $\bar{\partial}$ approach, we avoid all delicate estimates involving Cauchy projection operators in L^p spaces (which are central to the work in [8]). Instead it is only necessary to estimate certain double integrals, an exercise involving nothing more than calculus. Remarkably, this elementary approach also sharpens the result obtained in [8]. Our result is as follows.

---

[1]Since $q_0 \in H^{1,1}(\mathbb{R})$ implies that $(1 + |x|)q_0(x)$ is square-integrable, it follows by Cauchy-Schwarz that $H^{1,1}(\mathbb{R}) \subset L^1(\mathbb{R})$, which in turn implies that the reflection coefficient $r(z)$ is well-defined for each $z \in \mathbb{R}$.

Theorem 1.1 *The Cauchy problem* (1)–(2) *with initial data q_0 in the weighted Sobolev space $H^{1,1}(\mathbb{R})$ defined by* (7) *has a unique weak solution having the form* (3)–(5) *in which $r(z)$ is the reflection coefficient associated with q_0 and where the error term satisfies*

$$\sup_{x \in \mathbb{R}} |\mathcal{E}(x,t)| = \mathcal{O}\left(t^{-\frac{3}{4}}\right), \quad t \to +\infty. \tag{10}$$

The main features of this result are as follows.

- The error estimate is an improvement over the one reported in [8], i.e., we prove that the endpoint case $\kappa = \frac{1}{4}$ holds in (9). Our methods also suggest that the improved estimate (10) on the error is sharp.
- As with the result (9) obtained in [8], the improved estimate (10) only requires the condition $r \in H_1^{1,0}(\mathbb{R})$, i.e., it is not necessary that $zr(z) \in L^2(\mathbb{R})$, but only that r lies in the classical Sobolev space $H^1(\mathbb{R})$ and satisfies $|r(z)| \leq \rho$ for some $\rho < 1$. Dropping the weighted L^2 condition on r corresponds to admitting rougher initial data q_0. For such data, the solution of the Cauchy problem is of a weaker nature, as discussed at the end of [8].
- The new $\bar{\partial}$ method which is used to derive the estimate (10) affords a considerably less technical proof than previous results.
- The method used to establish the estimate (10) is readily extended to derive a more detailed asymptotic expansion, beyond the leading term (see the remark at the end of the paper).

Given the reflection coefficient $r \in H_1^{1,1}(\mathbb{R})$ associated with initial data $q_0 \in H^{1,1}(\mathbb{R})$ via the spectral transform $\mathcal{R}$ for the Zakharov-Shabat operator $\mathcal{L}$, the solution of the Cauchy problem for the nonlinear Schrödinger equation (1) may be described as follows. For full details, we again refer the reader to [17, Section 2]. Consider the following Riemann-Hilbert problem:

Riemann-Hilbert Problem 1 *Given parameters $(x,t) \in \mathbb{R}^2$, find $\mathbf{M} = \mathbf{M}(z) = \mathbf{M}(z; x, t)$, a 2×2 matrix, satisfying the following conditions:*

Analyticity $\mathbf{M}$ *is an analytic function of z in the domain $\mathbb{C} \setminus \mathbb{R}$. Moreover, $\mathbf{M}$ has a continuous extension to the real axis from the upper (lower) half-plane denoted $\mathbf{M}_+(z)$ ($\mathbf{M}_-(z)$) for $z \in \mathbb{R}$.*

Jump Condition *The boundary values satisfy the jump condition*

$$\mathbf{M}_+(z) = \mathbf{M}_-(z)\mathbf{V}_\mathbf{M}(z), \quad z \in \mathbb{R}, \tag{11}$$

where the jump matrix $\mathbf{V}_\mathbf{M}(z)$ is defined by

$$\mathbf{V_M}(z) := \begin{pmatrix} 1 - |r(z)|^2 & -\overline{r(z)}e^{-2it\theta(z;z_0)} \\ r(z)e^{2it\theta(z;z_0)} & 1 \end{pmatrix}, \quad z \in \mathbb{R}, \quad \theta(z;z_0) := 2z^2 - 4z_0 z,$$

$$z_0 := -\frac{x}{4t}. \tag{12}$$

Normalization *There is a matrix* $\mathbf{M}_1(x, t)$ *such that*

$$\mathbf{M}(z) = \mathbb{I} + z^{-1}\mathbf{M}_1(x, t) + o(z^{-1}), \quad z \to \infty. \tag{13}$$

From the solution of this Riemann-Hilbert problem, one defines a function $q(x, t)$, $(x, t) \in \mathbb{R}^2$, by

$$q(x, t) := 2iM_{1,12}(x, t). \tag{14}$$

The fact of the matter is then that $q(x, t)$ is the solution of the Cauchy problem (1)–(2).

Recent studies of the long-time behavior of the solution of the NLS initial-value problem (1)–(2) have involved the detailed analysis of the solution $\mathbf{M}$ to Riemann-Hilbert Problem 1. As regularity assumptions on the initial data q_0 are relaxed, this analysis becomes more involved, technically. The purpose of this manuscript is to carry out a complete analysis of the long-time asymptotic behavior of $\mathbf{M}$ under the assumption that $r \in H_1^{1,1}(\mathbb{R})$ (or really, $r \in H_1^{1,0}(\mathbb{R})$), as in [6], but via a $\bar{\partial}$ approach which replaces technical harmonic analysis estimates involving Cauchy projection operators with very straightforward estimates involving some explicit double integrals.

The proof of Theorem 1.1 using the methodology of [13, 14] was originally obtained by the first two authors in 2008 [9]. Since then the technique has been used successfully to study many other related problems of large-time behavior for various integrable equations. In [2], the authors used the methods of [9] to analyze the stability of multi-dark-soliton solutions of (1). In [1], the method of [9] was used to confirm the soliton resolution conjecture for the focusing version of the NLS equation under generic conditions on the discrete spectrum. In [12], the large-time behavior of solutions of the derivative NLS equation was studied using $\bar{\partial}$ methods, and in [11] the same techniques were used to establish a form of the soliton resolution conjecture for this equation. Similar $\bar{\partial}$ methods more based on the original approach of [13, 14] have also been useful in studying some problems of nonlinear wave theory not necessarily in the realm of large time asymptotics, for instance [15], which deals with boundary-value problems for (1) in the semiclassical limit. Based on this continued interest in $\bar{\partial}$ methods, we decided to write this review paper containing all of the results and arguments of [9], some in a new form, as well as some additional expository material which we hope the reader might find helpful.

2 An Unorthodox Approach to the Corresponding Linear Problem

In order to motivate the $\bar{\partial}$ steepest descent method, we first consider the Cauchy problem for the linear equation corresponding to (1), namely

$$i\frac{\partial q}{\partial t} + \frac{\partial^2 q}{\partial x^2} = 0, \tag{15}$$

with initial condition (2) for which $q_0 \in H^{1,1}(\mathbb{R})$. By Fourier transform theory, if

$$\hat{q}_0(z) := \int_{\mathbb{R}} q_0(x) e^{2izx} \, dx, \quad z \in \mathbb{R} \tag{16}$$

is the Fourier transform of the initial data, then $\hat{q}_0$ as a function of $z \in \mathbb{R}$ also lies in the weighted Sobolev space $H^{1,1}(\mathbb{R})$, and the solution of the Cauchy problem is given in terms of $\hat{q}_0$ by the integral

$$q(x, t) = \frac{1}{\pi} \int_{\mathbb{R}} \hat{q}_0(z) e^{-2it\theta(z;z_0)} \, dz, \tag{17}$$

where $\theta(z; z_0)$ and z_0 are as defined in (12). It is worth noticing that this formula is exactly what arises from Riemann-Hilbert Problem 1 via the formula (14) if only the jump matrix $\mathbf{V_M}(z)$ in (12) is replaced with the triangular form

$$\mathbf{V_M}(z) := \begin{pmatrix} 1 & -\hat{q}_0(z) e^{-2it\theta(z;z_0)} \\ 0 & 1 \end{pmatrix}, \quad z \in \mathbb{R} \tag{18}$$

in which case the solution of Riemann-Hilbert Problem 1 is explicitly given by

$$\mathbf{M}(z; x, t) = \mathbb{I} - \frac{1}{2\pi i} \int_{\mathbb{R}} \frac{\hat{q}_0(\zeta) e^{-2it\theta(\zeta;z_0)}}{\zeta - z} \, d\zeta \begin{pmatrix} 0 & 1 \\ 0 & 0 \end{pmatrix}. \tag{19}$$

This shows that the reflection coefficient $r(z)$ is a nonlinear analogue of (the complex conjugate of) the Fourier transform $\hat{q}_0(z)$.

Assuming that $z_0 \in \mathbb{R}$ is fixed, the method of stationary phase applies to deduce an asymptotic expansion of the integral in (17). The only point of stationary phase is $z = z_0$, and the classical formula of Stokes and Kelvin yields

$$q(x, t) = \frac{1}{\pi} \sqrt{\frac{2\pi}{t|-2\theta''(z_0; z_0)|}} \hat{q}_0(z_0) e^{-2it\theta(z_0;z_0) - i\pi/4} + \mathcal{E}(x, t)$$

$$= t^{-1/2} \frac{\hat{q}_0(z_0) e^{-i\pi/4}}{2\sqrt{\pi}} e^{ix^2/(4t)} + \mathcal{E}(x, t), \tag{20}$$

where the error term $\mathcal{E}(x,t)$ is of order $t^{-3/2}$ as $t \to +\infty$ under the best assumptions on $\hat{q}_0$, assumptions that guarantee that the error has a complete asymptotic expansion in terms proportional via explicit oscillatory factors to descending half-integer powers of t. To derive this expansion from first principles consists of several steps as follows.

- One introduces a smooth partition of unity to separate contributions to the integral from points z close to z_0 and far from z_0.
- One uses integration by parts to estimate the contributions from points z far from z_0. This requires having sufficiently many derivatives of $\hat{q}_0(z)$, which corresponds to having sufficient decay of $q_0(x)$.
- One approximates $\hat{q}_0(z)$ locally near z_0 by an *analytic function* with an accuracy related to the size of t and the number of terms of the expansion that are desired.
- One uses Cauchy's theorem to deform the path of integration for the approximating integrand to a diagonal path over the stationary phase point. The slope of the diagonal path produces the phase factor of $e^{-i\pi/4}$, and the path integral of the leading term $\hat{q}_0(z_0)e^{-2it\theta(z;z_0)}$ in the local approximation of $\hat{q}_0(z)e^{-2it\theta(z;z_0)}$ is a Gaussian integral that produces the factor of $\sqrt{\pi}$.

It is possible to implement all steps of this method assuming, say, that q_0 (and hence also $\hat{q}_0$) is a Schwartz-class function. However, as one reduces the regularity of q_0 it becomes impossible to obtain an expansion to all orders. More to the point, even in the presence of Schwartz-class regularity, the proof of the stationary phase expansion by the traditional methods outlined above is complicated, perhaps more so than necessary as we hope to convince the reader.

To explain an alternative approach that bears fruit in the case $q_0 \in H^{1,1}(\mathbb{R})$ that is of interest here, let Ω denote a simply-connected region in the complex plane with counter-clockwise oriented piecewise-smooth boundary $\partial\Omega$. If $f: \Omega \to \mathbb{C}$ is differentiable (as a function of two real variables $u = \text{Re}(z)$ and $v = \text{Im}(z)$) and extends continuously to $\partial\Omega$, then it follows from Stokes' theorem that

$$\oint_{\partial\Omega} f(u,v)\,dz = \iint_\Omega 2i\bar{\partial} f(u,v)\,dA(u,v) \qquad (21)$$

where $dA(u,v)$ denotes area measure in the plane and where $\bar{\partial}$ is the Cauchy-Riemann operator:

$$\bar{\partial} := \frac{1}{2}\left(\frac{\partial}{\partial u} + i\frac{\partial}{\partial v}\right), \quad z = u + iv, \qquad (22)$$

which annihilates all analytic functions of $z = u + iv$. Now consider the diagram shown in Fig. 1. We define a function $E(u,v)$ on $\Omega_+ \cup \Omega_-$ as follows:

$$E(u,v) := \cos(2\arg(u+iv-z_0))\hat{q}_0(u) + (1 - \cos(2\arg(u+iv-z_0)))\,\hat{q}_0(z_0),$$
$$u + iv \in \Omega_+ \cup \Omega_-. \qquad (23)$$

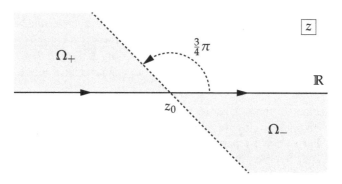

Fig. 1 The integration contour $\mathbb{R}$ in (17) and the unbounded domains Ω_+ and Ω_- in the $z = u+iv$ plane

Observe that:

- On the boundary $v = 0$ (i.e., $z \in \mathbb{R}$), we have $\cos(2\arg(u + iv - z_0)) \equiv 1$, so $E(u, 0) = \hat{q}_0(u)$.
- On the boundary $v = z_0 - u$, we have $\cos(2\arg(u + iv - z_0)) \equiv 0$, so $E(u, z_0 - u) = \hat{q}_0(z_0)$ which is independent of u.

The first point shows that $E(u, v)$ is an *extension* of the function $\hat{q}_0(z)$ from the real z-axis into the domain $\Omega_+ \cup \Omega_-$. The second point shows that the extension evaluates to a constant on the diagonal part of the boundary of $\Omega_+ \cup \Omega_-$. In the interior of $\Omega_+ \cup \Omega_-$, $E(u, v)$ inherits smoothness properties from $\hat{q}_0(u)$. In particular, under the assumption $\hat{q}_0 \in H^{1,1}(\mathbb{R})$, we may apply Stokes' theorem in the form (21) to the functions $\pm E(u, v)e^{-2it\theta(u+iv;z_0)}$ on the domains $\Omega_\pm$ and add up the results to obtain the formula

$$q(x,t) = \frac{1}{\pi}\int_{z_0+\infty e^{3\pi i/4}}^{z_0+\infty e^{-i\pi/4}} \hat{q}_0(z_0)e^{-2it\theta(z;z_0)}\,dz$$
$$+ \frac{1}{\pi}\iint_{\Omega_+ - \Omega_-} 2i\overline{\partial}\left(E(u,v)e^{-2it\theta(u+iv;z_0)}\right) dA(u,v). \quad (24)$$

The first term on the right-hand side originates from the diagonal boundary of $\Omega_+ \cup \Omega_-$ and because E is constant there it is an exact Gaussian integral evaluating to the explicit leading term on the right-hand side of (20). Therefore, the remaining term on the right-hand side of (24) is an exact double-integral representation of the error term $\mathcal{E}(x, t)$ in the formula (20). Since $q_0 \in H^{1,1}(\mathbb{R})$ implies $\hat{q}_0 \in H^{1,1}(\mathbb{R})$ which in turn implies that $\hat{q}_0(z)$ is defined for all $z \in \mathbb{R}$, the leading term in (20) certainly makes sense.

To estimate the error term we will only use the fact that $\hat{q}_0' \in L^2(\mathbb{R})$, i.e., that $\hat{q}_0$ lies in the (classical, unweighted) Sobolev space $H^1(\mathbb{R})$. First note that since

$e^{-2it\theta(z;z_0)}$ is an entire function of z, $\bar{\partial} e^{-2it\theta(z;z_0)} \equiv 0$, so by the product rule it suffices to have suitable estimates of $\bar{\partial} E(u, v)$ for $u + iv \in \Omega_\pm$. Indeed,

$$\left| \iint_{\Omega_\pm} 2i\bar{\partial} \left(E(u, v) e^{-2it\theta(u+iv;z_0)} \right) dA(u, v) \right|$$
$$\leq 2 \iint_{\Omega_\pm} |\bar{\partial} E(u, v)| e^{2t \operatorname{Im}(\theta(u+iv;z_0))} dA(u, v) \qquad (25)$$
$$= 2 \iint_{\Omega_\pm} |\bar{\partial} E(u, v)| e^{8t(u-z_0)v} dA(u, v).$$

A direct computation using (22) gives

$$\bar{\partial} E(u, v) = \bar{\partial} \left[\hat{q}_0(z_0) + \cos(2 \arg(u + iv - z_0)) \left(\hat{q}_0(u) - \hat{q}_0(z_0) \right) \right]$$
$$= \cos(2 \arg(u + iv - z_0)) \bar{\partial} \hat{q}_0(u)$$
$$+ \left(\hat{q}_0(u) - \hat{q}_0(z_0) \right) \bar{\partial} \cos(2 \arg(u + iv - z_0)) \qquad (26)$$
$$= \frac{1}{2} \cos(2 \arg(u + iv - z_0)) \hat{q}_0'(u)$$
$$+ \left(\hat{q}_0(u) - \hat{q}_0(z_0) \right) \bar{\partial} \cos(2 \arg(u + iv - z_0)).$$

In polar coordinates (ρ, ϕ) centered at the point $z_0 \in \mathbb{R}$ and defined by $u = z_0 + \rho \cos(\phi)$ and $v = \rho \sin(\phi)$, the Cauchy-Riemann operator (22) takes the equivalent form

$$\bar{\partial} = \frac{e^{i\phi}}{2} \left(\frac{\partial}{\partial \rho} + \frac{i}{\rho} \frac{\partial}{\partial \phi} \right), \qquad (27)$$

so as $\arg(u + iv - z_0) = \phi$ we have

$$\bar{\partial} \cos(2 \arg(u + iv - z_0)) = \frac{i e^{i\phi}}{2\rho} \frac{d}{d\phi} \cos(2\phi) = -\frac{i e^{i\phi}}{\rho} \sin(2\phi). \qquad (28)$$

Therefore we easily obtain the inequality

$$|\bar{\partial} E(u, v)| \leq \frac{1}{2} |\hat{q}_0'(u)| + \frac{|\hat{q}_0(u) - \hat{q}_0(z_0)|}{\sqrt{(u - z_0)^2 + v^2}}, \quad u + iv \in \Omega_+ \cup \Omega_-. \qquad (29)$$

Note that by the fundamental theorem of calculus and the Cauchy-Schwarz inequality,

$$|\hat{q}_0(u) - \hat{q}_0(z_0)| \le \int_{z_0}^u |\hat{q}_0'(w)| \, |dw| \le \sqrt{\int_{z_0}^u |dw|} \sqrt{\int_{z_0}^u |\hat{q}_0'(w)|^2 \, |dw|}$$

$$\le \|\hat{q}_0'\|_{L^2(\mathbb{R})} \sqrt{|u - z_0|} \le \|\hat{q}_0'\|_{L^2(\mathbb{R})} \left[(u - z_0)^2 + v^2\right]^{1/4}, \tag{30}$$

so (29) implies that also

$$|\bar{\partial} E(u, v)| \le \frac{1}{2} |\hat{q}_0'(u)| + \frac{\|\hat{q}_0'\|_{L^2(\mathbb{R})}}{[(u - z_0)^2 + v^2]^{1/4}}, \quad u + iv \in \Omega_+ \cup \Omega_-. \tag{31}$$

Therefore, using (31) in (25) gives

$$\left| \iint_{\Omega_\pm} 2i\bar{\partial}\left(E(u,v) e^{-2it\theta(u+iv, z_0)}\right) dA(u,v) \right| \le I^\pm(x,t) + 2\|\hat{q}_0'\|_{L^2(\mathbb{R})} J^\pm(x,t), \tag{32}$$

where

$$I^\pm(x,t) := \iint_{\Omega_\pm} |\hat{q}_0'(u)| e^{8t(u-z_0)v} \, dA(u,v) \quad \text{and}$$

$$J^\pm(x,t) := \iint_{\Omega_\pm} \frac{e^{8t(u-z_0)v}}{[(u-z_0)^2 + v^2]^{1/4}} \, dA(u,v). \tag{33}$$

The key point is that for $t > 0$, the exponential factors are bounded by 1 and decaying at infinity in $\Omega_\pm$. So, by iterated integration, Cauchy-Schwarz, and the change of variable $w = t^{1/2}(u - z_0)$,

$$I^+(x,t) = \int_{-\infty}^{z_0} du \int_0^{z_0 - u} dv \, |\hat{q}_0'(u)| e^{8t(u-z_0)v}$$

$$= \int_{-\infty}^{z_0} du \, |\hat{q}_0'(u)| \frac{1 - e^{-8t(u-z_0)^2}}{8t(z_0 - u)}$$

$$\le \|\hat{q}_0'\|_{L^2(\mathbb{R})} \sqrt{\int_{-\infty}^{z_0} \left[\frac{1 - e^{-8t(u-z_0)^2}}{8t(z_0 - u)}\right]^2 du} \tag{34}$$

$$= K \|\hat{q}_0'\|_{L^2(\mathbb{R})} t^{-3/4}, \quad K := \sqrt{\int_{-\infty}^0 \left[\frac{1 - e^{-8w^2}}{8w}\right]^2 dw} < \infty.$$

In exactly the same way, we also get $I^-(x,t) \le K \|\hat{q}_0'\|_{L^2(\mathbb{R})} t^{-3/4}$. Note that K is an absolute constant. The integrals $J^\pm(x,t)$ are independent of q_0 and by translation of z_0 to the origin and reflection through the origin, the integrals are also independent of x and are obviously equal. To calculate them we introduce rescaled polar coordinates by $u = z_0 + t^{-1/2}\rho \cos(\phi)$ and $v = t^{-1/2}\rho \sin(\phi)$ to get

$$J^{\pm}(x,t) = Lt^{-3/4}, \quad L := \int_0^{\infty} \rho d\rho \int_{3\pi/4}^{\pi} d\phi \, \rho^{-1/2} e^{8\rho^2 \sin(\phi)\cos(\phi)} \tag{35}$$

It is a calculus exercise to show that the above double integral is convergent and hence defines L as a second absolute constant.

It follows from these elementary calculations that if only $\hat{q}_0' \in L^2(\mathbb{R})$, then the error term $\mathcal{E}(x,t)$ in (20) obeys the estimate

$$\sup_{x \in \mathbb{R}} |\mathcal{E}(x,t)| \le \frac{2}{\pi}(K + 2L)\|\hat{q}_0'\|_{L^2(\mathbb{R})} t^{-3/4} \tag{36}$$

which decays as $t \to +\infty$ at exactly the same rate as in the claimed result for the nonlinear problem as formulated in Theorem 1.1. The same method can be used to obtain higher-order corrections under additional hypotheses of smoothness for the Fourier transform $\hat{q}_0$. One simply needs to integrate by parts with respect to $u = \text{Re}(z)$ in the double integral on the right-hand side of (24).

In the rest of the paper we will show that almost exactly the same elementary estimates suffice to prove the nonlinear analogue of this result, namely Theorem 1.1.

3 Proof of Theorem 1.1

We will prove Theorem 1.1 in several systematic steps. After some preliminary observations involving the jump matrix $\mathbf{V}_{\mathbf{M}}(z)$ in Riemann-Hilbert Problem 1 in Sects. 3.1 and 3.2, we shall see that the subsequent analysis of Riemann-Hilbert Problem 1 parallels our study of the associated linear problem detailed in Sect. 2. In particular we find natural analogues of the nonanalytic extension method (Sect. 3.3), of the Gaussian integral giving the leading term in the stationary phase formula (Sect. 3.4), and of the simple double integral estimates leading to the proof of its accuracy (Sect. 3.5). Finally, in Sect. 3.6 we assemble the ingredients to arrive at the formula (3) with the improved error estimate, completing the proof of Theorem 1.1.

3.1 Jump Matrix Factorization

The jump matrix $\mathbf{V}_{\mathbf{M}}(z)$ of Riemann-Hilbert Problem 1 defined in (12) can be factored in two ways that are useful in different intervals of the jump contour $\mathbb{R}$ as indicated:

$$\mathbf{V}_{\mathbf{M}}(z) = \begin{pmatrix} 1 & -\overline{r(z)}e^{-2it\theta(z;z_0)} \\ 0 & 1 \end{pmatrix} \begin{pmatrix} 1 & 0 \\ r(z)e^{2it\theta(z;z_0)} & 1 \end{pmatrix}, \quad z > z_0, \tag{37}$$

and

$$\mathbf{V}_M(z) = \begin{pmatrix} 1 & 0 \\ \dfrac{r(z)e^{2it\theta(z;z_0)}}{1-|r(z)|^2} & 1 \end{pmatrix} (1-|r(z)|^2)^{\sigma_3} \begin{pmatrix} 1 & -\dfrac{\overline{r(z)}e^{-2it\theta(z;z_0)}}{1-|r(z)|^2} \\ 0 & 1 \end{pmatrix}, \quad z < z_0.$$
(38)

The importance of these factorizations is that they provide an *algebraic* separation of the oscillatory exponential factors $e^{\pm 2it\theta(z;z_0)}$. Indeed, if the reflection coefficient $r(z)$ is an analytic function of $z \in \mathbb{R}$, then in each case the left-most (right-most) factor has an analytic continuation into the lower (upper) half-plane near the indicated half-line that is exponentially decaying to the identity matrix as $t \to +\infty$ due to z_0 being a simple critical point of $\theta(z;z_0)$. This observation is the basis for the steepest descent method for Riemann-Hilbert problems as first formulated in [4]. In the more realistic case that $r(z)$ is nowhere analytic, this analytic continuation method must be supplemented with careful approximation arguments that are quite detailed [8]. We will proceed differently in Sect. 3.3 below. But first we need to deal with the central diagonal factor in the factorization (38) to be used for $z < z_0$.

3.2 Modification of the Diagonal Jump

We now show how the diagonal factor $(1-|r(z)|^2)^{\sigma_3}$ in the jump matrix factorization (38) can be replaced with a constant diagonal matrix. Consider the complex scalar function defined by the formula

$$\delta(z;z_0) := \exp\left(\frac{1}{2\pi i}\int_{-\infty}^{z_0}\frac{\ln(1-|r(s)|^2)}{s-z}\,ds\right), \quad z \in \mathbb{C}\setminus(-\infty, z_0]. \quad (39)$$

This function is important because according to the Plemelj formula, it satisfies the scalar jump conditions $\delta_+(z;z_0) = \delta_-(z;z_0)(1-|r(z)|^2)$ for $z < z_0$ and $\delta_+(z;z_0) = \delta_-(z;z_0)$ for $z > z_0$. Hence the diagonal matrix $\delta(z;z_0)^{\sigma_3}$ is typically used in steepest descent theory to deal with the diagonal factor in (38). However, $\delta(z;z_0)$ has a mild singularity at $z = z_0$:

$$\delta(z;z_0) = K(z-z_0)^{i\nu(z_0)}(1+o(1)), \quad z \to z_0, \quad K = K(z_0) = \text{constant}, \quad (40)$$

where $\nu(z_0)$ is defined in (4) and the power function is interpreted as the principal branch. The use of $\delta(z;z_0)$ introduces this singularity unnecessarily into the Riemann-Hilbert analysis. In our approach we will therefore use a related function:

$$f(z;z_0) := c(z_0)\delta(z;z_0)(z-z_0)^{-i\nu(z_0)}, \quad (41)$$

where the constant $c(z_0)$ is defined by

$$c(z_0) := \exp\left(-\frac{1}{2\pi i}\left[\int_{-\infty}^{z_0-1}\frac{\ln(1-|r(s)|^2)}{s-z_0}\,ds\right.\right.$$
$$\left.\left.+\int_{z_0-1}^{z_0}\frac{\ln(1-|r(s)|^2)-\ln(1-|r(z_0)|^2)}{s-z_0}\,ds\right]\right) \quad (42)$$
$$=\exp\left(\frac{1}{2\pi i}\int_{-\infty}^{z_0}\ln(z_0-s)\,d\ln(1-|r(s)|^2)\right).$$

The function $f(z;z_0)$ has numerous useful properties that we summarize here.

Lemma 3.1 (Properties of $f(z;z_0)$) *Suppose that $r \in H^1(\mathbb{R})$ and there exists $\rho < 1$ such that $|r(z)| \le \rho$ holds for all $z \in \mathbb{R}$ (as is implied by $r \in H_1^{1,1}(\mathbb{R})$ which follows from $q_0 \in H^{1,1}(\mathbb{R})$). Then*

- *The functions $f(z;z_0)^{\pm 1}$ are well-defined and analytic in z for $\arg(z-z_0) \in (-\pi, \pi)$.*
- *The functions $f(z;z_0)^{\pm 1}$ are uniformly bounded independently of $z_0 \in \mathbb{R}$:*

$$\sup_{\substack{z_0\in\mathbb{R}\\ \arg(z-z_0)\in(-\pi,\pi)}}|f(z;z_0)|^{\pm 1} \le \frac{1}{1-\rho^2}. \quad (43)$$

- *The function $f(z;z_0)$ satisfies the following asymptotic condition:*

$$\lim_{\substack{z\to\infty\\ -\pi<\arg(z-z_0)<\pi}} f(z;z_0)(z-z_0)^{i\nu(z_0)} = c(z_0). \quad (44)$$

- *The functions $f(z;z_0)^{\pm 2}$ are Hölder continuous with exponent $1/2$. In particular, $f(z;z_0)^{\pm 2} \to 1$ as $z \to z_0$ and there is a constant $K = K(\rho) > 0$ such that $|f(z;z_0)^{\pm 2} - 1| \le K|z-z_0|^{1/2}$ holds whenever $\arg(z-z_0) \in (-\pi,\pi)$.*
- *The continuous boundary values $f_\pm(z;z_0)$ taken by $f(z;z_0)$ on $\mathbb{R}$ for $z < z_0$ from $\pm\text{Im}(z) > 0$ satisfy the jump condition*

$$f_+(z;z_0) = f_-(z;z_0)\frac{1-|r(z)|^2}{1-|r(z_0)|^2}, \quad z < z_0. \quad (45)$$

Proof The assumptions imply in particular that $\ln(1-|r(\cdot)|^2) \in L^1(\mathbb{R})$, so for z in a small neighborhood of each point disjoint from the integration contour, the integral in (39) is absolutely convergent and so $\delta(z;z_0)$ and $\delta(z;z_0)^{-1}$ are analytic functions of z on that neighborhood. The same argument shows that the first integral in the exponent of the expression (42) for $c(z_0)$ is convergent. Since $r \in H^1(\mathbb{R})$ implies that $r(\cdot)$ is Hölder continuous with exponent $1/2$, the condition $|r(\cdot)| \le \rho < 1$ further implies that $\ln(1-|r(s)|^2)$ is also Hölder continuous with exponent $1/2$, from which it follows that the second integral in the exponent of the expression (42) is also convergent. Therefore $c(z_0)$ exists, and clearly $|c(z_0)| = 1$. Since the principal branch of $(z - z_0)^{\mp i\nu(z_0)}$ is analytic for $\arg(z - z_0) \in (-\pi, \pi)$, the analyticity of $f(z; z_0)^{\pm 1}$ in the same domain follows. This proves the first statement.

In [8, Proposition 2.12] it is asserted that under the hypothesis $|r(z)| \leq \rho < 1$, the function $\delta(z; z_0)$ defined by (39) satisfies the uniform estimates $(1 - \rho^2)^{1/2} \leq |\delta(z; z_0)|^{\pm 1} \leq (1 - \rho^2)^{-1/2}$ whenever $\arg(z - z_0) \in (-\pi, \pi)$. If $\arg(z - z_0) = 0$, then obviously $|\delta(z; z_0)| = 1$, so it remains to prove the estimates hold for $\mathrm{Im}(z) \neq 0$. Following [12], since $\ln(1 - \rho^2) \leq \ln(1 - |r(s)|^2) \leq 0$, if $u = \mathrm{Re}(z)$ and $v = \mathrm{Im}(z)$ we have $\mathrm{Im}((s - z)^{-1}) = v/((s - u)^2 + v^2)$, so assuming $v > 0$,

$$\exp\left(\frac{v \ln(1 - \rho^2)}{2\pi} \int_{-\infty}^{z_0} \frac{ds}{(s - u)^2 + v^2}\right) \leq |\delta(u + iv; z_0)|. \tag{46}$$

Bounding the left-hand side below by extending the integration to $\mathbb{R}$ (using $v \ln(1 - \rho^2) < 0$) gives the lower bound $(1 - \rho^2)^{1/2} \leq |\delta(z; z_0)|$, and by taking reciprocals, the upper bound $|\delta(z; z_0)|^{-1} \leq (1 - \rho^2)^{-1/2}$ for $\mathrm{Im}(z) > 0$. The corresponding result for $\mathrm{Im}(z) < 0$ follows by the exact symmetry $\delta(\bar{z}; z_0)^{-1} = \overline{\delta(z; z_0)}$. Combining these bounds with $|c(z_0)| = 1$ and the elementary inequalities $(1 - \rho^2)^{1/2} \leq (1 - |r(z_0)|^2)^{1/2} = e^{-\pi \nu(z_0)} \leq |(z - z_0)^{i\nu(z_0)}| \leq e^{\pi \nu(z_0)} = (1 - |r(z_0)|^2)^{-1/2} \leq (1 - \rho^2)^{-1/2}$ holding for $\arg(z - z_0) \in (-\pi, \pi)$ then proves the second statement.

Since $\ln(1 - |r(\cdot)|^2) \in L^1(\mathbb{R})$, from (39) a dominated convergence argument shows that $\delta(z; z_0) \to 1$ as $z \to \infty$ provided only that the limit is taken in such a way that for some given $\epsilon > 0$, $\mathrm{dist}(z, [-\infty, z_0)) \geq \epsilon$. Combining this fact with (41) proves the third statement.

Analyticity implies Hölder continuity, so provided z is bounded away from the half-line $(-\infty, z_0]$, Hölder-$1/2$ continuity of $f(z; z_0)^{\pm 2}$ is obvious. But, since $\ln(1 - |r(\cdot)|^2)$ is Hölder continuous on $\mathbb{R}$ with exponent $1/2$, by the Plemelj-Privalov theorem [16, §19] and a related classical result [16, §22], the functions $\delta(z; z_0)^{\pm 1}$ are uniformly Hölder continuous with exponent $1/2$ in any neighborhood of the integration contour *except* for the endpoint $z = z_0$, and hence the same is true for the functions $f(z; z_0)^{\pm 2}$. However, the latter functions are better-behaved near $z = z_0$. To see this, note that since

$$(z - z_0)^{\mp i\nu(z_0)} = (z - (z_0 - 1))^{\mp i\nu(z_0)} \left[\frac{z - z_0}{z - (z_0 - 1)}\right]^{\mp i\nu(z_0)}$$

$$= (z - (z_0 - 1))^{\mp i\nu(z_0)} \exp\left(\mp \frac{1}{2\pi i} \int_{z_0 - 1}^{z_0} \frac{\ln(1 - |r(z_0)|^2)}{s - z} ds\right),$$

$$z \in \mathbb{C} \setminus (-\infty, z_0],$$

(47)

we have from (39) and (41) that

$$f(z;z_0)^{\pm 2} = c(z_0)^{\pm 2}(z-(z_0-1))^{\mp 2i\nu(z_0)}\exp\left(\pm\frac{1}{\pi i}\int_{-\infty}^{z_0-1}\frac{\ln(1-|r(s)|^2)}{s-z}ds\right)$$
$$\cdot \exp\left(\pm\frac{1}{\pi i}\int_{z_0-1}^{+\infty}\frac{h(s)\,ds}{s-z}\right) \tag{48}$$

where $h(s) := \ln(1-|r(s)|^2) - \ln(1-r(z_0)|^2)$ for $s < z_0$ and $h(s) := 0$ for $s \geq z_0$. As the first three factors are analytic at $z = z_0$ while $h(s)$ is Hölder continuous with exponent $1/2$ in a neighborhood of $s = z_0$, the same arguments cited above apply and yield the desired Hölder continuity of $f(z;z_0)^{\pm 2}$ near $z = z_0$. It only remains to show that $f(z_0;z_0)^{\pm 2} = 1$, but this follows immediately from (42) and (48). This proves the fourth statement.

Finally, the fifth statement follows from the definition (41) of $f(z;z_0)$ and the jump condition $\delta_+(z;z_0) = \delta_-(z;z_0)(1-|r(z)|^2)$ for $z < z_0$. □

Using the diagonal matrix $f(z;z_0)^{\sigma_3}$ to conjugate the unknown $\mathbf{M}(z)$ of Riemann-Hilbert Problem 1 by introducing

$$\mathbf{N}(z) = \mathbf{N}(z;x,t) := e^{i\omega(z_0)\sigma_3/2}e^{it\theta(z_0;z_0)\sigma_3} \cdot c(z_0)^{\sigma_3}\mathbf{M}(z;x,t)f(z;z_0)^{-\sigma_3}$$
$$\cdot e^{-it\theta(z_0;z_0)\sigma_3}e^{-i\omega(z_0)\sigma_3/2}, z \in \mathbb{C}\setminus\mathbb{R}, \tag{49}$$

where

$$\omega(z_0) := \arg(r(z_0)), \tag{50}$$

it is easy to check that $\mathbf{N}(z)$ satisfies several conditions explicitly related to those of $\mathbf{M}(z)$ according to Riemann-Hilbert Problem 1. Indeed, $\mathbf{N}(z)$ must be a solution of the following equivalent problem.

Riemann-Hilbert Problem 2 *Given parameters* $(x,t) \in \mathbb{R}^2$, *find* $\mathbf{N} = \mathbf{N}(z) = \mathbf{N}(z;x,t)$, *a* 2×2 *matrix, satisfying the following conditions:*

Analyticity $\mathbf{N}$ *is an analytic function of z in the domain* $\mathbb{C}\setminus\mathbb{R}$. *Moreover, $\mathbf{N}$ has a continuous extension to the real axis from the upper (lower) half-plane denoted* $\mathbf{N}_+(z)$ ($\mathbf{N}_-(z)$) *for* $z \in \mathbb{R}$.

Jump Condition *The boundary values satisfy the jump condition*

$$\mathbf{N}_+(z) = \mathbf{N}_-(z)\mathbf{V}_\mathbf{N}(z), \quad z \in \mathbb{R}, \tag{51}$$

where the jump matrix $\mathbf{V}_\mathbf{N}(z)$ *may be written in the alternate forms*

$$\mathbf{V_N}(z) = \begin{pmatrix} 1 & -f(z;z_0)^2 \overline{r(z)} e^{i\omega(z_0)} e^{-2it[\theta(z;z_0)-\theta(z_0;z_0)]} \\ 0 & 1 \end{pmatrix}$$

$$\cdot \begin{pmatrix} 1 & 0 \\ f(z;z_0)^{-2} r(z) e^{-i\omega(z_0)} e^{2it[\theta(z;z_0)-\theta(z_0;z_0)]} & 1 \end{pmatrix}, \quad z > z_0, \qquad (52)$$

$$\mathbf{V_N}(z) := \begin{pmatrix} 1 & 0 \\ \dfrac{f_-(z;z_0)^{-2} r(z) e^{-i\omega(z_0)} e^{2it[\theta(z;z_0)-\theta(z_0;z_0)]}}{1-|r(z)|^2} & 1 \end{pmatrix}$$

$$\cdot (1-|r(z_0)|^2)^{\sigma_3} \begin{pmatrix} 1 & -\dfrac{f_+(z;z_0)^2 \overline{r(z)} e^{i\omega(z_0)} e^{-2it[\theta(z;z_0)-\theta(z_0;z_0)]}}{1-|r(z)|^2} \\ 0 & 1 \end{pmatrix}, \quad z < z_0, \qquad (53)$$

where $f_+(z;z_0)$ ($f_-(z;z_0)$) is the boundary value taken by $f(z;z_0)$ from the upper (lower) half-plane.

Normalization *There is a matrix* $\mathbf{N}_1(x,t)$ *such that*

$$\mathbf{N}(z)(z-z_0)^{-i\nu(z_0)\sigma_3} = \mathbb{I} + z^{-1}\mathbf{N}_1(x,t) + o(z^{-1}), \quad z \to \infty. \qquad (54)$$

Note that the matrix coefficient $\mathbf{N}_1(x,t)$ is necessarily related to the coefficient $\mathbf{M}_1(x,t)$ in Riemann-Hilbert Problem 1 by a diagonal conjugation:

$$\mathbf{M}_1(x,t) = e^{-i\omega(z_0)\sigma_3/2} e^{-it\theta(z_0;z_0)\sigma_3} c(z_0)^{-\sigma_3} \mathbf{N}_1(x,t) c(z_0)^{\sigma_3} e^{it\theta(z_0;z_0)\sigma_3} e^{i\omega(z_0)\sigma_3/2}. \qquad (55)$$

Therefore, the reconstruction formula (14) can be written in terms of $\mathbf{N}_1(x,t)$ as

$$q(x,t) := 2i e^{-i\omega(z_0)} e^{-2it\theta(z_0;z_0)} c(z_0)^{-2} N_{1,12}(x,t). \qquad (56)$$

The net effect of this step is therefore to replace the non-constant diagonal central factor in (38) with its constant value at $z = z_0$ and to introduce power-law asymptotics at $z = \infty$ at the cost of slight modifications of the left-most and right-most factors in (37)–(38). In the formula (49) we have also taken the opportunity to conjugate off the constant value of $\theta(z;z_0)$ and the phase of $r(z)$ at the critical point $z = z_0$.

3.3 Nonanalaytic Extensions and $\bar{\partial}$ Steepest Descent

The key to the steepest descent method, both in its classical analytic framework and in the $\bar{\partial}$ setting, is to get the oscillatory factors $e^{\pm 2it\theta(z;z_0)}$ off the real axis and into appropriate sectors of the complex z-plane where they decay as $t \to +\infty$. We will accomplish this by exactly the same means as in the linear case, namely by defining

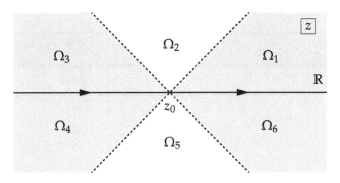

Fig. 2 The jump contour $\mathbb{R}$ in Riemann-Hilbert Problem 2 and the sectors Ω_j, $j = 1, \ldots, 6$ in the $z = u + iv$ plane

non-analytic extensions of the non-oscillatory coefficients of $e^{\pm 2it\theta(z;z_0)}$ in the left-most and right-most jump matrix factors in (37)–(38) by a slight generalization of the formula (23). In reference to the diagram in Fig. 2, we define sectors

$$\begin{aligned}
\Omega_1 &: \quad 0 < \arg(z - z_0) < \frac{1}{4}\pi \\
\Omega_2 &: \quad \frac{1}{4}\pi < \arg(z - z_0) < \frac{3}{4}\pi \\
\Omega_3 &: \quad \frac{3}{4}\pi < \arg(z - z_0) < \pi \\
\Omega_4 &: \quad -\pi < \arg(z - z_0) < -\frac{3}{4}\pi \\
\Omega_5 &: \quad -\frac{3}{4}\pi < \arg(z - z_0) < -\frac{1}{4}\pi \\
\Omega_6 &: \quad -\frac{1}{4}\pi < \arg(z - z_0) < 0.
\end{aligned} \quad (57)$$

Note that $\Omega_3 = \Omega_+$ and $\Omega_6 = \Omega_-$ in reference to Fig. 1. Now we define extensions on the domains shaded in Fig. 2 by following a very similar approach as in Sect. 2:

$$E_1(u, v) := \cos(2\arg(u + iv - z_0)) f(u + iv; z_0)^{-2} r(u) e^{-i\omega(z_0)}$$
$$+ (1 - \cos(2\arg(u + iv - z_0))) |r(z_0)|, \quad z = u + iv \in \Omega_1$$

$$E_3(u, v) := -\left[\cos(2\arg(u + iv - z_0)) f(u + iv; z_0)^2 \frac{\overline{r(u)} e^{i\omega(z_0)}}{1 - |r(u)|^2}\right.$$
$$\left. + (1 - \cos(2\arg(u + iv - z_0))) \frac{|r(z_0)|}{1 - |r(z_0)|^2}\right], \quad z = u + iv \in \Omega_3$$

$$E_4(u, v) := \cos(2\arg(u + iv - z_0)) f(u + iv; z_0)^{-2} \frac{r(u)e^{-i\omega(z_0)}}{1 - |r(u)|^2}$$

$$+ (1 - \cos(2\arg(u + iv - z_0))) \frac{|r(z_0)|}{1 - |r(z_0)|^2}, \quad z = u + iv \in \Omega_4$$

$$E_6(u, v) := -\left[\cos(2\arg(u + iv - z_0)) f(u + iv; z_0)^2 \overline{r(u)} e^{i\omega(z_0)}\right.$$

$$\left. + (1 - \cos(2\arg(u + iv - z_0)))|r(z_0)|\right], \quad z = u + iv \in \Omega_6.$$
(58)

It is easy to check that:

- $E_1(u, v)$ evaluates to $f(z; z_0)^{-2} r(z) e^{-i\omega(z_0)}$ for $z \in \mathbb{R}$ on the boundary of Ω_1.
- $E_3(u, v)$ evaluates to $-f_+(z; z_0)^2 \overline{r(z)} e^{i\omega(z_0)}/(1 - |r(z)|^2)$ for $z \in \mathbb{R}$ on the boundary of Ω_3.
- $E_4(u, v)$ evaluates to $f_-(z; z_0)^{-2} r(z) e^{-i\omega(z_0)}/(1 - |r(z)|^2)$ for $z \in \mathbb{R}$ on the boundary of Ω_4.
- $E_6(u, v)$ evaluates to $-f(z; z_0)^2 \overline{r(z)} e^{i\omega(z_0)}$ for $z \in \mathbb{R}$ on the boundary of Ω_6.

Thus exactly as in Sect. 2 these formulæ represent extensions of their values on the real sector boundaries into the complex plane that become constant on the diagonal sector boundaries (see (60) below), with the constant chosen in each case to ensure continuity of the extension along the interior boundary of each sector. The only essential difference between the extension formulæ (58) and the formula (23) from Sect. 2 is the way that the factors $f(z; z_0)^{\pm 2}$ are treated differently from the factors involving $r(z)$; the reason for using $f(u + iv; z_0)^{\pm 2}$ in (58) rather than $f(u; z_0)^{\pm 2}$ will become clearer in Sect. 3.5 when we compute $\overline{\partial} E_j(u, v)$, $j = 1, 3, 4, 6$, and take advantage of the fact (see Lemma 3.1) that $\overline{\partial} f(u + iv; z_0)^{\pm 2} \equiv 0$ in the interior of each sector.

We use the extensions to "open lenses" about the intervals $z < z_0$ and $z > z_0$ by making another substitution:

$$\mathbf{O}(u, v; x, t) := \begin{cases} \mathbf{N}(z; x, t) \begin{pmatrix} 1 & 0 \\ E_1(u, v) e^{2it[\theta(u+iv;z_0) - \theta(z_0;z_0)]} & 1 \end{pmatrix}^{-1}, & z = u + iv \in \Omega_1 \\ \mathbf{N}(z; x, t), & z = u + iv \in \Omega_2 \\ \mathbf{N}(z; x, t) \begin{pmatrix} 1 & E_3(u, v) e^{-2it[\theta(u+iv;z_0) - \theta(z_0;z_0)]} \\ 0 & 1 \end{pmatrix}^{-1}, & z = u + iv \in \Omega_3 \\ \mathbf{N}(z; x, t) \begin{pmatrix} 1 & 0 \\ E_4(u, v) e^{2it[\theta(u+iv;z_0) - \theta(z_0;z_0)]} & 1 \end{pmatrix}, & z = u + iv \in \Omega_4 \\ \mathbf{N}(z; x, t), & z = u + iv \in \Omega_5 \\ \mathbf{N}(z; x, t) \begin{pmatrix} 1 & E_6(u, v) e^{-2it[\theta(u+iv;z_0) - \theta(z_0;z_0)]} \\ 0 & 1 \end{pmatrix}, & z = u + iv \in \Omega_6. \end{cases}$$
(59)

Our notation $\mathbf{O}(u, v; x, t)$ reflects the viewpoint that unlike $\mathbf{N}(z; x, t)$, $z = u + iv$, $\mathbf{O}(u, v; x, t)$ is not a piecewise-analytic function in the complex plane due to the non-analytic extensions $E_j(u, v)$, $j = 1, 3, 4, 6$. The exponential factors in (59) all have modulus less than 1 and decay exponentially to zero as $t \to +\infty$ pointwise in the interior of each of the indicated sectors, a fact that suggests that (59) is a near-identity transformation in the limit $t \to +\infty$. We also have the following property.

Lemma 3.2 (Relation Between N and O for Large $z \in \mathbb{C}$) *Let $z_0 \in \mathbb{R}$ be fixed, and suppose that $r \in H^1(\mathbb{R})$ and that there exists a constant $\rho < 1$ such that $|r(z)| \le \rho$ holds for all $z \in \mathbb{R}$ (conditions that are true for $r \in H_1^{1,1}(\mathbb{R})$ as follows from $q_0 \in H^{1,1}(\mathbb{R})$). Then $\mathbf{O}(u, v; x, t) = \mathbf{N}(u + iv; x, t)(\mathbb{I} + o(1))$ holds as $z = u + iv \to \infty$ where the decay of the error term is uniform with respect to direction in each sector Ω_j, $j = 1, \ldots, 6$.*

Proof The exponential factors in (59) also decay as $z = u + iv \to \infty$ provided that $v \to \infty$. Since $r, r' \in L^2(\mathbb{R})$ means that $(1 + |\cdot|)\hat{r}(\cdot)$ is square-integrable where $\hat{r}$ denotes the Fourier transform of r, the Cauchy-Schwarz inequality implies that also $\hat{r} \in L^1(\mathbb{R})$. Hence by the Riemann-Lebesgue Lemma, $r(u)$ is bounded, continuous, and tends to zero as $u \to \infty$. As $1 - |r(u)|^2 \ge 1 - \rho^2 > 0$, the same properties hold for $r(u)/(1 - |r(u)|^2)$. Since the hypotheses of Lemma 3.1 hold, $f(u + iv; z_0)^{\pm 2}$ are bounded functions, so the desired result follows from using extension formulæ (58) in (59). $\square$

Despite the non-analyticity of the extensions, the above proof shows also that each of the extensions $E_j(u, v)$, $j = 1, 3, 4, 6$, is continuous on the relevant sector and therefore $\mathbf{O}(u, v; x, t)$ is a piecewise-continuous function of $(u, v) \in \mathbb{R}^2$ with jump discontinuities across the sector boundaries. We address these jump discontinuities next.

3.4 The Isomonodromy Problem of Its

Although $\mathbf{O}(u, v; x, t)$ is not analytic in the sectors shaded in Fig. 2 for essentially the same reason that the double integral error term in (24) does not vanish identically, the fact that the extensions $E_j(u, v)$, $j = 1, 3, 4, 6$, evaluate to constants on the diagonals:

$$E_1(u - z_0, u) = |r(z_0)| \quad \text{and} \quad E_6(u - z_0, -u) = -|r(z_0)|, \quad u > z_0,$$

$$E_3(u - z_0, -u) = -\frac{|r(z_0)|}{1 - |r(z_0)|^2} \quad \text{and} \quad E_4(u - z_0, u) = \frac{|r(z_0)|}{1 - |r(z_0)|^2}, \quad u < z_0,$$
(60)

implies that if we introduce the recentered and rescaled independent variable $\zeta := 2t^{1/2}(z - z_0)$, the jump conditions satisfied by $\mathbf{O}(u, v; x, t)$ across the sector boundaries are exactly the same as those satisfied by the matrix function $\mathbf{P}(\zeta; |r(z_0)|)$ solving the following Riemann-Hilbert problem.

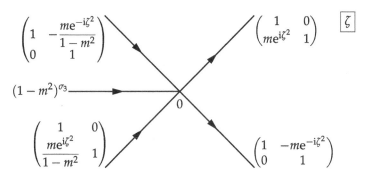

Fig. 3 The jump contour $\Sigma_{\mathbf{P}}$ and jump matrix $\mathbf{V}_{\mathbf{P}}(\zeta; m)$ for Riemann-Hilbert Problem 3

Riemann-Hilbert Problem 3 *Let $m \in [0, 1)$ be a parameter, and seek a 2×2 matrix function $\mathbf{P} = \mathbf{P}(\zeta) = \mathbf{P}(\zeta; m)$ with the following properties:*

Analyticity $\mathbf{P}(\zeta)$ *is an analytic function of ζ in the sectors $|\arg(\zeta)| < \frac{1}{4}\pi$, $\frac{1}{4}\pi < \pm\arg(\zeta) < \frac{3}{4}\pi$, and $\frac{3}{4}\pi < \pm\arg(\zeta) < \pi$. It admits a continuous extension from each of these five sectors to its boundary.*

Jump Conditions *Denoting by $\mathbf{P}_+(\zeta)$ (resp., $\mathbf{P}_-(\zeta)$) the boundary value taken on any one of the rays of the jump contour $\Sigma_{\mathbf{P}}$ from the left (resp., right) according to the orientation shown in Fig. 3, the boundary values are related by $\mathbf{P}_+(\zeta; m) = \mathbf{P}_-(\zeta; m)\mathbf{V}_{\mathbf{P}}(\zeta; m)$, where the jump matrix $\mathbf{V}_{\mathbf{P}}(\zeta; m)$ is defined on the five rays of $\Sigma_{\mathbf{P}}$ by*

$$\mathbf{V}_{\mathbf{P}}(\zeta; m) := \begin{cases} \begin{pmatrix} 1 & 0 \\ me^{i\zeta^2} & 1 \end{pmatrix}, & \arg(\zeta) = \frac{1}{4}\pi \\ \begin{pmatrix} 1 & -me^{-i\zeta^2} \\ 0 & 1 \end{pmatrix}, & \arg(\zeta) = -\frac{1}{4}\pi \\ \begin{pmatrix} 1 & -\dfrac{me^{-i\zeta^2}}{1-m^2} \\ 0 & 1 \end{pmatrix}, & \arg(\zeta) = \frac{3}{4}\pi \\ \begin{pmatrix} 1 & 0 \\ \dfrac{me^{i\zeta^2}}{1-m^2} & 1 \end{pmatrix}, & \arg(\zeta) = -\frac{3}{4}\pi \\ (1-m^2)^{\sigma_3}, & \arg(-\zeta) = 0. \end{cases} \quad (61)$$

Normalization $\mathbf{P}(\zeta; m)\zeta^{-\ln(1-m^2)\sigma_3/(2\pi i)} \to \mathbb{I}$ *as* $\zeta \to \infty$.

This Riemann-Hilbert problem is essentially the isomonodromy problem identified by Its [10], and it is the analogue in the nonlinear setting of the Gaussian integral that is the leading term of the stationary phase expansion (24) in the linear case. Although the jump conditions for $\mathbf{O}(u, v; x, t)$ correspond exactly to those of $\mathbf{P}(\zeta; |r(z_0)|)$, the scaling $z \mapsto \zeta = 2t^{1/2}(z - z_0)$ introduces an extra factor into the asymptotics as $z \to \infty$; the fact of the matter is that the matrix $(2t^{1/2})^{i\nu(z_0)\sigma_3}\mathbf{O}(u, v; x, t)$ satisfies the normalization condition of $\mathbf{P}(\zeta; |r(z_0)|)$, and the constant pre-factor has no effect on the jump conditions. Hence in Sect. 3.5 below we shall use the latter as a parametrix for the former.

However, we first develop the explicit solution of Riemann-Hilbert Problem 3. The first step is to consider the related unknown $\mathbf{U}(\zeta; m) := \mathbf{P}(\zeta; m)e^{-i\zeta^2\sigma_3/2}$ and observe that from the conditions of Riemann-Hilbert Problem 3 that $\mathbf{U}(\zeta; m)$ is analytic exactly in the same five sectors where $\mathbf{P}(\zeta; m)$ is, and that it satisfies jump conditions of exactly the form (61) except that the factors $e^{\pm i\zeta^2}$ are everywhere replaced by 1; in other words, the jump matrix for $\mathbf{U}(\zeta; m)$ on each jump ray is constant along the ray. It follows that the ζ-derivative $\mathbf{U}'(\zeta; m)$ satisfies the same "raywise constant" jump conditions as does $\mathbf{U}(\zeta; m)$ itself. Then, since it is easy to prove by Liouville's theorem that any solution $\mathbf{P}(\zeta; m)$ of Riemann-Hilbert Problem 3 has unit determinant, it follows that $\mathbf{U}(\zeta; m)$ is invertible and a calculation shows that the function $\mathbf{U}'(\zeta; m)\mathbf{U}(\zeta; m)^{-1}$ is continuous and hence by Morera's theorem analytic in the whole ζ-plane possibly excepting $\zeta = 0$. We will assume analyticity at the origin as well and show later that this is consistent. As an entire function of ζ, the product $\mathbf{U}'(\zeta; m)\mathbf{U}(\zeta; m)^{-1}$ is potentially determined by its asymptotic behavior as $\zeta \to \infty$. Assuming further that the normalization condition in Riemann-Hilbert Problem 3 means both that for some matrix coefficient $\mathbf{P}_1(m)$ to be determined,

$$\mathbf{P}(\zeta; m) = \left(\mathbb{I} + \zeta^{-1}\mathbf{P}_1(m) + \mathcal{O}(\zeta^{-2})\right)\zeta^{\ln(1-m^2)\sigma_3/(2\pi i)} \quad \text{and}$$

$$\mathbf{P}'(\zeta; m) = \left(\frac{\ln(1-m^2)}{2\pi i}\zeta^{-1}\sigma_3 + \mathcal{O}(\zeta^{-2})\right)\zeta^{\ln(1-m^2)\sigma_3/(2\pi i)} \tag{62}$$

hold as $\zeta \to \infty$, such as would arise from term-by-term differentiation, it follows also that

$$\mathbf{U}(\zeta; m) = \left(\mathbb{I} + \zeta^{-1}\mathbf{P}_1(m) + \mathcal{O}(\zeta^{-2})\right)\zeta^{\ln(1-m^2)\sigma_3/(2\pi i)}e^{-i\zeta^2\sigma_3/2} \quad \text{and}$$

$$\mathbf{U}'(\zeta; m) = \left(-i\zeta\sigma_3 - i\mathbf{P}_1(m)\sigma_3 + \mathcal{O}(\zeta^{-1})\right)\zeta^{\ln(1-m^2)\sigma_3/(2\pi i)}e^{-i\zeta^2\sigma_3/2} \tag{63}$$

as $\zeta \to \infty$. Therefore the entire function is determined by Liouville's theorem to be a linear polynomial:

$$\mathbf{U}'(\zeta; m)\mathbf{U}(\zeta; m)^{-1} = -i\zeta\sigma_3 + i[\sigma_3, \mathbf{P}_1(m)], \tag{64}$$

where $[\mathbf{A}, \mathbf{B}] := \mathbf{AB} - \mathbf{BA}$ is the matrix commutator. In other words, $\mathbf{U}(\zeta; m)$ satisfies the first-order system of linear differential equations:

$$\frac{d\mathbf{U}}{d\zeta}(\zeta; m) = \begin{pmatrix} -i\zeta & 2iP_{1,12}(m) \\ -2iP_{1,21}(m) & i\zeta \end{pmatrix} \mathbf{U}(\zeta; m). \tag{65}$$

Now, another easy consequence of Liouville's theorem is that there is at most one solution of Riemann-Hilbert Problem 3. Using the fact that $m \in [0, 1)$, it is not difficult to show that if $\mathbf{P}(\zeta; m)$ is a solution of Riemann-Hilbert Problem 3, then so is

$$\sigma_1 \overline{\mathbf{P}(\bar{\zeta}; m)} \sigma_1, \quad \text{where} \quad \sigma_1 := \begin{pmatrix} 0 & 1 \\ 1 & 0 \end{pmatrix}, \tag{66}$$

so by uniqueness it follows that $\mathbf{P}(\zeta; m) = \sigma_1 \overline{\mathbf{P}(\bar{\zeta}; m)} \sigma_1$. Combining this symmetry with the first expansion in (62) shows that $P_{1,21}(m) = \overline{P_{1,12}(m)}$, so the differential equations can be written in the form

$$\frac{d\mathbf{U}}{d\zeta}(\zeta; m) = \begin{pmatrix} -i\zeta & \beta \\ \bar{\beta} & i\zeta \end{pmatrix} \mathbf{U}(\zeta; m), \quad \beta = \beta(m) := 2i P_{1,12}(m). \tag{67}$$

The constant $\beta \in \mathbb{C}$ is unknown, but if it is considered as a parameter, then eliminating the second row shows that the elements U_{1j}, $j = 1, 2$, of the first row satisfy Weber's equation for parabolic cylinder functions in the form:

$$\frac{d^2 U_{1j}}{dy^2} - \left(\frac{1}{4}y^2 + a\right) U_{1j} = 0, \quad a := \frac{1}{2}(1+i|\beta|^2), \quad y := \sqrt{2} e^{-i\pi/4} \zeta, \quad j = 1, 2. \tag{68}$$

The solutions of this equation are well-documented in the Digital Library of Mathematical Functions [18, §12]. Equation (68) has particular solutions denoted $U(a, \pm y)$ and $U(-a, \pm iy)$, where $U(\cdot, \cdot)$ is a special function[2] with well-known integral representations, asymptotic expansions, and connection formulæ.

The second step is to represent the elements U_{1j} as linear combinations of a fundamental pair of so-called numerically satisfactory solutions specially adapted to each of the five sectors of analyticity for Riemann-Hilbert Problem 3. Thus, we write

$U_{1j}(\zeta; m)$
$$= \begin{cases} \beta A_j^{(0)} U(a, y) + \beta B_j^{(0)} U(-a, iy), & |\arg(\zeta)| < \frac{1}{4}\pi, \\ \beta A_j^{(1)} U(a, y) + \beta B_j^{(1)} U(-a, -iy), & \frac{1}{4}\pi < \arg(\zeta) < \frac{3}{4}\pi, \\ \beta A_j^{(-1)} U(a, -y) + \beta B_j^{(-1)} U(-a, iy), & -\frac{3}{4}\pi < \arg(\zeta) < -\frac{1}{4}\pi, \\ \beta A_j^{(2)} U(a, -y) + \beta B_j^{(2)} U(-a, -iy), & \frac{3}{4}\pi < \arg(\zeta) < \pi, \\ \beta A_j^{(-2)} U(a, -y) + \beta B_j^{(-2)} U(-a, -iy), & -\pi < \arg(\zeta) < -\frac{3}{4}\pi, \end{cases} \tag{69}$$

---

[2] In many works on long-time asymptotics for the Cauchy problem (1)–(2) written before the Digital Library of Mathematical Functions was freely available (e.g., [8, 9]), the solution of Riemann-Hilbert Problem 3 was developed in terms of the related function $D_\nu(y) := U(-\frac{1}{2} - \nu, y)$. Since most formulæ in [18, §12] are phrased in terms of $U(\cdot, \cdot)$, we favor the latter.

and then using the first row of (67) along with identities allowing the elimination of derivatives of U [18, Eqs. 12.8.2–12.8.3] we get the following representation of the elements of the second row of $\mathbf{U}(\zeta; m)$:

$$U_{2j}(\zeta; m)$$

$$= \sqrt{2}e^{-i\pi/4} \begin{cases} -A_j^{(0)}U(a-1, y)+i(a-\tfrac{1}{2})B_j^{(0)}U(1-a, iy), & |\arg(\zeta)|<\tfrac{1}{4}\pi, \\ -A_j^{(1)}U(a-1, y)-i(a-\tfrac{1}{2})B_j^{(1)}U(1-a, -iy), & \tfrac{1}{4}\pi<\arg(\zeta)<\tfrac{3}{4}\pi, \\ A_j^{(-1)}U(a-1, -y)+i(a-\tfrac{1}{2})B_j^{(-1)}U(1-a, iy), & -\tfrac{3}{4}\pi<\arg(\zeta)<-\tfrac{1}{4}\pi, \\ A_j^{(2)}U(a-1, -y)-i(a-\tfrac{1}{2})B_j^{(2)}U(1-a, -iy), & \tfrac{3}{4}\pi<\arg(\zeta)<\pi, \\ A_j^{(-2)}U(a-1, -y)-i(a-\tfrac{1}{2})B_j^{(-2)}U(1-a, -iy), & -\pi<\arg(\zeta)<-\tfrac{3}{4}\pi. \end{cases}$$

(70)

Finally, we determine the coefficients $A_j^{(i)}$ and $B_j^{(i)}$ for $j = 1, 2$ and $i = 0, \pm 1, \pm 2$, as well as the value of $\beta = \beta(m)$ so that all of the conditions of Riemann-Hilbert Problem 3 are satisfied by $\mathbf{P}(\zeta; m) = \mathbf{U}(\zeta; m)e^{i\zeta\sigma_3/2}$. The advantage of using numerically satisfactory fundamental pairs is that the asymptotic expansion [18, Eq. 12.9.1]

$$U(a, y) \sim e^{-\tfrac{1}{4}y^2} y^{-a-\tfrac{1}{2}} \sum_{k=0}^{\infty} (-1)^k \frac{\left(\tfrac{1}{2}+a\right)_{2k}}{k!(2y^2)^k}, \quad y \to \infty, \quad |\arg(y)| < \frac{3}{4}\pi$$

(71)

can be used to determine from (69)–(70) the asymptotic behavior of $\mathbf{U}(\zeta; m)$ in each sector for the purposes of comparison with the first formula in (63). This immediately shows that for consistency it is necessary to take $A_1^{(i)} = 0$ and $B_2^{(i)} = 0$ for $i = 0, \pm 1, \pm 2$. Next, it is useful to consider the trivial jump conditions for the first column of $\mathbf{U}(\zeta; m)$ (across $\arg(\zeta) = -\tfrac{1}{4}\pi$ and $\arg(\zeta) = \tfrac{3}{4}\pi$) and for the second column of $\mathbf{U}(\zeta; m)$ (across $\arg(\zeta) = \tfrac{1}{4}\pi$ and $\arg(\zeta) = -\tfrac{3}{4}\pi$). These imply the identities $B_1^{(0)} = B_1^{(-1)}$, $B_1^{(1)} = B_1^{(2)}$ (from matching the first column) and $A_2^{(0)} = A_2^{(1)}$, $A_2^{(-2)} = A_2^{(-1)}$ (from matching the second column). The diagonal jump condition satisfied by $\mathbf{U}(\zeta; m)$ across the negative real axis then yields the additional identities $B_1^{(-2)} = (1-m^2)^{-1}B_1^{(2)}$ and $A_2^{(2)} = (1-m^2)^{-1}A_2^{(-2)}$. With this information, we have found that $\mathbf{U}(\zeta; m)$ necessarily has the form

$$\mathbf{U}(\zeta; m) = \begin{pmatrix} \beta B_1^{(0)} U(-a, iy) & \beta A_2^{(0)} U(a, y) \\ \sqrt{2}e^{i\pi/4}(a-\tfrac{1}{2})B_1^{(0)} U(1-a, iy) & \sqrt{2}e^{3\pi i/4} A_2^{(0)} U(a-1, y) \end{pmatrix},$$

$$|\arg(\zeta)| < \frac{1}{4}\pi,$$

(72)

$$\mathbf{U}(\zeta; m) = \begin{pmatrix} \beta B_1^{(1)} U(-a, -iy) & \beta A_2^{(0)} U(a, y) \\ \sqrt{2}e^{-3\pi i/4}(a-\tfrac{1}{2})B_1^{(1)} U(1-a, -iy) & \sqrt{2}e^{3\pi i/4} A_2^{(0)} U(a-1, y) \end{pmatrix},$$

$$\frac{1}{4}\pi < \arg(\zeta) < \frac{3}{4}\pi,$$

(73)

$$\mathbf{U}(\zeta;m) = \begin{pmatrix} \beta B_1^{(0)} U(-a, iy) & \beta A_2^{(-1)} U(a, -y) \\ \sqrt{2} e^{i\pi/4}(a-\tfrac{1}{2}) B_1^{(0)} U(1-a, iy) & \sqrt{2} e^{-i\pi/4} A_2^{(-1)} U(a-1, -y) \end{pmatrix},$$
$$-\tfrac{3}{4}\pi < \arg(\zeta) < -\tfrac{1}{4}\pi, \tag{74}$$

$\mathbf{U}(\zeta;m)$
$$= \begin{pmatrix} \beta B_1^{(1)} U(-a, -iy) & \beta(1-m^2)^{-1} A_2^{(-1)} U(a, -y) \\ \sqrt{2} e^{-3\pi i/4}(a-\tfrac{1}{2}) B_1^{(1)} U(1-a, -iy) & \sqrt{2} e^{-i\pi/4}(1-m^2)^{-1} A_2^{(-1)} U(a-1, -y) \end{pmatrix},$$
$$\tfrac{3}{4}\pi < \arg(\zeta) < \pi, \tag{75}$$

and

$\mathbf{U}(\zeta;m)$
$$= \begin{pmatrix} \beta(1-m^2)^{-1} B_1^{(1)} U(-a, -iy) & \beta A_2^{(-1)} U(a, -y) \\ \sqrt{2} e^{-3\pi i/4}(a-\tfrac{1}{2})(1-m^2)^{-1} B_1^{(1)} U(1-a, -iy) & \sqrt{2} e^{-i\pi/4} A_2^{(-1)} U(a-1, -y) \end{pmatrix},$$
$$-\pi < \arg(\zeta) < -\tfrac{3}{4}\pi. \tag{76}$$

Appealing again to (71) now shows that $\mathbf{U}(\zeta;m)$ agrees with the first formula in (63) up to the leading term only if the parameter a in Weber's equation (68) satisfies

$$a - \frac{1}{2} = \frac{1}{2\pi i} \ln(1-m^2) \quad \Longrightarrow \quad |\beta|^2 = -\frac{1}{\pi} \ln(1-m^2) > 0, \tag{77}$$

and the remaining constants $A_2^{(0)}$, $A_2^{(-1)}$, $B_1^{(0)}$, and $B_1^{(1)}$, are given in terms of β by

$$B_1^{(0)} = \beta^{-1}(1-m^2)^{-1/8} \exp\left(i\frac{1}{4\pi} \ln(2) \ln(1-m^2)\right)$$
$$A_2^{(0)} = \frac{1}{\sqrt{2}}(1-m^2)^{-1/8} e^{-3\pi i/4} \exp\left(-i\frac{1}{4\pi} \ln(2) \ln(1-m^2)\right)$$
$$B_1^{(1)} = \beta^{-1}(1-m^2)^{3/8} \exp\left(i\frac{1}{4\pi} \ln(2) \ln(1-m^2)\right)$$
$$A_2^{(-1)} = \frac{1}{\sqrt{2}}(1-m^2)^{3/8} e^{i\pi/4} \exp\left(-i\frac{1}{4\pi} \ln(2) \ln(1-m^2)\right). \tag{78}$$

Only $\arg(\beta)$ remains to be determined, and for this we recall the nontrivial jump conditions for the first (second) column of $\mathbf{U}(\zeta;m)$ across the rays $\arg(\zeta) = \tfrac{1}{4}\pi, -\tfrac{3}{4}\pi$ (the rays $\arg(\zeta) = -\tfrac{1}{4}\pi, \tfrac{3}{4}\pi$). Actually all four of these jump conditions contain equivalent information due to the fact that the cyclic product of the jump matrices in Riemann-Hilbert Problem 3 about the origin is the identity, so we just examine the transition of the first column across the ray $\arg(\zeta) = \frac{1}{4}\pi$ implied by the jump conditions in Riemann-Hilbert Problem 3. Using all available information, the jump condition matches the connection formula [18, Eq. 12.2.18] if and only if

$$\arg(\beta) = \frac{\pi}{4} + \frac{1}{2\pi}\ln(2)\ln(1-m^2) - \arg\left(\Gamma\left(i\frac{1}{2\pi}\ln(1-m^2)\right)\right). \quad (79)$$

Combining this with (77) determines $\beta=\beta(m)$ and then using (78) in (72)–(76) fully determines $\mathbf{U}(\zeta;m)$ and hence also $\mathbf{P}(\zeta;m) = \mathbf{U}(\zeta;m)e^{i\zeta^2\sigma_3/2}$. This completes the construction of the necessarily unique solution of Riemann-Hilbert Problem 3. One can easily check directly that $\mathbf{U}'(\zeta;m)\mathbf{U}(\zeta;m)^{-1}$ is analytic at $\zeta = 0$, and using (71) (which is known to be a formally differentiable expansion) one confirms the asymptotic expansions (62)–(63), justifying after the fact all assumptions made to arrive at the explicit solution.

We note that for each $m \in [0,1)$, $\mathbf{P}(\zeta;m)$ is uniformly bounded with respect to $\zeta \in \mathbb{C}$, since it is locally bounded and the normalization factor in the asymptotics as $\zeta \to \infty$ satisfies

$$(1-m^2)^{1/2} < |\zeta^{-\ln(1-m^2)/(2\pi i)}| < (1-m^2)^{-1/2}, \quad \arg(\zeta) \in (-\pi,\pi). \quad (80)$$

Since $\det(\mathbf{P}(\zeta;m)) = 1$, the same holds for $\mathbf{P}(\zeta;m)^{-1}$. Moreover, it is not difficult to see that if $\|\cdot\|$ is a matrix norm, then $\sup_{\zeta\in\mathbb{C}\setminus\Sigma_\mathbf{P}}\|\mathbf{P}(\zeta;m)\|$ is a continuous function of $m \in [0,1)$. Therefore the estimates on $\mathbf{P}(\zeta;m)$ and $\mathbf{P}(\zeta;m)^{-1}$ hold uniformly with respect to $m \in [0,\rho]$ for any $\rho < 1$.

3.5 The Equivalent $\bar{\partial}$ Problem and Its Solution for Large t

The next part of the proof of Theorem 1.1 is the nonlinear analogue of the estimation of the error $\mathcal{E}(x,t)$ in the stationary phase formula (20) by double integrals in the z-plane. Here instead of a double integral we will have a double-integral equation arising from a $\bar{\partial}$-problem. To arrive at this problem, we simply define a matrix function $\mathbf{E}(u,v;x,t)$ by comparing the "open lenses" matrix $(2t^{1/2})^{i\nu(z_0)\sigma_3}\mathbf{O}(u,v;x,t)$ with its parametrix $\mathbf{P}(2t^{1/2}(z-z_0);|r(z_0)|)$:

$$\mathbf{E}(u,v;x,t) := (2t^{1/2})^{i\nu(z_0)\sigma_3}\mathbf{O}(u,v;x,t)\mathbf{P}(2t^{1/2}(u+iv-z_0);|r(z_0)|)^{-1}. \quad (81)$$

We claim that $\mathbf{E}(u,v;x,t)$ satisfies the following problem.

$\bar{\partial}$ **Problem 1** *Let $(x,t) \in \mathbb{R}^2$ be parameters. Find a 2×2 matrix function $\mathbf{E} = \mathbf{E}(u,v) = \mathbf{E}(u,v;x,t)$, $(u,v) \in \mathbb{R}^2$ with the following properties:*

Continuity $\mathbf{E}$ *is a continuous function of $(u,v) \in \mathbb{R}^2$.*

Nonanalyticity $\mathbf{E}$ *is a (weak) solution of the partial differential equation* $\bar{\partial}\mathbf{E}(u,v) = \mathbf{E}(u,v)\mathbf{W}(u,v)$, *where* $\mathbf{W}(u,v) = \mathbf{W}(u,v;x,t)$ *is defined by*

$$\mathbf{W}(u,v;x,t) := \mathbf{P}(2t^{1/2}(u+iv-z_0); |r(z_0)|)\mathbf{\Delta}(u,v;x,t)$$
$$\cdot \mathbf{P}(2t^{1/2}(u+iv-z_0); |r(z_0)|)^{-1}, \quad (82)$$

and

$$\mathbf{\Delta}(u,v;x,t) := \begin{cases} \begin{pmatrix} 0 & 0 \\ -\bar{\partial}E_1(u,v)e^{2it(\theta(u+iv;z_0)-\theta(z_0;z_0))} & 0 \end{pmatrix}, & u+iv \in \Omega_1 \\ \mathbf{0}, & u+iv \in \Omega_2 \\ \begin{pmatrix} 0 & -\bar{\partial}E_3(u,v)e^{-2it(\theta(u+iv;z_0)-\theta(z_0;z_0))} \\ 0 & 0 \end{pmatrix}, & u+iv \in \Omega_3 \\ \begin{pmatrix} 0 & 0 \\ \bar{\partial}E_4(u,v)e^{2it(\theta(u+iv;z_0)-\theta(z_0;z_0))} & 0 \end{pmatrix}, & u+iv \in \Omega_4 \\ \mathbf{0}, & u+iv \in \Omega_5 \\ \begin{pmatrix} 0 & \bar{\partial}E_6(u,v)e^{-2it(\theta(u+iv;z_0)-\theta(z_0;z_0))} \\ 0 & 0 \end{pmatrix}, & u+iv \in \Omega_6. \end{cases}$$
(83)

Note that $\mathbf{W}(u,v;x,t)$ *has jump discontinuities across the sector boundaries in general.*
Normalization $\mathbf{E}(u,v) \to \mathbb{I}$ *as* $(u,v) \to \infty$.

To show the continuity, first note that in each of the six sectors Ω_j, $j = 1, \ldots, 6$, $\mathbf{E}(u,v;x,t)$ is continuous as a function of (u,v) up to the sector boundary. Indeed, the first factor in (81) is independent of (u,v), and the second factor in (81) has the claimed continuity because this is a property of the solution $\mathbf{N}(u+iv;x,t)$ of Riemann-Hilbert Problem 2 and of the change-of-variables formula (59). Finally, $\mathbf{P}(\zeta;m)$ has unit determinant and its explicit formula in terms of parabolic cylinder functions shows that its restriction to each sector is an entire function of ζ, which guarantees the asserted continuity of the third factor in (81). Moreover, the matrices $(2t^{1/2})^{i\nu(z_0)\sigma_3}\mathbf{O}(u,v;x,t)$ and $\mathbf{P}(2t^{1/2}(u+iv-z_0); |r(z_0)|)$ satisfy exactly the same jump conditions across the six rays that form the common boundaries of neighboring sectors, from which it follows that $\mathbf{E}_+(u,v;x,t) = \mathbf{E}_-(u,v;x,t)$ holds across each of these rays and therefore $\mathbf{E}(u,v;x,t)$ may be regarded as a continuous function of $(u,v) \in \mathbb{R}^2$.

To show that $\bar{\partial}\mathbf{E} = \mathbf{EW}$ holds, one simply differentiates $\mathbf{E}(u,v;x,t)$ in each of the six sectors, using the fact that $\mathbf{O}(u,v;x,t)$ is related to $\mathbf{N}(u+iv;x,t)$ explicitly by (59) and that both $\mathbf{N}(u+iv;x,t)$ and the unit-determinant matrix function $\mathbf{P}(2t^{1/2}(u+iv-z_0); |r(z_0)|)$ are analytic functions of $u+iv$ in each sector, and hence are annihilated by $\bar{\partial}$. The region of non-analyticity of $\mathbf{E}$ is therefore the union of shaded sectors shown in Fig. 2.

Finally to show the normalization condition, we recall Lemma 3.2. Therefore, comparing the normalization conditions of Riemann-Hilbert Problem 2 for $\mathbf{N}(z; x, t)$ and of Riemann-Hilbert Problem 3 for $\mathbf{P}(\zeta; m)$ shows that $\mathbf{E}(u, v; x, t) \to \mathbb{I}$ as $(u, v) \to \infty$ in $\mathbb{R}^2$.

The rest of this section is devoted to the proof of the following result.

Proposition 3.3 *Suppose that $r \in H^1(\mathbb{R})$ with $|r(z)| \le \rho$ for some $\rho < 1$. If $t > 0$ is sufficiently large, then for all $x \in \mathbb{R}$ there exists a unique solution $\mathbf{E}(\cdot, \cdot; x, t) \in L^\infty(\mathbb{R}^2)$ of $\overline{\partial}$-Problem 1 with the property that*

$$\mathbf{E}_1(x, t) := \lim_{\substack{(u,v) \to \infty \\ u=0}} (u + iv) \left[\mathbf{E}(u, v; x, t) - \mathbb{I} \right] \tag{84}$$

exists and satisfies

$$\sup_{x \in \mathbb{R}} \|\mathbf{E}_1(x, t)\| = \mathcal{O}(t^{-3/4}), \quad t \to +\infty. \tag{85}$$

Proof To show that $\overline{\partial}$-Problem 1 has a unique solution for $t > 0$ sufficiently large, and simultaneously obtain estimates for the solution $\mathbf{E}(u, v; x, t)$, we formulate a weakly-singular integral equation whose solution is that of $\overline{\partial}$-Problem 1:

$$\mathbf{E}(u, v; x, t) = \mathbb{I} + \mathcal{J}\mathbf{E}(u, v; x, t),$$

$$\mathcal{J}\mathbf{F}(u, v) := -\frac{1}{\pi} \iint_{\mathbb{R}^2} \frac{\mathbf{F}(U, V) \mathbf{W}(U, V; x, t)}{(U - u) + i(V - v)} \, dA(U, V), \tag{86}$$

in which the identity matrix $\mathbb{I}$ is viewed as a constant function on $\mathbb{R}^2$. Indeed, this is a consequence of the distributional identity $\overline{\partial} z^{-1} = -\pi \delta$ where δ denotes the Dirac mass at the origin. We will solve the integral equation (86) in the space $L^\infty(\mathbb{R}^2)$, by computing the corresponding operator norm[3] of $\mathcal{J} : L^\infty(\mathbb{R}^2) \to L^\infty(\mathbb{R}^2)$ and showing that for large $t > 0$ it is less than 1. Thus, we begin with the elementary estimate

$$\|\mathcal{J}\mathbf{F}(u, v)\| \le \frac{1}{\pi} \|\mathbf{F}\|_{L^\infty(\mathbb{R}^2)} \iint_{\mathbb{R}^2} \frac{\|\mathbf{W}(U, V; x, t)\| \, dA(U, V)}{\sqrt{(U - u)^2 + (V - v)^2}}. \tag{87}$$

Using the uniform boundedness of $\mathbf{P}(\zeta; m)$ and its inverse with respect to ζ i.e., there exists $C > 0$ such that $\|\mathbf{P}(\zeta; m)\| \le C$ and $\|\mathbf{P}(\zeta; m)^{-1}\| \le C$ for all $\zeta \in \mathbb{C} \setminus \Sigma_{\mathbf{P}}$ and all $m \in [0, \rho]$ with $\rho < 1$, the assumption $|r(z)| \le \rho < 1$ gives that

---

[3] All L^p norms of matrix-valued functions in this section depend on the choice of matrix norm, which we always take to be induced by a norm on $\mathbb{C}^2$.

$$\|\mathbf{W}(u,v;x,t)\| \leq C^2 \begin{cases} e^{-8t(u-z_0)v}|\overline{\partial}E_1(u,v)|, & z = u+iv \in \Omega_1 \\ e^{8t(u-z_0)v}|\overline{\partial}E_3(u,v)|, & z = u+iv \in \Omega_3 \\ e^{-8t(u-z_0)v}|\overline{\partial}E_4(u,v)|, & z = u+iv \in \Omega_4 \\ e^{8t(u-z_0)v}|\overline{\partial}E_6(u,v)|, & z = u+iv \in \Omega_6, \end{cases} \quad (88)$$

and of course $\mathbf{W}(u,v;x,t) \equiv 0$ on $\Omega_2 \cup \Omega_5$. By direct computation using (58) along with the analyticity of $f(z;z_0)^{\pm 2}$ provided by Lemma 3.1 and straightforward estimates of $\cos(2\arg(u+iv-z_0))$ and its $\overline{\partial}$-derivative as in Sect. 2, we have the following analogues of (29):

$$|\overline{\partial}E_1(u,v)| \leq \frac{1}{2}|f(u+iv;z_0)^{-2}||r'(u)| + \frac{|f(u+iv;z_0)^{-2}r(u) - r(z_0)|}{\sqrt{(u-z_0)^2+v^2}},$$
$$z = u+iv \in \Omega_1, \quad (89)$$

$$|\overline{\partial}E_3(u,v)| \leq \frac{1}{2}|f(u+iv;z_0)^2|\left|\frac{d}{du}\frac{\overline{r(u)}}{1-|r(u)|^2}\right|$$
$$+ \left|\frac{f(u+iv;z_0)^2\overline{r(u)}}{1-|r(u)|^2} - \frac{\overline{r(z_0)}}{1-|r(z_0)|^2}\right|\frac{1}{\sqrt{(u-z_0)^2+v^2}}, \quad z = u+iv \in \Omega_3,$$
$$(90)$$

$$|\overline{\partial}E_4(u,v)| \leq \frac{1}{2}|f(u+iv;z_0)^{-2}|\left|\frac{d}{du}\frac{r(u)}{1-|r(u)|^2}\right|$$
$$+ \left|\frac{f(u+iv;z_0)^{-2}r(u)}{1-|r(u)|^2} - \frac{r(z_0)}{1-|r(z_0)|^2}\right|\frac{1}{\sqrt{(u-z_0)^2+v^2}}, \quad z = u+iv \in \Omega_4,$$
$$(91)$$

and

$$|\overline{\partial}E_6(u,v)| \leq \frac{1}{2}|f(u+iv;z_0)^2||r'(u)| + \frac{|f(u+iv;z_0)^2\overline{r(u)} - \overline{r(z_0)}|}{\sqrt{(u-z_0)^2+v^2}},$$
$$z = u+iv \in \Omega_6. \quad (92)$$

Note that

$$\left|\frac{d}{du}\frac{\overline{r(u)}}{1-|r(u)|^2}\right| = \left|\frac{d}{du}\frac{r(u)}{1-|r(u)|^2}\right| \leq \frac{1+\rho^2}{(1-\rho^2)^2}|r'(u)| \quad (93)$$

holds under the condition $|r(u)| \le \rho < 1$. Also, under the same condition,

$$\begin{aligned}|f(u+iv;z_0)^{-2}r(u)-r(z_0)| &= |(f(u+iv;z_0)^{-2}-1)r(u)+r(u)-r(z_0)|\\ &\le \rho|f(u+iv;z_0)^{-2}-1|+|r(u)-r(z_0)|\\ &\le \left(K\rho+\|r'\|_{L^2(\mathbb{R})}\right)[(u-z_0)^2+v^2]^{1/4},\end{aligned} \quad (94)$$

where we used Lemma 3.1 and (30), and $K > 0$ depends on ρ but not on z_0. Exactly the same estimate holds for $|f(u+iv;z_0)^2\overline{r(u)}-\overline{r(z_0)}|$. In the same way, but also using (93),

$$\left|\frac{f(u+iv;z_0)^2\overline{r(u)}}{1-|r(u)|^2}-\frac{\overline{r(z_0)}}{1-|r(z_0)|^2}\right| \le \left(\frac{K\rho}{1-\rho^2}+\frac{1+\rho^2}{(1-\rho^2)^2}\|r'\|_{L^2(\mathbb{R})}\right)[(u-z_0)^2+v^2]^{1/4}$$

$$\left|\frac{f(u+iv;z_0)^{-2}r(u)}{1-|r(u)|^2}-\frac{r(z_0)}{1-|r(z_0)|^2}\right| \le \left(\frac{K\rho}{1-\rho^2}+\frac{1+\rho^2}{(1-\rho^2)^2}\|r'\|_{L^2(\mathbb{R})}\right)[(u-z_0)^2+v^2]^{1/4}.$$
(95)

Therefore again using Lemma 3.1, we see that there are constants L and M depending only on the upper bound $\rho < 1$ for $\|r\|_{L^\infty(\mathbb{R})}$, on $\|r\|_{L^2(\mathbb{R})}$, and on $\|r'\|_{L^2(\mathbb{R})}$ such that

$$|\partial E_j(u,v)| \le L|r'(u)| + \frac{M}{[(u-z_0)^2+v^2]^{1/4}}, \quad z = u+iv \in \Omega_j, \quad j=1,3,4,6. \tag{96}$$

Note that (96) is the nonlinear analogue of the estimate (31).

Combining (96) with (87)–(88) shows that for some constant D independent of $(x,t) \in \mathbb{R}^2$,

$$\|\mathcal{J}\mathbf{F}(u,v;x,t)\| \le D\left[I^{[1,4]}(u,v;x,t) + J^{[1,4]}(u,v;x,t) + I^{[3,6]}(u,v;x,t)\right.$$
$$\left. + J^{[3,6]}(u,v;x,t)\right]\|\mathbf{F}\|_{L^\infty(\mathbb{R}^2)}, \tag{97}$$

where the four terms are analogues in the nonlinear case of the double integrals defined in (33) for the linear case:

$$I^{[1,4]}(u,v;x,t) := \iint_{\Omega_1 \cup \Omega_4} \frac{|r'(U)|e^{-8t(U-z_0)V}\,dA(U,V)}{\sqrt{(U-u)^2+(V-v)^2}},$$

$$I^{[3,6]}(u,v;x,t) := \iint_{\Omega_3 \cup \Omega_6} \frac{|r'(U)|e^{8t(U-z_0)V}\,dA(U,V)}{\sqrt{(U-u)^2+(V-v)^2}},$$

$$J^{[1,4]}(u, v; x, t) := \iint_{\Omega_1 \cup \Omega_4} \frac{e^{-8t(U-z_0)V} \, dA(U, V)}{[(U - z_0)^2 + V^2]^{1/4} \sqrt{(U - u)^2 + (V - v)^2}}, \quad \text{and}$$

$$J^{[3,6]}(u, v; x, t) := \iint_{\Omega_3 \cup \Omega_6} \frac{e^{8t(U-z_0)V} \, dA(U, V)}{[(U - z_0)^2 + V^2]^{1/4} \sqrt{(U - u)^2 + (V - v)^2}}. \tag{98}$$

Estimation of the integrals $I^{[1,4]}(u, v; x, t)$ and $J^{[1,4]}(u, v; x, t)$ requires nearly identical steps as estimation of $I^{[3,6]}(u, v; x, t)$ and $J^{[3,6]}(u, v; x, t)$ (just note that the sign of the exponent always corresponds to decay in the sectors of integration). So for brevity we just deal with $I^{[3,6]}(u, v; x, t)$ and $J^{[3,6]}(u, v; x, t)$.

To estimate $I^{[3,6]}(u, v; x, t)$, by iterated integration we have

$$I^{[3,6]}(u, v; x, t)$$

$$= \left[\int_0^{+\infty} dV \int_{-\infty}^{z_0-V} dU + \int_{-\infty}^0 dV \int_{z_0-V}^{+\infty} dU \right] \frac{|r'(U)| e^{8t(U-z_0)V}}{\sqrt{(U-u)^2 + (V-v)^2}}$$

$$\leq \left[\int_0^{+\infty} dV \int_{-\infty}^{z_0-V} dU + \int_{-\infty}^0 dV \int_{z_0-V}^{+\infty} dU \right] \frac{|r'(U)| e^{-8tV^2}}{\sqrt{(U-u)^2 + (V-v)^2}}. \tag{99}$$

The inner integrals can be estimated by Cauchy-Schwarz, using the fact that $r' \in L^2(\mathbb{R})$:

$$\pm \int_{\mp\infty}^{z_0-V} \frac{|r'(U)| \, dU}{\sqrt{(U-u)^2 + (V-v)^2}} \leq \int_{\mathbb{R}} \frac{|r'(U)| \, dU}{\sqrt{(U-u)^2 + (V-v)^2}}$$

$$\leq \|r'\|_{L^2(\mathbb{R})} \sqrt{\int_{\mathbb{R}} \frac{dU}{(U-u)^2 + (V-v)^2}} = \frac{\|r'\|_{L^2(\mathbb{R})} \sqrt{\pi}}{\sqrt{|V-v|}}. \tag{100}$$

Thus,

$$I^{[3,6]}(u, v; x, t) \leq \|r'\|_{L^2(\mathbb{R})} \sqrt{\pi} \int_{\mathbb{R}} \frac{e^{-8tV^2} \, dV}{\sqrt{|V-v|}}. \tag{101}$$

Without loss of generality, suppose that $v > 0$. Then

$$\int_{\mathbb{R}} \frac{e^{-8tV^2} \, dV}{\sqrt{|V-v|}} = \int_{-\infty}^0 \frac{e^{-8tV^2} \, dV}{\sqrt{v-V}} + \int_0^v \frac{e^{-8tV^2} \, dV}{\sqrt{v-V}} + \int_v^{+\infty} \frac{e^{-8tV^2} \, dV}{\sqrt{V-v}}. \tag{102}$$

Using monotonicity of $\sqrt{v-V}$ on $V < 0$ and the rescaling $V = t^{-1/2}w$, we get for the first term:

$$\int_{-\infty}^{0} \frac{e^{-8tV^2} dV}{\sqrt{v-V}} \leq \int_{-\infty}^{0} \frac{e^{-8tV^2} dV}{\sqrt{-V}} = t^{-1/4} \int_{-\infty}^{0} \frac{e^{-8w^2} dw}{\sqrt{-w}} = \mathcal{O}(t^{-1/4}). \tag{103}$$

For the second term, we use the inequality $e^{-b} \leq Cb^{-1/4}$ for $b > 0$ and the rescaling $V = vw$ to get

$$\int_{0}^{v} \frac{e^{-8tV^2} dV}{\sqrt{v-V}} \leq C(8t)^{-1/4} \int_{0}^{v} \frac{dV}{\sqrt{V(v-V)}} = C(8t)^{-1/4} \int_{0}^{1} \frac{dw}{\sqrt{w(1-w)}} = \mathcal{O}(t^{-1/4}). \tag{104}$$

Using monotonicity of e^{-8tV^2} on $V > v$ and the change of variable $V - v = t^{-1/2}w$ we get for the third term:

$$\int_{v}^{+\infty} \frac{e^{-8tV^2} dV}{\sqrt{V-v}} \leq \int_{v}^{+\infty} \frac{e^{-8t(V-v)^2} dV}{\sqrt{V-v}} = t^{-1/4} \int_{0}^{+\infty} \frac{e^{-8w^2} dw}{\sqrt{w}} = \mathcal{O}(t^{-1/4}). \tag{105}$$

The upper bounds in (103)–(104) are all independent of v (and u), so combining them with (101)–(102) gives

$$\sup_{(u,v) \in \mathbb{R}^2} I^{[3,6]}(u, v; x, t) \leq C \|r'\|_{L^2(\mathbb{R})} t^{-1/4}, \tag{106}$$

where C denotes an absolute constant.

To estimate $J^{[3,6]}(u, v; x, t)$ we again introduce iterated integrals in the same way as in (99) to obtain the inequality

$$J^{[3,6]}(u, v; x, t) \leq \left[\int_{0}^{+\infty} dV \int_{-\infty}^{z_0-V} dU + \int_{-\infty}^{0} dV \int_{z_0-V}^{+\infty} dU \right]$$
$$\cdot \frac{e^{-8tV^2}}{[(U-z_0)^2 + V^2]^{1/4} \sqrt{(U-u)^2 + (V-v)^2}}. \tag{107}$$

Now, to estimate the inner U-integrals we will use Hölder's inequality with conjugate exponents $p > 2$ and $q < 2$. Thus,

$$\pm \int_{\mp\infty}^{z_0-V} \frac{dU}{[(U-z_0)^2 + V^2]^{1/4} \sqrt{(U-u)^2 + (V-v)^2}}$$

$$\leq \left(\pm \int_{\mp\infty}^{z_0-V} \frac{dU}{[(U-z_0)^2 + V^2]^{p/4}} \right)^{1/p} \left(\pm \int_{\mp\infty}^{z_0-V} \frac{dU}{[(U-u)^2 + (V-v)^2]^{q/2}} \right)^{1/q}$$

$$\leq \left(\int_{\mathbb{R}} \frac{dU}{[(U-z_0)^2 + V^2]^{p/4}} \right)^{1/p} \left(\int_{\mathbb{R}} \frac{dU}{[(U-u)^2 + (V-v)^2]^{q/2}} \right)^{1/q}. \tag{108}$$

Now, by the change of variable $U - z_0 = |V|w$,

$$\left(\int_{\mathbb{R}} \frac{dU}{[(U-z_0)^2 + V^2]^{p/4}} \right)^{1/p} = |V|^{1/p - 1/2} \left(\int_{\mathbb{R}} \frac{dw}{[w^2 + 1]^{p/4}} \right)^{1/p}, \tag{109}$$

where the integral on the right-hand side is convergent as long as $p > 2$. Similarly, by the change of variable $U - u = |V - v|w$,

$$\left(\int_{\mathbb{R}} \frac{dU}{[(U-u)^2 + (V-v)^2]^{q/2}} \right)^{1/q} = |V-v|^{1/q - 1} \left(\int_{\mathbb{R}} \frac{dw}{[w^2 + 1]^{q/2}} \right)^{1/q}, \tag{110}$$

where the integral on the right-hand side is convergent as long as $q > 1$. Hence for any conjugate exponents $1 < q < 2 < p < \infty$ with $p^{-1} + q^{-1} = 1$, we have for some constant $C = C(p, q)$,

$$J^{[3,6]}(u, v; x, t) \leq C \int_{\mathbb{R}} e^{-8tV^2} |V|^{1/p - 1/2} |V - v|^{1/q - 1} dV. \tag{111}$$

As before, assume without loss of generality that $v > 0$. Then

$$\int_{\mathbb{R}} e^{-8tV^2} |V|^{1/p - 1/2} |V - v|^{1/q - 1} dV = \int_{-\infty}^{0} e^{-8tV^2} (-V)^{1/p - 1/2} (v - V)^{1/q - 1} dV$$

$$+ \int_{0}^{v} e^{-8tV^2} V^{1/p - 1/2} (v - V)^{1/q - 1} dV + \int_{v}^{+\infty} e^{-8tV^2} V^{1/p - 1/2} (V - v)^{1/q - 1} dV. \tag{112}$$

Using $q > 1$ and monotonicity of $(v - V)^{1/q - 1}$ on $V < 0$ along with $1/p + 1/q = 1$ and the rescaling $V = t^{-1/2}w$ gives for the first integral

$$\int_{-\infty}^{0} e^{-8tV^2} (-V)^{1/p - 1/2} (v - V)^{1/q - 1} dV \leq \int_{-\infty}^{0} e^{-8tV^2} (-V)^{1/p - 1/2 + 1/q - 1} dV$$

$$= \int_{-\infty}^{0} e^{-8tV^2} (-V)^{-1/2} dV$$

$$= t^{-1/4} \int_{-\infty}^{0} e^{-8w^2} (-w)^{-1/2} dw = \mathcal{O}(t^{-1/4}). \tag{113}$$

For the second integral, we again recall $e^{-b} \leq Cb^{-1/4}$ for $b > 0$ and rescale by $V = vw$ to get

$$\int_0^v e^{-8tV^2} V^{1/p-1/2}(v-V)^{1/q-1} dV \leq C(8t)^{-1/4} \int_0^v V^{1/p-1}(v-V)^{1/q-1} dV$$

$$= C(8t)^{-1/4} \int_0^1 w^{1/p-1}(1-w)^{1/q-1} dw$$

$$= \mathcal{O}(t^{-1/4}), \quad (114)$$

using also $q, p < \infty$. Finally, for the third integral, we use monotonicity of e^{-8tV^2} and $V^{1/p-1/2}$ (for $p > 2$) on $V > v$ and make the substitution $V - v = t^{-1/2}w$ to get

$$\int_v^{+\infty} e^{-8tV^2} V^{1/p-1/2}(V-v)^{1/q-1} dV \leq \int_v^{+\infty} e^{-8t(V-v)^2} (V-v)^{1/p-1/2}$$

$$\cdot (V-v)^{1/q-1} dV$$

$$= \int_v^{+\infty} e^{-8t(V-v)^2} (V-v)^{-1/2} dV$$

$$= t^{-1/4} \int_0^{+\infty} e^{-8w^2} w^{-1/2} dw = \mathcal{O}(t^{-1/4}). \quad (115)$$

Since the upper bounds in (113)–(115) are all independent of $(u,v) \in \mathbb{R}^2$, combining them with (111)–(112) gives

$$\sup_{(u,v) \in \mathbb{R}^2} J^{[3,6]}(u,v;x,t) \leq Ct^{-1/4}, \quad (116)$$

where C denotes an absolute constant.

Returning to (97) and taking a supremum over $(u,v) \in \mathbb{R}^2$, we see that

$$\|\mathcal{J}\mathbf{F}\|_{L^\infty(\mathbb{R}^2)} \leq Dt^{-1/4} \|\mathbf{F}\|_{L^\infty(\mathbb{R}^2)}, \quad \text{i.e.,} \quad \|\mathcal{J}\|_{L^\infty(\mathbb{R}^2)_\circlearrowleft} \leq Dt^{-1/4} \quad (117)$$

holds where D is a constant depending only on the upper bound $\rho < 1$ for $\|r\|_{L^\infty(\mathbb{R})}$, on $\|r\|_{L^2(\mathbb{R})}$, and on $\|r'\|_{L^2(\mathbb{R})}$, and where $\|\mathcal{J}\|_{L^\infty(\mathbb{R}^2)_\circlearrowleft}$ denotes the norm of the weakly-singular integral operator $\mathcal{J}$ acting in $L^\infty(\mathbb{R}^2)$. It is a consequence of (117) that the integral equation (86) is uniquely solvable in $L^\infty(\mathbb{R}^2)$ by convergent Neumann series for sufficiently large $t > 0$:

$$\mathbf{E}(u,v;x,t) = (\mathbb{I} - \mathcal{J})^{-1}\mathbb{I} = \mathbb{I} + \mathcal{J}\mathbb{I} + \mathcal{J}^2\mathbb{I} + \mathcal{J}^3\mathbb{I} + \cdots, \quad t > D^{-4}, \quad (118)$$

where $\mathcal{I}$ denotes the identity operator and $\mathbb{I}$ the constant function on $\mathbb{R}^2$, and that the solution satisfies

$$\|\mathbf{E} - \mathbb{I}\|_{L^\infty(\mathbb{R}^2)} \leq \frac{Dt^{-1/4}}{1 - Dt^{-1/4}} = \mathcal{O}(t^{-1/4}), \quad t \to +\infty, \tag{119}$$

an estimate that is uniform with respect to $x \in \mathbb{R}$. This proves the first assertion in Proposition 3.3.

To prove the existence of the limit $\mathbf{E}_1(x, t)$ in (84), note that from the integral equation (86) we have

$$(u + iv)\left[\mathbf{E}(u, v; x, t) - \mathbb{I}\right]$$
$$= \frac{1}{\pi} \iint_{\mathbb{R}^2} \mathbf{E}(U, V; x, t)\mathbf{W}(U, V; x, t)\,dA(U, V)$$
$$- \frac{1}{\pi} \iint_{\mathbb{R}^2} \frac{U + iV}{(U - u) + i(V - v)} \mathbf{E}(U, V; x, t)\mathbf{W}(U, V; x, t)\,dA(U, V). \tag{120}$$

The second term satisfies

$$\left\| \iint_{\mathbb{R}^2} \frac{U + iV}{(U - u) + i(V - v)} \mathbf{E}(U, V; x, t)\mathbf{W}(U, V; x, t)\,dA(U, V) \right\|$$
$$\leq \|\mathbf{E}\|_{L^\infty(\mathbb{R}^2)} \iint_{\mathbb{R}^2} \sqrt{\frac{U^2 + V^2}{(U - u)^2 + (V - v)^2}} \|\mathbf{W}(U, V; x, t)\|\,dA(U, V). \tag{121}$$

Now, following [12], let us examine the resulting double integral for $u = 0$, i.e., for $z = u + iv$ restricted to the imaginary axis. Some simple trigonometry shows that

$$\sup_{(U,V) \in \mathrm{supp}(\mathbf{W}(\cdot,\cdot;x,t))} \sqrt{\frac{U^2 + V^2}{U^2 + (V - v)^2}} = 1 + \sqrt{2} \frac{|v|}{|v| - |z_0|}, \quad |v| > |z_0|. \tag{122}$$

Therefore, if $u = 0$, the double integral on the right-hand side of (121) will tend to zero as $|v| \to \infty$ by the Lebesgue dominated convergence theorem provided that $\mathbf{W}(\cdot, \cdot; x, t) \in L^1(\mathbb{R}^2)$. Using (88) and (96), we have

$$\iint_{\mathbb{R}^2} \|\mathbf{W}(U, V; x, t)\|\,dA(U, V)$$
$$\leq D\left[\tilde{I}^{[1,4]}(x, t) + \tilde{J}^{[1,4]}(x, t) + \tilde{I}^{[3,6]}(x, t) + \tilde{J}^{[3,6]}(x, t)\right], \tag{123}$$

where (compare with (98), or better yet, (33))

$$\tilde{I}^{[1,4]}(x,t) := \iint_{\Omega_1 \cup \Omega_4} |r'(U)| e^{-8t(U-z_0)V} \, dA(U,V),$$

$$\tilde{I}^{[3,6]}(x,t) := \iint_{\Omega_3 \cup \Omega_6} |r'(U)| e^{8t(U-z_0)V} \, dA(U,V),$$

$$\tilde{J}^{[1,4]}(x,t) := \iint_{\Omega_1 \cup \Omega_4} \frac{e^{-8t(U-z_0)V} \, dA(U,V)}{[(U-z_0)^2 + V^2]^{1/4}}, \quad \text{and} \quad (124)$$

$$\tilde{J}^{[3,6]}(x,t) := \iint_{\Omega_3 \cup \Omega_6} \frac{e^{8t(U-z_0)V} \, dA(U,V)}{[(U-z_0)^2 + V^2]^{1/4}}.$$

Noting the resemblance with the double integrals (33) analyzed in Sect. 2, we can immediately obtain the estimate

$$\iint_{\mathbb{R}^2} \|\mathbf{W}(U,V;x,t)\| \, dA(U,V) \leq Ct^{-3/4} < \infty \quad (125)$$

for some constant C independent of x. Therefore, the second term on the right-hand side of (120) tends to zero as $v \to \infty$ if $u = 0$ (the limit is not uniform with respect to x since v is compared with z_0 in (122)). Comparing with (84), we obtain from (120) the formula

$$\mathbf{E}_1(x,t) := \frac{1}{\pi} \iint_{\mathbb{R}^2} \mathbf{E}(U,V;x,t) \mathbf{W}(U,V;x,t) \, dA(U,V), \quad (126)$$

and exactly the same argument shows that $\mathbf{E}_1(x,t)$ is finite and uniformly decaying as $t \to +\infty$:

$$\|\mathbf{E}_1(x,t)\| \leq \frac{1}{\pi} \|\mathbf{E}\|_{L^\infty(\mathbb{R}^2)} \|\mathbf{W}\|_{L^1(\mathbb{R}^2)}$$

$$\leq \frac{1}{\pi} \left(\|\mathbb{I}\|_{L^\infty(\mathbb{R}^2)} + \|\mathbf{E} - \mathbb{I}\|_{L^\infty(\mathbb{R}^2)} \right) \|\mathbf{W}\|_{L^1(\mathbb{R}^2)}$$

$$\leq \frac{C}{\pi} \left(1 + \frac{Dt^{-1/4}}{1 - Dt^{-1/4}} \right) t^{-3/4} = \mathcal{O}(t^{-3/4}), \quad (127)$$

where we have used (119) and (125) and noted that the constants C and D are independent of x. This proves the second assertion in Proposition 3.3. □

3.6 The Solution of the Cauchy Problem (1)–(2) for $t > 0$ Large

Now we complete the proof of Theorem 1.1 by combining our previous results. The matrix function $\mathbf{N}(u + iv; x, t)$ agrees with $\mathbf{O}(u, v; x, t)$ for $u = 0$ and $|v|$ sufficiently large given $z_0 = -x/(4t)$. Since according to (81),

$$\mathbf{O}(u, v; x, t) = (2t^{1/2})^{-iv(z_0)\sigma_3} \mathbf{E}(u, v; x, t) \mathbf{P}(2t^{1/2}(u+iv-z_0); |r(z_0)|), \quad (128)$$

we compute the matrix coefficient $\mathbf{N}_1(x, t)$ appearing in (56) by taking a limit along the imaginary axis in (54). Thus, we obtain $\mathbf{N}_1(x, t) = (2t^{1/2})^{-iv(z_0)\sigma_3} \mathbf{Q}(x, t)(2t^{1/2})^{iv(z_0)\sigma_3}$, where (using $z = u + iv$)

$$\mathbf{Q}(x, t) = (2t^{1/2})^{iv(z_0)\sigma_3} \left\{ \lim_{\substack{(u,v) \to \infty \\ u=0}} z \left[\mathbf{N}(z; x, t)(z - z_0)^{-iv(z_0)\sigma_3} - \mathbb{I} \right] \right\} (2t)^{-iv(z_0)\sigma_3}$$

$$= \lim_{\substack{(u,v) \to \infty \\ u=0}} z$$

$$\cdot \left[\mathbf{E}(u, v; x, t) \mathbf{P}(2t^{1/2}(z - z_0); |r(z_0)|)(2t^{1/2}(z - z_0))^{-iv(z_0)\sigma_3} - \mathbb{I} \right].$$

(129)

Using (62) and Proposition 3.3 yields

$$\mathbf{Q}(x, t) = \mathbf{E}_1(x, t) + \frac{1}{2}t^{-1/2}\mathbf{P}_1(|r(z_0)|). \quad (130)$$

Therefore, using (56) gives the following formula for the solution of the Cauchy problem (1)–(2):

$$q(x, t) = 2i e^{-i\omega(z_0)} e^{-2it\theta(z_0; z_0)} c(z_0)^{-2} (2t^{1/2})^{-2iv(z_0)} Q_{12}(x, t)$$

$$= e^{-i\omega(z_0)} e^{-2it\theta(z_0; z_0)} c(z_0)^{-2} (2t^{1/2})^{-2iv(z_0)} \left[2i E_{1,12}(x, t) + \frac{1}{2} t^{-1/2} 2i P_{1,12}(|r(z_0)|) \right]$$

$$= e^{-i\omega(z_0)} e^{-2it\theta(z_0; z_0)} c(z_0)^{-2} (2t^{1/2})^{-2iv(z_0)} \left[2i E_{1,12}(x, t) + \frac{1}{2} t^{-1/2} \beta(|r(z_0)|) \right],$$

(131)

where we recall $\omega(z_0) = \arg(r(z_0))$, $\theta(z_0; z_0) = -2z_0^2$, the definition (4) of $v(z_0)$, the definition (42) of $c(z_0)$, and the definitions (77) and (79) of $|\beta(m = |r(z_0)|)|^2$ and $\arg(\beta(m = |r(z_0)|))$ respectively. Since the factors to the left of the square brackets have unit modulus, from Proposition 3.3 it follows that $q(x, t)$ has exactly the representation (3) in which $|\mathcal{E}(x, t)| = |E_{1,12}(x, t)| = \mathcal{O}(t^{-3/4})$ as $t \to +\infty$, uniformly with respect to x. This completes the proof of Theorem 1.1.

Remark The use of truncations of the Neumann series (118) for $\mathbf{E}(u, v; x, t)$ yields a corresponding asymptotic expansion of $q(x, t)$ as $t \to +\infty$. In other words, it is straightforward (but tedious) to compute explicit corrections to the leading term in the asymptotic formula (3) by expanding $\mathcal{E}(x, t)$. For instance, the formula (126) gives

$$\mathbf{E}_1(x, t) = \frac{1}{\pi} \iint_{\mathbb{R}^2} \mathbf{W}(U, V; x, t) \, dA(U, V)$$

$$+ \frac{1}{\pi} \iint_{\mathbb{R}^2} (\mathbf{E}(U, V; x, t) - \mathbb{I}) \mathbf{W}(U, V; x, t) \, dA(U, V), \quad (132)$$

i.e., an explicit double integral plus a remainder. Using the estimates (119) and (125) we find that the remainder term satisfies

$$\sup_{x \in \mathbb{R}} \left\| \frac{1}{\pi} \iint_{\mathbb{R}^2} (\mathbf{E}(U, V; x, t) - \mathbb{I}) \mathbf{W}(U, V; x, t) \, dA \right\|$$

$$\leq \frac{1}{\pi} \sup_{x \in \mathbb{R}} \|\mathbf{E}(\cdot, \cdot; x, t)\|_{L^\infty(\mathbb{R}^2)} \|\mathbf{W}(\cdot, \cdot; x, t)\|_{L^1(\mathbb{R})} \quad (133)$$

$$= \mathcal{O}(t^{-1/4} t^{-3/4}) = \mathcal{O}(t^{-1}), \quad t \to +\infty.$$

Using this result in (131) gives in place of (3) the corrected asymptotic formula

$$q(x, t) = q^{(0)}(x, t) + q^{(1)}(x, t) + \mathcal{E}^{(1)}(x, t) \quad (134)$$

where

$$q^{(0)}(x, t) := t^{-1/2} \alpha(z_0) e^{ix^2/(4t) - i\nu(z_0) \ln(8t)} \quad (135)$$

is the leading term in (3),

$$q^{(1)}(x, t) := \frac{2i}{\pi} e^{-i\omega(z_0)} e^{-2it\theta(z_0; z_0)} c(z_0)^{-2} (2t^{1/2})^{-2i\nu(z_0)}$$

$$\cdot \iint_{\mathbb{R}^2} W_{12}(U, V; x, t) \, dA(U, V) \quad (136)$$

is an explicit correction (see (82)–(83)), and where $\mathcal{E}^{(1)}(x, t)$ is error term satisfying $\mathcal{E}^{(1)}(x, t) = \mathcal{O}(t^{-1})$ as $t \to +\infty$ uniformly with respect to $x \in \mathbb{R}$. Theorem 1.1 implies that the correction satisfies $\|q^{(1)}(\cdot, t)\|_{L^\infty(\mathbb{R})} = \mathcal{O}(t^{-3/4})$ as $t \to +\infty$, but the explicit formula (136) allows for a complete analysis of the correction. For instance, we are in a position to seek reflection coefficients $r(z)$ in the Sobolev space $H^1(\mathbb{R})$ with $|r(z)| \leq \rho < 1$ for which the correction saturates the upper bound of $\mathcal{O}(t^{-3/4})$, or to determine under which conditions on $r(z)$ the correction term can be smaller. Under additional hypotheses the expansion (134) can be carried out to higher order, with subsequent corrections involving iterated double integrals of $\mathbf{W}$, which in turn involve $\bar{\partial}$-derivatives of the extensions E_j, $j = 1, 3, 4, 6$, and the parabolic cylinder functions contained in the matrix $\mathbf{P}(\zeta; m)$ solving Riemann-Hilbert Problem 3.

Acknowledgements The first two authors were supported in part by NSF grants DMS-0451495, DMS-0800979, and the second author was supported by NSF Grant DMS-1733967. The third author was supported in part by NSF grant DMS-1812625.

References

1. M. Borghese, R. Jenkins, and K. D. T.-R. McLaughlin, "Long time asymptotic behavior of the focusing nonlinear Schrödinger equation," *Ann. Inst. H. Poincaré Anal. Non Linéaire* **35**, no. 4, 887–920, 2018.
2. S. Cuccagna and R. Jenkins, "On the asymptotic stability of N-soliton solutions of the defocusing nonlinear Schrödinger equation," *Commun. Math. Phys.* **343**, 921–969, 2016.
3. P. Deift, A. Its, and X. Zhou, "Long-time asymptotic for integrable nonlinear wave equations," in A. S. Fokas and V. E. Zakharov, editors, *Important Developments in Soliton Theory 1980–1990*, 181–204, Springer-Verlag, Berlin, 1993.
4. P. Deift and X. Zhou, "A steepest descent method for oscillatory Riemann–Hilbert problems. Asymptotics for the mKdV equation," *Ann. Math.* **137**, 295–368, 1993.
5. P. Deift and X. Zhou, "Long-time asymptotics for integrable systems. Higher order theory," *Comm. Math. Phys.* **165**, 175–191, 1994.
6. P. Deift and X. Zhou, *Long-time behavior of the non-focusing nonlinear Schrödinger equation — A case study*, volume 5 of *New Series: Lectures in Math. Sci.*, University of Tokyo, 1994.
7. P. Deift and X. Zhou, "Perturbation theory for infinite-dimensional integrable systems on the line. A case study," *Acta Math.* **188**, no. 2, 163–262, 2002.
8. P. Deift and X. Zhou, "Long-time asymptotics for solutions of the NLS equation with initial data in a weighted Sobolev space," *Comm. Pure Appl. Math.* **56**, 1029–1077, 2003.
9. M. Dieng and K. D. T.-R. McLaughlin, "Long-time asymptotics for the NLS equation via $\bar{\partial}$ methods," `arXiv:0805.2807`, 2008.
10. A. R. Its, "Asymptotic behavior of the solutions to the nonlinear Schrödinger equation, and isomonodromic deformations of systems of linear differential equations," *Dokl. Akad. Nauk SSSR* **261**, 14–18, 1981. (In Russian.)
11. R. Jenkins, J. Liu, P. Perry, and C. Sulem, "Soliton resolution for the derivative nonlinear Schrödinger equation," *Commun. Math. Phys.*, `doi.org/10.1007/s00220-018-3138-4`, 2018.
12. J. Liu, P. A. Perry, and C. Sulem, "Long-time behavior of solutions to the derivative nonlinear Schrödinger equation for soliton-free initial data," *Ann. Inst. H. Poincaré Anal. Non Linéaire* **35**, no. 1, 217–265, 2018.
13. K. D. T.-R. McLaughlin and P. D. Miller, "The $\bar{\partial}$ steepest descent method and the asymptotic behavior of polynomials orthogonal on the unit circle with fixed and exponentially varying nonanalytic weights," *Intern. Math. Res. Papers* **2006**, Article ID 48673, 1–77, 2006.
14. K. D. T.-R. McLaughlin and P. D. Miller, "The $\bar{\partial}$ steepest descent method for orthogonal polynomials on the real line with varying weights," *Intern. Math. Res. Notices* **2008**, Article ID rnn075, 1–66, 2008.
15. P. D. Miller and Z.-Y. Qin, "Initial-boundary value problems for the defocusing nonlinear Schrödinger equation in the semiclassical limit," *Stud. Appl. Math.* **134**, no. 3, 276–362, 2015.

16. N. I. Muskhelishvili, *Singular Integral Equations, Boundary Problems of Function Theory and Their Application to Mathematical Physics*, Second edition, Dover Publications, New York, 1992.
17. P. A. Perry, "Inverse scattering and global well-posedness in one and two space dimensions," in P. D. Miller, P. Perry, and J.-C. Saut, editors, *Nonlinear Dispersive Partial Differential Equations and Inverse Scattering, Fields Institute Communications*, volume 83, 161–252, Springer, New York, 2019. https://doi.org/10.1007/978-1-4939-9806-7_4.
18. F. W. J. Olver, A. B. Olde Daalhuis, D. W. Lozier, B. I. Schneider, R. F. Boisvert, C. W. Clark, B. R. Miller, and B. V. Saunders, eds., NIST Digital Library of Mathematical Functions, http://dlmf.nist.gov/, Release 1.0.17, 2017.
19. H. Segur and M. J. Ablowitz, "Asymptotic solutions and conservation laws for the nonlinear Schrödinger equation," *J. Math. Phys.* **17**, 710–713 (part I) and 714–716 (part II), 1976.
20. V. E. Zakharov and S. V. Manakov, "Asymptotic behavior of nonlinear wave systems integrated by the inverse method," *Sov. Phys. JETP* **44**, 106–112, 1976.
21. X. Zhou, "The L^2-Sobolev space bijectivity of the scattering and inverse-scattering transforms," *Comm. Pure Appl. Math.* **51**, 697–731, 1989.

Part II
Research Papers

Instability of Solitons in the 2d Cubic Zakharov-Kuznetsov Equation

Luiz Gustavo Farah, Justin Holmer, and Svetlana Roudenko

2010 Mathematics Subject Classification Primary: 35Q53, 37K40, 37K45, 37K05

1 Introduction

In this paper we consider the generalized Zakharov-Kuznetsov equation:

(gZK) $$u_t + \partial_{x_1}\left(\Delta u + u^p\right) = 0, \quad x = (x_1, \ldots, x_N) \in \mathbb{R}^N, \; t \in \mathbb{R}, \tag{1.1}$$

in two dimensions ($N = 2$) and with a specific power of nonlinearity $p = 3$. This equation is the higher-dimensional extension of the well-studied model describing, for example, the weakly nonlinear waves in shallow water, the Korteweg-de Vries (KdV) equation:

L. G. Farah
Department of Mathematics, UFMG, Belo Horizonte, Brazil
e-mail: farah@mat.ufmg.br

J. Holmer
Department of Mathematics, Brown University, Providence, RI, USA
e-mail: holmer@math.brown.edu

S. Roudenko (✉)
Department of Mathematics & Statistics, Florida International University, Miami, FL, USA
e-mail: sroudenko@fiu.edu

(KdV) $\quad u_t + u_{xxx} + (u^p)_x = 0, \quad p = 2, \quad x \in \mathbb{R}, \quad t \in \mathbb{R}.$ (1.2)

When other integer powers $p \neq 2$ are considered, it is referred to as the *generalized KdV* (gKdV) equation, possibly with one exception of $p = 3$, which is also referred to as the *modified* KdV (mKdV) equation. Despite its apparent universality, the gKdV equation is limited as a spatially one-dimensional model. While there are several higher dimensional generalizations of it, in this paper we are interested in the gZK equation (1.1). In the three dimensional setting and quadratic power ($N = 3$ and $p = 2$), Eq. (1.1) was originally derived by Zakharov and Kuznetsov to describe weakly magnetized ion-acoustic waves in a strongly magnetized plasma [33], thus, the name of the equation. In two dimensions, it is also physically relevant; for example, with $p = 2$, it governs the behavior of weakly nonlinear ion-acoustic waves in a plasma comprising cold ions and hot isothermal electrons in the presence of a uniform magnetic field [28, 29]. Melkonian and Maslowe [25] showed that Eq. (1.1) is the amplitude equation for two-dimensional long waves on the free surface of a thin film flowing down a vertical plane with moderate values of the surface fluid tension and large viscosity. Lannes et al. in [19] made the first rigorous derivation of Eq. (1.1) from the Euler-Poisson system with magnetic field in the long wave limit. Yet another derivation was carried by Han-Kwan in [16] from the Vlasov-Poisson system in a combined cold ions and long wave limit.

In this paper we consider the Cauchy problem of the 2d cubic ZK equation (sometimes it is referred as the modified ZK, mZK, or the generalized ZK, gZK) with initial data u_0:

$$\begin{cases} u_t + \partial_{x_1}\left(\Delta_{(x_1,x_2)}u + u^3\right) = 0, & (x_1, x_2) \in \mathbb{R}^2, \, t > 0, \\ u(0, x_1, x_2) = u_0(x_1, x_2) \in H^1(\mathbb{R}^2). \end{cases}$$ (1.3)

During their lifespan, the solutions $u(t, x_1, x_2)$ to (1.3) conserve the mass and energy:

$$M[u(t)] = \int_{\mathbb{R}^2} [u(t, x_1, x_2)]^2 \, dx_1 dx_2 = M[u(0)]$$ (1.4)

and

$$E[u(t)] = \frac{1}{2}\int_{\mathbb{R}^2} |\nabla u(t, x_1, x_2)|^2 \, dx_1 dx_2 - \frac{1}{4}\int_{\mathbb{R}^2} [u(t, x_1, x_2)]^4 \, dx_1 dx_2 = E[u(0)].$$ (1.5)

There is one more conserved quantity of L^1-type, which we omit as it is not needed in this paper. We also mention that unlike the KdV and mKdV, which are completely integrable, the gZK equations do not exhibit complete integrability for any p.

One of the useful symmetries in the evolution equations is *the scaling invariance*, which states that an appropriately rescaled version of the original solution is also a solution of the equation. For Eq. (1.1) the rescaled solution is

$$u_\lambda(t, x_1, x_2) = \lambda^{\frac{2}{p-1}} u(\lambda^3 t, \lambda x_1, \lambda x_2).$$

This scaling makes a specific Sobolev $\dot{H}^s$-norm invariant, i.e.,

$$\|u(0, \cdot, \cdot)\|_{\dot{H}^s(\mathbb{R}^N)} = \lambda^{\frac{2}{p-1} + s - \frac{N}{2}} \|u_0\|_{\dot{H}^s(\mathbb{R}^N)},$$

and the index s gives rise to the critical-type classification of equations. For the gKdV equation (1.2) the critical index is $s = \frac{1}{2} - \frac{2}{p-1}$, and for the two dimensional ZK equation (1.1) the index is $s = 1 - \frac{2}{p-1}$. When $s = 0$ (this corresponds to $p = 3$), Eq. (1.3) is referred to as the L^2-critical equation. The gZK equation has other invariances such as translation and dilation.

The generalized Zakharov-Kuznetsov equation has a family of travelling waves (or solitary waves, which sometimes are referred to as solitons), and observe that they travel only in x_1 direction

$$u(t, x_1, x_2) = Q_c(x_1 - ct, x_2) \tag{1.6}$$

with $Q_c(x_1, x_2) \to 0$ as $|x| \to +\infty$. Here, Q_c is the dilation of the ground state Q:

$$Q_c(\vec{x}) = c^{1/p-1} Q(c^{1/2} \vec{x}), \quad \vec{x} = (x_1, x_2),$$

with Q being a radial positive solution in $H^1(\mathbb{R}^2)$ of the well-known nonlinear elliptic equation $-\Delta Q + Q - Q^p = 0$. Note that $Q \in C^\infty(\mathbb{R}^2)$, $\partial_r Q(r) < 0$ for any $r = |x| > 0$, and for any multi-index α

$$|\partial^\alpha Q(\vec{x})| \leq c(\alpha) e^{-|\vec{x}|} \quad \text{for any} \quad \vec{x} \in \mathbb{R}^2. \tag{1.7}$$

In this work, we are interested in stability properties of traveling waves in the critical gZK equation (1.3), i.e., in the behavior of solutions close to the ground state Q (perhaps, up to translations). We begin with the precise concept of stability and instability used in this paper. For $\alpha > 0$, the neighborhood (or "tube") of radius α around Q (modulo translations) is defined by

$$U_\alpha = \left\{ u \in H^1(\mathbb{R}^2) : \inf_{\vec{y} \in \mathbb{R}^2} \|u(\cdot) - Q(\cdot + \vec{y})\|_{H^1} \leq \alpha \right\}. \tag{1.8}$$

Definition 1.1 (Stability of Q) We say that Q is stable if for all $\alpha > 0$, there exists $\delta > 0$ such that if $u_0 \in U_\delta$, then the corresponding solution $u(t)$ is defined for all $t \geq 0$ and $u(t) \in U_\alpha$ for all $t \geq 0$.

Definition 1.2 (Instability of Q) We say that Q is unstable if Q is not stable, in other words, there exists $\alpha > 0$ such that for all $\delta > 0$ the following holds: if $u_0 \in U_\delta$, then there exists $t_0 = t_0(u_0)$ such that $u(t_0) \notin U_\alpha$.

The main goal of this paper is to show that in the two dimensional case $p = 3$, the traveling waves are *unstable*. This result is also a necessary first step towards understanding whether the generalized ZK equation exhibits any blow-up behavior. We recall that unlike other dispersive models (such as the nonlinear Schrödinger or wave equations), the gKdV equation does not have a suitable virial quantity which would imply existence of blow-up via convexity-type arguments. Hence, proving the existence of blow-up must be done differently; for example, constructing explicit blow-up solutions. In the one-dimensional gKdV case the blow up behavior in the critical case was constructed by Merle [26] and Martel and Merle [24] via first obtaining the instability of solitary waves in [23]. We will address this question in the gZK context in a subsequent paper (see [11]).

In her study of dispersive solitary waves in higher dimensions, de Bouard [6] showed (her result holds in dimensions 2 and 3) that the traveling waves of the form (1.6) are stable for $p < p_c$ and unstable for $p > p_c$, where $p_c = 3$ in 2d. She followed the ideas developed for the gKdV equation by Bona et al. [2] for the instability, and Grillakis et al. [14] for the stability arguments. Here, we prove the instability of the traveling wave solutions of the form (1.6) for the $p = 3$ case in a spirit of Martel and Merle [23], thus, completing the stability picture for the two-dimensional ZK equation. We also note that the more delicate questions about different types of stability have been previously studied; in particular, Côte, Muñoz, Pilod and Simpson in [5] obtained the asymptotic stability of solitary waves in 2d for[1] $2 \leq p < p^* < 3$ by methods of Martel and Merle for the gKdV equation. The upper bound p^* in their restriction of nonlinearity comes from having a certain bilinear form positive-definite, which is needed for the linear Liouville property, see [5, Theorem 1.3], and numerically they show that the proper sign holds only for powers p up to $p^* \approx 2.15$. It would be interesting to investigate if asymptotic stability holds for all $p < 3$.

In this paper we prove the instability of the soliton $u(t, x_1, x_2) = Q(x_1 - t, x_2)$. Our main result reads as follows.

Theorem 1.3 (H^1**-instability of** Q **in the 2d Critical ZK Equation**) *There exists $\alpha_0 > 0$ such that for any $\delta > 0$ there exists $u_0 \in H^1(\mathbb{R}^2)$ satisfying*

$$\|u_0 - Q\|_{H^1} \leq \delta \quad \text{and} \quad u_0 - Q \perp \{Q_{x_1}, Q_{x_2}, \chi_0\}, \tag{1.9}$$

there exists a time $t_0 = t_0(u_0) < \infty$ with $u(t_0) \notin U_{\alpha_0}$, or equivalently,

$$\inf_{\vec{x} \in \mathbb{R}^2} \|u(t_0, \cdot) - Q(\cdot - \vec{x})\|_{H^1} \geq \alpha_0.$$

Here, χ_0 is the eigenfunction corresponding to the (unique) negative eigenvalue of the linearized operator $\mathcal{L}$, for details see Theorem 3.1.

---

[1] The nonlinearity in such gZK equation should be understood as $\partial_{x_1}(|u|^{p-1}u)$.

Remark 1.4 It suffices to prove Theorem 1.3 for the following initial data: define $a = -\frac{\int \chi_0 Q}{\|\chi_0\|_{L^2}^2}$ and let $\delta > 0$. Fix $n_0 = \left(1 + \frac{\|\chi_0\|_{H^1}}{\|\chi_0\|_{L^2}}\right) \|Q\|_{H^1} \cdot \delta^{-1}$. For any $n \geq n_0$ define

$$\varepsilon_0 = \frac{1}{n}(Q + a\chi_0).$$

Then $u_0 = Q + \varepsilon_0$ satisfies the hypotheses of the theorem, i.e., the conditions (1.9).

The main strategy follows the approach introduced by Martel and Merle [23] in their study of the same question in the critical gKdV model. For that they worked out pointwise decay estimates on the shifted linear equation and applied them to the nonlinear equation (bootstraping twice in time). In [9] we revisited that proof and showed that instead of pointwise decay estimates, it is possible to consider monotonicity properties of the solution, then apply them to the decomposition around the soliton and conclude the instability in the critical gKdV equation. In our two dimensional case of the critical ZK equation, we cannot obtain the instability result just relying on monotonicity (and truncation when needed), because there is a new term appearing in the virial-type quantity which is truly two-dimensional, see Lemma 6.1 and the last term in (6.51). We are forced to consider something else besides the monotonicity (since it would only give the boundedness of the new term, not the smallness), and thus, we develop new 2d pointwise decay estimates, see Sects. 8–12. These estimates by themselves are important results for the two-dimensional Airy-type kernel with the applications to the shifted linear equation as well as to the nonlinear equation. This part is a completely new development in the higher dimensional setting, and we believe that it will be useful in other contexts as well.

We note that we could prove the conditional instability with α being a multiple of δ in Definition 1.2, i.e., there exists a universal constant $c > 0$ such that for any $\delta > 0$ if $u_0 \in U_\delta$, then there exists $t_0 = t_0(u_0)$ such that $u(t_0) \notin U_{c\delta}$, using only monotonicity. However, we emphasize that in order to show the instability with α independent of δ, we need to use the pointwise decay estimates.

The paper is organized as follows. In Sect. 2 we provide the background information on the well-posedness of the generalized ZK equation in two dimensions. In Sect. 3 we discuss the properties of the linearized operator L around the ground state Q and exhibit the three sets of orthogonality conditions which make it positive-definite; the last one, which involves χ_0, is the one we use in the sequel to control various parameters in modulation theory and smallness of ε. Section 4 contains the canonical decomposition of a solution u around Q (thus, introducing ε and the equation for it), then the modulation theory and control of parameters coming from such a decomposition are described in Sect. 5. In Sect. 6 we introduce the key player, the virial-type functional, and make the first attempt to estimate it. Truncation helps us to obtain an upper bound, however, to proceed with the time derivative estimates, we need to develop more machinery, which we do in subsequent sections. In Sect. 7 we discuss the concept of monotonicity, which allows us to control several terms in the virial functional, but as mentioned above, not all terms. The next few sections, starting from Sect. 8 contain the new pointwise decay estimates. We state the main result in Sect. 8 and also re-examine H^1 well-posedness in the same section, then we develop the pointwise decay estimates on the 2d Airy-type kernel and its derivative in Sect. 9, after that we proceed with the application of them to the linear equation in Sect. 10, then for the Duhamel term in Sect. 11, and finally, for the nonlinear equation in Sect. 12. After all the tools are developed, we return to the virial-type functional estimates and obtain the lower bound on its time derivative, which allows us to conclude the instability result in Sect. 13.

2 Background on the Generalized ZK Equation

In this section we review the known results on the local and global well-posedness of the generalized ZK equation. To follow the notation in the literature, in this section we denote the power of nonlinearity as u^{k+1} (instead of u^p) and consider the Cauchy problem for the generalized ZK equation as follows:

$$\begin{cases} u_t + \partial_{x_1} \Delta u + \partial_{x_1} (u^{k+1}) = 0, & (x_1, x_2) \in \mathbb{R}^2, \ t > 0, \\ u(0, x_1, x_2) = u_0(x_1, x_2) \in H^s(\mathbb{R}^2). \end{cases} \quad (2.10)$$

Faminskii [8] showed the local well-posedness of the Cauchy problem (2.10) for the $k = 1$ case considering H^1 data (to be precise, he obtained the local well-posedness in H^m, for any integer $m \geq 1$). The current results on the local well-posedness are gathered in the following statement.

Theorem 2.1 *The local well-posedness in* (2.10) *holds in the following cases:*

- $k = 1$: *for* $s > \frac{1}{2}$, *see Grünrock and Herr* [15] *and Molinet and Pilod* [27],
- $k = 2$: *for* $s > \frac{1}{4}$, *see Ribaud and Vento* [30],
- $k = 3$: *for* $s > \frac{5}{12}$, *see Ribaud and Vento* [30],
- $k = 4, 5, 6, 7$: *for* $s > 1 - \frac{2}{k}$, *see Ribaud and Vento* [30],
- $k = 8$, $s > \frac{3}{4}$, *see Linares and Pastor* [21]
- $k > 8$, $s > s_k = 1 - 2/k$, *see Farah et al.* [12].

Note that in the last three cases (i.e., for $k \geq 4$), the bound on $s > s_k$ is optimal from the scaling conjecture. For previous results on the local well-posedness for $2 \leq k \leq 8$ for $s > 3/4$ see [20] and [21].

Following the approach of Holmer and Roudenko for the L^2-supercritical nonlinear Schrödinger (NLS) equation, see [17] and [7], the first author together with F. Linares and A. Pastor obtained the global well-posedness result for the nonlinearities $k \geq 3$ and under a certain mass-energy threshold, see [12].

Theorem 2.2 ([12]) *Let* $k \geq 3$ *and* $s_k = 1 - 2/k$. *Assume* $u_0 \in H^1(\mathbb{R}^2)$ *and suppose that*

$$E(u_0)^{s_k} M(u_0)^{1-s_k} < E(Q)^{s_k} M(Q)^{1-s_k}, \quad E(u_0) \geq 0. \tag{2.11}$$

If

$$\|\nabla u_0\|_{L^2}^{s_k} \|u_0\|_{L^2}^{1-s_k} < \|\nabla Q\|_{L^2}^{s_k} \|Q\|_{L^2}^{1-s_k}, \tag{2.12}$$

then for any t from the maximal interval of existence

$$\|\nabla u(t)\|_{L^2}^{s_k} \|u_0\|_{L^2}^{1-s_k} = \|\nabla u(t)\|_{L^2}^{s_k} \|u(t)\|_{L^2}^{1-s_k} < \|\nabla Q\|_{L^2}^{s_k} \|Q\|_{L^2}^{1-s_k},$$

where Q is the unique positive radial solution of

$$\Delta Q - Q + Q^{k+1} = 0.$$

In particular, this implies that H^1 solutions, satisfying (2.11) and (2.12) exist globally in time.

Remark 2.3 In the limit case $k = 2$ (or $p = 3$, the modified ZK equation), conditions (2.11) and (2.12) reduce to one condition, which is

$$\|u_0\|_{L^2} < \|Q\|_{L^2}.$$

Such a condition was already used in [20] and [21] to show the existence of global solutions, respectively, in $H^1(\mathbb{R}^2)$ and $H^s(\mathbb{R}^2)$, $s > 53/63$, see also [13] for another approach. Recently in [1] the global well-posedness was extended to $s > 3/4$.

We conclude this section with a note that while it would be important to obtain the local well-posedness down to the scaling index in the gZK equation for $1 \leq k \leq 3$, and the global well-posedness in the subcritical cases for $s < 3/4$ (ideally, all the way down to the L^2 level), for the purpose of this paper, it is sufficient to have the well-posedness theory in $H^1(\mathbb{R}^2)$.

3 The Linearized Operator L

The operator L, which is obtained by linearizing around the ground state Q, is defined by

$$L := -\Delta + 1 - p\, Q^{p-1}. \tag{3.13}$$

We first state the properties of this operator L (see Kwong [18] for all dimensions, Weinstein [32] for dimension 1 and 3, also Maris [22] and Chang et al. [3]).

Theorem 3.1 (Properties of L) *The following holds for an operator L defined in (3.13)*

- L is a self-adjoint operator and $\sigma_{ess}(L) = [\lambda_{ess}, +\infty)$ for some $\lambda_{ess} > 0$
- $\ker L = span\{Q_{x_1}, Q_{x_2}\}$
- L has a unique single negative eigenvalue $-\lambda_0$ (with $\lambda_0 > 0$) associated to a positive radially symmetric eigenfunction χ_0. Moreover, there exists $\delta > 0$ such that

$$|\chi_0(x)| \lesssim e^{-\delta |x|} \quad \text{for all } x \in \mathbb{R}^2. \tag{3.14}$$

We also define the generator Λ of the scaling symmetry as

$$\Lambda f = \frac{2}{p-1} f + \vec{x} \cdot \nabla f, \quad (x_1, x_2) \in \mathbb{R}^2. \tag{3.15}$$

The following identities are useful to have

Lemma 3.2 *The following identities hold*

(1) $L(\Lambda Q) = -2Q$
(2) $\int Q \Lambda Q = 0$ if $p = 3$ and $\int Q \Lambda Q = \frac{3-p}{p-1} \int Q^2$ for $p \neq 3$.

The proof is a direct simple computation and can be found in [10].

In general, the operator L is not positive-definite, however, on certain subspaces one can expect some positivity properties. We now consider only the L^2-critical case and power $p = 3$. First, we summarize known positivity estimates for the operator L (see Chang et al. [3] and Weinstein [32]):

Lemma 3.3 *The following conditions hold for L:*

(i) $(LQ, Q) = -2 \int Q^4 < 0$,
(ii) $L|_{\{Q^3\}^\perp} \geq 0$,
(iii) $L|_{\{Q\}^\perp} \geq 0$,
(iv) $L|_{\{Q, xQ, |x|^2 Q\}^\perp} > 0$.

The last property provides us with the orthogonality conditions that keep the quadratic form, generated by L, positive-definite (see Weinstein [32, Prop. 2.9]):

Lemma 3.4 *For any $f \in H^1(\mathbb{R}^2)$ such that*

$$(f, Q) = (f, x_j Q) = (f, |x|^2 Q) = 0, \quad j = 1, 2, \tag{3.16}$$

there exists a positive constant $C > 0$ such that

$$(Lf, f) \geq C (f, f).$$

While it shows that eliminating directions from (3.16) would make the bilinear form (Lf, f) positive, these directions are not quite suitable for our case. An alternative for the orthogonality conditions (3.16) would be to consider the kernel of L from Theorem 3.1 and Lemma 3.3(ii), which we do next.

Lemma 3.5 *For any $f \in H^1(\mathbb{R}^2)$ such that*

$$(f, Q^3) = (f, Q_{x_j}) = 0, \quad j = 1, 2, \tag{3.17}$$

there exists a positive constant $C > 0$ such that

$$(Lf, f) \geq C(f, f).$$

Proof From Chang et al. [3] (Lemma 2.2 (2.7)) we have

$$\inf_{(f, Q^3) = 0} (Lf, f) \geq 0. \tag{3.18}$$

Let $C_1 = \{(L\varepsilon, \varepsilon) : \|\varepsilon\|_{L^2} = 1, \ (f, Q^3) = (f, Q_{x_j}) = 0, \ j = 1, 2.\}$, then $C_1 \geq 0$ by (3.18). Assume, by contradiction, that $C_1 = 0$. In this case, as in [32, Proposition 2.9], we can find a function $\varepsilon^* \in H^1$ satisfying

(i) $(L\varepsilon^*, \varepsilon^*) = 0$
(ii) $(L - \alpha)\varepsilon^* = \beta Q^3 + \gamma Q_{x_1} + \delta Q_{x_2}$
(iii) $\|\varepsilon^*\|_{L^2} = 1$ and $(\varepsilon^*, Q^3) = (\varepsilon^*, Q_{x_j}) = 0, \quad j = 1, 2$.

Taking the scalar product of (ii) with ε^*, we deduce from (iii) that $(L\varepsilon^*, \varepsilon^*) = \alpha$, and thus, $\alpha = 0$ by (i). Now, taking the scalar product with Q_{x_1}, integrating by parts and recalling Theorem 3.1, we have

$$0 = (\varepsilon^*, LQ_{x_1}) = (L\varepsilon^*, Q_{x_1}) = \gamma \int Q_{x_1}^2 + \delta \int Q_{x_1} Q_{x_2}.$$

Since $Q_{x_1} \perp Q_{x_2}$, we deduce $\gamma = 0$. In a similar way (taking the scalar product with Q_{x_2}), we also have $\delta = 0$. Therefore, $L\varepsilon^* = \beta Q^3$, which implies

$$\varepsilon^* = -\frac{\beta}{2} Q + \theta_1 Q_{x_1} + \theta_2 Q_{x_2}, \tag{3.19}$$

where we have used Theorem 3.1 and Lemma 3.2. Taking the scalar product of (3.19) with Q^3, from (iii) and integration by parts, we get

$$0 = (\varepsilon^*, Q^3) = -\frac{\beta}{2} \int Q^4,$$

which implies $\beta = 0$.

Finally, using $Q_{x_j}, j = 1, 2$, we obtain $\theta_1 = \theta_2 = 0$. Thus, $\varepsilon^* = 0$, which is a contradiction with (iii). □

We deduce yet another set of orthogonality conditions, see (3.20), to keep the quadratic form, generated by L, positive-definite. This is the set, which we will use in this paper.

Lemma 3.6 *Let χ_0 be the positive radially symmetric eigenfunction associated to the unique single negative eigenvalue $-\lambda_0$ (with $\lambda_0 > 0$). Then, there exists $\sigma_0 > 0$ such that for any $f \in H^1(\mathbb{R}^2)$ satisfying*

$$(f, \chi_0) = (f, Q_{x_j}) = 0, \quad j = 1, 2, \tag{3.20}$$

one has

$$(Lf, f) \geq \sigma_0 (f, f).$$

Proof The result follows directly from Schechter [31, Chapter 8, Lemma 7.10] (see also [31, Chapter 1, Lemma 7.17]) □

In a sense, the last lemma shows that if we exclude the zero eigenvalue and negative eigenvalue directions, then only the "positive" directions are left, and thus, the positivity property of L must hold.

4 The Linearized Equation Around Q

In this section we decompose our solutions $u(t, \vec{x})$ around the soliton Q. Since we consider the L^2-critical problem, we must also incorporate the scaling parameter (besides the translation as we did in the supercritical case in [10]). We use the following canonical decomposition of u around Q:

$$v(t, y_1, y_2) = \lambda(t) u(t, \lambda(t) y_1 + x_1(t), \lambda(t) y_2 + x_2(t)). \tag{4.21}$$

Our next task is to examine the difference $\varepsilon = v - Q$, more precisely,

$$\varepsilon(t, \vec{y}) = v(t, \vec{y}) - Q(\vec{y}), \quad \vec{y} = (y_1, y_2). \tag{4.22}$$

4.1 Equation for ε

After we rescale time $t \mapsto s$ by $\frac{ds}{dt} = \frac{1}{\lambda^3}$, we obtain the equation for ε.

Lemma 4.1 *For all $s \geq 0$, we have*

$$\varepsilon_s = (L\varepsilon)_{y_1} + \frac{\lambda_s}{\lambda}\Lambda Q + \left(\frac{(x_1)_s}{\lambda} - 1\right) Q_{y_1} + \frac{(x_2)_s}{\lambda} Q_{y_2}$$

$$+ \frac{\lambda_s}{\lambda}\Lambda \varepsilon + \left(\frac{(x_1)_s}{\lambda} - 1\right)\varepsilon_{y_1} + \frac{(x_2)_s}{\lambda}\varepsilon_{y_2}$$

$$- 3(Q\varepsilon^2)_{y_1} - (\varepsilon^3)_{y_1}, \tag{4.23}$$

where $\Lambda f = f + \vec{y} \cdot \nabla f$ and L is the linearized operator around Q:

$$L\varepsilon = -\Delta\varepsilon + \varepsilon - 3Q^2\varepsilon.$$

Proof Using (4.21), we obtain

$$v_t = \lambda_t u + \lambda u_t + \lambda u_{x_1}(\lambda_t y_1 + (x_1)_t) + \lambda u_{x_2}(\lambda_t y_2 + (x_2)_t),$$

and for $i = 1, 2$

$$v_{y_i} = \lambda^2 u_{x_i}, \quad v_{y_i y_i} = \lambda^3 u_{x_i x_i}.$$

Substituting the above into $u_t + \partial_{x_1}(\Delta u + u^3) = 0$, we obtain

$$v_t = \lambda^{-1}\lambda_t v + \lambda^{-1}\lambda_t (\vec{y} \cdot \nabla v) + \lambda^{-1}\left(v_{y_1}(x_1)_t + v_{y_2}(x_2)_t\right) - \lambda^{-3}\partial_{y_1}\left(\Delta v + v^3\right).$$

Recalling that $\frac{ds}{dt} = \frac{1}{\lambda^3}$, we change the time variable $t \mapsto s$

$$\lambda^{-3}v_s = \lambda^{-4}\lambda_s v + \lambda^{-4}\lambda_s(\vec{y}\cdot\nabla v) + \lambda^{-4}\left(v_{y_1}(x_1)_s + v_{y_2}(x_2)_s\right) - \lambda^{-3}\partial_{y_1}\left(\Delta v + v^3\right),$$

Simplifying, we get

$$v_s = \frac{\lambda_s}{\lambda}(v + \vec{y}\cdot\nabla v) + \frac{((x_1)_s, (x_2)_s)}{\lambda}\cdot\nabla v - \partial_{y_1}\left(\Delta v + v^3\right).$$

Next, we use (4.22) and the fact that $\Delta Q = Q - Q^3$ to obtain the equation for ε:

$$\varepsilon_s = \frac{\lambda_s}{\lambda}(\Lambda Q + \Lambda\varepsilon) + \frac{((x_1)_s, (x_2)_s)}{\lambda}\cdot(\nabla Q + \nabla\varepsilon) - \partial_{y_1}\left(Q + \Delta\varepsilon + 3Q^2\varepsilon + 3Q\varepsilon^2 + \varepsilon^3\right).$$

Simplifying, we get Eq. (4.23). $\square$

4.2 Mass and Energy Relations

Our next task is to derive the basic mass and energy conservations for ε. First, denote

$$M_0 = 2\int_{\mathbb{R}^2} Q(\vec{y})\varepsilon(0, \vec{y})\,d\vec{y} + \int_{\mathbb{R}^2}\varepsilon^2(0, \vec{y})\,d\vec{y}. \tag{4.24}$$

For any $s \geq 0$ by the L^2 scaling invariance and mass conservation, we have

$$\int_{\mathbb{R}^2} v^2(s, \vec{y}) \, d\vec{y} = \int_{\mathbb{R}^2} \lambda^2(t) \, u^2(t, \lambda \vec{y} + \vec{x}(t)) \, d\vec{y} = \int_{\mathbb{R}^2} u^2(t) \, d\vec{x} = M[u(t)] \equiv M[u(0)].$$

On the other hand,

$$\begin{aligned}
\int_{\mathbb{R}^2} v^2(s, \vec{y}) \, d\vec{y} &= \int_{\mathbb{R}^2} (Q(\vec{y}) + \varepsilon(s, \vec{y}))^2 \, d\vec{y} \\
&= \int_{\mathbb{R}^2} Q^2(\vec{y}) \, d\vec{y} + 2 \int_{\mathbb{R}^2} Q(\vec{y}) \, \varepsilon(s, \vec{y}) \, d\vec{y} + \int_{\mathbb{R}^2} \varepsilon^2(s, \vec{y}) \, d\vec{y} \\
&= \int_{\mathbb{R}^2} u_0^2(\vec{x}) \, d\vec{x} = \int_{\mathbb{R}^2} Q^2(\vec{y}) \, d\vec{y} + 2 \int_{\mathbb{R}^2} Q(\vec{y}) \, \varepsilon(0, \vec{y}) \, d\vec{y} \\
&\quad + \int_{\mathbb{R}^2} \varepsilon^2(0, \vec{y}) \, d\vec{y},
\end{aligned}$$

and thus,

$$M[\varepsilon(s)] := 2 \int_{\mathbb{R}^2} Q(\vec{y}) \, \varepsilon(s, \vec{y}) \, d\vec{y} + \int_{\mathbb{R}^2} \varepsilon^2(s, \vec{y}) \, d\vec{y} = M_0. \tag{4.25}$$

Next, we examine the energy conservation for v, where a straightforward calculation gives

$$E[v(s)] = \lambda^2(s) \, E[u(t)] = \lambda^2(s) \, E[u_0]. \tag{4.26}$$

Since $v = Q + \varepsilon$, we also obtain

$$\begin{aligned}
E[Q + \varepsilon] &= \frac{1}{2} \int |\nabla(Q + \varepsilon)|^2 - \frac{1}{4} \int (Q + \varepsilon)^4 \\
&= \frac{1}{2} \left(\int |\nabla \varepsilon|^2 + \varepsilon^2 - 3 Q^2 \varepsilon^2 \right) + \int \left(\nabla Q \nabla \varepsilon - Q^3 \varepsilon \right) \\
&\quad - \frac{1}{2} \int \varepsilon^2 - \int Q \varepsilon^3 - \frac{1}{4} \int \varepsilon^4 \\
&= \frac{1}{2} (L\varepsilon, \varepsilon) - \left(\int Q\varepsilon + \frac{1}{2} \int \varepsilon^2 \right) - \frac{1}{4} \left[4 \int Q\varepsilon^3 + \int \varepsilon^4 \right],
\end{aligned} \tag{4.27}$$

where in the second line the integration by parts is used as well as $2\|\nabla Q\|_{L^2}^2 = \|Q\|_{L^4}^4$ (since Q is a solution of $\Delta Q + Q^3 = Q$). By Gagliardo-Nirenberg inequality, we can bound the last term as

$$4 \int Q\varepsilon^3 + \int \varepsilon^4 \leq c_1 \|\nabla \varepsilon\|_{L^2} \|\varepsilon\|_{L^2}^2 + c_2 \|\nabla \varepsilon\|_{L^2}^2 \|\varepsilon\|_{L^2}^2, \tag{4.28}$$

and if $\|\varepsilon\|_{H^1} \leq 1$, we get

$$\left| E[Q+\varepsilon] + \left(\int Q\varepsilon + \frac{1}{2}\int \varepsilon^2\right) - \frac{1}{2}(L\varepsilon, \varepsilon) \right| \leq c_0 \|\nabla\varepsilon\|_{L^2} \|\varepsilon\|_{L^2}^2.$$

Putting together (4.25), (4.26) and (4.28), we have the following

Lemma 4.2 *For any $s \geq 0$ we have mass and energy conservations for ε*

$$M[\varepsilon(s)] = M_0, \quad \text{and} \quad E[Q+\varepsilon(s)] = \lambda^2(s)\, E[u_0]. \tag{4.29}$$

Moreover, the energy linearization is

$$E[Q+\varepsilon] + \left(\int Q\varepsilon + \frac{1}{2}\int \varepsilon^2\right) = \frac{1}{2}(L\varepsilon, \varepsilon) - \frac{1}{4}\left(4\int Q\varepsilon^3 + \int \varepsilon^4\right), \tag{4.30}$$

and if $\|\varepsilon\|_{H^1} \leq 1$, then there exists a $c_0 > 0$ such that

$$\left| E[Q+\varepsilon] + \left(\int Q\varepsilon + \frac{1}{2}\int \varepsilon^2\right) - \frac{1}{2}(L\varepsilon, \varepsilon) \right| \leq c_0 \|\nabla\varepsilon\|_{L^2} \|\varepsilon\|_{L^2}^2. \tag{4.31}$$

5 Modulation Theory and Parameter Estimates

In this section, recalling the definition (1.8), we show that it is possible to choose parameters $\lambda(s) \in \mathbb{R}$ and $\vec{x}(s) = (x_1(s), x_2(s)) \in \mathbb{R}^2$ such that $\varepsilon(s) \perp \chi_0$ and $\varepsilon(s) \perp Q_{x_j}$, $j = 1, 2$. Moreover, assuming an additional symmetry, we can assume $x_2(s) = 0$.

Proposition 5.1 (Modulation Theory I) *There exists $\overline{\alpha}, \overline{\lambda} > 0$ and a unique C^1 map*

$$(\lambda_1, \vec{x}_1) : U_{\overline{\alpha}} \to (1 - \overline{\lambda}, 1 + \overline{\lambda}) \times \mathbb{R}^2$$

such that if $u \in U_{\overline{\alpha}}$ and $\varepsilon_{\lambda_1, \vec{x}_1}$ is given by

$$\varepsilon_{\lambda_1, \vec{x}_1}(y_1, y_2) = \lambda_1 u(\lambda_1 y_1 + (x_1)_1, \lambda_1 y_2 + (x_1)_2) - Q(y_1, y_2), \tag{5.32}$$

then

$$\varepsilon_{\lambda_1, \vec{x}_1} \perp \chi_0 \quad \text{and} \quad \varepsilon_{\lambda_1, \vec{x}_1} \perp Q_{y_j}, \quad j = 1, 2. \tag{5.33}$$

Moreover, there exists a constant $C_1 > 0$, such that if $u \in U_\alpha$ with $0 < \alpha < \overline{\alpha}$, then

$$\|\varepsilon_{\lambda_1, \vec{x}_1}\|_{H^1} \leq C_1 \alpha \quad \text{and} \quad |\lambda_1 - 1| \leq C_1 \alpha. \tag{5.34}$$

Proof Let $\varepsilon_{\lambda_1,\vec{x}_1}$ be defined as in (5.32). Differentiating and recalling the definition (3.15), we have

$$\frac{\partial \varepsilon_{\lambda_1,\vec{x}_1}}{\partial (x_1)_j}\bigg|_{\lambda_1=1,\vec{x}_1=0} = u_{y_j}, \quad j=1,2 \qquad (5.35)$$

and

$$\frac{\partial \varepsilon_{\lambda_1,\vec{x}_1}}{\partial \lambda_1}\bigg|_{\lambda_1=1,\vec{x}_1=0} = \Lambda u. \qquad (5.36)$$

Next, consider the following functionals

$$\rho^j_{\lambda_1,\vec{x}_1}(u) = \int \varepsilon_{\lambda_1,\vec{x}_1} Q_{y_j}, \quad j=1,2, \quad \text{and} \quad \rho^3_{\lambda_1,\vec{x}_1}(u) = \int \varepsilon_{\lambda_1,\vec{x}_1} \chi_0,$$

and define the function $S: \mathbb{R}^3 \times H^1 \to \mathbb{R}^3$ such that

$$S(\lambda_1, \vec{x}_1, u) = (\rho^1_{\lambda_1,\vec{x}_1}(u), \rho^2_{\lambda_1,\vec{x}_1}(u), \rho^3_{\lambda_1,\vec{x}_1}(u)).$$

From (5.35) and (5.36), we deduce

$$\frac{\partial \rho^j_{\lambda_1,\vec{x}_1}(u)}{\partial \lambda_1}\bigg|_{\lambda_1=1,\vec{x}_1=0,u=Q} = \int \Lambda Q Q_{y_j};$$

$$\frac{\partial \rho^1_{\lambda_1,\vec{x}_1}(u)}{\partial (x_1)_1}\bigg|_{\lambda_1=1,\vec{x}_1=0,u=Q} = \int Q_{y_1} Q_{y_1} = \int Q^2_{y_1} > 0;$$

$$\frac{\partial \rho^2_{\lambda_1,\vec{x}_1}(u)}{\partial (x_1)_1}\bigg|_{\lambda_1=1,\vec{x}_1=0,u=Q} = \int Q_{y_1} Q_{y_2} = 0;$$

$$\frac{\partial \rho^1_{\lambda_1,\vec{x}_1}(u)}{\partial (x_1)_2}\bigg|_{\lambda_1=1,\vec{x}_1=0,u=Q} = \int Q_{y_2} Q_{y_1} = 0;$$

$$\frac{\partial \rho^2_{\lambda_1,\vec{x}_1}(u)}{\partial (x_1)_2}\bigg|_{\lambda_1=1,\vec{x}_1=0,u=Q} = \int Q_{y_2} Q_{y_2} = \int Q^2_{y_2} > 0.$$

Moreover, since $L(\chi_0) = -\lambda_0 \chi_0$ (with $\lambda_0 > 0$), $L(\Lambda Q) = -2Q$, χ_0 and Q are positive functions and $\chi_0 \perp \text{span}\{Q_{y_1}, Q_{y_2}\}$, we also have

$$\frac{\partial \rho^3_{\lambda_1,\vec{x}_1}(u)}{\partial \lambda_1}\bigg|_{\lambda_1=1,\vec{x}_1=0,u=Q} = \int \Lambda Q \chi_0 = -\frac{1}{\lambda_0} \int \Lambda Q L(\chi_0) = \frac{2}{\lambda_0} \int Q \chi_0 > 0;$$

$$\left.\frac{\partial \rho^3_{\lambda_1,\vec{x}_1}(u)}{\partial (x_1)_1}\right|_{\lambda_1=1,\vec{x}_1=0,u=Q} = \int Q_{y_1}\chi_0 = 0;$$

$$\left.\frac{\partial \rho^3_{\lambda_1,\vec{x}_1}(u)}{\partial (x_1)_2}\right|_{\lambda_1=1,\vec{x}_1=0,u=Q} = \int Q_{y_2}\chi_0 = 0.$$

Noting that $S(1,0,0,Q) = (0,0,0)$, we can apply the Implicit Function Theorem to obtain the existence of $\bar{\beta} > 0$, a neighborhood V of $(1,0,0)$ in $\mathbb{R}^3$ and a unique C^1 map

$$(\lambda_1, \vec{x}_1) : \left\{u \in H^1(\mathbb{R}^2) : \|u - Q\|_{H^1} < \bar{\beta}\right\} \to V$$

such that $S((\lambda_1, \vec{x}_1)(u), u) = 0$, in other words, the orthogonality conditions (5.33) are satisfied.

Also note that there exists $C > 0$ such that if $\|u - Q\|_{H^1} < \alpha \leq \bar{\beta}$ then $|\lambda_1 - 1| + |\vec{x}_1| \leq C\alpha$. Moreover, by (4.22) we also have $\|\varepsilon_{\lambda_1,\vec{x}_1}\|_{H^1} \leq C\alpha$, for some $C > 0$.

It is straightforward to extend the map $(\lambda_1, \vec{x}_1)$ to the region U_α. Indeed, applying again the Implicit Function Theorem, there exists $\bar{\alpha} < \bar{\beta}$ and a unique C^1 map $r : U_{\bar{\alpha}} \to \mathbb{R}^2$, such that

$$\|u(\cdot) - Q(\cdot - r)\|_{H^1} = \inf_{r \in \mathbb{R}^2} \|u(\cdot) - Q(\cdot - r)\|_{H^1} < \bar{\alpha} < \bar{\beta},$$

for all $u \in U_{\bar{\alpha}}$.

Finally, defining $\lambda_1 = \lambda_1(u(\cdot + r(u)))$ and $\vec{x}_1 = \vec{x}_1(u(\cdot + r(u))) + r(u)$, we have that (5.33) and (5.34) are satisfied. □

Note that solitary waves (1.6) are traveling only in the x_1-direction, so it should be reasonable to consider a path $\vec{x}(t) = (x_1(t), x_2(t))$ so that $x_1(t) \approx ct$ and $x_2(t) \approx 0$. Inspired by the work of de Bouard [6], if we assume an additional symmetry, we can consider exactly that, and thus simplify the choice of parameters:

Proposition 5.2 (Modulation Theory II) *If we assume that u from Proposition 5.1 is cylindrically symmetric (i.e., $u(x_1, x_2) = u(x_1, |x_2|)$), then, reducing $\bar{\alpha} > 0$ if necessary, we have $(x_1)_2 \equiv 0$.*

Proof We first define $\varepsilon_{\lambda_1, x_1}$ by

$$\varepsilon_{\lambda_1, x_1}(y_1, y_2) = \lambda_1 u(\lambda_1 y_1 + x_1, \lambda_1 y_2) - Q(y_1, y_2), \quad (5.37)$$

and then the functionals

$$\rho^1_{\lambda_1, x_1}(u) = \int \varepsilon_{\lambda_1, x_1} Q_{y_j}, \quad j = 1, 2, \quad \text{and} \quad \rho^3_{\lambda_1, x_1}(u) = \int \varepsilon_{\lambda_1, x_1} \chi_0,$$

and the function $S: \mathbb{R}^2 \times H^1 \to \mathbb{R}^2$ such that

$$S(\lambda_1, x_1, u) = (\rho^1_{\lambda_1, \vec{x}_1}(u), \rho^3_{\lambda_1, \vec{x}_1}(u)).$$

Arguing as in the proof of Proposition 5.1, we have

$$\frac{\partial \rho^1_{\lambda_1, x_1}(u)}{\partial \lambda_1}\bigg|_{\lambda_1=1, x_1=0, u=Q} = \int \Lambda Q Q_{y_j};$$

$$\frac{\partial \rho^3_{\lambda_1, x_1}(u)}{\partial \lambda_1}\bigg|_{\lambda_1=1, x_1=0, u=Q} = \int \Lambda Q \chi_0 = -\frac{1}{\lambda_0} \int \Lambda Q L(\chi_0) = \frac{2}{\lambda_0} \int Q \chi_0 > 0;$$

$$\frac{\partial \rho^1_{\lambda_1, x_1}(u)}{\partial x_1}\bigg|_{\lambda_1=1, x_1=0, u=Q} = \int Q_{y_1} Q_{y_1} = \int Q^2_{y_1} > 0;$$

$$\frac{\partial \rho^3_{\lambda_1, x_1}(u)}{\partial x_1}\bigg|_{\lambda_1=1, \vec{x}_1=0, u=Q} = \int Q_{y_1} \chi_0 = 0.$$

Since $S(1, 0, Q) = (0, 0)$, we again apply the Implicit Function Theorem to obtain the existence of $\alpha_1 > 0$, a neighborhood V of $(1, 0)$ in $\mathbb{R}^2$ and a unique C^1 map

$$(\lambda_1, x_1) : \left\{ u \in H^1(\mathbb{R}^2) : \|u - Q\|_{H^1} < \alpha_1 \right\} \to V$$

such that $\varepsilon_{\lambda_1, x_1} \perp \chi_0$ and $\varepsilon_{\lambda_1, \vec{x}_1} \perp Q_{y_1}$.

Now, using the expression for $\varepsilon_{\lambda_1, x_1}$ in (5.37), we also deduce

$$\int \varepsilon_{\lambda_1, x_1} Q_{y_2} = \int \lambda_1 u(\lambda_1 y_1 + x_1, \lambda_1 y_2) Q_{y_2}(y_1, y_2) \, dy_1 dy_2 = 0,$$

if $u(x_1, x_2) = u(x_1, |x_2|)$, since $Q_{y_2} = \partial_r Q \cdot \frac{y_2}{r}$.

Finally, the uniqueness, which follows from the Implicit Function Theorem, yields (taking a smaller α_1 if necessary) $\vec{x}_1 = (x_1, 0)$ in Proposition 5.1, hence, completing the proof. □

Now, assume that $u(t) \in U_{\overline{\alpha}}$ for all $t \geq 0$. We define the functions $\lambda(t)$ and $x(t)$ as follows.

Definition 5.3 For all $t \geq 0$, let $\lambda(t)$ and $x(t)$ be such that $\varepsilon_{\lambda(t), x(t)}$, defined according to Eq. (5.37), satisfy

$$\varepsilon_{\lambda(t), x(t)} \perp \chi_0 \quad \text{and} \quad \varepsilon_{\lambda(t), x(t)} \perp Q_{y_j}, \quad j = 1, 2. \tag{5.38}$$

In this case we also define

$$\varepsilon(t) = \varepsilon_{\lambda(t), x(t)} = \lambda(t) u(t, \lambda(t) y_1 + x(t), \lambda(t) y_2) - Q(y_1, y_2). \quad (5.39)$$

We rescale time $t \mapsto s$ by $\frac{ds}{dt} = \frac{1}{\lambda^3}$ to better understand these parameters, which are now $\lambda(s)$ and $x(s)$. Indeed, the next proposition provides us with the equations and estimates for $\frac{\lambda_s}{\lambda}$ and $\left(\frac{x_s}{\lambda} - 1\right)$.

Lemma 5.4 (Modulation Parameters) *There exists $0 < \alpha_1 < \bar{\alpha}$ such that if for all $t \geq 0$, $u(t) \in U_{\alpha_1}$, then λ and x are C^1 functions of s and they satisfy the following equations:*

$$-\frac{\lambda_s}{\lambda} \int (\vec{y} \cdot \nabla Q_{y_1}) \varepsilon + \left(\frac{x_s}{\lambda} - 1\right) \left(\int |Q_{y_1}|^2 - \int Q_{y_1 y_1} \varepsilon \right)$$
$$= 6 \int Q Q_{y_1}^2 \varepsilon - 3 \int Q_{y_1 y_1} \varepsilon^2 Q - \int Q_{y_1 y_1} \varepsilon^3, \quad (5.40)$$

and

$$\frac{\lambda_s}{\lambda} \left(\frac{2}{\lambda_0} \int \chi_0 Q - \int (\vec{y} \cdot \nabla \chi_0) \varepsilon \right) - \left(\frac{x_s}{\lambda} - 1\right) \int (\chi_0)_{y_1} \varepsilon$$
$$= \int L((\chi_0)_{y_1}) \varepsilon - 3 \int (\chi_0)_{y_1} Q \varepsilon^2 - \int (\chi_0)_{y_1} \varepsilon^3. \quad (5.41)$$

Moreover, there exists a universal constant $C_2 > 0$ such that if $\|\varepsilon(s)\|_2 \leq \alpha$, for all $s \geq 0$, where $\alpha < \alpha_1$, then

$$\left|\frac{\lambda_s}{\lambda}\right| + \left|\frac{x_s}{\lambda} - 1\right| \leq C_2 \|\varepsilon(s)\|_2. \quad (5.42)$$

Proof Let χ be a smooth function with an exponential decay. We want to calculate $\frac{d}{ds} \int \chi \varepsilon(s)$. Indeed, we have

$$\frac{d}{ds} \int \chi u(s) = \lambda^3 \frac{d}{dt} \int \chi u(t) = -\lambda^3 \int \chi (\partial_x \Delta u + \partial_x (u^3))$$
$$= \lambda^3 \left[\int \partial_x \Delta \chi u + \int \chi_x u^3 \right].$$

Therefore, recalling the definition of v in (4.21), we get

$$\frac{d}{ds} \int \chi v(s) = \frac{d}{ds} \int \chi(\vec{y}) \lambda u(s, \lambda \vec{y} + \vec{x}(s)) d\vec{y}$$
$$= \frac{d}{ds} \left(\lambda^{-1} \int \chi(\lambda^{-1}(\vec{x} - \vec{x}(s))) u(s, \vec{x}) d\vec{x} \right)$$

$$= -\lambda^{-2}\lambda_s \int \chi(\lambda^{-1}(\vec{x} - \vec{x}(s)))u(s,\vec{x})d\vec{x}$$
$$+ \lambda^{-1} \int \left(\frac{d}{ds}\chi(\lambda^{-1}(\vec{x} - \vec{x}(s)))\right) u(s,\vec{x})d\vec{x}$$
$$+ \lambda^{-1}\lambda^3 \int \chi(\lambda^{-1}(\vec{x} - \vec{x}(s)))(\partial_x \Delta u + \partial_x(u^3))d\vec{x}$$
$$\equiv (A) + (B) + (C),$$

where

$$(A) = -\frac{\lambda_s}{\lambda} \int \chi v d\vec{y},$$

$$(B) = \lambda^{-1} \int \nabla \chi \cdot \frac{d}{ds}(\lambda^{-1}(\vec{x} - \vec{x}(s)))u(s,\vec{x})d\vec{x}$$
$$= -\frac{\lambda_s}{\lambda} \int (\nabla \chi \cdot \vec{y})v d\vec{y} - \int \left(\nabla \chi \cdot \frac{\vec{x}_s}{\lambda}\right) v d\vec{y},$$

$$(C) = \lambda^{-1} \int (\partial_{x_1} \Delta \chi)(\lambda^{-1}(\vec{x} - \vec{x}(s)))u d\vec{x} + \lambda \int \chi_{x_1}(\lambda^{-1}(\vec{x} - \vec{x}(s)))u d\vec{x}$$
$$= \int (\partial_{y_1} \Delta \chi)v d\vec{y} + \int \chi_{y_1} v^3 d\vec{y}.$$

Next, using $v = Q + \varepsilon$ and the definition of Λ in (3.15), we obtain

$$\frac{d}{ds}\int \chi v(s) = -\frac{\lambda_s}{\lambda} \int (\Lambda \chi)(Q + \varepsilon) - \left(\frac{(x_1)_s}{\lambda} - 1\right)\int \chi_{y_1}(Q + \varepsilon) - \frac{(x_2)_s}{\lambda}\int \chi_{y_2}(Q + \varepsilon)$$
$$- \int \chi_{y_1}(Q + \varepsilon) + \int (\partial_{y_1}\Delta \chi)(Q + \varepsilon) + \int \chi_{y_1}(Q + \varepsilon)^3.$$

Recalling $L\chi_{y_1} = -\partial_{y_1}\Delta \chi + \chi_{y_1} - 3Q^2 \chi_{y_1}$ and $-\Delta Q + Q - Q^3 = 0$, we deduce

$$\frac{d}{ds}\int \chi v(s) = -\frac{\lambda_s}{\lambda}\left(\int (\Lambda \chi)Q + \int (\Lambda \chi)\varepsilon\right)$$
$$- \left(\frac{(x_1)_s}{\lambda} - 1\right)\left(\int \chi_{y_1}Q + \int \chi_{y_1}\varepsilon\right)$$
$$- \frac{(x_2)_s}{\lambda}\left(\int \chi_{y_2}Q + \int \chi_{y_2}\varepsilon\right)$$
$$- \int (L\chi_{y_1})\varepsilon + 3\int \chi_{y_1}Q\varepsilon^2 + \int \chi_{y_1}\varepsilon^3.$$

Recall that we can assume $x_2 \equiv 0$ in view of Proposition 5.2. Now, setting $x_1 = x$ and taking $\chi = Q_{y_1}$ and using that $\int (\Lambda Q_{y_1})Q = 0$, $\int Q_{y_1 y_2} Q = \int Q_{y_1} Q_{y_2} = 0$, $L(Q_{y_1 y_1}) = 6QQ_{y_1}^2$ and $\int Q_{y_1} \varepsilon = 0$, we obtain (5.40).

Finally, fix $\chi = \chi_0$ and observe that

$$\int (\Lambda \chi_0) Q = \int \chi_0 Q + \int y_1 (\chi_0)_{y_1} Q + \int y_2 (\chi_0)_{y_2} Q$$

$$= -\int \chi_0 (\Lambda Q) = \frac{1}{\lambda_0} \int (L\chi_0)(\Lambda Q)$$

$$= -\frac{2}{\lambda_0} \int \chi_0 Q \neq 0.$$

Since $\int \chi_0 \varepsilon = 0$ and $\int (\chi_0)_{y_1} Q = -\int \chi_0 Q_{y_1} = 0$, we have (5.41).

Observe that there exists $\alpha_1 > 0$ such that

$$\left(\frac{2}{\lambda_0} \int \chi_0 Q - \int (\vec{y} \cdot \nabla \chi_0) \varepsilon \right) \left(\int |Q_{y_1}|^2 - \int Q_{y_1 y_1} \varepsilon \right)$$

$$- \left(\int (\vec{y} \cdot \nabla Q_{y_1}) \varepsilon \right) \left(\int (\chi_0)_{y_1} \varepsilon \right)$$

$$\geq \frac{1}{\lambda_0} \left(\int \chi_0 Q \right) \left(\int |Q_{y_1}|^2 \right),$$

if $\|\varepsilon(s)\| \leq \alpha < \alpha_1$, for all $s \geq 0$. Also, without loss of generality, we can assume $\alpha_1 < 1$.

Hence, we can solve the system of equations given by (5.40) and (5.41) and obtain a universal constant (depending only on powers of Q and its partial derivatives) $C_2 > 0$ such that (5.42) holds. In particular, if $\alpha < \frac{1}{C_2}$, we have

$$\left| \frac{\lambda_s}{\lambda} \right| + \left| \frac{x_s}{\lambda} - 1 \right| \leq 1. \tag{5.43}$$

□

6 Virial-Type Estimates

Our next step is to produce a virial-type functional which will help us to study the stability properties of the solutions close to Q. We first define a quantity depending on the ε variable, which incorporates the scaling generator Λ. This can be compared with the functional we created for the supercritical case, see [10, Section 5], where the eigenfunction χ_0 of L for the negative eigenvalue was also used (with the coefficient β), and it was possible to find $\beta \neq 0$. However, such a functional does not work in the critical case, since β becomes zero (due to $\int Q \Lambda Q = 0$).

We first start with defining a truncation function: let $\varphi \in C_0^\infty(\mathbb{R})$ be a function with

$$\varphi(y_1) = \begin{cases} 1, & \text{if } y_1 \leq 1 \\ 0, & \text{if } y_1 \geq 2. \end{cases}$$

For $A \geq 1$ we also define

$$\varphi_A(y_1) = \varphi\left(\frac{y_1}{A}\right).$$

Note that

$$\varphi_A(y_1) = \begin{cases} 1, & \text{if } y_1 \leq A \\ 0, & \text{if } y_1 \geq 2A. \end{cases} \tag{6.44}$$

Moreover

$$\varphi_A'(y_1) = \frac{1}{A}\varphi'\left(\frac{y_1}{A}\right).$$

We next define the function (note that we are integrating only in the first variable)

$$F(y_1, y_2) = \int_{-\infty}^{y_1} \Lambda Q(z, y_2)\,dz. \tag{6.45}$$

From the properties of Q, see (1.7), there exists a constant $c > 0$ such that

$$|F(y_1, y_2)| \leq c e^{-\frac{1}{2}|y_2|} \int_{-\infty}^{y_1} e^{-\frac{1}{2}|z|}\,dz,$$

which implies boundedness in y_1 and exponential decay in y_2:

$$\sup_{y_1 \in \mathbb{R}} |F(y_1, y_2)| \leq c e^{-\frac{1}{2}|y_2|} \quad \text{for all} \quad y_2 \in \mathbb{R} \tag{6.46}$$

as well as

$$|F(y_1, y_2)| \leq c e^{-\frac{1}{2}|y_2|} e^{\frac{1}{2}y_1} \quad \text{for all} \quad y_1 < 0. \tag{6.47}$$

Hence, F is a bounded function on $\mathbb{R}^2$, i.e., $F \in L^\infty(\mathbb{R}^2)$. We also note that $y_2 F_{y_2} \in L^\infty(\mathbb{R}^2)$.

We next define the virial-type functional

$$J_A(s) = \int_{\mathbb{R}^2} \varepsilon(s, y_1, y_2) F(y_1, y_2) \varphi_A(y_1)\,dy_1 dy_2. \tag{6.48}$$

It is clear that $J_A(s)$ is well-defined if $\varepsilon(s) \in L^2(\mathbb{R}^2)$. Indeed, since $\|\varphi_A\|_\infty = 1$, we can use the relation (6.44) and the properties of F to deduce

$$|J_A(s)| \leq \int_\mathbb{R} \int_{y_1<0} |\varepsilon(s) F(y_1, y_2)| \, dy_1 dy_2 + \int_\mathbb{R} \int_0^{2A} |\varepsilon(s) F(y_1, y_2)| \, dy_1 dy_2$$

$$\leq c \|\varepsilon(s)\|_2 \left(\int_\mathbb{R} \int_{y_1<0} e^{-|y_2|} e^{y_1} dy_1 dy_2 \right)^{1/2}$$

$$+ cA^{1/2} \int_\mathbb{R} \sup_{y_1} |F(y_1, y_2)| \left(\int_0^{2A} |\varepsilon(s)|^2 dy_1 \right)^{1/2} dy_2$$

$$\leq c \left(\int_\mathbb{R} e^{-|y_2|} dy_2 \right)^{1/2} \left(\int_{y_1<0} e^{y_1} dy_1 \right)^{1/2}$$

$$\|\varepsilon(s)\|_2 + cA^{1/2} \left(\int_\mathbb{R} e^{-|y_2|} dy_2 \right)^{1/2} \|\varepsilon(s)\|_2.$$

Therefore, we obtain the boundedness of J_A from above

$$|J_A(s)| \leq c(1 + A^{1/2}) \|\varepsilon(s)\|_2. \tag{6.49}$$

Next, we compute the derivative of $J_A(s)$.

Lemma 6.1 *Suppose that $\varepsilon(s) \in H^1(\mathbb{R}^2)$ for all $s \geq 0$. Then the function $s \mapsto J_A(s)$ is C^1 and*

$$\frac{d}{ds} J_A = -\frac{\lambda_s}{\lambda} (J_A - \kappa) + 2 \left(1 - \frac{1}{2} \left(\frac{x_s}{\lambda} - 1 \right) \right) \int \varepsilon Q + R(\varepsilon, A),$$

where

$$\kappa = \frac{1}{2} \int y_2^2 \left(\int Q_{y_2}(y_1, y_2) dy_1 \right)^2 dy_2, \tag{6.50}$$

and, there exists a universal constant $C_3 > 0$ such that, for $A \geq 1$, we have

$$|R(\varepsilon, A)| \leq C_3 \left(\|\varepsilon\|_2^2 + \|\varepsilon\|_2^2 \|\varepsilon\|_{H^1} + A^{-1/2} \|\varepsilon\|_2 \right.$$

$$+ \left| \frac{x_s}{\lambda} - 1 \right| (A^{-1} + \|\varepsilon\|_2)$$

$$+ \left. \left| \frac{\lambda_s}{\lambda} \right| \left(A^{-1} + \|\varepsilon\|_2 + A^{1/2} \|\varepsilon\|_{L^2(y_1 \geq A)} + \left| \int_{\mathbb{R}^2} y_2 F_{y_2} \varepsilon \varphi_A \right| \right) \right). \tag{6.51}$$

Remark 6.2 Note that by the decay properties of Q, see (1.7), the value of κ, defined in (6.50), is a finite number.

Proof First, arguing as in the proof of Lemma 4.1, we have that the function $\varepsilon(s)$, defined in (5.39), satisfy the following equation

$$\varepsilon_s = (L\varepsilon)_{y_1} + \frac{\lambda_s}{\lambda}(\Lambda Q + \Lambda \varepsilon) + \left(\frac{x_s}{\lambda} - 1\right)(Q_{y_1} + \varepsilon_{y_1}) - 3(Q\varepsilon^2)_{y_1} - (\varepsilon^3)_{y_1}. \quad (6.52)$$

Then, setting $R(\varepsilon) = 3Q\varepsilon^2 + \varepsilon^3$, we have

$$\frac{d}{ds}J_A = \int \varepsilon_s F \varphi_A$$

$$= \int \left((L\varepsilon)_{y_1} + \frac{\lambda_s}{\lambda}\Lambda\varepsilon + \left(\frac{x_s}{\lambda} - 1\right)\varepsilon_{y_1}\right) F \varphi_A$$

$$+ \int \left(\frac{\lambda_s}{\lambda}\Lambda Q + \left(\frac{x_s}{\lambda} - 1\right) Q_{y_1}\right) F \varphi_A$$

$$- \int R(\varepsilon)_{y_1} F \varphi_A$$

$$\equiv (I) + (II) + (III).$$

Now, since $\|\varphi_A\|_\infty \leq 1$ and $\varphi_{y_1} \in L^\infty$, we have

$$(III) = \int R(\varepsilon) \Lambda Q \varphi_A + \int R(\varepsilon) F \frac{1}{A} \varphi'\left(\frac{y_1}{A}\right)$$

$$\leq \|\Lambda Q\|_\infty \int |R(\varepsilon)| + \frac{\|F\|_\infty \|\varphi'\|_\infty}{A} \int |R(\varepsilon)|$$

$$\leq c_0 \left(\|\Lambda Q\|_\infty + \frac{\|F\|_\infty \|\varphi'\|_\infty}{A}\right)\left(\|\varepsilon\|_2^2 + \|\varepsilon\|_2^2\|\varepsilon\|_{H^1}\right), \quad (6.53)$$

by Gagliardo-Nirenberg inequality with

$$c_0 = \|3Q\|_\infty + C_{GN},$$

here, C_{GN} is the best constant for the cubic Gagliardo-Nirenberg inequality.
Furthermore,

$$(II) = \frac{\lambda_s}{\lambda}\int \Lambda Q F \varphi_A + \left(\frac{x_s}{\lambda} - 1\right)\int Q_{y_1} F \varphi_A$$

$$\equiv \frac{\lambda_s}{\lambda}(II.1) + \left(\frac{x_s}{\lambda} - 1\right)(II.2),$$

where, since $F_{y_1} = \Lambda Q$

$$(II.1) = \frac{1}{2}\int (F^2)_{y_1}\varphi_A$$

$$= \frac{1}{2}\int (F^2)_{y_1} + \frac{1}{2}\int (F^2)_{y_1}(\varphi_A - 1)$$

$$= \frac{1}{2}\int \left(\int \Lambda Q(y_1, y_2)dy_1\right)^2 dy_2 + R_1(A),$$

where in the last line we used (6.45).

Now, integration by parts yields

$$\int \Lambda Q dy_1 = \int Q dy_1 + \int y_1 Q_{y_1} dy_1 + \int y_2 Q_{y_2} dy_1$$

$$= \int y_2 Q_{y_2} dy_1,$$

and thus,

$$\int \left(\int \Lambda Q(y_1, y_2)dy_1\right)^2 dy_2 = \int y_2^2 \left(\int Q_{y_2}(y_1, y_2)dy_1\right)^2 dy_2 < +\infty.$$

Moreover, the error term can be estimated as follows

$$|R_1(A)| = \left|\int_{\mathbb{R}^2} \Lambda Q F(\varphi_A - 1)\right| \tag{6.54}$$

$$\leq \int_{\mathbb{R}} \sup_{y_1} |F(y_1, y_2)| \left(\int_{\mathbb{R}} |\Lambda Q(\varphi_A - 1)|dy_1\right) dy_2 \tag{6.55}$$

$$\leq 2\int_{\mathbb{R}} \sup_{y_1} |F(y_1, y_2)| \left(\int_{y_1 \geq A} \frac{|\Lambda Q||y_1|}{|y_1|}dy_1\right) dy_2 \tag{6.56}$$

$$\leq \frac{2\|F\|_\infty \|y_1 \Lambda Q\|_1}{A}. \tag{6.57}$$

On the other hand, since $\Lambda Q \perp Q$

$$(II.2) = -\int Q\Lambda Q\varphi_A - \int QF\frac{1}{A}\varphi'\left(\frac{y_1}{A}\right)$$

$$= -\int Q\Lambda Q(\varphi_A - 1) - \int QF\frac{1}{A}\varphi'\left(\frac{y_1}{A}\right)$$

$$\equiv R_2(A),$$

where, using again the definition of φ_A, we have

$$|R_2(A)| \leq 2 \int_{\mathbb{R}} \int_{y_1 \geq A} |Q \Lambda Q| \frac{|y_1|}{|y_1|} + \frac{\|F\|_\infty \|\varphi'\|_\infty \|Q\|_1}{A}$$

$$\leq \frac{1}{A} \left(2\|\Lambda Q\|_2 \|y_1 Q\|_2 + \|F\|_\infty \|\varphi'\|_\infty \|Q\|_1 \right). \qquad (6.58)$$

Next we estimate the term (I). Applying integration by parts, we get

$$(I) = -\int (L\varepsilon) \Lambda Q \varphi_A - \int (L\varepsilon) F \frac{1}{A} \varphi'\left(\frac{y_1}{A}\right)$$

$$+ \frac{\lambda_s}{\lambda} \int \Lambda \varepsilon F \varphi_A$$

$$- \left(\frac{x_s}{\lambda} - 1\right) \left(\int \varepsilon \Lambda Q \varphi_A + \int \varepsilon F \frac{1}{A} \varphi'\left(\frac{y_1}{A}\right) \right)$$

$$\equiv (I.1) + \frac{\lambda_s}{\lambda}(I.2) - \left(\frac{x_s}{\lambda} - 1\right)(I.3).$$

Let us first consider the term $(I.3)$. Using the definition (3.15), we have

$$(I.3) = \int \varepsilon Q + \int \varepsilon Q (\varphi_A - 1) + \int \varepsilon y_1 Q_{y_1} \varphi_A + \int \varepsilon y_2 Q_{y_2} \varphi_A + \frac{1}{A} \int \varepsilon F \varphi'\left(\frac{y_1}{A}\right)$$

$$\equiv \int \varepsilon Q + R_3(\varepsilon, A).$$

Next, it is easy to see that

$$\int \varepsilon Q (\varphi_A - 1) \leq 2 \int_{\mathbb{R}} \int_{y_1 \geq A} |\varepsilon Q| \frac{|y_1|}{|y_1|} \leq \frac{2}{A} \|\varepsilon\|_2 \|Q\|_2,$$

and, for $j = 1, 2$, we get

$$\int \varepsilon y_j Q_{y_j} \varphi_A \leq \|\varepsilon\|_2 \|y_j Q_{y_j}\|_2.$$

Moreover,

$$\int \varepsilon F \varphi'\left(\frac{y_1}{A}\right) \leq \|\varphi'\|_\infty \int_{\mathbb{R}} \left(\sup_{y_1} |F(y_1, y_2)| \int_A^{2A} |\varepsilon| dy_1 \right) dy_2$$

$$\leq A^{1/2} \|\varphi'\|_\infty \left(\int_{\mathbb{R}} \sup_{y_1} |F(y_1, y_2)|^2 dy_2 \right)^{1/2} \|\varepsilon\|_2.$$

Collecting the last three inequalities and using (6.46), we deduce

$$|R_3(\varepsilon, A)| \leq c\left(1 + \frac{1}{A} + \frac{1}{A^{1/2}}\right)\|\varepsilon\|_2, \qquad (6.59)$$

where the constant $c > 0$ is independent of ε and A.

Next, we turn to the term $(I.2)$. Integration by parts yields

$$\begin{aligned}(I.2) &= \int \varepsilon F \varphi_A + \int y_1 \varepsilon_{y_1} F \varphi_A + \int y_2 \varepsilon_{y_2} F \varphi_A \\ &= -\int y_1 \varepsilon \Lambda Q \varphi_A - \int y_1 \varepsilon F \frac{1}{A} \varphi'\left(\frac{y_1}{A}\right) \\ &\quad - \int \varepsilon F \varphi_A - \int y_2 \varepsilon F_{y_2} \varphi_A \\ &\equiv -J_A + R_4(\varepsilon, A),\end{aligned}$$

where in the last line we used definition (6.48). Let us first estimate the terms in $R_4(\varepsilon, A)$. Indeed, it is clear that

$$\int y_1 \varepsilon \Lambda Q \varphi_A \leq \|y_1 \Lambda Q\|_2 \|\varepsilon\|_2.$$

Furthermore,

$$\begin{aligned}\int y_1 \varepsilon F \frac{1}{A} \varphi'\left(\frac{y_1}{A}\right) &\leq \frac{1}{A}\|\varphi'\|_\infty \int_\mathbb{R} \left(\sup_{y_1} |F(y_1, y_2)| \int_A^{2A} |y_1 \varepsilon| dy_1\right) dy_2 \\ &\leq 2\|\varphi'\|_\infty \int_\mathbb{R} \left(\sup_{y_1} |F(y_1, y_2)| \int_A^{2A} |\varepsilon| dy_1\right) dy_2 \\ &\leq 2A^{1/2}\|\varphi'\|_\infty \left(\int_\mathbb{R} \sup_{y_1} |F(y_1, y_2)|^2 dy_2\right)^{1/2} \|\varepsilon\|_{L^2(y_1 \geq A)} \\ &\leq cA^{1/2}\|\varphi'\|_\infty \|\varepsilon\|_{L^2(y_1 \geq A)},\end{aligned}$$

where in the last line we used the inequality (6.46).

Collecting the last two estimates, we obtain

$$|R_4(\varepsilon, A)| \leq c(\|\varepsilon\|_2 + A^{1/2}\|\varepsilon\|_{L^2(y_1 \geq A)}) + \left|\int y_2 \varepsilon F_{y_2} \varphi_A\right|, \qquad (6.60)$$

where $c > 0$ is again independent of ε and A.

To estimate (I.1), we recall the definition of the operator L to get

$$L(fg) = -(\Delta f)g - f(\Delta g) - 2f_{y_1}g_{y_1} - 2f_{y_2}g_{y_2} + fg - 3Q^2 fg$$
$$= (Lf)g - 2f_{y_1}g_{y_1} - 2f_{y_2}g_{y_2} - f(\Delta g).$$

Hence,

$$L(\Lambda Q \varphi_A) = (L\Lambda Q)\varphi_A - 2(\Lambda Q)_{y_1}\frac{1}{A}\varphi'\left(\frac{y_1}{A}\right) - \Lambda Q\frac{1}{A^2}\varphi''\left(\frac{y_1}{A}\right)$$
$$\equiv L(\Lambda Q)\varphi_A + G_A,$$

and

$$L\left(F\frac{1}{A}\varphi'\left(\frac{y_1}{A}\right)\right) = \frac{1}{A}\left[L\left(\varphi'\left(\frac{y_1}{A}\right)\right)F - 2\Lambda Q\frac{1}{A^2}\varphi''\left(\frac{y_1}{A}\right)\right.$$
$$\left. -\varphi'\left(\frac{y_1}{A}\right)((\Lambda Q)_{y_1} + F_{y_2 y_2})\right]$$
$$\equiv \frac{1}{A}H_A.$$

Using the fact that L is a self-adjoint operator and $L(\Lambda Q) = -2Q$, we get

$$(I.1) = -\int \varepsilon L(\Lambda Q \varphi_A) - \int \varepsilon L\left(F\frac{1}{A}\varphi'\left(\frac{y_1}{A}\right)\right)$$
$$= 2\int \varepsilon Q \varphi_A - \int \varepsilon \left(G_A + \frac{1}{A}H_A\right)$$
$$= 2\int \varepsilon Q + 2\int \varepsilon Q(\varphi_A - 1) - \int \varepsilon \left(G_A + \frac{1}{A}H_A\right)$$
$$\equiv 2\int \varepsilon Q + R_5(\varepsilon, A).$$

Again, we estimate the terms in $R_5(\varepsilon, A)$ separately. First, we observe that

$$\int \varepsilon Q(\varphi_A - 1) \leq \int_{\mathbb{R}}\int_A^{+\infty} |\varepsilon Q|\frac{|y_1|}{|y_1|}dy_1 dy_2 \leq \frac{1}{A}\|y_1 Q\|_2 \|\varepsilon\|_2.$$

Moreover,

$$\int \varepsilon G_A \leq \frac{2}{A}\|\varphi'\|_\infty \|(\Lambda Q)_{y_1}\|_2 \|\varepsilon\|_2 + \frac{1}{A^2}\|\varphi''\|_\infty \|\Lambda Q\|_2 \|\varepsilon\|_2.$$

Now, note that $\|H_A\|_\infty \leq c$ (independent of $A \geq 1$) and

$$\operatorname{supp}(H_A) \subset \{A \leq y_1 \leq 2A\}.$$

Then, it is easy to see that (using that $F_{y_2 y_2}$ also satisfies similar estimates as the ones in (6.46) and (6.47))

$$\frac{1}{A} \int \varepsilon H_A \leq \frac{c}{A^{1/2}} \|\varepsilon\|_2.$$

Finally, for $A \geq 1$, we obtain

$$|R_5(\varepsilon, A)| \leq \frac{c}{A^{1/2}} \|\varepsilon\|_2, \qquad (6.61)$$

where, once again, $c > 0$ is independent of ε and A.

Collecting all the above estimates, we finally obtain

$$\begin{aligned}\frac{d}{ds} J_A &= 2 \int \varepsilon Q + R_5(\varepsilon, A) + \frac{\lambda_s}{\lambda}(-J_A + R_4(\varepsilon, A)) \\ &\quad - \left(\frac{x_s}{\lambda} - 1\right)\left(\int \varepsilon Q + R_3(\varepsilon, A)\right) \\ &\quad + \frac{\lambda_s}{\lambda}(\kappa + R_1(A)) + \left(\frac{x_s}{\lambda} - 1\right) R_2(A) \\ &\quad - (III) \\ &= -\frac{\lambda_s}{\lambda}(J_A - \kappa) + 2\left(1 - \frac{1}{2}\left(\frac{x_s}{\lambda} - 1\right)\right) \int \varepsilon Q + R(\varepsilon, A),\end{aligned}$$

where κ is given by (6.50) and

$$R(\varepsilon, A) = (III) + R_5(\varepsilon, A) + \left(\frac{x_s}{\lambda} - 1\right)(R_2(A) - R_3(\varepsilon, A)) + \frac{\lambda_s}{\lambda}(R_1(A) + R_4(\varepsilon, A)).$$

Furthermore, there exists a universal constant $C_3 > 0$ (independent of ε and A) such that, in view of (6.53), (6.54), and (6.58)–(6.61), for $A \geq 1$ the inequality (6.51) holds. □

6.1 Control of Parameters

Before we proceed with examining further properties of J_A, we need to understand how various parameters are interconnected and controlled by the initial time values, especially $\varepsilon(s)$. We proceed with the following two lemmas.

Lemma 6.3 (Comparison Between M_0, ε_0 and $\int \varepsilon_0 Q$) *There exists a universal constant $C_4 > 0$ such that, if $\|\varepsilon_0\|_{H^1} \leq 1$, then*

$$\left| M_0 - 2 \int \varepsilon_0 Q \right| + \left| E_0 + \int \varepsilon_0 Q \right| + \left| E_0 + \frac{1}{2} M_0 \right| \leq C_4 \|\varepsilon_0\|_{H^1}^2.$$

Proof First, observe that from the definition (4.24), we have

$$M_0 - 2 \int \varepsilon_0 Q = \int \varepsilon_0^2,$$

and thus, $\left| M_0 - 2 \int \varepsilon_0 Q \right| = \|\varepsilon_0\|_2^2$. Next, from (4.27), we obtain

$$E_0 = E[Q + \varepsilon_0] = \frac{1}{2} (L\varepsilon_0, \varepsilon_0) - \frac{1}{2} M_0 - \frac{1}{4} \left[4 \int Q\varepsilon_0^3 + \int \varepsilon_0^4 \right],$$

which implies, for some universal constant $c > 0$, that

$$\left| E_0 + \frac{1}{2} M_0 \right| \leq c \|\varepsilon_0\|_{H^1}^2,$$

by the definition of L, the Gagliardo-Nirenberg inequality (4.28) and the fact that $\|\varepsilon_0\|_{H^1} \leq 1$. Finally,

$$\left| E_0 + \int \varepsilon_0 Q \right| \leq \left| E_0 + \frac{1}{2} M_0 \right| + \frac{1}{2} \left| M_0 - 2 \int \varepsilon_0 Q \right| \leq \left(c + \frac{1}{2} \right) \|\varepsilon_0\|_{H^1}^2,$$

and setting $C_4 = c + \frac{1}{2}$, we conclude the proof. $\square$

Lemma 6.4 (Control of $\|\varepsilon(s)\|_{H^1}$) *There exists $\alpha_2 > 0$ such that, if $\|\varepsilon(s)\|_{H^1} < \alpha$, $|\lambda(s) - 1| < \alpha$ and $\varepsilon(s) \perp \{Q_{y_1}, Q_{y_2}, \chi_0\}$ for all $s \geq 0$, where $\alpha < \alpha_2$, then there exists a universal constant $C_5 > 0$ such that*

$$(L\varepsilon(s), \varepsilon(s)) \leq \|\varepsilon(s)\|_{H^1}^2 \leq C_5 \left(\alpha \left| \int \varepsilon_0 Q \right| + \|\varepsilon_0\|_{H^1}^2 \right).$$

Proof From (4.27) we have

$$(L\varepsilon(s), \varepsilon(s)) = 2E[Q + \varepsilon(s)] + M_0 + \frac{1}{4} \left[4 \int Q\varepsilon^3(s) + \int \varepsilon^4(s) \right]. \quad (6.62)$$

Therefore, from the Gagliardo-Nirenberg inequality (4.28), there exists a universal constant $c > 0$ such that if $\|\varepsilon(s)\|_{H^1} \leq 1$

$$(L\varepsilon(s), \varepsilon(s)) \leq 2E[Q + \varepsilon(s)] + M_0 + c \|\varepsilon(s)\|_{H^1} \|\varepsilon(s)\|_2^2$$

$$\leq 2E[Q + \varepsilon(s)] + M_0 + \frac{c}{\sigma_0} \|\varepsilon(s)\|_{H^1} (L\varepsilon(s), \varepsilon(s)), \quad (6.63)$$

where in the last line we used the coercivity of the quadratic form $(L\cdot,\cdot)$, provided $\varepsilon(s) \perp \{Q_{y_1}, Q_{y_2}, \chi_0\}$, which was obtained in Lemma 3.6.

Now, there exists $\alpha_2 > 0$ such that if $\|\varepsilon(s)\|_{H^1} < \alpha$ for all $s \geq 0$, where $\alpha < \alpha_2$, then
$$\frac{c}{\sigma_0}\|\varepsilon(s)\|_{H^1} \leq \frac{1}{2}.$$

Therefore, the last term in the RHS of (6.63) may be absorbed by the left-hand term, and we get
$$(L\varepsilon(s), \varepsilon(s)) \leq 4E[Q + \varepsilon(s)] + 2M_0$$
$$\leq 4\lambda^2(s)E_0 + 2M_0,$$

where in the last line we have used relation (4.26).

Next, we use the last estimate to control the H^1-norm of $\varepsilon(s)$ as well. Indeed, from the definition of L we have
$$\|\varepsilon(s)\|_{H^1}^2 = \int \varepsilon^2(s) + \int |\nabla \varepsilon(s)|^2 = (L\varepsilon(s), \varepsilon(s)) + 3\int Q^2 \varepsilon^2(s)$$
$$\leq (L\varepsilon(s), \varepsilon(s)) + \|3Q^2\|_\infty \|\varepsilon(s)\|_2^2$$
$$\leq \left(1 + \frac{\|3Q^2\|_\infty}{\sigma_0}\right)(L\varepsilon(s), \varepsilon(s))$$
$$\leq \left(1 + \frac{\|3Q^2\|_\infty}{\sigma_0}\right)(4\lambda^2(s)E_0 + 2M_0)$$
$$\leq 4\left(1 + \frac{\|3Q^2\|_\infty}{\sigma_0}\right)\left((\lambda(s) - 1)(\lambda(s) + 1)|E_0| + \left|E_0 + \frac{1}{2}M_0\right|\right).$$

Finally, since $|\lambda(s) - 1| < \alpha$, choosing $\alpha < 1$, we get $|\lambda(s) + 1| \leq 3$, and applying Lemma 6.3, we deduce
$$(\lambda(s) - 1)(\lambda(s) + 1)|E_0| + \left|E_0 + \frac{1}{2}M_0\right| \leq 3\alpha\left(\left|E_0 + \int \varepsilon_0 Q\right| + \left|\int \varepsilon_0 Q\right|\right) + C_4\|\varepsilon_0\|_{H^1}^2$$
$$\leq 3\alpha\left(C_4\|\varepsilon_0\|_{H^1}^2 + \left|\int \varepsilon_0 Q\right|\right) + C_4\|\varepsilon_0\|_{H^1}^2$$
$$\leq 4C_4\|\varepsilon_0\|_{H^1}^2 + 3\alpha\left|\int \varepsilon_0 Q\right|,$$

which implies the existence of a universal constant $C_5 > 0$ such that

$$(L\varepsilon(s), \varepsilon(s)) \leq \|\varepsilon(s)\|_{H^1}^2 \leq C_5 \left(\alpha \left| \int \varepsilon_0 Q \right| + \|\varepsilon_0\|_{H^1}^2 \right).$$

$\square$

Our next task is to bound the time derivative of J_A that we obtained in Lemma 6.1 from below. The main concern is to estimate the remainder $R(\varepsilon, A)$ from Lemma 6.1, and in particular, the last line (6.51). Via truncation we can always choose A to be large, so that the terms, which involve negative powers of A, would be controlled, however, the third term in (6.51) involves a positive power of A and the tail of the L^2 norm of ε, and hence, needs a delicate estimate. This can be done via monotonicity property, which we discuss in the next section. This is similar to our analysis of the supercritical gZK, see [10]. However, together with truncation and monotonicity the last term in (6.51) is still troublesome, and it is possible to bound it, but more is needed, namely, the smallness of that term. Thus, we need to develop another tool, the pointwise decay estimates, which we do in Sects. 8–12.

7 Monotonicity

For $M \geq 4$, define

$$\psi(x_1) = \frac{2}{\pi} \arctan(e^{\frac{x_1}{M}}).$$

The following properties hold for ψ:

1. $\psi(0) = \frac{1}{2}$,
2. $\lim_{x_1 \to -\infty} \psi(x_1) = 0$ and $\lim_{x_1 \to +\infty} \psi(x_1) = 1$,
3. $1 - \psi(x_1) = \psi(-x_1)$,
4. $\psi'(x_1) = \left(\pi M \cosh\left(\frac{x_1}{M}\right) \right)^{-1}$,
5. $|\psi'''(x_1)| \leq \frac{1}{M^2} \psi'(x_1) \leq \frac{1}{16} \psi'(x_1)$.

Let $(x_1(t), x_2(t)) \in C^1(\mathbb{R}, \mathbb{R}^2)$ and for $x_0, t_0 > 0$ and $t \in [0, t_0]$ define

$$I_{x_0, t_0}(t) = \int u^2(t, x_1, x_2) \psi(x_1 - x_1(t_0) + \frac{1}{2}(t_0 - t) - x_0) dx_1 dx_2, \qquad (7.64)$$

where $u \in C(\mathbb{R}, H^1(\mathbb{R}^2))$ is a solution of the gZK equation (1.1), satisfying

$$\|u(t, x_1 + x_1(t), x_2 + x_2(t)) - Q(x_1, x_2)\|_{H^1} \leq \alpha, \quad \text{for some} \quad \alpha > 0. \qquad (7.65)$$

While the functional $I_{x_0,t_0}(t)$, which localizes the mass of the solution respectively to the moving soliton, is a concept similar to the one originated in works of Martel and Merle, and is used to study the decay of the mass of the solution to the right of the soliton, and can be applied to a variety of questions for the gKdV equations (see also our review of the instability of the critical gKdV case via monotonicity [9]), we note that the integration in the definition (7.64) is two dimensional. Note that the function ψ is defined only in one variable x_1, this is similar to [5]. We next study the behavior of I_{x_0,t_0} in time and we have the following monotonicity-type result.

Lemma 7.1 (Almost Monotonicity) *Let $M \geq 4$ fixed and assume that $x_1(t)$ is an increasing function satisfying $x_1(t_0) - x_1(t) \geq \frac{3}{4}(t_0 - t)$ for every $t_0, t \geq 0$ with $t \in [0, t_0]$. Then there exist $\alpha_0 > 0$ and $\theta = \theta(M) > 0$ such that, if $u \in C(\mathbb{R}, H^1(\mathbb{R}^2))$ verify (7.65) with $\alpha < \alpha_0$, then for all $x_0 > 0$, $t_0, t \geq 0$ with $t \in [0, t_0]$, we have*

$$I_{x_0,t_0}(t_0) - I_{x_0,t_0}(t) \leq \theta e^{-\frac{x_0}{M}}.$$

Proof Using the equation and the fact that $|\psi'''(x)| \leq \frac{1}{M^2}\psi'(x) \leq \frac{1}{16}\psi'(x)$, we deduce

$$\frac{d}{dt} I_{x_0,t_0}(t) = 2 \int u u_t \psi - \frac{1}{2} \int u^2 \psi'$$

$$= - \int \left(3 u_{x_1}^2 + u_{x_2}^2 - \frac{3}{2} u^4 \right) \psi' + \int u^2 \psi''' - \frac{1}{2} \int u^2 \psi'$$

$$\leq - \int \left(3 u_{x_1}^2 + u_{x_2}^2 + \frac{1}{4} u^2 \right) \psi' + \frac{3}{2} \int u^4 \psi'. \quad (7.66)$$

We start with the estimate of the last term in (7.66), by using its closeness to Q,

$$\int u^4 \psi' = \int Q(\cdot - \vec{x}(t)) u^3 \psi' + \int (u - Q(\cdot - \vec{x}(t))) u^3 \psi', \quad (7.67)$$

where $\vec{x}(t) = (x_1(t), x_2(t))$. To estimate the second term, we use the Sobolev embedding $H^1(\mathbb{R}^2) \hookrightarrow L^q(\mathbb{R}^2)$, for all $2 \leq q < +\infty$, to obtain

$$\int (u - Q(\cdot - \vec{x}(t))) u^3 \psi' \leq \|(u - Q(\cdot - \vec{x}(t))) u\|_{4/3} \|u^2 \psi'\|_4$$

$$\leq c \|u - Q(\cdot - \vec{x}(t))\|_2 \|u\|_4 \|u \sqrt{\psi'}\|_8^2$$

$$\leq c \alpha \|Q\|_{H^1} \int (|\nabla u|^2 + |u|^2) \psi'. \quad (7.68)$$

For the first term on the right hand side of (7.67), we divide the integration into two regions $|\vec{x} - \vec{x}(t)| > R_0$ and $|\vec{x} - \vec{x}(t)| \leq R_0$, where R_0 is a positive number to be chosen later. Since $|Q(\vec{x})| \leq c e^{-|\vec{x}|}$, we obtain

$$\int_{|\vec{x}-\vec{x}(t)|>R_0} Q(\cdot-\vec{x}(t))u^3\psi' \leq c\,e^{-R_0}\|u\|_3\|u\sqrt{\psi'}\|_3^2$$

$$\leq c\,e^{-R_0}\|Q\|_{H^1}\int(|\nabla u|^2+|u|^2)\psi'. \qquad (7.69)$$

Next, when $|\vec{x}-\vec{x}(t)| \leq R_0$, we have

$$\left|x_1-x_1(t_0)+\frac{1}{2}(t_0-t)-x_0\right| \geq (x_1(t_0)-x_1(t)+x_0)-\frac{1}{2}(t_0-t)-|x_1-x_1(t)|$$

$$\geq \frac{1}{4}(t_0-t)+x_0-R_0,$$

where in the first inequality we used that $x_1(t)$ is increasing, $t_0 \geq t$ and $x_0 > 0$ to compute the modulus of the first term, and in the second line we used the assumption $x_1(t_0)-x_1(t) \geq \frac{3}{4}(t_0-t)$.

Since $\psi'(z) \leq \frac{2}{M\pi}e^{-\frac{|z|}{M}}$, we can use again the Sobolev embedding $H^1(\mathbb{R}^2) \hookrightarrow L^q(\mathbb{R}^2)$, for all $2 \leq q < +\infty$, to deduce that

$$\int_{|\vec{x}-\vec{x}(t)|\leq R_0} Q(\cdot-\vec{x}(t))u^3\psi' \leq \frac{2}{M\pi}\|Q\|_\infty e^{\frac{R_0}{M}}e^{-\frac{\left(\frac{1}{4}(t_0-t)+x_0\right)}{M}}\|u\|_{H^1}^3$$

$$\leq \frac{2}{M\pi}\|Q\|_\infty\|Q\|_{H^1}^3 e^{\frac{R_0}{M}}e^{-\frac{\left(\frac{1}{4}(t_0-t)+x_0\right)}{M}}. \qquad (7.70)$$

Therefore, choosing $\alpha > 0$ such that $c\alpha\|Q\|_{H^1} < \frac{2}{3}\cdot\frac{1}{16}$ and R_0 such that $c\,e^{-R_0}\|Q\|_{H^1} < \frac{2}{3}\cdot\frac{1}{16}$, collecting (7.68)–(7.70) together, we have

$$\frac{3}{2}\int u^4\psi' \leq \frac{1}{8}\int(|\nabla u|^2+|u|^2)\psi'+\frac{3}{M\pi}\|Q\|_\infty\|Q\|_{H^1}^3 e^{\frac{R_0}{M}}e^{-\frac{\left(\frac{1}{4}(t_0-t)+x_0\right)}{M}}.$$

Inserting the previous estimate in (7.66), we get that there exists a universal constant $c > 0$ such that

$$\frac{d}{dt}I_{x_0,t_0}(t) \leq -\int\left(\frac{3}{2}u_{x_1}^2+\frac{1}{2}u_{x_2}^2+\frac{1}{8}u^2\right)\psi'+\frac{c}{M}\|Q\|_\infty\|Q\|_{H^1}^3 e^{\frac{R_0}{M}}e^{-\frac{x_0}{M}}\cdot e^{-\frac{1}{4M}(t_0-t)}$$

$$\leq \frac{c}{M}\|Q\|_\infty\|Q\|_{H^1}^3 e^{\frac{R_0-x_0}{M}}\cdot e^{-\frac{1}{4M}(t_0-t)}.$$

Finally, integrating in time on $[t,t_0]$, we obtain the desired inequality for

$$\theta = \theta(M) = 4c\,\|Q\|_\infty\|Q\|_{H^1}^3 e^{\frac{c}{M}} > 0.$$

$\square$

The next lemma will be used to control several terms in the virial-type functional from Sect. 6 (see also Combet [4] for a similar result for the gKdV equation).

Lemma 7.2 *Let $x_1(t)$ satisfying the assumptions of Lemma 7.1. Also assume that $x_1(t) \geq \frac{1}{2}t$ and $x_2(t) = 0$ for all $t \geq 0$. Moreover, let $u \in C(\mathbb{R}, H^1(\mathbb{R}^2))$ be a solution of the gZK equation (1.3) satisfying (7.65) with $\alpha < \alpha_0$ (where α_0 is given in Lemma 7.1) and with the initial data u_0 verifying $\int |u_0(x_1, x_2)|^2 dx_2 \leq c e^{-\delta|x_1|}$ for some $c > 0$ and $\delta > 0$. Fix $M \geq \max\{4, \frac{2}{\delta}\}$, then there exists $C = C(M, \delta) > 0$ such that for all $t \geq 0$ and $x_0 > 0$*

$$\int_{\mathbb{R}} \int_{x_1 > x_0} u^2(t, x_1 + x_1(t), x_2) \, dx_1 dx_2 \leq C e^{-\frac{x_0}{M}}. \tag{7.71}$$

Proof From Lemma 7.1 with $t = 0$ and replacing t_0 by t, we deduce that for all $t \geq 0$

$$I_{x_0, t}(t) - I_{x_0, t}(0) \leq \theta e^{-\frac{x_0}{M}}.$$

This is equivalent to

$$\int u^2(t, x_1, x_2) \psi(x_1 - x_1(t) - x_0) \, dx_1 dx_2$$
$$\leq \int u_0^2(x_1, x_2) \psi(x_1 - x_1(t) + \frac{1}{2}t - x_0) \, dx_1 dx_2 + \theta e^{-\frac{x_0}{M}}.$$

On the other hand,

$$\int u^2(t, x_1, x_2) \psi(x_1 - x_1(t) - x_0) \, dx_1 dx_2 = \int u^2(t, x_1 + x_1(t), x_2) \psi(x_1 - x_0) \, dx_1 dx_2$$
$$\geq \frac{1}{2} \int_{\mathbb{R}} \int_{x_1 > x_0} u^2(t, x_1 + x_1(t), x_2) \, dx_1 dx_2,$$

where in the last inequality we used the fact that ψ is increasing and $\psi(0) = 1/2$.

Now, since $-x_1(t) + \frac{1}{2}t \leq 0$ and ψ is increasing, we get

$$\int u_0^2(x_1, x_2) \psi(x_1 - x_1(t) + \frac{1}{2}t - x_0) \, dx_1 dx_2 \leq \int u_0^2(x_1, x_2) \psi(x_1 - x_0) \, dx_1 dx_2.$$

Moreover, the assumption $\int |u_0(x_1, x_2)|^2 dx_2 \leq c e^{-\delta|x_1|}$ and the fact that $\psi(x_1) \leq c e^{\frac{x_1}{M}}$ for all $x_1 \in \mathbb{R}$, yield

$$\int u_0^2(x_1, x_2) \psi(x_1 - x_0) \, dx_1 dx_2 \le c \int e^{-\delta|x_1|} e^{\frac{x_1 - x_0}{M}} \, dx_1$$

$$\le c e^{-\frac{x_0}{M}} \int e^{-\left(\delta - \frac{1}{M}\right)|x_1|} \, dx_1$$

$$\le c e^{-\frac{x_0}{M}} \int e^{-\frac{\delta}{2}|x_1|} \, dx_1,$$

where in the last inequality we have used the fact that

$$\delta - \frac{1}{M} \ge \frac{\delta}{2} \iff M \ge \frac{2}{\delta}.$$

Therefore, the desired inequality (7.71) holds by taking

$$C = 4c \, \delta^{-1} e^{\frac{\delta|x_0|}{2}}.$$

$\square$

8 Pointwise Decay for ε and Review of the H^1 Well-Posedness Theory

We start this section with the main statement on the pointwise decay of $\varepsilon(x, y)$ for $x > 0$.

Lemma 8.1 (Pointwise Decay) *There exists $\sigma_0 > 0$ (large), $\delta_0 > 0$ (small) and $K > 0$ (large) such that the following holds for any $0 < \delta \le \delta_0$ and $\sigma \ge \sigma_0$.*
Recall $\varepsilon(s, x, y)$ solving Eq. (4.23) (with $y = 0$), i.e.,

$$\partial_s \varepsilon = (L\varepsilon)_x + \frac{\lambda_s}{\lambda}(\Lambda Q + \Lambda \varepsilon) + (\frac{x_s}{\lambda} - 1)(Q_x + \varepsilon_x) - 3(Q\varepsilon^2)_x - (\varepsilon^3)_x, \quad (8.72)$$

and suppose there exists $\delta > 0$ such that

$$\|\varepsilon(s)\|_{H^1_{xy}} + \left|\frac{\lambda_s(s)}{\lambda(s)}\right| + |\lambda(s) - 1| + \left|\frac{x_s(s)}{\lambda(s)} - 1\right| \lesssim \delta \qquad (8.73)$$

for all times $s \ge 0$.
Moreover, assume that for $x > K$ and $y \in \mathbb{R}$,

$$|\varepsilon(0, x, y)| \lesssim \delta \langle x \rangle^{-\sigma}. \qquad (8.74)$$

Then for $x > K$ and $y \in \mathbb{R}$, we have

$$|\varepsilon(s,x,y)| \lesssim \delta \begin{cases} s^{-7/12}\langle x\rangle^{-\sigma+\frac{7}{4}} & \text{if } 0 \le s \le 1 \\ \langle x\rangle^{-\frac{2}{3}\sigma+\frac{3}{4}} & \text{if } s \ge 1. \end{cases} \quad (8.75)$$

Remark 8.2 Note that we can fix K in Lemma 8.1 so that

$$\langle K\rangle^{-1} \le \delta_0, \quad (8.76)$$

which also implies that $e^{-K/2} \le \delta_0$.

Remark 8.3 Rescale the time s back to t via $\dfrac{ds}{dt} = \lambda^{-3}$, and define

$$\eta(t,x,y) = \lambda^{-1}\varepsilon(s(t), \lambda^{-1}(x+K), \lambda^{-1}y),$$

where $x_t = \frac{d}{dt}x(t)$ with $x(t)$ being the spatial shift. Then η solves

$$\partial_t \eta - \partial_x[(-\Delta + x_t)\eta] = F, \qquad F = f_1 + \partial_x f_2, \quad (8.77)$$

where

$$\begin{aligned} f_1 &= -(\lambda^{-1})_t \partial_{\lambda^{-1}}\tilde{Q}, \\ f_2 &= +(x_t - 1)\tilde{Q} - 3\tilde{Q}^2 \eta - 3\tilde{Q}\eta^2 - \eta^3. \end{aligned} \quad (8.78)$$

Here, $\tilde{Q}(x,y) = \lambda^{-1}Q(\lambda^{-1}(x+K), \lambda^{-1}y)$. Note that since $|Q(y_1,y_2)| \lesssim \langle \vec{y}\rangle^{-1/2}e^{-|\vec{y}|}$, we have $|\tilde{Q}(x,y)| \le \delta_0$ for $x > 0$ by (8.76). Also for $x > 0$ and $y \in \mathbb{R}$, we have

$$|\eta(0,x,y)| \lesssim \delta\langle x\rangle^{-\sigma}.$$

To show (8.75), it suffices to prove, for $x > 0$ and $y \in \mathbb{R}$,

$$|\eta(t,x,y)| \lesssim \delta \begin{cases} t^{-7/12}\langle x\rangle^{-\sigma+\frac{7}{4}} & \text{if } 0 \le t \le 1 \\ \langle x\rangle^{-\frac{2}{3}\sigma+\frac{3}{4}} & \text{if } t \ge 1. \end{cases} \quad (8.79)$$

Let $S(t,t_0)\phi$ be the solution $\rho(t,x,y)$ to the homogeneous problem

$$\begin{cases} \partial_t \rho - \partial_x(-\Delta + x_t)\rho = 0 \\ \rho(t_0,x,y) = \phi(x,y). \end{cases}$$

Then

$$\eta(t,x,y) = [S(t,0)\phi](x,y,t) + \int_0^t S(t,t')F(\bullet,\bullet,t')(x,y)\,dt'.$$

Moreover,

$$S(t, t_0)\phi(x, y) = \int A(x', y', t - t_0)\phi(x + x(t) - x', y - y')\, dx'\, dy', \quad (8.80)$$

where

$$A(x, y, t) = \iint_{\mathbb{R}^2} e^{i(t\xi^3 + t\xi\eta^2 + x\xi + y\eta)}\, d\xi\, d\eta.$$

We use the notation L_T^p as shorthand for $L_{[0,T]}^p$.

Theorem 8.4 (Following Faminskii [8], Linares-Pastor [20]) *For given functions $x(t)$, $\lambda(t)$, initial data $\eta_0(x, y) \in H_{xy}^1$ and $T > 0$, there exists a unique solution η to (8.77), (8.78) such that $\eta \in C([0, T]; H_{xy}^1)$ and $\eta(x + x(t), y, t) \in L_x^4 L_{yT}^\infty$.*

This type of uniqueness is called *conditional*, since it is only known to hold with the auxiliary condition $\eta(x + x(t), y, t) \in L_x^4 L_{yT}^\infty$.

Let $u(t, x, y) = \eta(t, x + x(t), y) + \tilde{Q}(x, y)$. Then u solves

$$\partial_t u + \partial_x(\Delta u + u^3) = 0. \quad (8.81)$$

For existence of η, it suffices to prove the existence of u solving (8.81) such that $u \in C([0, T]; H_{xy}^1) \cap L_x^4 L_{yT}^\infty$. Moreover, given two solutions η_1 and η_2, we can define corresponding u_1 and u_2 as above. Provided we have proved the uniqueness of solutions u to (8.81), we have $u_1 = u_2$, which implies $\eta_1 = \eta_2$. The existence and uniqueness of solutions u to (8.81) in the function class $C([0, T]; H_{xy}^1) \cap L_x^4 L_{yT}^\infty$ was established by Linares-Pastor [20]. It is proved by a contraction argument using the following estimates. Let $U(t)\phi$ denote the solution to the linear homogenous problem

$$\begin{cases} \partial_t \rho + \partial_x \Delta \rho = 0 \\ \rho(t_0, x, y) = \phi(x, y). \end{cases}$$

Then

$$u(t) = U(t)\phi + \int_0^t U(t - t')\partial_x[u(t')^3]\, dt'.$$

Lemma 8.5 (Linear Homogeneous Estimates) *We have*

(1) $\|U(t)\phi\|_{L_T^\infty H_{xy}^1} \lesssim \|\phi\|_{H_{xy}^1}$,
(2) $\|\partial_x U(t)\phi\|_{L_x^\infty L_{yT}^2} \lesssim \|\phi\|_{L_{xy}^2}$.

For $0 < T \leq 1$,

(3) $\|U(t)\phi\|_{L_x^4 L_{yT}^\infty} \lesssim \|\phi\|_{H_{xy}^1}$.

Proof The first estimate is a standard consequence of Plancherel and Fourier representation of the solution. The second estimate (local smoothing) is Faminskii [8, Theorem 2.2 on p. 1004]. The third estimate (maximal function estimate) is a special case ($s = 1$) of Faminskii [8, Theorem 2.4 on p. 1007]. All of these estimates are used by Linares-Pastor [20], and quoted as Lemma 2.7 on p. 1326 of that paper. □

Lemma 8.6 (Linear Inhomogeneous Estimates) *For $0 < T \leq 1$,*

(1) $\left\| \int_0^t U(t-t') \partial_x f(t') \, dt' \right\|_{L_T^\infty H_{xy}^1 \cap L_x^4 L_{yT}^\infty} \lesssim \|\partial_x f\|_{L_x^1 L_{yT}^2} + \|\partial_y f\|_{L_x^1 L_{yT}^2}$,

(2) $\left\| \int_0^t U(t-t') f(t') \, dt' \right\|_{L_T^\infty H_{xy}^1 \cap L_x^4 L_{yT}^\infty} \lesssim \|f\|_{L_T^1 H_{xy}^1}$.

Proof These follow from Lemma 8.5 by duality, T^*T, and the Christ-Kiselev lemma. □

Let us now summarize the proof of H_{xy}^1 local well-posedness following from these estimates. We note that Linares-Pastor [20], in fact, achieved local well-posedness in H_{xy}^s for $s > \frac{3}{4}$, although we only need the $s = 1$ case. Let X be the R-ball in the Banach space $C([0, T]; H_{xy}^1) \cap L_x^4 L_{yT}^\infty$, for T and R yet to be chosen. Consider the mapping Λ defined for $u \in X$ by

$$\Lambda u = U(t)\phi + \int_0^t U(t-t') \partial_x [u(t')^3] \, dt'.$$

Then we claim that for suitably chosen $R > 0$ and $T > 0$, we have $\Lambda : X \to X$ and Λ is a contraction. Indeed, by the estimates in Lemmas 8.5 and 8.6, we have

$$\|\Lambda u\|_X \lesssim \|\phi\|_{H_{xy}^1} + \|\partial_x(u^3)\|_{L_x^1 L_{yT}^2} + \|\partial_y(u^3)\|_{L_x^1 L_{yT}^2}.$$

We estimate

$$\|u_x u^2\|_{L_x^1 L_{yT}^2} \lesssim \|u_x\|_{L_x^2 L_{yT}^2} \|u\|_{L_x^4 L_{yT}^\infty}^2 \leq T^{1/2} \|u_x\|_{L_T^\infty L_{xy}^2} \|u\|_{L_x^4 L_{yT}^\infty}^2,$$

and similarly, for the x derivative replaced by the y-derivative. Consequently,

$$\|\Lambda u\|_X \leq C\|\phi\|_{H_{xy}^1} + CT^{1/2}\|u\|_X^3$$

for some constant $C > 0$. By similar estimates,

$$\|\Lambda u_2 - \Lambda u_1\|_X \leq CT^{1/2} \|u_2 - u_1\|_X \max(\|u_1\|_X, \|u_2\|_X)^2.$$

We can thus take $R = 2C\|\phi\|_{H_{xy}^1}$ and $T > 0$ such that $CR^2 T^{1/2} = \frac{1}{2}$ to obtain that $\Lambda : X \to X$ and is a contraction. The fixed point is the desired solution.

For the uniqueness statement, we can take $R \geq 2C\|\phi\|_{H^1_{xy}}$ large enough so that the two given solutions u_1, u_2 lie in X, and then take T so that $CR^2 T^{1/2} = \frac{1}{2}$. Then u_1 and u_2 are both fixed points of Λ in X, and since fixed points of a contraction are unique, $u_1 = u_2$.

This gives the local well-posedness in H^1_{xy}. Global well-posedness follows from the energy conservation.

9 Fundamental Solution Estimates

Recall the solution (8.80) and its kernel A. The first basic step is to get the estimates on this kernel, which is given in the following statement.

Proposition 9.1 (Fundamental Solution Estimate) *Let $t > 0$ and consider*

$$A(x, y, t) = \iint_{\mathbb{R}^2} e^{i(t\xi^3 + t\xi\eta^2 + x\xi + y\eta)} \, d\xi\, d\eta.$$

Let $\lambda = |x|^{3/2} t^{-1/2} > 0$ and $z = y|x|^{-1}$. Then for $x > 0$

$$|A(x, y, t)| \lesssim_{\alpha,\beta} t^{-2/3} \begin{cases} \langle\lambda\rangle^{-\alpha} & \text{if } |z| \leq 4, \text{ for any } \alpha \geq 0 \\ \langle\lambda |z|^{3/2}\rangle^{-\beta} & \text{if } |z| \geq 4, \text{ for any } \beta \geq 0. \end{cases}$$

If $x < 0$, then

$$|A(x, y, t)| \lesssim_\beta t^{-2/3} \begin{cases} \langle\lambda\rangle^{-1/6} & \text{if } |z| \leq 4 \\ \langle\lambda |z|^{3/2}\rangle^{-\beta} & \text{if } |z| \geq 4, \text{ for any } \beta \geq 0. \end{cases}$$

We give a proof of Proposition 9.1 based on factoring of A into the product of two Airy functions, which is possible in two dimensions. It is a short proof and gives estimates actually sharper than those claimed in the proposition statement. We remark that it is possible to obtain Proposition 9.1 by direct oscillatory integral methods (non-stationary and stationary phase), although it involves tedious calculations. The advantage of the oscillatory integral approach would be that such a proof could be generalized to higher dimensions.

Proof of Proposition 9.1 Making the change of variable $\xi \mapsto \frac{|x|^{1/2}}{t^{1/2}}\xi$ and $\eta \mapsto \frac{|x|^{1/2}}{t^{1/2}}\eta$, we obtain

$$A(x, y, t) = |x| t^{-1} \iint_{\xi,\eta} e^{i\lambda(\xi^3 + \xi\eta^2 + (\text{sgn } x)\xi + z\eta)} \, d\xi\, d\eta,$$

where $\lambda = |x|^{3/2} t^{-1/2} > 0$ and $z = y|x|^{-1} \in \mathbb{R}$, as given in the proposition statement. Rewriting $|x| t^{-1} = \lambda^{2/3} t^{-2/3}$, this becomes

$$A(x, y, t) = t^{-2/3} \lambda^{2/3} B_{\operatorname{sgn} x}(\lambda, z),$$

where

$$B_\pm(\lambda, z) = \iint_{\xi,\eta} e^{i\lambda(\xi^3 + \xi\eta^2 \pm \xi + z\eta)}\, d\xi\, d\eta.$$

To obtain symmetric conditions in the estimates, it is convenient to change variable $\xi \mapsto \frac{\xi}{\sqrt{3}}$, and make the inconsequential replacements $\frac{\lambda}{\sqrt{3}} \mapsto \lambda$ and $\sqrt{3} z \mapsto z$. This gives

$$B_\pm(\lambda, z) = \iint_{\xi,\eta} e^{i\lambda\phi(\xi,\eta;z)}\, d\xi\, d\eta,$$

where

$$\phi(\xi, \eta; z) = \tfrac{1}{3}\xi^3 + \xi\eta^2 \pm \xi + z\eta.$$

Now the goal is to prove the estimates

$$|B_+(\lambda, z)| \lesssim_{\alpha,\beta} \lambda^{-2/3} \begin{cases} \langle\lambda\rangle^{-\alpha} & \text{if } |z| \le 4, \text{ for any } \alpha \ge 0 \\ \langle\lambda|z|^{3/2}\rangle^{-\beta} & \text{if } |z| \ge 4, \text{ for any } \beta \ge 0 \end{cases} \quad (9.82)$$

and

$$|B_-(\lambda, z)| \lesssim_\beta \lambda^{-2/3} \begin{cases} \langle\lambda\rangle^{-1/6} & \text{if } |z| \le 4 \\ \langle\lambda|z|^{3/2}\rangle^{-\beta} & \text{if } |z| \ge 4, \text{ for any } \beta \ge 0. \end{cases} \quad (9.83)$$

Next, make the change of variable $\xi \mapsto \tfrac{1}{2}(\xi + \eta)$, $\eta \mapsto \tfrac{1}{2}(\xi - \eta)$, and replace $\tfrac{\lambda}{2} \mapsto \lambda$, which factors the exponential to obtain the splitting

$$B_\pm(\lambda, z) = \int_\xi e^{i\lambda(\tfrac{1}{3}\xi^3 + (z\pm 1)\xi)}\, d\xi \int_\eta e^{i\lambda(\tfrac{1}{3}\eta^3 + (-z\pm 1)\eta)}\, d\eta.$$

In terms of the Airy function $\mathcal{A}(x) = \int e^{i(\tfrac{1}{3}\xi^3 + x\xi)}\, d\xi$, this is

$$B_\pm(\lambda, z) = \lambda^{-2/3} \mathcal{A}(\lambda^{2/3}(z \pm 1)) \mathcal{A}(\lambda^{2/3}(-z \pm 1)). \quad (9.84)$$

We note that in either $+$ or $-$ case, if $|z| > 2$, then either $(z \pm 1) \geq \frac{1}{2}|z|$ or $(-z \pm 1) \geq \frac{1}{2}|z|$. Hence, by the strong decay of the Airy function on the right, we obtain

$$|B_\pm(\lambda, z)| \lesssim \lambda^{-2/3} \langle \lambda^{2/3}|z|\rangle^{-k}$$

for any $k \geq 0$. This gives the second half of the estimates (9.82) and (9.83).

For $|z| < 4$, first consider the $+$ case. If $0 \leq z < 4$, then $z + 1 > 1$, so we can use $|\mathcal{A}(\lambda^{2/3}(z+1))| \lesssim \langle \lambda^{2/3}\rangle^{-k}$ for any $k \geq 0$, together with the simple estimate $|\mathcal{A}(\lambda^{2/3}(-z+1))| \lesssim 1$, to achieve the first part of (9.82). If $-4 < z \leq 0$, then $-z + 1 > 1$, so we can use $|\mathcal{A}(\lambda^{2/3}(-z+1))| \lesssim \langle \lambda^{2/3}\rangle^{-k}$ for any $k \geq 0$, together with the simple estimate $|\mathcal{A}(\lambda^{2/3}(z+1))| \lesssim 1$, to achieve the first part of (9.82).

For $|z| < 4$, now consider the $-$ case. When $-1 \leq z \leq 1$, both $z - 1 \leq 0$ and $-z - 1 \leq 0$, so the amount of decay we can obtain from the Airy functions is limited. The worst situation is when $z = \pm 1$. For example, when $z = 1$, we have $\mathcal{A}(\lambda^{2/3}(z-1)) = A(0)$ and $|\mathcal{A}(\lambda^{2/3}(-z-1))| \lesssim \langle \lambda^{2/3}\rangle^{-1/4}$. When applied in (9.84), this gives the bound $|B_-(\lambda, z)| \lesssim \lambda^{-5/6}$ for $\lambda > 1$. □

Because of the form of the equation, we also need the x-derivative estimate.

Proposition 9.2 (x-**Derivative Fundamental Solution Estimate**) *Let $t > 0$ and consider*

$$A_x(x, y, t) = \iint_{\mathbb{R}^2} i\xi e^{i(t\xi^3 + t\xi\eta^2 + x\xi + y\eta)} \, d\xi \, d\eta.$$

Let $\lambda = |x|^{3/2} t^{-1/2} > 0$ and $z = y|x|^{-1}$. Then for $x > 0$

$$|A_x(x, y, t)| \lesssim_{\alpha,\beta} t^{-1} \begin{cases} \langle\lambda\rangle^{-\alpha} & \text{if } |z| \leq 4, \text{ for any } \alpha \geq 0 \\ \langle\lambda|z|^{3/2}\rangle^{-\beta} & \text{if } |z| \geq 4, \text{ for any } \beta \geq 0. \end{cases}$$

If $x < 0$, then

$$|A_x(x, y, t)| \lesssim_\beta t^{-1} \begin{cases} \langle\lambda\rangle^{1/6} & \text{if } |z| \leq 4 \\ \langle\lambda|z|^{3/2}\rangle^{-\beta} & \text{if } |z| \geq 4, \text{ for any } \beta \geq 0. \end{cases}$$

Proof Rescale $\xi \mapsto \frac{|x|^{1/2}}{(3t)^{1/2}}\xi$, $\eta \mapsto \frac{|x|^{1/2}}{t^{1/2}}\eta$ to obtain

$$A_x(x, y, t) = \lambda t^{-1} B_\pm(\lambda, z),$$

where $\lambda = |x|^{3/2}(3t)^{-1/2}$ and $z = \sqrt{3}y|x|^{-1}$, and

$$B_\pm(\lambda, z) = \iint i\xi e^{i\lambda\phi_\pm(\xi,\eta,z)} \, d\eta \, d\eta, \qquad \phi_\pm(\xi, \eta; z) = \tfrac{1}{3}\xi^3 + \xi\eta^2 \pm \xi + z\eta.$$

It suffices to prove

$$|B_+(\lambda, z)| \lesssim_{\alpha,\beta} \lambda^{-1} \begin{cases} \langle\lambda\rangle^{-\alpha} & \text{if } |z| \leq 4, \text{ for any } \alpha \geq 0 \\ \langle\lambda|z|^{3/2}\rangle^{-\beta} & \text{if } |z| \geq 4, \text{ for any } \beta \geq 0 \end{cases} \quad (9.85)$$

and

$$|B_-(\lambda, z)| \lesssim_\beta \lambda^{-1} \begin{cases} \langle\lambda\rangle^{1/6} & \text{if } |z| \leq 4 \\ \langle\lambda|z|^{3/2}\rangle^{-\beta} & \text{if } |z| \geq 4, \text{ for any } \beta \geq 0. \end{cases} \quad (9.86)$$

Changing variables again as

$$\xi \mapsto \tfrac{1}{2}(\xi + \eta), \qquad \eta \mapsto \tfrac{1}{2}(\xi - \eta)$$

and replacing $\tfrac{1}{2}\lambda$ by λ (thus, redefining $\lambda = \tfrac{1}{2}|x|^{3/2}(3t)^{-1/2}$), we obtain the factorization

$$8B_\pm(\lambda, z) = \iint i(\xi + \eta) e^{i\lambda(\frac{1}{3}\xi^3 + (\pm 1 + z)\xi)} e^{i\lambda(\frac{1}{3}\eta^3 + (\pm 1 - z)\eta)} d\xi\, d\eta$$

$$= \int i\xi e^{i\lambda(\frac{1}{3}\xi^3 + (\pm 1 + z)\xi)} d\xi \int e^{i\lambda(\frac{1}{3}\eta^3 + (\pm 1 - z)\eta)} d\eta$$

$$+ \int e^{i\lambda(\frac{1}{3}\xi^3 + (\pm 1 + z)\xi)} d\xi \int i\eta e^{i\lambda(\frac{1}{3}\eta^3 + (\pm 1 - z)\eta)} d\eta.$$

Change variables $\xi \mapsto \lambda^{-1/3}\xi$, $\eta \mapsto \lambda^{-1/3}\eta$ to obtain

$$8B_\pm(\lambda, z) = \lambda^{-1} \mathcal{A}'(\lambda^{2/3}(\pm 1 + z)) \mathcal{A}(\lambda^{2/3}(\pm 1 - z))$$

$$+ \lambda^{-1} \mathcal{A}(\lambda^{2/3}(\pm 1 + z)) \mathcal{A}'(\lambda^{2/3}(\pm 1 - z)),$$

where $\mathcal{A}(x) = \int e^{i(\frac{1}{3}\xi^3 + x\xi)} d\xi$ is the Airy function.

If $z > 2$, then $(\pm 1 + z) > \tfrac{1}{2}|z|$ and if $z < -2$, then $(\pm 1 - z) > \tfrac{1}{2}|z|$, so in either case, the strong rightward decay of $\mathcal{A}$ and $\mathcal{A}'$ gives $|B_\pm(\lambda, z)| \lesssim \lambda^{-1}\langle\lambda^{2/3}|z|\rangle^{-k}$ for all $k \geq 0$. Thus, the second parts of the claimed estimates in (9.85) and (9.86) hold.

In the $+$ case, if $0 \leq z < 2$, then $(1 + z) \geq 1$, so we use that $|\mathcal{A}'(\lambda^{2/3}(1 + z))| \lesssim \langle\lambda^{2/3}\rangle^{-k}$, $|\mathcal{A}(\lambda^{2/3}(1 - z))| \lesssim 1$, $|\mathcal{A}(\lambda^{2/3}(1 + z))| \lesssim \langle\lambda^{2/3}\rangle^{-k}$, and $|\mathcal{A}'(\lambda^{2/3}(1 - z))| \lesssim 1$ to achieve $|B_+(\lambda, z)| \lesssim \lambda^{-1}\langle\lambda^{2/3}\rangle^{-k}$ for any $k \geq 0$. On the other hand, if $-2 < z \leq 0$, then $1 - z \geq 1$, so we use that $|\mathcal{A}'(\lambda^{2/3}(1 + z))| \lesssim 1$, $|\mathcal{A}(\lambda^{2/3}(1 - z))| \lesssim \langle\lambda^{2/3}\rangle^{-k}$, $|\mathcal{A}(\lambda^{2/3}(1 + z))| \lesssim 1$, and $|\mathcal{A}'(\lambda^{2/3}(1 - z))| \lesssim \langle\lambda^{2/3}\rangle^{-k}$ to achieve $|B_+(\lambda, z)| \lesssim \lambda^{-1}\langle\lambda^{2/3}\rangle^{-k}$ for any $k \geq 0$. Thus, the first part of the estimate (9.85) holds.

In the $-$ case, if $-1 \leq z \leq 1$, then both $-1 + z \leq 0$ and $-1 - z \leq 0$, and we only have access to the weaker leftward decay estimates for the Airy function. The worst case arises when $z = \pm 1$. For example, if $z = 1$, then $-1 + z = 0$ and $-1 - z = -2$, so $|\mathcal{A}'(\lambda^{2/3}(-1+z))| \lesssim 1$, $|\mathcal{A}(\lambda^{2/3}(-1-z))| \lesssim \langle \lambda^{2/3} \rangle^{-1/4}$, $|\mathcal{A}(\lambda^{2/3}(-1+z))| \lesssim 1$, and $|\mathcal{A}'(\lambda^{2/3}(-1-z))| \lesssim \langle \lambda^{2/3} \rangle^{1/4}$, which gives $|B_-(\lambda, z)| \lesssim \langle \lambda^{2/3} \rangle^{1/4} \sim \langle \lambda \rangle^{1/6}$. The case of $z = -1$ is similar. $\square$

10 Linear Solution Decay Estimates

For this section, we will need the following estimates. Let

$$[\mu] = \begin{cases} \mu & \text{if } \mu > 0 \\ 0+ & \text{if } \mu = 0 \\ 0 & \text{if } \mu < 0. \end{cases}$$

We shall employ the two basic integral estimates (10.87), (10.88) below. Note that (10.87) requires $x > 0$ and restricts the integration to $x' < 0$, but then yields a stronger bound than (10.88) when $\sigma \gg \mu$. In fact, (10.87) even allows $\mu < 0$.

For any $\sigma \in \mathbb{R}$, $\mu \in \mathbb{R}$, $1 - \sigma < \mu$ and $x > 0$

$$\int_{x'=-\infty}^{0} \langle x - x' \rangle^{-\sigma} \langle x' \rangle^{-\mu} \, dx' \lesssim \langle x \rangle^{-\sigma + [1-\mu]}. \tag{10.87}$$

For any $\sigma > 1$, $\mu \geq 0$, and $x \in \mathbb{R}$,

$$\int_{x'=-\infty}^{+\infty} \langle x - x' \rangle^{-\sigma} \langle x' \rangle^{-\mu} \, dx' \lesssim \langle x \rangle^{-\min(\sigma - [1-\mu], \mu)}. \tag{10.88}$$

Let

$$\Phi(x, y, t) = \int A(x', y', t) \phi(x + t - x', y - y') \, dx' \, dy' = (S(t)\phi)(x, y), \tag{10.89}$$

i.e., the unique solution to $\partial_t \Phi + \partial_x (1 - \Delta_{xy}) \Phi = 0$ with initial condition $\Phi(x, y, 0) = \phi(x, y)$. For simplicity we have taken $x(t) = t$.

The proposition below gives rightward x-decay estimates for this linear solution.

Proposition 10.1 (Linear Solution Estimates) *Let $\sigma > \frac{9}{4}$, and suppose that*

$$\text{for } x > 0, \quad |\phi(x, y)| \leq C_1 \langle x \rangle^{-\sigma} \tag{10.90}$$

and

$$\|\langle x \rangle^{-1} \phi(x, y)\|_{L^2_{y \in \mathbb{R}, x < 0}} \leq C_1. \tag{10.91}$$

Then for t > 0,

$$\text{for } x > 0, \quad |\Phi(x, y, t)| \lesssim C_1 \begin{cases} t^{-7/12} \langle x \rangle^{-\sigma + \frac{7}{4}} & \text{if } t < 1 \\ t^{-13/12} \langle x \rangle^{-\tilde{\sigma}} & \text{if } t > 1, \end{cases}$$

where $\tilde{\sigma} = \min(\frac{2}{3}\sigma - \frac{3}{4}, \sigma - \frac{9}{4})$.

Note that for $\sigma > \frac{9}{2}$, we have $\sigma - \frac{9}{4} > \frac{2}{3}\sigma - \frac{3}{4}$, and thus, for $\tilde{\sigma} = \frac{2}{3}\sigma - \frac{3}{4}$. We also remark that the time decay factor $t^{-13/12}$ for $t > 1$ can be replaced by any negative power of t, provided the definition of $\tilde{\sigma}$ is suitably altered. We chose $t^{-13/12}$, since it is < -1, thus, integrable (over $t \geq 1$), and this integrability is needed in the Duhamel estimates.

Proof By linearity, it suffices to assume that $C_1 \leq 1$. Recall that we are assuming $x > 0$ and $t > 0$. From Proposition 9.1, we have

$$|A(x', y', t)| \lesssim t^{-2/3} \lambda^{-\alpha} (\lambda |z|^{3/2})^{-\beta} = t^{-\frac{2}{3} + \frac{1}{2}\alpha + \frac{1}{2}\beta} |x'|^{-\frac{3}{2}\alpha} |y'|^{-\frac{3}{2}\beta} \qquad (10.92)$$

with different constraints on the allowed values (and optimal values) of α and β depending upon whether

- $|z| < 4$ or $|z| > 4$,
- $\lambda < 1$ or $\lambda > 1$,
- $x' < 0$ or $x' > 0$

This is summarized in the following two tables:

$x' > 0$	$\lambda < 1$	$\lambda > 1$	$x' < 0$	$\lambda < 1$	$\lambda > 1$
$|z| < 4$	$\alpha = 0, \beta = 0$	$\alpha \geq 0, \beta = 0$	$|z| < 4$	$\alpha = 0, \beta = 0$	$\alpha = \frac{1}{6}, \beta = 0$
$|z| > 4$	$\alpha = 0, \beta \geq 0$	$\alpha \geq 0, \beta \geq 0$	$|z| > 4$	$\alpha = 0, \beta \geq 0$	$\alpha \geq 0, \beta \geq 0$

From (10.89), we see that we need to further subdivide according to $-\infty < x' < x + t$ and $x' > x + t$. When $-\infty < x' < x + t$, we have $x + t - x' > 0$, and we can use (10.90), and when $x' > x + t$, we have $x + t - x' < 0$, and we can only use that $\phi \in L^2_{xy}$.

Below our decomposition of the (x', y') integration space is given as 9 different regions. Each of the regions can be further divided according to whether $|x'|$ and $|y'|$ are < 1 or > 1. We label the corresponding subregions as $--$, $-+$, $+-$, and $++$, as follows:

- $--$ corresponds to $|x'| < 1$ and $|y'| < 1$
- $-+$ corresponds to $|x'| < 1$ and $|y'| > 1$
- $+-$ corresponds to $|x'| > 1$ and $|y'| < 1$
- $++$ corresponds to $|x'| > 1$ and $|y'| > 1$.

Thus,
$$\Phi = \Phi_1 + \cdots + \Phi_9,$$

where Φ_j denotes the convolution integral in (10.89) restricted to the region under consideration. We further use the decompositions

$$\Phi_j = \Phi_{j--} + \Phi_{j-+} + \Phi_{j+-} + \Phi_{j++}$$

as needed.

In Regions 1–5, we begin as follows: From (10.89), (10.90), and (10.92)

$$|\Phi_j(x, y, t)| \lesssim t^{-\frac{2}{3}+\frac{1}{2}\alpha+\frac{1}{2}\beta} \iint_{(x',y')\in R} |x'|^{-\frac{3}{2}\alpha}\langle x+t-x'\rangle^{-\sigma}|y'|^{-\frac{3}{2}\beta}\,dx'\,dy', \tag{10.93}$$

where R denotes the subregion of (x', y') space under consideration.

In Regions 6–9, the decay hypothesis (10.90) is not available, so we start from (10.89) with Cauchy-Schwarz in (x', y')

$$|\Phi_j(x, y, t)| \leq \left(\iint_{(x',y')\in R} |A(x', y', t)|^2\langle x+t-x'\rangle^2\,dx'\,dy'\right)^{1/2} \|\langle x\rangle^{-1}\phi(x, y)\|_{L^2_{y\in\mathbb{R},x<0}}.$$

Since $\|\langle x\rangle^{-1}\phi\|_{L^2_{y\in\mathbb{R},x<0}} \leq 1$, this term can be dropped above and (10.92) yields

$$|\Phi_j(x, y, t)| \leq t^{-\frac{2}{3}+\frac{1}{2}\alpha+\frac{1}{2}\beta} \left(\iint_{(x',y')\in R} |x'|^{-3\alpha}|y'|^{-3\beta}\langle x+t-x'\rangle^2\,dx'\,dy'\right)^{1/2}. \tag{10.94}$$

An argument used repeatedly below is

$$|x'| < t^{1/3} \implies \langle x+t-x'\rangle \sim \langle x+t\rangle. \tag{10.95}$$

Indeed, if $t < 8$, then $|x'| < t^{1/3} < 2$, so $\langle x+t-x'\rangle \sim \langle x+t\rangle$. On the other hand, if $t > 8$, then $|x'| < t^{1/3} \ll t \leq x+t$, so again $\langle x+t-x'\rangle \sim \langle x+t\rangle$.

Finally, we remark that it is Region 2 below that seems to limit $t^{-7/12}$ as the least singular power of t for $0 < t < 1$.

1. **Region** $x' < x+t$, $|x'| < t^{1/3}$, $|y'| < 4|x'|$. Here, $\lambda < 1$, $|z| < 4$. We take $\alpha = 0$, $\beta = 0$. From (10.95), we have $\langle x+t-x'\rangle \sim \langle x+t\rangle$. Starting with (10.93), we get

$$|\Phi_1(x, y, t)| \lesssim t^{-2/3}\langle x+t\rangle^{-\sigma} \int_{x',|x'|<t^{1/3}} \int_{y',|y'|<4|x'|} dy'\,dx' \lesssim \langle x+t\rangle^{-\sigma}.$$

2. **Region** $x' < 0$, $|x'| > t^{1/3}$, $|y'| < 4|x'|$. Here, $x' < 0$, $\lambda > 1$, $|z| < 4$. We take $\beta = 0$, and are limited to $\alpha = \frac{1}{6}$. Starting with (10.93),

$$|\Phi_2(x,y,t)| \lesssim t^{-7/12} \int_{x', -\infty < x' < 0} \langle x+t-x' \rangle^{-\sigma} |x'|^{-1/4} \int_{y', |y'| < 4|x'|} dy'\, dx'$$

$$\lesssim t^{-7/12} \int_{x', -\infty < x' < 0} \langle x+t-x' \rangle^{-\sigma} |x'|^{3/4}\, dx'.$$

By (10.87) with $\mu = -\frac{3}{4}$, provided $\sigma > \frac{7}{4}$, we have

$$\lesssim t^{-7/12} \langle x+t \rangle^{-\sigma + \frac{7}{4}}.$$

For $t \geq 1$, we note that $t^{-7/12} \langle x+t \rangle^{-\sigma + \frac{7}{4}} \leq t^{-13/12} \langle x \rangle^{-\sigma + \frac{9}{4}}$.

3. **Region** $0 < x' < x+t$, $|x'| > t^{1/3}$, $|y'| < 4|x'|$. Here, $x' > 0$, $\lambda > 1$, $|z| < 4$. We take $\beta = 0$. For $|x'| < 1$ (the $-*$ subregion), we take $\alpha = 0$, but for $|x'| > 1$ (the $+*$ subregion), we take $\alpha \gg 1$.

For $|x'| < 1$, we take $\alpha = \frac{1}{6}$ and use that $\langle x+t-x' \rangle \sim \langle x+t \rangle$ to obtain, starting with (10.93),

$$|\Phi_{3-*}(x,y,t)| \lesssim t^{-\frac{7}{12}} \langle x+t \rangle^{-\sigma} \int_{x', |x'| < 1} |x'|^{-1/4} \int_{y', |y'| < 4|x'|} dy'\, dx' \sim t^{-7/12} \langle x+t \rangle^{-\sigma}.$$

In the case $t \geq 1$, we note that $t^{-7/12} \langle x+t \rangle^{-\sigma} \leq t^{-13/12} \langle x \rangle^{-\sigma + \frac{1}{2}}$.

For $|x'| > 1$, we take $\alpha \gg 1$, starting with (10.93),

$$|\Phi_{3+*}(x,y,t)| \lesssim t^{-\frac{2}{3} + \frac{1}{2}\alpha} \int_{x', |x'| > 1} |x'|^{-\frac{3}{2}\alpha} \langle x+t-x' \rangle^{-\sigma} \int_{y', |y'| < 4|x'|} dy'\, dx'$$

$$\lesssim t^{-\frac{2}{3} + \frac{1}{2}\alpha} \int_{x', |x'| > 1} |x'|^{1 - \frac{3}{2}\alpha} \langle x+t-x' \rangle^{-\sigma}\, dx'.$$

For $\sigma > 1$, taking $\alpha = \frac{2}{3}(\sigma + 1)$ gives by (10.88)

$$\lesssim t^{-\frac{1}{3} + \frac{1}{3}\sigma} \langle x+t \rangle^{-\sigma}.$$

For $t \geq 1$, we decompose the exponent as $\langle x+t \rangle^{-\sigma} = \langle x+t \rangle^{\frac{1}{3} - \frac{1}{3}\sigma - \frac{13}{12}} \langle x+t \rangle^{\frac{3}{4} - \frac{2}{3}\sigma}$, and use $t \geq 1$ and $x \geq 0$ to obtain the bound by $t^{-13/12} \langle x \rangle^{\frac{3}{4} - \frac{2}{3}\sigma}$.

4. **Region** $-\infty < x' < x+t$, $|x'| < t^{1/3}$, **and** $|y'| > 4|x'|$. Here, $|z| > 4$ and $\lambda < 1$. By (10.95), we have $\langle x+t-x' \rangle \sim \langle x+t \rangle$. We take $\alpha = 0$ and any $\frac{2}{3} < \beta < \frac{4}{3}$ (so that $-2 < -\frac{3}{2}\beta < -1$) and starting from (10.93), we have

$$|\Phi_4(x,y,t)| \lesssim t^{-\frac{2}{3} + \frac{1}{2}\beta} \langle x+t \rangle^{-\sigma} \int_{x', |x'| < t^{1/3}} \int_{y', |y'| > |x'|} |y'|^{-\frac{3}{2}\beta}\, dy'\, dx'$$

$$\lesssim t^{-\frac{2}{3}+\frac{1}{2}\beta}\langle x+t\rangle^{-\sigma}\int_{x',\,|x'|<t^{1/3}}|x'|^{-\frac{3}{2}\beta+1}\,dx'\lesssim \langle x+t\rangle^{-\sigma},$$

where, in the last step, we used that $-\frac{3}{2}\beta+1>-1$.

5. **Region** $-\infty<x'<x+t$, $|x'|>t^{1/3}$, **and** $|y'|>4|x'|$. Here, $|z|>4$ and $\lambda>1$. Any choice of $\alpha,\beta\geq 0$ is permitted in (10.92).

For $|x'|<1$ and $|y'|<1$ (the $--$ subregion), we take $\alpha=\frac{1}{6}$ and $\beta=0$. Since $|x'|<1$, we have $\langle x+t-x'\rangle\sim\langle x+t\rangle$. Starting from (10.93), we get

$$|\Phi_{5--}(x,y,t)|\lesssim t^{-\frac{7}{12}}\langle x+t\rangle^{-\sigma}\int_{x',\,|x'|<1}|x'|^{-1/4}\int_{y',\,|y'|<1}dy'\,dx'\lesssim t^{-7/12}\langle x+t\rangle^{-\sigma}.$$

For $|x'|<1$ and $|y'|>1$ (the $-+$ subregion), we take $\alpha=0$ and $\beta=\frac{2}{3}+$. Since $|x'|<1$, we have $\langle x+t-x'\rangle\sim\langle x+t\rangle$. Starting from (10.93),

$$|\Phi_{5-+}(x,y,t)|\lesssim t^{-\frac{2}{3}+\frac{1}{2}\beta}\langle x+t\rangle^{-\sigma}\int_{x',\,|x'|<1}dx'\int_{y',\,|y'|>1}|y'|^{-\frac{3}{2}\beta}\,dy'\lesssim t^{-\frac{1}{3}+}\langle x+t\rangle^{-\sigma}.$$

For $|x'|>1$ and $|y'|<1$ (the $+-$ subregion), we take $\alpha\gg 1$ and $\beta=0$. Starting from (10.93),

$$|\Phi_{5+-}(x,y,t)|\lesssim t^{-\frac{2}{3}+\frac{1}{2}\alpha}\int_{x',\,|x'|>1}|x'|^{-\frac{3}{2}\alpha}\langle x+t-x'\rangle^{-\sigma}\,dx'\int_{y',\,|y'|<1}dy'.$$

For $\sigma>1$, take $\alpha=\frac{2}{3}\sigma$ and apply (10.88) to obtain

$$\lesssim t^{-\frac{2}{3}+\frac{1}{3}\sigma}\langle x+t\rangle^{-\sigma}.$$

For $t\geq 1$, we decompose as $\langle x+t\rangle^{-\sigma}=\langle x+t\rangle^{\frac{2}{3}-\frac{1}{3}\sigma-\frac{13}{12}}\langle x+t\rangle^{\frac{5}{12}-\frac{2}{3}\sigma}$, and using $x\geq 0$, we obtain the bound $t^{-13/12}\langle x\rangle^{\frac{5}{12}-\frac{2}{3}\sigma}$.

For $|x'|>1$ and $|y'|>1$ (the $++$ subregion), we take $\alpha\gg 1$ and any $\frac{2}{3}<\beta<\frac{4}{3}$ (so that $-2<-\frac{3}{2}\beta<-1$ and $-1<-\frac{3}{2}\beta+1<0$). Starting from (10.93), we obtain

$$|\Phi_{5++}(x,y,t)|\lesssim t^{-\frac{2}{3}+\frac{1}{2}\alpha+\frac{1}{2}\beta}\int_{x',\,|x'|>1}|x'|^{-\frac{3}{2}\alpha}\langle x+t-x'\rangle^{-\sigma}$$

$$\int_{y',\,|y'|>\max(1,|x'|)}|y'|^{-\frac{3}{2}\beta}\,dy'\,dx'$$

$$\lesssim t^{-\frac{2}{3}+\frac{1}{2}\alpha+\frac{1}{2}\beta}\int_{x',\,|x'|>1}|x'|^{1-\frac{3}{2}\alpha-\frac{3}{2}\beta}\langle x+t-x'\rangle^{-\sigma}\,dx'.$$

Take α so that $\alpha+\beta = \frac{2}{3}(\sigma+1)$ (and hence $1-\frac{3}{2}\alpha-\frac{3}{2}\beta = -\sigma$) and apply (10.88) to obtain

$$\lesssim t^{-\frac{1}{3}+\frac{1}{3}\sigma}\langle x+t\rangle^{-\sigma}.$$

For $t \geq 1$, decompose $\langle x+t\rangle^{-\sigma} = \langle x+t\rangle^{\frac{1}{3}-\frac{1}{3}\sigma-\frac{13}{12}}\langle x+t\rangle^{\frac{3}{4}-\frac{2}{3}\sigma}$ to obtain the bound $t^{-13/12}\langle x\rangle^{\frac{3}{4}-\frac{2}{3}\sigma}$.

Recall that for Regions 6–9 below, we must use (10.94). Thus, we need to recover the $\langle x+t\rangle^{-\sigma}$ decay factor from the constraint $x' > x+t$ and the decay on $A(x', y', t)$.

6. **Region $x' > x+t$, $|x'| < t^{1/3}$, and $|y'| < 4|x'|$.** Here, $|z| < 4$ and $\lambda < 1$, so $\alpha = 0$ and $\beta = 0$. Note that the constraints imply that $x+t < t^{1/3}$. Since $x+t < t^{1/3}$, it follows that $t < t^{1/3}$, from which we conclude that $t < 1$. Also from $x+t < t^{1/3}$, we conclude that $x < t^{1/3}$, and since $t < 1$, this implies $x < 1$. Consequently, $\langle x+t-x'\rangle \sim 1$.

Starting from (10.94), we obtain

$$|\Phi_6(x,y,t)| \lesssim t^{-2/3}\left(\int_{x', |x'|<t^{1/3}}\int_{y', |y'|<4|x'|} dy' dx'\right)^{1/2} \lesssim t^{-1/3}.$$

7. **Region $x' > x+t$, $|x'| > t^{1/3}$, and $|y'| < 4|x'|$.** Here, $|z| < 4$ and $\lambda > 1$, so we take $\beta = 0$, and we are allowed any $\alpha \geq 0$.

If $|x'| < 1$, we take $\alpha = \frac{1}{6}$. Since $|x'| < 1$, we have $x+t < 1$, so $\langle x+t-x'\rangle \sim 1$, $x \leq 1$, and $t \leq 1$. By (10.94),

$$|\Phi_{7-*}(x,y,t)| \lesssim t^{-\frac{7}{12}}\left(\int_{x', |x'|<1}|x'|^{-1/2}\int_{y', |y'|<4|x'|} dy' dx'\right)^{1/2} \lesssim t^{-7/12}$$

If $|x'| > 1$, we take $\alpha \gg 1$. Starting from (10.94), we get

$$|\Phi_{7+*}(x,y,t)| \lesssim t^{-\frac{2}{3}+\frac{1}{2}\alpha}\left(\int_{x', x'>\max(1,x+t)}|x'|^{-3\alpha}\langle x+t-x'\rangle^2\int_{y', |y'|<4|x'|} dy' dx'\right)^{1/2}$$

By changing variable $\tilde{x} = x+t-x'$,

$$\lesssim t^{-\frac{2}{3}+\frac{1}{2}\alpha}\left(\int_{\tilde{x}<0}\langle x+t-\tilde{x}\rangle^{-3\alpha+1}\langle \tilde{x}\rangle^2 d\tilde{x}\right)^{1/2}$$

By (10.87),

$$\lesssim t^{-\frac{2}{3}+\frac{1}{2}\alpha}\langle x+t\rangle^{-\frac{3}{2}\alpha+2}.$$

With $\alpha = \frac{2}{3}\sigma + \frac{4}{3}$, this becomes

$$\lesssim t^{\frac{1}{3}\sigma}\langle x+t\rangle^{-\sigma},$$

which is suitable for $t < 1$. For $t > 1$, first decompose $\langle x+t \rangle^{-\frac{3}{2}\alpha+2} = \langle x+t\rangle^{\frac{2}{3}-\frac{1}{2}\alpha-\frac{13}{12}}\langle x+t\rangle^{2+\frac{5}{12}-\alpha}$, which gives a bound by $t^{-13/12}\langle x\rangle^{2+\frac{5}{12}-\alpha}$. We can then take $\alpha = \sigma + 2 + \frac{5}{12}$.

8. **Region $x' > x+t$, $|x'| < t^{1/3}$, and $|y'| > 4|x'|$.** Here, $|z| > 4$ and $\lambda < 1$, so we take $\alpha = 0$. Note that the constraints imply that $x+t < t^{1/3}$. Since $x+t < t^{1/3}$, it follows that $t < t^{1/3}$, from which we conclude that $t < 1$. Also from $x+t < t^{1/3}$, we conclude that $x < t^{1/3}$ and since $t < 1$, this implies $x < 1$. Consequently, $\langle x+t-x'\rangle \sim 1$.

For $|y'| < 1$ we take $\beta = 0$. From (10.94), we have

$$|\Phi_{8*-}(x,y,t)| \lesssim t^{-\frac{2}{3}}\left(\int_{x',\,|x'|<t^{1/3}}\int_{y',\,|y'|<1} dy'\,dx'\right)^{1/2} \lesssim t^{-1/2}.$$

For $|y'| > 1$, we take any $\frac{1}{3} < \beta < \frac{2}{3}$ (so that $-3\beta < -1$ and $-1 < -3\beta + 1$). From (10.94), we have

$$|\Phi_{8*+}(x,y,t)| \lesssim t^{-\frac{2}{3}+\frac{1}{2}\beta}\left(\int_{x',\,|x'|<t^{1/3}}\int_{y',\,|y'|>\max(1,|x'|)} |y'|^{-3\beta}\,dy'\,dx'\right)^{1/2}$$

$$\lesssim t^{-\frac{2}{3}+\frac{1}{2}\beta}\left(\int_{x',\,|x'|<t^{1/3}} |x'|^{-3\beta+1}\,dx'\right)^{1/2} \lesssim t^{-\frac{1}{3}}.$$

9. **Region $x' > x+t$, $|x'| > t^{1/3}$, and $|y'| > 4|x'|$.** Here, $|z| > 4$ and $\lambda > 1$.

For $|x'| < 1$ and $|y'| < 1$ (the $--$ subregion), we take $\alpha = \frac{1}{6}$ and $\beta = 0$. Since $|x'| < 1$, it follows that $x+t \leq 1$, and thus, $\langle x+t-x'\rangle \sim 1$. From (10.94), we have

$$|\Phi_{9++}(x,y,t)| \lesssim t^{-7/12}\left(\int_{x',\,|x'|<1} |x'|^{-1/2}\,dx' \int_{y',\,|y'|<1} dy'\right)^{1/2} \lesssim t^{-7/12}.$$

For $|x'| < 1$ and $|y'| > 1$ (the $-+$ subregion), we take $\alpha = 0$ and $\beta = \frac{1}{3}+$. Since $|x'| < 1$, it follows that $x+t \leq 1$ and hence $\langle x+t-x'\rangle \sim 1$. From (10.94), we have

$$|\Phi_{9-+}(x,y,t)| \lesssim t^{-\frac{2}{3}+\frac{1}{2}\beta}\left(\int_{x',\,|x'|<1} dx' \int_{y',\,|y'|>1} |y'|^{-3\beta}\,dy'\right)^{1/2} \lesssim t^{-\frac{2}{3}+\frac{1}{2}\beta} = t^{-\frac{1}{2}+}.$$

For $|x'| > 1$ and $|y'| < 1$ (the $+-$ subregion), we take $\alpha \gg 1$ and $\beta = 0$. From (10.94), we have

$$|\Phi_{9+-}(x,y,t)| \lesssim t^{-\frac{2}{3}+\frac{1}{2}\alpha}\left(\int_{x',\,x'>\max(x+t,1)} |x'|^{-3\alpha}\langle x+t-x'\rangle^2\,dx' \int_{y',\,|y'|<1} dy'\right)^{1/2}$$

By the change of variable $\tilde{x} = x+t-x'$,

$$\lesssim t^{-\frac{2}{3}+\frac{1}{2}\alpha}\left(\int_{\tilde{x}<0} \langle x+t-\tilde{x}\rangle^{-3\alpha}\langle\tilde{x}\rangle^2\,d\tilde{x}\right)^{1/2}$$

By (10.87),

$$\lesssim t^{-\frac{2}{3}+\frac{1}{2}\alpha}\langle x+t\rangle^{-\frac{3}{2}\alpha+\frac{3}{2}}.$$

For $t \le 1$, we take $\alpha = \frac{2}{3}\sigma + 1$ to obtain

$$\lesssim t^{-\frac{1}{6}+\frac{1}{3}\sigma}\langle x+t\rangle^{-\sigma}.$$

For $t > 1$, we decompose the exponent $-\frac{3}{2}\alpha+\frac{3}{2} = (-\frac{1}{2}\alpha+\frac{2}{3}-\frac{13}{12})+(-\alpha+\frac{23}{12})$ and use $x \ge 0$ to obtain the bound $t^{-13/12}\langle x\rangle^{-\alpha+\frac{23}{12}}$. Then set $\alpha = \sigma + \frac{23}{12}$ to obtain the bound $t^{-13/12}\langle x\rangle^{-\sigma}$.

For $|x'| > 1$ and $|y'| > 1$ (the ++ subregion), we take $\alpha \gg 1$ and $\beta = \frac{1}{3}+$. From (10.94), we have

$$|\Phi_{9++}(x,y,t)| \lesssim t^{-\frac{2}{3}+\frac{1}{2}\alpha+\frac{1}{2}\beta}\left(\int_{x',\,x'>\max(x+t,1)} |x'|^{-3\alpha}\langle x+t-x'\rangle^2\right.$$

$$\left.\int_{y',\,|y'|>\max(1,|x'|)} |y'|^{-3\beta}\,dy'\,dx'\right)^{1/2}$$

$$\lesssim t^{-\frac{2}{3}+\frac{1}{2}\alpha+\frac{1}{2}\beta}\left(\int_{x',\,x'>\max(x+t,1)} |x'|^{-3\alpha-3\beta+1}\langle x+t-x'\rangle^2\,dx'\right)^{1/2}$$

By changing variable $\tilde{x} = x+t-x'$,

$$\lesssim t^{-\frac{2}{3}+\frac{1}{2}\alpha+\frac{1}{2}\beta}\left(\int_{\tilde{x}<0}\langle x+t-\tilde{x}\rangle^{-3\alpha-3\beta+1}\langle\tilde{x}\rangle^2\,d\tilde{x}\right)^{1/2}$$

By (10.87),

$$\lesssim t^{-\frac{2}{3}+\frac{1}{2}\alpha+\frac{1}{2}\beta}\langle x+t\rangle^{-\frac{3}{2}\alpha-\frac{3}{2}\beta+2}.$$

For $t < 1$, take α such that $\frac{3}{2}\alpha+\frac{3}{2}\beta - 2 = \sigma$, which gives

$$\lesssim t^{\frac{1}{3}\sigma}\langle x+t\rangle^{-\sigma}.$$

For $t > 1$, we first decompose the exponent as $-\frac{3}{2}\alpha - \frac{3}{2}\beta + 2 = (\frac{2}{3} - \frac{1}{2}\alpha - \frac{1}{2}\beta - \frac{13}{12}) + (-\alpha - \beta + \frac{29}{12})$, and use $x \geq 0$ to obtain the bound $t^{-13/12}\langle x\rangle^{-\alpha-\beta+\frac{29}{12}}$. Then select α so that $\alpha + \beta - \frac{29}{12} = \sigma$ to obtain the bound by $t^{-13/12}\langle x\rangle^{-\sigma}$.

□

Now we consider

$$\Phi_x(x, y, t) = \int A_x(x', y', t)\phi(x + t - x', y - y')\,dx'\,dy' = (\partial_x S(t)\phi)(x, y), \quad (10.96)$$

i.e., the x-derivative of the unique solution to $\partial_t \Phi + \partial_x(1 - \Delta_{xy})\Phi = 0$ with initial condition $\Phi(x, y, 0) = \phi(x, y)$.

Proposition 10.2 (Derivative Linear Solution Estimates) *Let $\sigma > \frac{9}{4}$, and suppose that*

$$\text{for } x > 0, \quad |\phi(x, y)| \leq C_1 \langle x \rangle^{-\sigma} \quad (10.97)$$

and

$$\|\langle x\rangle^{-1}\phi(x, y)\|_{L^2_{y\in\mathbb{R}, x<0}} \leq C_1.$$

Then for $t > 0$,

$$\text{for } x > 0, \quad |\Phi_x(x, y, t)| \lesssim C_1 t^{-13/12} \begin{cases} \langle x\rangle^{-\sigma + \frac{9}{4}} & \text{if } t < 1 \\ \langle x\rangle^{-\tilde{\sigma}} & \text{if } t > 1 \end{cases}$$

Here, $\tilde{\sigma} = \min(\sigma - \frac{9}{4}, \frac{2}{3}\sigma - \frac{5}{12})$. Note that if $\sigma > \frac{11}{2}$, then $\sigma - \frac{9}{4} > \tilde{\sigma}$.

Proof By linearity, it suffices to assume that $C_1 \leq 1$ and $\|\phi\|_{L^2_{xy}} \leq 1$. Recall that we are assuming $x > 0$ and $t > 0$. From Proposition 9.2, we have

$$|A_x(x', y', t)| \lesssim t^{-1}\lambda^{-\alpha}(\lambda|z|^{3/2})^{-\beta} = t^{-1+\frac{1}{2}\alpha+\frac{1}{2}\beta}|x'|^{-\frac{3}{2}\alpha}|y'|^{-\frac{3}{2}\beta} \quad (10.98)$$

with different constraints on the allowed values (and optimal values) of α and β depending upon whether

- $|z| < 4$ or $|z| > 4$,
- $\lambda < 1$ or $\lambda > 1$,
- $x' < 0$ or $x' > 0$.

This is summarized in the following two tables:

$x' > 0$	$\lambda < 1$	$\lambda > 1$	$x' < 0$	$\lambda < 1$	$\lambda > 1$
$|z| < 4$	$\alpha = 0, \beta = 0$	$\alpha \geq 0, \beta = 0$	$|z| < 4$	$\alpha = 0, \beta = 0$	$\alpha = -\frac{1}{6}, \beta = 0$
$|z| > 4$	$\alpha = 0, \beta \geq 0$	$\alpha \geq 0, \beta \geq 0$	$|z| > 4$	$\alpha = 0, \beta \geq 0$	$\alpha \geq 0, \beta \geq 0$

From (10.89), we see that we need to further subdivide according to $-\infty < x' < x + t$ and $x' > x + t$. When $-\infty < x' < x + t$, we have $x + t - x' > 0$ and we can use (10.90), and when $x' > x + t$, we have $x + t - x' < 0$ and we can only use that $\phi \in L^2_{xy}$.

Below our decomposition of the (x', y') integration space is given as nine different regions. Each of the regions can be further divided according to whether $|x'|$ and $|y'|$ are < 1 or > 1. We label the corresponding subregions as $--$, $-+$, $+-$, and $++$, as follows:

- $--$ corresponds to $|x'| < 1$ and $|y'| < 1$
- $-+$ corresponds to $|x'| < 1$ and $|y'| > 1$
- $+-$ corresponds to $|x'| > 1$ and $|y'| < 1$
- $++$ corresponds to $|x'| > 1$ and $|y'| > 1$.

Thus,

$$\Phi = \Phi_1 + \cdots + \Phi_9,$$

where Φ_j denotes the convolution integral in (10.89) restricted to the region under consideration. We further use the decompositions

$$\Phi_j = \Phi_{j--} + \Phi_{j-+} + \Phi_{j+-} + \Phi_{j++}$$

as needed.

In Regions 1–5, we begin as follows: From (10.89), (10.90), and (10.92)

$$|\Phi_j(x, y, t)| \lesssim t^{-1+\frac{1}{2}\alpha+\frac{1}{2}\beta} \iint_{(x',y') \in R} |x'|^{-\frac{3}{2}\alpha} \langle x + t - x' \rangle^{-\sigma} |y'|^{-\frac{3}{2}\beta} \, dx' \, dy', \tag{10.99}$$

where R denotes the subregion of (x', y') space under consideration.

In Regions 6–9, the decay hypothesis (10.90) is not available, so we start from (10.89) with Cauchy-Schwarz in (x', y')

$$|\Phi_j(x, y, t)| \leq \left(\iint_{(x',y') \in R} |A(x', y', t)|^2 \langle x + t - x' \rangle^2 \, dx' \, dy' \right)^{1/2} \|\langle x \rangle^{-1} \phi\|_{L^2_{xy}}.$$

Since $\|\langle x \rangle^{-1} \phi\|_{L^2_{xy}} \leq 1$, this term can be dropped above and (10.92) yields

$$|\Phi_j(x, y, t)| \leq t^{-1+\frac{1}{2}\alpha+\frac{1}{2}\beta} \left(\iint_{(x',y') \in R} |x'|^{-3\alpha} |y'|^{-3\beta} \langle x + t - x' \rangle^2 \, dx' \, dy' \right)^{1/2}. \tag{10.100}$$

An argument used repeatedly below is

$$|x'| < t^{1/3} \implies \langle x+t-x' \rangle \sim \langle x+t \rangle. \tag{10.101}$$

Indeed, if $t < 8$, then $|x'| < t^{1/3} < 2$, so $\langle x+t-x' \rangle \sim \langle x+t \rangle$. On the other hand, if $t > 8$, then $|x'| < t^{1/3} \ll t \leq x+t$, so again $\langle x+t-x' \rangle \sim \langle x+t \rangle$.

Another frequently employed inequality is for $\mu, \mu_1, \mu_2 \geq 0$ with $\mu_1 + \mu_2 = \mu$,

$$\langle x+t \rangle^{-\mu} \lesssim \langle x \rangle^{-\mu_1} \langle t \rangle^{-\mu_2}, \tag{10.102}$$

which is straightforward, since we are assuming $x \geq 0$ and $t \geq 0$.

1. **Region** $x' < x+t$, $|x'| < t^{1/3}$, $|y'| < 4|x'|$. Here, $\lambda < 1$, $|z| < 4$. We take $\alpha = 0$, $\beta = 0$. From (10.101), we have $\langle x+t-x' \rangle \sim \langle x+t \rangle$. Starting with (10.99), we get

$$|\Phi_1(x,y,t)| \lesssim t^{-1} \langle x+t \rangle^{-\sigma} \int_{x', |x'|<t^{1/3}}$$

$$\int_{y', |y'|<4|x'|} dy' \, dx' \lesssim t^{-1/3} \langle x+t \rangle^{-\sigma} \lesssim t^{-13/12} \langle x \rangle^{-\sigma + \frac{3}{4}}.$$

2. **Region** $x' < 0$, $|x'| > t^{1/3}$, $|y'| < 4|x'|$. Here, $x' < 0$, $\lambda > 1$, $|z| < 4$. We take $\beta = 0$, and we are limited to $\alpha = -\frac{1}{6}$. Starting with (10.99),

$$|\Phi_2(x,y,t)| \lesssim t^{-13/12} \int_{x', -\infty < x' < 0} \langle x+t-x' \rangle^{-\sigma} |x'|^{1/4} \int_{y', |y'|<4|x'|} dy' \, dx'$$

$$\lesssim t^{-13/12} \int_{x', -\infty < x' < 0} \langle x+t-x' \rangle^{-\sigma} |x'|^{5/4} \, dx'.$$

By (10.87) with $\mu = -\frac{5}{4}$, provided $\sigma > \frac{9}{4}$, we have

$$\lesssim t^{-13/12} \langle x+t \rangle^{-\sigma + \frac{9}{4}}.$$

3. **Region** $0 < x' < x+t$, $|x'| > t^{1/3}$, $|y'| < 4|x'|$. Here, $x' > 0$, $\lambda > 1$, $|z| < 4$. We take $\beta = 0$. For $|x'| < 1$ (the $-*$ subregion), we take $\alpha = 0$, but for $|x'| > 1$ (the $+*$ subregion), we take $\alpha \gg 1$.

For $|x'| < 1$, we take $\alpha = 0$ and use that $\langle x+t-x' \rangle \sim \langle x+t \rangle$ to obtain, starting with (10.99),

$$|\Phi_{3-*}(x,y,t)| \lesssim t^{-1} \langle x+t \rangle^{-\sigma} \int_{x', |x'|<1}$$

$$\int_{y', |y'|<4|x'|} dy' \, dx' \sim t^{-1} \langle x+t \rangle^{-\sigma} \lesssim t^{-\frac{13}{12}} \langle x \rangle^{-\sigma + \frac{1}{12}}.$$

For $|x'| > 1$, we take $\alpha \gg 1$, and starting with (10.99), we get

$$|\Phi_{3+*}(x,y,t)| \lesssim t^{-1+\frac{1}{2}\alpha} \int_{x', |x'|>1} |x'|^{-\frac{3}{2}\alpha} \langle x+t-x'\rangle^{-\sigma} \int_{y', |y'|<4|x'|} dy'\, dx'$$

$$\lesssim t^{-1+\frac{1}{2}\alpha} \int_{x', |x'|>1} |x'|^{1-\frac{3}{2}\alpha} \langle x+t-x'\rangle^{-\sigma}\, dx'.$$

For $\sigma > 1$, taking $\alpha = \frac{2}{3}(\sigma + 1)$ gives by (10.88)

$$\lesssim t^{-\frac{2}{3}+\frac{1}{3}\sigma} \langle x+t\rangle^{-\sigma}.$$

For $t \geq 1$, we apply (10.102) with $\mu = \sigma$, $\mu_1 = \frac{2}{3}\sigma - \frac{5}{12}$, and $\mu_2 = \frac{1}{3}\sigma + \frac{5}{12}$, we obtain

$$\lesssim t^{-13/12} \langle x\rangle^{-\frac{2}{3}\sigma+\frac{5}{12}}.$$

4. **Region** $-\infty < x' < x+t$, $|x'| < t^{1/3}$, **and** $|y'| > 4|x'|$. Here, $|z| > 4$ and $\lambda < 1$. By (10.101), we have $\langle x+t-x'\rangle \sim \langle x+t\rangle$. We take $\alpha = 0$ and any $\frac{2}{3} < \beta < \frac{4}{3}$ (so that $-2 < -\frac{3}{2}\beta < -1$) and starting from (10.99), we have

$$|\Phi_4(x,y,t)| \lesssim t^{-1+\frac{1}{2}\beta} \langle x+t\rangle^{-\sigma} \int_{x', |x'|<t^{1/3}} \int_{y', |y'|>|x'|} |y'|^{-\frac{3}{2}\beta}\, dy'\, dx'$$

$$\lesssim t^{-1+\frac{1}{2}\beta} \langle x+t\rangle^{-\sigma} \int_{x', |x'|<t^{1/3}} |x'|^{-\frac{3}{2}\beta+1}\, dx' \lesssim t^{-1/3} \langle x+t\rangle^{-\sigma},$$

where, in the last step, we used that $-\frac{3}{2}\beta + 1 > -1$. By (10.102) with $\mu = \sigma$, $\mu_1 = \sigma - \frac{3}{4}$, $\mu_2 = \frac{3}{4}$, we obtain

$$\lesssim t^{-13/12} \langle x\rangle^{-\sigma+\frac{3}{4}}.$$

5. **Region** $-\infty < x' < x+t$, $|x'| > t^{1/3}$, **and** $|y'| > 4|x'|$. Here, $|z| > 4$ and $\lambda > 1$. Any choice of $\alpha, \beta \geq 0$ is permitted in (10.98).

For $|x'| < 1$ and $|y'| < 1$ (the $--$ subregion), we take $\alpha = 0$ and $\beta = 0$. Since $|x'| < 1$, we have $\langle x+t-x'\rangle \sim \langle x+t\rangle$. Starting from (10.99), we obtain

$$|\Phi_{5--}(x,y,t)| \lesssim t^{-1} \langle x+t\rangle^{-\sigma} \int_{x', |x'|<1}$$

$$\int_{y', |y'|<1} dy'\, dx' \lesssim t^{-1} \langle x+t\rangle^{-\sigma} \lesssim t^{-13/12} \langle x\rangle^{-\sigma+\frac{1}{12}}.$$

For $|x'| < 1$ and $|y'| > 1$ (the $-+$ subregion), we take $\alpha = 0$ and $\beta = \frac{2}{3}+$. Since $|x'| < 1$, we have $\langle x+t-x'\rangle \sim \langle x+t\rangle$. Starting from (10.99), we have

$$|\Phi_{5-+}(x,y,t)| \lesssim t^{-1+\frac{1}{2}\beta}\langle x+t\rangle^{-\sigma}\int_{x',|x'|<1}dx'\int_{y',|y'|>1}|y'|^{-\frac{3}{2}\beta}dy' \lesssim t^{-\frac{2}{3}+}\langle x+t\rangle^{-\sigma}.$$

By (10.102) with $\mu = \sigma$, $\mu_1 = \sigma - \frac{5}{12}-$, $\mu_2 = \frac{5}{12}+$, we obtain

$$\lesssim t^{-13/12}\langle x\rangle^{-\sigma-\frac{5}{12}-}.$$

For $|x'| > 1$ and $|y'| < 1$ (the $+-$ subregion), we take $\alpha \gg 1$ and $\beta = 0$. Starting from (10.99), we get

$$|\Phi_{5+-}(x,y,t)| \lesssim t^{-1+\frac{1}{2}\alpha}\int_{x',|x'|>1}|x'|^{-\frac{3}{2}\alpha}\langle x+t-x'\rangle^{-\sigma}dx'\int_{y',|y'|<1}dy'.$$

For $\sigma > 1$, take $\alpha = \frac{2}{3}\sigma$ and apply (10.88) to obtain

$$\lesssim t^{-1+\frac{1}{3}\sigma}\langle x+t\rangle^{-\sigma}.$$

For $t \geq 1$, we apply (10.102) with $\mu = \sigma$, $\mu_1 = \frac{2}{3}\sigma - \frac{1}{12}$, $\mu_2 = \frac{1}{3}\sigma + \frac{1}{12}$ to obtain

$$\lesssim t^{-13/12}\langle x\rangle^{-\frac{2}{3}\sigma+\frac{1}{12}}.$$

For $|x'| > 1$ and $|y'| > 1$ (the $++$ subregion), we take $\alpha \gg 1$ and any $\frac{2}{3} < \beta < \frac{4}{3}$ (so that $-2 < -\frac{3}{2}\beta < -1$ and $-1 < -\frac{3}{2}\beta + 1 < 0$). Starting from (10.99), we get

$$|\Phi_{5++}(x,y,t)| \lesssim t^{-1+\frac{1}{2}\alpha+\frac{1}{2}\beta}\int_{x',|x'|>1}|x'|^{-\frac{3}{2}\alpha}\langle x+t-x'\rangle^{-\sigma}$$

$$\int_{y',|y'|>\max(1,|x'|)}|y'|^{-\frac{3}{2}\beta}dy'\,dx'$$

$$\lesssim t^{-1+\frac{1}{2}\alpha+\frac{1}{2}\beta}\int_{x',|x'|>1}|x'|^{1-\frac{3}{2}\alpha-\frac{3}{2}\beta}\langle x+t-x'\rangle^{-\sigma}dx'.$$

Take α so that $\alpha+\beta = \frac{2}{3}(\sigma+1)$ (and hence $1-\frac{3}{2}\alpha-\frac{3}{2}\beta = -\sigma$) and apply (10.88) to obtain

$$\lesssim t^{-\frac{2}{3}+\frac{1}{3}\sigma}\langle x+t\rangle^{-\sigma}.$$

For $t \geq 1$, apply (10.102) with $\mu = \sigma$, $\mu_1 = \frac{2}{3}\sigma - \frac{5}{12}$, $\mu_2 = \frac{1}{3}\sigma + \frac{5}{12}$ to obtain the bound of

$$\lesssim t^{-13/12}\langle x\rangle^{-\frac{2}{3}\sigma+\frac{5}{12}}.$$

Instability of Solitons in 2d Cubic ZK   349

Recall that for Regions 6–9 below, we must use (10.94). Thus, we need to recover the $\langle x+t \rangle^{-\sigma}$ decay factor from the constraint $x' > x+t$ and the decay on $A(x', y', t)$.

6. **Region $x' > x+t$, $|x'| < t^{1/3}$, and $|y'| < 4|x'|$.** Here, $|z| < 4$ and $\lambda < 1$, so $\alpha = 0$ and $\beta = 0$. Note that the constraints imply that $x+t < t^{1/3}$. Since $x+t < t^{1/3}$, it follows that $t < t^{1/3}$, from which we conclude that $t < 1$. Also from $x+t < t^{1/3}$, we conclude that $x < t^{1/3}$, and since $t < 1$, this implies $x < 1$. Consequently, $\langle x+t-x' \rangle \sim 1$.

    Starting from (10.100), we obtain

$$|\Phi_6(x,y,t)| \lesssim t^{-1}\left(\int_{x', |x'|<t^{1/3}} \int_{y', |y'|<4|x'|} dy' dx'\right)^{1/2} \lesssim t^{-2/3}.$$

7. **Region $x' > x+t$, $|x'| > t^{1/3}$, and $|y'| < 4|x'|$.** Here, $|z| < 4$ and $\lambda > 1$, so we take $\beta = 0$, and we are allowed any $\alpha \geq 0$.

    If $|x'| < 1$, we take $\alpha = 0$. Since $|x'| < 1$, we have $x+t < 1$, so $\langle x+t-x' \rangle \sim 1$. By (10.100),

$$|\Phi_{7-*}(x,y,t)| \lesssim t^{-1}\left(\int_{x', |x'|<1} \int_{y', |y'|<4|x'|} dy' dx'\right)^{1/2} \lesssim t^{-1}.$$

    If $|x'| > 1$, we take $\alpha \gg 1$. Starting from (10.100), we get

$$|\Phi_{7+*}(x,y,t)| \lesssim t^{-1+\frac{1}{2}\alpha}\left(\int_{x', x'>\max(1,x+t)} |x'|^{-3\alpha}\langle x+t-x'\rangle^2 \int_{y', |y'|<4|x'|} dy' dx'\right)^{1/2}$$

By the change of variable $\tilde{x} = x+t-x'$,

$$\lesssim t^{-1+\frac{1}{2}\alpha}\left(\int_{\tilde{x}<0} \langle x-t-\tilde{x}\rangle^{-3\alpha+1}\langle\tilde{x}\rangle^2 dx'\right)^{1/2}$$

By (10.87),

$$\lesssim t^{-1+\frac{1}{2}\alpha}\langle x+t\rangle^{-\frac{3}{2}\alpha+2}.$$

By (10.102) with $\mu = \frac{3}{2}\alpha - 2$, $\mu_1 = \sigma$ and $\mu_2 = \frac{3}{2}\alpha - 2 - \sigma$, we obtain

$$\lesssim t^{\sigma-\alpha+1}\langle x\rangle^{-\sigma}.$$

We thus take $\alpha = \sigma + \frac{25}{12}$.

8. **Region $x' > x+t$, $|x'| < t^{1/3}$, and $|y'| > 4|x'|$.** Here, $|z| > 4$ and $\lambda < 1$, so we take $\alpha = 0$. Note that the constraints imply that $x+t < t^{1/3}$. Since $x+t < t^{1/3}$, it follows that $t < t^{1/3}$, from which we conclude that $t < 1$. Also from $x+t < t^{1/3}$, we conclude that $x < t^{1/3}$, and since $t < 1$, this implies $x < 1$. Consequently, $\langle x+t-x' \rangle \sim 1$.

For $|y'| < 1$ we take $\beta = 0$. From (10.100), we have

$$|\Phi_{8*-}(x, y, t)| \lesssim t^{-1} \left(\int_{x', |x'|<t^{1/3}} \int_{y', |y'|<1} dy' \, dx' \right)^{1/2} \lesssim t^{-5/6}.$$

For $|y'| > 1$, we take any $\frac{1}{3} < \beta < \frac{2}{3}$ (so that $-3\beta < -1$ and $-1 < -3\beta + 1$). From (10.100), we have

$$|\Phi_{8*+}(x, y, t)| \lesssim t^{-1+\frac{1}{2}\beta} \left(\int_{x', |x'|<t^{1/3}} \int_{y', |y'|>\max(1,|x'|)} |y'|^{-3\beta} dy' \, dx' \right)^{1/2}$$

$$\lesssim t^{-1+\frac{1}{2}\beta} \left(\int_{x', |x'|<t^{1/3}} |x'|^{-3\beta+1} dx' \right)^{1/2} \lesssim t^{-\frac{1}{3}-\beta}.$$

9. **Region $x' > x + t$, $|x'| > t^{1/3}$, and $|y'| > 4|x'|$**. Here, $|z| > 4$ and $\lambda > 1$.

For $|x'| < 1$ and $|y'| < 1$ (the $--$ subregion), we take $\alpha = 0$ and $\beta = 0$. Since $|x'| < 1$, it follows that $x+t \leq 1$, and thus, $\langle x+t-x' \rangle \sim 1$. From (10.100), we have

$$|\Phi_{9++}(x, y, t)| \lesssim t^{-1} \left(\int_{x', |x'|<1} dx' \int_{y', |y'|<1} dy' \right)^{1/2} \lesssim t^{-1}.$$

For $|x'| < 1$ and $|y'| > 1$ (the $-+$ subregion), we take $\alpha = 0$ and $\beta = \frac{1}{3}+$. Since $|x'| < 1$, it follows that $x+t \leq 1$, and thus, $\langle x+t-x' \rangle \sim 1$. From (10.100), we have

$$|\Phi_{9-+}(x, y, t)| \lesssim t^{-1+\frac{1}{2}\beta} \left(\int_{x', |x'|<1} dx' \int_{y', |y'|>1} |y'|^{-3\beta} dy' \right)^{1/2} \lesssim t^{-1+\frac{1}{2}\beta} = t^{-\frac{5}{6}+}.$$

For $|x'| > 1$ and $|y'| < 1$ (the $+-$ subregion), we take $\alpha \gg 1$ and $\beta = 0$. From (10.100), we have

$$|\Phi_{9+-}(x, y, t)| \lesssim t^{-1+\frac{1}{2}\alpha} \left(\int_{x', x'>\max(x+t,1)} |x'|^{-3\alpha} (x+t-x')^2 dx' \int_{y', |y'|<1} dy' \right)^{1/2}$$

By the change of variable $\tilde{x} = x + t - x'$,

$$|\Phi_{9+-}(x, y, t)| \lesssim t^{-1+\frac{1}{2}\alpha} \left(\int_{\tilde{x}<0} \langle x+t-\tilde{x} \rangle^{-3\alpha} \langle \tilde{x} \rangle^2 \right)^{1/2}$$

$$\lesssim t^{-1+\frac{1}{2}\alpha} \langle x+t \rangle^{-\frac{3}{2}\alpha+\frac{3}{2}}.$$

Apply (10.102) with $\mu = \frac{3}{2}\alpha - \frac{1}{2}$, $\mu_1 = \sigma$, and $\mu_2 = \frac{3}{2}\alpha - \frac{3}{2} - \sigma$ to obtain

$$|\Phi_{9+-}(x,y,t)| \lesssim t^{\frac{1}{2}-\alpha+\sigma}\langle x\rangle^{-\sigma}.$$

Taking $\alpha = \sigma + \frac{19}{12}$, we obtain the desired bound.

For $|x'| > 1$ and $|y'| > 1$ (the ++ subregion), we take $\alpha \gg 1$ and $\beta = \frac{1}{3}+$. From (10.100), we have

$$|\Phi_{9++}(x,y,t)| \lesssim t^{-1+\frac{1}{2}\alpha+\frac{1}{2}\beta} \left(\int_{x', x' > \max(x+t, 1)} |x'|^{-3\alpha} \langle x+t-x'\rangle^2 \right.$$
$$\left. \int_{y', |y'| > \max(1, |x'|)} |y'|^{-3\beta} \, dy' \, dx' \right)^{1/2}$$

$$\lesssim t^{-1+\frac{1}{2}\alpha+\frac{1}{2}\beta} \left(\int_{x', x' > \max(x+t, 1)} |x'|^{-3\alpha-3\beta+1} \langle x+t-x'\rangle^2 \, dx' \right)^{1/2}$$

By the change of variable $\tilde{x} = x + t - x'$, we obtain

$$\lesssim t^{-1+\frac{1}{2}\alpha+\frac{1}{2}\beta} \left(\int_{\tilde{x}<0} \langle x+t-\tilde{x}\rangle^{-3\alpha-3\beta+1} \langle\tilde{x}\rangle^2 \, dx' \right)^{1/2}$$

By (10.87),

$$\lesssim t^{-1+\frac{1}{2}\alpha+\frac{1}{2}\beta} \langle x+t\rangle^{-\frac{3}{2}\alpha-\frac{3}{2}\beta+2}.$$

By (10.102) with $\mu = \frac{3}{2}\alpha + \frac{2}{3}\beta - 2$, $\mu_1 = \sigma$, and $\mu_2 = \frac{3}{2}\alpha + \frac{3}{2}\beta - 2 - \sigma$, we obtain

$$\lesssim t^{1-\alpha-\beta+\sigma}\langle x\rangle^{-\sigma}.$$

Taking $\alpha = -\beta + \sigma + \frac{25}{12}$ gives $t^{-13/12}\langle x\rangle^{-\sigma}$.

□

11 Duhamel Estimate

We will now use Proposition 10.2 to prove a Duhamel estimate in Proposition 11.2 below. First, we need

Lemma 11.1 *Suppose $\mu > 1$ and $\nu > 1$, $0 < \sigma_2 \leq \sigma_1$, and thus, $\tilde{\sigma}_2 \leq \tilde{\sigma}_1$ (where $\tilde{\sigma}_j$ is given in terms of σ_j as in Proposition 10.2), $t > 0$ and that $f(t, t') \geq 0$ satisfies for $0 < t' < t$,*

$$f(t,t') \lesssim (t-t')^{-\mu} \begin{cases} (t')^{-\nu}\langle x\rangle^{-\sigma_1} & \text{if } t' < 1, \ t-t' < 1 \\ \langle x\rangle^{-\sigma_2} & \text{if } t' > \frac{1}{2}, \ t-t' < 1 \\ (t')^{-\nu}\langle x\rangle^{-\tilde\sigma_1} & \text{if } t' < 1, \ t-t' > \frac{1}{2} \\ \langle x\rangle^{-\tilde\sigma_2} & \text{if } t' > \frac{1}{2}, \ t-t' > \frac{1}{2}. \end{cases}$$

(Note that the regions are overlapping for convenience in application. The meaning is that $f(t,t')$ is bounded by the minimum of the bounds across all applicable regions.) Then for any $0 \le a, b \le \frac{1}{2}$ with $0 \le a+b \le t$, we have

$$\int_b^{t-a} f(t,t')\,dt' \lesssim \begin{cases} (t^{-\nu}a^{-\mu+1} + t^{-\mu}b^{-\nu+1})\langle x\rangle^{-\sigma_1} & \text{if } t < 1 \\ b^{-\nu+1}\langle x\rangle^{-\tilde\sigma_1} + \langle x\rangle^{-\tilde\sigma_2} + a^{-\mu+1}\langle x\rangle^{-\sigma_2} & \text{if } t > 1. \end{cases} \qquad (11.103)$$

If, instead, $0 \le \nu < 1$, then

$$\int_0^{t-a} f(t,t')\,dt' \lesssim \begin{cases} t^{-\nu}a^{-\mu+1}\langle x\rangle^{-\sigma_1} & \text{if } t < 1 \\ \langle x\rangle^{-\tilde\sigma_2} + a^{-\mu+1}\langle x\rangle^{-\sigma_2} & \text{if } t > 1. \end{cases} \qquad (11.104)$$

Proof First consider the case $\nu > 1$. If $t < 1$,

$$\int_b^{t-a} f(t,t')\,dt' \lesssim \langle x\rangle^{-\sigma_1} \int_b^{t-a} (t-t')^{-\mu}(t')^{-\nu}\,dt',$$

and changing variable to $s = t'/t$, we continue

$$= t^{-\mu-\nu+1}\langle x\rangle^{-\sigma_1} \int_{b/t}^{1-a/t} (1-s)^{-\mu}s^{-\nu}\,ds \lesssim (t^{-\nu}a^{-\mu+1} + t^{-\mu}b^{-\nu+1})\langle x\rangle^{-\sigma_1}.$$

If $t > 1$, then we split into three pieces

$$\int_b^{1/2} f(t,t')\,dt' \lesssim \langle x\rangle^{-\tilde\sigma_1} \int_b^{1/2} (t-t')^{-\mu}(t')^{-\nu}\,dt' \lesssim b^{-\nu+1}\langle x\rangle^{-\tilde\sigma_1},$$

$$\int_{1/2}^{t-\frac{1}{2}} f(t,t')\,dt' \lesssim \langle x\rangle^{-\tilde\sigma_2} \int_{1/2}^{t-\frac{1}{2}} (t-t')^{-\mu}\,dt' \lesssim \langle x\rangle^{-\tilde\sigma_2},$$

$$\int_{t-\frac{1}{2}}^{t-a} f(t,t')\,dt' \lesssim \langle x\rangle^{-\sigma_2} \int_{t-\frac{1}{2}}^{t-a} (t-t')^{-\mu}\,dt' \lesssim a^{-\mu+1}\langle x\rangle^{-\sigma_2}.$$

Now we turn to the case $0 \le \nu < 1$. If $t < 1$,

$$\int_a^{t-a} f(t,t')\,dt' \lesssim \langle x \rangle^{-\sigma_1} \int_0^{t-a} (t-t')^{-\mu}(t')^{-\nu}\,dt',$$

and changing variable to $s = t'/t$, we obtain

$$= t^{-\mu-\nu+1}\langle x \rangle^{-\sigma_1} \int_0^{1-a/t} (1-s)^{-\mu} s^{-\nu}\,ds \lesssim t^{-\nu} a^{-\mu+1} \langle x \rangle^{-\sigma_1}.$$

If $t > 1$, then we split into three pieces

$$\int_0^{1/2} f(t,t')\,dt' \lesssim \langle x \rangle^{-\tilde{\sigma}_1} \int_0^{1/2} (t-t')^{-\mu}(t')^{-\nu}\,dt' \lesssim \langle x \rangle^{-\tilde{\sigma}_1},$$

$$\int_{1/2}^{t-\frac{1}{2}} f(t,t')\,dt' \lesssim \langle x \rangle^{-\tilde{\sigma}_2} \int_{1/2}^{t-\frac{1}{2}} (t-t')^{-\mu}\,dt' \lesssim \langle x \rangle^{-\tilde{\sigma}_2},$$

$$\int_{t-\frac{1}{2}}^{t-a} f(t,t')\,dt' \lesssim \langle x \rangle^{-\sigma_2} \int_{t-\frac{1}{2}}^{t-a} (t-t')^{-\mu}\,dt' \lesssim a^{-\mu+1}\langle x \rangle^{-\sigma_2}.$$

$\square$

Proposition 11.2 (Duhamel Estimate) *Suppose that $F(x, y, t)$ satisfies, for some $\sigma \gg 1$ and $\nu \geq 0$,*

- *for $x > 0$, $|F(x,y,t)| \lesssim C_2 \begin{cases} t^{-\nu}\langle x \rangle^{-\sigma_1} & \text{if } t < 1 \\ \langle x \rangle^{-\sigma_2} & \text{if } t > \frac{1}{2} \end{cases}$, where $1 < \sigma_2 \leq \sigma_1$,*
- $\|\langle x \rangle^{-1} F(x,y,t)\|_{L_t^\infty L_{y\in\mathbb{R}, x<0}^2} \lesssim C_2$,
- $\|\langle x \rangle^{-1} \partial_x F(x,y,t)\|_{L_t^\infty L_{y\in\mathbb{R}, x<0}^1} \lesssim C_2$,
- $\|\partial_x F(x,y,t)\|_{L_t^\infty L_{y\in\mathbb{R}, x>0}^1} \lesssim C_2$.

Then, if $\nu > 1$, we have for $x > 0$,

$$\left| \int_0^t [\partial_x S(t-t') F(\bullet, \bullet, t')](x,y)\,dt' \right|$$

$$\lesssim C_2 \begin{cases} t^{1/3} & \text{if } 0 < t^{\nu+\frac{5}{12}} \lesssim \langle x \rangle^{-(\sigma_1-\frac{9}{4})} \\ t^{-\frac{4}{5}\nu}\langle x \rangle^{-\frac{4}{5}(\sigma_1-\frac{9}{4})} & \text{if } \langle x \rangle^{-(\sigma_1-\frac{9}{4})} \ll t^{\nu+\frac{5}{12}} < 1 \text{ and } \nu < \frac{5}{4} \\ t^{-\frac{2}{3}-\frac{5}{12\nu}}\langle x \rangle^{-\frac{1}{\nu}(\sigma_1-\frac{9}{4})} & \text{if } \langle x \rangle^{-(\sigma_1-\frac{9}{4})} \ll t^{\nu+\frac{5}{12}} < 1 \text{ and } \nu > \frac{5}{4} \\ \langle x \rangle^{-\min(\frac{\tilde{\sigma}_1}{\nu}, \tilde{\sigma}_2, \frac{4}{5}(\sigma_2-\frac{9}{4}))} & \text{if } t > 1, \end{cases}$$

where

$$\tilde{\sigma}_1 = \min(\sigma_1 - \tfrac{9}{4}, \tfrac{2}{3}\sigma_1 - \tfrac{5}{12}),$$

$$\tilde{\sigma}_2 = \min(\sigma_2 - \tfrac{9}{4}, \tfrac{2}{3}\sigma_2 - \tfrac{5}{12}).$$

Proof By linearity, we can take $C_2 = 1$. Let

$$f(x, y, t, t') = [\partial_x S(t - t')F(\cdot, \cdot, t')](x, y).$$

By the assumed pointwise decay on $F(x, y, t)$ for $x > 0$, and the assumption $\|F\|_{L_t^\infty L_{xy}^2} \lesssim 1$, Proposition 10.2 gives, for $x > 0$, the pointwise estimate

$$|f(x, y, t, t')| \lesssim (t - t')^{-13/12} \begin{cases} (t')^{-\nu}\langle x \rangle^{-\sigma_1 + \frac{9}{4}} & \text{if } t - t' < 1, \, t' < 1 \\ (t')^{-\nu}\langle x \rangle^{-\tilde{\sigma}_1} & \text{if } t - t' > \dfrac{1}{2}, \, t' < 1 \\ \langle x \rangle^{-\sigma_2 + \frac{9}{4}} & \text{if } t - t' < 1, \, t' > \dfrac{1}{2} \\ \langle x \rangle^{-\tilde{\sigma}_2} & \text{if } t - t' > \dfrac{1}{2}, \, t' > \dfrac{1}{2}. \end{cases}$$

By Lemma 11.1, with σ_1 replaced by $\sigma_1 - \tfrac{9}{4}$, σ_2 replaced by $\sigma_2 - \tfrac{9}{4}$, and $\tilde{\sigma}_1, \tilde{\sigma}_2$ as given above, we obtain, for $x > 0$,

$$\left| \int_b^{t-a} f(x, y, t, t') \, dt' \right| \lesssim \begin{cases} (t^{-\nu} a^{-1/12} + t^{-13/12} b^{-\nu+1}) \langle x \rangle^{-\sigma_1 + \frac{9}{4}} & \text{if } t < 1 \\ b^{-\nu+1} \langle x \rangle^{-\tilde{\sigma}_1} + \langle x \rangle^{-\tilde{\sigma}_2} + a^{-1/12} \langle x \rangle^{-\sigma_2 + \frac{9}{4}} & \text{if } t > 1. \end{cases}$$

We have

$$[S(t)\phi](x, y) = \iint A(x + t - x', y - y', t)\phi(x', y') \, dx' \, dy'$$

By (10.92) with $\beta = 0$

$$|[S(t)\phi](x, y)| \lesssim t^{-\frac{2}{3} + \frac{1}{2}\alpha} \iint |x + t - x'|^{-\frac{3}{2}\alpha} |\phi(x', y')| \, dx' \, dy'$$

with the corresponding restrictions on α. Splitting the integration into $x' > -1$, where we use $\alpha = 0$, and $x' < -1$, where we use $\alpha = 1$, we obtain

$$|[S(t)\phi](x, y)| \lesssim t^{-\frac{2}{3}} \iint_{x' > -1} |\phi(x', y')| \, dx' \, dy'$$

$$+ t^{-\frac{1}{6}} \iint_{x' < -1} \langle x + t - x' \rangle^{-3/2} |\phi(x', y')| \, dx' \, dy'$$

In the second integral, since $x' < -1$, we have that $\langle x + t - x'\rangle^{-3/2} \leq \langle t \rangle^{-\frac{1}{2}} \langle x' \rangle^{-1}$ and thus

$$\|S(t)\phi\|_{L_{xy}^\infty} \lesssim t^{-2/3}(\|\phi\|_{L_{y\in\mathbb{R},x>-1}^1} + \|\langle x \rangle^{-1}\phi\|_{L_{y\in\mathbb{R},x<-1}^1})$$

$$\approx t^{-2/3}(\|\phi\|_{L_{y\in\mathbb{R},x>0}^1} + \|\langle x \rangle^{-1}\phi\|_{L_{y\in\mathbb{R},x<0}^1})$$

Hence,

$$\left|\int_0^b f(x,y,t,t')\,dt'\right| \lesssim \int_0^b \|S(t-t')\partial_x F(\cdot,\cdot,t')(x,y)\|_{L_{xy}^\infty}\,dt'$$

$$\lesssim \left(\|\partial_x F(x,y,t)\|_{L_t^\infty L_{y\in\mathbb{R},x>0}^1} + \|\langle x \rangle^{-1}\partial_x F(x,y,t)\|_{L_t^\infty L_{y\in\mathbb{R},x<0}^1}\right)$$

$$\int_0^b (t-t')^{-2/3}\,dt'$$

$$\lesssim t^{-2/3} b,$$

where, in the last step, we assumed that $b \leq \frac{1}{2}t$. Similarly,

$$\left|\int_{t-a}^t f(x,y,t,t')\,dt'\right| \lesssim \int_{t-a}^t (t-t')^{-2/3}\,dt' \lesssim a^{1/3}.$$

Now, if we take

$$G(x,y,t) = \int_0^t f(x,y,t,t')\,dt',$$

then the above estimates give, for $x > 0$,

$$|G(x,y,t)| \lesssim t^{-2/3}b + a^{1/3} + \begin{cases} (t^{-\nu}a^{-1/12} + t^{-13/12}b^{-\nu+1})\langle x \rangle^{-\sigma_1 + \frac{9}{4}} & \text{if } t < 1 \\ b^{-\nu+1}\langle x \rangle^{-\tilde{\sigma}_1} + \langle x \rangle^{-\tilde{\sigma}_2} + a^{-1/12}\langle x \rangle^{-\sigma_2 + \frac{9}{4}} & \text{if } t > 1, \end{cases}$$
(11.105)

provided $a + b \leq t$ and $b \leq \frac{1}{2}t$.

Case 1. $t < 1$ and $\langle x \rangle^{-\sigma_1 + \frac{9}{4}} \ll t^{\nu + \frac{5}{12}}$ (corresponding to $a \ll t$ and $b \ll t$, where a and b are defined below) In this case, we obtain from the first component of (11.105), for $x > 0$,

$$|G(x,y,t)| \lesssim t^{-\frac{2}{3}}b + t^{-\frac{13}{12}}b^{-\nu+1}\langle x \rangle^{-\sigma_1 + \frac{9}{4}} + t^{-\nu}a^{-\frac{1}{12}}\langle x \rangle^{-\sigma_1 + \frac{9}{4}} + a^{\frac{1}{3}}.$$

The optimal values of a and b are $a = [t^{-\nu}\langle x\rangle^{-\sigma_1+\frac{9}{4}}]^{12/5}$ and $b = [t^{-\frac{5}{12}}\langle x\rangle^{-\sigma_1+\frac{9}{4}}]^{1/\nu}$. This furnishes the bound

$$|G(x,y,t)| \lesssim t^{-(\frac{2}{3}+\frac{5}{12\nu})}\langle x\rangle^{-\frac{1}{\nu}(\sigma_1-\frac{9}{4})} + t^{-\frac{4\nu}{5}}\langle x\rangle^{-\frac{4}{5}(\sigma_1-\frac{9}{4})}.$$

We consider two further subcases

Case 1A. $\nu < \frac{5}{4}$ Then $\frac{1}{\nu} - \frac{4}{5} > 0$, so by raising $\langle x\rangle^{-(\sigma_1-\frac{9}{4})} < t^{\nu+\frac{5}{12}}$ to the positive power $\frac{1}{\nu} - \frac{4}{5}$, we obtain

$$\langle x\rangle^{-(\sigma_1-\frac{9}{4})(\frac{1}{\nu}-\frac{4}{5})} < t^{(\nu+\frac{5}{12})(\frac{1}{\nu}-\frac{4}{5})} = t^{\frac{2}{3}+\frac{5}{12\nu}-\frac{4}{5}\nu}.$$

Hence,

$$t^{-(\frac{2}{3}+\frac{5}{12\nu})}\langle x\rangle^{-\frac{1}{\nu}(\sigma_1-\frac{9}{4})} \leq t^{-\frac{4}{5}\nu}\langle x\rangle^{-\frac{4}{5}(\sigma_1-\frac{9}{4})}.$$

Consequently, in this case, we have the bound

$$|G(x,y,t)| \lesssim t^{-\frac{4}{5}\nu}\langle x\rangle^{-\frac{4}{5}(\sigma_1-\frac{9}{4})}.$$

Case 1B. $\nu > \frac{5}{4}$ Then $\frac{4}{5} - \frac{1}{\nu} > 0$, so by raising $\langle x\rangle^{-(\sigma_1-\frac{9}{4})} < t^{\nu+\frac{5}{12}}$ to the positive power $\frac{4}{5} - \frac{1}{\nu}$, we obtain

$$\langle x\rangle^{-(\sigma_1-\frac{9}{4})(\frac{4}{5}-\frac{1}{\nu})} < t^{(\nu+\frac{5}{12})(\frac{4}{5}-\frac{1}{\nu})} = t^{-\frac{2}{3}-\frac{5}{12\nu}+\frac{4}{5}\nu}.$$

Hence,

$$t^{-\frac{4}{5}\nu}\langle x\rangle^{-\frac{4}{5}(\sigma_1-\frac{9}{4})} < t^{-\frac{2}{3}-\frac{5}{12\nu}}\langle x\rangle^{-\frac{1}{\nu}(\sigma_1-\frac{9}{4})}.$$

Consequently, in this case, we have the bound

$$|G(x,y,t)| \lesssim t^{-\frac{2}{3}-\frac{5}{12\nu}}\langle x\rangle^{-\frac{1}{\nu}(\sigma_1-\frac{9}{4})}.$$

We also remark that in this case ($\nu > \frac{5}{4}$ case), it follows that $\frac{2}{3} + \frac{5}{12\nu} < 1$, so

$$t^{-\frac{2}{3}-\frac{5}{12\nu}}\langle x\rangle^{-\frac{1}{\nu}(\sigma_1-\frac{9}{4})} < t^{-1}\langle x\rangle^{-\frac{1}{\nu}(\sigma_1-\frac{9}{4})}.$$

Case 2. $t < 1$ and $t^{\nu+\frac{5}{12}} \lesssim \langle x\rangle^{-\sigma_1+\frac{9}{4}}$ In this case, we just use

$$|G(x,y,t)| \lesssim t^{1/3}.$$

Instability of Solitons in 2d Cubic ZK

Case 3. $t > 1$ In this case, we apply the second component of (11.105) to obtain

$$|G(x, y, t)| \lesssim b + b^{-\nu+1} \langle x \rangle^{-\tilde{\sigma}_1} + \langle x \rangle^{-\tilde{\sigma}_2} + a^{-\frac{1}{12}} \langle x \rangle^{-\sigma_2 + \frac{9}{4}} + a^{\frac{1}{3}}.$$

In this case, the optimal choices of a and b are $b = \langle x \rangle^{-\frac{\tilde{\sigma}_1}{\nu}}$ and $a = \langle x \rangle^{-\frac{12}{5}(\sigma_2 - \frac{9}{4})}$. □

In the nonlinear argument in the next section, we will need the following consequence of Proposition 11.2, which we state as a corollary.

Corollary 11.3 (Duhamel Estimate) *Suppose that $F(x, y, t)$ satisfies, for some $\sigma \gg 1$ and $\nu \geq 0$,*

- *for $x > 0$, $|F(x, y, t)| \lesssim C_2 \begin{cases} t^{-\nu} \langle x \rangle^{-\sigma_1} & \text{if } t < 1 \\ \langle x \rangle^{-\sigma_2} & \text{if } t > 1 \end{cases}$, where $1 < \sigma_2 \leq \sigma_1$,*
- $\|\langle x \rangle^{-1} F(x, y, t)\|_{L_t^\infty L_{y \in \mathbb{R}, x < 0}^2} \lesssim C_2,$
- $\|\langle x \rangle^{-1} \partial_x F(x, y, t)\|_{L_t^\infty L_{y \in \mathbb{R}, x < 0}^1} \lesssim C_2,$
- $\|\partial_x F(x, y, t)\|_{L_t^\infty L_{y \in \mathbb{R}, x > 0}^1} \lesssim C_2.$

Then, if $\nu > \frac{5}{4}$ and $r \geq 0$, we have for $x > 0$

$$\left| \int_0^t [\partial_x S(t - t') F(\bullet, \bullet, t')](x, y) \, dt' \right| \lesssim C_2 \begin{cases} t^{-\nu/3} \langle x \rangle^{-\sigma_1/3} & \text{if } 0 \leq t \leq 1 \\ \langle x \rangle^{-\frac{1}{3}\sigma_2 - r} & \text{if } t > \frac{1}{2}, \end{cases} \quad (11.106)$$

provided $\sigma_j \geq \frac{11}{2}$ for $j = 1, 2$ and

$$\sigma_2 \geq \max\left(\frac{27}{7} + \frac{15}{7}r, \frac{5}{4} + 3r\right) \quad (11.107)$$

and

$$\sigma_1 \geq \frac{27}{7}(\nu + 1) \text{ and } \sigma_1 \geq \frac{\nu}{2}\sigma_2 + \frac{5}{8} + \frac{3}{2}\nu r. \quad (11.108)$$

Proof By Proposition 11.2, the first line of (11.106) will hold provided

$$t^{\nu + \frac{5}{12}} \lesssim \langle x \rangle^{-(\sigma_1 - \frac{9}{4})} \implies t^{1/3} \lesssim t^{-\nu/3} \langle x \rangle^{-\sigma_1/3} \quad (11.109)$$

$$\langle x \rangle^{-(\sigma_1 - \frac{9}{4})} \ll t^{\nu + \frac{5}{12}} < 1 \implies t^{-\frac{2}{3} - \frac{5}{12\nu}} \langle x \rangle^{-\frac{1}{\nu}(\sigma_1 - \frac{9}{4})} \lesssim t^{-\nu/3} \langle x \rangle^{-\sigma_1/3}. \quad (11.110)$$

For $0 \leq t \leq 1$, we will show that (11.109) and (11.110) hold provided the left inequality of (11.108) holds.

Regarding (11.109), the right side of the implication can be reexpressed as follows:

$$\text{RHS} \iff t^{\nu+1} \lesssim \langle x \rangle^{-\sigma_1} \iff (t^{\nu+1})^{\frac{\sigma_1 - \frac{9}{4}}{\sigma_1}} \lesssim \langle x \rangle^{-(\sigma_1 - \frac{9}{4})}.$$

Thus, the implication in (11.109) will be true if

$$(t^{\nu+1})^{\frac{\sigma_1 - \frac{9}{4}}{\sigma_1}} \lesssim t^{\nu + \frac{5}{12}}.$$

We can reexpress this as $t^{(\nu+1)(\sigma_1 - \frac{9}{4})} \lesssim t^{(\nu + \frac{5}{12})\sigma_1}$. Since $0 \leq t \leq 1$, this is equivalent to

$$(\nu + 1)\left(\sigma_1 - \frac{9}{4}\right) \geq \left(\nu + \frac{5}{12}\right)\sigma_1.$$

With some algebra, this reduces to the left inequality of (11.108).

Regarding (11.110), the right side of the implication can be reexpressed as follows:

$$\langle x \rangle^{-[(1 - \frac{\nu}{3})\sigma_1 - \frac{9}{4}]} \lesssim t^{\frac{2}{3}\nu - \frac{1}{3}\nu^2 + \frac{5}{12}},$$

and also equivalently, by exponentiating

$$\langle x \rangle^{-(\sigma_1 - \frac{9}{4})} \lesssim t^\mu, \quad \mu = \frac{(-\frac{\nu^2}{3} + \frac{2\nu}{3} + \frac{5}{12})(\sigma_1 - \frac{9}{4})}{(1 - \frac{\nu}{3})\sigma_1 - \frac{9}{4}},$$

where we have assumed that $(1 - \frac{\nu}{3})\sigma_1 - \frac{9}{4} > 0$. Thus, the implication in (11.110) is true provided that

$$t^{\nu + \frac{5}{12}} \lesssim t^\mu.$$

Since $0 \leq t \leq 1$, this is equivalent to

$$\nu + \frac{5}{12} \geq \mu,$$

which we reexpress as

$$\left(\nu + \frac{5}{12}\right)\left((1 - \frac{\nu}{3})\sigma_1 - \frac{9}{4}\right) \geq \left(-\frac{\nu^2}{3} + \frac{2\nu}{3} + \frac{5}{12}\right)\left(\sigma_1 - \frac{9}{4}\right).$$

Some algebra reduces this to the condition on the left in (11.108). Thus, we have established that the left inequality in (11.108) suffices to imply (11.109), (11.110), from which it follows from Proposition 11.2 that the first line of (11.106) holds.

By Proposition 11.2, we have the second line of (11.106) holds provided

$$\min\left(\frac{\tilde{\sigma}_1}{\nu}, \tilde{\sigma}_2, \frac{4}{5}\left(\sigma_2 - \frac{9}{4}\right)\right) \geq \frac{\sigma_2}{3} + r \tag{11.111}$$

holds, where

$$\tilde{\sigma}_j = \min\left(\sigma_j - \frac{9}{4}, \frac{2}{3}\sigma_j - \frac{5}{12}\right).$$

Since we assume that $\sigma_j \geq \frac{11}{2}$ for $j = 1, 2$ we have $\tilde{\sigma}_j = \frac{2}{3}\sigma_j - \frac{5}{12}$. We observe that (11.111) holds provided (11.107) and the second inequality of (11.108) holds. □

12 Nonlinear Estimate

Now we return to the problem of estimating η. Recall that η solves

$$\partial_t \eta - \partial_x[(-\Delta + x_t)\eta] = F, \qquad F = f_1 + \partial_x f_2, \tag{12.112}$$

where

$$\begin{aligned} f_1 &= -(\lambda^{-1})_t \partial_{\lambda^{-1}} \tilde{Q} \\ f_2 &= +(x_t - 1)\tilde{Q} - 3\tilde{Q}^2\eta - 3\tilde{Q}\eta^2 - \eta^3. \end{aligned} \tag{12.113}$$

Here, $\tilde{Q}(x, y) = \lambda^{-1} Q(\lambda^{-1}(x + K), \lambda^{-1} y)$. Note that since $|Q(y_1, y_2)| \lesssim \langle \vec{y} \rangle^{-1/2} e^{-|\vec{y}|}$, we have $|\tilde{Q}(x, y)| \leq e^{-K/2}$ for $x > 0$ (see Remark 8.2). We know that for all $t > 0$,

$$\|\eta(t)\|_{H^1_{xy}} \lesssim \delta$$

and

$|\lambda(t) - 1| \lesssim \delta, \qquad |\lambda_t| \lesssim \delta, \qquad |x_t - 1| \lesssim \delta \quad \text{or} \quad (1 - \delta)t \lesssim x(t) \lesssim (1 + \delta)t.$

Furthermore, we know that $\phi(x, y) = \eta(0, x, y)$ satisfies, for $x > 0$, $y \in \mathbb{R}$,

$$|\phi(x, y)| \leq \delta \langle x \rangle^{-\sigma}.$$

Finally, we know that for any $T > 0$, η is the *unique* solution of (12.112) in $C([0, T]; H^1_{xy})$ such that $\eta(t, x + x(t), y) \in L^4_x L^\infty_{yT}$. Our goal is to show that for $x > 0$ and $y \in \mathbb{R}$,

$$|\eta(t, x, y)| \lesssim \delta \begin{cases} t^{-7/12} \langle x \rangle^{-\sigma + \frac{7}{4}} & \text{if } 0 \leq t \leq 1 \\ \langle x \rangle^{-\frac{2}{3}\sigma + \frac{3}{4}} & \text{if } t \geq 1. \end{cases} \quad (12.114)$$

Proposition 12.1 *There exists $\delta_0 > 0$ (small), $K > 0$ (large), and $\sigma_0 > 0$ (large) such that the following holds true. Suppose that $\sigma \geq \sigma_0$, $0 < \delta \leq \delta_0$, $\phi \in H^1$ with $\|\phi\|_{H^1} \leq \delta$, and*

$$\text{for } x > 0, \qquad |\phi(x, y)| \leq \delta \langle x \rangle^{-\sigma}.$$

Then the unique solution $\eta(t, x, y)$ solving (12.112) for all t satisfies

$$\text{for } x > 0, \qquad |\eta(t, x, y)| \lesssim \delta \begin{cases} t^{-7/12} \langle x \rangle^{-\sigma + \frac{7}{4}} & \text{if } 0 < t \leq 1 \\ \langle x \rangle^{-\frac{2}{3}\sigma + \frac{3}{4}} & \text{if } t \geq 1. \end{cases} \quad (12.115)$$

The proof consists of the following steps. The following lemma provides a key short-time step result.

Lemma 12.2 *There exists $\sigma_0 > 0$ (large), $K > 0$ (large), and $\delta_0 > 0$ (small) such that if $\sigma \geq \sigma_0$, $0 < \delta \leq \delta_0$, then the unique solution $\eta(t, x, y)$ solving (12.112) satisfies, for all $0 \leq t \leq 1$,*

$$\text{for } x > 0, \qquad |\eta(t, x, y)| \lesssim \delta \, t^{-7/12} \langle x \rangle^{-\sigma + \frac{7}{4}}. \quad (12.116)$$

Proof This is done using a contraction argument and the available decay estimates, and the Duhamel estimate Corollary 11.3, as follows. Take $T = 1$ and define the Y norm as follows

$$\|\eta\|_Y = \|\eta(t, x, y)\|_{L^\infty_T H^1_{xy}} + \|\eta(t, x + x(t), y)\|_{L^4_x L^\infty_{yT}} + \|\eta(t, x, y) t^{7/12} \langle x \rangle^{\sigma - \frac{7}{4}}\|_{L^\infty_{Ty,x>0}}.$$

Let Λ be defined on Y by

$$\Lambda \eta = S(t, 0)\phi + \int_0^t S(t, t') f_1(\bullet, \bullet, t') \, dt' + \int_0^t \partial_x S(t, t') f_2(\bullet, \bullet, t') \, dt'.$$

By Proposition 10.1 with $C_1 = \delta$, we obtain that for $x > 0$ and $y \in \mathbb{R}$, $0 \leq t \leq 1$,

$$|[S(t, 0)\phi](x, y)| \leq \tfrac{1}{4} C_3 \delta t^{-7/12} \langle x \rangle^{-\sigma + \frac{7}{4}} \quad (12.117)$$

for some absolute constant C_3 (which for convenience in writing below, we will take ≥ 1).

Since $f_1 = \lambda^{-3}\lambda_t(Q + (x+K)Q_x + yQ_y)$, where Q, Q_x, and Q_y are evaluated at $(\lambda^{-1}(x+K), \lambda^{-1}y)$, we have that $|f_1(x,y,t)| \lesssim \delta\delta_0\langle x\rangle^{-\sigma}$ for all t and $x > 0$. Thus, by Proposition 10.1, for $x > 0$,

$$\left|\int_0^t S(t,t')f_1(\bullet,\bullet,t')\,dt'\right| \lesssim \delta\delta_0\langle x\rangle^{-\sigma+\frac{7}{4}} \le \tfrac{1}{4}C_3\delta t^{-7/12}\langle x\rangle^{-\sigma+\frac{7}{4}}.$$

Suppose that $\|\eta\|_Y \le C_3\delta$. Then, requiring K large enough so that $\langle K\rangle^{-1} \le \delta_0$ (which also implies that $e^{-K/2} \le \delta_0$, see our Remark 8.2), we have for $x > 0$, $y \in \mathbb{R}$,

$$|f_2(t,x,y)| \le |x_t - 1|\tilde{Q} + 3|\eta|\tilde{Q}^2 + 3|\eta|^2\tilde{Q} + |\eta|^3 \lesssim C_3^3\delta_0\delta\, t^{-7/4}\langle x\rangle^{-3(\sigma-\frac{7}{4})}.$$

Moreover, by Sobolev

$$\|\langle x\rangle^{-1}\partial_x f_2\|_{L_T^\infty L_{y\in\mathbb{R},x<0}^1} + \|\partial_x f_2\|_{L_T^\infty L_{y\in\mathbb{R},x>0}^1} + \|\langle x\rangle^{-1}f_2\|_{L_T^\infty L_{y\in\mathbb{R},x<0}^2} \lesssim C_3^3\delta_0\delta.$$

Thus, in the hypothesis of Corollary 11.3, we can take $C_2 = C_3^3\delta_0\delta$ and $\sigma_1 = 3(\sigma - \frac{7}{4})$, and conclude that for $x > 0$ and $y \in \mathbb{R}$,

$$\left|\int_0^t [\partial_x S(t,t')f_2(\bullet,\bullet,t')](x,y)\,dt'\right| \le C_4 C_3^3\delta_0\delta\, t^{-7/12}\langle x\rangle^{-\sigma+\frac{7}{4}} \qquad (12.118)$$

for some absolute constant $C_4 > 0$. Taking $\delta_0 > 0$ sufficiently small so that $C_4 C_3^2\delta_0 \le \frac{1}{2}$, we obtain from (12.117) and (12.118) that for $x > 0$ and $y \in \mathbb{R}$, $0 \le t \le 1$,

$$|(\Lambda\eta)(t,x,y)| \le \frac{1}{2}C_3\delta t^{-7/12}\langle x\rangle^{-\sigma+\frac{7}{4}}. \qquad (12.119)$$

Moreover, by the estimates in Lemmas 8.5 and 8.6, if $\|\eta\|_{L_T^\infty H_{xy}^1} \le \frac{1}{4}C_3\delta$ and $\|\eta(t, x+x(t), y)\|_{L_x^4 L_{yT}^\infty} \le \frac{1}{4}C_3\delta$, then

$$\|\Lambda\eta\|_{L_T^\infty H_{xy}^1} + \|\Lambda\eta(t, x+x(t), y)\|_{L_x^4 L_{yT}^\infty} \le \frac{1}{2}C_3\delta \qquad (12.120)$$

as in the discussion following Theorem 8.4 reviewing the local well-posedness (with a possible adjust to C_3 and δ_0, as required by the absolute constants in those estimates). Combining (12.119) and (12.120), we obtain

$$\|\Lambda\eta\|_Y \le C_3\delta. \qquad (12.121)$$

Moreover, it also follows similarly from these estimates that

$$\|\Lambda \eta_2 - \Lambda \eta_1\|_Y \leq \frac{1}{2}\|\eta_2 - \eta_1\|_Y \tag{12.122}$$

for two $\eta_1, \eta_2 \in Y$ such that $\eta_1(0, x, y) = \eta_2(0, x, y) = \phi$. Hence, Λ is a contraction and the fixed point solves (12.112). By the uniqueness in Theorem 8.4, this fixed point is the unique solution in the function class stated in that Theorem. □

Now the proof proceeds as follows:

- Let $T_* \geq 0$ be the sup of all times for which (12.115) holds.
- By Lemma 12.2, $T_* \geq 1$.
- If $T_* < \infty$, then we will obtain a contradiction in the following series of steps. First, we know that at $T_1 \stackrel{\text{def}}{=} T_* - \frac{1}{2}$,

$$\text{for } x > 0, \qquad |\eta(T_1, x, y)| \lesssim \delta \langle x \rangle^{-\frac{2}{3}\sigma + \frac{3}{4}}.$$

- Apply Lemma 12.2 with $t = 0$ replaced by $t = T_1$ to obtain that η satisfies, for all $T_* - \frac{1}{2} \leq t \leq T_* + \frac{1}{2}$, the estimate

$$\text{for } x > 0, \qquad |\eta(t, x, y)| \lesssim \delta (t - T_1)^{-7/12} \langle x \rangle^{-\frac{2}{3}\sigma + \frac{5}{2}}.$$

Restricting to $T_* \leq t \leq T_* + \frac{1}{2}$, this is simplifies to

$$\text{for } x > 0, \qquad |\eta(t, x, y)| \lesssim \delta \langle x \rangle^{-\frac{2}{3}\sigma + \frac{5}{2}}.$$

- Now we know that for $0 \leq t \leq T_* + \frac{1}{2}$,

$$\text{for } x > 0, \qquad |\eta(t, x, y)| \lesssim \delta \begin{cases} t^{-7/12} \langle x \rangle^{-\sigma + \frac{7}{4}} \\ \langle x \rangle^{-\frac{2}{3}\sigma + \frac{5}{2}} \end{cases} \tag{12.123}$$

holds, which is slightly weaker than (12.115).

- We know that, on $0 \leq t \leq T_* + \frac{1}{2}$, η satisfies

$$\eta = S(t, 0)\phi + \int_0^t S(t, t') f_1(\bullet, \bullet, t') \, dt' + \int_0^t \partial_x S(t, t') f_2(\bullet, \bullet, t') \, dt'.$$

Apply the estimates in Proposition 10.1 and Corollary 11.3 to show that (12.123) suffices to conclude (12.115) holds on $0 \leq t \leq T_* + \frac{1}{2}$, which is a contradiction to the definition of T_*. Indeed, we apply Corollary 11.3 with $\nu = \frac{7}{4}$, $\sigma_1 = 3(\sigma - \frac{7}{4})$, $\sigma_2 = 3(\frac{2}{3}\sigma - \frac{5}{2})$, and $r = \frac{7}{4}$. Then $\frac{1}{3}\sigma_2 + r = \frac{2}{3}\sigma - \frac{3}{4}$, so that (12.115) is obtained.

13 H^1-Instability of Q for the Critical gZK

We are now ready to prove our main result, Theorem 1.3.

Proof of Theorem 1.3 For $n \in \mathbb{N}$ to be chosen later, let

$$u_0^n = Q + \varepsilon_0^n,$$

where

$$\varepsilon_0^n = \frac{1}{n}(Q + a\chi_0), \tag{13.124}$$

and $a \in \mathbb{R}$ is such that $\varepsilon_0^n \perp \chi_0$, that is,

$$a = -\frac{\int \chi_0 Q}{\|\chi_0\|_2^2}.$$

From Theorem 3.1, we have that for every $n \in \mathbb{N}$

$$\varepsilon_0^n \perp \{Q_{y_1}, Q_{y_2}, \chi_0\}.$$

Denote by $u^n(t)$ the solution of (1.1) associated to u_0^n.

Assume by contradiction that Q is stable. Then, for $\alpha_0 < \overline{\alpha}$, where $\overline{\alpha} > 0$ is given by Proposition 5.1, if n is sufficiently large, we have $u^n(t) \in U_{\alpha_0}$ (recall (1.8)). Thus, from Definition 5.3, there exist functions $\lambda^n(t)$ and $x^n(t)$ such that $\varepsilon^n(t)$, defined in (5.39), satisfies

$$\varepsilon^n(t) \perp \{Q_{y_1}, Q_{y_j}, \chi_0\},$$

and also $\lambda^n(0) = 1$ and $x^n(0) = 0$.

To simplify the notation we drop the index n in what follows. Rescaling the time $t \mapsto s$ by $\frac{ds}{dt} = \frac{1}{\lambda^3}$ and taking $\alpha_0 < \alpha_1$, where $\alpha_1 > 0$ is given by Lemma 5.4, we have that $\lambda(s)$ and $x(s)$ are C^1 functions, and that $\varepsilon(s)$ satisfies Eq. (6.52). Moreover, from Proposition 5.1, since $u(t) \in U_{\alpha_0}$, we have

$$\|\varepsilon(s)\|_{H^1} \leq C_1 \alpha_0 \quad \text{and} \quad |\lambda(s) - 1| \leq C_1 \alpha_0, \tag{13.125}$$

thus, taking $\alpha_0 < (2C_1)^{-1}$, we obtain

$$\|\varepsilon(s)\|_{H^1} \leq 1 \quad \text{and} \quad \frac{1}{2} \leq \lambda(s) \leq \frac{3}{2}, \quad \text{for all} \quad s \geq 0. \tag{13.126}$$

Furthermore, in view of (5.42), if $\alpha_0 > 0$ is small enough, we deduce

$$\left|\frac{\lambda_s}{\lambda}\right| + \left|\frac{x_s}{\lambda} - 1\right| \leq C_2 \|\varepsilon(s)\|_2 \leq C_1 C_2 \alpha_0.$$

Since $x_t = x_s/\lambda^3$, we conclude that

$$\frac{1 - C_1 C_2 \alpha_0}{(1 + C_1 \alpha_0)^2} \leq \frac{1 - C_1 C_2 \alpha_0}{\lambda^2} \leq x_t \leq \frac{1 + C_1 C_2 \alpha_0}{\lambda^2} \leq \frac{1 + C_1 C_2 \alpha_0}{(1 - C_1 \alpha_0)^2}$$

Hence, we can choose $\alpha_0 > 0$, small enough, such that

$$\frac{3}{4} \leq x_t \leq \frac{5}{4}.$$

The last inequality implies that $x(t)$ is increasing and by the Mean Value Theorem

$$x(t_0) - x(t) \geq \frac{3}{4}(t_0 - t)$$

for every $t_0, t \geq 0$ with $t \in [0, t_0]$. Also, recalling $x(0) = 0$, another application of the Mean Value Theorem yields

$$x(t) \geq \frac{1}{2}t$$

for all $t \geq 0$. Finally, by assumption (13.124) and properties of Q, we have

$$|u_0(\vec{x})| \leq ce^{-\delta|\vec{x}|},$$

for some $c > 0$ and $\delta > 0$.

From the monotonicity properties in Sect. 7, we obtain the L^2 exponential decay on the right for $\varepsilon(s)$. □

Corollary 13.1 *Let $M \geq 4$. If $\alpha_0 > 0$ is sufficiently small, then there exists $C = C(M, \delta) > 0$ such that for every $s \geq 0$ and $y_0 > 0$*

$$\int_{\mathbb{R}} \int_{y_1 > y_0} \varepsilon^2(s, y_1, y_2) dy_1 dy_2 \leq Ce^{-\frac{y_0}{2M}}.$$

Proof Applying Lemma 7.2, for a fixed $M \geq 4$, there exists $C = C(M) > 0$ such that for all $t \geq 0$ and $x_0 > 0$ we have

$$\int_{\mathbb{R}} \int_{x_1 > x_0} u^2(t, x_1 + x(t), x_2) dx_1 dx_2 \leq Ce^{-\frac{x_0}{M}}.$$

From the definition of $\varepsilon(s)$, we have that

$$\frac{1}{\lambda(s)}\varepsilon\left(s, \frac{y_1}{\lambda(s)}, \frac{y_2}{\lambda(s)}\right) = u(s, y_1 + x(s), y_2) - \frac{1}{\lambda(s)}Q\left(\frac{y_1}{\lambda(s)}, \frac{y_2}{\lambda(s)}\right).$$

Moreover, if $\alpha_0 < (2C_1)^{-1}$, we have $1/2 \leq \lambda(s) \leq 3/2$, and using (13.127), we get

$$\frac{1}{\lambda(s)}Q\left(\frac{y_1}{\lambda(s)}, \frac{y_2}{\lambda(s)}\right) \leq \frac{c}{\lambda(s)}e^{-\frac{|\bar{y}|}{\lambda(s)}} \leq 2c\,e^{-\frac{2}{3}|\bar{y}|} \leq c\,e^{-\frac{|\bar{y}|}{M}}, \qquad (13.127)$$

since $M \geq 3/2$.

Therefore, we deduce that

$$\int_{\mathbb{R}}\int_{y_1>y_0} \frac{1}{\lambda^2(s)}\varepsilon^2\left(s, \frac{y_1}{\lambda(s)}, \frac{y_2}{\lambda(s)}\right) dy_1 dy_2 \leq 2\int_{\mathbb{R}}\int_{y_1>y_0} u^2(s, y_1 + x(s), y_2) dy_1 dy_2$$

$$+ 2\int_{\mathbb{R}}\int_{y_1>y_0} \frac{1}{\lambda^2(s)}Q^2\left(\frac{y_1}{\lambda(s)}, \frac{y_2}{\lambda(s)}\right) dy_1 dy_2$$

$$\leq 2ce^{-\frac{y_0}{M}} + 2c\int_{\mathbb{R}}\int_{y_1>y_0} e^{-\frac{|\bar{y}|}{M}} dy$$

$$\leq Ce^{-\frac{y_0}{M}}$$

for some $C = C(M) > 0$.

Finally, by the scaling invariance of the L^2-norm, we get

$$\int_{\mathbb{R}}\int_{y>y_0} \varepsilon^2(s, y_1, y_2) dy_1 dy_2 = \int_{\mathbb{R}}\int_{y>\lambda(s)y_0} \frac{1}{\lambda^2(s)}\varepsilon^2\left(s, \frac{y_1}{\lambda(s)}, \frac{y_2}{\lambda(s)}\right) dy_1 dy_2$$

$$\leq Ce^{-\frac{\lambda(s)y_0}{M}} \leq Ce^{-\frac{y_0}{2M}},$$

since $\lambda(s) \geq 1/2$. $\square$

Next, we define a rescaled and shifted quantity of the virial-type. Recall the definition of J_A in (6.48) and let

$$K_A(s) = \lambda(s)(J_A(s) - \kappa).$$

(We remark that this quantity is similar to the corresponding one in Martel and Merle [23].)

From (6.49) and (13.126), it is clear that

$$|K_A(s)| \leq c\left((1 + A^{1/2})\|\varepsilon(s)\|_2 + \kappa\right) < +\infty, \qquad (13.128)$$

for all $s \geq 0$.

Moreover, using Lemma 6.1, we also have

$$\frac{d}{ds}K_A = \lambda_s (J_A - \kappa) + \lambda \frac{d}{ds} J_A$$

$$= \lambda \left(\frac{d}{ds} J_A + \frac{\lambda_s}{\lambda} (J_A - \kappa) \right)$$

$$= \lambda \left(2 \left(1 - \frac{1}{2} \left(\frac{x_s}{\lambda} - 1\right)\right) \int \varepsilon Q + R(\varepsilon, A) \right). \tag{13.129}$$

In the next result we obtain a strictly positive lower bound for $\frac{d}{ds} K_A(s)$ for a certain choice of $\alpha_0 > 0$, $n \in \mathbb{N}$ and $A \geq 1$.

Theorem 13.2 *There exist $\alpha_0 > 0$ sufficiently small, $n_0 \in \mathbb{N}$ and $A \geq 1$ sufficiently large such that*

$$\frac{d}{ds} K_A(s) \geq \frac{b}{2n_0} > 0, \quad \text{for all} \quad s \geq 1, \tag{13.130}$$

where

$$b = \int (Q + a\chi_0) Q = \|Q\|_2^2 - \frac{(\int Q\chi_0)^2}{\|\chi_0\|_2^2}.$$

Remark 13.3 Note that $b > 0$, since $Q \notin \text{span}\{\chi_0\}$.

Proof In view of (13.125), let $\alpha_0 < \min\{\alpha_1(C_1)^{-1}, \alpha_2(C_1)^{-1}, (2C_1)^{-1}, 1/2\}$ so that we can apply Lemmas 5.4 and 6.4. From (13.129) and the definition of M_0 (see (4.24)), we have

$$\frac{d}{ds} K_A(s) = \lambda \left(2 \left(1 - \frac{1}{2} \left(\frac{x_s}{\lambda} - 1\right)\right) M_0 + \widetilde{R}(\varepsilon, A) \right), \tag{13.131}$$

where $\widetilde{R}(\varepsilon, A) = R(\varepsilon, A) - \left(1 - \frac{1}{2}\left(\frac{x_s}{\lambda} - 1\right)\right) \int \varepsilon^2$.

Since $\alpha_0 < (2C_1)^{-1}$, we have $1/2 \leq \lambda(s) \leq 3/2$, and using (5.43), we obtain

$$\lambda \left(1 - \frac{1}{2}\left(\frac{x_s}{\lambda} - 1\right)\right) \geq \frac{1}{2} \cdot \frac{1}{2} = \frac{1}{4}.$$

Moreover, from the definition of M_0, we also get

$$M_0 = 2 \int \varepsilon_0 Q + \int \varepsilon_0^2 \geq 2 \int \varepsilon_0 Q = \frac{2b}{n}.$$

Therefore,

$$2\lambda \left(1 - \frac{1}{2}\left(\frac{x_s}{\lambda} - 1\right)\right) M_0 \geq \frac{b}{n}. \tag{13.132}$$

On the other hand, by Lemma 5.4, we have

$$\left|\frac{\lambda_s}{\lambda}\right| + \left|\frac{x_s}{\lambda} - 1\right| \leq C_2 \|\varepsilon(s)\|_2.$$

Therefore, using the inequalities (6.51) and (13.126), there exists a universal constant $C_6 > 0$, such that for $A \geq 1$ we have

$$\lambda \widetilde{R}(\varepsilon, A) \leq C_6 \|\varepsilon(s)\|_2 \left(\|\varepsilon(s)\|_2 + A^{-1/2} + A^{1/2} \|\varepsilon(s)\|_{L^2(y_1 \geq A)} + \left|\int_{\mathbb{R}^2} y_2 F_{y_2} \varepsilon \varphi_A\right| \right). \tag{13.133}$$

Moreover, by Lemma 6.4, we deduce

$$\|\varepsilon(s)\|_{H^1}^2 \leq C_5 \left(C_1 \alpha_0 \left|\int \varepsilon_0 Q\right| + \|\varepsilon_0\|_{H^1}^2 \right),$$

and thus, the assumption (13.124) yields

$$\|\varepsilon(s)\|_{H^1}^2 \leq C_5 \left(C_1 \alpha_0 \left(\frac{b}{n}\right) + \frac{d}{n^2} \right)$$

$$\leq C_5 \left(C_1 + \frac{d}{b} \right) \left(\alpha_0 + \frac{1}{n} \right) \left(\frac{b}{n}\right), \tag{13.134}$$

where $d = \|Q + a\chi_0\|_{H^1}$.

Set $C_7 = C_5 \left(C_1 + \frac{d}{b} \right)$. Collecting (13.133) and (13.134), we obtain

$$\lambda \widetilde{R}(\varepsilon, A) \leq C_7 C_6 \left(\alpha_0 + \frac{1}{n} \right) \left(\frac{b}{n}\right) +$$

$$+ \sqrt{C_7} C_6 \left(A^{-1/2} + A^{1/2} \|\varepsilon(s)\|_{L^2(y_1 \geq A)} \right) \left(\alpha_0 + \frac{1}{n} \right)^{1/2} \left(\frac{b}{n}\right)^{1/2}$$

$$+ \sqrt{C_7} C_6 \left|\int_{\mathbb{R}^2} y_2 F_{y_2} \varepsilon \varphi_A\right| \left(\alpha_0 + \frac{1}{n} \right)^{1/2} \left(\frac{b}{n}\right)^{1/2}.$$

Let $K \geq 1$ satisfy (8.76) and split the integral on the right hand side of the last inequality into two parts

$$\int_{\mathbb{R}^2} y_2 F_{y_2} \varepsilon \varphi_A = \int_{\mathbb{R}} \int_{y_1 < K} y_2 F_{y_2} \varepsilon \varphi_A \, dy_1 dy_2 + \int_{\mathbb{R}} \int_{y_1 > K} y_2 F_{y_2} \varepsilon \varphi_A \, dy_1 dy_2.$$

From (6.46) and (6.47) we have, for every $A > K \geq 1$, that

$$\int_{\mathbb{R}} \int_{y_1 < K} y_2 F_{y_2} \varepsilon \varphi_A dy_1 dy_2 \leq c \left(\int_{\mathbb{R}} \int_{y_1 < K} |y_2 F_{y_2}|^2 dy_1 dy_2 \right)^{1/2} \|\varepsilon(s)\|_2$$

$$\leq c K^{1/2} \|\varepsilon(s)\|_2$$

$$\leq c \sqrt{C_7} \left(\alpha_0 + \frac{1}{n} \right)^{1/2} \left(\frac{b}{n} \right)^{1/2}, \qquad (13.135)$$

where in the last line we used (13.134).

To bound the second part we use Lemma 8.1 with n large such that

$$\delta = \sqrt{C_7} \left(\alpha_0 + \frac{1}{n} \right)^{1/2} \left(\frac{b}{n} \right)^{1/2} < \delta_0.$$

Indeed, since ε_0 given by (13.124) has an exponential decay, the relation (8.74) is satisfied for any $\sigma > \frac{21}{8}$. Therefore, $\sigma^* = -\frac{2}{3}\sigma + \frac{3}{4} < -1$ and for every $s \geq 1$ and $A > K \geq 1$, that

$$\int_{\mathbb{R}} \int_{y_1 > K} y_2 F_{y_2} \varepsilon \varphi_A dy_1 dy_2 \leq \int_{\mathbb{R}} \sup_{y_1} |y_2 F_{y_2}| \left(\int_{y_1 > K} |\varepsilon| dy_1 \right) dy_2$$

$$\leq c(\sqrt{C_7} + 1) \left(\frac{1}{n} + \left(\alpha_0 + \frac{1}{n} \right)^{1/2} \left(\frac{b}{n} \right)^{1/2} \right),$$

where we also used (6.46) in the last line. Now, there exists a constant $C_8 > 0$ such that

$$\int_{\mathbb{R}} \int_{y_1 > K} y_2 F_{y_2} \varepsilon \varphi_A dy_1 dy_2 \leq C_8 \left(\frac{1}{n} + \left(\alpha_0 + \frac{1}{n} \right)^{1/2} \left(\frac{b}{n} \right)^{1/2} \right). \qquad (13.136)$$

Collecting (13.135) and (13.136), for every $s \geq 1$ and $A \geq 1$, we have

$$\left| \int_{\mathbb{R}^2} y_2 F_{y_2} \varepsilon \varphi_A \right| \leq C_9 \left(\frac{1}{b^{1/2} n^{1/2}} + \left(\alpha_0 + \frac{1}{n} \right)^{1/2} \right) \left(\frac{b}{n} \right)^{1/2}$$

for some constant $C_9 > 0$.

Now, we choose $\alpha_0 > 0$ sufficiently small and $n_0 \in \mathbb{N}$ sufficiently large such that

$$\sqrt{C_7} C_6 \left(\alpha_0 + \frac{1}{n_0} \right)^{1/2} \max \left\{ \sqrt{C_7} \left(\alpha_0 + \frac{1}{n_0} \right)^{1/2}, 1, C_9 \left(\frac{1}{b^{1/2} n_0^{1/2}} + \left(\alpha_0 + \frac{1}{n_0} \right)^{1/2} \right) \right\} < 1/6.$$

For fixed α_0 and n_0, satisfying the previous inequality, we choose $A \geq 1$ such that

$$A^{-1/2} + A^{1/2} \|\varepsilon(s)\|_{L^2(y_1 \geq A)} \leq \left(\frac{b}{n_0} \right)^{1/2},$$

which is possible due to Corollary 13.1.

Therefore, we finally deduce that

$$\lambda \widetilde{R}(\varepsilon, A) \leq \frac{b}{2n_0},$$

which implies from (13.131) and (13.132) that

$$\frac{d}{ds} K_A(s) \geq \frac{b}{2n_0} > 0, \quad \text{for all} \quad s \geq 1.$$

◻

Now we have all the ingredients to finish the proof of our main result.

Last Step in the Proof of Theorem 1.3. Integrating in s variable both sides of inequality (13.130), we get

$$K_A(s) \geq s \left(\frac{b}{2n_0} \right) + K_A(0), \quad \text{for all} \quad s \geq 1.$$

Therefore,

$$\lim_{s \to \infty} K_A(s) = \infty,$$

which is a contradiction to (13.128). Hence, our original assumption that Q is stable is not valid and we conclude the proof of the theorem. ◻

Acknowledgements Most of this work was done when the first author was visiting GWU in 2016–2017 under the support of the Brazilian National Council for Scientific and Technological Development (CNPq/Brazil), for which all authors are very grateful as it boosted the energy into the research project. S.R. would like to thank IHES and the organizers for the excellent working conditions during the trimester program "Nonlinear Waves" in May–July 2016. L.G.F. was partially supported by CNPq, CAPES and FAPEMIG/Brazil. J.H. was partially supported by the NSF grant DMS-1500106. S.R. was partially supported by the NSF CAREER grant DMS-1151618/1929029.

References

1. D. Bhattacharya, L. G. Farah, and S. Roudento, *Global well-posedness for low regularity data in the 2d modified Zakharov-Kuznetsov equation*, arxiv.org preprint. arXiv:1906.05822.
2. J.L. Bona, P. Souganidis and W. Strauss, *Stability and instability of solitary waves of Korteweg-de Vries type*, Proc. Roy. Soc. London 411 (1987), 395–412.
3. S.-M. Chang, S. Gustafson, K. Nakanishi and T.-P. Tsai, *Spectra of linearized operators for NLS solitary waves*, SIAM J. Math. Anal. 39 (2007/08), no. 4, 1070–1111.
4. V. Combet, *Construction and characterization of solutions converging to solitons for supercritical gKdV equations*, Differential Integral Equations 23 (2010), no. 5–6, 513–568.

5. R. Côte, C. Muñoz, D. Pilod, and G. Simpson, *Asymptotic Stability of high-dimensional Zakharov-Kuznetsov solitons*, Arch. Ration. Mech. Anal. 220 (2016), no. 2, 639–710.
6. de Bouard, A. *Stability and instability of some nonlinear dispersive solitary waves in higher dimension*, Proc. Roy. Soc. Edinburgh Sect. A 126 (1996), no. 1, 89–112.
7. T. Duyckaerts, J. Holmer and S. Roudenko, *Scattering for the non-radial 3d cubic nonlinear Schrödinger equation*, Math. Res. Lett. 15 (2008), no. 6, 1233–1250.
8. A. V. Faminskii, *The Cauchy problem for the Zakharov-Kuznetsov equation*, Differ. Equ. 31 (1995), 1002–1012.
9. L. G. Farah, J. Holmer and S. Roudenko, *Instability of solitons - revisited, I: the critical gKdV equation*, to appear in Contemp. Math., Amer. Math. Soc., arxiv.org preprint arXiv:1711.03187.
10. L. G. Farah, J. Holmer and S. Roudenko, *Instability of solitons - revisited, II: the supercritical Zakharov-Kuznetsov equation*, to appear in Contemp. Math., Amer. Math. Soc., arxiv.org preprint arXiv:1711.03207.
11. L. G. Farah, J. Holmer, S. Roudenko and Kai Yang, *Blow-up in finite or infinite time of the 2D cubic Zakharov-Kuznetsov equation*, arxiv.org preprint arXiv:1810.05121.
12. L. G. Farah, F. Linares and A. Pastor, *A note on the 2d generalized Zakharov-Kuznetsov equation: Local, global, and scattering results*, J. Differential Equations 253 (2012), 2558–2571.
13. G. Fonseca and M. Pachón, *Well-posedness for the two dimensional generalized Zakharov-Kuznetsov equation in anisotropic weighted Sobolev spaces*, J. Math. Anal. Appl. 443 (2016), no. 1, 566–584.
14. M. Grillakis, J. Shatah, and W. Strauss, *Stability theory of solitary waves in the presence of symmetry*, J. Funct. Anal. 74 (1987), 160–197.
15. A. Grünrock and S. Herr, *The Fourier restriction norm method for the Zakharov-Kuznetsov equation*, Discrete Contin. Dyn. Syst. 34 (2014), no. 5, 2061–2068.
16. D. Han-Kwan, *From Vlasov-Poisson to Korteweg-de Vries and Zakharov-Kuznetsov*, Comm. Math. Phys., 324 (2013), no. 3, 961–993.
17. J. Holmer and S. Roudenko, *A sharp condition for scattering of the radial 3d cubic nonlinear Schrödinger equation*, Comm. Math. Phys. 282 (2008), no. 2, 435–467.
18. M.K. Kwong, *Uniqueness of positive solutions of $\Delta u - u + u^p = 0$ in $\mathbb{R}^n$*, Arch. Rational Mech. Anal. 105 (1989), no. 3, 243–266.
19. D. Lannes, F. Linares and J.-C. Saut, *The Cauchy problem for the Euler-Poisson system and derivation of the Zakharov-Kuznetsov equation*, Studies in phase space analysis with applications to PDEs, 181–213, Progr. Nonlinear Differential Equations Appl., 84, Birkhäuser/Springer, New York, 2013.
20. F. Linares and A. Pastor, *Well-posedness for the two-dimensional modified Zakharov–Kuznetsov equation*, SIAM J. Math Anal. 41 (2009), 1323–1339.
21. F. Linares and A. Pastor, *Local and global well-posedness for the 2d generalized Zakharov-Kuznetsov equation*, J. Funct. Anal. 260 (2011), 1060–1085.
22. M. Maris, *Existence of nonstationary bubbles in higher dimensions*, J. Math. Pures Appl. (9) 81 (2002), no. 12, 1207–1239.
23. Y. Martel and F. Merle, *Instability of solitons for the critical gKdV equation*, GAFA, Geom. Funct. Anal., 11 (2001), 74–123.
24. Y. Martel and F. Merle, *Blow up in finite time and dynamics of blow up solutions for the L^2-critical generalized KdV equation*, J. Amer. Math. Soc. 15 (2002), 617–664.
25. S. Melkonian and S. A. Maslowe, *Two dimensional amplitude evolution equations for nonlinear dispersive waves on thin films*, Phys. D 34 (1989), 255–269.
26. F. Merle, *Existence of blow-up solutions in the energy space for the critical generalized Korteweg-de Vries equation*, J. Amer. Math. Soc. 14 (2001), 555–578.
27. L. Molinet and D. Pilod, *Bilinear Strichartz estimates for the Zakharov-Kuznetsov equation and applications*, Ann. Inst. H. Poincaré An. Non Lin., 32 (2015), 347–371.
28. S. Monro and E. J. Parkes, *The derivation of a modified Zakharov-Kuznetsov equation and the stability of its solutions*, J. Plasma Phys. 62 (3) (1999), 305–317.

29. S. Monro and E. J. Parkes, *Stability of solitary-wave solutions to a modified Zakharov-Kuznetsov equation*, J. Plasma Phys. 64 (3) (2000), 411–426.
30. F. Ribaud and S. Vento, *A note on the Cauchy problem for the 2d generalized Zakharov-Kuznetsov equations*, C. R. Math. Acad. Sci. Paris 350 (2012), no. 9–10, 499–503.
31. M. Schechter, *Spectra of Partial Differential Operator*, North Holland, 1986.
32. M. Weinstein, *Modulational stability of ground states of nonlinear Schrödinger equations*, SIAM J. Math. Anal. 16 (1985), no. 3, 472–491.
33. Zakharov V. E. and Kuznetsov E. A., *On three dimensional solitons*, Zhurnal Eksp. Teoret. Fiz, 66, 594–597 [in Russian]; Sov. Phys JETP, vol. 39, no. 2 (1974), 285–286.

On the Nonexistence of Local, Gauge-Invariant Birkhoff Coordinates for the Focusing NLS Equation

Thomas Kappeler and Peter Topalov

1 Introduction

It is well known that the non-linear Schrödinger (NLS) equation

$$\begin{cases} \dot{\varphi}_1 = i\varphi_{1xx} - 2i\varphi_1^2\varphi_2, \\ \dot{\varphi}_2 = -i\varphi_{2xx} + 2i\varphi_1\varphi_2^2, \end{cases} \qquad (1)$$

on the torus $\mathbb{T} \equiv \mathbb{R}/\mathbb{Z}$ is a Hamiltonian PDE on the scale of Sobolev spaces $H_c^s = H_\mathbb{C}^s \times H_\mathbb{C}^s$, $s \geq 0$, with Poisson bracket

$$\{F, G\}(\varphi) := -i \int_0^1 \left((\partial_{\varphi_1} F)(\partial_{\varphi_2} G) - (\partial_{\varphi_1} G)(\partial_{\varphi_2} F) \right) dx, \qquad (2)$$

and Hamiltonian $\mathcal{H} : H_c^1 \to \mathbb{C}$, given by

$$\mathcal{H}(\varphi) = \int_0^1 \left(\varphi_{1x}\varphi_{2x} + \varphi_1^2\varphi_2^2 \right) dx, \quad \varphi = (\varphi_1, \varphi_2) \in H_c^1. \qquad (3)$$

Here, for any $s \geq 0$, $H_\mathbb{C}^s \equiv H^s(\mathbb{T}, \mathbb{C})$ denotes the Sobolev space of complex valued functions on $\mathbb{T}$ and the Poisson bracket (2) is defined for functionals F and G on H_c^s, provided that the pairing given by the integral in (2) is well-defined (cf. Sect. 2

T. Kappeler (✉)
Institute of Mathematics, University of Zurich, Zurich, Switzerland
e-mail: thomas.kappeler@math.uzh.ch

P. Topalov
Department of Mathematics, Northeastern University, Boston, USA

for more details on these matters). The NLS phase space H_c^s is a direct sum of two real subspaces $H_{\mathbb{C}}^s = H_r^s \oplus_{\mathbb{R}} i H_r^s$ where

$$H_r^s = \{\varphi \in H_c^s \,|\, \varphi_2 = \overline{\varphi_1}\} \quad \text{and} \quad i H_r^s = \{\varphi \in H_c^s \,|\, \varphi_2 = -\overline{\varphi_1}\}.$$

The Hamiltonian vector field, corresponding to (2) and (3),

$$X_{\mathcal{H}}(\varphi) = i\big(-\partial_{\varphi_2}\mathcal{H}, \partial_{\varphi_1}\mathcal{H}\big) = \big(i\varphi_{1xx} - 2i\varphi_1^2\varphi_2, -i\varphi_{2xx} + 2i\varphi_1\varphi_2^2\big)$$

is tangent to the real subspaces H_r^s and iH_r^s (cf. Sect. 2) and for any $s \geq 0$ the restrictions

$$X_{\mathcal{H}}\big|_{H_r^2} : H_r^2 \to H_r^0 \quad \text{and} \quad X_{\mathcal{H}}\big|_{iH_r^2} : iH_r^2 \to iH_r^0$$

are real analytic maps (cf. Sect. 2). The vector field $X_{\mathcal{H}}\big|_{H_r^2}$ corresponds to the *defocusing* NLS (dNLS) equation

$$iu_t = -u_{xx} + 2|u|^2 u$$

whereas $X_{\mathcal{H}}\big|_{iH_r^2}$ corresponds to the *focusing* NLS (fNLS) equation

$$iu_t = -u_{xx} - 2|u|^2 u.$$

Both equations are known to be well-posed on the Sobolev space H_r^s and respectively iH_r^s for any $s \geq 0$. Moreover, they are integrable PDEs: the dNLS equation can be brought into Birkhoff normal form on the *entire* phase space H_r^0 (cf. [2]) whereas for the fNLS equation, an Arnold-Liouville type theorem has been established in [3]. In broad terms, the latter can be described as follows: By [8], the fNLS equation admits a Lax pair representation $\partial_t L(\varphi) = P(\varphi)L(\varphi) - L(\varphi)P(\varphi)$ where for any $\varphi = (\varphi_1, -\overline{\varphi}_1) \in iH_r^0$, $L(\varphi) : iH_r^1 \to iH_r^0$ is the first order differential operator

$$L(\varphi) := i\begin{pmatrix} 1 & 0 \\ 0 & -1 \end{pmatrix}\partial_x + \begin{pmatrix} 0 & \varphi_1 \\ -\overline{\varphi}_1 & 0 \end{pmatrix}$$

and $P(\varphi)$ is a certain second order differential operator. As a consequence, the spectrum $\operatorname{spec} L(\varphi)$ of the operator $L(\varphi)$, considered on the interval $[0, 2]$ with periodic boundary conditions, is invariant with respect to the fNLS flow. Since the resolvent of $L(\varphi)$ is compact, $\operatorname{spec} L(\varphi)$ is discrete and hence $\operatorname{spec} L(\varphi)$ being invariant means that the periodic eigenvalues of $L(\varphi)$ are first integrals of the fNLS equation. In particular, for any potential $\varphi \in iH_r^0$, the isospectral set $\operatorname{Iso}(\varphi) := \{\psi \in iH_r^0 \,|\, \operatorname{spec} L(\psi) = \operatorname{spec} L(\varphi)\}$ and its connected component $\operatorname{Iso}_0(\varphi)$, containing φ, are invariant under the fNLS flow. The Arnold-Liouville type theorem for the fNLS equation, established in [3] (cf. also [4]), says that for any potential $\varphi \in i H_r^0$ with the property that all periodic eigenvalues of $L(\varphi)$ are simple, there exists a fNLS invariant neighborhood of $\mathrm{Iso}_0(\varphi)$ in $i H_r^0$, on which the fNLS equation can be brought into Birkhoff normal form. In fact, any Hamiltonian in the (local) Poisson algebra, defined by the action variables, can be brought into such a form. We note that $\mathrm{Iso}_0(\varphi)$ is homeomorphic to an infinite product of circles, but that in any invariant neighborhood of it there is a dense set of invariant tori of *finite* dimension. Unlike in the classical, finite dimensional Arnold-Liouville theorem, such a neighborhood can be described in terms of Birkhoff coordinates rather than action angle coordinates.

The aim of this paper is to study the local properties of the vector field $X_{\mathcal{H}}$ in small neighborhoods of the *constant potentials*

$$\varphi_c(x) = (c, -\bar{c}) \in i H_r^0, \quad c \in \mathbb{C} \setminus \{0\}.$$

It is straightforward to see that all, but finitely many, periodic eigenvalues of $L(\varphi_c)$ have algebraic (and geometric) multiplicity two. Hence the Arnold-Liouville type theorem in [3] does not apply. More specifically, for a given $c \in \mathbb{C}$, $c \neq 0$, consider the *re-normalized* NLS Hamiltonian

$$\mathcal{H}^c = \mathcal{H} - 2|c|^2 \mathcal{H}_1$$

where

$$\mathcal{H}_1(\varphi) = -\int_0^1 \varphi_1(x) \varphi_2(x) \, dx.$$

Note that the flow, corresponding to the Hamiltonian $-2|c|^2 \mathcal{H}_1$, is the gauge transformation, given by the phase shift $(\varphi_1, \varphi_2) \mapsto (\varphi_1 e^{-2i|c|^2 t}, \varphi_2 e^{2i|c|^2 t})$ and that φ_c is a stationary solution of the Hamiltonian vector field $X_{\mathcal{H}^c}$. One of our main results is the following instance of an infinite dimensional version of Williamson's classification theorem in finite dimension [7].

Theorem 1.1 Assume that $c \in \mathbb{C}$ and $|c| \notin \pi \mathbb{Z}$. Then there exists a Darboux basis $\{\alpha_k, \beta_k\}_{k \in \mathbb{Z}}$ in $i L_r^2$ such that the Hessian $d_{\varphi_c}^2 \mathcal{H}^c$, when viewed as a quadratic form represented in this basis, takes the form

$$d_{\varphi_c}^2 \mathcal{H}^c = 4|c|^2 dp_0^2 - \sum_{0 < \pi k < |c|} 4\pi k \sqrt[+]{|c|^2 - \pi^2 k^2} (dp_k dq_k + dp_{-k} dq_{-k}) \\ - \sum_{\pi |k| > |c|} 4\pi |k| \sqrt[+]{\pi^2 k^2 - |c|^2} (dp_k^2 + dq_k^2) \tag{4}$$

where $\{(dp_k, dq_k)\}_{k \in \mathbb{Z}}$ are the coordinates dual to the Darboux basis $\{\alpha_k, \beta_k\}_{k \in \mathbb{Z}}$.

We conjecture that Theorem 1.1 can be generalized to a small neighborhood of the constant potential φ_c. We refer to the end of Sect. 3 where the precise statement of such a generalization is given. An analog of Theorem 1.1, formulated in terms of the linearization of the Hamiltonian vector field $X_{\mathcal{H}^c}$ at the constant potential φ_c, is formulated in Sect. 3 (see Theorem 3.2). As a consequence of these results, we obtain

Theorem 1.2 *For any given $c \in \mathbb{C}$ with $|c| \notin \pi\mathbb{Z}$ and $|c| > \pi$, the focusing NLS equation does not admit gauge invariant local Birkhoff coordinates in any neighborhood of the constant potential φ_c.*

Remark 1.1 The exceptional case $c \in \pi\mathbb{Z}$ can be treated in a similar way. Since it requires some additional work, the treatment of this case is beyond the scope of this paper, which aims at describing in precise terms, but as briefly as possible a generic situation in which no local gauge-invariant Birkhoff coordinates exist.

We refer to Sect. 4 for the precise definition of gauge invariant local Birkhoff coordinates. In more general terms, Theorem 1.2 means that there is no neighborhood of the constant potential φ_c with $|c| \notin \pi\mathbb{Z}$ and $|c| > \pi$ where one can introduce action-angle coordinates for the fNLS equation so that the action variables commute with the Hamiltonian $\mathcal{H}_1$. Note that a similar result could be obtained using the Bäcklund transform and the existence of a homoclinic orbit in a neighborhood of the constant potential φ_c (cf. [6]). However, such a neighborhood of φ_c is *not* arbitrarily small since it contains the homoclinic solution of the fNLS equation.

Finally, note that the same results hold for the potentials

$$\varphi_{c,k} := \left(ce^{2\pi ikx}, -\bar{c}e^{-2\pi ikx}\right)$$

where $k \in \mathbb{Z}$ and $c \in \mathbb{C}$ with $|c| \notin \pi\mathbb{Z}$ and $|c| > \pi$. The only difference is that the renormalized Hamiltonian for these potentials is of the form $\mathcal{H}_{c,k} = \mathcal{H} + \alpha\mathcal{H}_1 + \beta\mathcal{H}_2$ where $\alpha, \beta \in \mathbb{C}$ are specifically chosen constants depending on the choice of c and k and $\mathcal{H}_2(\varphi) = i\int_0^1 \varphi_1(x)\varphi_{2x}(x)\,dx$. This easily follows from the fact that for any given $k \in \mathbb{Z}$ and for any $s \geq 0$ the transformation

$$\tau_k : iH_r^s \to iH_r^s, \quad (\varphi_1, \varphi_2) \mapsto \left(\varphi_1 e^{2\pi ikx}, \varphi_2 e^{-2\pi ikx}\right),$$

preserves the symplectic structure induced by (2) and transforms the Hamiltonian $\mathcal{H}$ into a linear sum of the Hamiltonians $\mathcal{H}$, $\mathcal{H}_1$, and $\mathcal{H}_2$.

Organization of the Paper The paper is organized as follows: In Sect. 2 we introduce the basic notions related to the symplectic phase space geometry of the NLS equation that are needed in this paper. Theorem 1.1 and related results are proven in Sect. 3. In Sect. 4 we prove Theorem 1.2.

2 Set-Up

1. *The NLS phase space.* It is well-known that the non-linear Schrödinger equation is a Hamiltonian system on the phase space $L_c^2 := L_\mathbb{C}^2 \times L_\mathbb{C}^2$ where $L_\mathbb{C}^2 \equiv L^2(\mathbb{T}, \mathbb{C})$ is the space of square integrable complex-valued functions on the torus $\mathbb{T}$. For any two elements $f, g \in L_c^2$, the Hilbert scalar product on L_c^2 is defined as $(f, g)_{L^2} := \int_0^1 (f_1 \overline{g_1} + f_2 \overline{g_2}) \, dx$ where $f = (f_1, f_2)$, $g = (g_1, g_2)$, and $\overline{g_1}$ and $\overline{g_2}$ denote the complex conjugates of g_1 and g_2 respectively. In addition to the scalar product we will also need the non-degenerate pairing

$$\langle f, g \rangle_{L^2} := \int_0^1 (f_1 g_1 + f_2 g_2) \, dx. \tag{5}$$

The symplectic structure on L_c^2 is

$$\omega(f, g) := -i \int_0^1 \det \begin{pmatrix} f_1 & g_1 \\ f_2 & g_2 \end{pmatrix} dx \tag{6}$$

(Note that $\omega(f, g)$ is *not* the Kähler form of the Hermitian scalar product $(\cdot, \cdot)_{L^2}$ in L_c^2.) Consider also the scale of Sobolev spaces $H_c^s := H_\mathbb{C}^s \times H_\mathbb{C}^s$ where $H_\mathbb{C}^s \equiv H^s(\mathbb{T}, \mathbb{C})$ is the Sobolev space of complex-valued distributions on $\mathbb{T}$ and $s \in \mathbb{R}$. For any given $s \in \mathbb{R}$ the pairing (5) induces an isomorphism $\iota_s : (H_c^s)' \to H_c^{-s}$ where $(H_c^s)'$ denotes the space of continuous linear functionals on H_c^s. In this way, for any given $s \in \mathbb{R}$ the symplectic structure extends to a bounded bilinear map $\omega : H^s \times H^{-s} \to \mathbb{C}$. The L^2-gradient $\partial_\varphi F = (\partial_{\varphi_1} F, \partial_{\varphi_2} F)$ of a C^1-function $F : H^s \to \mathbb{C}$ at $\varphi \in L_c^2$ is defined by $\partial_\varphi F := \iota_s(d_\varphi F) \in H_c^{-s}$ where $d_\varphi F \in (H_c^s)'$ is the differential of F at $\varphi \in L_c^2$. In particular, the Hamiltonian vector field X_F corresponding to a C^1-smooth function $F : H_c^s \to \mathbb{C}$ at $\varphi \in H_c^s$ defined by the relation $\omega(\cdot, X_F(\varphi)) = d_\varphi F(\cdot)$ is then given by

$$X_F(\varphi) = i\big(-\partial_{\varphi_2} F, \partial_{\varphi_1} F\big). \tag{7}$$

The vector field X_F is a continuous map $X_F : H_c^s \to H_c^{-s}$.

Remark 2.1 Since $X_F : H_c^s \to H_c^{-s}$ we see that strictly speaking X_F is a weak vector field on H_c^{-s}. However, for the sake of convenience in this paper we will call such maps *vector fields* on H_c^s.

Remark 2.2 The Poisson bracket of two C^1-smooth functions $F, G : H_c^s \to \mathbb{C}$ is then given by

$$\{F, G\}(\varphi) := d_\varphi F(X_G) = -i \int_0^1 \big((\partial_{\varphi_1} F)(\partial_{\varphi_2} G) - (\partial_{\varphi_1} G)(\partial_{\varphi_2} F)\big) \, dx \tag{8}$$

provided that the pairing given by the integral in (8) is well-defined.

The Hamiltonian $\mathcal{H} : H_c^1 \to \mathbb{C}$ of the NLS equation is

$$\mathcal{H}(\varphi) := \int_0^1 \left(\varphi_{1x}\varphi_{2x} + \varphi_1^2\varphi_2^2\right) dx. \tag{9}$$

By (7) the corresponding Hamiltonian vector field is

$$\begin{aligned} X_{\mathcal{H}}(\varphi) &= i\left(-\partial_{\varphi_2}\mathcal{H}, \partial_{\varphi_1}\mathcal{H}\right) \\ &= i\left(\varphi_{1xx} - 2\varphi_1^2\varphi_2, -\varphi_{2xx} + 2\varphi_1\varphi_2^2\right). \end{aligned} \tag{10}$$

Clearly, $X_{\mathcal{H}} : H_c^2 \to L_c^2$ is an analytic map. The NLS equation is then written as

$$\begin{cases} \dot{\varphi}_1 = i\varphi_{1xx} - 2i\varphi_1^2\varphi_2, \\ \dot{\varphi}_2 = -i\varphi_{2xx} + 2i\varphi_1\varphi_2^2. \end{cases} \tag{11}$$

The phase space L_c^2 has two real subspaces

$$L_r^2 := \left\{\varphi \in L_c^2 \,\middle|\, \varphi_2 = \overline{\varphi_1}\right\} \quad \text{and} \quad iL_r^2 := \left\{\varphi \in L_c^2 \,\middle|\, \varphi_2 = -\overline{\varphi_1}\right\}$$

so that $L_c^2 = L_r^2 \oplus_{\mathbb{R}} iL_r^2$. For any $s \in \mathbb{R}$ one also defines in a similar way the real subspaces H_r^s and iH_r^s in H_c^s so that $H_c^s = H_r^s \oplus_{\mathbb{R}} iH_r^s$. It follows from (9) that the Hamiltonian $\mathcal{H}$ is real valued when restricted to H_r^1 and iH_r^1. Moreover, one easily sees from (11) that the Hamiltonian vector field $X_{\mathcal{H}}$ is "tangent" to the real subspaces H_r^1 and iH_r^1 so that the restrictions

$$X_{\mathcal{H}}\big|_{H_r^2} : H_r^2 \to L_r^2 \quad \text{and} \quad X_{\mathcal{H}}\big|_{iH_r^2} : iH_r^2 \to iL_r^2$$

are well-defined, and hence real analytic maps. The vector field $X_{\mathcal{H}}\big|_{H_r^2}$ corresponds to the defocusing NLS equation and the vector field $X_{\mathcal{H}}\big|_{iH_r^2}$ corresponds to the focusing NLS equation. This is consistent with the fact that the restriction of the symplectic structure ω to L_r^2 and iL_r^2 is real valued. For the sake of convenience in what follows we drop the restriction symbols in $X_{\mathcal{H}}\big|_{iH_r^2}$ and $X_{\mathcal{H}}\big|_{H_r^2}$ and simply write $X_{\mathcal{H}}$ instead.

2. *Constant potentials.* For any given complex number $c \in \mathbb{C}$, $c \neq 0$, consider the constant potential

$$\varphi_c(x) := (c, -\overline{c}) \in iL_r^2 \cap iC_r^\infty.$$

It follows from (10) that

$$X_{\mathcal{H}}(\varphi_c) = 2i|c|^2(c, \overline{c}). \tag{12}$$

Since this vector does not vanish we see that φ_c is *not* a critical point of the NLS Hamiltonian (9) and hence $d_{\varphi_c}\mathcal{H} \neq 0$ in $(H_c^1)'$.

3. *The re-normalized Hamiltonian.* In addition to the NLS Hamiltonian (9) consider the Hamiltonian

$$\mathcal{H}_1(\varphi) := -\int_0^1 \varphi_1(x)\varphi_2(x)\,dx\,. \tag{13}$$

Note that this is the first Hamiltonian appearing in the NLS hierarchy—see e.g. [2]. The corresponding Hamiltonian vector field is

$$X_{\mathcal{H}_1}(\varphi) = i(\varphi_1, -\varphi_2). \tag{14}$$

For any $s \in \mathbb{R}$ we have that $X_{\mathcal{H}_1} : H_c^s \to H_c^s$ and hence $X_{\mathcal{H}_1}$ is a (regular) vector field on H_c^s. This vector field is tangent to the real submanifolds H_r^s and iH_r^s and induces the following one-parameter group of diffeomorphisms of iH_r^s,[1]

$$S^t : iH_r^s \to iH_r^s, \quad (\varphi_1^0, \varphi_2^0) \xmapsto{S^t} (\varphi_1^0 e^{it}, \varphi_2^0 e^{-it}). \tag{15}$$

The transformations (15) preserves the vector field $X_{\mathcal{H}}$, i.e. for any $t \in \mathbb{R}$ and for any $\varphi \in iH_r^2$

$$S^t(X_{\mathcal{H}}(\varphi)) = X_{\mathcal{H}}(S^t(\varphi))\,. \tag{16}$$

It follows from (12) and (14) that

$$X_{\mathcal{H}}(\varphi_c) = 2|c|^2 X_{\mathcal{H}_1}(\varphi_c). \tag{17}$$

We have the following

Lemma 2.1 *Let $c \in \mathbb{C} \setminus \{0\}$. Then one has:*

(i) *The re-normalized Hamiltonian $\mathcal{H}^c : iH_r^1 \to \mathbb{R}$,*

$$\mathcal{H}^c(\varphi) := \mathcal{H}(\varphi) - 2|c|^2 \mathcal{H}_1(\varphi), \quad \varphi \in iH_r^1, \tag{18}$$

has a critical point at φ_c.

(ii) *The curve $\gamma_c : \mathbb{R} \to iH_r^1$,*

$$\gamma_c : t \mapsto \left(ce^{2i|c|^2 t}, -\bar{c}e^{-2i|c|^2 t}\right), \quad t \in \mathbb{R}, \tag{19}$$

is a solution of the NLS equation (11) with initial data at φ_c. This is a time periodic solution with period $\pi/|c|^2$.

---

[1] In what follows we will restrict our attention to the real space iH_r^s, $s \in \mathbb{R}$.

(iii) The range of the curve γ_c consists of critical points of the Hamiltonian $\mathcal{H}^c$.

Proof of Lemma 2.1 Item (i) follows directly from (17). Since the symmetry (15) preserves both $X_{\mathcal{H}}$ and $X_{\mathcal{H}_1}$ we conclude from (17) that for any $t \in \mathbb{R}$,

$$X_{\mathcal{H}}\big(S^t(\varphi_c)\big) = 2|c|^2 X_{\mathcal{H}_1}\big(S^t(\varphi_c)\big). \tag{20}$$

This together with the fact that $S^t(\varphi_c)$ is the integral curve of $X_{\mathcal{H}_1}$ with initial data at φ_c we conclude that $\gamma(t) := S_{2|c|^2 t}(\varphi_c)$ is an integral curve of $X_{\mathcal{H}}$ with initial data at φ_c. This proves item (ii). Item (iii) follows from (20). □

3 The Linearization of $X_{\mathcal{H}^c}$ at φ_c and Its Normal Form

In this section we determine the spectrum and the normal form of the linearized Hamiltonian vector field $X_{\mathcal{H}^c} : i H_r^2 \to i L_r^2$ at $\varphi_c \in i H_r^2$. In view of Lemma 2.1 the constant potential φ_c is a singular point of the vector field $X_{\mathcal{H}^c}$, i.e. $X_{\mathcal{H}^c}(\varphi_c) = 0$. Moreover, by Lemma 2.1 (iii), the range of the periodic trajectory γ_c consists of singular points of $X_{\mathcal{H}^c}$. It follows from (10) and (14) that for any $\varphi \in i H_r^2$,

$$X_{\mathcal{H}^c}(\varphi) = i \begin{pmatrix} \varphi_{1xx} - 2\varphi_1^2 \varphi_2 - 2|c|^2 \varphi_1 \\ -\varphi_{2xx} + 2\varphi_1 \varphi_2^2 + 2|c|^2 \varphi_2 \end{pmatrix}.$$

Hence, the linearized vector field $(dX_{\mathcal{H}^c})\big|_{\varphi=\varphi_c} : i H_r^2 \to i L_r^2$ is given by

$$(dX_{\mathcal{H}^c})\big|_{\varphi=\varphi_c} \begin{pmatrix} \delta\varphi_1 \\ \delta\varphi_2 \end{pmatrix} = i \begin{pmatrix} (\delta\varphi_1)_{xx} + 2|c|^2 (\delta\varphi_1) - 2c^2(\delta\varphi_2) \\ -(\delta\varphi_2)_{xx} + 2\bar{c}^2(\delta\varphi_1) - 2|c|^2(\delta\varphi_2) \end{pmatrix} \tag{21}$$

where $(\delta\varphi_1, \delta\varphi_2) \in i H_r^2$. Since the symmetry (15) preserves $X_{\mathcal{H}^c}$ and since for any $t \in \mathbb{R}$,

$$S^t(\varphi_c) = \big(ce^{it}, -\bar{c}e^{-it}\big),$$

the map S^t conjugates the operator (21) computed at φ_c with the one computed at φ_{c_t} with $c_t := ce^{it}$. More specifically, one has the following commutative diagram

$$\begin{array}{ccc} i H_r^2 & \xrightarrow{\mathcal{L}_{c_t}} & i L_r^2 \\ {\scriptstyle S^t}\uparrow & & \uparrow{\scriptstyle S^t} \\ i H_r^2 & \xrightarrow{\mathcal{L}_c} & i L_r^2 \end{array} \tag{22}$$

where for simplicity of notation we denote

$$\mathcal{L}_c \equiv (dX_{\mathcal{H}^c})|_{\varphi = \varphi_c}.$$

By choosing $t = -\arg(c)$ in the diagram above we obtain

Lemma 3.1 *The operators $\mathcal{L}_c$ and $\mathcal{L}_{|c|}$ are conjugate.*

With this in mind, in what follows we will assume without loss of generality that c is real. In this case

$$\mathcal{L}_c = i \begin{pmatrix} 1 & 0 \\ 0 & -1 \end{pmatrix} \partial_x^2 + 2ic^2 \begin{pmatrix} 1 & -1 \\ 1 & -1 \end{pmatrix}, \quad c \in \mathbb{R}. \tag{23}$$

For any $k \in \mathbb{Z}$ consider the vectors

$$\xi_k := \begin{pmatrix} 1 \\ 0 \end{pmatrix} e^{2\pi i k x} \quad \text{and} \quad \eta_k := \begin{pmatrix} 0 \\ 1 \end{pmatrix} e^{2\pi i k x}. \tag{24}$$

The system of vectors $\{(\xi_k, \eta_k)\}_{k \in \mathbb{Z}}$ give an orthonormal basis in the complex Hilbert space L_c^2 so that for any $\varphi = (\varphi_1, \varphi_2) \in L_c^2$,

$$\varphi = \sum_{k \in \mathbb{Z}} (z_k \xi_k + w_k \eta_k),$$

where $z_k := \widehat{(\varphi_1)}_k$ and $w_k := \widehat{(\varphi_2)}_k$. Denote $\ell_c^2 := \ell_{\mathbb{C}}^2 \times \ell_{\mathbb{C}}^2$ where $\ell_{\mathbb{C}}^2 \equiv \ell^2(\mathbb{Z}, \mathbb{C})$ is the space of square summable sequences of complex numbers. In this way, $\{(z_k, w_k)\}_{k \in \mathbb{Z}} \in \ell_c^2$ are coordinates in L_c^2. In these coordinates, the real subspace iL_r^2 is characterized by the condition that $\forall k \in \mathbb{Z}$, $w_k = -\overline{(z_{-k})}$, and the real subspace L_r^2 is characterized by the condition that $\forall k \in \mathbb{Z}$, $w_k = \overline{(z_{-k})}$. We denote the corresponding spaces of sequences respectively by $i\ell_r^2$ and ℓ_r^2. It follows from (6) that for any $k, l \in \mathbb{Z}$ one has

$$\omega(\xi_k, \xi_l) = \omega(\eta_k, \eta_l) = 0 \quad \text{and} \quad \omega(\xi_k, \eta_{-l}) = -i\delta_{kl}$$

where δ_{kl} is the Kronecker delta. In addition to the vectors in (24) consider for $k \in \mathbb{Z}$ the vectors

$$\xi_k' := \frac{1}{\sqrt{2}}(\xi_k - \eta_{-k}) = \frac{1}{\sqrt{2}} \begin{pmatrix} e^{2\pi k i x} \\ -e^{-2\pi k i x} \end{pmatrix}, \quad \eta_k' := \frac{i}{\sqrt{2}}(\xi_k + \eta_{-k}) = \frac{i}{\sqrt{2}} \begin{pmatrix} e^{2\pi k i x} \\ e^{-2\pi k i x} \end{pmatrix}. \tag{25}$$

Note that the system of vectors $\{\xi'_k, \eta'_k\}_{k\in\mathbb{Z}}$ form an orthonormal basis in the real subspace iL_r^2. In addition, this is a *Darboux basis* in iL_r^2 with respect to the restriction of the symplectic structure (6) to iL_r^2, i.e. for any $k, l \in \mathbb{Z}$ one has

$$\omega(\xi'_k, \xi'_l) = \omega(\eta'_k, \eta'_l) = 0 \quad \text{and} \quad \omega(\xi'_k, \eta'_l) = \delta_{kl}.$$

Moreover, for any $\varphi \in L_c^2$,

$$\varphi = \sum_{k\in\mathbb{Z}} (x_k \xi'_k + y_k \eta'_k),$$

where $\{(x_k, y_k)\}_{k\in\mathbb{Z}}$ are coordinates in L_c^2 so that the real subspace iL_r^2 is characterized by the condition that $\{(x_k, y_k)\}_{k\in\mathbb{Z}} \in \ell^2(\mathbb{Z}, \mathbb{R}) \times \ell^2(\mathbb{Z}, \mathbb{R})$. For any $k \in \mathbb{Z}$,

$$x_k = \frac{1}{\sqrt{2}}(z_k - w_{-k}) \quad \text{and} \quad y_k = \frac{1}{i\sqrt{2}}(z_k + w_{-k}).$$

Finally, consider the 2-(complex)dimensional subspaces in L_c^2,

$$V_k^{\mathbb{C}} := \operatorname{span}_{\mathbb{C}} \langle \xi_k, \eta_k \rangle, \quad k \in \mathbb{Z}, \tag{26}$$

together with the 4-(real)dimensional *symplectic subspaces* in iL_r^2,

$$W_k^{\mathbb{R}} := \operatorname{span}_{\mathbb{R}} \langle \xi'_k, \eta'_k, \xi'_{-k}, \eta'_{-k} \rangle, \quad k \in \mathbb{Z}_{\geq 1}. \tag{27}$$

and

$$W_0^{\mathbb{R}} := \operatorname{span}_{\mathbb{R}} \langle \xi'_0, \eta'_0 \rangle. \tag{28}$$

It follows from (25) that for any $k \in \mathbb{Z}_{\geq 1}$,

$$W_k^{\mathbb{R}} \otimes \mathbb{C} = V_k^{\mathbb{C}} \oplus_{\mathbb{C}} V_{-k}^{\mathbb{C}} \quad \text{and} \quad W_0^{\mathbb{R}} \otimes \mathbb{C} = V_0^{\mathbb{C}}. \tag{29}$$

Since $\{\xi'_k, \eta'_k\}_{k\in\mathbb{Z}}$ is an orthonormal basis of iL_r^2,

$$iL_r^2 = \bigoplus_{k\in\mathbb{Z}_{\geq 0}} W_k^{\mathbb{R}}$$

is a decomposition of iL_r^2 into L^2-orthogonal real subspaces. We have the following

Theorem 3.1 *For any $c \in \mathbb{R}$, $c \notin \pi\mathbb{Z}$, the operator*

$$\mathcal{L}_c \equiv (dX_{\mathcal{H}^c})\big|_{\varphi=\varphi_c} : iH_r^2 \to iL_r^2,$$

has a compact resolvent. In particular, the spectrum of $\mathcal{L}_c$ is discrete and has the following properties:

(i) *The spectrum of $\mathcal{L}_c$ consists of $\lambda_0 = 0$ and*

$$\lambda_k = \begin{cases} 4\pi k \sqrt[+]{|c|^2 - \pi^2 k^2}, & 0 < \pi|k| < |c|, \\ 4\pi i k \sqrt[+]{\pi^2 k^2 - |c|^2}, & \pi|k| > |c|, \end{cases} \qquad (30)$$

for any integer $k \in \mathbb{Z} \setminus \{0\}$. The eigenvalue λ_0 has algebraic multiplicity two and geometric multiplicity one; for any $k \in \mathbb{Z} \setminus \{0\}$ the eigenvalue λ_k has algebraic multiplicity two and geometric multiplicity two.

(ii) *For any $k \in \mathbb{Z}$ the complex linear space $V_k^{\mathbb{C}}$ (see (26)) is an invariant space of $\mathcal{L}_c$ in L_c^2 and*

$$\mathcal{L}_c\big|_{V_k^{\mathbb{C}}} = i \begin{pmatrix} 2|c|^2 - 4\pi^2 k^2 & -2|c|^2 \\ 2|c|^2 & 4\pi^2 k^2 - 2|c|^2 \end{pmatrix} \qquad (31)$$

in the basis of $V_k^{\mathbb{C}}$ given by ξ_k and η_k. If $0 < \pi|k| < |c|$ the matrix (31) has two real eigenvalues $\pm 4\pi|k|\sqrt[+]{|c|^2 - \pi^2 k^2}$ and if $|c| < |k|\pi$ it has two purely imaginary complex eigenvalues $\pm 4\pi i |k| \sqrt[+]{\pi^2 k^2 - |c|^2}$.

(iii) *For any $k \in \mathbb{Z}_{\geq 1}$ the real symplectic space $W_k^{\mathbb{R}}$ (see (27)) is an invariant space of $\mathcal{L}_c$ in iL_r^2. When written in the basis $\{\xi_k, \eta_k, \xi_{-k}, \eta_{-k}\}$ of the complexification of $W_k^{\mathbb{R}}$ the matrix representation of the operator $\mathcal{L}_c\big|_{W_k^{\mathbb{R}}}$ consists of two diagonal square blocks of the form (31). The real symplectic space $W_0^{\mathbb{R}}$ is an invariant space of the operator $\mathcal{L}_c$ in iL_r^2 and*

$$\mathcal{L}_c\big|_{W_0^{\mathbb{R}}} = 2i|c|^2 \begin{pmatrix} 1 & -1 \\ 1 & -1 \end{pmatrix}$$

when written in the basis $\{\xi_0, \eta_0\}$ of the complexification of $W_0^{\mathbb{R}}$. Zero is a double eigenvalue of this matrix with geometric multiplicity one.

Proof of Theorem 3.1 The proof of this Theorem follows directly from the matrix representation of $\mathcal{L}_c$ when computed in the basis of $V_k^{\mathbb{C}}$ given by the vectors ξ_k and η_k. □

Theorem 3.1 implies that the linearized vector field $(dX_{\mathcal{H}^c})\big|_{\varphi=\varphi_c}$ has the following *normal form*.

Theorem 3.2 *Assume that $c \in \mathbb{R}$ and $c \notin \pi\mathbb{Z}$. We have:*

(i) *The vectors $\alpha_0 := \xi'_0$ and $\beta_0 := \eta'_0$ form a Darboux basis of $W_0^{\mathbb{R}} \subseteq iL_r^2$ such that*

$$\mathcal{L}_c\big|_{W_0^{\mathbb{R}}} = 4|c|^2 \begin{pmatrix} 0 & 0 \\ 1 & 0 \end{pmatrix}. \tag{32}$$

(ii) *For any $k \in \mathbb{Z}_{\geq 1}$ there exists a Darboux basis $\{\alpha_k, \beta_k, \alpha_{-k}, \beta_{-k}\}$ in $W_k^{\mathbb{R}} \subseteq iL_r^2$ such that[2] for $0 < \pi k < |c|$,*

$$\mathcal{L}_c\big|_{W_k^{\mathbb{R}}} = 4\pi k \sqrt[+]{|c|^2 - \pi^2 k^2} \begin{pmatrix} 1 & 0 & 0 & 0 \\ 0 & -1 & 0 & 0 \\ 0 & 0 & 1 & 0 \\ 0 & 0 & 0 & -1 \end{pmatrix}, \tag{33}$$

and for $\pi k > |c|$,

$$\mathcal{L}_c\big|_{W_k^{\mathbb{R}}} = 4\pi k \sqrt[+]{\pi^2 k^2 - |c|^2} \begin{pmatrix} 0 & 1 & 0 & 0 \\ -1 & 0 & 0 & 0 \\ 0 & 0 & 0 & 1 \\ 0 & 0 & -1 & 0 \end{pmatrix}. \tag{34}$$

In addition, one has the following uniform in $k \in \mathbb{Z}$ with $\pi k > |c|$ estimates

$$\begin{cases} \alpha_k = \xi'_k + O(1/k^2), & \beta_k = \eta'_k + O(1/k^2), \\ \alpha_{-k} = \xi'_{-k} + O(1/k^2), & \beta_{-k} = \eta'_{-k} + O(1/k^2), \end{cases} \tag{35}$$

where $\{\xi'_k, \eta'_k\}_{k\in\mathbb{Z}}$ is the orthonormal Darboux basis (25) in iL_r^2.

Recall that $\mathfrak{h}^s(\mathbb{Z}, \mathbb{R})$ with $s \in \mathbb{R}$ denotes the Hilbert space of sequences of real numbers $(a_k)_{k\in\mathbb{Z}}$ so that $\sum_{k\in\mathbb{Z}} \langle k \rangle^{2s} |a_k|^2 < \infty$ where $\langle k \rangle := \sqrt{1 + |k|^2}$. We will also need the Banach space $\ell_s^1(\mathbb{Z}, \mathbb{R})$ of sequences of real numbers $(a_k)_{k\in\mathbb{Z}}$ so that $\sum_{k\in\mathbb{Z}} \langle k \rangle^s |a_k| < \infty$.

Remark 3.1 Note that the asymptotics (35) imply (see e.g. [5, Section 22.5]) that for any $s \in \mathbb{R}$ the system $\{\alpha_k, \beta_k\}_{k\in\mathbb{Z}}$ is a basis in iH_r^s in the sense that for any $\varphi \in iH_r^s$ there exists a unique sequence $\{(p_k, q_k)\}_{k\in\mathbb{Z}}$ in $\mathfrak{h}^s(\mathbb{Z}, \mathbb{R}) \times \mathfrak{h}^s(\mathbb{Z}, \mathbb{R})$ such that $\varphi = \sum_{k\in\mathbb{Z}} (p_k \alpha_k + q_k \beta_k)$ where the series converges in iH_r^s and the mapping

$$iH_r^s \to \mathfrak{h}^s(\mathbb{Z}, \mathbb{R}) \times \mathfrak{h}^s(\mathbb{Z}, \mathbb{R}), \quad \varphi \mapsto \{(p_k, q_k)\}_{k\in\mathbb{Z}},$$

is an isomorphism. Note that $\{\alpha_k, \beta_k\}_{k\in\mathbb{Z}}$ is a Darboux basis that is *not* an orthonormal basis in iL_r^2.

---

[2] Here $\omega(\alpha_k, \beta_k) = \omega(\alpha_{-k}, \beta_{-k}) = 1$ while all other skew-symmetric products between these vectors vanish.

Proof of Theorem 3.2 Item (i) follows by a direct computation in the basis of $W_0^{\mathbb{R}}$ provided by the vectors $\alpha_0 := \xi_0'$ and $\beta_0 := \eta_0'$ (see (25)). Towards proving item (ii), we first consider the case when $\pi k > |c|$. Denote for simplicity

$$L_{\pm k} := \mathcal{L}_c\big|_{V_{\pm k}^{\mathbb{C}}}, \quad a_k := 4\pi^2 k^2 - 2|c|^2, \quad b := 2|c|^2,$$

and note that $a_k^2 - b^2 = 16\pi^2 k^2 (\pi^2 k^2 - |c|^2) > 0$. Then, in view of (31),

$$L_k = i \begin{pmatrix} -a_k & -b \\ b & a_k \end{pmatrix} \quad \text{and} \quad L_{-k} = L_k$$

in the basis of $V_m^{\mathbb{C}}$ given by $\{\xi_m, \eta_m\}$ for $m = \pm k$. Denote, by $\varkappa_k$ the positive square root of the quantity

$$\varkappa_k^2 = \sqrt[+]{a_k^2 - b^2}\left(a_k + \sqrt[+]{a_k^2 - b^2}\right).$$

It follows from Theorem 3.1 (ii) and (31) that

$$F_k := -\frac{1}{\varkappa_k}\left(\frac{-b}{a_k + \sqrt[+]{a_k^2 - b^2}}\right) e^{-2\pi i k x} \tag{36}$$

and

$$F_{-k} := -\frac{1}{\varkappa_k}\left(\frac{-b}{a_k + \sqrt[+]{a_k^2 - b^2}}\right) e^{2\pi i k x} \tag{37}$$

are linearly independent eigenfunctions of the restriction of $\mathcal{L}_c$ to the invariant space $V_k^{\mathbb{C}} \oplus_{\mathbb{C}} V_{-k}^{\mathbb{C}} = W_k^{\mathbb{R}} \otimes \mathbb{C}$ with eigenvalue

$$\lambda_k = i\sqrt[+]{a_k^2 - b^2} = 4\pi i k \sqrt[+]{\pi^2 k^2 - |c|^2}.$$

The eigenfunctions have been normalized in a way convenient for our purposes. Denote by σ the *complex conjugation* in L_c^2 corresponding to the real subspace $i L_r^2$,

$$\sigma : L_c^2 \to L_c^2, \quad (\varphi_1, \varphi_2) \mapsto (-\overline{\varphi_2}, -\overline{\varphi_1}). \tag{38}$$

Remark 3.2 One easily sees from (38) that $\sigma\big|_{iL_r^2} = \mathrm{id}_{iL_r^2}$ and $\sigma\big|_{L_r^2} = -\mathrm{id}_{L_r^2}$. This implies that for any $\alpha, \beta \in i L_r^2$,

$$\sigma(\alpha + i\beta) = \alpha - i\beta. \tag{39}$$

Since the operator $\mathcal{L}_c$ is real (i.e., $\mathcal{L}_c : iH_r^2 \to iL_r^2$) and complex-linear, we conclude from (39) that if $f \in H_c^2$ is an eigenfunction of $\mathcal{L}_c$ with eigenvalue $\lambda \in \mathbb{C}$ then $\sigma(f)$ is an eigenfunction of $\mathcal{L}_c$ with eigenvalue $\overline{\lambda}$. Moreover, one easily checks that for any $f, g \in H_c^2$,

$$\omega(\mathcal{L}_c f, g) = -\omega(f, \mathcal{L}_c g), \qquad (40)$$

which is consistent with the fact that $X_{\mathcal{H}^c}$ is a Hamiltonian vector field. Equation (40) implies that if $f, g \in H_c^2$ are eigenfunctions of $\mathcal{L}_c$ with the same eigenvalue $\lambda \in \mathbb{C} \setminus \{0\}$ then they are isotropic, i.e. $\omega(f, g) = 0$.

Since $L_c^2 = iL_r^2 \oplus_\mathbb{R} L_r^2$ we have that

$$F_k = \alpha_k + i\beta_k \quad \text{and} \quad F_{-k} = \alpha_{-k} + i\beta_{-k} \qquad (41)$$

where by (39)

$$\alpha_{\pm k} := \frac{F_{\pm k} + \sigma(F_{\pm k})}{2} \quad \text{and} \quad \beta_{\pm k} := \frac{F_{\pm k} - \sigma(F_{\pm k})}{2i}$$

are elements in iL_r^2. Moreover, in view of Remark 3.2,

$$G_k := \sigma(F_k) = \frac{1}{\varkappa_k} \begin{pmatrix} a_k + \sqrt[+]{a_k^2 - b^2} \\ -b \end{pmatrix} e^{2\pi i k x} \qquad (42)$$

and

$$G_{-k} := \sigma(F_{-k}) = \frac{1}{\varkappa_k} \begin{pmatrix} a_k + \sqrt[+]{a_k^2 - b^2} \\ -b \end{pmatrix} e^{-2\pi i k x} \qquad (43)$$

are linearly independent eigenfunctions of the restriction of $\mathcal{L}_c$ to the invariant space $V_k^\mathbb{C} \oplus_\mathbb{C} V_{-k}^\mathbb{C} = W_k^\mathbb{R} \otimes \mathbb{C}$ with eigenvalue

$$-\lambda_k = -i \sqrt[+]{a_k^2 - b^2} = -4\pi i k \sqrt[+]{\pi^2 k^2 - |c|^2}.$$

It follows from (6), (36), (37), (42), and (43) that

$$\omega(F_k, G_k) = \omega(F_{-k}, G_{-k}) = -2i \quad \text{and} \quad \omega(F_k, G_{-k}) = \omega(F_{-k}, G_k) = 0 \qquad (44)$$

while $\omega(F_k, F_{-k}) = \omega(G_k, G_{-k}) = 0$ in view of Remark 3.2. Hence,

$$\omega(\alpha_k, \beta_k) = \frac{1}{4i} \omega(F_k + \sigma(F_k), F_k - \sigma(F_k)) = 1$$

and similarly $\omega(\alpha_{-k}, \beta_{-k}) = 1$ whereas all other values of ω evaluated at pairs of vectors from the set $\{\alpha_k, \beta_k, \alpha_{-k}, \beta_{-k}\}$ vanish. This shows that the vectors $\{\alpha_k, \beta_k, \alpha_{-k}, \beta_{-k}\}$ form a Darboux basis. The matrix representation (34) of $\mathcal{L}_c$ in this basis then follows from (41) and the fact that F_k and F_{-k} are eigenfunctions of $\mathcal{L}_c$ with eigenvalue $i \sqrt[+]{a_k^2 - b^2}$,

$$\mathcal{L}_c(\alpha_k + i\beta_k) = i \sqrt[+]{a_k^2 - b^2} \,(\alpha_k + i\beta_k) \quad \text{and}$$

$$\mathcal{L}_c(\alpha_{-k} + i\beta_{-k}) = i \sqrt[+]{a_k^2 - b^2} \,(\alpha_{-k} + i\beta_{-k}).$$

The asymptotic relations in (35) follow from the explicit formulas for $F_{\pm k}$ and $G_{\pm k}$ above together with $\alpha_m = (F_m + G_m)/2$, $\beta_m = (F_m - G_m)/2i$ with $m = \pm k$, and $a_k = 4\pi^2 k^2 - 2|c|^2$.

The case when $0 < \pi|k| < |c|$ is treated in a similar way. In fact, take $k \in \mathbb{Z}$ with $0 < \pi k < |c|$ and denote by $\varkappa_k$ the branch of the square root of

$$\varkappa_k^2 = \sqrt[+]{b^2 - a_k^2}\left(a_k - i\sqrt{b^2 - a_k^2}\right)$$

that lies in the fourth quadrant of the complex plane $\mathbb{C}$. It follows from Theorem 3.1 (ii) and (31) that

$$F_k := \frac{1}{\varkappa_k}\begin{pmatrix} -b \\ a_k - i \sqrt[+]{b^2 - a_k^2} \end{pmatrix} e^{2\pi i k x} \tag{45}$$

and

$$F_{-k} := \sigma(F_k) = -\frac{1}{\overline{\varkappa_k}}\begin{pmatrix} a_k + i \sqrt[+]{b^2 - a_k^2} \\ -b \end{pmatrix} e^{-2\pi i k x} \tag{46}$$

are linearly independent eigenfunctions of the restriction of $\mathcal{L}_c$ to the invariant space $V_k^{\mathbb{C}} \oplus_{\mathbb{C}} V_{-k}^{\mathbb{C}} = W_k^{\mathbb{R}} \otimes \mathbb{C}$ with eigenvalue

$$\lambda_k = \sqrt[+]{b^2 - a_k^2} = 4\pi k \sqrt[+]{|c|^2 - \pi^2 k^2}.$$

By arguing in the same way as above, one sees that

$$G_k := -\frac{1}{\varkappa_k}\begin{pmatrix} -b \\ a_k + i \sqrt[+]{b^2 - a_k^2} \end{pmatrix} e^{2\pi i k x} \tag{47}$$

and

$$G_{-k} := \sigma(G_k) = \frac{1}{\varkappa_k} \begin{pmatrix} a_k - i \sqrt[+]{b^2 - a_k^2} \\ -b \end{pmatrix} e^{-2\pi i k x} \qquad (48)$$

are linearly independent eigenfunctions of the restriction of $\mathcal{L}_c$ to the invariant space $V_k^{\mathbb{C}} \oplus_{\mathbb{C}} V_{-k}^{\mathbb{C}} = W_k^{\mathbb{R}} \otimes \mathbb{C}$ with eigenvalue

$$-\lambda_k = -\sqrt[+]{b^2 - a_k^2} = -4\pi k \sqrt[+]{|c|^2 - \pi^2 k^2}.$$

Since $L_c^2 = iL_r^2 \oplus_{\mathbb{R}} L_r^2$ we have that

$$F_k = \alpha_k + i\alpha_{-k} \quad \text{and} \quad G_k = \beta_k + i\beta_{-k}, \qquad (49)$$

where

$$\alpha_{\pm k} := \frac{F_{\pm k} + \sigma(F_{\pm k})}{2} \quad \text{and} \quad \beta_{\pm k} := \frac{F_{\pm k} - \sigma(F_{\pm k})}{2i}$$

are elements in iL_r^2. By (39) this implies that

$$F_{-k} = \alpha_k - i\alpha_{-k} \quad \text{and} \quad G_{-k} = \beta_k - i\beta_{-k}. \qquad (50)$$

It follows from (6), (45)–(48) that

$$\omega(F_k, G_k) = \omega(F_{-k}, G_{-k}) = 0 \quad \text{and} \quad \omega(F_k, G_{-k}) = \omega(F_{-k}, G_k) = 2 \qquad (51)$$

while $\omega(F_k, F_{-k}) = \omega(G_k, G_{-k}) = 0$ in view of Remark 3.2. This together with (49) and (50) implies that the vectors $\{\alpha_k, \beta_k, \alpha_{-k}, \beta_{-k}\}$ form a Darboux basis in $W_k^{\mathbb{R}}$. The matrix representation (33) of $\mathcal{L}_c$ in this basis then follows from (49), (50), and the fact that F_k and G_k given by (45) and (47) are eigenfunctions of $\mathcal{L}_c$ with (real) eigenvalues $\pm\sqrt[+]{b^2 - a_k^2}$,

$$\mathcal{L}_c(\alpha_k + i\alpha_{-k}) = \sqrt[+]{b^2 - a_k^2} (\alpha_k + i\alpha_{-k}) \quad \text{and}$$

$$\mathcal{L}_c(\beta_k + i\beta_{-k}) = -\sqrt[+]{b^2 - a_k^2} (\beta_k + i\beta_{-k}).$$

This completes the proof of Theorem 3.2. □

Remark 3.3 In fact, the canonical form (32)–(34), of the restriction of the operator $\mathcal{L}_c$ to the invariant symplectic space $W_k^{\mathbb{R}}$ with $k \geq 0$ can be deduced from the description of the spectrum of $\mathcal{L}_c$ obtained in Theorem 3.1 (*i*), (*ii*), and the Williamson classification of linear Hamiltonian systems in $\mathbb{R}^{2n}$ (see [1, 7]). Instead of doing this, we choose to construct the normalizing Darboux basis directly. The reason is twofold: first, in this way we obtain explicit formulas for the normalizing basis, and second, we need the asymptotic relations (35) to conclude that the system of vectors $\{\alpha_0, \beta_0\}$ together with $\{\alpha_k, \beta_k, \alpha_{-k}, \beta_{-k}\}_{k \in \mathbb{Z}_{\geq 1}}$ form a Darboux basis in iL_r^2 in the sense described in Remark 3.1.

In this way, as a consequence of Theorem 3.2 we obtain the following instance of an infinite dimensional version of the Williamson classification of linear Hamiltonian systems in $\mathbb{R}^{2n}$ described in [7].

Theorem 3.3 *Assume that $c \in \mathbb{R}$ and $c \notin \pi\mathbb{Z}$. Then the Hessian $d_{\varphi_c}^2 \mathcal{H}^c$, when viewed as a quadratic form represented in the Darboux basis $\{\alpha_k, \beta_k\}_{k \in \mathbb{Z}}$ in iL_r^2 given by Theorem 3.2, takes the form*

$$d_{\varphi_c}^2 \mathcal{H}^c = 4|c|^2 dp_0^2 - \sum_{0 < \pi k < |c|} 4\pi k \sqrt[+]{|c|^2 - \pi^2 k^2} (dp_k dq_k + dp_{-k} dq_{-k})$$

$$- \sum_{\pi|k| > |c|} 4\pi |k| \sqrt[+]{\pi^2 k^2 - |c|^2} (dp_k^2 + dq_k^2) \tag{52}$$

where $\{(dp_k, dq_k)\}_{k \in \mathbb{Z}}$ are the dual coordinates in this basis.

For any $0 < \pi k < |c|$ denote

$$I_k := p_k q_k + p_{-k} q_{-k} \quad \text{and} \quad I_{-k} := p_k q_{-k} - p_{-k} q_k, \tag{53}$$

and for $\pi|k| > |c|$,

$$I_k := (p_k^2 + q_k^2)/2$$

whereas for $k = 0$

$$I_0 := p_0^2/2 \,.$$

Note that the functions in (53) are the commuting integrals characterizing the *focus-focus* singularity in the symplectic space $\mathbb{R}^4$—see e.g. [9]. We conjecture that the following holds: *There exists an open neighborhood U of φ_c in iL_r^2, an open neighborhood V of zero in $\ell^2(\mathbb{Z}, \mathbb{R}) \times \ell^2(\mathbb{Z}, \mathbb{R})$, and a canonical real analytic diffeomorphism $\Phi : U \to V$ such that for any $s \geq 0$,*

$$\Phi : U \cap i H_r^s \to V \cap \left(\mathfrak{h}^s(\mathbb{Z}, \mathbb{R}) \times \mathfrak{h}^s(\mathbb{Z}, \mathbb{R})\right), \quad \varphi \mapsto \{(p_k, q_k)\}_{k \in \mathbb{Z}},$$

and for any $(p, q) \in V \cap \left(\mathfrak{h}^1(\mathbb{Z}, \mathbb{R}) \times \mathfrak{h}^1(\mathbb{Z}, \mathbb{R})\right)$,

$$\mathcal{H}^c \circ \Phi^{-1}(p, q) = H^c\left(\{I_k\}_{k \in \mathbb{Z}}\right),$$

where $H^c : \ell_2^1(\mathbb{Z}, \mathbb{R}) \to \mathbb{R}$ is a real analytic map. We will discuss this conjecture in future work.

4 Nonexistence of Local Birkhoff Coordinates

First, we will discuss the notion of *local Birkhoff coordinates*. Let $\mathfrak{h}^s \equiv \mathfrak{h}^s(\mathbb{Z}, \mathbb{R})$ and $\ell^2 \equiv \ell^2(\mathbb{Z}, \mathbb{R})$.

Definition 4.1 We say that the focusing NLS equation has *local Birkhoff coordinates* in a neighborhood of $\varphi^\bullet \in iH_r^2$ if there exist an open connected neighborhood U of $\varphi^\bullet$ in iL_r^2, an open neighborhood V of $(p^\bullet, q^\bullet) \in \mathfrak{h}^2 \times \mathfrak{h}^2$ in $\ell^2 \times \ell^2$, and a canonical C^2-diffeomorphism $\Phi : U \to V$ such that for any $0 \le s \le 2$,

$$\Phi : U \cap iH_r^s \to V \cap (\mathfrak{h}^s \times \mathfrak{h}^s), \quad \varphi \mapsto \{(p_k, q_k)\}_{k \in \mathbb{Z}},$$

is a C^2-diffeomorphism and for any $k \in \mathbb{Z}$ the Poisson bracket $\{I_k, H\}$, where $H := \mathcal{H} \circ \Phi^{-1}$ and $I_k := (p_k^2 + q_k^2)/2$, vanishes on $V \cap (\mathfrak{h}^1 \times \mathfrak{h}^1)$.

The map $\Phi : U \to V$ being *canonical* means that

$$(\Phi^{-1})^* \omega = \sum_{k \in \mathbb{Z}} dp_k \wedge dq_k \tag{54}$$

where $\Phi^{-1} : V \to U$ is the inverse of $\Phi : U \to V$ and ω is the symplectic form (6) on iL_r^2. Assume that the focusing NLS equation has local Birkhoff coordinates in a neighborhood of $\varphi^\bullet \in iH_r^2$. Then, for any $k \in \mathbb{Z}$ and $0 \le s \le 2$ consider the action variable

$$\mathfrak{I}_k : U \cap iH_r^s \to \mathbb{R}, \quad \mathfrak{I}_k := I_k \circ \Phi.$$

Recall that $\{S^t\}_{t \in \mathbb{R}}$ denotes the Hamiltonian flow,

$$S^t : iH_r^s \to iH_r^s, \quad (\varphi_1, \varphi_2) \mapsto (\varphi_1 e^{it}, \varphi_2 e^{-it}),$$

generated by the Hamiltonian $\mathcal{H}_1(\varphi) = -\int_0^1 \varphi_1(x) \varphi_2(x) \, dx$ (see (13)) from the standard NLS hierarchy (see e.g. [2]).

Definition 4.2 The local Birkhoff coordinates are called *gauge invariant* if for any $k \in \mathbb{Z}$, $0 \le s \le 2$, and for any $\varphi \in U \cap iH_r^s$ and $t \in \mathbb{R}$ such that $S^t(\varphi) \in U \cap iH_r^s$ one has $\mathfrak{I}_k(S^t(\varphi)) = \mathfrak{I}_k(\varphi)$.

Remark 4.1 The gauge invariance of local Birkhoff coordinates means that the Hamiltonian $\mathcal{H}_1$ belongs to the Poisson algebra $\mathcal{A}_\mathfrak{I} := \{F \in C^1(U, \mathbb{C}) \,|\, \{F, \mathfrak{I}_k\} = 0 \,\forall k \in \mathbb{Z}\}$ generated by the local action variables $\{\mathfrak{I}_k\}_{k \in \mathbb{Z}}$. Note that, for example, $\mathcal{H}_1$ belongs to the Poisson algebra generated by the functionals $\{\Delta_\lambda\}_{\lambda \in \mathbb{C}}$ where $\Delta_\lambda : iL_r^2 \to \mathbb{C}$ is the discriminant $\Delta_\lambda(\varphi) \equiv \Delta(\lambda, \varphi) := \operatorname{tr} M(x, \lambda, \varphi)|_{x=1}$ and $M(x, \lambda, \varphi)$ is the fundamental 2×2-matrix solution of the Zakharov-Shabat system (see e.g. [2]).

The main result of this section is Theorem 1.2 stated in Sect. 1 which we recall for the convenience of the reader.

Theorem 4.1 *For any given $c \in \mathbb{C}$ with $|c| \notin \pi \mathbb{Z}$ and $|c| > \pi$, the focusing NLS equation does* not *admit gauge invariant local Birkhoff coordinates in a neighborhood of the constant potential φ_c.*

Consider the commutative diagram

$$\begin{array}{ccc} U \cap i H_r^2 & \xrightarrow{X_{\mathcal{H}^c}} & i L_r^2 \\ \Phi \downarrow & & \downarrow \Phi_* \\ V \cap (\mathfrak{h}^2 \times \mathfrak{h}^2) & \xrightarrow{\widetilde{X}_{\mathcal{H}^c}} & \ell^2 \times \ell^2 \end{array} \qquad (55)$$

where $X_{\mathcal{H}^c}$ is the Hamiltonian vector field of the re-normalized Hamiltonian $\mathcal{H}^c$, $\Phi_* \equiv (d\Phi)|_{\varphi=\varphi_c}$, and $\widetilde{X}_{\mathcal{H}^c}$ is defined by the diagram. By linearizing the maps in this diagram at φ_c we obtain

$$\begin{array}{ccc} i H_r^2 & \xrightarrow{\mathcal{L}_c} & i L_r^2 \\ \Phi_* \downarrow & & \downarrow \Phi_* \\ \mathfrak{h}^2 \times \mathfrak{h}^2 & \xrightarrow{\widetilde{\mathcal{L}}_c} & \ell^2 \times \ell^2 \end{array} \qquad (56)$$

where $\mathcal{L}_c$ is the linearization of $X_{\mathcal{H}^c}$ at the critical point φ_c and $\widetilde{\mathcal{L}}_c$ is the linearization of $\widetilde{X}_{\mathcal{H}^c}$ at the critical point $(p^\bullet, q^\bullet) = \Phi(\varphi_c)$. In particular, we see that the (unbounded) linear operator $\mathcal{L}_c$ on $i L_r^2$ with domain $i H_r^2$ is conjugated to the operator $\widetilde{\mathcal{L}}_c$ on $\ell^2 \times \ell^2$ with domain $\mathfrak{h}^2 \times \mathfrak{h}^2$. We have

Lemma 4.1 *Assume that for a given $c \in \mathbb{C}$ the focusing NLS equation has gauge invariant local Birkhoff coordinates in a neighborhood of the constant potential φ_c. Then the spectrum of the operator $\widetilde{\mathcal{L}}_c$ is discrete and lies on the imaginary axis.*

Proof of Lemma 4.1 Let $\{(p_k, q_k)\}_{k \in \mathbb{Z}}$ be the local Birkhoff coordinates on $V \cap (\ell^2 \times \ell^2)$ and let $H^c := \mathcal{H}^c \circ \Phi^{-1} : V \cap (\mathfrak{h}^1 \times \mathfrak{h}^1) \to \mathbb{R}$ be the Hamiltonian $\mathcal{H}^c$ in these coordinates. One easily concludes from (54) and (55) that in the open neighborhood $V \cap (\mathfrak{h}^2 \times \mathfrak{h}^2)$ of the critical point $z^\bullet := (p^\bullet, q^\bullet)$ one has

$$\widetilde{X}_{\mathcal{H}^c} = X_{H^c} = \sum_{n \in \mathbb{Z}} \left(\frac{\partial H^c}{\partial p_n} \partial_{q_n} - \frac{\partial H^c}{\partial q_n} \partial_{p_n} \right).$$

Since by Lemma 2.1 and (55), $z^\bullet$ is a critical point of $\widetilde{X}_{\mathcal{H}^c}$,

$$\left. \frac{\partial H^c}{\partial p_n} \right|_{z^\bullet} = \left. \frac{\partial H^c}{\partial q_n} \right|_{z^\bullet} = 0 \qquad (57)$$

for any $n \in \mathbb{Z}$. In addition, we obtain that the operator $\widetilde{\mathcal{L}}_c : \mathfrak{h}^2 \times \mathfrak{h}^2 \to \ell^2 \times \ell^2$ takes the form

$$\widetilde{\mathcal{L}}_c \equiv d_z \cdot \widetilde{X}_{\mathcal{H}^c}$$
$$= \sum_{n \in \mathbb{Z}} \partial_{q_n} \otimes \sum_{l \in \mathbb{Z}} \left(\frac{\partial^2 H^c}{\partial p_n \partial p_l} dp_l + \frac{\partial^2 H^c}{\partial p_n \partial q_l} dq_l \right) \Big|_{z^\bullet} \tag{58}$$
$$- \sum_{n \in \mathbb{Z}} \partial_{p_n} \otimes \sum_{l \in \mathbb{Z}} \left(\frac{\partial^2 H^c}{\partial q_n \partial p_l} dp_l + \frac{\partial^2 H^c}{\partial q_n \partial q_l} dq_l \right) \Big|_{z^\bullet}.$$

Note that for any $l \in \mathbb{Z}$,

$$X_{I_l} = p_l \partial_{q_l} - q_l \partial_{p_l}.$$

Since the local Birkhoff coordinates are assumed gauge invariant and since $dH(X_{I_k}) = \{H, I_k\} = 0$ for any $k \in \mathbb{Z}$, we obtain that for any $l \in \mathbb{Z}$,

$$0 = (dH^c)(X_{I_l}) = p_l \frac{\partial H^c}{\partial q_l} - q_l \frac{\partial H^c}{\partial p_l} \tag{59}$$

in the open neighborhood $V \cap (\mathfrak{h}^2 \times \mathfrak{h}^2)$ of $z^\bullet$. By taking the partial derivatives ∂_{p_n} and ∂_{q_n} of the equality above at $z^\bullet$ for $n \in \mathbb{Z}$ we obtain, in view of (57), that for any $n, l \in \mathbb{Z}$,

$$\left(\frac{\partial^2 H^c}{\partial p_n \partial q_l} p_l - \frac{\partial^2 H^c}{\partial p_n \partial p_l} q_l \right) \Big|_{z^\bullet} = 0 \quad \text{and} \quad \left(\frac{\partial^2 H^c}{\partial q_n \partial q_l} p_l - \frac{\partial^2 H^c}{\partial q_n \partial p_l} q_l \right) \Big|_{z^\bullet} = 0. \tag{60}$$

We split the set of indices $\mathbb{Z}$ in the sum above into two subsets

$$A := \{ l \in \mathbb{Z} \mid (p_l^\bullet, q_l^\bullet) \neq (0, 0) \} \quad \text{and} \quad B := \{ l \in \mathbb{Z} \mid (p_l^\bullet, q_l^\bullet) = (0, 0) \}.$$

Note that for $l \in B$ the relations (60) are trivial. More generally, by taking the partial derivatives ∂_{p_n} and ∂_{q_n} of (59) in $V \cap (\mathfrak{h}^2 \times \mathfrak{h}^2)$ for $n \neq l$ we see that for any $l \in \mathbb{Z}$ and for any $n \neq l$ we have

$$\frac{\partial^2 H^c}{\partial p_n \partial q_l} p_l - \frac{\partial^2 H^c}{\partial p_n \partial p_l} q_l = 0 \quad \text{and} \quad \frac{\partial^2 H^c}{\partial q_n \partial q_l} p_l - \frac{\partial^2 H^c}{\partial q_n \partial p_l} q_l = 0 \tag{61}$$

for any $(p, q) \in V \cap (\mathfrak{h}^2 \times \mathfrak{h}^2)$. This and Lemma 4.2 below, applied to I equal to $(p_l^2 + q_l^2)/2$ and F equal to $\frac{\partial H^c}{\partial p_n}$ and $\frac{\partial H^c}{\partial q_n}$ respectively, implies that for any $l \in B$ and for any $n \neq l$,

$$\frac{\partial^2 H^c}{\partial p_n \partial q_l} \Big|_{z^\bullet} = \frac{\partial^2 H^c}{\partial p_n \partial p_l} \Big|_{z^\bullet} = 0 \quad \text{and} \quad \frac{\partial^2 H^c}{\partial q_n \partial q_l} \Big|_{z^\bullet} = \frac{\partial^2 H^c}{\partial q_n \partial p_l} \Big|_{z^\bullet} = 0. \tag{62}$$

By combining (62) with (58) we obtain

$$\widetilde{\mathcal{L}}_c = \sum_{n \in A} \partial_{q_n} \otimes \sum_{l \in A} \left(\frac{\partial^2 H^c}{\partial p_n \partial p_l} dp_l + \frac{\partial^2 H^c}{\partial p_n \partial q_l} dq_l \right)\Big|_{z^\bullet}$$

$$- \sum_{n \in A} \partial_{p_n} \otimes \sum_{l \in A} \left(\frac{\partial^2 H^c}{\partial q_n \partial p_l} dp_l + \frac{\partial^2 H^c}{\partial q_n \partial q_l} dq_l \right)\Big|_{z^\bullet}$$

$$+ \sum_{n \in B} (\partial_{p_n}, \partial_{q_n}) \otimes \begin{pmatrix} -\frac{\partial^2 H^c}{\partial q_n \partial p_n} & -\frac{\partial^2 H^c}{\partial q_n \partial q_n} \\ \frac{\partial^2 H^c}{\partial p_n \partial p_n} & \frac{\partial^2 H^c}{\partial p_n \partial q_n} \end{pmatrix}\Big|_{z^\bullet} \begin{pmatrix} dp_n \\ dq_n \end{pmatrix}. \quad (63)$$

Since the local Birkhoff coordinates are assumed gauge invariant and since $dH(X_{I_k}) = \{H, I_k\} = 0$ for any $k \in \mathbb{Z}$, we conclude that the flow S_k^t of the vector field X_{I_k} preserves X_{H^c}, that is for any $t \in \mathbb{R}$ and for any $(p, q) \in V \cap (\mathfrak{h}^2 \times \mathfrak{h}^2)$ such that $S_k^t(p, q) \in V \cap (\mathfrak{h}^2 \times \mathfrak{h}^2)$ we have the following commutative diagram

$$\begin{array}{ccc} V \cap (\mathfrak{h}^2 \times \mathfrak{h}^2) & \xrightarrow{X_{H^c}} & \ell^2 \times \ell^2 \\ S_k^t \downarrow & & \downarrow S_k^t \\ V \cap (\mathfrak{h}^2 \times \mathfrak{h}^2) & \xrightarrow{X_{H^c}} & \ell^2 \times \ell^2 \end{array} \quad (64)$$

Remark 4.2 Note that for any $s \in \mathbb{R}$ and for any $k \in \mathbb{Z}$ we have that $X_{I_k} = p_k \partial_{q_k} - q_k \partial_{p_k}$ is a vector field in the proper sense (i.e. non-weak) on $\mathfrak{h}^s \times \mathfrak{h}^s$ and that $S_k^t : \mathfrak{h}^s \times \mathfrak{h}^s \to \mathfrak{h}^s \times \mathfrak{h}^s$ is a bounded linear map. In fact, if we introduce complex variables $z_k := p_k + iq_k, k \in \mathbb{Z}$, then

$$(S_k^t(z))_l = \begin{cases} z_l, & l \neq k, \\ e^{-it} z_k, & l = k. \end{cases}$$

In particular, we see from (64) that for any $k \in \mathbb{Z}$ and for any $t \in \mathbb{R}$ near zero we have that $\left(d_{S_k^t(z^\bullet)} X_{H^c}\right) \circ S_k^t = S_k^t \circ (d_{z^\bullet} X_{H^c})$. For $k \in B$ we have $S_k^t(z^\bullet) = z^\bullet$ and hence, for any $t \in \mathbb{R}$ near zero,

$$\widetilde{\mathcal{L}}_c \circ S_k^t = S_k^t \circ \widetilde{\mathcal{L}}_c.$$

By taking the t-derivative at $t = 0$ we obtain that for any $k \in B$,

$$[\widetilde{\mathcal{L}}_c, d_{z^\bullet} X_{I_k}] = 0 \quad \text{where} \quad d_{z^\bullet} X_{I_k} = \partial_{q_k} \otimes dp_k - \partial_{p_k} \otimes dq_k. \quad (65)$$

Formula (63) together with (65) and (60) then implies that

$$\widetilde{\mathcal{L}}_c = \sum_{n,k \in A} A_{nk} \left(X_{I_n}\big|_{z^\bullet}\right) \otimes \left(d_{z^\bullet} I_k\right) + \sum_{n \in B} B_n \left(\partial_{q_n} \otimes dp_n - \partial_{p_n} \otimes dq_n\right) \quad (66)$$

for some matrices $(A_{nk})_{n,k \in A}$ and $(B_n)_{n \in B}$ with constant elements. Note that in view of the commutative diagram (56) and Theorem 3.1, the unbounded operator $\widetilde{\mathcal{L}}_c$ on $\ell^2 \times \ell^2$ with domain $\mathfrak{h}^2 \times \mathfrak{h}^2$ has a compact resolvent. In particular, it has discrete spectrum. Moreover, by Theorem 3.1 (i), zero belongs to the spectrum of $\widetilde{\mathcal{L}}_c$ and has geometric multiplicity one. Since, in view of (66), the vectors $X_{I_k}\big|_{z^\bullet}$, $k \in A$, are eigenvectors of $\widetilde{\mathcal{L}}_c$ with eigenvalue zero, we conclude that A consists of one element $A = \{n_0\}$ and that $B_n \neq 0$ for any $n \in \mathbb{Z} \setminus \{n_0\}$. Hence, the spectrum of $\widetilde{\mathcal{L}}_c$ consists of $\{\pm i B_n\}_{n \in \mathbb{Z} \setminus \{n_0\}}$ and zero, which has algebraic multiplicity two and geometric multiplicity one. This completes the proof of Lemma 4.1. □

Let $\{(x, y)\}$ be the coordinates in $\mathbb{R}^2$ equipped with the canonical symplectic form $dx \wedge dy$ and let $I = (x^2 + y^2)/2$. The proof of the following Lemma is not complicated and thus omitted.

Lemma 4.2 *If $F : \mathbb{R}^2 \to \mathbb{R}$ is a C^1-map such that $\{F, I\} = 0$ in some open neighborhood of zero then $d_{(0,0)} F = 0$.*

Now, we are ready to prove Theorem 4.1.

Proof of Theorem 4.1 Take $c \in \mathbb{C}$ such that $|c| \notin \pi \mathbb{Z}$ and $|c| > \pi$, and assume that there exist gauge invariant local Birkhoff coordinates of the focusing NLS equation in a neighborhood of the constant potential φ_c. In view of Lemma 3.1 and Theorem 3.1 (i) the spectrum of $\mathcal{L}_c$ on iL_r^2 is discrete and contains non-zero real eigenvalues. On the other side, by Lemma 4.1, the spectrum of $\widetilde{\mathcal{L}}_c$ lies on the imaginary axis. This shows that the two operators are not conjugated and hence, contradicts the existence of local Birkhoff coordinates. □

Acknowledgements Both authors would like to thank the Fields Institute and its staff for the excellent working conditions and the organizers of the workshop "Inverse Scattering and Dispersive PDEs in One Space Dimension" for their efforts to make it such a successful and pleasant event. The second author also acknowledges the support of the Institute of Mathematics of the BAS.

T.K. is partially supported by the Swiss National Science Foundation. P.T. is partially supported by the Simons Foundation, Award #526907.

References

1. V. Arnold, Mathematical methods of classical mechanics, Graduate Texts in Mathematics, **60**, Springer-Verlag, New York, 1989
2. B. Grébert, T. Kappeler, *The defocusing NLS equation and its normal form*, EMS Series of Lectures in Mathematics, EMS, Zürich, 2014

3. T. Kappeler, P. Topalov, *Arnold-Liouville theorem for integrable PDEs: a case study of the focusing NLS equation*, in preparation
4. T. Kappeler, P. Topalov, *On a Arnold-Liouville type theorem for the focusing NLS and the focusing mKdV equations*, preprint 2018, to appear in Integrable Systems: A recognition of E. Previatos work. Vol. 1, Ed. R. Donagi, T. Shaska, Cambridge University Press, LMS Lecture Notes Series
5. P. Lax, Functional Analysis, Wiley, 2002
6. Li, Y., McLaughlin, D., *Morse and Melnikov functions for NLS PDEs*, Comm. Math. Phys., **162**(1994), no. 1, 175–214
7. J. Williamson, *On the algebraic problem concerning the normal forms of linear dynamical systems*, Amer. J. Math., **58**(1936), 141–163
8. Zakharov, V., Shabat, A., *A scheme for integrating nonlinear equations of mathematical physics by the method of the inverse scattering problem I*, Functional Anal. Appl., **8**(1974), 226–235
9. N. Zung, *A note on focus-focus singularities*, Differential Geometry and its Applications, **7**(1997), 123–130

Extended Decay Properties for Generalized BBM Equation

Chulkwang Kwak and Claudio Muñoz

2010 Mathematics Subject Classification 35Q35, 35Q51

1 Introduction and Main Results

1.1 Setting of the Problem

In this note we shall consider nonlinear scattering and decay properties for the one-dimensional generalized Benjamin, Bona and Mahony (gBBM) equation [5] (or regularized long wave equation) in the energy space:

$$(1 - \partial_x^2)u_t + \left(u + u^p\right)_x = 0, \quad (t,x) \in \mathbb{R} \times \mathbb{R}, \quad p = 2, 3, 4, \ldots \quad (1.1)$$

Here $u = u(t,x)$ is a real-valued scalar function. The original BBM equation, which is the case $p = 2$ above, was originally derived by Benjamin et al. [5] and Peregrine [33] as a model for the uni-directional propagation of long-crested, surface water waves. It also arises mathematically as a regularized version of the KdV equation, obtained by performing the standard "Boussinesq trick". This leads to simpler

---

C. Kwak
Facultad de Matemáticas, Pontificia Universidad Católica de Chile, Santiago, Chile
e-mail: chkwak@mat.uc.cl

C. Muñoz (✉)
CNRS and Departamento de Ingeniería Matemática and Centro de Modelamiento Matemático (UMI 2807 CNRS), Universidad de Chile, Santiago, Chile
e-mail: cmunoz@dim.uchile.cl

well-posedness and better dynamical properties compared with the original KdV equation. Moreover, BBM is not integrable, unlike KdV [11, 27].

It is well-known (see [12]) that (1.1) for $p = 2$ is globally well-posed in H^s, $s \geq 0$, and weakly ill-posed for $s < 0$. As for the remaining cases $p = 3, 4, \ldots$, gBBM is globally well-posed in H^1 [5], thanks to the preservation of the mass and energy

$$M[u](t) := \frac{1}{2} \int \left(u^2 + u_x^2 \right)(t, x) dx, \tag{1.2}$$

$$E[u](t) := \int \left(\frac{1}{2} u^2 + \frac{u^{p+1}}{p+1} \right)(t, x) dx. \tag{1.3}$$

Since now, we will identify H^1 as the standard *energy space* for (1.1).

1.2 Main Result

In this note we consider the problem of decay for small solutions to gBBM (1.1). Let $b > 0$ and $a > \frac{1}{8}$ be any positive numbers, and $I(t)$ be given by

$$I(t) := (-\infty, -at) \cup ((1+b)t, \infty), \quad t > 0. \tag{1.4}$$

Theorem 1.1 *Let $u_0 \in H^1$ be such that, for some $\varepsilon = \varepsilon(b) > 0$ small, one has*

$$\|u_0\|_{H^1} < \varepsilon. \tag{1.5}$$

Let $u \in C(\mathbb{R}, H^1)$ be the corresponding global (small) solution of (1.1) with initial data $u(t = 0) = u_0$. Then, for $I(t)$ as in (1.4), there is strong decay to zero:

$$\lim_{t \to \infty} \|u(t)\|_{H^1(I(t))} = 0. \tag{1.6}$$

Additionally, one has the mild rate of decay

$$\int_2^\infty \int e^{-c_0|x+\sigma t|} \left(u^2 + u_x^2 \right)(t, x) dx \, dt \lesssim_{c_0} \varepsilon^2, \tag{1.7}$$

where σ is fixed and such that $\sigma > \frac{1}{8}$ or $\sigma = -(1+b)$.

Remark 1.1 The case of decay inside the interval $((1+b)t, +\infty)$ is probably well-known in the literature, coming from arguments similar to those exposed by El Dika and Martel in [16]. However, decay for the left portion $\left(-\infty, -\frac{1}{8}^+ t\right)$ seems completely new as far as we understand, and it is in strong contrast with the similar decay problem for the KdV equation on the left, which has not been rigorously proved yet.

Remark 1.2 Note that our results also consider the cases $p = 2$ and $p = 3$, which are difficult to attain using standard scattering techniques because of very weak linear decay estimates, and the presence of long range nonlinearities. Recall that the standard linear decay estimates for BMM are $O(t^{-1/3})$ [1].

Remark 1.3 Theorem 1.1 is in concordance with the existence of solitary waves for (1.1) [16]. Indeed, for any $c > 1$,

$$u(t,x) := (c-1)^{1/(p-1)} Q\left(\sqrt{\frac{c-1}{c}}(x-ct)\right), \quad Q(s) := \left(\frac{p+1}{2\cosh^2(\frac{p-1}{2}s)}\right)^{1/(p-1)},$$

is a solitary wave solution of (1.1), moving to the right with speed $c > 1$. Small solitary waves in the energy space have $c \sim 1$ ($p < 5$), which explains the emergence of the coefficient b in (1.4). Also, (1.1) has solitary waves with negative speed: for $c > 0$ and p even,

$$u(t,x) := -(c+1)^{1/(p-1)} Q\left(\sqrt{\frac{c+1}{c}}(x+ct)\right),$$

is solitary wave for (1.1), but it is never small in the energy space. The stability problem for these solitary waves it is well-known: it was studied in [4, 7, 10, 34, 35]. Indeed, solitary waves are stable for $p = 2, 3, 4, 5$, and stable/unstable for $p > 5$, depending on the speed c. See also [27] for the study of the collision problem for $p = 2$.

Remark 1.4 The extension of this result to the case of perturbations of solitary waves is an interesting open problem, which will be treated elsewhere.

1.3 About the Literature

Albert [1] showed scattering in the L^∞ norm for solutions of (1.1) provided $p > 4$, with resulting global decay $O(t^{-1/3})$. Here the power 4 is important to close the nonlinear estimates, based in weighted Sobolev and Lebesgue spaces. Biler et al. [6] showed decay in several space dimensions, using similar techniques. Hayashi and Naumkin [18] considered BBM with a diffusion term, proving asymptotics for small solutions. Our result improves [1, 18] in the sense that it also considers the cases $p = 2$ and 3, which are not part of the standard scattering theory, and it does not requires a damping term to be valid.

Concerning asymptotic regimes around solitary waves, the fundamental work of Miller and Weinstein [29] showed asymptotic stability of the BBM solitary wave in exponentially weighted Sobolev spaces. El Dika [14, 15] proved asymptotic stability properties of the BBM solitary wave in the energy space. El Dika and Martel [16] showed stability and asymptotic stability for the sum of N solitary waves. See also Mizumachi [32] for similar results. All these results are proved on the right of the main part of the solution itself, and no information is given on the remaining left part. Theorem 1.1 is new in the sense that it also gives information on the left portion of the space.

1.4 About the Proof

In order to prove Theorem 1.1, we follow the ideas of the proof described in [23], where decay for an $abcd$-Boussinesq system [8, 9, 13] was considered. The main tool in [23] was the construction of a suitable virial functional for which the dynamics is converging to zero when integrated in time. See also [22] for further improvements. In this paper, this construction is somehow simpler but still interesting enough, because it allows to consider two different regions of the physical space, on the left (dispersive) and on the right (soliton region), unlike KdV for which virial estimates only reach the soliton region [24–26]. The virial that we use here is also partly inspired in the ones introduced in [19–21], and previously in [24, 28]. See also [2, 17, 30] for similar results.

2 Proof of Theorem 1.1

Let $L > 0$ be large, and $\varphi = \varphi(x)$ be a smooth, bounded weight function, to be chosen later. For each $t, \sigma \in \mathbb{R}$, we consider the following functionals (see [16] for similar choices):

$$\mathcal{I}(t) := \frac{1}{2} \int \varphi\left(\frac{x+\sigma t}{L}\right) \left(u^2 + u_x^2\right)(t, x) dx, \qquad (2.1)$$

and

$$\mathcal{J}(t) := \int \varphi\left(\frac{x+\sigma t}{L}\right) \left(\frac{1}{2}u^2 + \frac{u^{p+1}}{p+1}\right)(t, x) dx. \qquad (2.2)$$

Clearly each functional above is well-defined for H^1 functions. Using (1.1) and integration by parts, we have the following standard result (see also [16, 23] for similar computations).

Lemma 2.1 *For any $t \in \mathbb{R}$, we have*

$$\frac{d}{dt}\mathcal{J}(t) = \frac{\sigma}{2L} \int \varphi' u_x^2 + \frac{1}{2L}(\sigma - 1) \int \varphi' u^2 + \frac{1}{L} \int \varphi' u(1 - \partial_x^2)^{-1} u \qquad (2.3)$$
$$- \frac{1}{L(p+1)} \int \varphi' u^{p+1} + \frac{1}{L} \int \varphi' u(1 - \partial_x^2)^{-1}(u^p),$$

and if $v := (1 - \partial_x^2)^{-1}(u + u^p)$,

$$\frac{d}{dt}\mathcal{J}(t) = \frac{\sigma}{2L} \int \varphi' \left(u^2 + \frac{2}{p+1} u^{p+1} \right) + \frac{1}{2L} \int \varphi'(v^2 - v_x^2). \qquad (2.4)$$

Proof of (2.3) We compute:

$$\frac{d}{dt}\mathcal{J}(t) = \frac{\sigma}{2L} \int \varphi'(u^2 + u_x^2) + \int \varphi(uu_t + u_x u_{tx})$$
$$= \frac{\sigma}{2L} \int \varphi'(u^2 + u_x^2) + \int \varphi u(u_t - u_{txx}) - \frac{1}{L} \int \varphi' u u_{tx}.$$

Replacing (1.1), and integrating by parts, we get

$$\frac{d}{dt}\mathcal{J}(t) = \frac{\sigma}{2L} \int \varphi'(u^2 + u_x^2) - \int \varphi u(u + u^p)_x + \frac{1}{L} \int (\varphi' u)_{xx} (1 - \partial_x^2)^{-1}(u + u^p)$$
$$=: I_1 + I_2 + I_3.$$

I_1 is already done. On the other hand,

$$I_2 = -\int \varphi(uu_x + pu^p u_x) = \frac{1}{L} \int \varphi' \left(\frac{1}{2} u^2 + \frac{p}{p+1} u^{p+1} \right).$$

Finally,

$$I_3 = -\frac{1}{L} \int \varphi' u(u + u^p) + \frac{1}{L} \int \varphi' u(1 - \partial_x^2)^{-1}(u + u^p)$$
$$= -\frac{1}{L} \int \varphi'(u^2 + u^{p+1}) + \frac{1}{L} \int \varphi' u(1 - \partial_x^2)^{-1}(u + u^p).$$

We conclude that

$$\frac{d}{dt}\mathcal{J}(t) = \frac{\sigma}{2L} \int \varphi' u_x^2 + \frac{1}{2L}(\sigma - 1) \int \varphi' u^2 + \frac{1}{L} \int \varphi' u(1 - \partial_x^2)^{-1} u$$
$$- \frac{1}{L(p+1)} \int \varphi' u^{p+1} + \frac{1}{L} \int \varphi' u(1 - \partial_x^2)^{-1}(u^p).$$

This last equality proves (2.3). □

Proof of (2.4) We compute:

$$\frac{d}{dt}\mathcal{I}(t) = \frac{\sigma}{2L}\int \varphi'\left(u^2 + \frac{2}{p+1}u^{p+1}\right) + \int \varphi(u+u^p)u_t$$

$$= \frac{\sigma}{2L}\int \varphi'\left(u^2 + \frac{2}{p+1}u^{p+1}\right) - \int \varphi(u+u^p)\partial_x(1-\partial_x^2)^{-1}(u+u^p).$$

Recall that $v = (1-\partial_x^2)^{-1}(u+u^p)$. Then

$$-\int \varphi(u+u^p)\partial_x(1-\partial_x^2)^{-1}(u+u^p) = -\int \varphi(1-\partial_x^2)vv_x = \frac{1}{2L}\int \varphi'(v^2 - v_x^2).$$

Therefore,

$$\frac{d}{dt}\mathcal{I}(t) = \frac{\sigma}{2L}\int \varphi'\left(u^2 + \frac{2}{p+1}u^{p+1}\right) + \frac{1}{2L}\int \varphi'(v^2 - v_x^2).$$

This proves (2.4). □

For α real number, define the modified virial

$$\mathcal{H}(t) := \mathcal{H}_\alpha(t) := \mathcal{I}(t) + \alpha\mathcal{J}(t). \tag{2.5}$$

From Lemma 2.1, we get (recall that $v = (1-\partial_x^2)^{-1}(u+u^p)$)

$$\frac{d}{dt}\mathcal{H}(t) = \frac{\sigma}{2L}\int \varphi'u_x^2 + \frac{1}{2L}(\sigma(1+\alpha)-1)\int \varphi'u^2 + \frac{1}{L}\int \varphi'u(1-\partial_x^2)^{-1}u$$

$$+ \frac{\alpha}{2L}\int \varphi'(v^2-v_x^2) - \frac{1}{L(p+1)}(\alpha\sigma-1)\int \varphi'u^{p+1}$$

$$+ \frac{1}{L}\int \varphi'u(1-\partial_x^2)^{-1}(u^p). \tag{2.6}$$

Let also, for $u \in H^1$,

$$f := (1-\partial_x^2)^{-1}u \in H^3. \tag{2.7}$$

We have

$$\int \varphi'u^2 = \int \varphi'\left(f^2 + 2f_x^2 + f_{xx}^2\right) - \frac{1}{L^2}\int \varphi'''f^2,$$

$$\int \varphi'u_x^2 = \int \varphi'\left(f_x^2 + 2f_{xx}^2 + f_{xxx}^2\right) - \frac{1}{L^2}\int \varphi'''f_x^2,$$

and

$$\int \varphi' u(1-\partial_x^2)^{-1} u = \int \varphi'\left(f^2 + f_x^2\right) - \frac{1}{2L^2}\int \varphi''' f^2.$$

Additionally, we easily have

$$v^2 = f^2 + 2f(1-\partial_x^2)^{-1}(u^p) + ((1-\partial_x^2)^{-1} u^p)^2,$$

and similarly,

$$v_x = f_x^2 + 2f_x(1-\partial_x^2)^{-1}(u^p)_x + (\partial_x(1-\partial_x^2)^{-1} u^p)^2. \tag{2.8}$$

Replacing these values in (2.6) and rearranging terms, we get

$$\frac{d}{dt}\mathcal{H}(t) = \mathcal{Q}(t) + \mathcal{S}(t) + \mathcal{N}(t), \tag{2.9}$$

where

$$\mathcal{Q}(t) := \frac{1}{2L}(1+\sigma)(1+\alpha)\int \varphi' f^2 + \frac{1}{2L}(\sigma(3+2\alpha) - \alpha)\int \varphi' f_x^2 \\ + \frac{1}{2L}((3+\alpha)\sigma - 1)\int \varphi' f_{xx}^2 + \frac{\sigma}{2L}\int \varphi' f_{xxx}^2, \tag{2.10}$$

$$\mathcal{S}(t) := -\frac{1}{2L^3}(\sigma(1+\alpha) - 1)\int \varphi''' f^2 - \frac{\sigma}{2L^3}\int \varphi''' f_x^2 - \frac{1}{2L^3}\int \varphi''' f^2, \tag{2.11}$$

and

$$\mathcal{N}(t) := \frac{\alpha}{2L}\int \varphi'\left(2f(1-\partial_x^2)^{-1}(u^p) + ((1-\partial_x^2)^{-1} u^p)^2\right) \\ - \frac{\alpha}{2L}\int \varphi'\left(2f_x(1-\partial_x^2)^{-1}(u^p)_x + (\partial_x(1-\partial_x^2)^{-1} u^p)^2\right) \tag{2.12} \\ - \frac{1}{L(p+1)}(\alpha\sigma - 1)\int \varphi' u^{p+1} + \frac{1}{L}\int \varphi' u(1-\partial_x^2)^{-1}\left(u^p\right).$$

Now we consider two different cases.

Case $x > 0$ This is the simpler case. We choose $\varphi := \tanh$, $\alpha = 0$, and $\sigma = -(1+b) < 0$, for b any fixed positive number. Note that $\varphi' = \text{sech}^2 > 0$. Then

$$\mathcal{Q}(t) = -\frac{1}{2L}b\int \varphi' f^2 - \frac{3}{2L}(1+b)\int \varphi' f_x^2 - \frac{1}{2L}(4+3b)$$
$$\int \varphi' f_{xx}^2 - \frac{1}{2L}(1+b)\int \varphi' f_{xxx}^2. \tag{2.13}$$

Now we recall the following result.

Lemma 2.2 (Equivalence of Local H^1 Norms, [23]) *Let f be as in (2.7). Let ϕ be a smooth, bounded positive weight satisfying $|\phi''| \leq \lambda \phi$ for some small but fixed $0 < \lambda \ll 1$. Then, for any $a_1, a_2, a_3, a_4 > 0$, there exist $c_1, C_1 > 0$, depending on a_j and $\lambda > 0$, such that*

$$c_1 \int \phi (u^2 + u_x^2) \leq \int \phi \left(a_1 f^2 + a_2 f_x^2 + a_3 f_{xx}^2 + a_4 f_{xxx}^2 \right) \leq C_1 \int \phi (u^2 + u_x^2). \tag{2.14}$$

Thanks to this lemma, we get for this case

$$\mathcal{Q}(t) \lesssim_{b,L} -\int \varphi'(f^2 + f_x^2 + f_{xx}^2 + f_{xxx}^2) \sim -\int \varphi'(u_x^2 + u^2). \tag{2.15}$$

Case $x < 0$ Here we need different estimates. In (2.10), we will impose

$$\sigma = \frac{1}{8}(1+\tilde{\sigma}), \quad \tilde{\sigma} > 0, \quad \text{and} \quad \alpha = 1.$$

We choose now $\varphi := -\tanh$. Note that $\varphi' = -\operatorname{sech}^2 < 0$. Then we have

$$-16L\mathcal{Q}(t) = 2(9+\tilde{\sigma})\int |\varphi'|f^2 + (-3+5\tilde{\sigma})\int |\varphi'|f_x^2$$
$$+ 4(-1+\tilde{\sigma})\int |\varphi'|f_{xx}^2 + (1+\tilde{\sigma})\int |\varphi'|f_{xxx}^2.$$

Define $g := |\varphi'|^{1/2} f = \operatorname{sech}(\frac{x+\sigma t}{L}) f$. Then we have the following easy identities

$$g_x = \operatorname{sech}\left(\frac{x+\sigma t}{L}\right) f_x + \frac{1}{L}(\operatorname{sech})'\left(\frac{x+\sigma t}{L}\right) f$$
$$= \operatorname{sech}\left(\frac{x+\sigma t}{L}\right) f_x - \frac{1}{L}\tanh\left(\frac{x+\sigma t}{L}\right) g,$$

$$g_{xx} = \operatorname{sech}\left(\frac{x+\sigma t}{L}\right) f_{xx} + \frac{2}{L}(\operatorname{sech})'\left(\frac{x+\sigma t}{L}\right) f_x + \frac{1}{L^2}(\operatorname{sech})''\left(\frac{x+\sigma t}{L}\right) f,$$

and

$$g_{xxx} = \text{sech}\left(\frac{x+\sigma t}{L}\right) f_{xxx} + \frac{3}{L}(\text{sech})'\left(\frac{x+\sigma t}{L}\right) f_{xx}$$
$$+ \frac{3}{L^2}(\text{sech})''\left(\frac{x+\sigma t}{L}\right) f_x + \frac{1}{L^3}(\text{sech})'''\left(\frac{x+\sigma t}{L}\right) f.$$

Therefore,

$$g_{xx} = -\frac{1}{L^2}g - \frac{2}{L}\tanh\left(\frac{x+\sigma t}{L}\right) g_x + \text{sech}\left(\frac{x+\sigma t}{L}\right) f_{xx}$$

and

$$g_{xxx} = -\frac{1}{L^3}\tanh\left(\frac{x+\sigma t}{L}\right) g - \frac{3}{L^2}g_x - \frac{3}{L}\tanh\left(\frac{x+\sigma t}{L}\right) g_{xx} + \text{sech}\left(\frac{x+\sigma t}{L}\right) f_{xxx}.$$

Consequently, for L large enough,

$$-16L\mathcal{Q}(t) = 2(9+\tilde{\sigma})\int g^2 + (-3+5\tilde{\sigma})\int g_x^2$$
$$+ 4(-1+\tilde{\sigma})\int g_{xx}^2 + (1+\tilde{\sigma})\int g_{xxx}^2 + O\left(\frac{1}{L}\int (g^2 + g_x^2 + g_{xx}^2)\right).$$

Now we have for $g \in H^3$,

$$\int (g_{xxx} - \sqrt{2}g_{xx} + 3g_x - 3\sqrt{2}g)^2 \geq 0.$$

Expanding terms and integrating by parts,

$$\int g_{xxx}^2 - 4\int g_{xx}^2 - 3\int g_x^2 + 18\int g^2 \geq 0.$$

We conclude that for $\tilde{\sigma} > 0$ fixed and L large enough,

$$-16L\mathcal{Q}(t) \geq \tilde{\sigma}\int g^2 + 4\tilde{\sigma}\int g_x^2 + 3\tilde{\sigma}\int g_{xx}^2 + \frac{1}{2}\tilde{\sigma}\int g_{xxx}^2.$$

Coming back to the variable f, we obtain for L even larger if necessary,

$$-16L\mathcal{Q}(t) \geq \frac{1}{2}\tilde{\sigma}\int |\varphi'|f^2 + 3\tilde{\sigma}\int |\varphi'|f_x^2 + 2\tilde{\sigma}\int |\varphi'|f_{xx}^2 + \frac{1}{4}\tilde{\sigma}\int |\varphi'|f_{xxx}^2,$$

Then we have

$$Q(t) \lesssim_{\tilde{\sigma},L} -\int |\varphi'|(f^2 + f_x^2 + f_{xx}^2 + f_{xxx}^2) \sim -\int |\varphi'|(u_x^2 + u^2). \qquad (2.16)$$

From (2.15) and (2.16) we conclude that

$$Q(t) \lesssim -\int |\varphi'|(u_x^2 + u^2), \qquad (2.17)$$

provided $\sigma = -(1+b)$, $b > 0$, or $\sigma > \frac{1}{8}$. The terms in (2.11) can be absorbed by this last term using $L > 0$ large and the fact that $|\varphi'''| \lesssim |\varphi'|$. Finally, (2.12) can be absorbed by (2.17) using (1.5) (provided ε is small enough compared with b), just as in [15, 23]. See Appendix for more details. We get

$$\frac{d}{dt}\mathcal{H}(t) \lesssim -\int |\varphi'|(u_x^2 + u^2). \qquad (2.18)$$

Therefore, we conclude that

$$\int_2^\infty \int \operatorname{sech}^2\left(\frac{x + \sigma t}{L}\right)\left(u^2 + u_x^2\right)(t,x)\,dx\,dt \lesssim_L \varepsilon^2. \qquad (2.19)$$

This proves (1.7). As an immediate consequence, there exists an increasing sequence of time $t_n \to \infty$ as $n \to \infty$ such that

$$\int \operatorname{sech}^2\left(\frac{x + \sigma t_n}{L}\right)\left(u^2 + u_x^2\right)(t_n,x)\,dx \longrightarrow 0 \text{ as } n \to \infty. \qquad (2.20)$$

2.1 End of Proof of the Theorem 1.1

Consider $\mathcal{I}(t)$ in (2.1). Choose now $\varphi := \frac{1}{2}(1 + \tanh)$ (for the right side) and $\varphi := \frac{1}{2}(1 - \tanh)$ (for the left hand side) in (2.1). The conclusion (1.6) follows directly from the ideas in [26]. Indeed, for the right side (i.e. $((1+b)t, \infty)$, $b > 0$ fixed), we choose $\tilde{b} = \frac{b}{2}$ and fix $t_0 > 2$. For $2 < t \leq t_0$ and large $L \gg 1$ (to make all estimates above hold), we consider the functional $\mathcal{I}_{t_0}(t)$ by

$$\mathcal{I}_{t_0}(t) := \frac{1}{2}\int \varphi\left(\frac{x + \sigma t_0 - \tilde{\sigma}(t_0 - t)}{L}\right)\left(u^2 + u_x^2\right)(t,x)\,dx,$$

where $\sigma = -(1+b)$ and $\tilde{\sigma} = -(1+\tilde{b})$. From Lemma 2.1, (2.13) with $\tilde{b} > 0$ and the smallness condition (1.5), we have

$$\frac{d}{dt}\mathcal{J}_{t_0}(t) \lesssim_{\tilde{b},L} -\int \operatorname{sech}^2\left(\frac{x+\sigma t_0 - \tilde{\sigma}(t_0-t)}{L}\right)(u^2+u_x^2) \leq 0,$$

which shows that the new functional $\mathcal{J}_{t_0}(t)$ is decreasing on $[2, t_0]$. On the other hand, since $\lim_{x\to-\infty}\varphi(x) = 0$, we have

$$\limsup_{t\to\infty}\int \varphi\left(\frac{x-\beta t - \gamma}{L}\right)(u^2+u_x^2)(\delta, x)dx = 0,$$

for any fixed $\beta, \gamma, \delta > 0$. Together with all above, for any $2 < t_0$, we have

$$0 \leq \int \varphi\left(\frac{x-(1+b)t_0}{L}\right)\left(u^2+u_x^2\right)(t_0, x)dx$$

$$\leq \int \varphi\left(\frac{x-(b-\tilde{b})t_0 - 2(1+\tilde{b})}{L}\right)\left(u^2+u_x^2\right)(2, x)dx,$$

which implies

$$\limsup_{t\to\infty}\int \varphi\left(\frac{x-(1+b)t}{L}\right)\left(u^2+u_x^2\right)(t, x)dx = 0.$$

A analogous argument can be applied to the left side $((-\infty, -at))$, fixed $a > \frac{1}{8}$), thus we conclude (1.6).

Remark 2.1 The understanding of the decay procedure inside the interval $(-t/8, t)$ is an interesting open problem that we hope to consider in a forthcoming publication (see [3]), at least in the case $p = 2$. See also [31] for similar recent results in the KdV case.

Acknowledgements We thank F. Rousset and M. A. Alejo for many interesting discussions on this subject and the BBM equation. C. K. is supported by FONDECYT Postdoctorado 2017 Proyect N° 3170067. C. M. work was partly funded by Chilean research grants FONDECYT 1150202, Fondo Basal CMM-Chile, MathAmSud EEQUADD and Millennium Nucleus Center for Analysis of PDE NC130017.

Part of this work was carried out while the authors were part of the Focus Program on Nonlinear Dispersive Partial Differential Equations and Inverse Scattering (August 2017) held at Fields Institute, Canada. They would like to thank the Institute and the organizers for their warming support.

Appendix: About the Proof of (2.18)

In this section we estimate the nonlinear term

$$\mathcal{N}(t) = \frac{\alpha}{2}\int \varphi'\left(2f(1-\partial_x^2)^{-1}(u^p) + ((1-\partial_x^2)^{-1}u^p)^2\right)$$
$$-\frac{\alpha}{2}\int \varphi'\left(2f_x(1-\partial_x^2)^{-1}(u^p)_x + (\partial_x(1-\partial_x^2)^{-1}u^p)^2\right)$$
$$-\frac{1}{p+1}(\alpha\sigma-1)\int \varphi' u^{p+1} + \int \varphi' u(1-\partial_x^2)^{-1}(u^p).$$

Clearly,

$$\left|\frac{1}{p+1}(\alpha\sigma-1)\int \varphi' u^{p+1}\right| \lesssim \varepsilon^{p-1}\int |\varphi'|u^2,$$

which is enough. Now, recall the following results.

Lemma A.1 ([15]) *The operator $(1-\partial_x^2)^{-1}$ satisfies the following comparison principle: for any $u, v \in H^1$,*

$$v \leq w \implies (1-\partial_x^2)^{-1}v \leq (1-\partial_x^2)^{-1}w. \tag{A.1}$$

Also,

Lemma A.2 ([15, 23]) *Suppose that $\phi = \phi(x)$ is such that*

$$(1-\partial_x^2)^{-1}\phi(x) \lesssim \phi(x), \quad x \in \mathbb{R}, \tag{A.2}$$

for $\phi(x) > 0$ satisfying $|\phi^{(n)}(x)| \lesssim \phi(x)$, $n \geq 0$. Then, for $v, w \in H^1$, we have

$$\int \phi^{(n)} v(1-\partial_x^2)^{-1}(w^p) \lesssim \|v\|_{H^1}\|w\|_{H^1}^{p-2}\int \phi w^2 \tag{A.3}$$

and

$$\int \phi v_x(1-\partial_x^2)^{-1}(w^p)_x \lesssim \|v\|_{H^1}\|w\|_{H^1}^{p-2}\int \phi(w^2 + w_x^2). \tag{A.4}$$

Using (A.3) with $n = 0$,

$$\left|\int \varphi' u(1-\partial_x^2)^{-1}(u^p)\right| \lesssim \varepsilon \int |\varphi'||u^p| \lesssim \varepsilon^{p-1}\int |\varphi'|u^2.$$

Using (A.3) with $n=0$ and (A.4), we also have from $\|f\|_{L^\infty}, \|f_x\|_{L^\infty} \lesssim \|u\|_{H^1}$ that

$$\left|\int \varphi' f(1-\partial_x^2)^{-1}(u^p)\right| \lesssim \varepsilon \int |\varphi'||u^p| \lesssim \varepsilon^{p-1}\int |\varphi'| u^2$$

and

$$\left|\int \varphi' f_x(1-\partial_x^2)^{-1}(u^p)_x\right| \lesssim \varepsilon \int |\varphi'||(u^p)_x| \lesssim \varepsilon^{p-1}\int |\varphi'|(u^2+u_x^2).$$

For the rest terms, using (A.3) with $n=0$ and (A.4),

$$\int |\varphi'|((1-\partial_x^2)^{-1}(u^p))^2 \lesssim \|(1-\partial_x^2)^{-1}(u^p)\|_{H^1}\varepsilon^{p-2}\int |\varphi'|u^2$$

and

$$\int |\varphi'|((1-\partial_x^2)^{-1}\partial_x(u^p))^2 \lesssim \|(1-\partial_x^2)^{-1}(u^p)\|_{H^1}\varepsilon^{p-2}\int |\varphi'|(u^2+u_x^2).$$

Finally, $\|(1-\partial_x^2)^{-1}(u^p)\|_{H^1} \lesssim \|u^p\|_{H^{-1}} \lesssim \varepsilon^p$. Gathering these estimates, we get for some δ small enough,

$$|\mathcal{N}(t)| \lesssim \delta \int |\varphi'|(u^2+u_x^2).$$

References

1. J. Albert, *On the Decay of Solutions of the Generalized Benjamin-Bona-Mahony Equation*, J. Math. Anal. Appl. 141, 527–537 (1989).
2. M.A. Alejo, and C. Muñoz, *Almost sharp nonlinear scattering in one-dimensional Born-Infeld equations arising in nonlinear electrodynamics*, Proceedings of the AMS 146 (2018), no. 5, 2225–2237.
3. M. A. Alejo, M. F. Cortez, C. Kwak, and C. Muñoz, *On the dynamics of zero-speed solutions for Camassa-Holm type equations*, to appear in International Math. Research Notices, https://doi.org/10.1093/imrn/rnz038.
4. T. B. Benjamin, *The stability of solitary waves*. Proc. Roy. Soc. London Ser. A 328 (1972), 153–183.
5. Benjamin, T. B.; Bona, J. L.; Mahony, J. J. (1972), *Model Equations for Long Waves in Nonlinear Dispersive Systems*, Philosophical Transactions of the Royal Society of London. Series A, Mathematical and Physical Sciences, 272 (1220): 47–78.
6. P. Biler, J. Dziubanski, and W. Hebisch, *Scattering of small solutions to generalized Benjamin-Bona-Mahony equation in several space dimensions*, Comm. PDE Vol. 17 (1992) - Issue 9-10, pp. 1737–1758.
7. J. L. Bona, *On the stability theory of solitary waves*. Proc. Roy. Soc. London A 349, (1975), 363–374.

8. J. L. Bona, M. Chen, and J.-C. Saut, *Boussinesq equations and other systems for small-amplitude long waves in nonlinear dispersive media. I: Derivation and linear theory*, J. Nonlinear. Sci. Vol. 12: pp. 283–318 (2002).
9. J. L. Bona, M. Chen, and J.-C. Saut, *Boussinesq equations and other systems for small-amplitude long waves in nonlinear dispersive media. II: The nonlinear theory*, Nonlinearity 17 (2004) 925–952.
10. J. L. Bona, W.R. McKinney, J. M. Restrepo, *Stable and unstable solitary-wave solutions of the generalized regularized long wave equation*. J. Nonlinear Sci. 10 (2000), 603–638.
11. J. L. Bona, W. G. Pritchard, and L. R. Scott, *Solitary-wave interaction*, Physics of Fluids 23, 438 (1980).
12. J. L. Bona, and N. Tzvetkov, *Sharp well-posedness results for the BBM equation*, Discrete Contin. Dyn. Syst. 23 (2009), no. 4, 1241–1252.
13. J. Boussinesq, *Théorie des ondes et des remous qui se propagent le long d'un canal rectangulaire horizontal, en communiquant au liquide contenu dans ce canal des vitesses sensiblement pareilles de la surface au fond*, J. Math. Pure Appl. (2) 17 (1872), 55–108.
14. K. El Dika, *Asymptotic stability of solitary waves for the Benjamin-Bona-Mahony equation*, Discrete and Contin. Dyn. Syst. 2005, 13(3): pp. 583–622. https://doi.org/10.3934/dcds.2005.13.583.
15. K. El Dika, *Smoothing effect of the generalized BBM equation for localized solutions moving to the right*, Discrete and Contin. Dyn. Syst. 12 (2005), no 5, 973–982.
16. K. El Dika, and Y. Martel, *Stability of N solitary waves for the generalized BBM equations*, Dyn. Partial Differ. Equ. 1 (2004), no. 4, 401–437.
17. P. Germain, F. Pusateri, and F. Rousset, *Asymptotic stability of solitons for mKdV*, Adv. in Math. Vol. 299, 20 August 2016, pp. 272–330.
18. T. Hayashi, and P. Naumkin, *Large Time Asymptotics for the BBM-Burgers Equation*, Annales Henri Poincaré June 2007, Vol. 8, Issue 3, pp. 485–511.
19. M. Kowalczyk, Y. Martel, and C. Muñoz, *Kink dynamics in the ϕ^4 model: asymptotic stability for odd perturbations in the energy space*, J. Amer. Math. Soc. 30 (2017), 769–798.
20. M. Kowalczyk, Y. Martel, and C. Muñoz, *Nonexistence of small, odd breathers for a class of nonlinear wave equations*, Letters in Mathematical Physics, May 2017, Volume 107, Issue 5, pp 921–931.
21. M. Kowalczyk, Y. Martel, and C. Muñoz, *On asymptotic stability of nonlinear waves*, Laurent Schwartz seminar notes (2017), see url at http://slsedp.cedram.org/slsedp-bin/fitem?id=SLSEDP_2016-2017____A18_0.
22. C. Kwak, and C. Muñoz, *Asymptotic dynamics for the small data weakly dispersive one-dimensional hamiltonian ABCD system*, to appear in Trans. of the AMS. Preprint arXiv:1902.00454.
23. C. Kwak, C. Muñoz, F. Poblete and J.-C. Pozo, *The scattering problem for the hamiltonian abcd Boussinesq system in the energy space*, J. Math. Pures Appl. (9) 127 (2019), 121–159.
24. Y. Martel and F. Merle, *A Liouville theorem for the critical generalized Korteweg-de Vries equation*, J. Math. Pures Appl. (9) **79** (2000), no. 4, 339–425.
25. Y. Martel and F. Merle, *Asymptotic stability of solitons for subcritical generalized KdV equations*, Arch. Ration. Mech. Anal. **157** (2001), no. 3, 219–254.
26. Y. Martel and F. Merle, *Asymptotic stability of solitons for subcritical gKdV equations revisited*. Nonlinearity, 18 (2005), no. 1, 55–80.
27. Y. Martel, F. Merle, and T. Mizumachi, *Description of the inelastic collision of two solitary waves for the BBM equation*, Arch. Rat. Mech. Anal. May 2010, Volume 196, Issue 2, pp 517–574.
28. F. Merle and P. Raphaël, *The blow-up dynamic and upper bound on the blow-up rate for critical nonlinear Schrödinger equation*, Ann. of Math. (2) **161** (2005), no. 1, 157–222.
29. J. R. Miller, M. I. Weinstein, *Asymptotic stability of solitary waves for the regularized long-wave equation*. Comm. Pure Appl. Math. 49 (1996), no. 4, 399–441.
30. C. Muñoz, F. Poblete, and J. C. Pozo, *Scattering in the energy space for Boussinesq equations*, Comm. Math. Phys. 361 (2018), no. 1, 127–141.

31. C. Muñoz, and G. Ponce, *Breathers and the dynamics of solutions to the KdV type equations*, Comm. Math. Phys. 367 (2019), no. 2, 581–598.
32. T. Mizumachi, *Asymptotic stability of solitary wave solutions to the regularized long-wave equation*, J. Differential Equations, 200 (2004), no. 2, 312–341.
33. D. H. Peregrine, *Long waves on a beach*. J. Fluid Mechanics 27 (1967), 815–827.
34. P. E. Souganidis, W. A. Strauss, *Instability of a class of dispersive solitary waves*. Proc. Roy. Soc. Edinburgh Sect. A 114 (1990), no. 3–4, 195–212.
35. M. I. Weinstein, *Existence and dynamic stability of solitary wave solutions of equations arising in long wave propagation*. Comm. Partial Differential Equations 12 (1987), no. 10, 1133–1173.

Ground State Solutions of the Complex Gross Pitaevskii Equation

Slim Ibrahim

1 Introduction of the Model and Main Result

The possibility of condensation of bosons was predicted by Bose [4] and Einstein [9, 10]. It was obtained experimentally for the first time in Anderson et al. [2], Bradley et al. [5], and Davis et al. [8] in a system consisting of about half a million alkali atoms cooled down to nanokelvin-level temperatures.

Gross Pitaevskii equation (GP) was introduced by Gross [12], Pitaevskii [22], Pitaevskii and Stringari [23] and its variants are widely used to understand Bose Einstein Condensate (BEC) in various systems.

The principal interest in BEC lies in its nature as a macroscopic quantum system, and some of the dynamics of atomic BEC have been successfully described by the GP equation, a nonlinear Schrödinger equation:

$$i\hbar \frac{\partial \psi}{\partial t} = \left(-\frac{\hbar^2}{2m} \Delta + V(x) + \delta |\psi|^2 \right) \psi, \tag{GP}$$

derived from the mean field theory of weakly interacting bosons. Here, $\psi = \psi(x, t)$ is the wave function of the condensate, δ is a constant characterizing the strength of the boson-boson interactions, m is the mass of the particles, and $V(x)$ is the trapping potential.

A serious obstacle in the study of BEC in atomic systems is the extremely low temperatures required to create the condensate. Because of this difficulty, other, non-atomic systems are being explored that can undergo condensation at higher temperatures. One possible candidate is a system of exciton-polaritons, which are

---

S. Ibrahim (✉)
Department of Mathematics and Statistics, University of Victoria, Victoria, BC, Canada
e-mail: ibrahims@uvic.ca

quasi-particles that can be created in semiconductor cavities as a result of interaction between excitons and a laser field in the cavity (see Kasprzak et al. [16], Coldren and Corzine [6]). A two-dimensional quantum structure consisting of coupled quantum wells embedded in an optical microcavity is used, and Excitons are produced in the coupled quantum wells, that interact with the photons trapped inside the optical cavity by means of two highly reflective mirrors. Due to this confinement, the effective mass of the polaritons is very small: 10^{-4} times the free-electron mass (see [16]). Since the temperature of condensation is inversely proportional to the mass of the particles, the exciton-polariton systems afford relatively high temperatures of condensation.

Two drawbacks in this new condensate: The polaritons are highly unstable and exhibit strong interactions. The excitons disappear with the recombination of the electron-hole pairs through emission of photons. One way to deal with this problem is to introduce a polariton reservoir: Polaritons are "cooled" and "pumped" from this reservoir into the condensate to compensate for the decay.

Various mathematical models have been proposed for this new condensate (see Keeling and Berloff [18], Sanvitto et al. [25], Wouters and Carusotto [27]). One of them, called complex GP equation (see Keeling and Berloff [19]), is explored in this work. This CGP equation reflects the non-equilibrium dynamics described above by adding pumping and decaying terms to the GP equation. The Complex Gross-Pitaevskii Equation reads

$$i\frac{\partial \psi}{\partial t} = (-\Delta + V(x) + |\psi|^2)\psi + i(\sigma(x) - \alpha|\psi|^2)\psi, \quad \text{(GPPD)}$$

where $\psi = \psi(x,t)$ is a complex-valued function defined on $\mathbb{R}^2 \times \mathbb{R}$, Δ is the Laplace operator on $\mathbb{R}^2$, $V(x) = |x|^2$ is the harmonic potential, $\sigma(x) \geq 0$ and $\alpha \geq 0$. The complex GP equation can be thought of as a complex Ginzburg-Landau equation without viscous dissipation.

$$i\frac{\partial \psi}{\partial t} = (-\Delta + V(x) + |\psi|^2)\psi + i(\sigma(x) - \alpha|\psi|^2)\psi + i\Delta\psi, \quad \text{(CGL)}$$

The absence of this dissipation makes it difficult even to get time-uniform estimates of the solutions of this equation in an appropriate energy space.

Instabilities are most likely to occur: It has been observed in various experiments on polariton condensates (see, e.g., Manni et al. [20], Ohadi et al. [21], Amo et al. [1]) that as the pumping intensity increases, polariton-polariton interactions become stronger, resulting in higher polariton dispersion and instabilities. Furthermore, Ballarini et al. [3] have observed long-lived polariton states across a parametric threshold of the pumping intensity.

A numerical approach has been introduced by Sierra et al. [26] who have developed a numerical collocation method to (numerically) construct radially symmetric "ground state", and showed its linear stability. They used a "smooth" Heaviside function. The original complex GP equation proposed by Keeling and

Berloff [19] includes a Heaviside function for the pumping part. With the latter, it is seen that the radially symmetric solutions have a discontinuity in the second derivative. This discontinuity reduces the accuracy of the collocation method and produces the Gibbs phenomenon in the splitting method due to the spectral part.

They observed the emergence of complicated vortex lattices after symmetry breaking (as in Keeling and Berloff [19]). These lattices remain rotating with a constant angular velocity, becoming the stable solution of the system.

Recall that the mass $\mathcal{M}$, the Hamiltonian $\mathcal{H}$, the action $\mathcal{S}_\mu$ ($\mu > 0$) and the functional $\mathcal{K}$ associated to Eq. (GPPD) are given by:

$$\mathcal{M}(u) := \|u\|_{L^2}^2, \tag{1.1}$$

$$\mathcal{H}(u) := \frac{1}{2}(\|\nabla u\|_{L^2}^2 + \|xu\|_{L^2}^2) + \frac{1}{4}\|u\|_{L^4}^4 := \mathcal{H}_0(u) + \frac{1}{4}\|u\|_{L^4}^4 \tag{1.2}$$

$$\mathcal{S}_\mu(u) := -\frac{\mu}{2}\mathcal{M}(u) + \mathcal{H}(u), \tag{1.3}$$

$$\mathcal{K}(u) := \int_{\mathbb{R}^2} (\sigma(x) - \alpha|u(x)|^2)|u(x)|^2 \, dx, \tag{1.4}$$

respectively. Observe that

$$\frac{d}{dt}\mathcal{M}(\psi(t)) = \mathcal{K}(\psi(t)) \tag{1.5}$$

and

$$\frac{d}{dt}\mathcal{H}(\psi(t)) = \int_{\mathbb{R}^2} (\sigma - \alpha|\psi|^2)(|\psi|^4 + V|\psi|^2 + |\nabla\psi|^2) - 2\alpha(R(\psi\nabla\bar{\psi}))^2 \, dx. \tag{1.6}$$

Identity (1.5) shows that, at least formally, the mass and the energy are pumped into the system through the term $i\sigma\psi$ involving the parameter σ and they are nonlinearly damped by the term $-i\alpha|\psi|^2\psi$ involving the parameter α. Contrarily to the complex Ginzburg-Landau equation (when a dissipation term of the form $i\Delta\psi$ is added to the right hand side of (GPDP)), one cannot obtain time-uniform estimates of the solution in the energy space. The complex Gross-Pitaevskii equation reflects the non-equilibrium dynamics described above by adding pumping and decaying terms to the GP equation.

Before going any further, we recall a few results about the linear equation without dissipation and pumping. The equation then reads

$$i\frac{\partial\phi}{\partial t} = (-\Delta + V(x))\phi.$$

We define the energy space $\Sigma := H^1(\mathbb{R}^2) \cap \{u : xu \in L^2\}$, endowed with the L^2-scalar product $(u, v)_2 := \int_{\mathbb{R}^2} u(x)\bar{v}(x)\,dx$, by

$$(u, v)_\Sigma = (\nabla u, \nabla v)_2 + (xu, xv)_2 + (u, v)_2 : \|u\|_\Sigma^2 = \|\nabla u\|_2^2 + \left\|(1 + (|\cdot|^2)^{\frac{1}{2}} u\right\|_2^2.$$

Also, define the dual space Σ^* of Σ as follows. For any $v \in \Sigma^*$, there exists a unique $u \in \Sigma$ such that $H_0 u = v$ with the norm on Σ^* given by

$$\|H_0 u\|_{\Sigma^*} = \|v\|_{\Sigma^*} := \|u\|_\Sigma.$$

Recall that $\|\cdot\|_p$ is the norm in $L^p(\mathbb{R}^2)$. It is well known that the unbounded operator $H_0 := -\Delta + V$ defined on

$$D(H_0) := \{u \in \Sigma : H_0(u) \in L^2(\mathbb{R}^2)\}$$

is self-adjoint. Moreover, the lowest eigenvalue of H_0 denoted by $\omega_1 = 2$ is simple with eigenfunction $e_1(x) = \frac{1}{\sqrt{\pi}} e^{-|x|^2/2}$. Notice that (e_1, ω_1) can be constructed variationally as

$$\omega_1 = \min_{\|u\|_{L^2}=1} \frac{1}{2} \int_{\mathbb{R}^2} |\nabla u|^2 + |x|^2 |u|^2 \, dx := \min_{\|u\|_{L^2}=1} \mathcal{H}_0.$$

In particular, for any $u \in D(H_0)$, we have

$$2\|u\|_{L^2}^2 \leq \|xu\|_{L^2}^2 + \|\nabla u\|_{L^2}^2.$$

For more details, we refer for example to [17].

Now, suppose $\psi(x, t) = Q(x)e^{\mu t}$ is a solitary wave with a complex chemical potential $\mu = \mu_r + i\mu_i$, then

$$\psi(x, t) = Q(x)e^{t\mu_i} e^{-it\mu_r}$$

and the wave would grow exponentially fast as $|t| \to \infty$ if $\mu_i \neq 0$. To avoid this, we assume that $\mu = \mu_r$. Equation (GPPD) then yields the following stationary problem for Q:

$$\mu Q = (-\Delta + V(x) + |Q|^2)Q + i(\sigma(x) - \alpha|Q|^2)Q, \qquad Q \in \Sigma \setminus \{0\}. \quad (\mu\text{-SP})$$

Multiplying (μ-SP) by $\bar{Q}$ and integrating gives the following identity.

$$\mu \mathcal{M}(Q) = 2\mathcal{H}(Q) + 1/2 \|Q\|_{L^4}^4 + i\mathcal{K}(Q).$$

Ground State Solutions of the Complex Gross Pitaevskii Equation

The condition for the chemical potential μ of being real is then equivalent to the fact that Q is a zero of $\mathcal{K}$.

It is important to emphasize that due to the presence of the dissipation and pumping mechanisms, we find it hard to apply the standard variational or PDE methods to construct soliton-type solutions of (GPPD) (i.e. a solution Q of (μ-SP)). In this paper, our idea to construct a solution of (μ-SP) with real chemical potential μ goes along a perturbative way by introducing a small parameter factor in the dissipation and pumping term. More precisely, for all $\varepsilon > 0$, consider

$$i\frac{\partial \psi}{\partial t} = (-\Delta + V(x) + |\psi|^2)\psi + i\varepsilon(\sigma(x) - \alpha|\psi|^2)\psi, \quad \text{(GPPD}_\varepsilon)$$

and its corresponding stationary equation

$$\mu Q = (-\Delta + V(x) + |Q|^2)Q + i\varepsilon(\sigma(x) - \alpha|Q|^2)Q \quad Q \in \Sigma \setminus \{0\}. \quad (\mu\text{-SP}_\varepsilon)$$

The object is to construct a solution $(Q_\varepsilon, \mu_\varepsilon)$ in the form

$$Q_\varepsilon = Q_\varepsilon^a + \psi_\varepsilon, \quad \text{and} \quad \mu_\varepsilon = \mu_\varepsilon^a + \mu_\varepsilon,$$

where the approximate solution $(Q_\varepsilon^a, \mu_\varepsilon^a)$ will be given explicitly, and $(\psi_\varepsilon, \mu_\varepsilon)$ is an error term. The main result of this paper reads as follows.

Theorem 1.1 *Assume that $\sigma(x) \geq 0$ is the Heaviside function. There exist a positive ε_0 small enough and $\alpha_0 > 0$ such that, for any $0 < \varepsilon < \varepsilon_0$, and $\alpha < \alpha_0$, the complex Gross-Pitaevskii equation (GPPD$_\varepsilon$) has a solitary wave solution $\psi^\varepsilon(x, t) = e^{it\mu_\varepsilon} Q^\varepsilon(x)$ with $(Q, \mu_\varepsilon) \in \Sigma \times (2, \infty)$ solving μ-SP$_\varepsilon$.*

Remark 1.1 It would be very desirable to extend the branch of standing wave solutions we constructed for ε small to all values of ε. Unfortunately, so far we were not able to do so given the non-equilibrium structure of the model.

2 Idea of the Proof

The main idea of the proof goes as follows. Plugging the Ansatz

$$Q_\varepsilon = Q_0 + \varepsilon(Q_{1,r} + iQ_{1,i}) + \varepsilon^2(Q_{2,r} + iQ_{2,i}) + \varepsilon^3(Q_{3,r} + iQ_{3,i}) + \cdots$$

and

$$\mu_\varepsilon = \mu_0 + \varepsilon\mu_1 + \varepsilon^2\mu_2 + \varepsilon^3\mu_3 + \cdots$$

into (GPPD) and taking into consideration the constraint $\mathcal{K}(Q_\varepsilon) = 0$ gives the following relations

$$Q_{2k+1,r} = Q_{2k+2,i} = 0, \quad \text{and} \quad \mu_{2k+1} = 0$$

for all $k = 0, 1, 2, \cdots$. In addition, Q_0 has to solve

$$\mu Q = (-\Delta + V(x) + |Q|^2)Q, \quad Q \in \Sigma \setminus \{0\} \quad (\mu\text{-SP}_0)$$

with the constraint $\mathcal{K}(Q_0) = 0$. The first step then is to construct such a Q_0. Next, in order to define $Q_{2k+1,i}$ and $Q_{2k,r}$, we proceed as follows. Denote by

$$L_- := -\Delta + V + Q_0^2 - \mu_0,$$

and

$$L_+ := -\Delta + V + 3Q_0^2 - \mu_0.$$

A fundamental property satisfied by L_- is given by the following proposition.

Proposition 2.1 *Let $<Q_0>^\perp$ be the subspace of Σ consisting of all functions L^2-orthogonal to Q_0. Then we have*

$$\ker(L_-) = \{Q_0\}, \quad \text{and} \quad L_- :<Q_0>^\perp \to <Q_0>^\perp \quad \text{is bijective.}$$

and

$$L_+ : \Sigma \to \Sigma^* \quad \text{is bijective, for large } \alpha.$$

We refer to [14] for complete details. Now, since $\mathcal{K}(Q_0) = (Q_0, (\sigma - \alpha|Q_0|^2)Q_0)_2 = 0$, then thanks to Proposition 2.1, one can uniquely define Q_{1i} by

$$L_- Q_{1i} := (\alpha|Q_0|^2 - \sigma)Q_0.$$

Now, define Q_{2r} and Q_{3i} by

$$L_+ Q_{2r} = \mu_2 Q_0 + (\sigma - \alpha|Q_0|^2)Q_{1i} - Q_0 Q_{1i}^2, \tag{2.1}$$

and

$$L_- Q_{3i} = (2Q_{2r}Q_0 - Q_{1i}^2)Q_{1i} + \mu_2 Q_{1i} + ((2+\alpha)Q_0^2 - \sigma)Q_{2r} + Q_{1i}^2 Q_0. \tag{2.2}$$

The bijectivity of L_+ enables us to determine Q_{2r}, and again the regularity of Q_0 shows that $Q_{2r} \in \text{Dom}(L_+)$. Thus it only remains to determine the coefficient μ_2, and Q_{3i}. They are uniquely calculated by the orthogonality condition

$$(L_- Q_{3i}, Q_0)_2 = 0.$$

Indeed, substituting Q_{2r} (given by inverting (2.1)) into (2.2) gives

$$L_- Q_{3i} = \mu_2[Q_{1i} + ((2+\alpha)Q_0^2 - \sigma + 2Q_0 Q_{1i})L_+^{-1} Q_0] + Q_{1i}^2 Q_0 - Q_{1i}^3 \quad (2.3)$$

$$+((2+\alpha)Q_0^2 - \sigma + 2Q_0 Q_{1i})L_+^{-1}((\sigma - Q_0^2)Q_{1i} - Q_0 Q_{1i}^2). \quad (2.4)$$

Now since $(Q_0, Q_{1i})_2 = 0$, then clearly

$$(L_+^{-1}((2+\alpha)Q_0^2 - \sigma + 2Q_0 Q_{1i}), Q_0) = \|Q_0\|_{L^2}^2 \neq 0,$$

which ensures that μ_2 is uniquely determined in terms of Q_0, $Q_{1,i}$ which were already defined. Then Q_{3i} follows by inverting L_- using the orthogonality $(Q_{3i}, Q_0)_2 = 0$. To this end, a fixed point argument enables us to construct $(Q_\varepsilon, \mu_\varepsilon)$ as a perturbation of the approximate solution

$$Q_\varepsilon^a := Q_0 + i\varepsilon Q_{1i} + \varepsilon^2 Q_{2r} + i\varepsilon^3 Q_{3i}, \quad \text{and} \quad \mu_\varepsilon^a = \mu_0 + \varepsilon^2 \mu_2. \quad (2.5)$$

3 Sketch of the Proofs

3.1 The Case Without Energy Pumping and Dissipation

Here we focus on the problem without pumping and decay of the energy, that is when $\varepsilon = 0$. We start by recalling a few know fact about the space Σ, for which the proof can for example be found in Kavian and Weissler [17].

Lemma 3.1 *The Hilbert space Σ is compactly embedded in $L^p(\mathbb{R}^2)$ for any $p \in [2, \infty)$.*

Throughout this paper, we suppose that $\sigma \geq 0$ in $L^\infty(\mathbb{R}^2)$.

Lemma 3.2 *For any $M > 0$, there exists a unique $v_M \in \Sigma$ solving the following constrained variational problem:*

$$(V_M): \quad \mu_M = \inf\{\mathcal{H}(u) : \int u^2 = M\};$$

In addition, v_M is non-negative, radial and radially decreasing.

Proof It is sufficient to show the existence of a minimizer of (V_M). The uniqueness of the minimizer follows directly from the strict convexity of the functional $\mathcal{H}$.

Now let us fix $M > 0$, let (v_n) be a minimizing sequence of (V_M), i.e., $\lim_{n \to \infty} \mathcal{H}(v_n) = \mu_M$ and $\int v_n^2 = M$. Then

$$\mathcal{H}(v_n) \geq \frac{1}{2}\|\nabla v_n\|_2^2 + \frac{1}{2}\|x v_n\|_2^2.$$

Therefore, we can find $K_M > 0$ such that

$$\|\nabla v_n\|_2^2 + \|x v_n\|_2^2 \leq K_M.$$

This implies that

$$\|v_n\|_\Sigma^2 \leq M + K_M. \qquad (3.1)$$

Consequently, there exists $u \in \Sigma$ such that

$$v_n \rightharpoonup u \quad \text{in } \Sigma.$$

This implies, thanks to Lemma 3.1, that $v_n \to u$ in $L^2(\mathbb{R}^2)$ and $L^4(\mathbb{R}^2)$. Thus, we certainly have that $\int u^2 = M$ implying that u is non-trivial, and by the lower semi-continuity, we can write:

$$\mathcal{H}(u) \leq \liminf_n \mathcal{H}(v_n) = \mu_M.$$

Therefore, $\mathcal{H}(u) = \mu_M$. On the other hand, let u be the unique minimizer of (V_M), then u is a non-negative function in Σ since

$$\mathcal{H}(|u|) \leq \mathcal{H}(u), \quad \text{and} \quad M(|u|) = M(u).$$

Furthermore, by standard rearrangement inequalities [13], we have:

$$\mathcal{H}(|u|^*) \leq \mathcal{H}(|u|).$$

$\square$

The next Lemma, addresses the regularity of the Hamiltonian $\mathcal{H}$, as well as the map $M \to \mu_M$.

Lemma 3.3

$$\mathcal{H} \in C^1(\Sigma, \mathbb{R}). \qquad (i)$$

$$\|\mathcal{H}'(u)\|_{\Sigma^{-1}} \leq C\{\|u\|_\Sigma + \|u\|_\Sigma^3\} \quad \text{for all} \quad \in \Sigma. \qquad (ii)$$

$$\text{the function} \quad M \to \mu_M = \mathcal{H}(v_M), \quad \text{is continuous on} \quad (0, \infty). \qquad (iii)$$

Proof The proofs of (i) and (ii) follow from standard arguments. For example, we refer to reference [15], and we just prove (iii). Fix $M > 0$. Let $M_n \subset (0, \infty)$ be a sequence of positive real numbers such that $M_n \to M$. We will first prove that

$$\limsup_n \mu_{M_n} \leq \mu_M. \qquad (3.2)$$

Let (v_n) be a sequence such that $\int v_n^2 = M$ and $\mathcal{H}(v_n) \to \mu_M$. By (3.1), we can find $L > 0$ such that

$$\|v_n\|_\Sigma^2 \leq L.$$

Now let $w_n = \frac{M_n}{M} v_n$, then $\int w_n^2 = M_n$ and

$$\|v_n - w_n\|_\Sigma = \left|1 - \frac{M_n}{M}\right| \|v_n\|_\Sigma \leq \left|1 - \frac{M_n}{M}\right| L$$

for any $n \in \mathbb{N}$.

Therefore, we can find n_0 such that

$$\|v_n - w_n\|_\Sigma \leq L + 1$$

for any $n \geq n_0$.

It follows from (ii) that we can find a constant $K(L)$ such that $\|\mathcal{H}'(u)\|_{\Sigma^{-1}} \leq K(L)$ for all $u \in \Sigma$ such that $\|u\|_\Sigma \leq 2L + 1$.

Thus, for all $n \geq n_0$,

$$|\mathcal{H}(w_n) - \mathcal{H}(v_n)| = \left|\int_0^1 \frac{d}{dt} \mathcal{H}(tw_n + (1-t)v_n) dt\right|$$

$$\leq \sup_{\|u\|_\Sigma \leq 2L+1} \|\mathcal{H}'(u)\|_{\Sigma^{-1}} \|v_n - w_n\|_\Sigma$$

$$\leq K(L) L \left|1 - \frac{M_n}{M}\right|.$$

Consequently, $\mu_{M_n} \leq \mathcal{H}(w_n) \leq \mathcal{H}(v_n) + K(L) L |1 - \frac{M_n}{M}|$.
Then $\limsup \mu_{M_n} \leq \lim \mathcal{H}(v_n) = \mu_M$ and then

$$\limsup \mu_{M_n} \leq \mu_M. \tag{3.3}$$

Now let us prove that if $M_n \to M$, then

$$\mu_M \leq \liminf \mu_{M_n}. \tag{3.4}$$

For all $n \in \mathbb{N}$, there exists (v_n) a sequence of functions in Σ such that $\int v_n^2 = M_n$ and

$$\mu_{M_n} \leq \mathcal{H}(v_n) \leq \mu_{M_n} + \frac{1}{n}.$$

Combining the proof of (3.1) and (3.4), we can find $K > 0$ such that $\|v_n\|_\Sigma \leq K$ for all $n \in \mathbb{N}$. Setting $w_n = \frac{M}{M_n} v_n$, we have that $\int w_n^2 = M$ and

$$\|v_n - w_n\|_\Sigma \leq K \left|1 - \frac{M}{M_n}\right|.$$

Thus, following the proof of (3.4), we certainly get:

$$|\mathcal{H}(w_n) - \mathcal{H}(v_n)| \leq L(K)K \left|1 - \frac{M}{M_n}\right|.$$

Consequently, we have:

$$\mu_{M_n} \geq \mathcal{H}(v_n) - \frac{1}{n} \geq \mathcal{H}(w_n) - L(K)K|1 - \frac{M}{M_n}| - \frac{1}{n},$$

yielding $\liminf \mu_{M_n} \geq \mu_M$ as desired. □

Proposition 3.4 *Let $(M_n) \subset (0, \infty)$ be a sequence of positive real numbers such that $M_n \to M$. Denote by v_{M_n} the unique minimizer of (V_{M_n}), and v_M the unique minimizer of (V_M). Then*

$$\mathcal{K}(v_{M_n}) \to \mathcal{K}(v_M),$$

and

$$\mathcal{H}(v_{M_n}) \to \mathcal{H}(v_M).$$

Proof We will first prove that there exists $\bar{u} \in \Sigma$ such that v_{M_n} converges weakly in Σ to $\bar{u}$ ($v_{M_n} \rightharpoonup \bar{u}$ in Σ). First obviously $\|v_{M_n}\|_2^2 \leq A$. Now noticing that

$$\mu_{M_n} = \frac{1}{2}\|\nabla v_{M_n}\|_2^2 + \frac{1}{2}\|xv_{M_n}\|_2^2 + \frac{1}{4}\|v_{M_n}\|_4^4,$$

one has

$$\mu_{M_n} \geq \frac{1}{2}\|\nabla v_{M_n}\|_2^2 + \frac{1}{2}\|xv_{M_n}\|_2^2.$$

Therefore, using (3.4), there exists a constant $B > 0$ such that

$$\|v_{M_n}\|_\Sigma \leq B.$$

Thus, (up to a subsequence), there exists $\bar{u} \in \Sigma$ such that

$$v_{M_n} \rightharpoonup \bar{u} \text{ in } \Sigma.$$

Now using Lemma 3.1, we have that

$$v_{M_n} \to \bar{u} \quad \text{in } L^2(\mathbb{R}^2) \cap L^4(\mathbb{R}^2).$$

In particular, $\int \bar{u}^2 = M$. Thus,

$$\mu_M \leq \mathcal{H}(\bar{u}) \leq \liminf \mathcal{H}(v_{M_n}) = \liminf \mu_{M_n}$$

and then $\mathcal{H}(\bar{u}) = \mu_M$. This shows that $\bar{u}$ is the unique minimizer of (V_M). To end the proof, we need to show that

$$\int \sigma(x) v_{M_n}^2(x) \to \int \sigma(x) v_M^2(x) \qquad (3.5)$$

and

$$\int v_{M_n}^4(x) \to \int v_M^4(x). \qquad (3.6)$$

To prove (3.5), it is sufficient to notice that $\sigma \in L^\infty(\mathbb{R}^2)$ and $v_n \to v$ in $L^2(\mathbb{R}^2)$, while (3.6) follows from the fact that $v_n \to u$ in $L^4(\mathbb{R}^2)$. □

Always in the case $\varepsilon = 0$, and within the class of minimizers we have just constructed, we would like to intersect it with the co-dimension one manifold characterized by the zeros of the functional $\mathcal{K}$. Before doing so, let us first fix our assumptions on the decay and pumping parameters. Throughout this section, we suppose that:

(H_0) There exists $R > 0$ such that $\sigma(x) > 0$, $\forall\, |x| < R$.
(H_1) $\sigma \in L^\infty(\mathbb{R}^2)$.
(H_2) $\alpha > 0$.

The first preliminary result is the first iteration. We can choose a ground state with the following property.

Proposition 3.5 *There exists a non-negative radial function $Q_0 \in \Sigma$ and $\mu_0 > 2$ solving (μ-SP$_0$). Moreover, Q_0 satisfies*

$$\mathcal{K}(Q_0) = 0.$$

Remark 3.1 As explained in the section Idea of the proof, (Q_0, μ_0) will be the first approximate solution in the iteration process to construct the full solution $(Q_\varepsilon, \mu_\varepsilon)$ of (μ-SP$_\varepsilon$).

To construct a nonlinear solution to (μ-SP$_0$), one can use several techniques. Variationally, for any given amount of mass $M > 0$, we have shown that a radial positive solution (u_M, μ_M) to (μ-SP$_0$) can be constructed through the following minimizing problem

$$\mu_M = \mathcal{H}(u_M) := \min_{\|u\|_{L^2}^2 = M} \mathcal{H}(u).$$

Moreover, this family of solutions coincides also with the one constructed using bifurcation arguments pioneered by Crandall and Rabinowitz [7]. Indeed, (u, μ) is a solution to (μ-SP$_0$) if and only if $(I - \mu K)u = \mathcal{N}(u)$, where $K = A^{-1}B$, $\mathcal{N} = A^{-1}G'(u)$, and the operators A, B and G are defined by

$$A : \Sigma \to \Sigma^*, \quad \text{for any} \quad u, v \in \Sigma; \quad < Au, v > := (\nabla u, \nabla v)_2 + (xu, xv)_2,$$

$$B : \Sigma \to \Sigma^*, \quad \text{for any} \quad u, v \in \Sigma; \quad < Bu, v > := (u, v)_2,$$

and

$$G : \Sigma \to \mathbb{R}, \quad \text{for any} \quad u \in \Sigma; \quad G(u) = -\frac{1}{4}\|u\|_{L^4}^4.$$

The following proposition shows that a local branch of solutions of (μ-SP$_0$) emerging from the linear solution (e_1, ω_1) can be constructed. This is in particular helpful to show that the spectral assumption on L_+ is satisfied when the value of α is sufficiently large. We refer to [14] for its proof.

Proposition 3.6 *There exists $\eta_0 > 0$ such that for all $0 < \eta < \eta_0$, a solution $u(\eta) \in \Sigma$, $\mu(\eta) > 2$ of (μ-SP$_0$) exists such that*

$$u(\eta) = \eta e_1 + \eta z(\eta),$$

with $z \in \Sigma$, $z(0) = 0$ and $(z(\eta), e_1)_2 = 0$.

Now, we are ready to prove Proposition 3.5.

Proof of Proposition 3.5 It is sufficient to prove that the functional $\mathcal{K}$ changes sign when the mass varies. Then the conclusion will follow using Lemma 3.3. On the one hand, by the Gagliardo-Nirenberg inequality, there is a constant $C_* > 0$ such that for any $u \in H^1$, we have

$$\|u\|_{L^4}^4 \leq C_* \|\nabla u\|_{L^2}^2 \|u\|_{L^2}^2.$$

On the other hand, multiplying (μ-SP$_0$) by $\bar{u}$ and integrating shows that any solution u of (μ-SP$_0$) satisfies

$$\mu \|u\|_{L^2}^2 = \|\nabla u\|_{L^2}^2 + \|xu\|_{L^2}^2 + \|u\|_{L^4}^4.$$

Thus, if $\|u\|_{L^2}^2 = M$ we have

$$\|u\|_{L^4}^4 \lesssim M^2 \mu_M.$$

this shows that if $M \leq 1$ then $\mu_M \lesssim 1$ and thus $\mathcal{K}(u_M) \geq 0$ as $M \to 0$. Now, we just need to show that $\mathcal{K}(u_M)$ becomes negative for large masses. In fact, first we will show that

$$\mathcal{H}(u_M) \lesssim M^{\frac{3}{2}}, \quad \text{as} \quad M \to \infty. \tag{3.7}$$

If we let $\mathcal{H}_{int}(u) := \frac{1}{2}(\|xu\|_{L^2}^2 + \frac{1}{2}\|u\|_{L^4}^4)$, then clearly

$$\mathcal{H}_{int}(u_M) \leq \mathcal{H}(u_M).$$

Now, we will explicitly calculate

$$v_M := \inf_{\|u\|_{L^2}^2 = M} \mathcal{H}_{int}(u), \quad u \in \Sigma_{int},$$

where $\Sigma_{int} = \{u \in L^2(\mathbb{R}^2), u \in L^4(\mathbb{R}^2) : \int |x|^2 u^2 < \infty\}$ with the norm

$$\|u\|_{\Sigma_2^4} = \|u\|_2 + \|u\|_4 + \||x|u\|_2.$$

Let $(u)_n$ be a minimizing sequence of v_M that is

$$\|u_n\|_{L^2}^2 = M, \quad \text{and} \quad \frac{1}{2}(\|xu_n\|_{L^2}^2 + \frac{1}{2}\|u_n\|_{L^4}^4) \to v_M. \tag{3.8}$$

From the above bounds, let us just denote by u (instead of u_M), an L^2-weak limit of (u_n). Denote by $f_n := u_n^2$. First we show that $\|f\|_{L^1(\mathbb{R}^2)} = M$. Up to an extraction, we may assume that a subsequence of (f_n) (also denoted by (f_n)) converges weakly to f in the sense of distributions; that is for any $\varphi \in \mathcal{C}_0^\infty(\mathbb{R}^2)$ (smooth and compactly supported function), we have

$$\int_{\mathbb{R}^2} \varphi f_n \, dx \to \int_{\mathbb{R}^2} \varphi f \, dx.$$

To show strong convergence in L^1, we observe that (see for example [11])

$$\limsup_n \|f_n - f\|_{L^1} \leq C(\{f_n, \ n = 1, 2, \cdots\}),$$

where, for any subset $\mathcal{A} = \{f_n, \ n = 1, 2, 3, \cdots\} \subset L^1(\mathbb{R}^2)$, the function $C(\mathcal{A})$ introduced by Rosenthal [24] is given by

$$C(\mathcal{A}) = \inf_\varepsilon \sup_{|A|<\varepsilon} \sup_n \int_A f_n \, dx.$$

In our case, we take $\mathcal{A} = \{f_n, \ n = 1, 2 \cdots\}$. Using Hölder inequality and the above bounds (3.8), we have for any $R > 0$

$$\int_A f_n \, dx \leq \sqrt{|A|}\sqrt{\int_A f_n^2 \, dx} + \frac{1}{R^2}\int_{A \cap \{|x|>R\}} |x|^2 f_n \, dx$$

$$\lesssim \sqrt{\varepsilon} + \frac{1}{R^2},$$

which clearly shows that $C(\{f_n, \ n = 1, 2, \cdots\}) = 0$, and thus $\|u_n - u\|_{L^2} \to 0$ and $\|f\|_{L^1(\mathbb{R}^2)} = \|u\|_{L^2(\mathbb{R}^2)}^2 = M$, as desired. Moreover, by the lower semi-continuity of the norms, we have

$$\frac{1}{2}\left(\|xu\|_{L^2}^2 + \frac{1}{2}\|u\|_{L^4}^4\right) = \frac{1}{2}(\||x|^2 f\|_{L^1} + \frac{1}{2}\|f^2\|_{L^2}^2)$$

$$\leq \liminf_n \frac{1}{2}\left(\|xu_n\|_{L^2}^2 + \frac{1}{2}\|u_n\|_{L^4}^4\right) \leq \nu_M.$$

If the estimate was strict, that would contradict the minimality of ν_M. The convergence is therefore strong in $u_n \to u$, and at the minimum we have

$$|x|^2 u + u^3 = \nu u, \quad u^2 = (\nu - |x|^2)_+$$

yielding

$$M = \|u_M\|_{L^2}^2 = \int_{\mathbb{R}^2}(\nu - V)_+ \, dx = \int_{\{|x|^2 < \nu\}}(\nu - |x|^2)_+ \, dx = \frac{\pi}{2}\nu^2,$$

and

$$\|u_M\|_{L^4}^4 = \int_{\mathbb{R}^2}(\nu - |x|^2)_+ |u|^2 \, dx = \int_{\mathbb{R}^2}(\nu - |x|^2)_+^2 \, dx \leq \frac{\pi}{3}\nu^3 \sim M^{\frac{3}{2}}.$$

Now we mollify u_M in order to get an upper bound for ν_M. Set

$$\tilde{u}_M := \left((\nu - |x|^2)_+^2 + 1\right)^{\frac{1}{4}} - 1, \quad w_M := \sqrt{M}\frac{\tilde{u}_M}{\|\tilde{u}_M\|_{L^2}}.$$

Calculating $\|\tilde{u}_M\|_{L^2}^2$ shows that

$$\|\tilde{u}_M\|_{L^2}^2 = \int_0^\mu \left((s^2 + 1)^{\frac{1}{4}} - 1\right)^2 ds \sim \mu^2 = M \quad \text{as} \quad M \to \infty. \tag{3.9}$$

Moreover, similar calculation enables us to see that

$$\|\nabla \tilde{u}_M\|_{L^2}^2 \lesssim \nu^3 \quad \text{and} \quad \||x|\tilde{u}_M\|_{L^2}^2 \lesssim \nu^3. \tag{3.10}$$

In summary, in virtue of (3.9) and (3.10), we have

$$\|w_M\|_{L^2}^2 = M \quad \text{and} \quad \||x|w_M\|_{L^2}^2 \lesssim M^{\frac{3}{2}}, \tag{3.11}$$

which implies, thanks to the fact $\mathcal{H}(u_M) \leq \mathcal{H}(w_M)$,

$$\|u_M\|_{L^2}^2 = M, \quad \text{and} \quad \|xu_M\|_{L^2}^2 \lesssim M^{\frac{3}{2}}, \text{ as } M \to \infty.$$

The above estimates automatically imply

$$M^{\frac{3}{2}} \lesssim \|u_M\|_{L^4}^4. \tag{3.12}$$

Indeed, if (3.12) does not hold, then there would exists a sequence $M_n \to \infty$, and u_n satisfying

$$\|u_n\|_{L^2}^2 = M_n \quad \text{and} \quad \||x|u_n\|_{L^2}^2 \lesssim M_n^{\frac{3}{2}}$$

and

$$\|u_n\|_{L^4}^4 \leq \frac{M_n^{\frac{3}{2}}}{n}.$$

But the following estimate, true for all $R > 0$ and $n \in \mathbb{N}$

$$\|u_n\|_{L^2}^2 \lesssim \frac{M_n^{\frac{3}{2}}}{R^2} + R\|u_n\|_{L^4}^2$$

$$\lesssim \frac{M_n^{\frac{3}{2}}}{R^2} + R\frac{M_n^{\frac{3}{4}}}{n^{\frac{1}{2}}}.$$

Now choosing $R = M_n^{\frac{1}{4}} n^{\frac{1}{8}}$, gives the bound

$$1 \lesssim \frac{1}{n^{\frac{1}{4}}}$$

leading to a contradiction by taking $n \to \infty$. Clearly, (3.12) shows that $\mathcal{K}(u_M)$ becomes negative as $M \to \infty$ which finishes the proof. □

3.2 The Ground State

The main result of this section is the following.

Theorem 3.1 *There exists $\varepsilon_0 > 0$ such that for all $0 < \varepsilon < \varepsilon_0$, Eq. (GPPD$_\varepsilon$) has a solution $(Q_\varepsilon, \mu_\varepsilon) \in \Sigma \times (2, \infty)$ that can be decomposed as*

$$(Q_\varepsilon, \mu_\varepsilon) = (Q_\varepsilon^a + \psi_\varepsilon, \mu_\varepsilon^a + \kappa_\varepsilon), \tag{3.13}$$

with $\psi_\varepsilon = \psi_{\varepsilon,r} + i\psi_{\varepsilon,i}$ satisfying

$$|\kappa_\varepsilon| + \|\psi_{\varepsilon,r}\|_\Sigma \lesssim \varepsilon^4 \tag{3.14}$$

$$\|\psi_{\varepsilon,i}\|_\Sigma \lesssim \varepsilon^5. \tag{3.15}$$

Here, $(Q_\varepsilon^a, \mu_\varepsilon^a)$ is the approximate solution that we constructed.

Proof of Theorem 3.1 First, we write an equation for $(Q_\varepsilon, \mu_\varepsilon)$ being a solution of $(\mu - \mathrm{SP}_\varepsilon)$. We start by further decomposing $Q_\varepsilon^a = Q_{\varepsilon,r}^a + iQ_{\varepsilon,i}^a$ and observe that

$$|Q_\varepsilon|^2 = |Q_{\varepsilon,r}^a|^2 + |Q_{\varepsilon,i}^a|^2 + 2Q_{\varepsilon,r}^a \psi_{\varepsilon,r} + 2Q_{\varepsilon,i}^a \psi_{\varepsilon,i} + |\psi_{\varepsilon,r}|^2 + |\psi_{\varepsilon,i}|^2.$$

Substituting this in equation $(\mu\text{-SP}_\varepsilon)$ and splitting the real and imaginary parts, we obtain

$$(\mu_\varepsilon^a + \kappa_\varepsilon)(Q_{\varepsilon,r}^a + \psi_{\varepsilon,r}) = (-\Delta + V + |Q_\varepsilon|^2)(Q_{\varepsilon,r}^a + \psi_{\varepsilon,r})$$
$$-\varepsilon(\sigma - \alpha|Q_\varepsilon|^2)(Q_{\varepsilon,i}^a + \psi_{\varepsilon,i}), \tag{3.16}$$

and

$$(\mu_\varepsilon^a + \kappa_\varepsilon)(Q_{\varepsilon,i}^a + \psi_{\varepsilon,i}) = (-\Delta + V + |Q_\varepsilon|^2)(Q_{\varepsilon,i}^a + \psi_{\varepsilon,i})$$
$$+\varepsilon(\sigma - \alpha|Q_\varepsilon|^2)(Q_{\varepsilon,r}^a + \psi_{\varepsilon,r}), \tag{3.17}$$

respectively. The identity coming from the real part can be rewritten in the following way.

$$L_+ \psi_{\varepsilon,r} = \mu_\varepsilon^a Q_{\varepsilon,r}^a - (-\Delta + V + |Q_\varepsilon^a|^2)\psi_{\varepsilon,r} + \varepsilon(\sigma - \alpha|Q_\varepsilon^a|^2)Q_{\varepsilon,i}^a$$
$$+ \kappa_\varepsilon Q_{\varepsilon,r}^a + \varepsilon^2 \mu_2 \psi_{\varepsilon,r} - 2Q_{\varepsilon,i}^a Q_{\varepsilon,r}^a \psi_{\varepsilon,r} + \varepsilon(\sigma - \alpha|Q_\varepsilon^a|^2)\psi_{\varepsilon,i}^a$$
$$- 2|Q_{\varepsilon,r}^i|^2 \psi_{\varepsilon,i}^a + \kappa_\varepsilon \psi_{\varepsilon,r} + \psi_{\varepsilon,r}(2Q_{\varepsilon,r}^a \psi_{\varepsilon,r} + 2Q_{\varepsilon,i}^a \psi_{\varepsilon,i} + \psi_{\varepsilon,r}^2 + \psi_{\varepsilon,i}^2)$$
$$- \varepsilon\psi_{\varepsilon,r}(2Q_{\varepsilon,r}^a \psi_{\varepsilon,r} + 2Q_{\varepsilon,i}^a \psi_{\varepsilon,i} + \psi_{\varepsilon,r}^2 + \psi_{\varepsilon,i}^2)$$
$$:= \kappa_\varepsilon Q_0 + \varepsilon^4 \varphi_1 + F_\varepsilon(\psi_{\varepsilon,r}, \psi_{\varepsilon,i}, \kappa_\varepsilon)$$

Ground State Solutions of the Complex Gross Pitaevskii Equation

where φ_1 is given by

$$\varphi_1 := \mu_2 Q_{2r} - Q_{1i}^2 Q_{2r} - (Q_{2r}^2 + 2Q_{3i}Q_{1i})Q_0 + (\sigma - \alpha Q_0^2)Q_{3i} - (2Q_0 Q_{2r} + Q_{1i}^2)Q_{1i}$$

and F_ε can be easily be explicitly computed. In particular it satisfies

$$\|F_\varepsilon(\psi_{\varepsilon,r}, \psi_{\varepsilon,i}, \kappa_\varepsilon)\|_\Sigma \lesssim \varepsilon^6.$$

The identity coming from the imaginary part can be rewritten in the following way.

$$\begin{aligned}
L_-\psi_{\varepsilon,i} &= \mu_\varepsilon^a Q_{\varepsilon,i}^a - (-\Delta + V + |Q_\varepsilon^a|^2) Q_{\varepsilon,i}^a - \varepsilon(\sigma - \alpha |Q_\varepsilon^a|^2) Q_{\varepsilon,r}^a \\
&+ \kappa_\varepsilon Q_{\varepsilon,i}^a + \varepsilon^2 \mu_2 \psi_{\varepsilon,i} - 2Q_{\varepsilon,i}^a (Q_{\varepsilon,r}^a \psi_{\varepsilon,r} + Q_{\varepsilon,i}^a \psi_{\varepsilon,i}) - \varepsilon(\sigma - \alpha|Q_0|^2)\psi_{\varepsilon,r} \\
&+ 2\varepsilon Q_{\varepsilon,r}^a (Q_{\varepsilon,r}^a \psi_{\varepsilon,r} + Q_{\varepsilon,i}^a \psi_{\varepsilon,i}) \\
&- 2\psi_{\varepsilon,i}(Q_{\varepsilon,r}^a \psi_{\varepsilon,r} + Q_{\varepsilon,i}^a \psi_{\varepsilon,i}) + 2\varepsilon \psi_{\varepsilon,r}(Q_{\varepsilon,r}^a \psi_{\varepsilon,r} + Q_{\varepsilon,i}^a \psi_{\varepsilon,i}) \\
&+ \varepsilon Q_{\varepsilon,r}^a(\psi_{\varepsilon,r}^2 + \psi_{\varepsilon,i}^2) - \psi_{\varepsilon,i}(\psi_{\varepsilon,r}^2 + \psi_{\varepsilon,i}^2) + \varepsilon \psi_{\varepsilon,r}(\psi_{\varepsilon,r}^2 + \psi_{\varepsilon,i}^2) + \kappa_\varepsilon \psi_{\varepsilon,i} \\
&:= \varepsilon \big(\kappa_\varepsilon Q_{1i} + ((2+\alpha)Q_0^2 - \sigma - 2Q_0 Q_{1i})\psi_{\varepsilon,r}\big) \\
&+ \varepsilon^5 \varphi_2 + G_\varepsilon(Q_{\varepsilon,r}, Q_{\varepsilon,i}, \kappa_\varepsilon),
\end{aligned}$$

where φ_2 is given by

$$\begin{aligned}
\varphi_2 := &- (2Q_0 Q_{2r} + Q_{1i}^2)Q_{3i} - (Q_{2r}^2 + 2Q_{1i}Q_{3i})Q_{1i} \\
&+ (2Q_0 Q_{2r} + Q_{1i}^2)Q_{2r} + (Q_{2r}^2 + 2Q_{1i}Q_{3i})Q_0
\end{aligned}$$

and G_ε can be explicitly computed. In particular it satisfies

$$\|G_\varepsilon(\psi_{\varepsilon,r}, \psi_{\varepsilon,i}, \kappa_\varepsilon)\|_\Sigma \lesssim \varepsilon^7.$$

Now we define a map $\Phi_\varepsilon : \Sigma \times \Sigma \times (0, \infty) \to \Sigma \times \Sigma \times (0, \infty)$ by

$$\Phi_\varepsilon(\tilde\psi_{\varepsilon,r}, \tilde\psi_{\varepsilon,i}, \tilde\kappa_\varepsilon) = (\psi_{\varepsilon,r}, \psi_{\varepsilon,i}, \kappa_\varepsilon)$$

where, $(\psi_{\varepsilon,r}, \psi_{\varepsilon,i}, \kappa_\varepsilon)$ solves

$$\begin{cases}
L_+\psi_{\varepsilon,r} = \kappa_\varepsilon Q_0 + \varepsilon^4 \varphi_1 + F_\varepsilon(\tilde\psi_{\varepsilon,r}, \tilde\psi_{\varepsilon,i}, \tilde\kappa_\varepsilon) \\
L_-\psi_{\varepsilon,i} = \varepsilon\big(\kappa_\varepsilon Q_{1i} + ((2+\alpha)Q_0^2 - \sigma - 2Q_0 Q_{1i})\psi_{\varepsilon,r}\big) \\
\qquad\quad + \varepsilon^5 \varphi_2 + G_\varepsilon(\tilde\psi_{\varepsilon,r}, \tilde\psi_{\varepsilon,i}, \tilde\kappa_\varepsilon), \\
(L_-\psi_{\varepsilon,i}, Q_0)_2 = 0.
\end{cases} \quad (3.18)$$

Now the purpose is to show that there are positive constants C_1, C_2 and C_3 such that the above map is a contraction on the ball

$$B_\varepsilon := \{(\psi_{\varepsilon,r}, \psi_{\varepsilon,i}, \kappa_\varepsilon): \quad |\kappa_\varepsilon| \le C_1\varepsilon^4, \ \|\psi_{\varepsilon,r}\|_\Sigma \le C_2\varepsilon^4, \ \|\psi_{\varepsilon,i}\|_\Sigma \le C_3\varepsilon^5\},$$

for $\varepsilon > 0$ sufficiently small. The ball B_ε is endowed with the norm

$$\max\left\{\frac{|\kappa_\varepsilon|}{C_1\varepsilon^4}, \frac{\|\psi_{\varepsilon,r}\|_\Sigma}{C_2\varepsilon^4}, \frac{\|\psi_{\varepsilon,i}\|_\Sigma}{C_3\varepsilon^5}\right\}. \tag{3.19}$$

Thanks to the equation for $\psi_{\varepsilon,r}$ and the invertibility of L_+, we can write

$$\psi_{\varepsilon,r} = \kappa_\varepsilon Q'(\mu_0) + \varepsilon^4 L_+^{-1}(\varphi_1) + L_+^{-1}\big(F_\varepsilon(\tilde{\psi}_{\varepsilon,r}, \tilde{\psi}_{\varepsilon,i}, \tilde{\kappa}_\varepsilon)\big). \tag{3.20}$$

Plugging the above identity in the equation for $\psi_{\varepsilon,i}$ we obtain

$$L_-\psi_{\varepsilon,i} = \varepsilon\kappa_\varepsilon\big(Q_{1,i} + ((2+\alpha)Q_0^2 + 2Q_0 Q_{1,i} - \sigma)Q'(\mu_0)$$
$$+\varepsilon^5((2+\alpha)Q_0^2 + 2Q_0 Q_{1,i} - \sigma)L_+^{-1}(\varphi_1) + L_+^{-1}\big(F_\varepsilon(\tilde{\psi}_{\varepsilon,r}, \tilde{\psi}_{\varepsilon,i}, \tilde{\kappa}_\varepsilon)\big).$$

Since,

$$\big(Q_{1,i} + ((2+\alpha)Q_0^2 - 2Q_0 Q_{1,i} - \sigma)Q'(\mu_0), Q_0\big)_2 = \|Q_0\|_{L^2}^2$$

then the choice of

$$\kappa_\varepsilon = -\frac{\varepsilon^4}{\|Q_0\|_{L^2}^2}\int_{\mathbb{R}^2} Q_0\varphi_1\,dx - \frac{1}{\varepsilon}\int_{\mathbb{R}^2} Q_0 L_+^{-1}\big(F_\varepsilon(\tilde{\psi}_{\varepsilon,r}, \tilde{\psi}_{\varepsilon,i}, \tilde{\kappa}_\varepsilon)\big)\,dx$$

makes

$$(L_-\psi_{\varepsilon,i}, Q_0)_2 = 0$$

which enables us to invert L_- and thus calculate $\psi_{\varepsilon,i}$:

$$\psi_{\varepsilon,i} = \varepsilon\kappa_\varepsilon L_-^{-1}\big(Q_{1,i} + ((2+\alpha)Q_0^2 - 2Q_0 Q_{1,i} - \sigma)Q'(\mu_0)\big)$$
$$+ L_-^{-1}L_+^{-1}\big(F_\varepsilon(\tilde{\psi}_{\varepsilon,r}, \tilde{\psi}_{\varepsilon,i}, \tilde{\kappa}_\varepsilon)\big)$$
$$+ \varepsilon^5 L_-^{-1}\big(((2+\alpha)Q_0^2 - 2Q_0 Q_{1,i} - \sigma)L_+^{-1}(\varphi_1)\big) \tag{3.21}$$

Let

$$C_1 := \frac{2\|\varphi_1\|_{L^2}}{\|Q_0\|_{L^2}},$$

$$C_2 = 2\big(C_1 \|Q'(\mu_0)\|_\Sigma + \|L_+^{-1}(\varphi_1)\|_\Sigma\big),$$

and

$$C_3 := 2C_1 \|L_-^{-1}\big(Q_{1,i} + ((2+\alpha)Q_0^2 - 2Q_0 Q_{1,i} - \sigma)Q'(\mu_0)\big)\|_\Sigma$$
$$+ 2\|L_-^{-1}\big(((2+\alpha)Q_0^2 - 2Q_0 Q_{1,i} - \sigma)L_+^{-1}(\varphi_1)\big)\|_\Sigma$$

To show that Φ_ε is a contraction, consider $(\tilde{\psi}_{\varepsilon,r}^a, \tilde{\psi}_{\varepsilon,i}^a, \tilde{\kappa}_\varepsilon^a)$ and $(\tilde{\psi}_{\varepsilon,r}^b, \tilde{\psi}_{\varepsilon,i}^b, \tilde{\kappa}_\varepsilon^b)$ in the ball B_ε and denote by $(\psi_{\varepsilon,r}^a, \psi_{\varepsilon,i}^a, \kappa_\varepsilon^a)$ and $(\psi_{\varepsilon,r}^b, \psi_{\varepsilon,i}^b, \kappa_\varepsilon^b)$ their respective images through the map Φ_ε. We have

$$\kappa_\varepsilon := \kappa_\varepsilon^a - \kappa_\varepsilon^b = \frac{1}{\varepsilon} \int_{\mathbb{R}^2} Q_0 L_+^{-1}\big[F_\varepsilon(\tilde{\psi}_{\varepsilon,r}^b, \tilde{\psi}_{\varepsilon,i}^b, \tilde{\kappa}_\varepsilon^b) - F_\varepsilon(\tilde{\psi}_{\varepsilon,r}^a, \tilde{\psi}_{\varepsilon,i}^a, \tilde{\kappa}_\varepsilon^a)\big] dx,$$

$$\psi_{\varepsilon,r} := \psi_{\varepsilon,r}^a - \psi_{\varepsilon,r}^b = \kappa_\varepsilon Q'(\mu_0) - L_+^{-1}\big(F_\varepsilon(\tilde{\psi}_{\varepsilon,r}^b, \tilde{\psi}_{\varepsilon,i}^b, \tilde{\kappa}_\varepsilon^b) - F_\varepsilon(\tilde{\psi}_{\varepsilon,r}^a, \tilde{\psi}_{\varepsilon,i}^a, \tilde{\kappa}_\varepsilon^a)\big),$$

and

$$\psi_{\varepsilon,i} := \psi_{\varepsilon,r}^a - \psi_{\varepsilon,r}^b = \varepsilon \kappa_\varepsilon L_-^{-1}\big(Q_{1,i} + ((2+\alpha)Q_0^2 - 2Q_0 Q_{1,i} - \sigma)Q'(\mu_0)\big)$$
$$- L_-^{-1} L_+^{-1}\big(F_\varepsilon(\tilde{\psi}_{\varepsilon,r}^b, \tilde{\psi}_{\varepsilon,i}^b, \tilde{\kappa}_\varepsilon^b) - F_\varepsilon(\tilde{\psi}_{\varepsilon,r}^a, \tilde{\psi}_{\varepsilon,i}^a, \tilde{\kappa}_\varepsilon^a)\big).$$

Estimating κ_ε, $\psi_{\varepsilon,r}$ and $\psi_{\varepsilon,i}$ using the above bounds on F_ε and G_ε yields

$$|\kappa_\varepsilon| \lesssim \varepsilon^5, \quad \|\psi_{\varepsilon,r}\|_\Sigma \lesssim \varepsilon^5, \quad \|\psi_{\varepsilon,i}\|_\Sigma \lesssim \varepsilon^6$$

showing the contraction of the map Φ_ε. □

Acknowledgement S.I. was supported by NSERC grant (371637-2014).

References

1. Amo, A., Lefrère, J., Pigeon, S., Adrados, C., Ciuti, C., Carusotto, I., HoudréA, R., Giacobino, E., Bramati, A.: Superfluidity of polaritons in semiconductor microcavities. Nat. Phys. 5(11), 805–810 (2009)
2. Anderson, M., Ensher, J., Matthews, M., Wieman, C., Cornell, E., Observation of Bose-Einstein conden- sation in a dilute atomic vapor. Science 269(5221), 198–201 (1995)
3. Ballarini, D., Sanvitto, D., Amo, A., Viña, L., Wouters, M., Carusotto, I., Lemaitre, A., Bloch, J., Observation of long-lived polariton states in semiconductor microcavities across the parametric threshold. Phys. Rev. Lett. 102(5), 056402 (2009)
4. S. Bose, Plancks gesetz und lichtquantenhypothese. Z. phys 26(3), 178 (1924).

5. Bradley, C., Sackett, C., Tollett, J., Hulet, R., Evidence of Bose-Einstein condensation in an atomic gas with attractive interactions. Phys. Rev. Lett. 75(9), 1687 (1995)
6. Coldren, L., Corzine, S., Diode Lasers and Photonic Integrated Circuits, volume 218. Wiley Series in Microwave and Optical Engineering, New York (1995)
7. Crandall, M. G. and Rabinowitz, P., Bifurcation, perturbation of simple eigenvalues and linearized stability. Arch. Rational Mech. Anal. 52 (1973), 161–180.
8. Davis, K., Mewes, M., Andrews, M., van Druten, N., Durfee, D., Kurn, D., Ketterle, W., Bose-Einstein condensation in a gas of sodium atoms. Phys. Rev. Lett. 75(22), 3969–3973 (1995)
9. Einstein, A., Sitzungsberichte der preussischen akademie der wissenschaften. Physikalisch-mathematische Klasse 261(3) (1924)
10. Einstein, A., Quantum theory of the monoatomic ideal gas. Sitzungsber. Preuss. Akad. Wiss, page 261 (1925)
11. Florescu, L., Weak compactness results in L^1, Analele Stiitfcifice Ale Universitat II al.l.Cuza Iasi Tomul XLV, s.I a, Matematica, 1999, f.1.
12. Gross, E., Hydrodynamics of a superfluid condensate. J. Math. Phys. 4, 195 (1963)
13. H. Hajaiej, C. A. Stuart; Symmetrization inequalities for composition operators of Carathéodory type, Proc. London. Math. Soc., 87(2003), 396–418.
14. H. Hajaiej, S. Ibrahim, N. Masmoudi, Ground State Solutions of the Complex Gross Pitaevskii Equation Associated to Exciton-Polariton Bose-Einstein Condensates. https://arxiv.org/pdf/1905.07660.pdf
15. H. Hajaiej, C. A. Stuart; On the variational approach to the stability of standing waves for the nonlinear Schrdinger equation, Adv Nonlinear Studies, 4 (2004), 469–501.
16. Kasprzak, J., Richard, M., Kundermann, S., Baas, A., Jeambrun, P., Keeling, J., Marchetti, F., Szymanacute, M., Andre, R., Staehli, Bose–Einstein condensation of exciton polaritons. Nature 443(7110), 409–414. (2006)
17. Kavian, O. and Weissler, Fred B., Self-similar solutions of the pseudo-conformally invariant nonlinear Schrödinger equation. Michigan Math. J. **41** (1994), no. 1, 151–173.
18. Keeling, J., Berloff, N., Exciton-polariton condensation. Contemp. Phys. 52(2), 131–151 (2011)
19. Keeling, J., Berloff, N.G., Spontaneous rotating vortex lattices in a pumped decaying condensate. Phys. Rev. Lett. 100(25), 250401 (2008). ISSN 1079-7114
20. Manni, F., Liew, T., Lagoudakis, K., Ouellet-Plamondon, C., André, R., Savona, V., Deveaud, B.: Spon- taneous self-ordered states of vortex-antivortex pairs in a polariton condensate. Phys. Rev. B 88(20), 201303 (2013)
21. Ohadi, H., Kammann, E., Liew, T., Lagoudakis, K., Kavokin, A., Lagoudakis, P., Spontaneous symmetry breaking in a polariton and photon laser. Phys. Rev. Lett. 109(1), 016404 (2012)
22. Pitaevskii, L., Vortex lines in an imperfect Bose gas. Sov. Phys. JETP 13(2), 451–454 (1961)
23. Pitaevskii, L., Stringari, S., Bose–Einstein Condensation. Number 116. Oxford University Press, Oxford (2003)
24. Rosenthal, H.P., Sous-espaces de L^1, Lectures Univ. Paris VI, 1979.
25. Sanvitto, D., Marchetti, F.M., Szymanska, M.H., Tosi, G., Baudisch, M., Laussy, F.P., Krizhanovskii, D.N., Skolnick, M.S., Marrucci, L., Lemaitre, A., Bloch, J., Tejedor, C., Vina, L., Persistent currents and quantized vortices in a polariton superfluid. Nat. Phys. (2010). ISSN 1745-2473
26. Sierra, J., Kasimov, A., Markowich, P. Weishäupl, R.-M., On the Gross–Pitaevskii Equation with Pumping and Decay: Stationary States and Their Stability. J Nonlinear Sci (2015) 25:709739. https://doi.org/10.1007/s00332-015-9239-8
27. Wouters, M., Carusotto, I., Excitations in a nonequilibrium Bose–Einstein condensate of exciton polaritons. Phys. Rev. Lett. 99(14), 140–402 (2007)

The Phase Shift of Line Solitons for the KP-II Equation

Tetsu Mizumachi

2010 Mathematics Subject Classification Primary: 35B35, 37K40; Secondary 35Q35

1 Introduction

The KP-II equation

$$\partial_x(\partial_t u + \partial_x^3 u + 3\partial_x(u^2)) + 3\sigma \partial_y^2 u = 0 \quad \text{for } t > 0 \text{ and } (x, y) \in \mathbb{R}^2, \tag{1.1}$$

where $\sigma = 1$, is a generalization to two spatial dimensions of the KdV equation

$$\partial_t u + \partial_x^3 u + 3\partial_x(u^2) = 0, \tag{1.2}$$

and has been derived as a model to explain the transverse stability of solitary wave solutions to the KdV equation with respect to 2 dimensional perturbation when the surface tension is weak or absent. See [15] for the derivation of (1.1).

The global well-posedness of (1.1) in $H^s(\mathbb{R}^2)$ ($s \geq 0$) on the background of line solitons has been studied by Molinet et al. [28] whose proof is based on the work of Bourgain [3]. For the other contributions on the Cauchy problem of the KP-II equation, see e.g. [8–10, 13, 35–38] and the references therein.

T. Mizumachi (✉)
Division of Mathematical and Information Sciences, Hiroshima University, Kagamiyama, Hiroshima, Japan
e-mail: tetsum@hiroshima-u.ac.jp

© Springer Science+Business Media, LLC, part of Springer Nature 2019
P. D. Miller et al. (eds.), *Nonlinear Dispersive Partial Differential Equations and Inverse Scattering*, Fields Institute Communications 83,
https://doi.org/10.1007/978-1-4939-9806-7_10

Let

$$\varphi_c(x) \equiv c \operatorname{sech}^2\left(\sqrt{\frac{c}{2}}x\right), \quad c > 0.$$

Then $\varphi_c(x - 2ct)$ is a solitary wave solution of the KdV equation (1.2) and a line soliton solution of (1.1) as well. Transverse linear stability of line solitons for the KP-II equation was studied by Burtsev [4]. See also [1] for the spectral stability of KP line solitons. Recently, transverse spectral and linear stability of periodic waves for the KP-II equation has been studied in [11, 12, 14].

If $\sigma = -1$, then (1.1) is called KP-I which is a model for long waves in a media with positive dispersion, e.g. water waves with large surface tension. The KP-I equation has a stable ground state [7] and line solitons are unstable for the KP-I equation except for thin domains in $\mathbb{R}^2$ where the 2 dimensional nature of the equation is negligible (see [31–33, 40]).

Nonlinear stability of line solitons for the KP-II equation has been proved for localized perturbations as well as for perturbations which have 0-mean along all the lines parallel to the x-axis [22, 23].

Theorem 1.1 ([23, Theorem 1.1]) *Let $c_0 > 0$ and $u(t, x, y)$ be a solution of* (1.1) *satisfying $u(0, x, y) = \varphi_{c_0}(x) + v_0(x, y)$. There exist positive constants ε_0 and C satisfying the following: if $v_0 \in \partial_x L^2(\mathbb{R}^2)$ and $\|v_0\|_{L^2(\mathbb{R}^2)} + \||D_x|^{1/2}v_0\|_{L^2(\mathbb{R}^2)} + \||D_x|^{-1/2}|D_y|^{1/2}v_0\|_{L^2(\mathbb{R}^2)} < \varepsilon_0$ then there exist C^1-functions $c(t, y)$ and $x(t, y)$ such that for every $t \geqslant 0$ and $k \geqslant 0$,*

$$\|u(t, x, y) - \varphi_{c(t,y)}(x - x(t, y))\|_{L^2(\mathbb{R}^2)} \leqslant C\|v_0\|_{L^2}, \tag{1.3}$$

$$\|c(t, \cdot) - c_0\|_{H^k(\mathbb{R})} + \|\partial_y x(t, \cdot)\|_{H^k(\mathbb{R})} + \|x_t(t, \cdot) - 2c(t, \cdot)\|_{H^k(\mathbb{R})} \leqslant C\|v_0\|_{L^2}, \tag{1.4}$$

$$\lim_{t \to \infty} \left(\|\partial_y c(t, \cdot)\|_{H^k(\mathbb{R})} + \|\partial_y^2 x(t, \cdot)\|_{H^k(\mathbb{R})}\right) = 0, \tag{1.5}$$

and for any $R > 0$,

$$\lim_{t \to \infty} \|u(t, x + x(t, y), y) - \varphi_{c(t,y)}(x)\|_{L^2((x > -R) \times \mathbb{R}_y)} = 0. \tag{1.6}$$

Theorem 1.2 ([23, Theorem 1.2]) *Let $c_0 > 0$ and $s > 1$. Suppose that u is a solutions of* (1.1) *satisfying $u(0, x, y) = \varphi_{c_0}(x) + v_0(x, y)$. Then there exist positive constants ε_0 and C such that if $\|\langle x \rangle^s v_0\|_{H^1(\mathbb{R}^2)} < \varepsilon_0$, there exist $c(t, y)$ and $x(t, y)$ satisfying* (1.5), (1.6) *and*

$$\|u(t, x, y) - \varphi_{c(t,y)}(x - x(t, y))\|_{L^2(\mathbb{R}^2)} \leqslant C\|\langle x \rangle^s v_0\|_{H^1(\mathbb{R}^2)}, \tag{1.7}$$

$$\|c(t,\cdot) - c_0\|_{H^k(\mathbb{R})} + \|\partial_y x(t,\cdot)\|_{H^k(\mathbb{R})} + \|x_t(t,\cdot) - 2c(t,\cdot)\|_{H^k(\mathbb{R})}$$
$$\leqslant C\|\langle x\rangle^s v_0\|_{H^1(\mathbb{R}^2)} \tag{1.8}$$

for every $t \geqslant 0$ and $k \geqslant 0$.

Remark 1.1 The parameters $c(t_0, y_0)$ and $x(t_0, y_0)$ represent the local amplitude and the local phase shift of the modulating line soliton $\varphi_{c(t,y)}(x - x(t, y))$ at time t_0 along the line $y = y_0$ and that $x_y(t, y)$ represents the local orientation of the crest of the line soliton.

Remark 1.2 In view of Theorem 1.1,

$$\lim_{t\to\infty}\sup_{y\in\mathbb{R}}(|c(t, y) - c_0| + |x_y(t, y)|) = 0,$$

and as $t \to \infty$, the modulating line soliton $\varphi_{c(t,y)}(x - x(t, y))$ converges to a y-independent modulating line soliton $\varphi_{c_0}(x - x(t, 0))$ in $L^2((x > -R) \times (|y| \leqslant R))$ for any $R > 0$.

For the KdV equation as well as for the KP-II equation posed on $L^2(\mathbb{R}_x \times \mathbb{T}_y)$, the dynamics of a modulating soliton $\varphi_{c(t)}(x - x(t))$ is described by a system of ODEs

$$\dot{c} \simeq 0, \quad \dot{x} \simeq 2c.$$

See [30] for the KdV equation and [26] for the KP-II equation with the y-periodic boundary condition. However, to analyze transverse stability of line solitons for localized perturbation in $\mathbb{R}^2$, we need to study a system of PDEs for $c(t, y)$ and $x(t, y)$ in [22, 23] as is the case with the planar traveling waves for the heat equations (e.g. [16, 20, 39]) and planar kinks for the ϕ^4-model [5].

By analyzing modulation PDEs, it turns out the set of exact 1-line solitons

$$\mathcal{K} = \{\varphi_c(x + ky - (2c + 3k^2)t + \gamma) \mid c > 0, k, \gamma \in \mathbb{R}\}$$

is not stable in $L^2(\mathbb{R}^2)$.

Theorem 1.3 ([22, Theorem 1.4]) *Let $c_0 > 0$. Then for any $\varepsilon > 0$, there exists a solution of (1.1) such that $\|u(0, x, y) - \varphi_{c_0}(x)\|_{L^2(\mathbb{R}^2)} < \varepsilon$ and $\liminf_{t\to\infty} t^{-1/4} \inf_{v\in\mathcal{A}} \|u(t,\cdot) - v\|_{L^2(\mathbb{R}^2)} > 0$.*

Remark 1.3 Theorem 1.3 is a consequence of finite speed propagation of local phase shifts and the fact that the line solitons have infinite length in the $\mathbb{R}^2$ case. Indeed, the phase $x(t, y)$ has *jumps* around the points $y = \pm\sqrt{8c_0 t}$.

Such phenomena are observed for Boussinesq equations in the physics literature. See e.g. [29] and the reference therein.

The following result is an improvement of [22, Theorem 1.5].

Theorem 1.4 *Let $c_0 = 2$ and $u(t)$ be as in Theorem 1.2. There exist positive constants ε_0 and C such that if $\varepsilon := \|\langle x \rangle (\langle x \rangle + \langle y \rangle) v_0\|_{H^1(\mathbb{R}^2)} < \varepsilon_0$, then there exist C^1-functions $c(t, y)$ and $x(t, y)$ satisfying (1.3)–(1.6) and*

$$\left\| \begin{pmatrix} c(t, \cdot) - 2 \\ x_y(t, \cdot) \end{pmatrix} - \begin{pmatrix} 2 & 2 \\ 1 & -1 \end{pmatrix} \begin{pmatrix} u_B^+(t, y + 4t) \\ u_B^-(t, y - 4t) \end{pmatrix} \right\|_{L^2(\mathbb{R})} = o(\varepsilon t^{-1/4}) \tag{1.9}$$

as $t \to \infty$, where $u_B^\pm$ are self similar solutions of the Burgers equation

$$\partial_t u = 2\partial_y^2 u \pm 4\partial_y(u^2)$$

such that

$$u_B^\pm(t, y) = \frac{\pm m_\pm H_{2t}(y)}{2\left(1 + m_\pm \int_0^y H_{2t}(y_1) \, dy_1\right)}, \quad H_t(y) = (4\pi t)^{-1/2} e^{-y^2/4t},$$

and that $m_\pm$ are constants satisfying

$$\int_\mathbb{R} u_B^\pm(t, y) \, dy = \frac{1}{4} \int_\mathbb{R} (c(0, y) - 2) \, dy + O(\varepsilon^2).$$

Remark 1.4 Since (1.1) is invariant under the scaling $u \mapsto \lambda^2 u(\lambda^3 t, \lambda x, \lambda^2 y)$, we may assume that $c_0 = 2$ without loss of generality.

Remark 1.5 The linearized operator around the line soliton solution has resonant continuous eigenvalues near $\lambda = 0$ whose corresponding eigenmodes grow exponentially as $x \to -\infty$. See (2.1)–(2.3). The diffraction of the line soliton around $y = \pm 4t$ can be thought as a mechanism to emit energy from those resonant continuous eigenmodes.

If we disregard diffractions of waves propagating along the crest of line solitons, then time evolution of the phase shift is approximately described by the 1-dimensional wave equation

$$x_{tt} = 8c_0 x_{yy}.$$

It is natural to expect that $\sup_{t, y \in \mathbb{R}} |x(t, y) - 2c_0 t|$ remains small for localized perturbations although the $L^2(\mathbb{R}_y)$ norm of $x(t, y) - 2c_0 t$ grows as $t \to \infty$.

Our main result in the present paper is the following.

Theorem 1.5 *Let $u(t, x, y)$ and $x(t, y)$ be as in Theorem 1.2. There exist positive constants ε_0 and C such that if $\varepsilon := \|\langle x \rangle(\langle x \rangle + \langle y \rangle)v_0\|_{H^1(\mathbb{R}^2)} < \varepsilon_0$, then $\sup_{t \geq 0, y \in \mathbb{R}} |x(t, y) - 2c_0 t| \leq C\varepsilon$.*

Moreover, there exists an $h \in \mathbb{R}$ such that for any $\delta > 0$,

$$\begin{cases} \lim_{t \to \infty} \|x(t, \cdot) - 2c_0 t - h\|_{L^\infty(|y| \leq (\sqrt{8c_0} - \delta)t)} = 0, \\ \lim_{t \to \infty} \|x(t, \cdot) - 2c_0 t\|_{L^\infty(|y| \geq (\sqrt{8c_0} + \delta)t)} = 0. \end{cases} \tag{1.10}$$

In the case where $h \neq 0$ in (1.10), the $L^2(\mathbb{R}^2)$-distance between the solution u and the set of exact 1-line solitons grows like $t^{1/2}$ or faster.

Corollary 1.6 *Let $c_0 > 0$. Then for any $\varepsilon > 0$, there exists a solution of (1.1) such that $\|\langle x \rangle (\langle x \rangle + \langle y \rangle) \{u(0, x, y) - \varphi_{c_0}(x)\}\|_{H^1(\mathbb{R}^2)} < \varepsilon$ and $\liminf_{t \to \infty} t^{-1/2} \inf_{v \in \mathcal{A}} \|u(t, \cdot) - v\|_{L^2(\mathbb{R}^2)} > 0$.*

To investigate the large time behavior of $x(t, y)$, we derive estimates of fundamental solutions to the linearized equation of modulation equations for parameters $c(t, y)$ and $x(t, y)$ which is a system of 1-dimensional damped wave equations (see Sect. 2.2). As is the same with the 1-dimensional wave equation, we need integrability of the initial data of the modulation equation to prove the boundedness of the phase shift.

In our construction of modulation parameters, we impose a secular term condition on $c(t, y)$ and $x(t, y)$ only for y-frequencies in a small interval $[-\eta_0, \eta_0]$. This facilitates the estimates of modulation parameters because the truncation of Fourier modes turns the modulation equations into semilinear equations. On the other hand, it was not clear in [22] whether the initial data of modulation equations are integrable even if perturbations to line solitons are exponentially localized. We find that $c(0, y)$ can be decomposed into a sum of an integrable function and a derivative of a function that belongs to $\mathcal{F}^{-1} L^\infty(\mathbb{R})$ for polynomially localized perturbations in $\mathbb{R}^2$.

The decomposition of initial data also enables us to prove Theorem 1.4 which shows the large time asymptotic of the local amplitude and the local orientation of line solitons in $L^2(\mathbb{R})$ whereas the result in [22] shows large time asymptotics in a region $y = \pm\sqrt{8c_0} t + O(\sqrt{t})$.

In [25], we study the 2-dimensional linearized Benney-Luke equation around line solitary waves in the weak surface tension case and find that the time evolution of resonant continuous eigenmodes is similar to (1.10). We except our argument presented in this paper is useful to investigate phase shifts of modulating line solitary waves for the 2-dimensional Benney-Luke equation and the other long wave models for 3D water.

Finally, let us introduce several notations. Let $\mathbf{1}_A$ be the characteristic function of the set A. For Banach spaces V and W, let $B(V, W)$ be the space of all the linear continuous operators from V to W and $\|T\|_{B(V,W)} = \sup_{\|u\|_V = 1} \|Tu\|_W$ for $T \in B(V, W)$. We abbreviate $B(V, V)$ as $B(V)$. For $f \in \mathcal{S}(\mathbb{R}^n)$ and $m \in \mathcal{S}'(\mathbb{R}^n)$, let

$$(\mathcal{F} f)(\xi) = \hat{f}(\xi) = (2\pi)^{-n/2} \int_{\mathbb{R}^n} f(x) e^{-ix\xi} \, dx,$$

$$(\mathcal{F}^{-1} f)(x) = \check{f}(x) = \hat{f}(-x),$$

and $(m(D)f)(x) = (2\pi)^{-n/2} (\check{m} * f)(x)$.

The symbol $\langle x \rangle$ denotes $\sqrt{1 + x^2}$ for $x \in \mathbb{R}$. We use $a \lesssim b$ and $a = O(b)$ to mean that there exists a positive constant such that $a \leq Cb$. Various constants will be simply denoted by C and C_i ($i \in \mathbb{N}$) in the course of the calculations.

2 Preliminaries

2.1 Semigroup Estimates for the Linearized KP-II Equation

First, we recall decay estimates of the semigroup generated by the linearized operator around a 1-line soliton in exponentially weighted spaces.

Let

$$\varphi = \varphi_2, \quad \mathcal{L} = -\partial_x^3 + 4\partial_x - 3\partial_x^{-1}\partial_y^2 - 6\partial_x(\varphi \cdot).$$

We remark that $\mathcal{L}$ generates a C^0-semigroup on $X := L^2(\mathbb{R}^2; e^{2\alpha x}dxdy)$ for any $\alpha > 0$.

Let $\mathcal{L}(\eta) = -\partial_x^3 + 4\partial_x + 3\eta^2\partial_x^{-1} - 6\partial_x(\varphi \cdot)$ be an operator on $L^2(\mathbb{R}; e^{2\alpha x}dx)$ with its domain $D(\mathcal{L}(\eta)) = e^{-\alpha x}H^3(\mathbb{R})$. Obviously, we have $\mathcal{L}(u(x)e^{i y\eta}) = e^{i y\eta}\mathcal{L}(\eta)u(x)$ for any $\eta \in \mathbb{R}$. If $\eta \simeq 0$, then $\mathcal{L}(\eta)$ has two isolated eigenvalues near 0 and the rest of the spectrum is bounded away from the imaginary axis and lies in the stable half plane (see [22, Chapter 2]). We remark that that $\mathcal{L}(0)$ is the linearized KdV operator around φ which has an isolated 0 eigenvalue of multiplicity 2 in $L^2(\mathbb{R}; e^{2\alpha x}dx)$ with $\alpha \in (0, 2)$ (see [30]).

Let

$$\beta(\eta) = \sqrt{1+i\eta}, \quad \lambda(\eta) = 4i\eta\beta(\eta), \tag{2.1}$$

$$g(x,\eta) = \frac{-i}{2\eta\beta(\eta)}\partial_x^2(e^{-\beta(\eta)x}\operatorname{sech} x), \quad g^*(x,\eta) = \partial_x(e^{\beta(-\eta)x}\operatorname{sech} x). \tag{2.2}$$

Then

$$\mathcal{L}(\eta)g(x,\pm\eta) = \lambda(\pm\eta)g(x,\pm\eta), \quad \mathcal{L}(\eta)^*g^*(x,\pm\eta) = \lambda(\mp\eta)g^*(x,\pm\eta). \tag{2.3}$$

The continuous eigenvalues $\lambda(\eta)$ belongs to the stable half plane $\{\lambda \in \mathbb{C} \mid \Re\lambda < 0\}$ for $\eta \in \mathbb{R} \setminus \{0\}$ and $\lambda(\eta) \to \lambda(0) = 0$ as $\eta \to 0$.

Let $\nu(\eta) := \Re\beta(\eta) - 1$ and η_0 be a small positive number. Since $g(x,\eta) = O(e^{\nu(\eta)|x|})$ as $x \to -\infty$ and $\nu(\eta) = O(\eta^2)$ for small η, we choose α and so that $\alpha \geqslant \nu(\eta)$ and $g(x,\pm\eta) \in L^2(\mathbb{R}; e^{2\alpha x}dx)$ for $\eta \in [-\eta_0, \eta_0]$. The continuous eigenmodes $g(x,\eta)e^{iy\eta}$ grow exponentially as $x \to -\infty$. Nevertheless, they have to do with modulation of line solitons. See [4] and the references therein.

The spectral projection to the continuous eigenmodes $\{g(x, \pm\eta)e^{iy\eta}\}_{-\eta_0 \leqslant \eta \leqslant \eta_0}$ is given by

$$P_0(\eta_0)f(x,y) = \frac{1}{\sqrt{2\pi}} \sum_{k=1,2} \int_{-\eta_0}^{\eta_0} a_k(\eta)g_k(x,\eta)e^{iy\eta}\,d\eta,$$

$$a_k(\eta) = \int_{\mathbb{R}} (\mathcal{F}_y f)(x,\eta)g_k^*(x,\eta)\,dx,$$

where

$$g_1(x,\eta) = 2\Re g(x,\eta), \quad g_2(x,\eta) = -2\eta \Im g(x,\eta),$$
$$g_1^*(x,\eta) = \Re g^*(x,\eta), \quad g_2^*(x,\eta) = -\eta^{-1}\Im g^*(x,\eta).$$

We remark that for an $\alpha \in (0,2)$,

$$g_1(x,\eta) = \frac{1}{4}\varphi' + \frac{x}{4}\varphi' + \frac{1}{2}\varphi + O(\eta^2), \quad g_2(x,\eta) = -\frac{1}{2}\varphi' + O(\eta^2) \quad \text{in } L^2(\mathbb{R}; e^{2\alpha x}dx),$$

$$g_1^*(x,\eta) = \frac{1}{2}\varphi + O(\eta^2), \quad g_2^*(x,\eta) = \int_{-\infty}^{x} \partial_c\varphi\, dx + O(\eta^2) \quad \text{in } L^2(\mathbb{R}; e^{-2\alpha x}dx),$$

where $\partial_c\varphi = \partial_c\varphi_c|_{c=2}$. See [22, Chapter 3].

For η_0 and M satisfying $0 < \eta_0 \leqslant M \leqslant \infty$, let

$$P_1(\eta_0, M)u(x,y) := \frac{1}{2\pi} \int_{\eta_0 \leqslant |\eta| \leqslant M} \int_{\mathbb{R}} u(x,y_1) e^{i\eta(y-y_1)} dy_1 d\eta,$$

$$P_2(\eta_0, M) := P_1(0, M) - P_0(\eta_0).$$

The semigroup $e^{t\mathcal{L}}$ is exponentially stable on $(I - P_0(\eta_0))X$.

Proposition 2.1 ([22, proposition 3.2 and Corollary 3.3]) *Let $\alpha \in (0,2)$ and η_1 be a positive number satisfying $\nu(\eta_1) < \alpha$. Then there exist positive constants K and b such that for any $\eta_0 \in (0,\eta_1]$, $M \geqslant \eta_0$, $f \in X$ and $t \geqslant 0$,*

$$\|e^{t\mathcal{L}} P_2(\eta_0, M)f\|_X \leqslant K e^{-bt} \|f\|_X.$$

Moreover, there exist positive constants K' and b' such that for $t > 0$,

$$\|e^{t\mathcal{L}} P_2(\eta_0, M)\partial_x f\|_X \leqslant K' e^{-b't} t^{-1/2} \|e^{\alpha x} f\|_X,$$

$$\|e^{t\mathcal{L}} P_2(\eta_0, M)\partial_x f\|_X \leqslant K' e^{-b't} t^{-3/4} \|e^{\alpha x} f\|_{L_x^1 L_y^2}.$$

2.2 Decay Estimates for Linearized Modulation Equations

Time evolution of parameters $c(t,y)$ and $x(t,y)$ of a modulating line soliton $\varphi_{c(t,y)}(x - x(t,y))$ is described by a system of Burgers type equations. In this subsection, we introduce linear estimates which will be used to prove boundedness of the phase shift $x(t,y) - 2c_0 t$. The estimates are a substitute of d'Alembert's formula for the 1-dimensional wave equation.

Let $\omega(\eta) = \sqrt{16 + (8\mu_3 - 1)\eta^2}$, $\mu_3 > 1/8$, $\lambda_*^\pm(\eta) = -2\eta^2 \pm i\eta\omega(\eta)$ and

$$\mathcal{A}_*(\eta) = \begin{pmatrix} -3\eta^2 & -8\eta^2 \\ 2 + \mu_3\eta^2 & -\eta^2 \end{pmatrix}, \quad \mathcal{P}_*(\eta) = \frac{1}{4\eta} \begin{pmatrix} 8\eta & 8\eta \\ -\eta - i\omega(\eta) & -\eta + i\omega(\eta) \end{pmatrix}.$$

Then $\mathcal{P}_*(\eta)^{-1} \mathcal{A}_*(\eta) \mathcal{P}_*(\eta) = \operatorname{diag}(\lambda_*^+(\eta), \lambda_*^-(\eta))$ and

$$e^{t\mathcal{A}_*(\eta)} = e^{-2t\eta^2} \begin{pmatrix} \cos t\eta\omega(\eta) - \frac{\eta}{\omega(\eta)} \sin t\eta\omega(\eta) & -\frac{8\eta}{\omega(\eta)} \sin t\eta\omega(\eta) \\ \frac{\eta^2 + \omega(\eta)^2}{8\eta\omega(\eta)} \sin t\eta\omega(\eta) & \cos t\eta\omega(\eta) + \frac{\eta}{\omega(\eta)} \sin t\eta\omega(\eta) \end{pmatrix}. \tag{2.4}$$

Let η_0 be a positive number and let $\chi_1(\eta)$ be a nonnegative smooth function such that $0 \leq \chi_1(\eta) \leq 1$ for $\eta \in \mathbb{R}$, $\chi_1(\eta) = 1$ if $|\eta| \leq \frac{1}{2}\eta_0$ and $\chi_1(\eta) = 0$ if $|\eta| \geq \frac{3}{4}\eta_0$. Let $\chi_2(\eta) = 1 - \chi_1(\eta)$. Then

$$\|\chi_2(D_y) e^{t\mathcal{A}_*(D_y)}\|_{B(L^2(\mathbb{R}))} \lesssim e^{-\eta_0^2 t/2} \quad \text{for } t \geq 0. \tag{2.5}$$

Next, we will estimate the low frequency part of $e^{t\mathcal{A}_*(\eta)}$. Let

$$K_1(t, y) = \frac{1}{\sqrt{2\pi}} \mathcal{F}^{-1} \left(\chi_1(\eta) e^{-2t\eta^2} \cos t\eta\omega(\eta) \right),$$

$$K_2(t, y) = \frac{1}{\sqrt{2\pi}} \mathcal{F}^{-1} \left(e^{-2t\eta^2} \frac{\eta \chi_1(\eta)}{\omega(\eta)} \sin t\eta\omega(\eta) \right),$$

$$K_3(t, y) = \frac{1}{\sqrt{2\pi}} \mathcal{F}^{-1} \left(e^{-2t\eta^2} \frac{\chi_1(\eta) \omega(\eta)}{\eta} \sin t\eta\omega(\eta) \right).$$

Then

$$\chi_1(D_y) e^{t\mathcal{A}_*(D_y)} \delta = \begin{pmatrix} K_1(t, y) - K_2(t, y) & -8K_2(t, y) \\ \frac{1}{8}(K_2(t, y) + K_3(t, y)) & K_1(t, y) + K_2(t, y) \end{pmatrix}. \tag{2.6}$$

We have the following estimates for K_1, K_2 and K_3.

Lemma 2.2 *Let $j \in \mathbb{Z}_{\geq 0}$. Then*

$$\sup_{t>0} \|K_1(t, \cdot)\|_{L^1(\mathbb{R})} < \infty, \quad \|K_1(t, \cdot)\|_{L^2(\mathbb{R})} \lesssim \langle t \rangle^{-1/4}, \tag{2.7}$$

$$\|\partial_y^{j+1} K_1(t, \cdot)\|_{L^1(\mathbb{R})} + \|\partial_y^j K_2(t, \cdot)\|_{L^1(\mathbb{R})} + \|\partial_y^{j+2} K_3(t, \cdot)\|_{L^1(\mathbb{R})} \lesssim \langle t \rangle^{-(j+1)/2}, \tag{2.8}$$

$$\|\partial_y^{j+1} K_1(t, \cdot)\|_{L^2(\mathbb{R})} + \|\partial_y^j K_2(t, \cdot)\|_{L^2(\mathbb{R})} + \|\partial_y^{j+2} K_3(t, \cdot)\|_{L^2(\mathbb{R})} \lesssim \langle t \rangle^{-(2j+3)/4}, \tag{2.9}$$

$$\sup_{t>0} \|\partial_y K_3(t,\cdot)\|_{L^1(\mathbb{R})} < \infty, \quad \|\partial_y K_3(t,\cdot)\|_{L^2(\mathbb{R})} \lesssim \langle t \rangle^{-1/4}, \tag{2.10}$$

$$\sup_{t>0} \|K_3(t,\cdot) * f\|_{L^\infty(\mathbb{R})} \lesssim \|f\|_{L^1(\mathbb{R})}. \tag{2.11}$$

Proof Let $\tilde{\omega}(\eta) = \omega(\eta) - 4$ and

$$K_{1,\pm}(t,y) = \frac{1}{2\sqrt{2\pi}} \mathcal{F}^{-1}\left(\chi_1(\eta) e^{-(2\eta^2 \pm i\eta\tilde{\omega}(\eta))t}\right). \tag{2.12}$$

Since $K_1(t,y) = \sum_\pm K_{1,\pm}(t, y \mp 4t)$, it suffices to show that $\sup_{t>0} \|K_{1,\pm}(t,\cdot)\|_{L^1(\mathbb{R})} < \infty$ and $\|K_{1,\pm}(t,\cdot)\|_{L^2(\mathbb{R})} \lesssim \langle t \rangle^{-1/4}$ to prove (2.7). Using the Plancherel identity, we have

$$\|K_{1,\pm}(t,\cdot)\|_{L^2(\mathbb{R})} \lesssim \left\|\chi_1(\eta) e^{-2t\eta^2}\right\|_{L^2(\mathbb{R})} \lesssim \langle t \rangle^{-1/4},$$

$$\|y K_{1,\pm}(t,y)\|_{L^2(\mathbb{R})} \lesssim \left\|\partial_\eta \left(\chi_1(\eta) e^{-(2\eta^2 \pm i\tilde{\omega}(\eta))t}\right)\right\|_{L^2(\mathbb{R})}$$
$$\lesssim \|\chi_1'(\eta) e^{-2t\eta^2}\|_{L^2(\mathbb{R})} + t(\|\eta \chi_1(\eta) e^{-2t\eta^2}\|_{L^2(\mathbb{R})}$$
$$+ \|\tilde{\omega}'(\eta) \chi_1(\eta) e^{-2t\eta^2}\|_{L^2(\mathbb{R})})$$
$$\lesssim \langle t \rangle^{1/4}.$$

Note that

$$|\tilde{\omega}(\eta)| \lesssim \min\{1, \eta^2\}, \quad |\tilde{\omega}'(\eta)| \lesssim \min\{1, |\eta|\}. \tag{2.13}$$

Combining the above, we have

$$\|K_{1,\pm}(t,\cdot)\|_{L^1(\mathbb{R})} \lesssim t^{1/4} \|K_{1,\pm}(t,\cdot)\|_{L^2(|y|\leq \sqrt{t})}$$
$$+ \|y K_{1,\pm}(t,y)\|_{L^2(|y|\geq \sqrt{t})} \|y^{-1}\|_{L^2(|y|\geq \sqrt{t})} = O(1).$$

Thus we have (2.7). We can prove (2.8)–(2.10) in the same way.

Now we will prove (2.11). Let

$$K_{3,1}(t,y) = \frac{1}{2\sqrt{2\pi}} \mathcal{F}^{-1}\left(\omega(\eta) \chi_1(\eta) e^{-2t\eta^2} \cos t\eta \tilde{\omega}(\eta)\right),$$

$$K_{3,2}(t,y) = \frac{1}{2\sqrt{2\pi}} \mathcal{F}^{-1}\left(\omega(\eta) \chi_1(\eta) e^{-2t\eta^2} \frac{\sin t\eta \tilde{\omega}(\eta)}{\eta}\right).$$

Then

$$K_3(t, y) = K_{3,1}(t, \cdot) * \mathbf{1}_{[-4t,4t]} + K_{3,2}(t, y + 4t) + K_{3,2}(t, y - 4t). \tag{2.14}$$

We can prove that

$$\sup_{t>0} \|K_{3,1}(t, \cdot)\|_{L^1(\mathbb{R})} < \infty, \tag{2.15}$$

in the same way as (2.7). It follows from (2.13) that

$$\sup_{t>0} \|K_{3,2}(t, cdot)\|_{L^{infty}(\mathbb{R})} < \infty. \tag{2.16}$$

Combining (2.14)–(2.16), we have (2.11). This completes the proof of Lemma 2.2. □

Let Y and Z be closed subspaces of $L^2(\mathbb{R})$ defined by

$$Y = \mathcal{F}_\eta^{-1} Z \quad \text{and} \quad Z = \{f \in L^2(\mathbb{R}) \mid \operatorname{supp} f \subset [-\eta_0, \eta_0]\},$$

and let $X_1 = L^1(\mathbb{R}_y; L^2(\mathbb{R}; e^{\alpha x} dx))$, $Y_1 = \mathcal{F}_\eta^{-1} Z_1$ and $Z_1 = \{f \in Z \mid \|f\|_{Z_1} := \|f\|_{L^\infty} < \infty\}$.

Let $E_1 = \operatorname{diag}(1, 0)$ and $E_2 = \operatorname{diag}(0, 1)$ and let $\chi(\eta)$ be a smooth function such that $\chi(\eta) = 1$ if $\eta \in [-\frac{\eta_0}{4}, \frac{\eta_0}{4}]$ and $\chi(\eta) = 0$ if $\eta \notin [-\frac{\eta_0}{2}, \frac{\eta_0}{2}]$. We will use the following estimates to investigate large time behavior of modulation parameters.

Lemma 2.3 *For $t \geqslant 0$ and $k \geqslant 1$,*

$$\|\chi_1(D_y)e^{t\mathcal{A}_*}E_1\|_{B(L^1;L^\infty)} = O(1), \quad \|(I - \chi(D_y))e^{t\mathcal{A}_*}E_1\|_{B(Y;L^\infty)} = O(e^{-c_1 t}), \tag{2.17}$$

$$\|e^{t\mathcal{A}_*}E_2\|_{B(Y;L^\infty)} \lesssim \langle t \rangle^{-1/4}, \quad \|e^{t\mathcal{A}_*}E_2\|_{B(Y_1;L^\infty)} \lesssim \langle t \rangle^{-1/2}, \tag{2.18}$$

$$\|\partial_y^k e^{t\mathcal{A}_*}\|_{B(Y,L^\infty)} \lesssim \langle t \rangle^{-(2k-1)/4}, \quad \|\partial_y^k e^{t\mathcal{A}_*}\|_{B(Y_1;L^\infty)} \lesssim \langle t \rangle^{-k/2}, \tag{2.19}$$

where c_1 is a positive constant. Moreover,

$$\left\| e^{t\mathcal{A}_*} \begin{pmatrix} f_1 \\ f_2 \end{pmatrix} - \frac{1}{2} H_{2t} * W_{4t} * f_1 \mathbf{e}_2 \right\|_{L^\infty} \lesssim \langle t \rangle^{-1/2}(\|f_1\|_{Y_1} + \|f_2\|_{Y_1}), \tag{2.20}$$

$$\left\| \operatorname{diag}(1, \partial_y) e^{t\mathcal{A}_*} \begin{pmatrix} f_1 \\ f_2 \end{pmatrix} - \frac{1}{4} \begin{pmatrix} 2 & 2 \\ 1 & -1 \end{pmatrix} \begin{pmatrix} H_{2t}(\cdot + 4t) \\ H_{2t}(\cdot - 4t) \end{pmatrix} * f_1 \right\|_{L^\infty}$$
$$\lesssim \langle t \rangle^{-1}(\|f_1\|_{Y_1} + \|f_2\|_{Y_1}), \tag{2.21}$$

$$\left\| e^{t\mathcal{A}_*} \operatorname{diag}(\partial_y, 1) \begin{pmatrix} f_1 \\ f_2 \end{pmatrix} - \frac{1}{4} \sum_{\pm} H_{2t}(\cdot \pm 4t)(2f_2 \pm f_1) \mathbf{e}_2 \right\|_{L^\infty}$$
(2.22)
$$\lesssim \langle t \rangle^{-1}(\|f_1\|_{Y_1} + \|f_2\|_{Y_1}),$$

where $H_t(y) = (4\pi t)^{-1/2} \exp(-y^2/4t)$ and $W_t(y) = \frac{1}{2}\mathbf{1}_{[-t,t]}(y)$.

Proof Equations (2.17)–(2.19) follow immediately from Lemma 2.2, (2.5) and (2.6).

In view of (2.12) and (2.13),

$$\left\| 2K_{1,\pm}(t,\cdot) * f - \chi_1(D) e^{2t\partial_y^2} f \right\|_{L^\infty} \lesssim \left\| \chi_1(\eta) e^{-2t\eta^2}(e^{\pm it\eta\tilde{\omega}(\eta)} - 1) \right\|_{L^1} \|\hat{f}\|_{L^\infty}$$
$$\lesssim \min\left\{ t\|\eta^3 e^{-2t\eta^2}\|_{L^1}, \|\chi_1\|_{L^1} \right\} \|f\|_{Y_1}$$
$$\lesssim \langle t \rangle^{-1} \|f\|_{Y_1}.$$

Since $K_1(t,\cdot) = \sum_{\pm} K_{1,\pm}(t, \cdot \mp 4t)$,

$$\left\| K_1(t,\cdot) * f - \frac{1}{2} \sum_{\pm} H_{2t}(\cdot \pm 4t) * f \right\|_{L^\infty} \lesssim \langle t \rangle^{-1} \|f\|_{Y_1}. \quad (2.23)$$

We can prove

$$\left\| \partial_y K_3(t,\cdot) * f - 2H_{2t}(\cdot + 4t) * f + 2H_{2t}(\cdot - 4t) * f \right\|_{L^\infty(\mathbb{R})} \lesssim \langle t \rangle^{-1} \|f\|_{Y_1},$$
(2.24)
$$\|K_3(t,\cdot) * f - 4H_{2t} * W_{4t} * f\|_{L^\infty(\mathbb{R})} \lesssim \langle t \rangle^{-1/2} \|f\|_{Y_1} \quad (2.25)$$

in the same way. Combining (2.23)–(2.25) with Lemma 2.2 and (2.5), we obtain (2.20)–(2.22). Thus we complete the proof. □

To investigate the large time behavior of $x(t,y)$, we need the following.

Lemma 2.4 *Suppose that $f \in L^1(\mathbb{R}_+ \times \mathbb{R})$. Then for any $\delta > 0$,*

$$\lim_{t \to \infty} \sup_{|y| \leqslant (4-\delta)t} \left| \int_0^t H_{2(t-s)} * W_{4(t-s)} * f(s,\cdot)(y)\,ds - \frac{1}{2}\int_0^\infty \int_\mathbb{R} f(s,y)\,dy\,ds \right| = 0,$$
(2.26)

$$\lim_{t \to \infty} \sup_{|y| \geqslant (4+\delta)t} \left| \int_0^t H_{2(t-s)} * W_{4(t-s)} * f(s,\cdot)(y)\,ds \right| = 0. \quad (2.27)$$

Proof Let $K_t(y, y_1) = \{(s, y_2) \mid 0 \leq s < t, |y_2 - y + 2y_1(t-s)^{1/2}| \leq 4(t-s)\}$. Then

$$\int_0^t H_{2(t-s)} * W_{4(t-s)} * f(s, \cdot)(y)\, ds$$

$$= \frac{1}{4\sqrt{2\pi}} \int_0^t ds\, (t-s)^{-1/2} \int_{\mathbb{R}} dy_1\, e^{-(y-y_1)^2/8(t-s)} \int_{y_1-4(t-s)}^{y_1+4(t-s)} dy_2\, f(s, y_2)$$

$$= \frac{1}{2\sqrt{2\pi}} \int_{\mathbb{R}} dy_1\, e^{-y_1^2/2} \int_{K_t(y,y_1)} ds dy_2\, f(s, y_2).$$

Since $e^{-y_1^2/2} f(s, y_2)$ is integrable on $\mathbb{R}_+ \times \mathbb{R}^2$,

$$\left| \int_{|y_1| \geq \delta\sqrt{t}/4} dy_1\, e^{-y_1^2/2} \int_{K_t(y,y_1)} ds dy_2\, f(s, y_2) \right|$$

$$\leq \|f\|_{L^1(\mathbb{R}_+ \times \mathbb{R})} \int_{|y_1| \geq \delta\sqrt{t}/4} dy_1\, e^{-y_1^2/2} \to 0 \quad \text{as } t \to \infty.$$

Moreover,

$$\lim_{t \to \infty} \sup_{|y| \leq (4-\delta)t} \left| \frac{1}{\sqrt{2\pi}} \int_{|y_1| \leq \delta\sqrt{t}/4} e^{-y_1^2/2} \left(\int_{K_t(y,y_1)} f(s, y_2)\, dy_2 ds \right) dy_1 \right.$$

$$\left. - \int_{\mathbb{R}_+ \times \mathbb{R}} f(s, y)\, dy ds \right| = 0,$$

$$\lim_{t \to \infty} \sup_{|y| \geq (4+\delta)t} \left| \int_{|y_1| \leq \delta\sqrt{t}/4} e^{-y_1^2/2} \left(\int_{K_t(y,y_1)} f(s, y_2)\, dy_2 ds \right) dy_1 \right| = 0$$

because

$$\lim_{t \to \infty} \bigcap_{|y| \leq (4-\delta)t, |y_1| \leq \delta\sqrt{t}/4} K_t(y, y_1) = \mathbb{R}_+ \times \mathbb{R}, \quad \lim_{t \to \infty} \bigcup_{|y| \geq (4+\delta)t, |y_1| \leq \delta\sqrt{t}/4} K_t(y, y_1) = \emptyset.$$

$\square$

3 Decomposition of Solutions Around 1-Line Solitons

Following [22, 23], we decompose a solution around a line soliton $\varphi(x - 4t)$ into a sum of a modulating line soliton and a dispersive part plus a small wave which is caused by amplitude changes of the line soliton:

$$u(t, x, y) = \varphi_{c(t,y)}(z) - \psi_{c(t,y),L}(z + 3t) + v(t, z, y), \quad z = x - x(t, y), \quad (3.1)$$

where $\psi_{c,L}(x) = 2(\sqrt{2c} - 2)\psi(x + L)$, $\psi(x)$ is a nonnegative function such that $\psi(x) = 0$ if $|x| \geq 1$ and that $\int_\mathbb{R} \psi(x)\,dx = 1$ and $L > 0$ is a large constant to be fixed later. The modulation parameters $c(t_0, y_0)$ and $x(t_0, y_0)$ denote the maximum height and the phase shift of the modulating line soliton $\varphi_{c(t,y)}(x - x(t, y))$ along the line $y = y_0$ at the time $t = t_0$, and $\psi_{c,L}$ is an auxiliary smooth function such that

$$\int_\mathbb{R} \psi_{c,L}(x)\,dx = \int_\mathbb{R} (\varphi_c(x) - \varphi(x))\,dx. \tag{3.2}$$

Now we further decompose v into a small solution of (1.1) and an exponentially localized part as in [21, 23, 27]. If $v_0(x, y)$ is polynomially localized, then as in [23], we can decompose the initial data as a sum of an amplified line soliton and a remainder part $v_*(x, y)$ that satisfies $\int_\mathbb{R} v_*(x, y)\,dx = 0$ for every $y \in \mathbb{R}$. Let

$$c_1(y) = \left\{\sqrt{c_0} + \frac{1}{2\sqrt{2}}\int_\mathbb{R} v_0(x, y)\,dx\right\}^2, \tag{3.3}$$

$$v_*(x, y) = v_0(x, y) + \varphi_{c_0}(x) - \varphi_{c_1(y)}(x). \tag{3.4}$$

Then we have the following.

Lemma 3.1 *Let $c_0 > 0$ and $s > 1$. There exists a positive constant ε_0 such that if $\varepsilon := \|\langle x\rangle^{s/2}(\langle x\rangle + \langle y\rangle)^{s/2} v_0\|_{H^1(\mathbb{R}^2)} < \varepsilon_0$, then*

$$\left\|\langle y\rangle^{s/2}(c_1 - c_0)\right\|_{L^2(\mathbb{R})} + \left\|\langle y\rangle^{s/2}\partial_y c_1\right\|_{L^2(\mathbb{R})} \lesssim \left\|\langle x\rangle^{s/2}\langle y\rangle^{s/2} v_0\right\|_{H^1(\mathbb{R}^2)}, \tag{3.5}$$

$$\left\|\langle x\rangle^s v_*\right\|_{L^2(\mathbb{R}^2)} \lesssim \left\|\langle x\rangle^s v_0\right\|_{L^2(\mathbb{R}^2)}, \quad \left\|\langle x\rangle^{s/2}\langle y\rangle^{s/2} v_*\right\|_{L^2(\mathbb{R}^2)} \lesssim \left\|\langle x\rangle^{s/2}\langle y\rangle^{s/2} v_0\right\|_{L^2(\mathbb{R}^2)}, \tag{3.6}$$

$$\|\partial_x^{-1} v_*\|_{L^2} + \|\partial_x^{-1}\partial_y v_*\|_{L^2} + \|v_*\|_{H^1(\mathbb{R}^2)} \lesssim \|\langle x\rangle^s v_0\|_{H^1(\mathbb{R}^2)}. \tag{3.7}$$

Moreover, the mapping

$$\langle x\rangle^{-s/2}(\langle x\rangle + \langle y\rangle)^{-s/2} H^1(\mathbb{R}^2) \ni v_0 \mapsto (v_*, c_1 - c_0) \in H^1(\mathbb{R}^2) \times H^1(\mathbb{R}) \cap \langle y\rangle^{-s/2} L^2(\mathbb{R})$$

is continuous.

Proof By [23, (10.4)],

$$\sup_y \left|\sqrt{c_1(y)} - \sqrt{c_0}\right| \lesssim \|\langle x\rangle^{s/2} v_0\|_{L^2} + \|\langle x\rangle^{s/2}\partial_y v_0\|_{L^2}.$$

Hence it follows from (3.3) and (3.4) that for sufficiently small ε,

$$\left\|\langle y\rangle^{s/2}\partial_y^i(c_1-c_0)\right\|_{L^2(\mathbb{R})} \lesssim \left\|\langle y\rangle^{s/2}\partial_y^i\left(\sqrt{c_1}-\sqrt{c_0}\right)\right\|_{L^2(\mathbb{R})}$$

$$\lesssim \left\|\langle y\rangle^{s/2}\int_\mathbb{R} \partial_y^i v_0(x,y)\,dx\right\|_{L^2(\mathbb{R}_y)}$$

$$\lesssim \left\|\langle x\rangle^{s/2}\langle y\rangle^{s/2}\partial_y^i v_0\right\|_{L^2(\mathbb{R}^2)},$$

$$\left\|\langle x\rangle^{s/2}\langle y\rangle^{s/2} v_*\right\|_{L^2(\mathbb{R}^2)} \lesssim \left\|\langle x\rangle^{s/2}\langle y\rangle^{s/2} v_0\right\|_{L^2(\mathbb{R}^2)} + \left\|\langle x\rangle^{s/2}\langle y\rangle^{s/2}(\varphi_{c_1(y)}-\varphi_{c_0})\right\|_{L^2(\mathbb{R}^2)}$$

$$\lesssim \left\|\langle x\rangle^{s/2}\langle y\rangle^{s/2} v_0\right\|_{L^2(\mathbb{R}^2)}.$$

Using [23, (10.2)], we can prove $\|\langle x\rangle^s v_*\|_{L^2(\mathbb{R}^2)} \lesssim \|\langle x\rangle^s v_0\|_{L^2(\mathbb{R}^2)}$ in the same way. We have (3.7) from [23, Lemma 10.1] and its proof. Since the continuity of the mapping $v_0 \mapsto (v_*, c_1 - c_0)$ can be proved in the similar way, we omit the proof. Thus we complete the proof. □

Let $\tilde{v}_1$ be a solution of

$$\begin{cases} \partial_t \tilde{v}_1 + \partial_x^3 \tilde{v}_1 + 3\partial_x(\tilde{v}_1^2) + 3\partial_x^{-1}\partial_y^2 \tilde{v}_1 = 0, \\ \tilde{v}_1(0,x,y) = v_*(x,y). \end{cases} \quad (3.8)$$

Since $v_* \in H^1(\mathbb{R}^2)$ and $\partial_x^{-1}\partial_y v_* \in L^2(\mathbb{R}^2)$, we have $\tilde{v}_1(t) \in C(\mathbb{R}; H^1(\mathbb{R}^2))$ from [28]. Applying the Strichartz estimate in [34, Proposition 2.3] to

$$\partial_x^{-1}\partial_y \tilde{v}_1(t) = e^{tS}\partial_x^{-1}\partial_y v_* - 6\int_0^t e^{(t-s)S}(\tilde{v}_1 \partial_y \tilde{v}_1)(s)\,ds, \quad S = -\partial_x^3 - 3\partial_x^{-1}\partial_y^2,$$

we have $\partial_x^{-1}\partial_y \tilde{v} \in C(\mathbb{R}; L^2(\mathbb{R}))$. Suppose that v_0 satisfies the assumption of Lemma 3.1 and that $u(t)$ is a solution to (1.1) with $u(0,x,y) = \varphi(x) + v_0(x,y)$, where $\varphi = \varphi_2$. Then as [23, Lemmas 3.1 and 3.3], we can prove that $w(t,x,y) := u(t,x+4t,y) - \varphi(x) - \tilde{v}_1(t,x+4t,y)$ belongs to an exponentially weighted space $X = L^2(\mathbb{R}^2; e^{2\alpha x}dxdy)$ for an $\alpha \in (0,1)$, that

$$w \in C([0,\infty); X), \quad \partial_x w, \ \partial_x^{-1}\partial_y w \in L^2(0,T; X),$$

and that the mapping

$$\langle x\rangle^{-1}(\langle x\rangle + \langle y\rangle)^{-1}H^1(\mathbb{R}^2) \ni v_0 \mapsto w \in C([0,T]; X)$$

is continuous for any $T > 0$ by using by Lemma 3.1.

Now let

$$v_1(t, z, y) = \tilde{v}_1(t, z + x(t, y), y), \quad v_2(t, z, y) = v(t, z, y) - v_1(t, z, y). \quad (3.9)$$

To fix the decomposition (3.1), we impose the constraint that for $k = 1, 2$,

$$\int_{\mathbb{R}^2} v_2(t, z, y) g_k^*(z, \eta, c(t, y)) e^{-iy\eta} \, dz dy = 0 \quad \text{in } L^2(-\eta_0, \eta_0), \quad (3.10)$$

where $g_1^*(z, \eta, c) = c g_1^*(\sqrt{c/2}z, \eta)$ and $g_2^*(z, \eta, c) = \frac{c}{2} g_2^*(\sqrt{c/2}z, \eta)$.
Let

$$F_k[u, \tilde{c}, \gamma, L](\eta) := \mathbf{1}_{[-\eta_0, \eta_0]}(\eta) \int_{\mathbb{R}^2} \{u(x, y) + \varphi(x) - \varphi_{c(y)}(x - \gamma(y)) \\
+ \tilde{\psi}_{c(y), L}(x - \gamma(y))\} g_k^*(x - \gamma(y), \eta, c(y)) e^{-iy\eta} \, dx dy$$

for $k = 1, 2$, where $c(y) = 2 + \tilde{c}(y)$. Since $w(0) = \varphi_{c_1} - \varphi$ and

$$\|\varphi_{c_1} - \varphi\|_{X_1} \lesssim \|\langle y \rangle (c_1 - 2)\|_{L^2(\mathbb{R}_y)} \lesssim \|\langle x \rangle \langle y \rangle v_0\|_{L^2(\mathbb{R}^2)}$$

by Lemma 3.1, it follows from [22, Lemmas 5.2 and 5.4] that there exists $(\tilde{c}_*, x_*) \in Y_1 \times Y_1$ satisfying

$$F_1[w(0), \tilde{c}_*, x_*, L] = F_2[w(0), \tilde{c}_*, x_*, L] = 0,$$

$$\|\tilde{c}_*\|_Y + \|x_*\|_Y \lesssim \|w(0)\|_X \lesssim \|\langle x \rangle v_0\|_{L^2(\mathbb{R}^2)},$$

$$\|\tilde{c}_*\|_{Y_1} + \|x_*\|_{Y_1} \lesssim \|w(0)\|_{X_1} \lesssim \|\langle x \rangle \langle y \rangle v_0\|_{L^2(\mathbb{R}^2)}, \quad (3.11)$$

provided $\|\langle x \rangle \langle y \rangle v_0\|_{L^2(\mathbb{R}^2)}$ is sufficiently small. By the definitions,

$$\begin{cases} v_2(0, x, y) = v_{2,*}(x, y) := \varphi_{c_1(y)}(x) - \varphi_{c_*(y)}(x - x_*(y)) + \tilde{\psi}_{c_*(y), L}(x - x_*(y)), \\ \tilde{c}(0, y) = \tilde{c}_*(y), \quad x(0, y) = x_*(y), \end{cases} \quad (3.12)$$

where $c_* = 2 + \tilde{c}_*$ and it follows from Lemma 3.1 and (3.11) that

$$\|v_{2,*}\|_X \lesssim \|c_1 - 2\|_{L^2(\mathbb{R})} + \|\tilde{c}_*\|_Y \lesssim \|\langle x \rangle v_0\|_{L^2(\mathbb{R}^2)}. \quad (3.13)$$

Lemma 5.2 in [22] implies that there exist a $T > 0$, $\tilde{c}(t, \cdot) := c(t, \cdot) - 2 \in C([0, T]; Y)$ and $\tilde{x}(t, \cdot) := x(t, \cdot) - 4t \in C([0, T]; Y)$ satisfying

$$F_1[w(t), \tilde{c}(t), \tilde{x}(t), L] = F_2[w(t), \tilde{c}(t), \tilde{x}(t), L] = 0 \quad \text{for } t \in [0, T]$$

because $w \in C([0, \infty); X)$. If $(v_2(t), \tilde{c}(t))$ remains small in $X \times Y$ for $t \in [0, T]$, then the decomposition (3.1) and (3.9) satisfying (3.10) exists beyond $t = T$ thanks to the continuation argument.

Proposition 3.2 ([23, Proposition 3.9]) *There exists a $\delta_1 > 0$ such that if (3.1), (3.9) and (3.10) hold for $t \in [0, T)$ and*

$$(\tilde{c}, \tilde{x}) \in C([0, T); Y \times Y) \cap C^1((0, T); Y \times Y),$$

$$\sup_{t \in [0,T]} (\|v_2(t)\|_X + \|\tilde{c}(t)\|_Y) < \delta_1, \quad \sup_{t \in [0,T)} \|\tilde{x}(t)\|_Y < \infty,$$

then either $T = \infty$ or T is not the maximal time of the decomposition (3.1) and (3.9) satisfying (3.10).

4 Modulation Equations

By [23, Lemma 3.6 and Remark 3.7],

$$v_2(t) \in C([0, T); X), \quad (\tilde{c}, \tilde{x}) \in C([0, T); Y \times Y) \cap C^1((0, T); Y \times Y),$$

where T is the maximal time of the decomposition (3.1) and (3.9) satisfying (3.10). Substituting (3.1) and (3.9) into (1.1) with $\sigma = 1$ and (3.8), we have

$$\partial_t v_1 - 2c\partial_z v_1 + \partial_z^3 v_1 + 3\partial_z^{-1}\partial_y^2 v_1 = \partial_z(N_{1,1} + N_{1,2}) + N_{1,3}, \tag{4.1}$$

where $N_{1,1} = -3v_1^2$, $N_{1,2} = \{x_t - 2c - 3(x_y)^2\}v_1$ and $N_{1,3} = 6\partial_y(x_y v_1) - 3x_{yy}v_1$, and

$$\begin{cases} \partial_t v_2 = \mathcal{L}_c v_2 + \ell + \partial_z(N_{2,1} + N_{2,2} + N_{2,4}) + N_{2,3}, \\ v_2(0) = v_{2,*}, \end{cases} \tag{4.2}$$

where $\mathcal{L}_c v = -\partial_z(\partial_z^2 - 2c + 6\varphi_c)v - 3\partial_z^{-1}\partial_y^2$, $\ell = \sum_{k=1}^2 \ell_k$, $\ell_k = \sum_{j=1}^3 \ell_{kj}$ ($k = 1, 2$), $\tilde{\psi}_c(z) = \psi_{c,L}(z + 3t)$ and

$$\ell_{11} = (x_t - 2c - 3(x_y)^2)\varphi_c' - (c_t - 6c_y x_y)\partial_c\varphi_c, \quad \ell_{12} = 3x_{yy}\varphi_c,$$

$$\ell_{13} = 3c_{yy}\int_z^\infty \partial_c\varphi_c(z_1)\, dz_1 + 3(c_y)^2 \int_z^\infty \partial_c^2\varphi_c(z_1)\, dz_1,$$

$$\ell_{21} = (c_t - 6c_y x_y)\partial_c\tilde{\psi}_c - (x_t - 4 - 3(x_y)^2)\tilde{\psi}_c',$$

$$\ell_{22} = (\partial_z^3 - \partial_z)\tilde{\psi}_c - 3\partial_z(\tilde{\psi}_c^2) + 6\partial_z(\varphi_c\tilde{\psi}_c) - 3x_{yy}\tilde{\psi}_c,$$

$$\ell_{23} = -3c_{yy}\int_z^\infty \partial_c\tilde{\psi}_c(z_1)\,dz_1 - 3(c_y)^2\int_z^\infty \partial_c^2\tilde{\psi}_c(z_1)\,dz_1,$$

$$N_{2,1} = -3(2v_1v_2 + v_2^2), \quad N_{2,2} = \{x_t - 2c - 3(x_y)^2\}v_2 + 6\tilde{\psi}_c v_2,$$

$$N_{2,3} = 6\partial_y(x_y v_2) - 3x_{yy}v_2, \quad N_{2,4} = 6(\tilde{\psi}_c - \varphi_c)v_1.$$

Differentiating (3.10) with respect to t and substituting (4.2) into the resulting equation, we have in $L^2(-\eta_0, \eta_0)$

$$\frac{d}{dt}\int_{\mathbb{R}^2} v_2(t,z,y)g_k^*(z,\eta,c(t,y))e^{-iy\eta}\,dzdy$$
$$= \int_{\mathbb{R}^2}\ell g_k^*(z,\eta,c(t,y))e^{-iy\eta}\,dzdy + \sum_{j=1}^{6}II_k^j(t,\eta) = 0, \tag{4.3}$$

where

$$II_k^1 = \int_{\mathbb{R}^2} v_2(t,z,y)\mathcal{L}_{c(t,y)}^*(g_k^*(z,\eta,c(t,y))e^{-iy\eta})\,dzdy,$$

$$II_k^2 = -\int_{\mathbb{R}^2} N_{2,1}\partial_z g_k^*(z,\eta,c(t,y))e^{-iy\eta}\,dzdy,$$

$$II_k^3 = \int_{\mathbb{R}^2} N_{2,3}g_k^*(z,\eta,c(t,y))e^{-iy\eta}\,dzdy$$
$$+ 6\int_{\mathbb{R}^2} v_2(t,z,y)c_y(t,y)x_y(t,y)\partial_c g_k^*(z,\eta,c(t,y))e^{-iy\eta}\,dzdy,$$

$$II_k^4 = \int_{\mathbb{R}^2} v_2(t,z,y)\left(c_t - 6c_y x_y\right)(t,y)\partial_c g_k^*(z,\eta,c(t,y))e^{-iy\eta}\,dzy,$$

$$II_k^5 = -\int_{\mathbb{R}^2} N_{2,2}\partial_z g_k^*(z,\eta,c(t,y))e^{-iy\eta}\,dzdy,$$

$$II_k^6 = -\int_{\mathbb{R}^2} N_{2,4}\partial_z g_k^*(z,\eta,c(t,y))e^{-iy\eta}\,dzdy.$$

Using the fact that $g_1^*(z,\eta,c) \simeq \varphi_c(z)$ and $g_2^*(z,\eta,c) \simeq (c/2)^{3/2}\int_{-\infty}^z \partial_c\varphi_c$ for $\eta \simeq 0$, we derive the modulation equations for $c(t,y)$ and $x(t,y)$ (see [23, Section 4]).

To write down the modulation equation, let us introduce several notations. Let R^j, $\tilde{R}^j$, $\tilde{S}_j$, $\bar{S}_j$, $\tilde{\mathcal{A}}_1(t)$, B_j and $\tilde{\mathcal{C}}_j$ be the same as those in [23, pp. 165–168] except for the definitions of R^4 and R^5. We move a part of R^4 into R^5. See (B.1) and (B.2) in Appendix B.

Note that

$$b(t,\cdot) := \frac{1}{3}\tilde{P}_1\left\{\sqrt{2}c(t,\cdot)^{3/2} - 4\right\} = \tilde{c}(t,\cdot) + O(\tilde{c}^2), \tag{4.4}$$

where $\widetilde{P}_1 f = \mathcal{F}_\eta^{-1} \mathbf{1}_{[-\eta_0, \eta_0]} \mathcal{F}_y f$ and $\mathbf{1}_{[-\eta_0, \eta_0]}$ is a characteristic function of $[-\eta_0, \eta_0]$. We make use of (4.4) to translate the nonlinear term $c_y x_y$ into a divergence form.

Now let us introduce localized norms of $v(t)$. Let $p_\alpha(z) = 1 + \tanh \alpha z$ and $\|v\|_{W(t)} = \|p_\alpha(z + 3t + L)^{1/2} v\|_{L^2(\mathbb{R}^2)}$. Assuming the smallness of the following quantities, we can derive modulation equations of $b(t, y)$ and $x(t, y)$ for $t \in [0, T]$. Let $0 \leqslant T \leqslant \infty$ and

$$\mathbb{M}_{c,x}(T) = \sum_{k=0,1} \sup_{t \in [0,T]} \left\{ \langle t \rangle^{(2k+1)/4} (\|\partial_y^k \tilde{c}(t)\|_Y + \|\partial_y^{k+1} x(t)\|_Y) \right.$$
$$\left. + \langle t \rangle (\|\partial_y^2 \tilde{c}(t)\|_Y + \|\partial_y^3 x(t)\|_Y) \right\},$$

$$\mathbb{M}_v(T) = \sup_{t \in [0,T]} \|v(t)\|_{L^2},$$

$$\mathbb{M}_1(T) = \sup_{t \in [0,T]} \{\langle t \rangle^2 \|v_1(t)\|_{W(t)} + \langle t \rangle \|(1 + z_+) v_1(t)\|_{W(t)}\} + \|\mathcal{E}(v_1)^{1/2}\|_{L^2(0,T;W(t))},$$

$$\mathbb{M}'_1(\infty) = \sup_{t \geqslant 0} \|\mathcal{E}(\tilde{v}_1(t))^{1/2}\|_{L^2(\mathbb{R}^2)},$$

$$\mathbb{M}_2(T) = \sup_{0 \leqslant t \leqslant T} \langle t \rangle^{3/4} \|v_2(t)\|_X + \|\mathcal{E}(v_2)^{1/2}\|_{L^2(0,T;X)},$$

where $\mathcal{E}(v) = (\partial_x v)^2 + (\partial_x^{-1} \partial_y v)^2 + v^2$. We remark that by an anisotropic Sobolev inequality (see e.g. [2]),

$$\|v\|_{L^2(\mathbb{R}^2)} + \|v\|_{L^6(\mathbb{R}^2)} \lesssim \|\mathcal{E}(v)^{1/2}\|_{L^2(\mathbb{R}^2)}. \tag{4.5}$$

We can prove the following result exactly in the same way as [23, Proposition 3.9].

Proposition 4.1 *There exists a* $\delta_2 > 0$ *such that if* $\mathbb{M}_{c,x}(T) + \mathbb{M}_2(T) + \eta_0 + e^{-\alpha L} < \delta_2$ *for a* $T \geqslant 0$, *then*

$$\begin{pmatrix} b_t \\ \tilde{x}_t \end{pmatrix} = \mathcal{A}(t) \begin{pmatrix} b \\ \tilde{x} \end{pmatrix} + \sum_{i=1}^{6} \mathcal{N}^i, \tag{4.6}$$

$$b(0, \cdot) = b_*, \quad x(0, \cdot) = x_*, \tag{4.7}$$

where $b_* = 4/3 \widetilde{P}_1 \{(c_*/2)^{3/2} - 1\}$, $\mathcal{A}(t) = \mathcal{A}_* + B_4^{-1} \widetilde{\mathcal{A}}_1(t) + \partial_y^4 \mathcal{A}_1(t) + \partial_y^2 \mathcal{A}_2(t)$,

$$\mathcal{A}_1(t) = -B_4^{-1} (\widetilde{S}_1 B_1^{-1} B_2 + \widetilde{S}_0), \quad \mathcal{A}_2(t) = B_4^{-1} \widetilde{S}^3 B_1^{-1} B_2,$$

$$\mathcal{N}^1 = \widetilde{P}_1 \begin{pmatrix} 6(b\tilde{x}_y)_y \\ 2(\tilde{c} - b) + 3(\tilde{x}_y)^2 \end{pmatrix}, \quad \mathcal{N}^2 = \mathcal{N}^{2a} + \mathcal{N}^{2b},$$

$$\mathcal{N}^{2a} = B_3^{-1} \left(\widetilde{P}_1 R_1^7 \mathbf{e}_1 + \widetilde{R}^1 + \widetilde{R}^3 \right), \quad \mathcal{N}^{2b} = B_3^{-1} \widetilde{P}_1 R_2^7 \mathbf{e}_2, \quad \mathbf{e}_1 = \begin{pmatrix} 1 \\ 0 \end{pmatrix}, \quad \mathbf{e}_2 = \begin{pmatrix} 0 \\ 1 \end{pmatrix},$$

$$\mathcal{N}^3 = B_3^{-1}\partial_y(\widetilde{R}^2 + \widetilde{R}^4), \quad \mathcal{N}^4 = (B_3^{-1} - B_4^{-1})(B_2 - \partial_y^2\widetilde{S}_0)\partial_y\begin{pmatrix} b_y \\ x_y \end{pmatrix},$$

$$\mathcal{N}^5 = (B_3^{-1} - B_4^{-1})\widetilde{A}_1(t)\begin{pmatrix} b \\ \widetilde{x} \end{pmatrix}, \quad \mathcal{N}^6 = B_3^{-1}R^{v_1}.$$

We remark that $\mathcal{N}^6$ equals to $\mathcal{N}^5$ in [23] and that $\mathcal{A}(t) + \mathcal{N}^5$ equals to $\mathcal{A}(t)$ in [23]. The other terms are exactly the same.

To apply (2.17) and (2.20) to (4.6) in Sect. 9, we need to decompose b_* into a sum of an integrable function and a function that belongs to $\partial_y^2 Y_1$. Note that $\widetilde{P}_1 L^1(\mathbb{R}) \subset Y_1 \subset Y$ and $Y \subset \cap_{k\geqslant 0} H^k(\mathbb{R})$.

Lemma 4.2 *There exist $\overset{\bullet}{b} \in L^1(\mathbb{R})$ and $\overset{\circ}{b} \in Y_1$ such that $b_* = \widetilde{P}_1 \overset{\bullet}{b} + \partial_y^2 \overset{\circ}{b}$ and*

$$\|\overset{\bullet}{b}\|_{L^1(\mathbb{R})} + \|\overset{\circ}{b}\|_{Y_1} \lesssim \|\langle x\rangle\langle y\rangle v_0\|_{L^2(\mathbb{R}^2)}.$$

Proof Since $b_* - \tilde{c}_* = \frac{4}{3}\widetilde{P}_1 \left\{(c_*/2)^{3/2} - 1 - \frac{3}{4}\tilde{c}_*\right\}$ and $\left\|(c_*/2)^{3/2} - 1 - \frac{3}{4}c_*\right\|_{L^1(\mathbb{R})}$
$\lesssim \|\tilde{c}_*\|_Y^2 \lesssim \|\langle x\rangle v_0\|_{L^2(\mathbb{R}^2)}^2$, it suffices to show that there exist $\overset{\bullet}{c} \in L^1(\mathbb{R})$ and $\overset{\circ}{c} \in Y_1$ such that $\tilde{c}_* = \overset{\bullet}{c} + \partial_y^2 \overset{\circ}{c}$ and

$$\|\overset{\bullet}{c}\|_{L^1(\mathbb{R})} + \|\overset{\circ}{c}\|_{Y_1} \lesssim \|\langle x\rangle\langle y\rangle v_0\|_{L^2(\mathbb{R}^2)}.$$

Let

$$F_{10}[u, \tilde{c}, \gamma, L](\eta) := \frac{1}{2}\mathbf{1}_{[-\eta_0, \eta_0]}(\eta)\int_{\mathbb{R}^2} \{u(x, y) + \Phi[\tilde{c}, \gamma](x, y)\}\varphi_{c(y)}(x - \gamma(y))e^{-iy\eta}\,dxdy,$$

$$F_{11}[u, \tilde{c}, \gamma, L](\eta) := \mathbf{1}_{[-\eta_0, \eta_0]}(\eta)\int_{\mathbb{R}^2} \{u(x, y) + \Phi[\tilde{c}, \gamma](x, y)\}g_{k1}^*(x - \gamma(y), \eta, c(y))e^{-iy\eta}\,dxdy$$

where $c(y) = 2 + \tilde{c}(y)$ and $\Phi[\tilde{c}, \gamma](x, y) = \varphi(x) - \varphi_{c(y)}(x - \gamma(y)) + \psi_{c(y),L}(x - \gamma(y))$. Then

$$F_1[u, \tilde{c}, \gamma, L](\eta) = F_{10}[u, \tilde{c}, \gamma, L](\eta) + \eta^2 F_{11}[u, \tilde{c}, \gamma, L](\eta),$$

and we can prove

$$\|F_{11}[u, \tilde{c}, \gamma, L]\|_{Z_1} \lesssim \|u\|_{X_1} + \|\tilde{c}\|_{Y_1} + \|\gamma\|_{Y_1} + \|u\|_X(\|\tilde{c}\|_Y + \|\gamma\|_Y)$$

in exactly the same way as the proof of [22, Lemma 5.1].

Let $w_0(x, y) = \varphi_{c_1(y)}(x) - \varphi(x)$. Since $F_1[w_0, \tilde{c}_*, x_*, L] = 0$ and $(w_0, \tilde{c}_*, x_*) \in X_1 \times Y_1 \times Y_1$,

$$F_{10}[w_0, \tilde{c}_*, x_*, L](\eta) = -\eta^2 F_{11}[w_0, \tilde{c}_*, x_*, L](\eta), \tag{4.8}$$

$$F_{11}[w_0, \tilde{c}_*, x_*, L] \in Z_1. \tag{4.9}$$

Let

$$\Phi_0(x, y) = -\tilde{c}_*(y)\{\partial_c \varphi(x) - \psi(x + L)\} + x_*(y)\varphi'(x),$$

$$\Psi_1(x, y) = \frac{1}{2}\{\varphi_{c_*(y)}(x - x_*(y)) - \varphi(x)\}.$$

Then $\mathcal{F}_\eta^{-1} F_{10}[w_0, \tilde{c}_*, x_*, L](y) = (2\pi)^{1/2} \widetilde{P}_1(J_0 + J_1 + J_2 + J_3)$, where

$$J_0 = \frac{1}{2}\int_\mathbb{R} \Phi_0(x, y)\varphi(x)\,dx = \left(-1 + \frac{1}{2}\int_\mathbb{R} \varphi(x)\psi(x+L)\,dx\right)\tilde{c}_*,$$

$$J_1 = \frac{1}{2}\int_\mathbb{R} w_0(x, y)\varphi_{c_*(y)}(x - x_*(y))\,dx, \quad J_2 = \int_\mathbb{R} \Phi[\tilde{c}_*, x_*](x, y)\Psi_1(x, y)\,dx,$$

$$J_3 = \frac{1}{2}\int_\mathbb{R} \{\Phi[\tilde{c}_*, x_*](x, y) - \Phi_0(x, y)\}\varphi(x)\,dx,$$

and

$$\|J_1\|_{L^1(\mathbb{R})} \lesssim \|w_0\|_{L^1(\mathbb{R}^2)} \lesssim \|\langle y\rangle(c_1 - 2)\|_{L^2(\mathbb{R})},$$

$$\|J_2\|_{L^1(\mathbb{R})} \lesssim \|\Phi\|_{L^2(\mathbb{R}^2)} \|\Psi_1\|_{L^2(\mathbb{R}^2)} \lesssim (\|\tilde{c}_*\|_Y + \|x_*\|_Y)^2,$$

$$\|J_3\|_{L^1(\mathbb{R})} \lesssim \|\Phi - \Phi_0\|_{L^1(\mathbb{R}^2)} \lesssim (\|\tilde{c}_*\|_Y + \|x_*\|_Y)^2.$$

Combining the above with Lemma 3.1 and (3.11), we obtain Lemma 4.2. □

5 A Priori Estimates for Modulation Parameters

In this section, we will estimate $\mathbb{M}_{c,x}(T)$ assuming smallness of $\mathbb{M}_1(T)$, $\mathbb{M}_2(T)$, $\mathbb{M}_v(T)$, η_0 and $e^{-\alpha L}$.

Lemma 5.1 *There exist positive constants δ_3 and C such that if $\mathbb{M}_{c,x}(T) + \mathbb{M}_1(T) + \mathbb{M}_2(T) + \eta_0 + e^{-\alpha L} \leq \delta_3$, then*

$$\mathbb{M}_{c,x}(T) \leq C\|\langle x\rangle(\langle x\rangle + \langle y\rangle)v_0\|_{L^2} + C(\mathbb{M}_1(T) + \mathbb{M}_2(T)^2). \tag{5.1}$$

To prove Lemma 5.1, we need the following.

Claim 5.1 Let δ_2 be as in Proposition 4.1. Suppose $\mathbb{M}_{c,x}(T) + \mathbb{M}_1(T) + \mathbb{M}_2(T) + \eta_0 + e^{-\alpha L} < \delta_2$ for a $T \geq 0$. Then for $t \in [0, T]$,

$$\|c_t\|_Y + \|x_t - 2c - 3(x_y)^2\|_Y \lesssim (\mathbb{M}_{c,x}(T) + \mathbb{M}_1(T) + \mathbb{M}_2(T)^2)\langle t\rangle^{-3/4}.$$

Proof In view of (4.6),

$$\|c_t(t)\|_Y + \|x_t - 2c - 3(x_y)^2\|_Y \lesssim I + \sum_{2 \leq i \leq 6} \|\mathcal{N}^i\|_Y,$$

$$I = \|b_t - c_t\|_Y + \|\widetilde{\mathcal{A}}_1(t)(b, \tilde{x})\|_Y + \|b_{yy}\|_Y + \|x_{yy}\|_Y + \|(b\tilde{x}_y)_y\|_Y + \|(I - \widetilde{P}_1)x_y^2\|_Y,$$

and it follows from Claim A.5, [22, Claim D.6] and the definition of $\mathbb{M}_{c,x}(T)$ that

$$I \lesssim \mathbb{M}_{c,x}(T)\langle t\rangle^{-3/4} \quad \text{for } t \in [0, T].$$

See the proof of Lemma 5.2 in [23]. Following the line of [22, Chapter 7] and using (5.18), we can prove that for $t \in [0, T]$,

$$\sum_{i=2}^{5} \|\mathcal{N}_i(t)\|_Y \lesssim (\mathbb{M}_{c,x}(T) + \mathbb{M}_1(T) + \mathbb{M}_2(t))^2 \langle t\rangle^{-1}.$$

Combining the above with (5.21), we have Claim 5.1. □

To deal with $E_1 \mathcal{N}^6$, we decompose $\chi(D_y)B_3^{-1}$ and $\chi(D_y)B_4^{-1}$ into a sum of operators that belong to $B(L^1(\mathbb{R}))$ and operators that belong to $\partial_y^2 B(Y_1)$. Since

$$B_3 = B_1 + \widetilde{\mathcal{C}}_1 + \partial_y^2(\bar{S}_1 + \bar{S}_2) - \bar{S}_3 - \bar{S}_4 - \bar{S}_5,$$

$$B_4 = B_1 + \partial_y^2 \widetilde{S}_1 - \widetilde{S}_3 = B_3|_{\tilde{c}=0, v_2=0},$$

we have

$$\overset{\bullet}{B}_4^{-1} = \overset{\bullet}{B}_4 - \partial_y^2 \overset{\circ}{B}_{14}, \quad \overset{\bullet}{B}_3^{-1} - \overset{\bullet}{B}_4^{-1} = \overset{\bullet}{B}_{34} - \partial_y^2 \overset{\circ}{B}_{34}, \tag{5.2}$$

$$\overset{\bullet}{B}_4 = (B_1 - \widetilde{S}_{31})^{-1}, \quad \overset{\circ}{B}_{14} = B_4^{-1}(\widetilde{S}_1 + \widetilde{S}_{32})\overset{\bullet}{B}_4, \tag{5.3}$$

$$\overset{\bullet}{B}_{34} = -\overset{\bullet}{B}_4(\widetilde{\mathcal{C}}_1 + \widetilde{S}_{31} - \bar{S}_{31} - \bar{S}_{41} - \bar{S}_{51})B_3^{-1}, \tag{5.4}$$

$$\overset{\circ}{B}_{34} = \overset{\circ}{B}_{14}(B_4 - B_3)B_3^{-1} + \overset{\bullet}{B}_4(\bar{S}_1 - \widetilde{S}_1 + \bar{S}_2 + \widetilde{S}_{32} - \bar{S}_{32} + \bar{S}_{42} + \bar{S}_{52}) \tag{5.5}$$

where $\widetilde{S}_j$ and $\bar{S}_j$ are the same as those in [23, p. 167] and $\widetilde{S}_{j1}$, $\widetilde{S}_{j2}$, $\bar{S}_{j1}$ and $\bar{S}_{j2}$ are defined by (A.1)–(A.8) and (A.16)–(A.18) in Appendix A. We remark that $\widetilde{S}_j =$

$\widetilde{S}_{j1} - \partial_y^2 \widetilde{S}_{j2}$, that $\bar{S}_j = \bar{S}_{j1} - \partial_y^2 \bar{S}_{j2}$ and that $\widetilde{S}_{j1}$ is a time-dependent constant multiple of $\widetilde{P}_1$.

Claim 5.2 Let $0 \leqslant T \leqslant \infty$. There exist positive constants η_0, L_0 and C such that if $|\eta| \leqslant \eta_0$ and $L \geqslant L_0$, then for every $t \in [0, T]$,

$$\|\dot{B}_4\|_{B(Y) \cap B(Y_1)} + \|\overset{\circ}{B}_{14}\|_{B(Y) \cap B(Y_1)} \leqslant C, \tag{5.6}$$

$$\|\chi(D_y)\dot{B}_4\|_{B(L^1)} \leqslant C, \tag{5.7}$$

$$\|\dot{B}_4 - B_1^{-1}\|_{B(Y_1)} + \|\chi(D_y)(\dot{B}_4 - B_1^{-1})\|_{B(L^1)} \leqslant Ce^{-\alpha(3t+L)}. \tag{5.8}$$

Moreover, if $\mathbb{M}_{c,x}(T) + \mathbb{M}_2(T) \leqslant \delta$ is sufficiently small, then there exists a positive constant C_1 such that for $t \geqslant 0$,

$$\|\dot{B}_{34}\|_{B(Y,Y_1)} + \|\overset{\circ}{B}_{34}\|_{B(Y,Y_1)} \leqslant C_1(\mathbb{M}_{c,x}(T) + \mathbb{M}_2(T))\langle t \rangle^{-1/4}, \tag{5.9}$$

$$\|\chi(D_y)\dot{B}_{34}\|_{B(Y,L^1)} \leqslant C_1(\mathbb{M}_{c,x}(T) + \mathbb{M}_2(T))\langle t \rangle^{-1/4}, \tag{5.10}$$

$$\|[\partial_y, \dot{B}_{34}]\|_{B(Y,Y_1)} + \|\chi(D_y)[\partial_y, \dot{B}_{34}]\|_{B(Y,L^1)} \tag{5.11}$$
$$\leqslant C_1(\mathbb{M}_{c,x}(T) + \mathbb{M}_2(T))\langle t \rangle^{-3/4}.$$

Proof By [22, Claim 6.3] and [22, Claim B.1],

$$\sup_{t \geqslant 0} \|B_4^{-1}\|_{B(Y) \cap B(Y_1)} = O(1), \quad \|\widetilde{S}_1\|_{B(Y) \cap B(Y_1)} = O(1).$$

Combining the above with (A.20) and (A.22), we have (5.6).

Next, we will prove (5.7). Since $\chi(\eta)\chi_1(\eta) = \chi(\eta)$ and $[\chi_1(D_y), \widetilde{S}_{31}] = 0$,

$$\chi(D_y)\dot{B}_4 = \sum_{n \geqslant 0} \chi(D_y)(B_1^{-1}\chi_1(D_y)\widetilde{S}_{31})^n B^{-1}. \tag{5.12}$$

Hence it follows from (A.20) that

$$\|\chi(D_y)\dot{B}_4\|_{B(L^1)} \lesssim \|\check{\chi}\|_{L^1} \sum_{n \geqslant 0} \|B_1^{-1}\chi_1(D_y)\widetilde{S}_{31}\|_{B(L^1(\mathbb{R}))}^n = O(1).$$

We can prove (5.8) in the same way.

By [22, Claims 6.1 and 6.2], we have

$$\sup_{t \in [0,T]} \|B_3^{-1}\|_{B(Y)} < \infty \quad \text{and} \quad \|\widetilde{\mathcal{C}}_1\|_{B(Y,Y_1)} \lesssim \mathbb{M}_{c,x}(T)\langle t \rangle^{-1/4} \quad \text{for } t \in [0,T].$$

It follows from Claims 6.1, B.1–B.3 in [22] that

$$\|B_3 - B_4\|_{B(Y,Y_1)} + \|\bar{S}_1 - \widetilde{S}_1\|_{B(Y,Y_1)} \lesssim (\mathbb{M}_{c,x}(T) + \mathbb{M}_2(T))\langle t\rangle^{-1/4} \quad \text{for } t \in [0, T].$$

Combining the above with (5.6) and Claim A.4, we have (5.9).
Since

$$\chi(D_y)\dot{B}_{34} = -\chi(D_y)\dot{B}_4\chi_1(D_y)\left(\widetilde{\mathcal{C}}_1 + \widetilde{S}_{31} - \sum_{3\leqslant j\leqslant 5} \bar{S}_{j1}\right) B_3^{-1},$$

we have (5.10) from Claim A.4. Using the fact that

$$\|[\partial_y, \widetilde{\mathcal{C}}_1]\|_{B(Y,Y_1)} + \|\chi_1(D_y)[\partial_y, \widetilde{\mathcal{C}}_1]\|_{B(Y,L^2)} \lesssim \mathbb{M}_{c,x}(T)\langle t\rangle^{-3/4}$$

(see [22, Claim B.7]), we can prove (5.11) in the same way as (5.9) and (5.10). This completes the proof of Claim 5.2. □

Now we are in position to prove Lemma 5.1.

Proof of Lemma 5.1 Let $\mathcal{A}(t) = \mathrm{diag}(1, \partial_y)\mathcal{A}(t)\,\mathrm{diag}(1, \partial_y^{-1})$ and $\mathcal{U}(t, s)$ be the semigroup generated by $\mathcal{A}(t)$. Lemma 4.2 in [22] implies that there exists a $C = C(\eta_0, k) > 0$ such that

$$\|\partial_y^k \mathcal{U}(t,s) f\|_Y \leqslant C\langle t-s\rangle^{-k/2}\|f\|_Y \quad \text{for } t \geqslant s \geqslant 0, \tag{5.13}$$

$$\|\partial_y^k \mathcal{U}(t,s) f\|_Y \leqslant C\langle t-s\rangle^{-(2k+1)/4}\|f\|_{Y_1} \quad \text{for } t \geqslant s \geqslant 0, \tag{5.14}$$

provided η_0 is sufficiently small.

Multiplying (4.6) by $\mathrm{diag}(1, \partial_y)$ from the left, we have

$$\begin{cases} \partial_t \begin{pmatrix} b \\ x_y \end{pmatrix} = \mathcal{A}(t)\begin{pmatrix} b \\ x_y \end{pmatrix} + \sum_{i=1}^{6} \mathrm{diag}(1, \partial_y)\mathcal{N}^i, \\ b(0, \cdot) = b_*, \quad \partial_y x(0, \cdot) = \partial_y x_*. \end{cases} \tag{5.15}$$

Since $\|\mathcal{N}^6\|_{Y_1}$ does not necessarily decay as $t \to \infty$, we make use of the change of variable

$$k(t, y) = \frac{1}{2}\widetilde{P}_1\left(\int_{\mathbb{R}} v_1(t, z, y)\varphi_{c(t,y)}(z)\,dz\right), \quad S_{11}^3[\psi](t) = \frac{1}{2}\int_{\mathbb{R}^2} \psi(z+3t+L)\varphi(z)\,dz, \tag{5.16}$$

$$\widetilde{b}(t, y) = b(t, y) + \widetilde{k}(t, y), \quad \widetilde{k}(t, y) = (2 - S_{11}^3(t))^{-1}k(t, \cdot).$$

Then
$$\begin{cases} \partial_t \begin{pmatrix} \tilde{b} \\ x_y \end{pmatrix} = A(t) \begin{pmatrix} \tilde{b} \\ x_y \end{pmatrix} + \sum_{i=1}^{6} \operatorname{diag}(1, \partial_y) \mathcal{N}^i + \partial_t \tilde{k}(t) \mathbf{e_1} + A(t) \tilde{k}(t) \mathbf{e_1}, \\ \tilde{b}(0, \cdot) = b_* + \tilde{k}(0, \cdot), \quad \partial_y x(0, \cdot) = \partial_y x_*. \end{cases} \quad (5.17)$$

By (3.11) and (5.14),
$$\left\| \partial_y^k U(t, 0) \begin{pmatrix} \tilde{b}(0, \cdot) \\ x_y(0, \cdot) \end{pmatrix} \right\|_Y \lesssim \langle t \rangle^{-(2k+1)/4} (\|\tilde{b}(0, \cdot)\|_{Y_1} + \|\partial_y x_*\|_{Y_1}),$$

and
$$\|\tilde{b}(0)\|_{Y_1} + \|\partial_y x_*\|_{Y_1} \lesssim \|b_*\|_{Y_1} + \|\langle y \rangle v_*\|_{L^2} + \eta_0 \|x_*\|_{Y_1} + \|\varphi_{c_*(y)}(x - x_*(y)) v_*\|_{L^1(\mathbb{R}^2)}$$
$$\lesssim \|\langle x \rangle (\langle x \rangle + \langle y \rangle) v_0\|_{L^2}.$$

Except for $\mathcal{N}^6$, the term which includes v_1 in (5.15) and needs to be treated differently from [22, (6.16)] is $\mathcal{N}^{2a}$ because $\tilde{R}^1$ includes R^4 and R^4 includes the inverse Fourier transform of $\mathbf{1}_{[-\eta_0, \eta_0]}(\eta) I I_k^2(t, \eta)$. But thanks to the smallness of $\mathbb{M}_1(T)$,

$$\|II_k^2\|_{L^\infty[-\eta_0, \eta_0]} = \sup_{\eta \in [-\eta_0, \eta_0]} \left| \int_{\mathbb{R}^2} N_{2,1} \partial_z g_k^*(z, \eta, c(t, y)) e^{-iy\eta} \, dz dy \right|$$
$$\lesssim (\|p_\alpha(z) v_1\|_{L^2} \|v_2\|_X + \|v_2\|_X^2) \sup_{c \in [2-\delta, 2+\delta], \eta \in [-\eta_0, \eta_0]} \|e^{-2\alpha z} g_k^*(z, \eta, c)\|_{L_z^\infty}$$
$$\lesssim (\mathbb{M}_1(T) + \mathbb{M}_2(T))^2 \langle t \rangle^{-3/2}, \quad (5.18)$$

and $\|R^4\|_{Y_1}$ decays at the rate as in [22, Claim D.5]. Using (5.13), (5.14) and (5.18), we can prove that for $t \in [0, T]$ and $k = 0, 1, 2$,

$$\sum_{i=1}^{5} \int_0^t \|\partial_y^k U(t, s) \operatorname{diag}(1, \partial_y) \mathcal{N}^j(s)\|_Y \, ds \lesssim (\mathbb{M}_{c,x}(T) + \mathbb{M}_1(T) + \mathbb{M}_2(T))^2 \langle t \rangle^{-\min\{1, (2k+1)/4\}}.$$

in the same way as [22, Chapter 7].

Since $\operatorname{diag}(1, \partial_y) \mathcal{N}^6 = E_1 \mathcal{N}^6 + \partial_y E_2 \mathcal{N}^6$, it follows from (5.13)

$$\mathfrak{N}^k := \left\| \int_0^t \|\partial_y^k U(t, s) \partial_y E_2 \mathcal{N}^6(s)\|_Y \, ds \right\|_Y \lesssim \int_0^t \langle t - s \rangle^{-(k+1)/2} \|\mathcal{N}^6(s)\|_Y \, ds. \quad (5.19)$$

Using the fact that

$$\sup_{\eta\in[-\eta_0,\eta_0]} \Big(|\varphi_{c(t,y)}(z)\partial_z g_k^*(z,\eta,c(t,y)) - \varphi(z)\partial_z g_k^*(z,\eta)|$$
$$+|\tilde{\psi}_{c(t,y)}(z)\partial_\tau g_k^*(z,\eta,c(t,y))|\Big)$$
$$\lesssim e^{-2\alpha|z|}|\tilde{c}(t,y)|,$$

we see that

$$\|R^{v_1}(t)\|_Y$$
$$\lesssim \sum_{k=1,2}\left\|\int_{\mathbb{R}^2}\varphi(z)v_1(t,z,y)\partial_z g_k^*(z,\eta)e^{-iy\eta}\,dzdy\right\|_{L^2(-\eta_0,\eta_0)} \quad (5.20)$$
$$+ \|\tilde{c}\|_Y \|e^{-\alpha|z|}v_1(t)\|_{L^2(\mathbb{R}^2)}$$
$$\lesssim \|e^{-\alpha|z|}v_1(t)\|_{L^2(\mathbb{R}^2)} \lesssim \langle t\rangle^{-2}\mathbb{M}_1(T) \quad t\in[0,T],$$

and that

$$\|\mathcal{N}^6(t)\|_Y \lesssim \langle t\rangle^{-2}\mathbb{M}_1(T) \quad \text{for } t\in[0,T], \quad (5.21)$$

follows from the boundedness of B_3^{-1} (see [23, Claim 4.5]). Combining (5.19) and (5.21), we have $\mathfrak{N}^k \lesssim \mathbb{M}_1(T)\langle t\rangle^{-(k+1)/2}$ for $t\in[0,T]$ and $k=0,1,2$.

Now we investigate

$$E_1\mathcal{N}^6 = E_1 B_3^{-1} R^{v_1} = E_1\{\dot{B}_4 + \dot{B}_{34} - \partial_y^2(\overset{\circ}{B}_{14} + \overset{\circ}{B}_{34})\}R^{v_1} \quad (5.22)$$

more precisely. We remark that $\|\mathcal{N}^6\|_{Y_1}$ cannot be expected to decay like $\|\mathcal{N}^6\|_Y$ as $t\to\infty$ because

$$\left\|\int_{\mathbb{R}}\varphi(z)v_1(t,z,y)\,dz\right\|_{L^1(\mathbb{R}_y)}$$

does not necessarily decay as $t\to\infty$. By (5.20) and Claim 5.2,

$$\|\dot{B}_4 R^{v_1}\|_Y + \|(\overset{\circ}{B}_{14} + \overset{\circ}{B}_{34})R^{v_1}\|_Y \lesssim \mathbb{M}_1(T)\langle t\rangle^{-2}, \quad (5.23)$$

$$\|\chi(D_y)\dot{B}_{34}R^{v_1}\|_{L^1} + \|\dot{B}_{34}R^{v_1}\|_{Y_1} \lesssim \mathbb{M}_1(T)(\mathbb{M}_{c,x}(T) + \mathbb{M}_2(T))\langle t\rangle^{-9/4}. \quad (5.24)$$

In view of (5.13), (5.14) and (5.22)–(5.24), we see that for $k = 0, 1, 2$ and $t \in [0, T]$,

$$\left\| \int_0^t \partial_y^k U(t,s) E_1 \{ \dot{B}_{34} - \partial_y^2 (\mathring{B}_{14} + \mathring{B}_{34}) \} R^{v_1} \, ds \right\|_Y$$
$$\lesssim \int_0^t \langle t-s \rangle^{-(2k+1)/4} \| \dot{B}_{34} R^{v_1} \|_{Y_1} \, ds + \int_0^t \langle t-s \rangle^{-(k+2)/2} \| (\mathring{B}_{14} + \mathring{B}_{34}) \} R^{v_1} \|_Y \, ds$$
$$\lesssim \mathbb{M}_1(T) \langle t \rangle^{-(2k+1)/4},$$

and that $E_1 \dot{B}_4 R^{v_1}$ is the hazardous part of $E_1 \mathcal{N}^6$.

The worst part of $E_1 \dot{B}_4 R^{v_1}$ can be expressed as a time derivative of a decaying function as in [23]. The operator $B_1 - \widetilde{S}_{31}$ and its inverse $\dot{B}_4$ are lower triangular on $Y \times Y$ and

$$E_1 \dot{B}_4 R^{v_1} = (2 - S_{11}^3[\psi])^{-1} R_1^{v_1} \mathbf{e_1}. \tag{5.25}$$

In view of [23, pp. 175–176],

$$R_1^{v_1} = S_1^7 [\partial_c \varphi_c](c_t) - S_1^7 [\varphi_c'](x_t - 2c - 3(x_y)^2) - k_t + \dot{R}_{v_1} + \partial_y \mathring{R}_{v_1}, \tag{5.26}$$

where

$$S_1^7 [q_c](f)(t, y) = \frac{1}{2} \widetilde{P}_1 \left(\int_{\mathbb{R}} v_1(t, z, y) f(y) q_{c(t,y)}(z) \, dz \right),$$

and $\dot{R}_{v_1}$ and $\mathring{R}_{v_1}$ are chosen such that $\dot{R}_{v_1} + \partial_y \mathring{R}_{v_1} = R_{11}^{v_1} + \partial_y R_{12}^{v_1}$ and that

$$\chi(D_y) \dot{R}_{v_1} \in L^1(0, \infty; L^1(\mathbb{R})).$$

We give the precise definitions of $\dot{R}_{v_1}$ and $\mathring{R}_{v_1}$ later.

The term $\partial_t \widetilde{k} \mathbf{e_1}$ in (5.17) cancels out with a bad part of $E_1 \dot{B}_4 R^{v_1}$ which comes from $-k_t$ in (5.26). In fact,

$$\partial_t \widetilde{k}(t, y) - (2 - S_{11}^3[\psi](t))^{-1} \partial_t k(t, y) = (2 - S_{11}^3[\psi](t))^{-2} \partial_t S_{11}^3[\psi](t) k(t, y),$$

and by the definition,

$$|S_{11}^3[\psi](t)| + \left| \partial_t S_{11}^3[\psi](t) \right| \lesssim e^{-2(3t+L)} \quad \text{for } t \geq 0. \tag{5.27}$$

Combining Claim C.2 and (5.27) with (5.14), we have for $t \in [0, T]$ and $k \geqslant 0$,

$$\left\| \int_0^t \partial_y^k U(t, s) \left\{ \partial_t \tilde{k}(s, \cdot) - (2 - S_{11}^3[\psi](s))^{-1} \partial_t k(s, \cdot) \right\} \mathbf{e}_1 \, ds \right\|_Y$$
$$\lesssim \mathbb{M}_1(T) \int_0^t \langle t - s \rangle^{-(2k+1)/4} \langle s \rangle e^{-2(3s+L)} \, ds \lesssim \mathbb{M}_1(T) \langle t \rangle^{-(2k+1)/4}.$$

Next, we will investigate $\dot{R}_{v_1}$ and $\overset{\circ}{R}_{v_1}$. We write II_{13}^6 in [23, p. 175] as $II_{13}^6 = II_{131}^6 + \eta^2 II_{132}^6$,

$$II_{131}^6 = 3 \int_{\mathbb{R}^2} v_1(t, z, y) \tilde{\psi}_{c(t,y)}(z) \varphi_{c(t,y)}(z) e^{-iy\eta} \, dz dy,$$

$$II_{132}^6 = 6 \int_{\mathbb{R}^2} v_1(t, z, y) \tilde{\psi}_{c(t,y)}(z) \partial_z g_{11}^*(z, \eta, c(t, y)) e^{-iy\eta} \, dz dy,$$

$$g_{k1}^*(z, \eta, c) = \frac{g_k^*(z, \eta, c) - g_k^*(z, 0, c)}{\eta^2},$$

because $\partial_z g_1^*(z, 0, c(t, y)) = \frac{1}{2} \varphi_{c(t,y)}(z)$ and let

$$\dot{R}_{v_1} = \frac{1}{2\pi} \int_{-\eta_0}^{\eta_0} \left\{ II_{111}^6(t, \eta) - II_{131}^6(t, \eta) \right\} e^{iy\eta} \, d\eta$$
$$= R_{11}^{v_1} - \frac{\partial_y^2}{2\pi} \int_{-\eta_0}^{\eta_0} II_{132}^6(t, \eta) e^{iy\eta} \, d\eta,$$

$$\overset{\circ}{R}_{v_1} = \frac{1}{2\pi} \int_{-\eta_0}^{\eta_0} \left\{ II_{112}^6(t, \eta) - i\eta II_{12}^6(t, \eta) + i\eta II_{132}^6(t, \eta) \right\} e^{iy\eta} \, d\eta.$$

Then

$$\dot{R}_{v_1} = \frac{3}{2} \widetilde{P}_1 \int_{\mathbb{R}} v_1(t, z, y)^2 \varphi'_{c(t,y)}(z) \, dz - 3\widetilde{P}_1 \int_{\mathbb{R}} v_1(t, z, y) \tilde{\psi}_{c(t,y)}(z) \varphi'_{c(t,y)}(z) \, dz$$
$$+ \frac{3}{2} \widetilde{P}_1 \left[\int_{\mathbb{R}} v_1(t, z, y) \left\{ c_{yy}(t, y) \int_{-\infty}^z \partial_c \varphi_{c(t,y)}(z_1) \, dz_1 \right. \right.$$
$$\left. \left. + c_y(t, y)^2 \int_{-\infty}^z \partial_c^2 \varphi_{c(t,y)}(z_1) \, dz_1 \right\} dz \right]$$
$$- \frac{3}{2} \widetilde{P}_1 \int_{\mathbb{R}} v_1(t, z, y) \left\{ x_{yy}(t, y) \varphi_{c(t,y)}(z) + 2(c_y x_y)(t, y) \partial_c \varphi_{c(t,y)}(z) \right\} dz,$$

$\overset{\circ}{R}_{v_1} = \overset{\circ}{R}_{v_1,1} + \partial_y \overset{\circ}{R}_{v_1,2}$ and

$$\overset{\circ}{R}_{v_1,1} = -\frac{3}{2}\widetilde{P}_1 \int_{\mathbb{R}} v_1(t,z,y) c_y(t,y) \left(\int_{-\infty}^{z} \partial_c \varphi_{c(t,y)}(z_1)\, dz_1 \right) dz$$
$$+ 3\widetilde{P}_1 \int_{\mathbb{R}} v_1(t,z,y) x_y(t,y) \varphi_{c(t,y)}(z)\, dz,$$
$$\overset{\circ}{R}_{v_1,2} = \frac{3}{2}\widetilde{P}_1 \int_{\mathbb{R}} v_1(t,z,y) \left(\int_{-\infty}^{z} \varphi_{c(t,y)}(z_1)\, dz_1 \right) dz$$
$$- \frac{1}{2\pi} \int_{-\eta_0}^{\eta_0} \left\{ II_{12}^6(t,\eta) - II_{132}^6(t,\eta) \right\} e^{iy\eta}\, d\eta,$$
$$II_{12}^6(t,\eta) = 6\int_{\mathbb{R}^2} v_1(t,z,y)\varphi_{c(t,y)}(z)\partial_z g_{11}^*(z,\eta,c(t,y)) e^{-iy\eta}\, dzdy,$$

and we have

$$\|\dot{R}_{v_1}\|_{Y_1} + \|\chi(D_y)\dot{R}_{v_1}\|_{L^1(\mathbb{R})} \lesssim \mathbb{M}_1(T)(\mathbb{M}_{c,x}(T) + \mathbb{M}_1(T))\langle t \rangle^{-3/2}, \quad (5.28)$$

$$\|\overset{\circ}{R}_{v_1,1}\|_{Y_1} \lesssim \mathbb{M}_{c,x}(T)\mathbb{M}_1(T)\langle t \rangle^{-7/4}, \quad \|\overset{\circ}{R}_{v_1,2}\|_{Y} \lesssim \mathbb{M}_1(T)\langle t \rangle^{-1}. \quad (5.29)$$

Combining the above with (5.13) and (5.14), we have

$$\left\| \int_0^t \partial_y^k U(t,s)(2 - S_{11}^3[\psi](s))^{-1} \left(\dot{R}_{v_1} + \partial_y \overset{\circ}{R}_{v_1} \right) ds \right\|_Y$$
$$\lesssim \int_0^t \langle t-s \rangle^{-(2k+1)/4} \|\dot{R}_{v_1}\|_{Y_1} ds$$
$$+ \int_0^t \langle t-s \rangle^{-(2k+3)/2} \|\overset{\circ}{R}_{v_1,1}\|_{Y_1} ds + \int_0^t \langle t-s \rangle^{-(k+4)/2} \|\overset{\circ}{R}_{v_1,2}\|_Y ds$$
$$\lesssim \mathbb{M}_1(T)\langle t \rangle^{-\min\{1,(2k+1)/4\}} \quad \text{for } k=0,1,2 \text{ and } t \in [0,T].$$

Next, we will estimate $S_1^7[\partial_c \varphi_c](c_t)$ and $S_1^7[\varphi'_c](x_t - 2c - 3(x_y)^2)$. By Claim 5.1,

$$\|S_1^7[\partial_c \varphi_c](c_t)\|_{Y_1} + \|\chi(D_y)S_1^7[\partial_c \varphi_c](c_t)\|_{L^1}$$
$$+ \|S_1^7[\varphi'_c](x_t - 2c - 3(x_y)^2)\|_{Y_1} + \|\chi(D_y)S_1^7[\varphi'_c](x_t - 2c - 3(x_y)^2)\|_{L^1}$$
$$\lesssim \mathbb{M}_1(T)(\mathbb{M}_{c,x}(T) + \mathbb{M}_1(T) + \mathbb{M}_2(T)^2)\langle t \rangle^{-11/4}. \quad (5.30)$$

Finally, we will estimate $A(t)\tilde{k}\mathbf{e}_1$. Since $\mathcal{A}_1(t)E_2 = O$ and $[\mathcal{A}_i(t), \partial_y] = O$ for $i = 1, 2$,

$$A(t) = A_* + \mathrm{diag}(\partial_y^3, \partial_y^4) A_1(t) \,\mathrm{diag}(\partial_y, 1) + \mathrm{diag}(\partial_y, \partial_y^2) A_2(t) \,\mathrm{diag}(\partial_y, 1) + A_3(t),$$

$$A_* = \begin{pmatrix} 3\partial_y^2 & 8\partial_y \\ (2-\mu_3 \partial_y^2)\partial_y & \partial_y^2 \end{pmatrix}, \quad A_3(t) = \mathrm{diag}(1, \partial_y) B_4^{-1} \widetilde{A}_1(t) E_1,$$

$$\sup_{t \geqslant 0} \|\partial_y^{-1}(A(t) - A_3(t))\|_{B(Y)} < \infty, \quad \|A_3(t)\|_{B(Y_1)} \lesssim e^{-\alpha(3t+L)}.$$

Combining the above with (5.13), (5.14) and Claims C.1 and C.2, we have for $k = 0, 1, 2$,

$$\left\| \int_0^t \partial_y^k U(t,s) A(s) \tilde{k}(s) \, ds \right\|_Y$$

$$\leqslant \int_0^t \|\partial_y^{k+1} U(t,s)\|_{B(Y)} \left\| \partial_y^{-1}(A(s) - A_3(s)) \right\|_{B(Y)} \|\tilde{k}(s)\|_Y \, ds$$

$$+ \int_0^t \|\partial_y^k U(t,s)\|_{B(Y_1, Y)} \|A_3(s)\|_{B(Y_1)} \|\tilde{k}(s)\|_{L^1} \, ds$$

$$\lesssim \mathbb{M}_1(T) \left\{ \int_0^t \langle t-s \rangle^{-(k+1)/2} \langle s \rangle^{-2} \, ds + \int_0^t \langle t-s \rangle^{-(2k+1)/4} \langle s \rangle e^{-\alpha(3s+L)} \, ds \right\}$$

$$\lesssim \mathbb{M}_1(T) \langle t \rangle^{-(2k+1)/4}.$$

This completes the proof of Lemma 5.1. □

6 The $L^2(\mathbb{R}^2)$ Estimate

In this section, we will estimate $\mathbb{M}_v(T)$ assuming smallness of $\mathbb{M}_{c,x}(T)$, $\mathbb{M}_1(T)$ and $\mathbb{M}_2(T)$.

Lemma 6.1 *Let δ_3 be the same as in Lemma 5.1. Suppose that $\mathbb{M}_{c,x}(T) + \mathbb{M}_1(T) + \mathbb{M}_2(T) + \eta_0 + e^{-\alpha L} \leqslant \delta_3$. Then there exists a positive constant C such that*

$$\mathbb{M}_v(T) \leqslant C(\|v_0\|_{L^2(\mathbb{R}^2)} + \mathbb{M}_{c,x}(T) + \mathbb{M}_1(T) + \mathbb{M}_2(T)).$$

To prove Lemma 6.1, we use a variant of the L^2 conservation law on v as in [22, 23].

Lemma 6.2 ([23, Lemma 6.2]) *Let $0 \leqslant T \leqslant \infty$. Let $\tilde{v}_1$ be a solution of (3.8) and v_2 be a solution of (4.2). Suppose that $(v_2(t), c(t), \gamma(t))$ satisfy (3.1), (3.9) and (3.10). Then*

$$Q(t, v) := \int_{\mathbb{R}^2} \left\{ v(t,z,y)^2 - 2\psi_{c(t,y),L}(z + 3t) v(t,z,y) \right\} dz\,dy$$

satisfies for $t \in [0, T]$,

$$Q(t, v) = Q(0, v) + 2 \int_0^t \int_{\mathbb{R}^2} \left(\ell_{11} + \ell_{12} + 6\varphi'_{c(s,y)}(z)\tilde{\psi}_{c(s,y)}(z) \right) v(s, z, y) \, dz \, dy \, ds$$

$$- 2 \int_0^t \int_{\mathbb{R}^2} \ell \psi_{c(t,y),L}(z + 3s) \, dz \, dy \, ds - 6 \int_0^t \int_{\mathbb{R}^2} \varphi'_{c(s,y)}(z) v(s, z, y)^2 \, dz \, dy \, ds$$

$$- 6 \int_0^t \int_{\mathbb{R}^2} (\partial_z^{-1} \partial_y v)(s, z, y) c_y(s, y) \partial_c \varphi_{c(t,y)}(z) \, dz \, dy.$$
(6.1)

Proof of Lemma 6.1 We can estimate the right hand side of (6.1) in exactly the same way as in the proof of [22, Lemma 8.1] except for the last term. By the definition, we have for $t \in [0, T]$,

$$\left| \int_{\mathbb{R}^2} (\partial_z^{-1} \partial_y v)(s, z, y) c_y(s, y) \partial_c \varphi_{c(t,y)}(z) \, dz \, dy \right|$$

$$\lesssim \| e^{-\alpha |z|} \partial_z^{-1} \partial_y v \|_{L^2(0,T;L^2(\mathbb{R}^2))} \| c_y \|_{L^2(0,T;Y)}$$

$$\lesssim \mathbb{M}_{c,x}(T)(\mathbb{M}_1(T) + \mathbb{M}_2(T)).$$

and

$$Q(t, v) + 8 \| \psi \|_{L^2}^2 \| \sqrt{c(t)} - \sqrt{c_0} \|_{L^2(\mathbb{R})}^2$$

$$\lesssim \| v_0 \|_{L^2(\mathbb{R}^2)}^2 + (\mathbb{M}_1(T) + \mathbb{M}_2(T) + \mathbb{M}_{c,x}(T))^2 \int_0^t \langle s \rangle^{-5/4} \, dt$$

$$\lesssim \| v_0 \|_{L^2(\mathbb{R}^2)}^2 + (\mathbb{M}_1(T) + \mathbb{M}_2(T) + \mathbb{M}_{c,x}(T))^2.$$

Combining the above with the fact that $Q(t, v) = \| v(t) \|_{L^2}^2 + O(\| \tilde{c}(t) \|_Y \| v(t) \|_{L^2})$, we have Lemma 6.1. Thus we complete the proof. □

7 Estimates for Small Solutions for the KP-II Equation

In this section, we will give upper bounds of $\mathbb{M}_1(T)$ and $\mathbb{M}'_1(\infty)$. First, we will prove decay estimates for v_1 assuming that $v_0(x, y)$ is polynomially localized as $x \to \infty$.

Lemma 7.1 *Let $\tilde{v}_1$ be a solution of (3.8). There exist positive constants C and δ_4 such that if $\| \langle x \rangle^2 v_0 \|_{L^2} + \mathbb{M}_{c,x}(T) + \mathbb{M}_1(T) + \mathbb{M}_2(T) < \delta_4$, then $\mathbb{M}_1(T) \leq C \| \langle x \rangle^2 v_0 \|_{L^2}$ for $t \in [0, T]$.*

To prove Lemma 7.1, we make use of the virial identity for the KP-II equation that was shown in [6]. Let u be a solution of (1.1) with $u(0) \in L^2(\mathbb{R}^2)$ and

$$I(t) = \int_{\mathbb{R}^2} p_\alpha(x - x(t)) u(t, x, y)^2 \, dxdy.$$

Suppose that $\inf_{t \geqslant 0} x'(t) > 0$. There exist positive constants α_0 and δ such that if $\alpha \in (0, \alpha_0)$ and $\|v_0\|_{L^2} < \delta$, then

$$I(t) + \int_0^t \int_{\mathbb{R}^2} p'_\alpha(x - x(s)) \mathcal{E}(u)(s, x, y) \, dxdyds \lesssim I(0). \tag{7.1}$$

See e.g. [26, Lemma 5.3] for the proof.

If $u(0)$ is small in $L^2(\mathbb{R}^2)$ and polynomially localized, we can prove time decay estimates by using (7.1).

Lemma 7.2 *Let $u(t)$ be a solution of (1.1). Let $0 \leqslant T \leqslant \infty$ and let $x(t)$ be a C^1 function satisfying $x(0) = 0$ and $\inf_{t \in [0,T]} \dot{x}(t) > c_1$ for a $c_1 > 0$. Suppose that $(1 + x_+)^\rho u(0) \in L^2(\mathbb{R}^2)$ for a $\rho \geqslant 0$. Then there exist positive constants α_0 and δ such that if $\alpha \in (0, \alpha_0)$ and $\|u(0)\|_{L^2(\mathbb{R}^2)} < \delta$, then*

$$\int_{\mathbb{R}^2} p_\alpha(x - x(t)) u(t, x, y)^2 \, dxdy \lesssim \langle t \rangle^{-2\rho} \|(1 + x_+)^\rho u(0)\|^2_{L^2(\mathbb{R}^2)}, \tag{7.2}$$

$$\int_0^T \int_{\mathbb{R}^2} p_\alpha(x - x(t)) \mathcal{E}(u)(t, x, y) \, dxdydt \lesssim \|(1 + x_+)^{1/2} p_\alpha(x)^{1/2} u(0)\|^2_{L^2(\mathbb{R}^2)}, \tag{7.3}$$

$$\int_{\mathbb{R}^2} (1 + x_+)^{2\rho_1} p_\alpha(x) u(t, x + x(t), y)^2 \, dxdy$$

$$\lesssim \langle t \rangle^{-2(\rho_2 - \rho_1)} \|(1 + x_+)^{\rho_2} u(0)\|^2_{L^2(\mathbb{R}^2)}, \tag{7.4}$$

where ρ_1 and ρ_2 are positive constants satisfying $\rho_2 > \rho_1 > 0$.

Proof We can prove (7.2) in the same way as in [24, Section 14.1]. Since $\min(1, e^{2\alpha x}) \leqslant p_\alpha(x) \leqslant 2 \min(1, e^{2\alpha x})$ and $p'_\alpha(x) = \alpha \operatorname{sech}^2 \alpha x = O(e^{-2\alpha |x|})$, it follows that for $x \leqslant 0$,

$$\sum_{j \geqslant 0} p_\alpha(x - j) \lesssim \begin{cases} \sum_{j \geqslant 0} e^{-2\alpha(|x|+j)} \lesssim p_\alpha(x) & \text{for } x \leqslant 0, \\ \sum_{0 \leqslant j \leqslant [x]} 1 + \sum_{j \geqslant [x]+1} e^{-2\alpha|x-j|} \lesssim 1 + x & \text{for } x \geqslant 0. \end{cases}$$

Similarly, we have $p_\alpha(x) \lesssim \sum_{j \geqslant 0} p'_\alpha(x - j)$. Hence it follows from (7.1) that for $t \in [0, T]$,

$$\int_0^t \int_{\mathbb{R}^2} p_\alpha(x - x(s)) \mathcal{E}(u)(s, x, y) \, dx\,dy\,ds$$

$$\lesssim \sum_{j=0}^\infty \int_0^t \int_{\mathbb{R}^2} p'_\alpha(x - x(s) - j) \mathcal{E}(u)(s, x, y) \, dx\,dy\,ds$$

$$\lesssim \sum_{j=0}^\infty \int_{\mathbb{R}^2} p_\alpha(x - j) u(0, x, y)^2 \, dx\,dy \lesssim \int_{\mathbb{R}^2} (1 + x_+) p_\alpha(x) u(0, x, y)^2 \, dx\,dy.$$

Finally, we will prove (7.2). Let c_1 and c_2 be constants satisfying $0 < c_1 < c_2 \leq \inf_{0 \leq t \leq T} \dot{x}(t)$ and let $q_\ell(x) = (1 + x_+)^{2\rho_\ell} p_\alpha(x)$ for $\ell = 1, 2$. Since

$$q_\ell(x) \simeq \sum_{j \geq 0} (1 + j)^{2\rho_\ell - 1} p_\alpha(x - j),$$

it follows from (7.1) that for $t \in [0, T]$,

$$\int_{\mathbb{R}^2} q_2(x - c_1 t) u(t, x, y)^2 \, dx\,dy \lesssim \int_{\mathbb{R}^2} q_2(x) u(0, x, y)^2, \qquad (7.5)$$

provided $\|u(0)\|_{L^2}$ is sufficiently small. Combining (7.5) with the fact that

$$q_1(x - x(t)) \leq q_1(x - c_2 t) \lesssim \langle t \rangle^{2(\rho_1 - \rho_2)} q_2(x - c_1 t),$$

we have (7.2). Thus we complete the proof. □

Now we are in position to prove Lemma 7.1.

Proof of Lemma 7.1 By Claim 5.1, there exists a $c_1 > 0$ such that $x_t(t, y) \geq c_1$ for every $t \in [0, T]$ and $y \in \mathbb{R}$. Hence it follows from Lemmas 3.1 and 7.2 that $\mathbb{M}_1(T) \lesssim \|(1 + x_+)^2 v_*\|_{L^2(\mathbb{R}^2)} \lesssim \|\langle x \rangle^2 v_0\|_{L^2(\mathbb{R}^2)}$. Thus we complete the proof. □

The scattering result by Hadac et al. [10] which uses U^p and V^p spaces introduced by [18, 19] implies that higher order Sobolev norms of solutions to (3.8) remain small for all the time.

Lemma 7.3 *Let $\tilde{v}_1(t)$ be a solution of (3.8). There exists positive constants δ_5 and C such that if $\|\langle x \rangle^2 v_0\|_{H^1(\mathbb{R}^2)} \leq \delta_5$, then $\mathbb{M}'_1(\infty) \leq C \|\langle x \rangle^2 v_0\|_{H^1(\mathbb{R}^2)}$ for every $t \in \mathbb{R}$.*

Proof It follows from [10, Proposition 3.1 and Theorem 3.2] that

$$\|\partial_x \tilde{v}_1(t)\|_{L^2(\mathbb{R}^2)} + \left\| |D_x|^{-1/2} \langle D_y \rangle^{1/2} \tilde{v}_1(t) \right\|_{L^2(\mathbb{R}^2)}$$

$$\lesssim \|\partial_x v_*\|_{L^2(\mathbb{R}^2)} + \left\| |D_x|^{-1/2} \langle D_y \rangle^{1/2} v_* \right\|_{L^2(\mathbb{R}^2)}$$

$$\lesssim \|\mathcal{E}(v_*)^{1/2}\|_{L^2(\mathbb{R}^2)}.$$

See e.g. [23, Section 7.2] for an explanation. Combining the above with the L^2-conservation law $\|\tilde{v}_1(t)\|_{L^2(\mathbb{R}^2)} = \|v_*\|_{L^2(\mathbb{R}^2)}$ and the Sobolev inequality (4.5), we have

$$\|\tilde{v}_1(t)\|_{L^3(\mathbb{R}^2)} \lesssim \left\||D_x|^{1/2}\tilde{v}_1(t)\right\|_{L^2(\mathbb{R}^2)}$$
$$+ \left\||D_x|^{-1/2}\langle D_y\rangle^{1/2}\tilde{v}_1(t)\right\|_{L^2(\mathbb{R}^2)} \lesssim \|\mathcal{E}(v_*)^{1/2}\|_{L^2(\mathbb{R}^2)}.$$

Let

$$H(u) = \frac{1}{2}\int_{\mathbb{R}^2}\left\{(\partial_x u)^2 - 3(\partial_x^{-1}\partial_y u)^2 - 2u^3\right\}dxdy.$$

Since $H(u)$ is the Hamiltonian of the KP-II equation and $\tilde{v}_1$ is a solution of (3.8) satisfying $\tilde{v}_1 \in C(\mathbb{R}; H^1(\mathbb{R}^2))$ and $\partial_x^{-1}\partial_y\tilde{v}_1 \in C(\mathbb{R}; L^2(\mathbb{R}^2))$,

$$3\|\partial_x^{-1}\partial_y\tilde{v}_1(t)\|_{L^2(\mathbb{R}^2)}^2 \leqslant -2H(\tilde{v}_1(t)) + \|\partial_x\tilde{v}_1(t)\|_{L^2(\mathbb{R}^2)}^2 + 2\|\tilde{v}_1(t)\|_{L^3(\mathbb{R}^2)}^3$$
$$= -2H(v_*) + \|\partial_x\tilde{v}_1(t)\|_{L^2(\mathbb{R}^2)}^2$$
$$+ 2\|\tilde{v}_1(t)\|_{L^3(\mathbb{R}^2)}^3 \lesssim \|\mathcal{E}(v_*)^{1/2}\|_{L^2(\mathbb{R}^2)}^2.$$

Combining the above with Lemma 3.1, we have Lemma 7.3. □

8 Decay Estimates for the Exponentially Localized Part of Perturbations

In this section, we will estimate $v_2(t)$ following the line of [22].

Lemma 8.1 *Let η_0 be a small positive number and $\alpha \in (\nu(\eta_0), 2)$. Suppose that $\mathbb{M}'_1(\infty)$ is sufficiently small. Then there exist positive constants δ_6 and C such that if $\mathbb{M}_2(T) + \mathbb{M}_v(T) \leqslant \delta_6$,*

$$\mathbb{M}_2(T) \leqslant C\left(\|\langle x\rangle v_0\|_{L^2(\mathbb{R}^2)} + \mathbb{M}_{c,x}(T) + \mathbb{M}_1(T)\right). \tag{8.1}$$

First, we estimate the low frequency part of $v_2(t)$ assuming the smallness of $\mathbb{M}_{c,x}(T)$, $\mathbb{M}_2(t)$ and $\mathbb{M}_v(T)$.

Lemma 8.2 *Let η_0, α and M be positive constants satisfying $\nu(\eta_0) < \alpha < 2$ and $\nu(M) > \alpha$. Suppose that $v_2(t)$ is a solution of (4.2). Then there exist positive constants b_1, δ_6 and C such that if $\mathbb{M}_{c,x}(T) + \mathbb{M}_v(T) + \mathbb{M}_1(T) + \mathbb{M}_2(T) < \delta_6$, then for $t \in [0, T]$,*

$$\|P_1(0, M)v_2(t, \cdot)\|_X \leqslant Ce^{-bt}\|v_{2,*}\|_X$$
$$+ C\left\{\mathbb{M}_{c,x}(T) + \mathbb{M}_1(T) + \mathbb{M}_2(T)(\mathbb{M}_2(T) + \mathbb{M}_v(T))\right\} \langle t \rangle^{-3/4}. \tag{8.2}$$

Proof Let $\tilde{v}_2(t) = P_2(\eta_0, M)v_2(t)$ and $N'_{2,2} = \{2\tilde{c}(t, y) + 6(\varphi(z) - \varphi_{c(t,y)}(z))\}v_2(t, z, y)$. Applying Proposition 2.1 to (4.2), we have

$$\|\tilde{v}_2(t)\|_X \lesssim e^{-bt}\|v_{2,*}\|_X + \int_0^t e^{-b'(t-s)}(t-s)^{-3/4}\|e^{\alpha z}P_2 N_{2,1}(s)\|_{L_z^1 L_y^2}\,ds$$
$$+ \int_0^t e^{-b'(t-s)}(t-s)^{-1/2}(\|N_{2,2}(s)\|_X + \|N'_{2,2}(s)\|_X + \|N_{2,4}\|_X)\,ds$$
$$+ \int_0^t e^{-b(t-s)}(\|\ell(s)\|_X + \|P_2 N_{2,3}(s)\|_X)\,ds,$$
$$\tag{8.3}$$

where we abbreviate $P_2(\eta_0, M)$ as P_2. It follows from [22, Claim 9.1] that for $t \in [0, T]$,

$$\|e^{\alpha z}P_2 N_{2,1}\|_{L_z^1 L_y^2} \lesssim \sqrt{M}(\|v_1\|_{L^2} + \|v_2\|_{L^2})\|v_2\|_X$$
$$\lesssim \sqrt{M}(\mathbb{M}_1(T) + \mathbb{M}_v(T))\mathbb{M}_2(T)\langle t \rangle^{-3/4}. \tag{8.4}$$

By the definitions and Claim 5.1,

$$\|\ell_1\|_X \lesssim (\mathbb{M}_{c,x}(T) + \mathbb{M}_1(T) + \mathbb{M}_2(T)^2)\langle t \rangle^{-3/4},$$
$$\|\ell_2\|_X \lesssim e^{-\alpha(3t+L)}(\mathbb{M}_{c,x}(T) + \mathbb{M}_1(T) + \mathbb{M}_2(T)^2), \tag{8.5}$$

and

$$\|N_{2,2}\|_X \lesssim (\|x_t - 2c - 3(x_y)^2\|_{L^\infty} + \|\tilde{c}\|_{L^\infty})\|v_2\|_X,$$
$$\lesssim (\mathbb{M}_{c,x}(T) + \mathbb{M}_1(T) + \mathbb{M}_2(T)^2)\mathbb{M}_2(T)\langle t \rangle^{-5/4}. \tag{8.6}$$

in the same way as (8.6) and (8.7) in [23]. Since

$$\|\tilde{c}(t)\|_{L^\infty} + \sum_{k=1,2}\|\partial_y^k x(t)\|_{L^\infty} \lesssim \mathbb{M}_{c,x}(T)\langle t \rangle^{-1/2} \quad \text{for } t \in [0, T],$$

$$\|N'_{2,2}\|_X + \|P_2 N_{2,3}\|_X \lesssim (\|\tilde{c}\|_{L^\infty} + \|x_y\|_{L^\infty}$$
$$+ \|x_{yy}\|_{L^\infty})\|v_2\|_X \lesssim \mathbb{M}_{c,x}(T)\mathbb{M}_2(T)\langle t \rangle^{-5/4}. \tag{8.7}$$

Here we use the fact that $\|\partial_y P_2\|_{B(X)} \lesssim M$. Since $|e^{\alpha z}\{\varphi_c(z) - \tilde{\psi}_c(z)\}| \lesssim p_\alpha(z + 3t + L)$, we have

$$\|N_{2,4}\|_X \lesssim \mathbb{M}_1(T)\langle t\rangle^{-2} \quad \text{for } t \in [0, T]. \tag{8.8}$$

Combining (8.3)–(8.8), we have for $t \in [0, T]$,

$$\|\tilde{v}_2(t)\|_X \lesssim e^{-bt}\|v_{2,*}\|_X + \{\mathbb{M}_{c,x}(T) + \mathbb{M}_1(T) + (\mathbb{M}_v(T) + \mathbb{M}_2(T))\mathbb{M}_2(T)\}\langle t\rangle^{-3/4}.$$

As long as $v_2(t)$ satisfies the orthogonality condition (3.10) and $\tilde{c}(t, y)$ remains small, we have

$$\|\tilde{v}_2(t)\|_X \lesssim \|P_1(0, M)v_2(t)\|_X \lesssim \|\tilde{v}_2(t)\|_X$$

in exactly the same way as the proof of lemma 9.2 in [22]. Thus we have (8.2). This completes the proof of Lemma 8.2. □

Using a virial identity, we can estimate the exponentially weighted norm of $v_2(t)$ for high frequencies in y in the same way as [23, Lemma 8.3].

Lemma 8.3 *Let* $\alpha \in (0, 2)$ *and* $v_2(t)$ *be a solution of* (4.2). *Suppose* $\mathbb{M}'_1(\infty)$ *is sufficiently small. Then there exist positive constants* δ_6 *and* M_1 *such that if* $\mathbb{M}_{c,x}(T) + \mathbb{M}_1(T) + \mathbb{M}_2(T) + \mathbb{M}_v(T) < \delta_6$ *and* $M \geqslant M_1$, *then for* $t \in [0, T]$,

$$\|v_2(t)\|_X^2 \lesssim e^{-M\alpha t}\|v_{2,*}\|_X^2$$
$$+ \int_0^t e^{-M\alpha(t-s)}\left(\|\ell(s)\|_X^2 + \|P_1(0, M)v_2(s)\|_X^2 + \|N_{2,4}(s)\|_X^2\right) ds,$$

$$\|\mathcal{E}(v_2)^{1/2}\|_{L^2(0,T;X)}$$
$$\lesssim \|v_{2,*}\|_X + \|\ell\|_{L^2(0,T;X)} + \|P_1(0, M)v_2\|_{L^2(0,T);X)} + \|N_{2,4}\|_{L^2(0,T;X)}.$$

Now we are in position to prove Lemma 8.1.

Proof of Lemma 8.1 Combining Lemmas 8.2 and 8.3, (8.5) and (8.8) with (3.13), we have

$$\mathbb{M}_2(T) \lesssim \|v_{2,*}\|_X + \mathbb{M}_{c,x}(T) + \mathbb{M}_1(T) + \mathbb{M}_2(T)(\mathbb{M}_2(T) + \mathbb{M}_v(T))$$
$$\lesssim \|\langle x\rangle v_0\|_{L^2(\mathbb{R}^2)} + \mathbb{M}_1(T) + \mathbb{M}_2(T)(\mathbb{M}_2(T) + \mathbb{M}_v(T)).$$

Thus we obtain (8.1) provided $\mathbb{M}_2(T)$ and $\mathbb{M}_v(T)$ are sufficiently small. This completes the proof of Lemma 8.1. □

9 Large Time Behavior of the Phase Shift of Line Solitons

In this section, we will prove Theorem 1.5. To begin with, we remark that $\mathbb{M}_{c,x}(T)$, $\mathbb{M}_1(T)$, $\mathbb{M}_2(T)$ and $\mathbb{M}_v(T)$ remain small for every $T \in [0, \infty]$ provided the initial perturbation v_0 is sufficiently small. Combining Proposition 3.2, Lemmas 5.1, 6.1, 7.1, 7.3 and 8.1, we have the following.

Proposition 9.1 *There exist positive constants ε_0 and C such that if $\varepsilon := \|\langle x \rangle(\langle x \rangle + \langle y \rangle)v_0\|_{H^1(\mathbb{R}^2)} < \varepsilon_0$, then $\mathbb{M}_{c,x}(\infty) + \mathbb{M}_1(\infty) + \mathbb{M}_2(\infty) + \mathbb{M}_v(\infty) + \mathbb{M}_1'(\infty) \leqslant C\varepsilon$.*

When we estimate the L^∞ norm of $\tilde{x}$ by applying Lemma 2.3 to (4.6), two terms $\mathcal{N}^6$ and $B_4^{-1}\widetilde{\mathcal{A}}_1(t)^t(b, \tilde{x})$ are hazardous because they do not necessarily belong to $L^1(\mathbb{R})$.

We introduce a change of variable to eliminate $E_1 \dot{B}_4 k_t \mathbf{e_1}$ and a bad part of $B_4^{-1}\widetilde{\mathcal{A}}_1(t)$. Let

$$\begin{pmatrix} \tilde{a}_3(t, D_y) & 0 \\ \tilde{a}_4(t, D_y) & 0 \end{pmatrix} := B_4^{-1}\widetilde{\mathcal{A}}_1(t), \quad \tilde{a}_{31}(t) = \tilde{a}_3(t, 0), \quad \tilde{a}_{32}(t, \eta) = \frac{\tilde{a}_3(t, \eta) - \tilde{a}_3(t, 0)}{\eta^2},$$

and $\gamma(t) = e^{-\int_0^t \tilde{a}_{31}(s)\,ds}$. Note that $\tilde{a}_3(t, \eta)$ is even in η because $g_k^*(z, \eta)$ thus the symbols of B_4 and $\widetilde{\mathcal{A}}_1(t)$ are even in η. By [23, Claim 6.3], the operator B_4^{-1} is uniformly bounded in $B(Y)$ and we can prove that for $t \geqslant 0$,

$$|\tilde{a}_{31}(t)| + |\tilde{a}_{31}'(t)| + \|\tilde{a}_{32}(t, D_y)\|_{B(Y)} + \|\tilde{a}_3(t, D_y)\|_{B(Y)} + \|\tilde{a}_4(t, D_y)\|_{B(Y)} \lesssim e^{-\alpha(3t+L)} \tag{9.1}$$

in exactly the same way as [22, Claim D.3]. We need to replace $e^{-\alpha(4t+L)}$ in [22, Claim D.3] by $e^{-\alpha(3t+L)}$ because $\tilde{\psi}_c(z) = \psi_{c,L}(z + 3t)$ in our paper whereas $\tilde{\psi}_c(z) = \psi_{c,L}(z + 4t)$ in [22]. By the definitions of $\widetilde{S}_0$, $\widetilde{S}_1$ and $\widetilde{S}_3$ (see [22, pp. 40–41]) and [23, (A1), (A6)],

$$\|\mathcal{A}_1(t)\|_{B(Y) \cap B(Y_1)} = O(1), \quad \|\mathcal{A}_2(t)\|_{B(Y) \cap B(Y_1)} = O(e^{-\alpha(3t+L)}). \tag{9.2}$$

By (9.1),

$$0 < \inf_{t \geqslant 0} \gamma(t) \leqslant \sup_{t \geqslant 0} \gamma(t) < \infty, \quad \lim_{t \to \infty} \gamma(t) > 0.$$

Let $\mathbf{k}(t, y) = \gamma(t)\tilde{k}(t, y)\mathbf{e_1}$, $\tilde{b}(t, y) = b(t, y) + \tilde{k}(t, y)$ and

$$\mathbf{b}(t, y) = {}^t(b_1(t, y), b_2(t, y)) = \gamma(t){}^t(\tilde{b}(t, y), \tilde{x}(t, y)). \tag{9.3}$$

By Claim C.1 and (5.27),

$$\|\partial_y^k b_1(t)\|_Y \lesssim \|\partial_y^k b(t)\|_Y + \|k(t)\|_Y$$
$$\lesssim \langle t \rangle^{-(2k+1)/4} \mathbb{M}_{c,x}(\infty) + \mathbb{M}_1(\infty)\langle t \rangle^{-2} \quad \text{for } k \geq 0, \tag{9.4}$$

$$\|\partial_y^k b_2(t)\|_Y \lesssim \|\partial_y^k \tilde{x}(t)\|_Y \lesssim \langle t \rangle^{-(2k-1)/4} \mathbb{M}_{c,x}(\infty) \quad \text{for } k \geq 1, \tag{9.5}$$

$$\|\tilde{x}(t)\|_{L^\infty} \lesssim \|b_2(t)\|_{L^\infty} + \|k(t)\|_Y \lesssim \|b_2(t)\|_{L^\infty} + \mathbb{M}_1(\infty)\langle t \rangle^{-2}. \tag{9.6}$$

Note that $\|\partial_y^k k(t)\|_Y \lesssim \eta_0^k \|k(t)\|_Y$ for any $k \geq 1$.

Substituting (9.3) into (4.6) and using (5.22), (5.25) and (5.26), we have

$$\partial_t \mathbf{b} = \mathcal{A}_* \mathbf{b} + \gamma \left\{ \sum_{i=1}^{5} \mathcal{N}^i + \dot{\mathcal{N}}_6 + \partial_y \overset{\circ}{\mathcal{N}}_6 + \dot{\mathcal{N}}_7 + \partial_y^2 \overset{\circ}{\mathcal{N}}_7 \right\}, \tag{9.7}$$

where $\dot{\mathcal{N}}_6 = \sum_{1 \leq j \leq 3} \dot{\mathcal{N}}_{6j}$, $\overset{\circ}{\mathcal{N}}_6 = \overset{\circ}{\mathcal{N}}_{61} + \partial_y \overset{\circ}{\mathcal{N}}_{62}$,

$$\dot{\mathcal{N}}_{61} = \gamma^{-1} \partial_t \{\gamma(2 - S_{11}^3[\psi])^{-1}\} k \mathbf{e}_1, \quad \dot{\mathcal{N}}_{62} = E_2 \dot{B}_4 R^{v_1} - 2E_{21} \mathbf{k}, \quad E_{21} = \begin{pmatrix} 0 & 0 \\ 1 & 0 \end{pmatrix},$$

$$\dot{\mathcal{N}}_{63} = (2 - S_{11}^3[\psi])^{-1} \left\{ \dot{R}_{v_1} + S_1^7[\partial_c \varphi_c](c_t) - S_1^7[\varphi_c'](x_t - 2c - 3(x_y)^2) \right\} + \dot{B}_{34} R^{v_1},$$

$$\overset{\circ}{\mathcal{N}}_{61} = \dot{B}_4 \overset{\circ}{R}_{v_1,1} \mathbf{e}_1, \quad \overset{\circ}{\mathcal{N}}_{62} = \left\{ \dot{B}_4 \overset{\circ}{R}_{v_1,2} \mathbf{e}_1 - (\overset{\circ}{B}_{14} + \overset{\circ}{B}_{34}) R^{v_1} \right\},$$

$$\dot{\mathcal{N}}_7 = \{\tilde{a}_4(t, D_y) b(t, \cdot) - \tilde{a}_{31}(t) \tilde{x}(t, \cdot)\} \mathbf{e}_2,$$

$$\overset{\circ}{\mathcal{N}}_7 = \left(\partial_y^2 \mathcal{A}_1 + \mathcal{A}_2\right)(b(t, \cdot)\mathbf{e}_1 + \tilde{x}(t, \cdot)\mathbf{e}_2) - \tilde{a}_{32}(t, D_y) b(t, \cdot)\mathbf{e}_1$$
$$- \partial_y^{-2}(\mathcal{A}_* - 2E_{21}) \mathbf{k}(t, \cdot).$$

Now we start to prove Theorem 1.5.

Proof of Theorem 1.5 Using the variation of constants formula, we can translate (9.7) into

$$\mathbf{b}(t) = e^{t\mathcal{A}_*} \mathbf{b}(0)$$
$$+ \int_0^t e^{(t-s)\mathcal{A}_*} \gamma(s) \left(\sum_{i=1}^{5} \mathcal{N}^i(s) + \dot{\mathcal{N}}_6(s) + \partial_y \overset{\circ}{\mathcal{N}}_6(s) + \dot{\mathcal{N}}_7(s) + \partial_y^2 \overset{\circ}{\mathcal{N}}_7(s) \right) ds. \tag{9.8}$$

Now we will estimate the L^∞-norm of the right hand side of (9.8). By (4.7) and (9.3),

$$b_1(0, y) = b_*(y) + \frac{1}{2}\left(2 - S_{11}^3[\psi](0)\right)^{-1} \widetilde{P}_1\left(\int_{\mathbb{R}} v_*(x, y)\varphi_{c_*(y)}(x - x_*(y))\, dx\right),$$

$$b_2(0, y) = x_*(y),$$

and it follows from Lemmas 2.3 and 4.2 that

$$\left\|e^{t\mathcal{A}_*}\mathbf{b}(0)\right\|_{L^\infty(\mathbb{R})} \lesssim \|\overset{\bullet}{b}\|_{L^1} + \|\overset{\bullet}{b}\|_{Y_1} + \|x_*\|_{Y_1} + \|v_*\|_{L^1(\mathbb{R}^2)} \lesssim \varepsilon,$$

$$\left\|\mathbf{e}_2 \cdot e^{t\mathcal{A}_*}\mathbf{b}(0) - \frac{1}{2} H_{2t} * W_{4t} * b_1(0)\right\|_{L^\infty} \lesssim \langle t\rangle^{-1/2}\varepsilon, \qquad (9.9)$$

where $\varepsilon = \|\langle x\rangle\langle y\rangle v_0\|_{L^2(\mathbb{R}^2)}$.

Next, we will estimate $\mathcal{N}^1$. Let $\overset{\bullet}{\mathcal{N}}_1 = \widetilde{P}_1\{2(\tilde{c} - b) + 3(x_y)^2\}\mathbf{e}_2$ and $\overset{\circ}{\mathcal{N}}_1 = 6\widetilde{P}_1(b\tilde{x}_y)\mathbf{e}_1$. Then $E_1\mathcal{N}_1 = \partial_y\widetilde{\mathcal{N}}$, $E_2\mathcal{N}_1 = \overset{\bullet}{\mathcal{N}}_1$, $\mathcal{N}^1 = \text{diag}(\partial_y, 1)(\overset{\circ}{\mathcal{N}}_1 + \overset{\bullet}{\mathcal{N}}_1)$, and

$$III_1(t) := \|\overset{\bullet}{\mathcal{N}}_1\|_{Y_1} + \|\overset{\circ}{\mathcal{N}}_1\|_{Y_1} \lesssim \mathbb{M}_{c,x}(\infty)^2 \langle t\rangle^{-1/2}.$$

By (1.9) and the fact that

$$b - \tilde{c} = \frac{4}{3}\widetilde{P}_1\left\{\left(\frac{c}{2}\right)^{3/2} - 1 - \frac{3}{4}\tilde{c}\right\} = \frac{1}{8}\widetilde{P}_1\tilde{c}^2 + O(\tilde{c}^3) \qquad (9.10)$$

(see [22, Claim D.6]),

$$III_2(t) := \left\|bx_y - 2\{u_B^+(t, \cdot + 4t)^2 - u_B^-(t, \cdot - 4t)^2\}\right\|_{L^1} \lesssim \varepsilon^2 \delta(t)\langle t\rangle^{-1/2},$$

$$III_3(t) := \left\|2(\tilde{c} - b) + 3(x_y)^2 - 2\{u_B^+(t, \cdot + 4t)^2 + u_B^-(t, \cdot - 4t)^2\}\right\|_{L^1} \lesssim \varepsilon^2 \delta(t)\langle t\rangle^{-1/2},$$

where $\delta(t)$ is a function that tends to 0 as $t \to \infty$. Note that $\|u_B^+(t, \cdot + 4t)u_B^-(t, \cdot - 4t)\|_{L^1} = O(\varepsilon^2 t^{-1/2} e^{-8t})$. By Lemma 2.3 and [22, Claim 4.1],

$$\left\|\int_0^t e^{(t-s)\mathcal{A}_*}\gamma(s)\mathcal{N}^1(s)\, ds\right\|_{L^\infty} \lesssim \int_0^t \langle t-s\rangle^{-1/2} III_1(s)\, ds$$

$$\lesssim \mathbb{M}_{c,x}(\infty)^2 \int_0^t \langle t-s\rangle^{-1/2}\langle s\rangle^{-1/2}$$

$$\lesssim \mathbb{M}_{c,x}(\infty)^2,$$

$$\left\| \int_0^t e^{(t-s)\mathcal{A}_*} \gamma(s) \{\mathcal{N}^1(s) \, ds \right.$$

$$\left. - \sum_{\pm} H_{2t}(\cdot \pm 4(t-s)) * \{4u_B^{\pm}(s, \cdot \pm 4s)^2 - 2u_B^{\mp}(s, \cdot \mp 4s)^2 \} \, ds \mathbf{e}_2 \right\|_{L^\infty}$$

$$\lesssim \int_0^t \langle t-s \rangle^{-1} III_1(s) \, ds + \int_0^t \langle t-s \rangle^{-1/2} (III_2(s) + III_3(s)) \, ds$$

$$\lesssim \varepsilon^2 \langle t \rangle^{-1/2} \log(t+2) + \varepsilon^2 \int_0^t (t-s)^{-1/2} s^{-1/2} \delta(s) \, ds \to 0 \quad \text{as } t \to \infty.$$

For y satisfying $\min\{|y - 4t|, |y + 4t|\} \geq 2\delta(t)^{-1/2}\sqrt{t}$,

$$\int_0^t \int_{\mathbb{R}} H_{2(t-s)}(y - y_1 \pm 4t) \left\{ H_{2s}(y_1)^2 + H_{2s}(y_1 \mp 8s)^2 \right\} dy_1 ds$$

$$\lesssim \int_0^t (t-s)^{-1/2} s^{-1} e^{-\delta(t)^{-1}/8} \left(\int_{|y_1| \leq \delta(t)^{-1/2}\sqrt{t}} e^{-y_1^2/8s} \, dy_1 \right) ds$$

$$+ \int_0^t (t-s)^{-1/2} s^{-1} \left(\int_{|y_1| \geq \delta(t)^{-1/2}\sqrt{t}} e^{-y_1^2/8s} \, dy_1 \right) ds$$

$$\lesssim \exp\left(-\delta(t)^{-1}/8\right) \to 0 \quad \text{as } t \to \infty.$$

Combining the above with the fact that $|u_B^{\pm}(s, y)| \lesssim H_{2s}(y)$, we obtain

$$\lim_{t \to \infty} \left\| \int_0^t e^{(t-s)\mathcal{A}_*} \gamma(s) \mathcal{N}^1(s) \, ds \right\|_{L^\infty(|y \pm 4t| \geq \delta t)} = 0.$$

The other terms can be decomposed as

$$\sum_{i=2}^{5} \mathcal{N}^i(t) + \overset{\bullet}{\mathcal{N}_6} + \partial_y \overset{\circ}{\mathcal{N}_6} + \overset{\bullet}{\mathcal{N}_7} + \partial_y^2 \overset{\circ}{\mathcal{N}_7} = \overset{\bullet}{\mathcal{N}_a} + \overset{\bullet}{\mathcal{N}_b} + \partial_y (\overset{\circ}{\mathcal{N}_a} + \overset{\circ}{\mathcal{N}_b}) + \partial_y^2 \overset{\circ}{\mathcal{N}_c}, \qquad (9.11)$$

such that $\overset{\bullet}{\mathcal{N}_b} = E_2 \overset{\bullet}{\mathcal{N}_b}$ and

$$\|\chi(D_y) E_1 \overset{\bullet}{\mathcal{N}_a}\|_{L^1(\mathbb{R})} + \|\partial_y^{-1} (I - \chi(D_y)) E_1 \overset{\bullet}{\mathcal{N}_a}\|_{Y_1} \lesssim \left(e^{-\alpha L} \varepsilon + \varepsilon^2\right) \langle t \rangle^{-3/2}, \qquad (9.12)$$

$$\|E_2 \overset{\bullet}{\mathcal{N}_a}\|_{Y_1} + \|\overset{\circ}{\mathcal{N}_a}\|_{Y_1} \lesssim \varepsilon^2 \langle t \rangle^{-1}, \qquad (9.13)$$

$$\|\overset{\bullet}{\mathcal{N}_b}\|_Y + \|\overset{\circ}{\mathcal{N}_b}\|_Y \lesssim \varepsilon \langle t \rangle^{-7/4}, \quad \|\overset{\circ}{\mathcal{N}_c}\|_Y \lesssim \varepsilon \langle t \rangle^{-1}. \qquad (9.14)$$

Hence it follows from Proposition 9.1 and Lemma 2.3 that

$$\sup_{t\geq 0}\left\|\int_0^t e^{(t-s)\mathcal{A}_*}\gamma(s)\chi(D_y)E_1\overset{\bullet}{\mathcal{N}}_a(s)\,ds\right\|_{L^\infty} \lesssim e^{-\alpha L}\varepsilon + \varepsilon^2, \qquad (9.15)$$

$$\left\|\int_0^t e^{(t-s)\mathcal{A}_*}\gamma(s)(I-\chi(D_y))E_1\overset{\bullet}{\mathcal{N}}_a(s)\,ds\right\|_{L^\infty} \lesssim (e^{-\alpha L}\varepsilon + \varepsilon^2)\langle t\rangle^{-1/2}, \qquad (9.16)$$

$$\left\|\int_0^t e^{(t-s)\mathcal{A}_*}\gamma(s)\bigl[E_2\overset{\bullet}{\mathcal{N}}_a(s)+\overset{\bullet}{\mathcal{N}}_b(s) + \partial_y\{\overset{\circ}{\mathcal{N}}_a(s)+\overset{\circ}{\mathcal{N}}_b(s)\}\right.$$
$$\left. + \partial_y^2\overset{\circ}{\mathcal{N}}_c(s)\bigr]ds\right\|_{L^\infty} \lesssim \varepsilon\langle t\rangle^{-1/4}. \qquad (9.17)$$

By Lemma 2.4, (2.20) and (9.12), that for any $\delta > 0$,

$$\lim_{t\to\infty}\sup_{|y|\leq(4-\delta)t}\left|\int_0^t e^{(t-s)\mathcal{A}_*}\gamma(s)\chi(D_y)E_1\overset{\bullet}{\mathcal{N}}_a(s)\,ds - h_a\mathbf{e_2}\right| = 0, \qquad (9.18)$$

$$h_a = \frac{1}{2}\int_0^\infty\int_\mathbb{R}\gamma(s)\mathbf{e_1}\cdot\overset{\bullet}{\mathcal{N}}_a(s,y)\,dy\,ds, \quad |h_a| \lesssim \varepsilon e^{-\alpha L} + \varepsilon^2, \qquad (9.19)$$

$$\lim_{t\to\infty}\sup_{|y|\geq(4+\delta)t}\left|\int_0^t e^{(t-s)\mathcal{A}_*}\gamma(s)\chi(D_y)E_1\overset{\bullet}{\mathcal{N}}_a(s)\,ds\right| = 0.$$

Now, we will prove (9.11)–(9.14). First, we will estimate $\mathcal{N}^2$. As in [22, Claim D.7],

$$\|\widetilde{P}_1 R_1^7\|_{Y_1} + \|\chi(D_y)R_1^7\|_{L^1} \lesssim M_{c,x}(\infty)^2\langle t\rangle^{-3/2}, \qquad (9.20)$$

$$\|\widetilde{P}_1 R_2^7\|_{Y_1} \lesssim M_{c,x}(\infty)^2\langle t\rangle^{-1}, \quad \|\widetilde{P}_1 R_2^7\|_Y \lesssim M_{c,x}(\infty)^2\langle t\rangle^{-5/4}, \qquad (9.21)$$

where

$$R_1^7 = \{4\sqrt{2}c^{3/2} - 16 - 12b\}x_{yy} - 6(2b_y - (2c)^{1/2}c_y)x_y - 3c^{-1}(c_y)^2,$$

$$R_2^7 = 6\left\{\left(\frac{c}{2}\right)^{3/2} - 1\right\}x_{yy} + 3\left(\frac{c}{2}\right)^{1/2}c_yx_y - 3(bx_y)_y + \mu_2\frac{2}{c}(c_y)^2$$
$$+ \frac{3}{2}(c^2 - 4)(I - \widetilde{P}_1)(x_y)^2.$$

Let

$$\widetilde{R}^{11} = R^{31} + R^{41} + R^{61} + \widetilde{S}_{41}\begin{pmatrix}0\\2\tilde{c}\end{pmatrix}, \quad \widetilde{R}^{12} = R^{32} + R^{42} + R^{62} + \widetilde{S}_{42}\begin{pmatrix}0\\2\tilde{c}\end{pmatrix},$$

$$\widetilde{R}^{31} = R^{91} + R^{11,1}, \quad \widetilde{R}^{32} = R^{92} + R^{11,2}.$$

Then $\widetilde{R}^1 = \widetilde{R}^{11} - \partial_y^2 \widetilde{R}^{12}$ and $\widetilde{R}^3 = \widetilde{R}^{31} - \partial_y^2 \widetilde{R}^{32}$. See Appendix B for the definitions of R^{j1} and R^{j2} and see (A.1)–(A.4) and (A.17) for the definitions of $\widetilde{S}_{4\ell}$ ($\ell = 1, 2$).
By Claims A.1, A.2, B.2–B.4 and B.6,

$$\|\chi(D_y)\widetilde{R}^{11}\|_{L^1(\mathbb{R})} + \|\widetilde{R}^{11}\|_{Y_1} + \|\widetilde{R}^{12}\|_{Y_1}$$
$$\lesssim (\mathbb{M}_{c,x}(\infty)^2 + \mathbb{M}_2(\infty)^2 + \mathbb{M}_1(\infty)\mathbb{M}_2(\infty))\langle t\rangle^{-3/2}, \qquad (9.22)$$

$$\|\chi(D_y)\widetilde{R}^{31}\|_{L^1(\mathbb{R})} + \|\widetilde{R}^{31}\|_{Y_1} + \|\widetilde{R}^{32}\|_{Y_1}$$
$$\lesssim \mathbb{M}_{c,x}(\infty)(\mathbb{M}_{c,x}(\infty) + \mathbb{M}_2(\infty))\langle t\rangle^{-3/2}. \qquad (9.23)$$

Let $\overset{\bullet}{\mathcal{N}}_2 = \overset{\bullet}{\mathcal{N}}_{21} + \overset{\bullet}{\mathcal{N}}_{22} + \overset{\bullet}{\mathcal{N}}_{23}$ and

$$\overset{\bullet}{\mathcal{N}}_{21} = \overset{\bullet}{B}_4(\widetilde{P}_1 R_1^7 \mathbf{e_1} + \widetilde{R}^{11} + \widetilde{R}^{31}) + \overset{\bullet}{B}_{34}(\widetilde{P}_1 R^7 + \widetilde{R}^1 + \widetilde{R}^3),$$
$$\overset{\bullet}{\mathcal{N}}_{22} = B_1^{-1}\widetilde{P}_1 R_2^7 \mathbf{e_2}, \quad \overset{\bullet}{\mathcal{N}}_{23} = (\overset{\bullet}{B}_4 - B_1^{-1})\widetilde{P}_1 R_2^7 \mathbf{e_2},$$
$$\overset{\circ}{\mathcal{N}}_2 = \overset{\bullet}{B}_4(\widetilde{R}^{12} + \widetilde{R}^{32}) + (\overset{\circ}{B}_{14} + \overset{\circ}{B}_{34})(\widetilde{P}_1 R^7 + \widetilde{R}^1 + \widetilde{R}^3).$$

Since $B_3^{-1} = \overset{\bullet}{B}_4 + \overset{\bullet}{B}_{34} - \partial_y^2(\overset{\circ}{B}_{14} + \overset{\circ}{B}_{34})$ and $[\overset{\bullet}{B}_4, \partial_y] = O$, we have $\mathcal{N}^2 = \overset{\bullet}{\mathcal{N}}_2 - \partial_y^2 \overset{\circ}{\mathcal{N}}_2$. By (9.20)–(9.23) and Claim 5.2,

$$\|\overset{\bullet}{\mathcal{N}}_{21}\|_{Y_1} + \|\chi(D_y)\overset{\bullet}{\mathcal{N}}_{21}\|_{L^1(\mathbb{R})} \lesssim (\mathbb{M}_{c,x}(\infty) + \mathbb{M}_2(\infty))^2 \langle t\rangle^{-3/2},$$
$$\|\overset{\circ}{\mathcal{N}}_2\|_{Y_1} \lesssim (\mathbb{M}_{c,x}(\infty) + \mathbb{M}_2(\infty))^2 \langle t\rangle^{-1}.$$

By (5.8), (9.21) and the fact that $B_1^{-1}\mathbf{e_2} = \frac{1}{2}\mathbf{e_2}$,

$$\|\overset{\bullet}{\mathcal{N}}_{22}\|_{Y_1} \lesssim \mathbb{M}_{c,x}(\infty)^2 \langle t\rangle^{-1}, \quad \overset{\bullet}{\mathcal{N}}_{22} = E_2 \overset{\bullet}{\mathcal{N}}_{22},$$
$$\|\overset{\bullet}{\mathcal{N}}_{23}\|_{Y_1} + \|\chi(D_y)\overset{\bullet}{\mathcal{N}}_{23}\|_{L^1(\mathbb{R})} \lesssim e^{-\alpha(3t+L)} \langle t\rangle^{-1} \mathbb{M}_{c,x}(\infty)^2.$$

Next we will estimate $\mathcal{N}^3$. Let

$$\overset{\bullet}{\mathcal{N}}_3 = [\overset{\bullet}{B}_{34}, \partial_y](\widetilde{R}^2 + \widetilde{R}^4), \quad \overset{\circ}{\mathcal{N}}_3 = (B_4^{-1} + \overset{\bullet}{B}_{34} - \partial_y \overset{\circ}{B}_{34}\partial_y)(\widetilde{R}^2 + \widetilde{R}^4).$$

Then $\mathcal{N}^3 = \overset{\bullet}{\mathcal{N}}_3 + \partial_y \overset{\circ}{\mathcal{N}}_3$. By Claims B.1 and B.3,

$$\|\widetilde{R}^2\|_{Y_1} \lesssim \mathbb{M}_{c,x}(\infty)(\mathbb{M}_{c,x}(\infty) + \mathbb{M}_2(\infty))\langle t\rangle^{-1},$$
$$\|\widetilde{R}^2\|_Y \lesssim \mathbb{M}_{c,x}(\infty)(\mathbb{M}_{c,x}(\infty) + \mathbb{M}_2(\infty))\langle t\rangle^{-5/4}. \qquad (9.24)$$

By Claims B.4 and B.5,

$$\|\widetilde{R}^4\|_{Y_1} \lesssim \mathbb{M}_{c,x}(\infty)^2 \langle t \rangle^{-1}, \quad \|\widetilde{R}^4\|_Y \lesssim \mathbb{M}_{c,x}(\infty)^2 \langle t \rangle^{-5/4}. \qquad (9.25)$$

Combining (9) and (9.25) with Claim 5.2, we have

$$\|\overset{\bullet}{\mathcal{N}}_3\|_{Y_1} + \|\chi(D_y)\overset{\bullet}{\mathcal{N}}_3\|_{L^1} \lesssim \mathbb{M}_{c,x}(\infty)(\mathbb{M}_{c,x}(\infty) + \mathbb{M}_2(\infty))^2 \langle t \rangle^{-2},$$

$$\|\overset{\circ}{\mathcal{N}}_3\|_{Y_1} \lesssim \mathbb{M}_{c,x}(\infty)(\mathbb{M}_{c,x}(\infty) + \mathbb{M}_2(\infty)) \langle t \rangle^{-1}.$$

Next, we will estimate $\mathcal{N}^4$. Let $n_{41} = (B_2 - \partial_y^2 \widetilde{S}_0) b_y \mathbf{e}_1$, $n_{42} = (B_2 - \partial_y^2 \widetilde{S}_0) x_{yy} \mathbf{e}_2$ and

$$\overset{\bullet}{\mathcal{N}}_{41} = [\overset{\bullet}{B}_{34}, \partial_y] n_{41}, \quad \overset{\bullet}{\mathcal{N}}_{42} = E_2 \overset{\bullet}{B}_{34} n_{42}, \quad \overset{\bullet}{\mathcal{N}}_{43} = E_1 \overset{\bullet}{B}_{34} n_{42},$$

$$\overset{\circ}{\mathcal{N}}_{41} = \overset{\bullet}{B}_{34} n_{41}, \quad \overset{\circ}{\mathcal{N}}_{42} = \overset{\bullet}{B}_{34} (\partial_y n_{41} + n_{42})$$

By the definitions, $E_2 \overset{\bullet}{\mathcal{N}}_{42} = \overset{\bullet}{\mathcal{N}}_{42}$ and

$$\mathcal{N}^4 = (\overset{\bullet}{B}_{34} - \partial_y^2 \overset{\circ}{B}_{34})(\partial_y n_{41} + n_{42}) = \overset{\bullet}{\mathcal{N}}_{41} + \overset{\bullet}{\mathcal{N}}_{42} + \overset{\bullet}{\mathcal{N}}_{43} + \partial_y \overset{\circ}{\mathcal{N}}_{41} - \partial_y^2 \overset{\circ}{\mathcal{N}}_{42}.$$

Since $\|\widetilde{S}_0\|_{B(Y)} = O(1)$ by [22, Claim B.1], we have $\|n_{41}\|_Y + \|n_{42}\|_Y \lesssim \mathbb{M}_{c,x}(\infty) \langle t \rangle^{-3/4}$. It follows from Claim 5.2 that

$$\|\chi(D_y)\overset{\bullet}{\mathcal{N}}_{41}\|_{L^1} + \|\overset{\bullet}{\mathcal{N}}_{41}\|_{Y_1} \lesssim \mathbb{M}_{c,x}(\infty)(\mathbb{M}_{c,x}(\infty) + \mathbb{M}_2(\infty)) \langle t \rangle^{-3/2},$$

$$\|\overset{\bullet}{\mathcal{N}}_{42}\|_{Y_1} + \|\overset{\circ}{\mathcal{N}}_{41}\|_{Y_1} + \|\overset{\circ}{\mathcal{N}}_{42}\|_{Y_1} \lesssim \mathbb{M}_{c,x}(\infty)(\mathbb{M}_{c,x}(\infty) + \mathbb{M}_2(\infty)) \langle t \rangle^{-1}.$$

Since $E_1 B_1^{-1} \widetilde{\mathcal{C}}_1 = O$,

$$\overset{\bullet}{\mathcal{N}}_{43} = E_1 \left(\overset{\bullet}{B}_{34} + B_1^{-1} \widetilde{\mathcal{C}}_1 B_3^{-1} \right) n_{42}.$$

By Claim A.4 and (5.4),

$$\|\overset{\bullet}{B}_{34} + \overset{\bullet}{B}_4 \widetilde{\mathcal{C}}_1 B_3^{-1}\|_{B(Y,Y_1)} = \|\overset{\bullet}{B}_4 (\widetilde{S}_{31} - \bar{S}_{31} - \bar{S}_{41} - \bar{S}_{51}) B_3^{-1}\|_{B(Y,Y_1)}$$

$$\lesssim (\mathbb{M}_{c,x}(\infty) + \mathbb{M}_2(\infty)) \langle t \rangle^{-3/4}.$$

By (5.8) and the above,

$$\|\overset{\bullet}{B}_{34} + B_1^{-1} \widetilde{\mathcal{C}}_1 B_3^{-1}\|_{B(Y,Y_1)}$$

$$\leq \|\overset{\bullet}{B}_{34} + \overset{\bullet}{B}_4 \widetilde{\mathcal{C}}_1 B_3^{-1}\|_{B(Y,Y_1)} + \|B_1^{-1} - \overset{\bullet}{B}_4\|_{B(Y_1)} \|\widetilde{\mathcal{C}}_1 B_3^{-1}\|_{B(Y,Y_1)} \qquad (9.26)$$

$$\lesssim \langle t \rangle^{-3/4} (\mathbb{M}_{c,x}(\infty) + \mathbb{M}_2(\infty)).$$

Using Claim 5.2 and (5.12), we can prove

$$\|\chi(D_y)\overset{\bullet}{B}_{34}+\chi(D_y)B_1^{-1}\widetilde{\mathscr{C}}_1 B_3^{-1}\|_{B(Y,L^1)} \lesssim \langle t\rangle^{-3/4}(\mathbb{M}_{c,x}(\infty)+\mathbb{M}_2(\infty)) \quad (9.27)$$

in the same way. By (9.26) and (9.27),

$$\|\overset{\bullet}{\mathcal{N}}_{43}\|_{Y_1} + \|\chi(D_y)\overset{\bullet}{\mathcal{N}}_{43}\|_{L^1} \lesssim \langle t\rangle^{-3/2}\mathbb{M}_{c,x}(\infty)(\mathbb{M}_{c,x}(\infty) + \mathbb{M}_2(\infty)). \quad (9.28)$$

Secondly, we will estimate $\mathcal{N}^5$. By (5.2),

$$\mathcal{N}^5 = \overset{\bullet}{\mathcal{N}}_5 - \partial_y^2 \overset{\circ}{\mathcal{N}}_5, \quad \overset{\bullet}{\mathcal{N}}_5 = \overset{\bullet}{B}_{34}\widetilde{\mathscr{A}}_1(t)\binom{b}{\tilde{x}}, \quad \overset{\circ}{\mathcal{N}}_5 = \overset{\circ}{B}_{34}\widetilde{\mathscr{A}}_1(t)\binom{b}{\tilde{x}}.$$

Since $\|\widetilde{\mathscr{A}}_1(t)^t(b,\tilde{x})\|_Y \lesssim \mathbb{M}_{c,x}(\infty)e^{-\alpha(3t+L)}$, it follows from Claim 5.2 that

$$\|\overset{\bullet}{\mathcal{N}}_5\|_{Y_1} + \|\chi(D_y)\overset{\bullet}{\mathcal{N}}_5\|_{L^1} + \|\overset{\circ}{\mathcal{N}}_5\|_{Y_1}$$
$$\lesssim e^{-\alpha(3t+L)}\langle t\rangle^{-1/4}\mathbb{M}_{c,x}(\infty)(\mathbb{M}_{c,x}(\infty) + \mathbb{M}_2(\infty)). \quad (9.29)$$

Next, we will estimate $\overset{\bullet}{\mathcal{N}}_6$ and $\overset{\circ}{\mathcal{N}}_6$. By Claim C.2, (5.27) and (9.1),

$$\|\overset{\bullet}{\mathcal{N}}_{61}\|_{Y_1} + \|\chi(D_y)\overset{\bullet}{\mathcal{N}}_{61}\|_{L^1} \lesssim e^{-\alpha(3t+L)}\langle t\rangle\varepsilon.$$

We see that $\overset{\bullet}{\mathcal{N}}_{62} = E_2\overset{\bullet}{\mathcal{N}}_{62}$ and that

$$\|\overset{\bullet}{\mathcal{N}}_{62}\|_Y \lesssim \|R^{v_1}\|_Y + \|k\|_Y \lesssim \mathbb{M}_1(\infty)\langle t\rangle^{-2}$$

follows from Claim C.1 and (5.20). By (5.24), (5.28) and (5.30),

$$\|\chi(D_y)\overset{\bullet}{\mathcal{N}}_{63}\|_{L^1} + \|\overset{\bullet}{\mathcal{N}}_{63}\|_{Y_1} \lesssim \mathbb{M}_1(\infty)(\mathbb{M}_{c,x}(\infty) + \mathbb{M}_1(\infty) + \mathbb{M}_2(\infty))\langle t\rangle^{-3/2}.$$

By Claims 5.2, (5.23) and (5.29),

$$\|\overset{\circ}{\mathcal{N}}_{61}\|_Y \lesssim \|\overset{\circ}{R}_{v_1,1}\|_Y \lesssim \mathbb{M}_{c,x}(\infty)\mathbb{M}_1(\infty)\langle t\rangle^{-7/4},$$

$$\|\overset{\circ}{\mathcal{N}}_{62}\|_Y \lesssim \|\overset{\circ}{R}_{v_1,2}\|_Y + \|(\overset{\circ}{B}_{14} + \overset{\circ}{B}_{34})R^{v_1}\|_Y \lesssim \mathbb{M}_1(\infty)\langle t\rangle^{-1}.$$

Finally, we will estimate $\overset{\bullet}{\mathcal{N}}_7$ and $\overset{\circ}{\mathcal{N}}_7$. By the definition and (9.1), we have $\overset{\bullet}{\mathcal{N}}_7 = E_2\overset{\bullet}{\mathcal{N}}_7$ and

$$\|\overset{\bullet}{\mathcal{N}}_7(t)\|_Y \lesssim \left(\mathbb{M}_{c,x}(\infty) + \mathbb{M}_1(\infty) + \mathbb{M}_2(\infty)^2\right)e^{-\alpha(3t+L)}\langle t\rangle^{1/4}.$$

In view of Claim 5.1,

$$\|\tilde{x}(t)\|_Y \lesssim (\mathbb{M}_{c,x}(\infty) + \mathbb{M}_1(\infty) + \mathbb{M}_2(\infty)^2)\langle t \rangle^{1/4}.$$

Combining the above with Claim C.1, (9.1) and (9.2),

$$\|\overset{\circ}{\mathcal{N}}_7(t)\|_Y \lesssim \left\{(\eta_0 + e^{-\alpha L})\mathbb{M}_{c,x}(\infty) + \mathbb{M}_1(\infty) + \mathbb{M}_2(\infty)^2\right\}\langle t \rangle^{-1}.$$

This completes the proof of Theorem 1.5. □

Next, we will prove Corollary 1.6.

Proof of Corollary 1.6 Let $\zeta \in C_0^\infty(-\eta_0, \eta_0)$ such that $\zeta(0) = 1$ and let

$$u(0, x, y) = \varphi_{2+c_*(y)}(x) - \psi_{2+\tilde{c}_*(y), L}(x), \quad \tilde{c}_*(y) = \varepsilon(\mathcal{F}_\eta^{-1}\zeta)(y).$$

Then it follows from [22, Lemma 5.2] that

$$\tilde{c}(0, y) = \tilde{c}_*(y), \quad \tilde{x}(0, y) \equiv 0, \quad b_1(0, y) = b_*(y), \quad v_* = v_{2,*} = 0.$$

Since $\|b_* - \tilde{c}_*\|_{L^1} \lesssim \|\tilde{c}_*(0)\|_Y^2$, we see that $b_* \in L^1(\mathbb{R})$ and that

$$\int_\mathbb{R} b_1(0, y) \, dy \geq \int_\mathbb{R} \tilde{c}_*(y) \, dy - \|b_* - \tilde{c}_*\|_{L^1} = \frac{\varepsilon}{\sqrt{2\pi}} + O(\varepsilon^2).$$

If ε and $e^{-\alpha L}$ are sufficiently small, then it follows from (9.8), (9.9), (9.11), (9.15)–(9.17) and the above that

$$h \gtrsim \liminf_{t \to \infty} \tilde{x}(t, 0) \gtrsim \liminf_{t \to \infty} b_2(t, 0) \gtrsim \varepsilon, \tag{9.30}$$

where h is a constant in (1.10). Corollary 1.6 follows immediately from (9.30) and Theorem 1.5. Thus we complete the proof. □

10 Behavior of the Local Amplitude and the Local Inclination of Line Solitons

In this section, we will prove Theorem 1.4 following a compactness argument in [17].

Let $\mathbf{b}(t, \cdot) = \gamma(t)\mathcal{P}_*(D_y)e^{4t\sigma_3\partial_y}\mathbf{d}(t, \cdot)$ and

$$\Pi_*(\eta) = \frac{1}{4i}\begin{pmatrix} 8i & 8i \\ \eta + i\omega(\eta) & \eta - i\omega(\eta) \end{pmatrix} = \begin{pmatrix} 1 & 0 \\ 0 & i\eta \end{pmatrix}\mathcal{P}_*(\eta).$$

Then (9.7) is translated to

$$\partial_t \mathbf{d} = \{2\partial_y^2 I + \partial_y \tilde{\omega}(D_y)\sigma_3\}\mathbf{d} + \overset{\bullet}{\tilde{\mathcal{N}}}_a + \partial_y(\tilde{\mathcal{N}} + \tilde{\mathcal{N}}') + \partial_y^2 \tilde{\mathcal{N}}'', \qquad (10.1)$$

where $\sigma_3 = \text{diag}(1, -1)$, $\overset{\bullet}{\tilde{\mathcal{N}}}_a = e^{-4t\sigma_3 \partial_y} \Pi_*(D_y)^{-1} E_1 \chi(D_y) \overset{\bullet}{\mathcal{N}}_a$ and

$$\tilde{\mathcal{N}} = e^{-4t\sigma_3 \partial_y} \Pi_*(D_y)^{-1} \begin{pmatrix} 6bx_y \\ 2(\tilde{c} - b) + 3(x_y)^2 \end{pmatrix},$$

$$\tilde{\mathcal{N}}' = e^{-4t\sigma_3 \partial_y} \Pi_*(D_y)^{-1}$$
$$\left\{ \partial_y^{-1}(I - \chi(D_y)) E_1 \overset{\bullet}{\mathcal{N}}_a + E_2 \overset{\bullet}{\mathcal{N}}_a + \overset{\bullet}{\mathcal{N}}_b + \text{diag}(1, \partial_y)\left(\overset{\circ}{\mathcal{N}}_a + \overset{\circ}{\mathcal{N}}_b \right) \right\},$$

$$\tilde{\mathcal{N}}'' = e^{-4t\sigma_3 \partial_y} \Pi_*(D_y)^{-1} \text{diag}(1, \partial_y) \overset{\circ}{\mathcal{N}}_c.$$

Note that $\chi(\eta) = 1$ for $\eta \in [-\eta_0/4, \eta_0/4]$ and that $\text{diag}(1, \partial_y)\overset{\bullet}{\mathcal{N}}_b = \partial_y \overset{\bullet}{\mathcal{N}}_b$. By (2.13), we have for $\eta \in [-\eta_0, \eta_0]$,

$$\left| \Pi_*(\eta) - \begin{pmatrix} 2 & 2 \\ 1 & -1 \end{pmatrix} \right| + \left| \Pi_*(\eta)^{-1} - \frac{1}{4} \begin{pmatrix} 1 & 2 \\ 1 & -2 \end{pmatrix} \right| \lesssim |\eta|. \qquad (10.2)$$

If η_0 is sufficiently small, then $\Pi_*(D_y)$, $\Pi_*^{-1}(D_y) \in B(Y)$ and it follows from Claim C.1 and the definitions of $\mathbf{b}$ and $\mathbf{d}$ that

$$\left\| \begin{pmatrix} b(t, \cdot) \\ x_y(t, \cdot) \end{pmatrix} - \begin{pmatrix} 2 & 2 \\ 1 & -1 \end{pmatrix} e^{4t\sigma_3 \partial_y} \mathbf{d}(t, \cdot) \right\|_Y \lesssim \|k(t, \cdot)\|_Y + \|\partial_y \mathbf{d}(t, \cdot)\|_Y \lesssim \varepsilon \langle t \rangle^{-3/4}. \qquad (10.3)$$

To investigate the asymptotic behavior of solutions, we consider the rescaled solution $\mathbf{d}_\lambda(t, y) = \lambda \mathbf{d}(\lambda^2 t, \lambda y)$. We will show that for any t_1 and t_2 satisfying $0 < t_1 < t_2 < \infty$,

$$\lim_{\lambda \to \infty} \sup_{t \in [t_1, t_2]} \|\mathbf{d}_\lambda(t, y) - \mathbf{d}_\infty(t, y)\|_{L^2(\mathbb{R})} = 0, \qquad (10.4)$$

where $\mathbf{d}_\infty(t, y) = {}^t(d_{\infty,+}(t, y), d_{\infty,-}(t, y))$ and $d_{\infty,\pm}(t, y)$ are self-similar solutions of Burgers equations

$$\begin{cases} \partial_t d_+ = 2\partial_y^2 d_+ + 4\partial_y(d_+^2), \\ \partial_t d_- = 2\partial_y^2 d_- - 4\partial_y(d_-)^2. \end{cases} \qquad (10.5)$$

satisfying

$$\lambda \mathbf{d}_\infty(\lambda^2 t, \lambda y) = \mathbf{d}_\infty(t, y) \quad \text{for every } \lambda > 0. \qquad (10.6)$$

First, we will show uniform boundedness of $\mathbf{d}_\lambda$ with respect to $\lambda \geq 1$.

Lemma 10.1 *Let ε be as in Theorem 1.4. Then there exists a positive constants C such that for any $\lambda \geqslant 1$ and $t \in (0, \infty)$,*

$$\sum_{k=0,1} \|\partial_y^k \mathbf{d}_\lambda(t, \cdot)\|_{L^2} \leqslant C \varepsilon t^{-(2k+1)/4}, \quad \|\partial_y^2 \mathbf{d}_\lambda(t, \cdot)\|_{L^2} \leqslant C \varepsilon \lambda^{1/2} t^{-1}, \qquad (10.7)$$

$$\|\partial_t \mathbf{d}_\lambda(t, \cdot)\|_{H^{-2}} \leqslant C(t^{-1/4} + t^{-3/2})\varepsilon. \qquad (10.8)$$

Proof The proof follows the line of the proof of [22, Lemma 12.1]. By Proposition 9.1 and (10.3),

$$\sum_{k=0,1} \langle t \rangle^{(2k+1)/4} \|\partial_y^k \mathbf{d}(t)\|_Y + \langle t \rangle \|\partial_y^2 \mathbf{d}(t)\|_Y \lesssim \varepsilon \quad \text{for every } t \geqslant 0, \qquad (10.9)$$

and (10.7) follows immediately from (10.9).

Next, we will prove (10.8). By (10.1),

$$\partial_t \mathbf{d}_\lambda = 2\partial_y^2 \mathbf{d}_\lambda + \lambda \sigma_3 \partial_y \tilde{\omega}(\lambda^{-1} D_y)\mathbf{d}_\lambda + \widetilde{\mathcal{N}}_{a,\lambda} + \partial_y(\widetilde{\mathcal{N}}_\lambda + \widetilde{\mathcal{N}}'_\lambda) + \partial_y^2 \widetilde{\mathcal{N}}''_\lambda,$$

where $\widetilde{\mathcal{N}}_{a,\lambda}(t, y) = \lambda^3 \widetilde{\mathcal{N}}_a(\lambda^2 t, \lambda y)$ and

$$\widetilde{\mathcal{N}}_\lambda(t, y) = \lambda^2 \widetilde{\mathcal{N}}(\lambda^2 t, \lambda y), \quad \widetilde{\mathcal{N}}'_\lambda(t, y) = \lambda^2 \widetilde{\mathcal{N}}'(\lambda^2 t, \lambda y), \quad \widetilde{\mathcal{N}}''_\lambda(t, y) = \lambda \widetilde{\mathcal{N}}''(\lambda^2 t, \lambda y).$$

In view of (2.13) and (10.7),

$$\|\partial_y^2 \mathbf{d}_\lambda(t, \cdot)\|_{H^{-2}} + \|\lambda \partial_y \tilde{\omega}(\lambda^{-1} D_y)\mathbf{d}_\lambda(t, \cdot)\|_{H^{-2}} \lesssim \|\mathbf{d}_\lambda(t, \cdot)\|_{L^2} \lesssim \varepsilon t^{-1/4}.$$

Using (9.12)–(9.14), (10.2) and the fact that $Y_1 \subset Y$, we have

$$\|\widetilde{\mathcal{N}}_a\|_{L^1} \lesssim \left(e^{-\alpha L} \varepsilon + \varepsilon^2\right) \langle t \rangle^{-3/2}, \qquad (10.10)$$

$$\|\widetilde{\mathcal{N}}\|_Y \lesssim \varepsilon^2 \langle t \rangle^{-3/4}, \quad \|\widetilde{\mathcal{N}}'\|_Y \lesssim \varepsilon^2 \langle t \rangle^{-1}, \quad \|\widetilde{\mathcal{N}}''\|_Y \lesssim \varepsilon \langle t \rangle^{-1},$$

and

$$\|\widetilde{\mathcal{N}}_{a,\lambda}(t, \cdot)\|_{L^1} = \lambda^2 \|\widetilde{\mathcal{N}}_a(\lambda^2 t, \cdot)\|_Y \lesssim \left(e^{-\alpha L} \varepsilon + \varepsilon^2\right) \lambda^{-1} t^{-3/2}, \qquad (10.11)$$

$$\|\widetilde{\mathcal{N}}_\lambda(t, \cdot)\|_{L^2} = \lambda^{3/2} \|\widetilde{\mathcal{N}}(\lambda^2 t, \cdot)\|_Y \lesssim \varepsilon^2 t^{-3/4}, \qquad (10.12)$$

$$\|\widetilde{\mathcal{N}}'_\lambda(t, \cdot)\|_{L^2} = \lambda^{3/2} \|\widetilde{\mathcal{N}}'_1(\lambda^2 t, \cdot)\|_Y \lesssim \varepsilon \lambda^{3/2}(1 + \lambda^2 t)^{-1} \lesssim \varepsilon \lambda^{-1/4} t^{-7/8}, \qquad (10.13)$$

$$\|\widetilde{\mathcal{N}}''_\lambda(t, \cdot)\|_{L^2} = \lambda^{1/2} \|\widetilde{\mathcal{N}}''(\lambda^2 t, \cdot)\|_Y \lesssim \varepsilon \lambda^{1/2}(1 + \lambda^2 t)^{-1} \lesssim \varepsilon \lambda^{-1/2} t^{-1/2}. \qquad (10.14)$$

Combining the above, we have (10.8). Thus we complete the proof. □

Using a standard compactness argument along with the Aubin-Lions lemma, we have the following.

Corollary 10.2 *There exist a sequence $\{\lambda_n\}_{n\geqslant 1}$ satisfying $\lim_{n\to\infty}\lambda_n = \infty$ and $\mathbf{d}_\infty(t, y)$ such that*

$$\mathbf{d}_{\lambda_n}(t, \cdot) \to \mathbf{d}_\infty(t, \cdot) \quad \text{weakly star in } L^\infty_{loc}((0, \infty); H^1(\mathbb{R})),$$

$$\partial_t \mathbf{d}_{\lambda_n}(t, \cdot) \to \partial_t \mathbf{d}_\infty(t, \cdot) \quad \text{weakly star in } L^\infty_{loc}((0, \infty); H^{-2}(\mathbb{R})),$$

$$\sup_{t>0} t^{1/4} \|\mathbf{d}_\infty(t)\|_{L^2} \leqslant C\varepsilon, \tag{10.15}$$

where C is a constant given in Lemma 10.1. Moreover, for any $R > 0$ and t_1, t_2 with $0 < t_1 \leqslant t_2 < \infty$,

$$\lim_{n\to\infty} \sup_{t\in[t_1, t_2]} \|\mathbf{d}_{\lambda_n}(t, \cdot) - \mathbf{d}_\infty(t, \cdot)\|_{L^2(|y|\leqslant R)} = 0. \tag{10.16}$$

Next, we will show that $\mathbf{d}_\infty(t, y)$ tends to a constant multiple of the delta function as $t \downarrow 0$. To find initial data of $\mathbf{d}_\infty(t, y)$, we transform (10.1) into a conservative system. Let

$$\tilde{\mathbf{d}}(t, y) = \begin{pmatrix} \tilde{d}_+(t, y) \\ \tilde{d}_-(t, y) \end{pmatrix} := \mathbf{d}(t, y) - \bar{\mathbf{d}}(t, y), \quad \bar{\mathbf{d}}(t, y) = -\int_t^\infty \widetilde{\mathcal{N}}_a(s, \cdot) \, ds.$$

Then

$$\partial_t \tilde{\mathbf{d}} = 2\partial_y^2 \tilde{\mathbf{d}} + \partial_y(\widetilde{\mathcal{N}} + \widetilde{\mathcal{N}}') + \partial_y^2(\widetilde{\mathcal{N}}'' + \widetilde{\mathcal{N}}'''), \tag{10.17}$$

where $\widetilde{\mathcal{N}}''' = 2\bar{\mathbf{d}} + \partial_y^{-1}\tilde{\omega}(D_y)\sigma_3 \mathbf{d}$.

Lemma 10.3

$$\lim_{t\downarrow 0} \int_\mathbb{R} \mathbf{d}_\infty(t, y) h(y) \, dy = h(0) \int_\mathbb{R} \tilde{\mathbf{d}}(0, y) \, dy \quad \text{for any } h \in H^2(\mathbb{R}). \tag{10.18}$$

Proof Let $\tilde{\mathbf{d}}_\lambda(t, y) = \lambda \tilde{\mathbf{d}}(\lambda^2 t, \lambda y)$ and $\bar{\mathbf{d}}_\lambda(t, y) = \lambda \bar{\mathbf{d}}(\lambda^2 t, \lambda y)$. By (10.10), (10.11) and the fact that $\|\bar{\mathbf{d}}(t, \cdot)\|_Y \lesssim \|\bar{\mathbf{d}}(t, \cdot)\|_{L^1}$,

$$\|\bar{\mathbf{d}}(t)\|_{L^1} + \|\bar{\mathbf{d}}(t)\|_Y \lesssim \left(e^{-\alpha L}\varepsilon + \varepsilon^2\right)\langle t\rangle^{-1/2}, \tag{10.19}$$

$$\|\bar{\mathbf{d}}_\lambda(t, \cdot)\|_{L^2} = \lambda^{1/2}\|\bar{\mathbf{d}}(\lambda^2 t, \cdot)\|_Y \lesssim \left(e^{-\alpha L}\varepsilon + \varepsilon^2\right)\lambda^{-1/2} t^{-1/2}. \tag{10.20}$$

Hence the limiting profile of $\mathbf{d}_\lambda(t)$ and $\tilde{\mathbf{d}}_\lambda(t)$ as $\lambda \to \infty$ are the same for every $t > 0$.

By (10.17),
$$\partial_t \tilde{\mathbf{d}}_\lambda = 2\partial_y^2 \tilde{\mathbf{d}}_\lambda + \partial_y(\widetilde{\mathcal{N}}_\lambda + \widetilde{\mathcal{N}}'_\lambda) + \partial_y^2(\widetilde{\mathcal{N}}''_\lambda + \widetilde{\mathcal{N}}'''_\lambda),$$

where $\widetilde{\mathcal{N}}'''_\lambda = 2\bar{\mathbf{d}}_\lambda + \lambda \partial_y^{-1}\tilde{\omega}(\lambda^{-1}D_y)\mathbf{d}_\lambda$. By Lemma 10.1 and (2.13),

$$\left\| 2\tilde{\mathbf{d}}_\lambda + \widetilde{\mathcal{N}}'''_\lambda \right\|_{L^2} \lesssim \|\mathbf{d}_\lambda\|_{L^2} \lesssim \varepsilon t^{-1/4}. \tag{10.21}$$

Combining the above with (10.12)–(10.14), we have

$$\sup_{\lambda \geq 1} \|\partial_t \tilde{\mathbf{d}}_\lambda(t, \cdot)\|_{H^{-2}} \lesssim \varepsilon(t^{-1/4} + t^{-7/8}).$$

Thus for $t > s > 0$ and $h \in H^2(\mathbb{R})$,

$$\left| \int_\mathbb{R} \tilde{\mathbf{d}}_\lambda(t, y) h(y)\, dy - \int_\mathbb{R} \tilde{\mathbf{d}}_\lambda(s, y) h(y)\, dy \right| \leq C \left\{ (t-s)^{3/4} + (t-s)^{1/8} \right\},$$

where C is a constant independent of λ. Passing to the limit as $s \downarrow 0$ in the above, we obtain for $t > 0$,

$$\left| \int_\mathbb{R} \tilde{\mathbf{d}}_\lambda(t, y) h(y)\, dy - \int_\mathbb{R} \tilde{\mathbf{d}}_\lambda(0, y) h(y)\, dy \right| \leq C(t^{3/4} + t^{1/8}). \tag{10.22}$$

Since $\tilde{\mathbf{d}}(0, \cdot) \in L^1(\mathbb{R}) + \partial_y Y_1$, it follows from Lebesgue's dominated convergence theorem that as $\lambda \to \infty$,

$$\int_\mathbb{R} \tilde{\mathbf{d}}_\lambda(0, y) h(y)\, dy = \int_\mathbb{R} \mathcal{F}_y \tilde{\mathbf{d}}(0, \lambda^{-1}\eta) \mathcal{F}_y^{-1} h(\eta)\, d\eta \to \sqrt{2\pi} \mathcal{F}_y \tilde{\mathbf{d}}(0, 0) h(0).$$

On the other hand, Corollary 10.2 and (10.20) imply that for any $t > 0$ and $h \in L^2(\mathbb{R})$,

$$\lim_{n \to \infty} \int_\mathbb{R} \tilde{\mathbf{d}}_{\lambda_n}(t, y) h(y)\, dy = \int_\mathbb{R} \mathbf{d}_\infty(t, y) h(y)\, dy.$$

This completes the proof of Lemma 10.3. □

Now we will improve (10.16) to show (10.4).

Lemma 10.4 *Suppose that ε is sufficiently small. Then for every t_1 and t_2 satisfying $0 < t_1 \leq t_2 < \infty$, there exist a positive constant C and a function $\tilde{\delta}(R)$ satisfying $\lim_{R \to \infty} \tilde{\delta}(R) = 0$ such that*

$$\sup_{t \in [t_1, t_2]} \|\mathbf{d}_\lambda(t, \cdot)\|_{L^2(|y| \geq R)} \leq C(\tilde{\delta}(R) + \lambda^{-1/4}) \quad \text{for } \lambda \geq 1.$$

Proof Let ζ be a smooth function such that $\zeta(y) = 0$ if $|y| \leqslant 1/2$ and $\zeta(y) = 1$ if $|y| \geqslant 1$ and $\zeta_R(y) = \zeta(y/R)$. Multiplying (10.17) by ζ_R, we have

$$(\partial_t - 2\partial_y^2)(\zeta_R \tilde{\mathbf{d}}_\lambda) = \partial_y\{\zeta_R(\widetilde{\mathcal{N}}_\lambda + \widetilde{\mathcal{N}}'_\lambda)\} + \partial_y^2\{\zeta_R(\widetilde{\mathcal{N}}''_\lambda + \widetilde{\mathcal{N}}'''_\lambda)\} - \widetilde{\mathcal{N}}_R, \quad (10.23)$$

where $\widetilde{\mathcal{N}}_R = \widetilde{\mathcal{N}}_{R,1} + \widetilde{\mathcal{N}}_{R,2}$, $\widetilde{\mathcal{N}}_{R,1} = [\partial_y, \zeta_R](\widetilde{\mathcal{N}}_\lambda + \widetilde{\mathcal{N}}'_\lambda)$ and $\widetilde{\mathcal{N}}_{R,2} = [\partial_y^2, \zeta_R](2\tilde{\mathbf{d}}_\lambda + \widetilde{\mathcal{N}}''_\lambda + \widetilde{\mathcal{N}}'''_\lambda)$. Using the variation of constants formula, we have

$$\zeta_R \tilde{\mathbf{d}}_\lambda(t) = e^{2t\partial_y^2}\zeta_R \tilde{\mathbf{d}}(0) + \sum_{j=1}^{6} IV_j,$$

$$IV_1 = \int_0^t e^{2(t-\tau)\partial_y^2}\partial_y(\zeta_R \widetilde{\mathcal{N}}_\lambda(\tau))\,d\tau, \quad IV_2 = \int_0^t e^{2(t-\tau)\partial_y^2}\partial_y(\zeta_R \widetilde{\mathcal{N}}'_\lambda(\tau))\,d\tau,$$

$$IV_3 = \int_0^t e^{2(t-\tau)\partial_y^2}\partial_y^2(\zeta_R \widetilde{\mathcal{N}}''_\lambda(\tau))\,d\tau, \quad IV_4 = \int_0^t e^{2(t-\tau)\partial_y^2}\partial_y^2(\zeta_R \widetilde{\mathcal{N}}'''_\lambda(\tau))\,d\tau,$$

$$IV_5 = -\int_0^t e^{2(t-\tau)\partial_y^2}\widetilde{\mathcal{N}}_{R,1}(\tau)\,d\tau, \quad IV_6 = -\int_0^t e^{2(t-\tau)\partial_y^2}\widetilde{\mathcal{N}}_{R,2}(\tau)\,d\tau.$$

By Lemma 4.2, (3.11), (4.7) and (10.19), we can decompose $\tilde{\mathbf{d}}(0)$ as

$$\tilde{\mathbf{d}}(0) = \dot{\mathbf{d}}_0 + \partial_y \overset{\circ}{\mathbf{d}}_0, \quad \|\dot{\mathbf{d}}_0\|_{L^1(\mathbb{R})} + \|\overset{\circ}{\mathbf{d}}_0\|_{Y_1} \lesssim \varepsilon. \quad (10.24)$$

Let $\dot{\mathbf{d}}_{0,\lambda}(y) = \lambda \dot{\mathbf{d}}_0(\lambda y)$ and $\overset{\circ}{\mathbf{d}}_{0,\lambda}(y) = \overset{\circ}{\mathbf{d}}_0(\lambda y)$. Then $\tilde{\mathbf{d}}_\lambda(0, y) = \dot{\mathbf{d}}_{0,\lambda}(y) + \partial_y \overset{\circ}{\mathbf{d}}_{0,\lambda}(y)$ and

$$\|e^{2t\partial_y^2}\zeta_R \tilde{\mathbf{d}}_\lambda(0)\|_{L^2} \lesssim t^{-1/4}\|\zeta_R \dot{\mathbf{d}}_{0,\lambda}\|_{L^1} + t^{-1/2}\|\overset{\circ}{\mathbf{d}}_{0,\lambda}\|_{L^2} + \|[\partial_y, \zeta_R]\overset{\circ}{\mathbf{d}}_{0,\lambda}\|_{L^2}$$

$$\lesssim t^{-1/4}\|\dot{\mathbf{d}}_0\|_{L^1(|y|\geqslant \lambda R)} + \{R^{-1} + (t\lambda)^{-1/2}\}\|\overset{\circ}{\mathbf{d}}_0\|_{L^2}.$$

By Lemma 10.1 and (10.20),

$$\|IV_1\|_{L^2} \lesssim \int_0^t (t-\tau)^{-3/4}\|\zeta_R \mathbf{d}_\lambda(\tau)\|_{L^2}\|\mathbf{d}_\lambda(\tau)\|_{L^2}\,d\tau$$

$$\lesssim \varepsilon \int_0^t (t-\tau)^{-3/4}\tau^{-1/4}\|\zeta_R \tilde{\mathbf{d}}_\lambda(\tau)\|_{L^2}\,d\tau + \varepsilon^2 \lambda^{-1/2} \int_0^t (t-\tau)^{-3/4}\tau^{-3/4}\,d\tau$$

$$\lesssim \varepsilon \int_0^t (t-\tau)^{-3/4}\tau^{-1/4}\|\zeta_R \tilde{\mathbf{d}}_\lambda(\tau)\|_{L^2}\,d\tau + \varepsilon^2 \lambda^{-1/2} t^{-1/2}.$$

By (10.13),

$$\|IV_2\|_{L^2} \lesssim \int_0^t (t-\tau)^{-1/2} \|\zeta_R \widetilde{\mathcal{N}}'_\lambda(\tau)\|_{L^2} d\tau$$

$$\lesssim \varepsilon \lambda^{-1/4} \int_0^t (t-\tau)^{-1/2} \tau^{-7/8} d\tau \lesssim \varepsilon \lambda^{-1/4} t^{-3/8}.$$

Since $\mathcal{F}_y \widetilde{\mathcal{N}}''_\lambda(t, \eta) = 0$ for $\eta \notin [-\lambda\eta_0, \lambda\eta_0]$, it follows from (10.14) that

$$\|\widetilde{\mathcal{N}}''_\lambda(\tau, \cdot)\|_{H^{1/4}} \lesssim \lambda^{1/4} \|\widetilde{\mathcal{N}}''_\lambda(\tau, \cdot)\|_{L^2} \lesssim \varepsilon \lambda^{-1/4} \tau^{-1/2},$$

$$\|IV_3\|_{L^2} \lesssim \varepsilon \lambda^{-1/4} \int_0^t (t-\tau)^{-7/8} \tau^{-1/2} ds \lesssim \varepsilon \lambda^{-1/4} t^{-3/8}.$$

Using Lemma 10.1, (2.13) and (10.20), we have

$$\|\widetilde{\mathcal{N}}'''_\lambda\|_{H^{1/4}} \lesssim \|\bar{\mathbf{d}}_\lambda\|_{H^{1/4}} + \lambda \|\partial_y^{-1} \tilde{\omega}(\lambda^{-1} D_y) \mathbf{d}_\lambda\|_{H^{1/4}}$$

$$\lesssim \lambda^{1/4} \|\bar{\mathbf{d}}_\lambda\|_{L^2} + \lambda^{-1/4} \|\mathbf{d}_\lambda\|_{H^{1/2}} \lesssim \varepsilon \lambda^{-1/4} t^{-1/2},$$

and

$$\|IV_4\|_{L^2} \lesssim \int_0^t (t-\tau)^{-7/8} \|\zeta_R \widetilde{\mathcal{N}}'''_\lambda(\tau)\|_{H^{1/4}} d\tau$$

$$\lesssim \varepsilon \lambda^{-1/4} \int_0^t (t-\tau)^{-7/8} \tau^{-1/2} d\tau \lesssim \varepsilon \lambda^{-1/4} t^{-3/8}.$$

By (10.12) and (10.13),

$$\|IV_5\|_{L^2} \lesssim \int_0^t \|\partial_y \zeta_R\|_{L^\infty} \left(\|\widetilde{\mathcal{N}}_\lambda(\tau)\|_{L^2} + \|\widetilde{\mathcal{N}}'_\lambda(\tau)\|_{L^2}\right) d\tau \lesssim \frac{\varepsilon}{R}(t^{1/4} + t^{1/8}).$$

By (10.7), (10.14), (10.20) and (10.21),

$$\|IV_6\|_{L^2} \lesssim \int_0^t \{\|\partial_y^2 \zeta_r\|_{L^\infty} + (t-\tau)^{-1/2} \|\partial_y \zeta_R\|_{L^\infty}\}$$

$$\left(\|\widetilde{\mathcal{N}}''_\lambda(\tau)\|_{L^2} + \left\|2\tilde{\mathbf{d}}_\lambda(\tau) + \widetilde{\mathcal{N}}'''_\lambda(\tau)\right\|_{L^2}\right) d\tau$$

$$\lesssim \frac{\varepsilon}{R} \int_0^t \{1 + (t-\tau)^{-1/2}\}\{(\lambda\tau)^{-1/2} + \tau^{-1/4}\} d\tau \lesssim \frac{\varepsilon}{R} \langle t \rangle^{3/4}.$$

Combining the above, we have for $t \in (0, t_2)$,

$$\|\zeta_R \mathbf{d}_\lambda(t)\|_{L^2} \lesssim t^{-1/4} \|\dot{\mathbf{d}}_0\|_{L^1(|y| \geq \lambda R)} + \frac{\varepsilon}{R} + \varepsilon \lambda^{-1/4} t^{-1/2}$$
$$+ \varepsilon \int_0^t (t-\tau)^{-3/4} \tau^{-1/4} \|\zeta_R \tilde{\mathbf{d}}_\lambda(\tau)\| \, d\tau,$$

and if ε is sufficiently small,

$$\sup_{t \in (0,t_2)} t^{1/2} \|\zeta_R \mathbf{d}_\lambda(t)\|_{L^2} \lesssim C(t_1, t_2) \left(\|\dot{\mathbf{d}}_0\|_{L^1(|y| \geq \lambda R)} + \frac{\varepsilon}{R} + \varepsilon \lambda^{-1/4} \right),$$

where $C(t_2)$ is a constant depending only on t_2. Thus we complete the proof. □

Now we are in position to prove Theorem 1.4

Proof of Theorem 1.4 Corollary 10.2 and Lemma 10.4 imply

$$\lim_{n \to \infty} \sup_{t \in [t_1, t_2]} \|\mathbf{d}_{\lambda_n}(t, y) - \mathbf{d}_\infty(t, y)\|_{L^2(\mathbb{R})} = 0,$$

and that $\mathbf{d}_\infty(t, y)$ is a solutions of (10.5) satisfying $\|\mathbf{d}_\infty(t, \cdot)\|_{L^2} \leq C \varepsilon t^{-1/4}$ for every $t > 0$.

Let $m_\pm \in (-2\sqrt{2}, 2\sqrt{2})$ be constants satisfying

$$\frac{1}{2} \log \left(\frac{2\sqrt{2} \pm m_\pm}{2\sqrt{2} \mp m_\pm} \right) = \int_\mathbb{R} \tilde{d}_\pm(0, y) \, dy.$$

Then for every $h \in H^1(\mathbb{R})$,

$$\lim_{t \downarrow 0} \int_\mathbb{R} u_B^\pm(t, y) h(y) \, dy = h(0) \int_\mathbb{R} \tilde{d}_\pm(0, y) \, dy.$$

If ε is sufficiently small, then solutions of (10.5) satisfying (10.15) and (10.18) are unique (see e.g. [22, pp. 74–75]). Hence it follows that

$$\mathbf{d}_\infty(t, y) = \begin{pmatrix} u_B^+(t, y + 4t) \\ u_B^-(t, y - 4t) \end{pmatrix}, \tag{10.25}$$

and that $\mathbf{d}_\infty(t, y)$ satisfies (10.6). Thanks to the uniqueness of the limiting profile $\mathbf{d}_\infty(t, y)$, we have (10.4).

By (10.4) and (10.6),

$$t^{1/4} \|\mathbf{d}(t, \cdot) - \mathbf{d}_\infty(t, \cdot)\|_{L^2(\mathbb{R})} = \|\mathbf{d}_{\sqrt{t}}(1, \cdot) - \mathbf{d}_\infty(1, \cdot)\|_{L^2(\mathbb{R})} \to 0 \quad \text{as } t \to \infty, \tag{10.26}$$

and Theorem 1.4 follows immediately from (10.3), (10.25) and (10.26). This completes the proof of Theorem 1.4. □

Acknowledgement This research is supported by JSPS KAKENHI Grant Number 17K05332.

Appendix A: Operator Norms of S_k^j

First, we recall the definitions of operators S_k^j and $\widetilde{S}^j$ used in [22, 23]. For $q_c = \varphi_c$, φ_c', $\partial_c \varphi_c$ and $\partial_z^{-1} \partial_c^m \varphi_c(z) = -\int_z^\infty \partial_c^m \varphi_c(z_1)\,dz_1$ ($m \geq 1$), let $S_k^1[q_c]$ and $S_k^2[q_c]$ be operators defined by

$$S_k^1[q_c](f)(t,y) = \frac{1}{2\pi} \int_{-\eta_0}^{\eta_0} \int_{\mathbb{R}^2} f(y_1) q_2(z) g_{k1}^*(z, \eta, 2) e^{i(y-y_1)\eta}\, dy_1 dz d\eta,$$

$$S_k^2[q_c](f)(t,y) = \frac{1}{2\pi} \int_{-\eta_0}^{\eta_0} \int_{\mathbb{R}^2} f(y_1) \tilde{c}(t, y_1) g_{k2}^*(z, \eta, c(t,y_1)) e^{i(y-y_1)\eta}\, dy_1 dz d\eta,$$

where

$$\delta q_c(z) = \frac{q_c(z) - q_2(z)}{c - 2},$$

$$g_{k2}^*(z, \eta, c) = g_{k1}^*(z, \eta, 2) \delta q_c(z) + \frac{g_{k1}^*(z, \eta, c) - g_{k1}^*(z, \eta, 2)}{c - 2} q_c(z),$$

$$\widetilde{S}_0 = 3 \begin{pmatrix} -S_1^1[\partial_z^{-1} \partial_c \varphi_c] & S_1^1[\varphi_c] \\ -S_2^1[\partial_z^{-1} \partial_c \varphi_c] & S_2^1[\varphi_c] \end{pmatrix}, \quad \widetilde{S}_j = \begin{pmatrix} -S_1^j[\partial_c \varphi_c] & S_1^j[\varphi_c'] \\ -S_2^j[\partial_c \varphi_c] & S_2^j[\varphi_c'] \end{pmatrix} \text{ for } j = 1, 2.$$

Let $S_k^3[p]$ and $S_k^4[p]$ be operators defined by

$$S_k^3[p](f)(t,y) = \frac{1}{2\pi} \int_{-\eta_0}^{\eta_0} \int_{\mathbb{R}^2} f(y_1) p(z + 3t + L) g_k^*(z, \eta) e^{i(y-y_1)\eta}\, dy_1 dz d\eta,$$

$$S_k^4[p](f)(t,y) = \frac{1}{2\pi} \int_{-\eta_0}^{\eta_0} \int_{\mathbb{R}^2} f(y_1) \tilde{c}(t, y_1) p(z + 3t + L)$$

$$\times g_{k3}^*(z, \eta, c(t,y_1)) e^{i(y-y_1)\eta}\, dy_1 dz d\eta,$$

where $g_{k3}^*(z, \eta, c) = (c - 2)^{-1}(g_k^*(z, \eta, c) - g_k^*(z, \eta))$ and $p(z) \in C_0^\infty(\mathbb{R})$. Let S_k^5 and S_k^6 be operators defined by

$$S_k^5(f)(t,y) = \frac{1}{2\pi} \int_{-\eta_0}^{\eta_0} \int_{\mathbb{R}^2} v_2(t, z, y_1) f(y_1) \partial_c g_k^*(z, \eta, c(t,y_1)) e^{i(y-y_1)\eta}\, dz dy_1 d\eta,$$

$$S_k^6(f)(t,y) = -\frac{1}{2\pi} \int_{-\eta_0}^{\eta_0} \int_{\mathbb{R}^2} v_2(t, z, y_1) f(y_1) \partial_z g_k^*(z, \eta, c(t,y_1)) e^{i(y-y_1)\eta}\, dz dy_1 d\eta.$$

Let $\widetilde{S}_3 = S_1^3[\psi]E_1 + S_2^3[\psi]E_{21}$,

$$\widetilde{S}_4 = \begin{pmatrix} S_1^3[\psi]((\sqrt{2/c}-1)\cdot) + S_1^4[\psi](\sqrt{2/c}\cdot) & -2(S_1^3[\psi'] + S_1^4[\psi'])((\sqrt{2c}-2)\cdot) \\ S_2^3[\psi]((\sqrt{2/c}-1)\cdot) + S_2^4[\psi](\sqrt{2/c}\cdot) & -2(S_2^3[\psi'] + S_2^4[\psi'])((\sqrt{2c}-2)\cdot) \end{pmatrix},$$

$$\widetilde{S}_5 = \begin{pmatrix} S_1^5 & S_1^6 \\ S_2^5 & S_2^6 \end{pmatrix},$$

and $\bar{S}_j = \widetilde{S}_j(I + \widetilde{\mathcal{C}}_2)^{-1}$ for $1 \leq j \leq 5$, where

$$\mathcal{C}_2 = \widetilde{P}_1\left\{(c(t,\cdot)/2)^{1/2} - 1\right\}\widetilde{P}_1, \quad \widetilde{\mathcal{C}}_2 = \mathcal{C}_2 E_1.$$

Now we decompose the operator S_k^j ($1 \leq j \leq 6$, $k = 1, 2$) into a sum of a time-dependent constant multiple of $\widetilde{P}_1$ and an operator which belongs to $\partial_y^2 B(Y)$. Let

$$S_{k1}^3[p](t)f(y) = \frac{1}{2\pi}\int_{-\eta_0}^{\eta_0}\int_{\mathbb{R}^2} f(y_1)p(z+3t+L)g_k^*(z,0)e^{i(y-y_1)\eta}\,dy_1\,dz\,d\eta$$

$$= \left(\int_{\mathbb{R}^2} p(z+3t+L)g_k^*(z,0)\,dz\right)\widetilde{P}_1 f, \tag{A.1}$$

$$S_{k2}^3[p](f)(t,y) = \frac{1}{2\pi}\int_{-\eta_0}^{\eta_0}\int_{\mathbb{R}^2} f(y_1)p(z+3t+L)g_{k1}^*(z,\eta)e^{i(y-y_1)\eta}\,dy_1\,dz\,d\eta, \tag{A.2}$$

$$S_{k1}^4[p](f)(t,y) = \widetilde{P}_1\left\{\tilde{c}(t,\cdot)f\int_{\mathbb{R}} p(z+3t+L)g_{k3}^*(z,0,c(t,\cdot))\,dz\right\}, \tag{A.3}$$

$$S_{k2}^4[p](f)(t,y) = \frac{1}{2\pi}\int_{-\eta_0}^{\eta_0}\int_{\mathbb{R}^2} f(y_1)\tilde{c}(t,y_1)p(z+3t+L)$$
$$\times g_{k4}^*(z,\eta,c(t,y_1))e^{i(y-y_1)\eta}\,dy_1\,dz\,d\eta, \tag{A.4}$$

where $g_{k4}^*(z,\eta,c) = \eta^{-2}\{g_{k3}^*(z,\eta,c) - g_{k3}^*(z,0,c)\}$ and

$$S_{k1}^5(f)(t,y) = \frac{1}{2\pi}\int_{-\eta_0}^{\eta_0}\int_{\mathbb{R}^2} v_2(t,z,y_1)f(y_1)\partial_c g_k^*(z,0,c(t,y_1))e^{i(y-y_1)\eta}\,dz\,dy_1\,d\eta, \tag{A.5}$$

$$S_{k2}^5(f)(t,y) = \frac{1}{2\pi}\int_{-\eta_0}^{\eta_0}\int_{\mathbb{R}^2} v_2(t,z,y_1)f(y_1)\partial_c g_{k1}^*(z,\eta,c(t,y_1))e^{i(y-y_1)\eta}\,dz\,dy_1\,d\eta, \tag{A.6}$$

$$S_{k1}^6(f)(t,y) = -\frac{1}{2\pi}\int_{-\eta_0}^{\eta_0}\int_{\mathbb{R}^2} v_2(t,z,y_1)f(y_1)\partial_z g_k^*(z,0,c(t,y_1))e^{i(y-y_1)\eta}\,dz\,dy_1\,d\eta, \tag{A.7}$$

$$S_{k2}^6(f)(t,y) = -\frac{1}{2\pi}\int_{-\eta_0}^{\eta_0}\int_{\mathbb{R}^2} v_2(t,z,y_1)f(y_1)\partial_z g_{k1}^*(z,\eta,c(t,y_1))e^{i(y-y_1)\eta}\,dzdy_1d\eta.$$
(A.8)

Then $S_k^j = S_{k1}^j - \partial_y^2 S_{k2}^j$ for $j = 3, 4, 5, 6$.

Claim A.1 Let $\alpha \in (0, 2)$. There exist positive constants C and η_1 such that for $\eta \in (0, \eta_1]$, $k = 1, 2$ and $t \geq 0$,

$$\|\chi(D_y)S_{k1}^3[p](f)(t,\cdot)\|_{L^1} \leq Ce^{-\alpha(3t+L)}\|e^{\alpha z}p\|_{L^2}\|f\|_{L^1(\mathbb{R})},$$
(A.9)

$$\|S_{k1}^3[p](f)(t,\cdot)\|_Y + \|S_{k2}^3[p](f)(t,\cdot)\|_Y \leq Ce^{-\alpha(3t+L)}\|e^{\alpha z}p\|_{L^2}\|f\|_{L^2(\mathbb{R})},$$
(A.10)

$$\|S_{k1}^3[p](f)(t,\cdot)\|_{Y_1} + \|S_{k2}^3[p](f)(t,\cdot)\|_{Y_1} \leq Ce^{-\alpha(3t+L)}\|e^{\alpha z}p\|_{L^2}\|\widetilde{P}_1 f\|_{Y_1}.$$
(A.11)

Claim A.2 There exist positive constants η_1, δ and C such that if $\eta_0 \in (0, \eta_1]$ and $\mathbb{M}_{c,x}(T) \leq \delta$, then for $k = 1, 2$, $t \in [0, T]$ and $f \in L^2$,

$$\|\chi(D_y)S_{k1}^4[p](f)(t,\cdot)\|_{L^1(\mathbb{R})} \leq Ce^{-\alpha(3t+L)}\|e^{\alpha z}p\|_{L^2}\|\tilde{c}\|_Y\|f\|_{L^2},$$
(A.12)

$$\|S_{k1}^4[p](f)(t,\cdot)\|_{Y_1} + \|S_{k2}^4[p](f)(t,\cdot)\|_{Y_1} \leq Ce^{-\alpha(3t+L)}\|e^{\alpha z}p\|_{L^2}\|\tilde{c}\|_Y\|f\|_{L^2}.$$
(A.13)

Claim A.3 There exist positive constants η_1, δ and C such that if $\eta_0 \in (0, \eta_1]$ and $\mathbb{M}_{c,x}(T) \leq \delta$, then for $k = 1, 2$, $t \in [0, T]$ and $f \in L^2$,

$$\|\chi(D_y)S_{k1}^5(f)(t,\cdot)\|_{L^1(\mathbb{R})} + \|\chi(D_y)S_{k1}^6(f)(t,\cdot)\|_{L^1(\mathbb{R})} \leq C\|v_2(t)\|_X\|f\|_{L^2(\mathbb{R})},$$
(A.14)

$$\sum_{j=5,6}\left(\|S_{k1}^j(f)(t,\cdot)\|_{Y_1} + \|S_{k2}^j(f)(t,\cdot)\|_{Y_1}\right) \leq C\|v_2(t)\|_X\|f\|_{L^2}.$$
(A.15)

Proof of Claims A.1–A.3 Since $\chi(D_y)\widetilde{P}_1 = \chi(D_y)$,

$$\|\chi(D_y)S_{k1}^3[p](f)(t,\cdot)\|_{L^1(\mathbb{R})}$$
$$= \frac{1}{\sqrt{2\pi}}\left|\int_{\mathbb{R}} p(z + 3t + L)g_k^*(z, 0)\,dz\right|\|\check{\chi}*f\|_{L^1(\mathbb{R})}$$
$$\leq \|\check{\chi}\|_{L^1(\mathbb{R})}\|f\|_{L^1(\mathbb{R})}\|e^{\alpha z}p(z + 3t + L)\|_{L^2(\mathbb{R})}\|e^{-\alpha z}g_k^*(z,0)\|_{L^2(\mathbb{R})}$$
$$\lesssim e^{-\alpha(3t+L)}\|f\|_{L^1(\mathbb{R})}.$$

Using Young's inequality, we have

$$\|\chi(D_y)S_{k1}^5(f)(t,\cdot)\|_{L^1(\mathbb{R})}$$

$$= \left\|\int_\mathbb{R} \check{\chi}(y-y_1)f(y_1)\left\{\int_\mathbb{R} v_2(t,z,y_1)\partial_c g_k^*(z,0,c(t,y_1))\,dz\right\}dy_1\right\|_{L^1(\mathbb{R})}$$

$$\lesssim \|\check{\chi}\|_{L^1}\|f\|_{L^2(\mathbb{R})}\left\|\int_\mathbb{R} v_2(t,z,\cdot)\partial_c g_k^*(z,0,c(t,\cdot))\,dz\right\|_{L^2(\mathbb{R})}$$

$$\lesssim \|f\|_{L^2(\mathbb{R})}\|v_2(t)\|_X \sup_{c\in[2-\delta,2+\delta]}\|e^{-\alpha z}\partial_c g_k^*(z,0,c)\|_{L^2(\mathbb{R})}$$

$$\lesssim \|f\|_{L^2(\mathbb{R})}\|v_2(t)\|_X.$$

Similarly, we have (A.12) and $\|\chi(D_y)S_{k1}^6(f)(t,\cdot)\|_{L^1(\mathbb{R})} \lesssim \|f\|_{L^2(\mathbb{R})}\|v_2(t)\|_X$.

We can prove (A.10), (A.11), (A.13) and (A.15) in exactly the same way as the proof of Claims B.3–B.5 in [22]. Thus we complete the proof. □

For $\ell = 1, 2$, let

$$\widetilde{S}_{3\ell} = \begin{pmatrix} S_{1\ell}^3[\psi] & 0 \\ S_{2\ell}^3[\psi] & 0 \end{pmatrix}, \tag{A.16}$$

$$\widetilde{S}_{4\ell} = \begin{pmatrix} S_{1\ell}^3[\psi]((\sqrt{2/c}-1)\cdot) + S_{1\ell}^4[\psi](\sqrt{2/c}\cdot) & -2(S_{1\ell}^3[\psi'] + S_{1\ell}^4[\psi'])((\sqrt{2c}-2)\cdot) \\ S_{1\ell}^4[\psi]((\sqrt{2/c}-1)\cdot) + S_{2\ell}^4[\psi](\sqrt{2/c}\cdot) & -2(S_{2\ell}^3[\psi'] + S_{2\ell}^4[\psi'])((\sqrt{2c}-2)\cdot) \end{pmatrix},$$
$$\tag{A.17}$$

$$\widetilde{S}_{5\ell} = \begin{pmatrix} S_{1\ell}^5 & S_{1\ell}^6 \\ S_{2\ell}^5 & S_{2\ell}^6 \end{pmatrix}, \tag{A.18}$$

and $\bar{S}_{j\ell} = \widetilde{S}_{j\ell}(1+\widetilde{\mathcal{C}}_2)^{-1}$. Then $\widetilde{S}_j = \widetilde{S}_{j1} - \partial_y^2 \widetilde{S}_{j2}$ and $\bar{S}_j = \bar{S}_{j1} - \partial_y^2 \bar{S}_{j2}$ for $j = 3, 4, 5$.

Claim A.4 *There exist positive constants η_1, δ and C such that if $\eta_0 \in (0,\eta_1]$ and $\mathbb{M}_{c,x}(T) \leqslant \delta$, then for $k = 1, 2$ and $t \in [0,T]$,*

$$\|\chi(D_y)\mathcal{C}_k\|_{B(L^2;L^1)} \leqslant C\mathbb{M}_{c,x}(T)\langle t\rangle^{-1/4}, \tag{A.19}$$

$$\|\chi(D_y)\widetilde{S}_{31}\|_{B(L^1(\mathbb{R}))} + \|\widetilde{S}_{31}\|_{B(Y_1)} \leqslant Ce^{-\alpha(3t+L)}, \tag{A.20}$$

$$\|\chi(D_y)(\bar{S}_{31}-\widetilde{S}_{31})\|_{B(L^2(\mathbb{R}),L^1(\mathbb{R}))} + \|\bar{S}_{31} - \widetilde{S}_{31}\|_{B(Y,Y_1)} \leqslant C\mathbb{M}_{c,x}(T)\langle t\rangle^{-1/4}e^{-\alpha(3t+L)}, \tag{A.21}$$

$$\sum_{k=1,2}\left(\|\widetilde{S}_{3k}\|_{B(Y)\cap B(Y_1)} + \|\bar{S}_{3k}\|_{B(Y)\cap B(Y_1)}\right) \leqslant Ce^{-\alpha(3t+L)}, \tag{A.22}$$

$$\|\chi(D_y)\widetilde{S}_{41}\|_{B(L^2(\mathbb{R}),L^1(\mathbb{R}))} + \|\chi(D_y)\bar{S}_{41}\|_{B(L^2(\mathbb{R}),L^1(\mathbb{R}))} \leqslant C\langle t\rangle^{-1/4}e^{-\alpha(3t+L)}\mathbb{M}_{c,x}(T), \tag{A.23}$$

$$\sum_{k=1,2} \left(\|\widetilde{S}_{4k}\|_{B(L^2(\mathbb{R}),Y_1)} + \|\bar{S}_{4k}\|_{B(L^2(\mathbb{R}),Y_1)} \right) \lesssim C\langle t\rangle^{-1/4} e^{-\alpha(3t+L)} \mathbb{M}_{c,x}(T),$$
(A.24)

$$\|\chi(D_y)\widetilde{S}_{51}\|_{B(L^2(\mathbb{R}),L^1(\mathbb{R}))} + \|\chi(D_y)\bar{S}_{51}\|_{B(L^2(\mathbb{R})),L^1(\mathbb{R}))} \lesssim C\langle t\rangle^{-3/4} \mathbb{M}_2(T),$$
(A.25)

$$\sum_{k=1,2} \left(\|\widetilde{S}_{5k}\|_{B(L^2(\mathbb{R}),Y_1)} + \|\bar{S}_{5k}\|_{B(L^2(\mathbb{R}),Y_1)} \right) \lesssim C\langle t\rangle^{-3/4} \mathbb{M}_2(T).$$
(A.26)

Proof By the definition, we have for $f \in L^2(\mathbb{R})$,

$$\|\chi(D_y)\mathcal{C}_1 f\|_{L^1} = \frac{1}{2\sqrt{2\pi}} \left\| \check{\chi}_1 * (c^2 - 4)\widetilde{P}_1 f \right\|_{L^1(\mathbb{R})} \lesssim \langle t\rangle^{-1/4} \mathbb{M}_{c,x}(T) \|f\|_{L^2}.$$

We can prove $\|\chi(D_y)\mathcal{C}_2 f\|_{L^1} \lesssim \langle t\rangle^{-1/4} \mathbb{M}_{c,x}(T)\|f\|_{L^2}$ in the same way.

Equation (A.20) follows from Claim A.1. Let $f \in L^2(\mathbb{R})$ and $f_1 = \widetilde{\mathcal{C}}_2 f$. Since $\chi(\eta) = \chi(\eta)\chi_1(\eta)$,

$$\chi(D_y)S^3_{k1}[p](f_1) = \frac{1}{\sqrt{2\pi}} \int_\mathbb{R} \chi(\eta) \hat{f}_1(\eta) e^{iy\eta} d\eta \left(\int_\mathbb{R} p(z+3t+L)g_k^*(z,0) dz \right)$$
$$= \chi(D_y)S^3_{k1}[p](\chi_1(D_y)f_1),$$

and it follow from Claim A.1 that

$$\|\chi(D_y)S^3_{k1}[p](f_1)\|_{L^1(\mathbb{R})} \lesssim e^{-\alpha(3t+L)} \|e^{\alpha z}p\|_{L^2(\mathbb{R})} \|\chi_1(D_y)f_1\|_{L^1(\mathbb{R})}.$$

Combining the above and (A.19) with χ replaced by χ_1, we have for $k = 1, 2$ and $t \in [0,T]$,

$$\|\chi(D_y)S^3_{k1}[p](\widetilde{\mathcal{C}}_2 f)\|_{L^1(\mathbb{R})} \lesssim \langle t\rangle^{-1/4} \mathbb{M}_{c,x}(T) \|f\|_{L^2(\mathbb{R})}.$$
(A.27)

Since $\widetilde{S}_{31} - \bar{S}_{31} = \widetilde{S}_{31}\widetilde{\mathcal{C}}_2(I + \widetilde{\mathcal{C}}_2)^{-1}$ and $I + \widetilde{\mathcal{C}}_2$ has a bounded inverse on $L^2(\mathbb{R})$, (A.21) follows immediately from Claim A.1 and (A.27). We can prove (A.23) and (A.25) in the same way.

Equations (A.22)–(A.26) follow from Claims A.1–A.3. □

Let $\ell_{2,lin}$ be the linear part of $\ell_{22} + \ell_{23}$ in $\tilde{c}$ (see [23, p. 166]),

$$\tilde{a}_k(t, D_y)\tilde{c} := \frac{1}{2\pi} \int_{-\eta_0}^{\eta_0} \int_{\mathbb{R}^2} \ell_{2,lin}(t, z, y_1) g_k^*(z, \eta) e^{i(y-y_1)\eta} dy_1 dz d\eta \quad \text{for } k = 1, 2,$$

and $\widetilde{\mathcal{A}}_1(t) = \tilde{a}_1(t, D_y)E_1 + \tilde{a}_2(t, D_y)E_{21}$. More precisely,

$$\tilde{a}_k(t,\eta) = \left[\int_{\mathbb{R}} \left\{\partial_z\left(\partial_z^2 - 1 + 6\varphi(z)\right)\psi(z + 3t + L)\right\} g_k^*(z,\eta)\,dz\right.$$
$$\left. + 3\eta^2 \int_{\mathbb{R}}\left(\int_z^\infty \psi(z_1 + 3t + L)\,dz_1\right) g_k^*(z,\eta)\,dz\right]\mathbf{1}_{[-\eta_0,\eta_0]}(\eta),$$

Let $\tilde{\mathcal{A}}_{1j}(t) = \tilde{a}_{1j}(t)E_1 + \tilde{a}_{2j}E_{21}$ for $j = 1, 2$, where

$$\tilde{a}_{k1}(t) = \int_{\mathbb{R}} \left\{\partial_z\left(\partial_z^2 - 1 + 6\varphi(z)\right)\psi(z + 3t + L)\right\} g_k^*(z,0)\,dz,$$

$$\tilde{a}_{k2}(t,\eta) = \left[\int_{\mathbb{R}} \left\{\partial_z\left(\partial_z^2 - 1 + 6\varphi(z)\right)\psi(z + 3t + L)\right\} g_{k1}^*(z,\eta)\,dz\right.$$
$$\left. + 3\int_{\mathbb{R}}\left(\int_z^\infty \psi(z_1 + 3t + L)\,dz_1\right) g_k^*(z,\eta)\,dz\right]\mathbf{1}_{[-\eta_0,\eta_0]}(\eta).$$

Then $\tilde{\mathcal{A}}_1(t) = \tilde{\mathcal{A}}_{11}(t) - \partial_y^2 \tilde{\mathcal{A}}_{12}(t)$ and we have the following.

Claim A.5 There exist positive constants C and L_0 such that if $L \geq L_0$, then for every $t \geq 0$,

$$\|\chi(D_y)\tilde{\mathcal{A}}_{11}(t)\|_{B(L^1(\mathbb{R}))} + \|\tilde{\mathcal{A}}_{11}(t)\|_{B(Y_1)} \leq Ce^{-\alpha(3t+L)}, \tag{A.28}$$

$$\|\tilde{\mathcal{A}}_{12}(t)\|_{B(Y)} + \|\tilde{\mathcal{A}}_{12}(t)\|_{B(Y_1)} \leq Ce^{-\alpha(3t+L)}. \tag{A.29}$$

Since $\chi(D_y)\tilde{P}_1 = \chi(D_y)$, $\chi \in C_0^\infty$ and $\check{\chi}$ is integrable, we have (A.28). Equation (A.29) can be shown in exactly the same way as [22, Claims D.3].

Appendix B: Estimates of R^j

Let R_k^2 be as in [22, p. 39], $R_k^3 = R_k^{31} - \partial_y^2 R_k^{32}$ and

$$R_k^{31}(t, y) := \tilde{P}_1 \int_{\mathbb{R}} (\ell_{22} + \ell_{23}) g_k^*(z, 0, c(t, y_1))\,dz - \tilde{P}_1 \int_{\mathbb{R}} \ell_{2,lin} g_k^*(z, 0)\,dz,$$

$$R_k^{32}(t, y) := \frac{1}{2\pi} \int_{-\eta_0}^{\eta_0} \int_{\mathbb{R}^2} (\ell_{22} + \ell_{23}) g_{k1}^*(z, \eta, c(t, y_1)) e^{i(y-y_1)\eta}\,dzdy_1 d\eta$$
$$- \frac{1}{2\pi}\int_{-\eta_0}^{\eta_0}\int_{\mathbb{R}^2} \ell_{2,lin} g_{k1}^*(z,\eta) e^{i(y-y_1)\eta}\,dzdy_1 d\eta.$$

Then we have the following.

Claim B.1 ([22, Claim D.1]) There exist positive constants δ and C such that if $\mathbb{M}_{c,x}(T) \leqslant \delta$, then for $t \in [0, T]$,

$$\|R_k^2(t, \cdot)\|_{Y_1} \leqslant C\mathbb{M}_{c,x}(T)^2 \langle t \rangle^{-1}, \quad \|\partial_y R_k^2(t, \cdot)\|_{Y_1} \leqslant C\mathbb{M}_{c,x}(T)^2 \langle t \rangle^{-5/4}.$$

Claim B.2 There exist positive constants δ and C such that if $\mathbb{M}_{c,x}(T) \leqslant \delta$, then for $t \in [0, T]$,

$$\|R_k^3(t, \cdot)\|_{Y_1} + \|R_{k2}^3(t, \cdot)\|_{Y_1} \leqslant Ce^{-\alpha(3t+L)}\mathbb{M}_{c,x}(T)^2,$$

$$\|R_{k1}^3(t, \cdot)\|_{Y_1} + \|\chi(D_y)R_{k1}^3(t, \cdot)\|_{L^1(\mathbb{R})} \leqslant Ce^{-\alpha(3t+L)}\mathbb{M}_{c,x}(T)^2.$$

We can prove Claim B.2 in exactly the same way as Claim D.2 in [22]. Note that $\chi(D_y)\widetilde{P}_1 = \chi(D_y)$ and $\chi(D_y) \in B(L^1(\mathbb{R}))$.

In this paper, we slightly modify the definitions of R_k^4 and R_k^5 from [22, 23]. We move II_{k1}^1 into R_k^5 from R_k^4. Let

$$R_k^4(t, y) = \frac{1}{2\pi} \int_{-\eta_0}^{\eta_0} \left\{ II_{k2}^1(t, \eta) + II_{k3}^1(t, \eta) + II_k^2(t, \eta) + II_{k1}^3(t, \eta) \right\} e^{iy\eta} d\eta, \tag{B.1}$$

$$R_k^5(t, y) = \frac{1}{2\pi} \int_{-\eta_0}^{\eta_0} \left\{ II_{k1}^1(t, \eta) + II_{k2}^3(t, \eta) \right\} e^{iy\eta} d\eta. \tag{B.2}$$

See [23, p. 166] for the definitions of II_{kj}^3. For the definitions of II_{kj}^1, replace $v(t, z, y)$ by $v_2(t, z, y)$ in II_{kj}^1 defined in the proof of [22, Claim D.5]. We decompose R_k^4 further. For $j, k, \ell = 1, 2$,

$$h_{jk1}(t, y) = \int_{\mathbb{R}} v_2(t, z, y) \left(\int_{-\infty}^z \partial_c^j g_k^*(z_1, 0, c(t, y)) dz_1 \right) dz,$$

$$h_{jk2}(t, y, \eta) = \int_{\mathbb{R}} v_2(t, z, y) \left(\int_{-\infty}^z \partial_c^j g_{k1}^*(z_1, \eta, c(t, y)) dz_1 \right) dz,$$

$$II_{k2\ell}^1(t, \eta) = 3 \int_{\mathbb{R}} c_{yy}(t, y) h_{1k\ell}(t, y) e^{-iy\eta} dy,$$

$$II_{k3\ell}^1(t, \eta) = 3 \int_{\mathbb{R}} c_y(t, y)^2 h_{2k\ell}(t, y) e^{-iy\eta} dy,$$

Then $II_{k2}^1 + II_{k3}^1 = \overset{\bullet}{II}_{1,k} + \eta^2 \overset{\circ}{II}_{1,k}$, $\overset{\bullet}{II}_{1,k} = II_{k21}^1 + II_{k31}^1$, $\overset{\circ}{II}_{1,k} = II_{k22}^1 + II_{k32}^1$ and

$$\|\mathcal{F}^{-1}\overset{\bullet}{II}_{1,k}(t, \cdot)\|_{L^1(\mathbb{R})} + \|\overset{\circ}{II}_{1,k}(t, \cdot)\|_{Z_1} \lesssim \langle t \rangle^{-3/2} \mathbb{M}_{c,x}(T) \mathbb{M}_2(T) \quad \text{for } t \in [0, T].$$

Let

$$II_{k1}^2 = -\int_{\mathbb{R}^2} N_{2,1} \partial_z g_k^*(z, 0, c(t, y)) e^{-iy\eta} \, dzdy,$$

$$II_{k2}^2 = -\int_{\mathbb{R}^2} N_{2,1} \partial_z g_{k1}^*(z, \eta, c(t, y)) e^{-iy\eta} \, dzdy,$$

$$II_{k11}^3 = -3 \int_{\mathbb{R}^2} v_2(t, z, y) x_{yy}(t, y) g_k^*(z, 0, c(t, y)) e^{-iy\eta} \, dzdy,$$

$$II_{k12}^3 = -3 \int_{\mathbb{R}^2} v_2(t, z, y) x_{yy}(t, y) g_{k1}^*(z, \eta, c(t, y)) e^{-iy\eta} \, dzdy.$$

Let

$$R_k^{41}(t, y) = \frac{1}{2\pi} \int_{-\eta_0}^{\eta_0} \left\{ \overset{\bullet}{II}_k^1(t, \eta) + II_{k1}^2(t, \eta) + II_{k11}^3(t, \eta) \right\} e^{iy\eta} \, d\eta,$$

$$R_k^{42}(t, y) = \frac{1}{2\pi} \int_{-\eta_0}^{\eta_0} \left\{ \overset{\circ}{II}_{k1}^1(t, \eta) + II_{k2}^2(t, \eta) + II_{k12}^3(t, \eta) \right\} e^{iy\eta} \, d\eta.$$

Then $R_k^4 = R_k^{41} - \partial_y^2 R_k^{42}$. Let $R_k^6 = R_k^{61} - \partial_y^2 R_k^{62}$ and

$$R_k^{61} = -6\widetilde{P}_1 \left(\int_{\mathbb{R}} \psi_{c(t,y_1),L}(z + 3t) v_2(t, z, y_1) \partial_z g_k^*(z, 0, c(t, y_1)) \, dz \right),$$

$$R_k^{62} = -\frac{3}{\pi} \int_{-\eta_0}^{\eta_0} \int_{\mathbb{R}^2} \psi_{c(t,y_1),L}(z + 3t) v_2(t, z, y_1) \partial_z g_{k1}^*(z, \eta, c(t, y_1)) e^{i(y-y_1)\eta} \, dy_1 dz d\eta.$$

We can prove the following in the same way as [22, Claim D.5].

Claim B.3 Suppose $\alpha \in (0, 1)$ and $\mathbb{M}_{c,x}(T) \leq \delta$. If δ is sufficiently small, then there exists a positive constant C such that for $t \in [0, T]$,

$$\|\chi(D_y) R_k^{41}(t)\|_{L^1} + \|R_k^{41}(t)\|_{Y_1} + \|R_k^{42}(t)\|_{Y_1}$$
$$\leq C \langle t \rangle^{-3/2} (\mathbb{M}_{c,x}(T) + \mathbb{M}_1(T) + \mathbb{M}_2(T)) \mathbb{M}_2(T),$$

$$\|R_k^5(t)\|_{Y_1} \leq C \langle t \rangle^{-1} \mathbb{M}_{c,x}(T) \mathbb{M}_2(T), \quad \|R_k^5(t)\|_Y \leq C \langle t \rangle^{-5/4} \mathbb{M}_{c,x}(T) \mathbb{M}_2(T),$$

$$\|\chi(D_y) R_k^{61}\|_{L^1(\mathbb{R})} + \|R_k^{61}\|_{Y_1} + \|R_k^{62}\|_{Y_1} \leq C \langle t \rangle^{-1} e^{-\alpha(3t+L)} \mathbb{M}_{c,x}(T) \mathbb{M}_2(T).$$

Let $R^9 = R^{91} - \partial_y^2 R^{92}$ and

$$R^{91} = -6 \sum_{3 \leq j \leq 5} \bar{S}_{j1} \{(I + \mathcal{C}_2)(c_y x_y) - (bx_y)_y\} \mathbf{e}_1,$$

$$R^{92} = -6 \sum_{3 \leq j \leq 5} \bar{S}_{j2} \{(I + \mathcal{C}_2)(c_y x_y) - (bx_y)_y\} \mathbf{e}_1.$$

Using Claims A.1–A.3 and boundedness of operators ∂_y, $\bar{S}_1$, $\bar{S}_2$ and $\widetilde{\mathcal{C}}_2$ ([22, pp. 83–84], [22, Claims 6.1, B.6]), we have the following.

Claim B.4 There exist positive constants C and δ such that if $\mathbb{M}_{c,x}(T) \leqslant \delta$, then for $t \in [0, T]$,

$$\|R^8(t)\|_{Y_1} \leqslant C\langle t\rangle^{-1}\mathbb{M}_{c,x}(T)^2, \quad \|R^8(t)\|_Y \leqslant C\langle t\rangle^{-5/4}\mathbb{M}_{c,x}(T)^2,$$

$$\|\chi(D_y)R^{91}(t)\|_{L^1} + \|R^{91}(t)\|_{Y_1} + \|R^{92}(t)\|_{Y_1} \leqslant C(e^{-\alpha L} + \mathbb{M}_2(T))\mathbb{M}_{c,x}(T)^2\langle t\rangle^{-2}.$$

For $R^{10} = (\partial_y^2 \widetilde{S}_0 - B_2)(b_y - c_y)\mathbf{e_1}$, we have the following from [22, Claim D.6] and the fact that $\widetilde{S}_0 \in B(Y) \cap B(Y_1)$.

Claim B.5 There exist positive constants C and δ such that if $\mathbb{M}_{c,x}(T) \leqslant \delta$, then for $t \in [0, T]$,

$$\|R^{10}(t)\|_{Y_1} \leqslant C\langle t\rangle^{-1}\mathbb{M}_{c,x}(T)^2, \quad \|R^{10}(t)\|_Y \leqslant C\langle t\rangle^{-5/4}\mathbb{M}_{c,x}(T)^2.$$

Let $R^{11} = R^{11,1} - \partial_y^2 R^{11,2}$ and

$$R^{11,1} = \widetilde{A}_{11}(t)(\tilde{c} - b)\mathbf{e_1}, \quad R^{11,2} = \widetilde{A}_{12}(t)(\tilde{c} - b)\mathbf{e_1}.$$

Claim B.6 There exist positive constants C and δ such that if $\mathbb{M}_{c,x}(T)$, then for $t \in [0, T]$,

$$\|\chi(D_y)R^{11,1}\|_{L^1} + \|R^{11,1}\|_{Y_1} + \|R^{11,2}\|_{Y_1} \leqslant Ce^{-\alpha(3t+L)}\langle t\rangle^{-1/2}\mathbb{M}_{c,x}(T)^2.$$

Proof By the definition,

$$\chi(D_y)R^{11,1}(t) = \chi(D_y)\{\tilde{c}(t,\cdot) - b(t,\cdot)\}(\tilde{a}_{11}(t)\mathbf{e_1} + \tilde{a}_{21}(t)\mathbf{e_2}).$$

Claim A.5 and (9.10) imply

$$\|\chi(D_y)R^{11,1}(t)\|_{L^1} \lesssim (|\tilde{a}_{11}(t)| + |\tilde{a}_{12}(t)|)\|\tilde{c}(t)\|_Y^2 \lesssim e^{-\alpha(3t+L)}\langle t\rangle^{-1/2}\mathbb{M}_{c,x}(T)^2.$$

We can prove the rest in the similar manner by using Claim A.5. Thus we complete the proof. □

Appendix C: Estimates for $k(t, y)$

By Lemmas 3.1 and 7.2, the L^2-norm of $k(t, y)$ decays like t^{-2} as $t \to \infty$.

Claim C.1 Suppose that $\inf_{t \geqslant 0, y \in \mathbb{R}} x_t(t, y) \geqslant c_1$ for a $c_1 > 0$. Then there exist positive constants δ and C such that if $\|\langle x\rangle^2 v_0\|_{H^1(\mathbb{R}^2)} < \delta$, then

$$\|k(t,y)\|_{L^2} \leqslant C\langle t\rangle^{-2}\|\langle x\rangle^2 v_0\|_{L^2(\mathbb{R}^2)}.$$

Next, we will give an upper bound of the growth rate of $\|k(t,y)\|_{L^1}$ when $v_*(x,y)$ is polynomially localized in $\mathbb{R}^2$.

Claim C.2 Let $\tilde{v}_1$ be a solution of (3.8). There exist positive constant C and ε_0 such that if $\|\langle x\rangle(\langle x\rangle+\langle y\rangle)v_0\|_{H^1(\mathbb{R}^2)} \leqslant \varepsilon_0$, then for every $t \geqslant 0$,

$$\|\langle y\rangle k(t,\cdot)\|_{L^2(\mathbb{R})} \leqslant C\langle t\rangle\,\|\langle x\rangle(\langle x\rangle+\langle y\rangle)v_0\|_{H^1(\mathbb{R}^2)}.$$

Proof Multiplying (3.8) by $2(1+y^2)\tilde{v}_1$ and integrating the resulting equation over $\mathbb{R}^2$, we have after some integration by parts,

$$\frac{d}{dt}\int_{\mathbb{R}^2}(1+y^2)\tilde{v}_1(t,x,y)^2\,dxdy = 12\int_{\mathbb{R}^2} y\tilde{v}_1(t,x,y)(\partial_x^{-1}\partial_y\tilde{v}_1)(t,x,y)\,dxdy.$$

By Lemmas 3.1 and 7.3 and the definition of $\mathbb{M}'_1(\infty)$,

$$\|\langle y\rangle\tilde{v}_1(t)\|_{L^2} \leqslant \|\langle y\rangle v_*\|_{L^2} + 6\int_0^t \|\partial_x^{-1}\partial_y\tilde{v}_1(s)\|_{L^2}\,ds$$

$$\lesssim \|\langle x\rangle\langle y\rangle v_0\|_{L^2(\mathbb{R}^2)} + \mathbb{M}'_1(\infty)t$$

$$\lesssim \|\langle x\rangle\langle y\rangle v_0\|_{L^2(\mathbb{R}^2)} + \left(\left\|\langle x\rangle^2 v_0\right\|_{H^1(\mathbb{R}^2)} + \|\langle x\rangle\langle y\rangle v_0\|_{L^2(\mathbb{R}^2)}\right)t.$$

Thus we complete the proof. $\square$

References

1. J. C. Alexander, R. L. Pego and R. L. Sachs, *On the transverse instability of solitary waves in the Kadomtsev-Petviashvili equation*, Phys. Lett. A **226** (1997), 187–192.
2. O. V. Besov, V. P. Ilín and S. M. Nikolskii. *Integral representations of functions and imbedding theorems* Vol. I (New York-Toronto: J. Wiley & Sons, 1978).
3. J. Bourgain, *On the Cauchy problem for the Kadomtsev-Petviashvili equation*, GAFA **3** (1993), 315–341.
4. S. P. Burtsev, *Damping of soliton oscillations in media with a negative dispersion law*, Sov. Phys. JETP **61** (1985).
5. S. Cuccagna, *On asymptotic stability in 3D of kinks for the ϕ^4 model*, Trans. Amer. Math. Soc. **360** (2008), 2581–2614.
6. A. de Bouard and Y. Martel, *Non existence of L^2-compact solutions of the Kadomtsev-Petviashvili II equation*, Math. Ann. **328** (2004) 525–544.
7. A. de Bouard and J. C. Saut, *Remarks on the stability of generalized KP solitary waves*, Mathematical problems in the theory of water waves, 75–84, Contemp. Math. **200**, Amer. Math. Soc., Providence, RI, 1996.
8. A. Grünrock, M. Panthee and J. Drumond Silva, *On KP-II equations on cylinders*, Ann. IHP Analyse non linéaire **26** (2009), 2335–2358.

9. M. Hadac, *Well-posedness of the KP-II equation and generalizations*, Trans. Amer. Math. Soc. **360** (2008), 6555–6572.
10. M. Hadac, S. Herr and H. Koch, *Well-posedness and scattering for the KP-II equation in a critical space*, Ann. IHP Analyse non linéaire **26** (2009), 917–941.
11. M. Haragus, *Transverse spectral stability of small periodic traveling waves for the KP equation*, Stud. Appl. Math. **126** (2011), 157–185.
12. M. Haragus, Jin Li and D. E. Pelinovsky, *Counting Unstable Eigenvalues in Hamiltonian Spectral Problems via Commuting Operators*, Comm. Math. Phys. **354** (2017), 247–268.
13. P. Isaza and J. Mejia, *Local and global Cauchy problems for the Kadomtsev-Petviashvili (KP-II) equation in Sobolev spaces of negative indices*, Comm. Partial Differential Equations **26** (2001), 1027–1057.
14. M. A. Johnson and K. Zumbrun, *Transverse instability of periodic traveling waves in the generalized Kadomtsev-Petviashvili equation* SIAM J. Math. Anal. **42** (2010), 2681–2702.
15. B. B. Kadomtsev and V. I. Petviashvili, *On the stability of solitary waves in weakly dispersive media*, Sov. Phys. Dokl. **15** (1970), 539–541.
16. T. Kapitula, *Multidimensional stability of planar traveling waves*, Trans. Amer. Math. Soc. **349** (1997), 257–269.
17. G. Karch, Self-similar large time behavior of solutions to Korteweg-de Vries-Burgers equation. Nonlinear Anal. Ser. A: Theory Methods **35** (1999), 199–219.
18. H. Koch and D. Tataru, *Dispersive estimates for principally normal pseudodifferential operators*, Comm. Pure Appl. Math. **58** (2005), 217–284.
19. H. Koch and D. Tataru, *A priori bounds for the 1D cubic NLS in negative Sobolev spaces*, Int. Math. Res. Not. (2007), Art. ID rnm053.
20. C. D. Levermore and J. X. Xin, *Multidimensional stability of traveling waves in a bistable reaction-diffusion equation, II*. Comm. Partial Differential Equations **17** (1992), 1901–1924.
21. T. Mizumachi, *Asymptotic stability of lattice solitons in the energy space*, Comm. Math. Phys. **288** (2009), 125–144.
22. T. Mizumachi, *Stability of line solitons for the KP-II equation in* $\mathbb{R}^2$, Mem. of AMS **238** (2015), 1125.
23. T. Mizumachi, *Stability of line solitons for the KP-II equation in* $\mathbb{R}^2$. *II*, Proc. Roy. Soc. Edinburgh Sect. A. **148** (2018), 149–198.
24. T. Mizumachi, R. L. Pego and J. R. Quintero, *Asymptotic stability of solitary waves in the Benney-Luke model of water waves*, Differential Integral Equations **26** (2013), 253–301.
25. T. Mizumachi and Y. Shimabukuro, *Asymptotic linear stability of Benney-Luke line solitary waves in 2D*, Nonlinearity **30** (2017), 3419–3465.
26. T. Mizumachi and N. Tzvetkov, *Stability of the line soliton of the KP-II equation under periodic transverse perturbations*, Mathematische Annalen **352** (2012), 659–690.
27. T. Mizumachi and N. Tzvetkov, L^2-*stability of solitary waves for the KdV equation via Pego and Weinstein's method*, RIMS Kôkyûroku Bessatsu B49 (2014): Harmonic Analysis and Nonlinear Partial Differential Equations, eds. M. Sugimoto and H. Kubo, pp.33–63.
28. L. Molinet, J. C. Saut and N. Tzvetkov, *Global well-posedness for the KP-II equation on the background of a non-localized solution*, Ann. Inst. H. Poincaré Anal. Non Linéaire **28** (2011), 653–676.
29. G. Pedersen, *Nonlinear modulations of solitary waves*, J. Fluid Mech. **267** (1994), 83–108.
30. R. L. Pego and M. I. Weinstein, *Asymptotic stability of solitary waves*, Comm. Math. Phys. **164** (1994), 305–349.
31. F. Rousset and N. Tzvetkov, *Transverse nonlinear instability for two-dimensional dispersive models*, Ann. IHP, Analyse Non Linéaire **26** (2009), 477–496.
32. F. Rousset and N. Tzvetkov, *Transverse nonlinear instability for some Hamiltonian PDE's*, J. Math. Pures Appl. **90** (2008), 550–590.
33. F. Rousset and N. Tzvetkov, *Stability and instability of the KDV solitary wave under the KP-I flow*, Commun. Math. Phys. **313** (2012), 155–173.
34. J. C. Saut, *Remarks on the generalized Kadomtsev-Petviashvili equations*, Indiana Univ. Math. J. **42** (1993), 1011–1026.

35. H. Takaoka, *Global well-posedness for the Kadomtsev-Petviashvili II equation*, Discrete Contin. Dynam. Systems **6** (2000), 483–499.
36. H. Takaoka and N. Tzvetkov, *On the local regularity of Kadomtsev-Petviashvili-II equation*, IMRN **8** (2001), 77–114.
37. N. Tzvetkov, *Global low regularity solutions for Kadomtsev-Petviashvili equation*, Diff. Int. Eq. **13** (2000), 1289–1320.
38. S. Ukai, *Local solutions of the Kadomtsev-Petviashvili equation*, J. Fac. Sc. Univ. Tokyo Sect. IA Math. **36** (1989), 193–209.
39. J. X. Xin, *Multidimensional stability of traveling waves in a bistable reaction-diffusion equation, I*. Comm. Partial Differential Equations **17** (1992), 1889–1899.
40. V. Zakharov, *Instability and nonlinear oscillations of solitons*, JEPT Lett. **22**(1975), 172–173.

Inverse Scattering for the Massive Thirring Model

Dmitry E. Pelinovsky and Aaron Saalmann

1 Introduction

The massive Thirring model (MTM) was derived by Thirring in 1958 [33] in the context of general relativity. It represents a relativistically invariant nonlinear Dirac equation in the space of one dimension. Another relativistically invariant one-dimensional Dirac equation is given by the Gross–Neveu model [12] also known as the massive Soler model [32] when it is written in the space of three dimensions.

It was discovered in 1970s by Mikhailov [24], Kuznetsov and Mikhailov [21], Orfanidis [25], Kaup and Newell [18] that the MTM is integrable with the inverse scattering transform method in the sense that it admits a Lax pair, countably many conserved quantities, the Bäcklund transformation, and other common features of integrable models. We write the MTM system in the laboratory coordinates by using the normalized form:

$$\begin{cases} i(u_t + u_x) + v + |v|^2 u = 0, \\ i(v_t - v_x) + u + |u|^2 v = 0. \end{cases} \tag{1.1}$$

The MTM system (1.1) appears as the compatibility condition in the Lax representation

$$L_t - A_x + [L, A] = 0, \tag{1.2}$$

D. E. Pelinovsky (✉)
Department of Mathematics, McMaster University, Hamilton, ON, Canada
e-mail: dmpeli@math.mcmaster.ca

A. Saalmann
Mathematisches Institut, Universität zu Köln, Köln, Germany
e-mail: asaalman@math.uni-koeln.de

where the 2×2-matrices L and A are given by

$$L = \frac{i}{4}(|u|^2 - |v|^2)\sigma_3 - \frac{i\lambda}{2}\begin{pmatrix} 0 & \bar{v} \\ v & 0 \end{pmatrix} + \frac{i}{2\lambda}\begin{pmatrix} 0 & \bar{u} \\ u & 0 \end{pmatrix} + \frac{i}{4}\left(\lambda^2 - \frac{1}{\lambda^2}\right)\sigma_3 \quad (1.3)$$

and

$$A = -\frac{i}{4}(|u|^2 + |v|^2)\sigma_3 - \frac{i\lambda}{2}\begin{pmatrix} 0 & \bar{v} \\ v & 0 \end{pmatrix} - \frac{i}{2\lambda}\begin{pmatrix} 0 & \bar{u} \\ u & 0 \end{pmatrix} + \frac{i}{4}\left(\lambda^2 + \frac{1}{\lambda^2}\right)\sigma_3. \quad (1.4)$$

Other forms of L and A with nonzero trace have also been introduced by Barashenkov and Getmanov [1]. The traceless representation of L and A in (1.3) and (1.4) is more useful for inverse scattering.

Formal inverse scattering results for the linear operators (1.3) and (1.4) can be found in [21]. The first steps towards rigorous developments of the inverse scattering transform for the MTM system (1.1) were made in 1990s by Villarroel [34] and Zhou [38]. In the former work, the treatment of the Riemann–Hilbert problems is sketchy, whereas in the latter work, an abstract method to solve Riemann–Hilbert problems with rational spectral dependence is developed with applications to the sine-Gordon equation in the laboratory coordinates. Although the MTM system (1.1) does not appear in the list of examples in [38], one can show that the abstract method of Zhou is also applicable to the MTM system.

The present paper relies on recent progress in the inverse scattering transform method for the derivative NLS equation [27, 29]. The key element of our technique is a transformation of the spectral plane λ for the operator L in (1.3) to the spectral plane $z = \lambda^2$ for a different spectral problem. This transformation can be performed uniformly in the λ plane for the Kaup–Newell spectral problem related to the derivative NLS equation [19]. In the contrast, one needs to consider separately the subsets of the λ plane near the origin and near infinity for the operator L in (1.3) due to its rational dependence on λ. Therefore, two Riemann–Hilbert problems are derived for the MTM system (1.1) with the components (u, v): the one near $\lambda = 0$ recovers u and the other one near $\lambda = \infty$ recovers v.

Let $\dot{L}^{2,m}(\mathbb{R})$ denote the space of square integrable functions with the weight $|x|^m$ for $m \in \mathbb{Z}$ so that $L^{2,m}(\mathbb{R}) \equiv \dot{L}^{2,m}(\mathbb{R}) \cap L^2(\mathbb{R})$. Let $\dot{H}^{n,m}(\mathbb{R})$ denote the Sobolev space of functions, the n-th derivative of which is square integrable with the weight $|x|^m$ for $n \in \mathbb{N}$ and $m \in \mathbb{Z}$ so that $H^{n,m}(\mathbb{R}) \equiv \dot{H}^{n,m}(\mathbb{R}) \cap \dot{L}^{2,m}(\mathbb{R}) \cap H^n(\mathbb{R})$ with $H^n(\mathbb{R}) \equiv \dot{H}^n(\mathbb{R}) \cap L^2(\mathbb{R})$. Norms on any of these spaces are introduced according to the standard convention.

The inverse scattering transform for the linear operators (1.3) and (1.4) can be controlled when the potential (u, v) belongs to the function space

$$X_{(u,v)} := H^2(\mathbb{R}) \cap H^{1,1}(\mathbb{R}). \quad (1.5)$$

Transformations of the spectral plane employed here allow us to give a sharp requirement on the L^2-based Hilbert spaces, for which the Riemann–Hilbert problem can be solved by using the technique from Deift and Zhou [11, 37]. Note that both the direct and inverse scattering transforms for the NLS equation are solved in function space $H^1(\mathbb{R}) \cap L^{2,1}(\mathbb{R})$, which is denoted by the same symbol $H^{1,1}(\mathbb{R})$ in the previous works [11, 37]. Compared to this space, the reflection coefficients (r_+, r_-) introduced in our paper for the linear operators (1.3) and (1.4) belong to the function space

$$X_{(r_+,r_-)} := \dot{H}^1(\mathbb{R}\setminus[-1,1]) \cap \dot{H}^{1,1}([-1,1]) \cap \dot{L}^{2,1}(\mathbb{R}) \cap \dot{L}^{2,-2}(\mathbb{R}). \tag{1.6}$$

In the application of the inverse scattering transform to the derivative NLS equation, alternative methods were recently developed [16, 22] based on a different (gauge) transformation of the Kaup–Newell spectral problem to the spectral problem for the Gerdjikov–Ivanov equation. Both the potentials and the reflection coefficients were controlled in the same function space $H^2(\mathbb{R}) \cap L^{2,2}(\mathbb{R})$ [16, 22]. These function spaces are more restrictive compared to the function spaces for the potential and the reflection coefficients used in [27, 29].

Unlike the recent literature on the derivative NLS equation, our interest to the inverse scattering for the MTM system (1.1) is not related to the well-posedness problems. Indeed, the local and global existence of solutions to the Cauchy problem for the MTM system (1.1) in the L^2-based Sobolev spaces $H^m(\mathbb{R})$, $m \in \mathbb{N}$ can be proven with the standard contraction and energy methods, see review of literature in [26]. Low regularity solutions in $L^2(\mathbb{R})$ were already obtained for the MTM system by Selberg and Tesfahun [31], Candy [5], Huh [13–15], and Zhang [35, 36]. The well-posedness results can be formulated as follows.

Theorem 1 ([5, 15]) *For every $(u_0, v_0) \in H^m(\mathbb{R})$, $m \in \mathbb{N}$, there exists a unique global solution $(u, v) \in C(\mathbb{R}, H^m(\mathbb{R}))$ such that $(u, v)|_{t=0} = (u_0, v_0)$ and the solution (u, v) depends continuously on the initial data (u_0, v_0). Moreover, for every $(u_0, v_0) \in L^2(\mathbb{R})$, there exists a global solution $(u, v) \in C(\mathbb{R}, L^2(\mathbb{R}))$ such that $(u, v)|_{t=0} = (u_0, v_0)$. The solution (u, v) is unique in a certain subspace of $C(\mathbb{R}, L^2(\mathbb{R}))$ and it depends continuously on the initial data (u_0, v_0).*

The inverse scattering transform and the reconstruction formulas for the global solutions (u, v) to the MTM system (1.1) can be used to solve other interesting analytical problems such as long-range scattering to zero [6], orbital and asymptotic stability of the Dirac solitons [9, 28], and an analytical proof of the soliton resolution conjecture. Similar questions have been recently addressed in the context of the cubic NLS equation [8, 10, 30] and the derivative NLS equation [17, 23].

The goal of our paper is to explain how the inverse scattering transform for the linear operators (1.3) and (1.4) can be developed by using the Riemann–Hilbert problem. For simplicity of presentation, we assume that the initial data to the MTM system (1.1) admit no eigenvalues and resonances in the sense of Definition 1 given in Sect. 3. Note that eigenvalues can be easily added by using Bäcklund transformation for the MTM system [9], whereas resonances can be removed by perturbations of initial data [3] (see relevant results in [27]). The following theorem represents the main result of this paper.

Theorem 2 *For every $(u_0, v_0) \in X_{(u,v)}$ admitting no eigenvalues or resonances in the sense of Definition 1, there is a direct scattering transform with the spectral data (r_+, r_-) defined in $X_{(r_+,r_-)}$. The unique solution $(u, v) \in C(\mathbb{R}, X_{(u,v)})$ to the MTM system (1.1) can be uniquely recovered by means of the inverse scattering transform for every $t \in \mathbb{R}$.*

The paper is organized as follows. Section 2 describes Jost functions obtained after two transformations of the differential operator L given by (1.3). Section 3 is used to set up scattering coefficients (r_+, r_-) and to introduce the scattering relations between the Jost functions. Section 4 explains how the Riemann–Hilbert problems can be solved and how the potentials (u, v) can be recovered in the time evolution of the MTM system (1.1). Section 5 concludes the paper with a review of open questions.

2 Jost Functions

The linear operator L in (1.3) can be rewritten in the form:

$$L = Q(\lambda; u, v) + \frac{i}{4}\left(\lambda^2 - \frac{1}{\lambda^2}\right)\sigma_3,$$

where

$$Q(\lambda; u, v) = \frac{i}{4}(|u|^2 - |v|^2)\sigma_3 - \frac{i\lambda}{2}\begin{pmatrix} 0 & \bar{v} \\ v & 0 \end{pmatrix} + \frac{i}{2\lambda}\begin{pmatrix} 0 & \bar{u} \\ u & 0 \end{pmatrix}.$$

Here we freeze the time variable t and drop it from the list of arguments. Assuming fast decay of (u, v) to $(0, 0)$ as $|x| \to \infty$, solutions to the spectral problem

$$\psi_x = L\psi \tag{2.1}$$

can be defined by the following asymptotic behavior:

$$\psi_1^{(-)}(x; \lambda) \sim \begin{pmatrix} 1 \\ 0 \end{pmatrix} e^{ix(\lambda^2 - \lambda^{-2})/4}, \quad \psi_2^{(-)}(x; \lambda) \sim \begin{pmatrix} 0 \\ 1 \end{pmatrix} e^{-ix(\lambda^2 - \lambda^{-2})/4} \quad \text{as } x \to -\infty$$

and

$$\psi_1^{(+)}(x; \lambda) \sim \begin{pmatrix} 1 \\ 0 \end{pmatrix} e^{ix(\lambda^2 - \lambda^{-2})/4}, \quad \psi_2^{(+)}(x; \lambda) \sim \begin{pmatrix} 0 \\ 1 \end{pmatrix} e^{-ix(\lambda^2 - \lambda^{-2})/4} \quad \text{as } x \to +\infty.$$

The *normalized Jost functions*

$$\varphi_\pm(x; \lambda) = \psi_1^{(\pm)}(x; \lambda)e^{-ix(\lambda^2 - \lambda^{-2})/4}, \; \phi_\pm(x; \lambda) = \psi_2^{(\pm)}(x; \lambda)e^{ix(\lambda^2 - \lambda^{-2})/4} \tag{2.2}$$

satisfy the constant boundary conditions at infinity:

$$\lim_{x \to \pm \infty} \varphi_\pm(x; \lambda) = e_1 \quad \text{and} \quad \lim_{x \to \pm \infty} \phi_\pm(x; \lambda) = e_2, \qquad (2.3)$$

where $e_1 = (1, 0)^T$ and $e_2 = (0, 1)^T$. The normalized Jost functions are solutions to the following Volterra integral equations:

$$\varphi_\pm(x; \lambda) = e_1 \qquad (2.4a)$$
$$+ \int_{\pm\infty}^x \begin{pmatrix} 1 & 0 \\ 0 & e^{-\frac{i}{2}(\lambda^2 - \lambda^{-2})(x-y)} \end{pmatrix} Q(\lambda; u(y), v(y)) \, \varphi_\pm(y; \lambda) dy,$$

$$\phi_\pm(x; \lambda) = e_2 \qquad (2.4b)$$
$$+ \int_{\pm\infty}^x \begin{pmatrix} e^{\frac{i}{2}(\lambda^2 - \lambda^{-2})(x-y)} & 0 \\ 0 & 1 \end{pmatrix} Q(\lambda; u(y), v(y)) \, \phi_\pm(y; \lambda) dy.$$

A standard assumption in analyzing Volterra integral equations is $Q(\lambda; u(\cdot), v(\cdot)) \in L^1(\mathbb{R})$ for fixed $\lambda \neq 0$ which is equivalent to $(u, v) \in L^1(\mathbb{R}) \cap L^2(\mathbb{R})$ by the definition of Q. In this case, for every $\lambda \in (\mathbb{R} \cup i\mathbb{R}) \setminus \{0\}$, Volterra integral equations (2.4) admit unique solutions $\varphi_\pm(\cdot; \lambda)$ and $\phi_\pm(\cdot; \lambda)$ in the space $L^\infty(\mathbb{R})$. However, even if $(u, v) \in L^1(\mathbb{R}) \cap L^2(\mathbb{R})$ the L^1-norm of $Q(\lambda; u(\cdot), v(\cdot))$ is not controlled uniformly in λ as $\lambda \to 0$ and $|\lambda| \to \infty$. This causes difficulties in studying the behaviour of $\varphi_\pm(\cdot; \lambda)$ and $\phi_\pm(\cdot; \lambda)$ as $\lambda \to 0$ and $|\lambda| \to \infty$ and thus we need to transform the spectral problem (2.1) to two equivalent forms. These two transformations generalize the exact transformation of the Kaup–Newell spectral problem to the Zakharov–Shabat spectral problem, see [19, 29].

2.1 Transformation of the Jost Functions for Small λ

Assume $u \in L^\infty(\mathbb{R})$, $\lambda \neq 0$, and define the transformation matrix by

$$T(u; \lambda) := \begin{pmatrix} 1 & 0 \\ u & \lambda^{-1} \end{pmatrix}. \qquad (2.5)$$

Let ψ be a solution of the spectral problem (2.1) and define $\Psi := T\psi$. Straightforward computations show that Ψ satisfies the equivalent linear equation

$$\Psi_x = \mathcal{L}\Psi, \qquad (2.6)$$

with new linear operator

$$\mathcal{L} = Q_1(u, v) + \lambda^2 Q_2(u, v) + \frac{i}{4}\left(\lambda^2 - \frac{1}{\lambda^2}\right)\sigma_3 \qquad (2.7)$$

where

$$Q_1(u,v) = \begin{pmatrix} -\frac{i}{4}(|u|^2+|v|^2) & \frac{i}{2}\overline{u} \\ u_x - \frac{i}{2}u|v|^2 - \frac{i}{2}v & \frac{i}{4}(|u|^2+|v|^2) \end{pmatrix}$$

and

$$Q_2(u,v) = \frac{i}{2}\begin{pmatrix} u\overline{v} & -\overline{v} \\ u+u^2\overline{v} & -u\overline{v} \end{pmatrix}.$$

Let us define $z := \lambda^2$ and introduce the partition $\mathbb{C} = B_0 \cup \mathbb{S}^1 \cup B_\infty$ with

$$B_0 := \{z \in \mathbb{C} : |z| < 1\}, \mathbb{S}^1 := \{z \in \mathbb{C} : |z| = 1\}, B_\infty := \{z \in \mathbb{C} : |z| > 1\}. \quad (2.8)$$

The second term in (2.7) is bounded if $z \in B_0$. The normalized Jost functions associated to the spectral problem (2.6) denoted by $\{m_\pm, n_\pm\}$ can be obtained from the original Jost functions $\{\varphi_\pm, \psi_\pm\}$ by the transformation formulas:

$$m_\pm(x;z) = T(u(x);\lambda)\varphi_\pm(x;\lambda), \quad n_\pm(x;z) = \lambda T(u(x);\lambda)\phi_\pm(x;\lambda), \quad (2.9)$$

subject to the constant boundary conditions at infinity:

$$\lim_{x\to\pm\infty} m_\pm(x;\lambda) = e_1 \quad \text{and} \quad \lim_{x\to\pm\infty} n_\pm(x;\lambda) = e_2. \quad (2.10)$$

The transformed Jost functions are solutions to the following Volterra integral equations:

$$m_\pm(x;z) = e_1 \quad (2.11a)$$

$$+ \int_{\pm\infty}^x \begin{pmatrix} 1 & 0 \\ 0 & e^{-\frac{i}{2}(z-z^{-1})(x-y)} \end{pmatrix} [Q_1(u(y),v(y)) + zQ_2(u(y),v(y))] m_\pm(y;z)dy,$$

$$n_\pm(x;z) = e_2 \quad (2.11b)$$

$$+ \int_{\pm\infty}^x \begin{pmatrix} e^{\frac{i}{2}(z-z^{-1})(x-y)} & 0 \\ 0 & 1 \end{pmatrix} [Q_1(u(y),v(y)) + zQ_2(u(y),v(y))] n_\pm(y;z)dy.$$

Compared to [29], we have an additional term $\frac{i}{2}z(x-y)$ in the argument of the oscillatory kernel and the additional term $zQ_2(u,v)$ under the integration sign. However, both additional terms are bounded in B_0 where $|z| < 1$. Therefore, the same analysis as in the proof of Lemmas 1 and 2 in [29] yields the following.

Lemma 1 *Let* $(u,v) \in L^1(\mathbb{R}) \cap L^\infty(\mathbb{R})$ *and* $u_x \in L^1(\mathbb{R})$. *For every* $z \in (-1,1)$, *there exist unique solutions* $m_\pm(\cdot;z) \in L^\infty(\mathbb{R})$ *and* $n_\pm(\cdot;z) \in L^\infty(\mathbb{R})$ *satisfying* the integral equations (2.11). For every $x \in \mathbb{R}$, $m_\pm(x, \cdot)$ and $n_\mp(x, \cdot)$ are continued analytically in $\mathbb{C}^\pm \cap B_0$. There exist a positive constant C such that

$$\|m_\pm(\cdot; z)\|_{L^\infty} + \|n_\mp(\cdot; z)\|_{L^\infty} \leq C, \quad z \in \mathbb{C}^\pm \cap B_0. \tag{2.12}$$

Lemma 2 *Under the conditions of Lemma 1, for every $x \in \mathbb{R}$ the normalized Jost functions $m_\pm$ and $n_\pm$ satisfy the following limits as $\operatorname{Im}(z) \to 0$ along a contour in the domains of their analyticity:*

$$\lim_{z \to 0} \frac{m_\pm(x; z)}{m_\pm^\infty(x)} = e_1, \quad \lim_{z \to 0} \frac{n_\pm(x; z)}{n_\pm^\infty(x)} = e_2, \tag{2.13}$$

where

$$m_\pm^\infty(x) = e^{-\frac{i}{4}\int_{\pm\infty}^x (|u|^2 + |v|^2)dy}, \quad n_\pm^\infty(x) = e^{\frac{i}{4}\int_{\pm\infty}^x (|u|^2 + |v|^2)dy}.$$

If in addition $u \in C^1(\mathbb{R})$, then

$$\lim_{z \to 0} \frac{1}{z} \left[\frac{m_\pm(x; z)}{m_\pm^\infty(x)} - e_1 \right] = \begin{pmatrix} -\int_{\pm\infty}^x \left[\bar{u}(u_x - \frac{i}{2}u|v|^2 - \frac{i}{2}v) - \frac{i}{2}u\bar{v} \right] dy \\ 2iu_x + u|v|^2 + v \end{pmatrix}, \tag{2.14a}$$

$$\lim_{z \to 0} \frac{1}{z} \left[\frac{n_\pm(x; z)}{n_\pm^\infty(x)} - e_2 \right] = \begin{pmatrix} \bar{u} \\ \int_{\pm\infty}^x \left[\bar{u}(u_x - \frac{i}{2}u|v|^2 - \frac{i}{2}v) - \frac{i}{2}u\bar{v} \right] dy \end{pmatrix}. \tag{2.14b}$$

Remark 1 By Sobolev's embedding of $H^1(\mathbb{R})$ into the space of continuous, bounded, and decaying at infinity functions, if $u \in H^1(\mathbb{R})$, then $u \in C(\mathbb{R}) \cap L^\infty(\mathbb{R})$ and $u(x) \to 0$ as $|x| \to \infty$. By the embedding of $L^{2,1}(\mathbb{R})$ into $L^1(\mathbb{R})$, if $u \in H^{1,1}(\mathbb{R})$, then $u \in L^1(\mathbb{R})$ and $u_x \in L^1(\mathbb{R})$. Thus, requirements of Lemma 1 are satisfied if $(u, v) \in H^{1,1}(\mathbb{R})$. The additional requirement $u \in C^1(\mathbb{R})$ of Lemma 2 is satisfied if $u \in H^2(\mathbb{R})$. Hence, $X_{(u,v)}$ in (1.5) is an optimal L^2-based Sobolev space for direct scattering of the MTM system (1.1).

Remark 2 Notations $(m_\pm, n_\pm)$ for the Jost functions used here are different from notations $(m_\pm, n_\pm)$ used in [29], where an additional transformation was used to generate $n_\pm$ (denoted by $p_\pm$ in [29]). This additional transformation is not necessary for our further work.

2.2 Transformation of the Jost Functions for Large λ

Assume $v \in L^\infty(\mathbb{R})$ and define the transformation matrix by

$$\widehat{T}(v; \lambda) := \begin{pmatrix} 1 & 0 \\ v & \lambda \end{pmatrix}. \tag{2.15}$$

Let ψ be a solution of the spectral problem (2.1) and define $\widehat{\Psi} := \widehat{T}\psi$. Straightforward computations show that $\widehat{\Psi}$ satisfies the equivalent linear equation

$$\widehat{\Psi}_x = \widehat{\mathcal{L}}\widehat{\Psi}, \tag{2.16}$$

with new linear operator

$$\widehat{\mathcal{L}} = \widehat{Q}_1(u, v) + \frac{1}{\lambda^2}\widehat{Q}_2(u, v) + \frac{i}{4}\left(\lambda^2 - \frac{1}{\lambda^2}\right)\sigma_3 \tag{2.17}$$

where

$$\widehat{Q}_1(u, v) = \begin{pmatrix} \frac{i}{4}(|u|^2 + |v|^2) & -\frac{i}{2}\bar{v} \\ v_x + \frac{i}{2}|u|^2 v + \frac{i}{2}u & -\frac{i}{4}(|u|^2 + |v|^2) \end{pmatrix}$$

and

$$\widehat{Q}_2(u, v) = -\frac{i}{2}\begin{pmatrix} \bar{u}v & -\bar{u} \\ v + \bar{u}v^2 & -\bar{u}v \end{pmatrix}.$$

We introduce the same variable $z := \lambda^2$ and note that the second term in (2.17) is now bounded for $z \in B_\infty$. The normalized Jost functions associated to the spectral problem (2.16) denoted by $\{\widehat{m}_\pm, \widehat{n}_\pm\}$ can be obtained from the original Jost functions $\{\varphi_\pm, \psi_\pm\}$ by the transformation formulas:

$$\widehat{m}_\pm(x; z) = \widehat{T}(v(x); \lambda)\varphi_\pm(x; \lambda), \quad \widehat{n}_\pm(x; z) = \lambda^{-1}\widehat{T}(v(x); \lambda)\phi_\pm(x; \lambda), \tag{2.18}$$

subject to the constant boundary conditions at infinity:

$$\lim_{x \to \pm\infty} \widehat{m}_\pm(x; \lambda) = e_1 \quad \text{and} \quad \lim_{x \to \pm\infty} \widehat{n}_\pm(x; \lambda) = e_2. \tag{2.19}$$

The transformed Jost functions are solutions to the following Volterra integral equations:

$$\widehat{m}_\pm(x; z) = e_1 + \int_{\pm\infty}^x \begin{pmatrix} 1 & 0 \\ 0 & e^{-\frac{i}{2}(z-z^{-1})(x-y)} \end{pmatrix} \tag{2.20a}$$

$$\left[\widehat{Q}_1(u(y), v(y)) + z^{-1}\widehat{Q}_2(u(y), v(y))\right]\widehat{m}_\pm(y; z)dy,$$

$$\widehat{n}_\pm(x; z) = e_2 + \int_{\pm\infty}^x \begin{pmatrix} e^{\frac{i}{2}(z-z^{-1})(x-y)} & 0 \\ 0 & 1 \end{pmatrix} \tag{2.20b}$$

$$\left[\widehat{Q}_1(u(y), v(y)) + z^{-1}\widehat{Q}_2(u(y), v(y))\right]\widehat{n}_\pm(y; z)dy.$$

Again, we have an additional term $\frac{i}{2}z^{-1}(x-y)$ in the argument of the oscillatory kernel and the additional term $z^{-1}\widetilde{Q}_2(u,v)$ under the integration sign. However, both additional terms are bounded in B_∞ where $|z|>1$. The following two lemmas contain results analogous to Lemmas 1 and 2.

Lemma 3 *Let $(u,v) \in L^1(\mathbb{R}) \cap L^\infty(\mathbb{R})$ and $v_x \in L^1(\mathbb{R})$. For every $z \in \mathbb{R}\setminus[-1,1]$, there exist unique solutions $\widehat{m}_\pm(\cdot;z) \in L^\infty(\mathbb{R})$ and $\widehat{n}_\pm(\cdot;z) \in L^\infty(\mathbb{R})$ satisfying the integral equations (2.20). For every $x \in \mathbb{R}$, $\widehat{m}_\pm(x,\cdot)$ and $\widehat{n}_\mp(x,\cdot)$ are continued analytically in $\mathbb{C}^\pm \cap B_\infty$. There exist a positive constant C such that*

$$\|\widehat{m}_\pm(\cdot;z)\|_{L^\infty} + \|\widehat{n}_\mp(\cdot;z)\|_{L^\infty} \leq C, \quad z \in \mathbb{C}^\pm \cap B_\infty. \qquad (2.21)$$

Lemma 4 *Under the conditions of Lemma 3, for every $x \in \mathbb{R}$ the normalized Jost functions $\widehat{m}_\pm$ and $\widehat{n}_\pm$ satisfy the following limits as $\mathrm{Im}(z) \to \infty$ along a contour in the domains of their analyticity:*

$$\lim_{|z|\to\infty} \frac{\widehat{m}_\pm(x;z)}{\widehat{m}_\pm^\infty(x)} = e_1, \quad \lim_{|z|\to\infty} \frac{\widehat{n}_\pm(x;z)}{\widehat{n}_\mp^\infty(x)} = e_2, \qquad (2.22)$$

where

$$\widehat{m}_\pm^\infty(x) = e^{\frac{i}{4}\int_{\pm\infty}^x (|u|^2+|v|^2)dy}, \quad \widehat{n}_\pm^\infty(x) = e^{-\frac{i}{4}\int_{\pm\infty}^x (|u|^2+|v|^2)dy}.$$

If in addition $v \in C^1(\mathbb{R})$, then

$$\lim_{|z|\to\infty} z\left[\frac{\widehat{m}_\pm(x;z)}{\widehat{m}_\pm^\infty(x)} - e_1\right] = \begin{pmatrix} -\int_{\pm\infty}^x \left[\overline{v}(v_x + \frac{i}{2}|u|^2 v + \frac{i}{2}u) + \frac{i}{2}\overline{u}v\right]dy \\ -2iv_x + |u|^2 v + u \end{pmatrix}, \qquad (2.23a)$$

$$\lim_{|z|\to\infty} z\left[\frac{\widehat{n}_\pm(x;z)}{\widehat{n}_\mp^\infty(x)} - e_2\right] = \begin{pmatrix} \overline{v} \\ \int_{\pm\infty}^x \left[\overline{v}(v_x + \frac{i}{2}|u|^2 v + \frac{i}{2}u) + \frac{i}{2}\overline{u}v\right]dy \end{pmatrix}. \qquad (2.23b)$$

2.3 Continuation of the Transformed Jost Functions Across $\mathbb{S}^1$

In Lemmas 1 and 3 we showed the existence of the transformed Jost functions

$$\{m_\pm(\cdot;z), n_\pm(\cdot;z)\}, \quad z \in B_0, \quad \text{and} \quad \{\widehat{m}_\pm(\cdot;z), \widehat{n}_\pm(\cdot;z)\}, \quad z \in B_\infty,$$

respectively, where the partition (2.8) is used. Because both sets of the transformed Jost functions are connected to the set $\{\varphi_\pm, \phi_\pm\}$ of the original Jost functions by the transformation formulas (2.9) and (2.18), respectively, we find the following connection formulas for every $z \in \mathbb{S}^1$:

$$m_{\pm}(x;z) = \begin{pmatrix} 1 & 0 \\ u(x) - z^{-1}v(x) & z^{-1} \end{pmatrix} \widehat{m}_{\pm}(x;z), \tag{2.24a}$$

$$n_{\pm}(x;z) = \begin{pmatrix} z & 0 \\ u(x)z - v(x) & 1 \end{pmatrix} \widehat{n}_{\pm}(x;z), \tag{2.24b}$$

or in the opposite direction,

$$\widehat{m}_{\pm}(x;z) = \begin{pmatrix} 1 & 0 \\ v(x) - zu(x) & z \end{pmatrix} m_{\pm}(x;z), \tag{2.25a}$$

$$\widehat{n}_{\pm}(x;z) = \begin{pmatrix} z^{-1} & 0 \\ v(x)z^{-1} - u(x) & 1 \end{pmatrix} n_{\pm}(x;z). \tag{2.25b}$$

By Lemmas 3 and 4, the right-hand sides of (2.24a) and (2.24b) yield analytic continuations of $m_{\pm}(x;\cdot)$ and $n_{\mp}(x;\cdot)$ in $\mathbb{C}^{\pm} \cap B_{\infty}$ respectively with the following limits as $\mathrm{Im}(z) \to \infty$ along a contour in the domains of their analyticity:

$$\lim_{|z|\to\infty} \frac{m_{\pm}(x;z)}{m_{\pm}^{\infty}(x)} = e_1 + u(x)e_2, \quad \lim_{|z|\to\infty} \frac{n_{\pm}(x;z)}{n_{\pm}^{\infty}(x)} = \bar{v}(x)e_1 + (1 + u(x)\bar{v}(x))e_2. \tag{2.26}$$

Analogously, by Lemmas 1 and 2, the right-hand sides of (2.25a) and (2.25b) yield analytic continuations of $\widehat{m}_{\pm}(x;\cdot)$ and $\widehat{n}_{\mp}(x;\cdot)$ in $\mathbb{C}^{\pm} \cap B_0$ respectively with the following limits as $\mathrm{Im}(z) \to 0$ along a contour in the domains of their analyticity:

$$\lim_{z\to 0} \frac{\widehat{m}_{\pm}(x;z)}{m_{\pm}^{\infty}(x)} = e_1 + v(x)e_2, \quad \lim_{z\to 0} \frac{\widehat{n}_{\pm}(x;z)}{n_{\pm}^{\infty}(x)} = \bar{u}(x)e_1 + (1 + \bar{u}(x)v(x))e_2. \tag{2.27}$$

By Lemmas 1–4, and the continuation formulas (2.24), (2.25), we obtain the following result.

Lemma 5 *Let* $(u,v) \in L^1(\mathbb{R}) \cap L^{\infty}(\mathbb{R})$ *and* $(u_x, v_x) \in L^1(\mathbb{R})$. *For every* $x \in \mathbb{R}$ *the Jost functions defined by the integral equations (2.11) and (2.20) can be continued such that* $m_{\pm}(x;\cdot)$, $n_{\mp}(x;\cdot)$, $\widehat{m}_{\pm}(x;\cdot)$, *and* $\widehat{n}_{\mp}(x;\cdot)$ *are analytic in* $\mathbb{C}^{\pm}$ *and continuous in* $\mathbb{C}^{\pm} \cup \mathbb{R}$ *with bounded limits as* $z \to 0$ *and* $|z| \to \infty$ *given by (2.13), (2.22), (2.26), (2.27).*

3 Scattering Coefficients

In order to define the scattering coefficients between the transformed Jost functions $\{m_{\pm}, n_{\pm}\}$ and $\{\widehat{m}_{\pm}, \widehat{n}_{\pm}\}$, we go back to the original Jost functions $\{\varphi_{\pm}, \phi_{\pm}\}$. For every $\lambda \in (\mathbb{R} \cup i\mathbb{R}) \setminus \{0\}$, we define the standard form of the scattering relation by

$$\begin{pmatrix} \varphi_-(x;\lambda)e^{ix(\lambda^2-\lambda^{-2})/4} \\ \phi_-(x;\lambda)e^{-ix(\lambda^2-\lambda^{-2})/4} \end{pmatrix} = \begin{pmatrix} \alpha(\lambda) & \beta(\lambda) \\ \gamma(\lambda) & \delta(\lambda) \end{pmatrix} \begin{pmatrix} \varphi_+(x;\lambda)e^{ix(\lambda^2-\lambda^{-2})/4} \\ \phi_+(x;\lambda)e^{-ix(\lambda^2-\lambda^{-2})/4} \end{pmatrix}. \quad (3.1)$$

Since the operator L in (1.3) admits the symmetry

$$\overline{\phi_\pm(x;\lambda)} = \pm \begin{pmatrix} 0 & -1 \\ 1 & 0 \end{pmatrix} \varphi_\pm(x;\overline{\lambda}),$$

we obtain

$$\gamma(\lambda) = -\overline{\beta(\overline{\lambda})}, \quad \delta(\lambda) = \overline{\alpha(\overline{\lambda})}, \quad \lambda \in (\mathbb{R} \cup i\mathbb{R}) \setminus \{0\}. \quad (3.2)$$

Since the matrix operator L in (1.3) has zero trace, the Wronskian determinant W of any two solutions to the spectral problem (2.1) for any $\lambda \in \mathbb{C}$ is independent of x. By computing the Wronskian determinants of the solutions $\{\varphi_-, \phi_+\}$ and $\{\varphi_+, \varphi_-\}$ as $x \to +\infty$ and using the scattering relation (3.1) and the asymptotic behavior of the Jost functions $\{\varphi_\pm, \psi_\pm\}$, we obtain

$$\begin{cases} \alpha(\lambda) = W\left(\varphi_-(x;\lambda)e^{ix(\lambda^2-\lambda^{-2})/4}, \phi_+(x;\lambda)e^{-ix(\lambda^2-\lambda^{-2})/4}\right), \\ \beta(\lambda) = W\left(\varphi_+(x;\lambda)e^{ix(\lambda^2-\lambda^{-2})/4}, \varphi_-(x;\lambda)e^{ix(\lambda^2-\lambda^{-2})/4}\right). \end{cases} \quad (3.3)$$

It follows from the asymptotic behavior of $\{\varphi_-, \phi_-\}$ as $x \to -\infty$ that $W(\varphi_-, \phi_-) = 1$. Substituting (3.1) and using the asymptotic behavior of $\{\varphi_+, \phi_+\}$ as $x \to +\infty$ yield the following constraint on the scattering data:

$$\alpha(\lambda)\delta(\lambda) - \beta(\lambda)\gamma(\lambda) = 1, \quad \lambda \in (\mathbb{R} \cup i\mathbb{R}) \setminus \{0\}. \quad (3.4)$$

In view of the constraints (3.2), the constraint (3.4) can be written as

$$\alpha(\lambda)\overline{\alpha(\overline{\lambda})} + \beta(\lambda)\overline{\beta(\overline{\lambda})} = 1, \quad \lambda \in (\mathbb{R} \cup i\mathbb{R}) \setminus \{0\}. \quad (3.5)$$

By using the transformation formulas (2.9) we can rewrite the scattering relation (3.1) in terms of the transformed Jost functions $\{m_\pm, n_\pm\}$. In particular, we apply $T(u;\lambda)$ to the first equation in (3.1) and $\lambda T(u;\lambda)$ to the second equation in (3.1), so that we obtain for $z \in \mathbb{R}\setminus\{0\}$,

$$\begin{pmatrix} m_-(x;z)e^{ix(z-z^{-1})/4} \\ n_-(x;z)e^{-ix(z-z^{-1})/4} \end{pmatrix} = \begin{pmatrix} a(z) & b_+(z) \\ -\overline{b_-(z)} & \overline{a(z)} \end{pmatrix} \begin{pmatrix} m_+(x;z)e^{ix(z-z^{-1})/4} \\ n_+(x;z)e^{-ix(z-z^{-1})/4} \end{pmatrix}, \quad (3.6)$$

where we have recalled $z = \lambda^2$ and defined the scattering coefficients:

$$a(z) := \alpha(\lambda), \quad b_+(z) := \lambda^{-1}\beta(\lambda), \quad b_-(z) := \lambda\beta(\lambda), \quad z \in \mathbb{R}\setminus\{0\}. \quad (3.7)$$

Since $m_\pm(x;z)$ and $n_\pm(x;z)$ depend on $z = \lambda^2$, we deduce that α is even in λ and β is odd in λ for $\lambda \in (\mathbb{R} \cup i\mathbb{R}) \setminus \{0\}$. The latter condition yields $\bar{\lambda}\beta(\bar{\lambda}) = \lambda\beta(\lambda)$, which has been used already in the expression (3.7) for $b_-(z)$. Thanks to the relation (3.5), we have the following constraints

$$\begin{cases} |\alpha(\lambda)|^2 + |\beta(\lambda)|^2 = 1, & \lambda \in \mathbb{R}\setminus\{0\}, \\ |\alpha(\lambda)|^2 - |\beta(\lambda)|^2 = 1, & \lambda \in i\mathbb{R}\setminus\{0\}. \end{cases} \quad (3.8)$$

Since the matrix operator $\mathcal{L}$ in (2.7) has zero trace, the Wronskian determinant W of any two solutions to the spectral problem (2.6) is also independent of x. As a result, by computing the Wronskian determinant as $x \to +\infty$ and using the asymptotic behavior of the Jost functions $\{m_\pm, n_\pm\}$, we obtain from the scattering relation (3.6) for $z \in \mathbb{R}\setminus\{0\}$:

$$\begin{cases} a(z) = W\left(m_-(x;z)e^{ix(z-z^{-1})/4}, n_+(x;z)e^{-ix(z-z^{-1})/4}\right), \\ b_+(z) = W\left(m_+(x;z)e^{ix(z-z^{-1})/4}, m_-(x;z)e^{ix(z-z^{-1})/4}\right), \\ \overline{b_-(z)} = W\left(n_+(x;z)e^{-ix(z-z^{-1})/4}, n_-(x;z)e^{-ix(z-z^{-1})/4}\right), \end{cases} \quad (3.9)$$

in accordance with the representation (3.3).

Analogously, by using the transformation formulas (2.18) we can rewrite the scattering relation (3.1) in terms of the transformed Jost functions $\{\widehat{m}_\pm, \widehat{n}_\pm\}$. In particular, we apply $\widehat{T}(u;\lambda)$ to the first equation in (3.1) and $\lambda^{-1}\widehat{T}(u;\lambda)$ to the second equation in (3.1), so that we obtain for $z \in \mathbb{R} \setminus \{0\}$,

$$\begin{pmatrix} \widehat{m}_-(x;z)e^{ix(z-z^{-1})/4} \\ \widehat{n}_-(x;z)e^{-ix(z-z^{-1})/4} \end{pmatrix} = \begin{pmatrix} \widehat{a}(z) & \widehat{b}_+(z) \\ -\overline{\widehat{b}_-(z)} & \overline{\widehat{a}(z)} \end{pmatrix} \begin{pmatrix} \widehat{m}_+(x;z)e^{ix(z-z^{-1})/4} \\ \widehat{n}_+(x;z)e^{-ix(z-z^{-1})/4} \end{pmatrix}, \quad (3.10)$$

where we have recalled $z = \lambda^2$ and defined the scattering coefficients

$$\widehat{a}(z) := \alpha(\lambda), \quad \widehat{b}_+(z) := \lambda\beta(\lambda), \quad \widehat{b}_-(z) := \lambda^{-1}\beta(\lambda), \quad z \in \mathbb{R}\setminus\{0\}. \quad (3.11)$$

Since the matrix operator $\widehat{\mathcal{L}}$ in (2.17) has zero trace, we obtain from the scattering relation (3.10) for $z \in \mathbb{R}\setminus\{0\}$:

$$\begin{cases} \widehat{a}(z) = W\left(\widehat{m}_-(x;z)e^{ix(z-z^{-1})/4}, \widehat{n}_+(x;z)e^{-ix(z-z^{-1})/4}\right), \\ \widehat{b}_+(z) = W\left(\widehat{m}_+(x;z)e^{ix(z-z^{-1})/4}, \widehat{m}_-(x;z)e^{ix(z-z^{-1})/4}\right), \\ \overline{\widehat{b}_-(z)} = W\left(\widehat{n}_+(x;z)e^{-ix(z-z^{-1})/4}, \widehat{n}_-(x;z)e^{-ix(z-z^{-1})/4}\right), \end{cases} \quad (3.12)$$

in accordance with the representation (3.3).

It follows from (3.7) and (3.11) that the two sets of scattering data are actually related by

$$\widehat{a}(z) = a(z), \quad \widehat{b}_+(z) = b_-(z), \quad \widehat{b}_-(z) = b_+(z), \quad z \in \mathbb{R}\setminus\{0\}. \tag{3.13}$$

These relations are in agreement with the continuation formulas (2.24) and (2.25). By using the representations (3.9) and (3.12), as well as Lemma 2, 4, and 5, we obtain the following.

Lemma 6 *Let* $(u, v) \in L^1(\mathbb{R}) \cap L^\infty(\mathbb{R})$ *and* $(u_x, v_x) \in L^1(\mathbb{R})$. *Then,* $a = \widehat{a}$ *is continued analytically into* $\mathbb{C}^-$ *with the following limits in* $\mathbb{C}^-$:

$$\lim_{z \to 0} a(z) = e^{-\frac{i}{4}\int_\mathbb{R}(|u|^2+|v|^2)dy} =: a_0 \tag{3.14}$$

and

$$\lim_{|z| \to \infty} a(z) = e^{\frac{i}{4}\int_\mathbb{R}(|u|^2+|v|^2)dy} =: a_\infty. \tag{3.15}$$

On the other hand, $b_\pm = \widehat{b}_\mp$ *are not continued analytically beyond the real line and satisfy the following limits on* $\mathbb{R}$:

$$\lim_{z \to 0} b_\pm(z) = \lim_{|z| \to \infty} b_\pm(z) = 0. \tag{3.16}$$

To simplify the inverse scattering transform, we consider the case of no eigenvalues or resonances in the spectral problem (2.1), where eigenvalues and resonances are defined as follows.

Definition 1 We say that the potential (u, v) admits an eigenvalue at $z_0 \in \mathbb{C}^-$ if $a(z_0) = 0$ and a resonance at $z_0 \in \mathbb{R}$ if $a(z_0) = 0$.

By taking the limit $x \to +\infty$ in the Volterra integral equations (2.11a) and (2.20a) for m_- and $\widehat{m}_-$ respectively and comparing it with the first equations in the scattering relations (3.6) and (3.10), we obtain the following equivalent representations for $a = \widehat{a}$:

$$a(z) = 1 - \frac{i}{4}\int_\mathbb{R}\left[(|u|^2 + |v|^2)m_-^{(1)} - 2\bar{u}m_-^{(2)} - 2z\bar{v}(um_-^{(1)} - m_-^{(2)})\right]dx,$$
$$z \in B_0 \cap \mathbb{C}^-, \tag{3.17a}$$

$$a(z) = 1 + \frac{i}{4}\int_\mathbb{R}\left[(|u|^2 + |v|^2)\widehat{m}_-^{(1)} - 2\bar{v}\widehat{m}_-^{(2)} - 2z^{-1}\bar{u}(v\widehat{m}_-^{(1)} - \widehat{m}_-^{(2)})\right]dx,$$
$$z \in B_\infty \cap \mathbb{C}^-, \tag{3.17b}$$

where the superscripts denote components of the Jost functions. If $(u, v) \in H^{1,1}(\mathbb{R})$ are defined in the ball of radius δ for some $\delta \in (0, 1)$, then constants C in (2.12) and (2.21) are independent of δ. Then, it follows from (3.17) that if δ is sufficiently small, then the integrals can be made as small as needed for every $z \in \mathbb{C}^- \cup \mathbb{R}$. This implies the following.

Lemma 7 *Let* $(u, v) \in L^1(\mathbb{R}) \cap L^\infty(\mathbb{R})$ *and* $(u_x, v_x) \in L^1(\mathbb{R})$ *be sufficiently small. Then* (u, v) *does not admit eigenvalues or resonances in the sense of Definition 1.*

Remark 3 The result of Lemma 7 was first obtained in Theorem 6.1 in [26]. No transformation of the spectral problem (2.1) was employed in [26]. Transformations similar to those we are using here were employed later in [29] in the context of the derivative NLS equation.

Remark 4 The result of Lemma 7 is useful for the study of long-range scattering from small initial data. Eigenvalues can always be included by using Bäcklund transformation for the MTM system [9, 27]. Resonances are structurally unstable and can be removed by perturbations of initial data [3, 27].

4 Riemann–Hilbert Problems

We will derive two Riemann–Hilbert problems. The first problem is formulated for the transformed Jost functions $\{m_\pm, n_\pm\}$, whereas the second problem is formulated for the transformed Jost functions $\{\widehat{m}_\pm, \widehat{n}_\pm\}$. Thanks to the asymptotic representations (2.14) and (2.23), the first problem is useful for reconstruction of the component u as $z \to 0$, whereas the second problem is useful for reconstruction of the component v as $|z| \to \infty$, both components satisfy the MTM system (1.1). This pioneering idea has first appeared on a formal level in [34]. The following assumption is used to simplify solutions to the Riemann–Hilbert problems.

Assumption 1 *Assume that the scattering coefficient a admits no zeros in* $\mathbb{C}^- \cup \mathbb{R}$.

Assumption 1 corresponds to the initial data (u_0, v_0) which admit no eigenvalues or resonances in the sense of Definition 1. By Lemma 7, the assumption is satisfied if the $H^{1,1}(\mathbb{R})$ norm on the initial data is sufficiently small. Since a is continued analytically into $\mathbb{C}^-$ by Lemma 6 with nonzero limits (3.14) and (3.15), zeros of a lie in a compact set. Therefore, if a admits no zeros in $\mathbb{C}^- \cup \mathbb{R}$ by Assumption 1, then there is $A > 0$ such that $|a(z)| \geq A$ for every $z \in \mathbb{R}$.

4.1 Riemann-Hilbert Problem for the Potential u

The asymptotic limit (2.26) presents a challenge to use $\{m_\pm, n_\pm\}$ for reconstruction of (u, v) as $|z| \to \infty$. On the other hand, the reconstruction formula for (u, v) in terms of $\{m_\pm, n_\pm\}$ is available from the asymptotic limit (2.14) as $z \to 0$. In order to avoid this complication, we use the inversion transformation $\omega = 1/z$, which maps 0 to ∞ and vice versa. The analyticity regions swap under the inversion transformation so that $\{m_-, n_+\}$ become analytic in $\mathbb{C}^+$ for ω and $\{m_+, n_-\}$ become analytic in $\mathbb{C}^-$ for ω.

Let us define matrices $P_\pm(x; \omega) \in \mathbb{C}^{2\times 2}$ for every $x \in \mathbb{R}$ and $\omega \in \mathbb{R}$ by

$$P_+(x; \omega) := \left[\frac{m_-(x; \omega^{-1})}{a(\omega^{-1})}, n_+(x; \omega^{-1})\right], \quad P_-(x; \omega) := \left[m_+(x; \omega^{-1}), \frac{n_-(x; \omega^{-1})}{a(\omega^{-1})}\right], \quad (4.1)$$

and two reflection coefficients

$$r_\pm(\omega) = \frac{b_\pm(\omega^{-1})}{a(\omega^{-1})}, \quad \omega \in \mathbb{R}, \quad (4.2)$$

The scattering relation (3.6) can be rewritten as the following jump condition for the Riemann–Hilbert problem:

$$P_+(x; \omega) = P_-(x; \omega) \begin{bmatrix} 1 + r_+(\omega)\overline{r_-(\omega)} & \overline{r_-(\omega)}e^{-\frac{i}{2}(\omega-\omega^{-1})x} \\ r_+(\omega)e^{\frac{i}{2}(\omega-\omega^{-1})x} & 1 \end{bmatrix}$$

If the scattering coefficient a satisfies Assumption 1, then $P_\pm(x; \cdot)$ for every $x \in \mathbb{R}$ are continued analytically in $\mathbb{C}^\pm$ by Lemmas 5 and 6. We denote these continuations by the same letters. Asymptotic limits (2.13) and (3.14) yield the following behavior of $P_\pm(x; \omega)$ for large $|\omega|$ in the domains of their analyticity:

$$P_\pm(x; \omega) \to \begin{bmatrix} m_+^\infty(x) & 0 \\ 0 & n_+^\infty(x) \end{bmatrix} =: P^\infty(x) \quad \text{as } |\omega| \to \infty.$$

Since we prefer to work with x-independent boundary conditions, we normalize the boundary conditions by defining

$$M_\pm(x; \omega) := [P^\infty(x)]^{-1} P_\pm(x; \omega), \quad \omega \in \mathbb{C}^\pm. \quad (4.3)$$

The following Riemann-Hilbert problem is formulated for the function $M(x; \cdot)$.

> **Riemann-Hilbert Problem 1** *For each $x \in \mathbb{R}$, find a 2×2-matrix valued function $M(x;\cdot)$ such that*
>
> (1) $M(x;\cdot)$ *is piecewise analytic in $\mathbb{C} \setminus \mathbb{R}$ with continuous boundary values*
>
> $$M_{\pm}(x;\omega) = \lim_{\varepsilon \downarrow 0} M(x; \omega \pm i\varepsilon), \quad z \in \mathbb{R}.$$
>
> (2) $M(x;\omega) \to I$ as $|\omega| \to \infty$.
> (3) *The boundary values $M_{\pm}(x;\cdot)$ on $\mathbb{R}$ satisfy the jump relation*
>
> $$M_+(x;\omega) - M_-(x;\omega) = M_-(x;\omega) R(x;\omega), \quad \omega \in \mathbb{R},$$
>
> *where*
>
> $$R(x;\omega) := \begin{bmatrix} r_+(\omega)\overline{r_-(\omega)} & \overline{r_-(\omega)} e^{-\frac{i}{2}(\omega-\omega^{-1})x} \\ r_+(\omega) e^{\frac{i}{2}(\omega-\omega^{-1})x} & 0 \end{bmatrix}.$$

It follows from the asymptotic limits (2.14) and the normalization (4.3) that the components (u, v) of the MTM system (1.1) are related to the solution of the Riemann–Hilbert problem 1 by using the following reconstruction formulas:

$$\left[2iu'(x) + u(x)|v(x)|^2 + v(x) \right] e^{\frac{i}{2}\int_x^{+\infty}(|u|^2+|v|^2)dy} = \lim_{|\omega| \to \infty} \omega[M(x;\omega)]_{21} \quad (4.4)$$

and

$$\overline{u}(x) e^{-\frac{i}{2}\int_x^{+\infty}(|u|^2+|v|^2)dy} = \lim_{|\omega| \to \infty} \omega[M(x;\omega)]_{12}, \quad (4.5)$$

where the subscript denotes the element of the 2×2 matrix M.

Remark 5 The gauge factors in (4.4)–(4.5) appear because of the normalization (4.3) and the asymptotic limits (2.14). A different approach was utilized in [16, 22] to avoid these gauge factors. The inverse scattering transform was developed to a different spectral problem, which was obtained from the Kaup–Newell spectral problem after a gauge transformation.

4.2 Riemann-Hilbert Problem for the Potential v

Let us define matrices $\widehat{P}_{\pm}(x; z) \in \mathbb{C}^{2 \times 2}$ for every $x \in \mathbb{R}$ and $z \in \mathbb{R}$ by

$$\widehat{P}_+(x;z) := \left[\widehat{m}_+(x;z), \frac{\widehat{n}_-(x;z)}{\overline{\widehat{a}(z)}} \right], \quad \widehat{P}_-(x;z) := \left[\frac{\widehat{m}_-(x;z)}{\widehat{a}(z)}, \widehat{n}_+(x;z) \right], \quad (4.6)$$

and two reflection coefficients by

$$\widehat{r}_\pm(z) = \frac{\widehat{b}_\pm(z)}{\widehat{a}(z)} = \frac{b_\mp(z)}{a(z)}, \quad z \in \mathbb{R}, \tag{4.7}$$

where the relations (3.13) have been used. The scattering relation (3.10) can be rewritten as the following jump condition for the Riemann–Hilbert problem:

$$\widehat{P}_+(x;z) = \widehat{P}_-(x;z) \begin{bmatrix} 1 & -\overline{\widehat{r}_-(z)}e^{\frac{i}{2}(z-z^{-1})x} \\ -\widehat{r}_+(z)e^{-\frac{i}{2}(z-z^{-1})x} & 1 + \widehat{r}_+(z)\overline{\widehat{r}_-(z)} \end{bmatrix}$$

If the scattering coefficient a satisfies Assumption 1, then $\widehat{P}_\pm(x;\cdot)$ for every $x \in \mathbb{R}$ are continued analytically in $\mathbb{C}^\pm$ by Lemmas 5 and 6. We denote these continuations by the same letters. Asymptotic limits (2.22) and (3.15) yield the following behavior of $\widehat{P}(x;z)$ for large $|z|$ in the domains of their analyticity:

$$\widehat{P}_\pm(x;z) \to \begin{bmatrix} \widehat{m}_+^\infty(x) & 0 \\ 0 & \widehat{n}_+^\infty(x) \end{bmatrix} =: \widehat{P}^\infty(x), \quad \text{as } |z| \to \infty.$$

In order to normalize the boundary conditions, we define

$$\widehat{M}_\pm(x;z) := \left[\widehat{P}^\infty(x)\right]^{-1} \widehat{P}_\pm(x;z), \quad z \in \mathbb{C}^\pm. \tag{4.8}$$

The following Riemann-Hilbert problem is formulated for the function $\widehat{M}(x;\cdot)$.

Riemann-Hilbert Problem 2 *For each $x \in \mathbb{R}$, find a 2×2-matrix valued function $\widehat{M}(x;\cdot)$ such that*

(1) $\widehat{M}(x;\cdot)$ *is piecewise analytic in $\mathbb{C} \setminus \mathbb{R}$ with continuous boundary values*

$$\widehat{M}_\pm(x;z) = \lim_{\varepsilon \downarrow 0} \widehat{M}(x;z \pm i\varepsilon), \quad z \in \mathbb{R}.$$

(2) $\widehat{M}(x;z) \to I$ *as $|z| \to \infty$.*
(3) *The boundary values $\widehat{M}_\pm(x;\cdot)$ on $\mathbb{R}$ satisfy the jump relation*

$$\widehat{M}_+(x;z) - \widehat{M}_-(x;z) = \widehat{M}_-(x;z)\widehat{R}(x;z),$$

where

$$\widehat{R}(x;z) := \begin{bmatrix} 0 & -\overline{\widehat{r}_-(z)}e^{\frac{i}{2}(z-z^{-1})x} \\ -\widehat{r}_+(z)e^{-\frac{i}{2}(z-z^{-1})x} & \widehat{r}_+(z)\overline{\widehat{r}_-(z)} \end{bmatrix}.$$

It follows from the asymptotic limit (2.23) and the normalization (4.8) that the components (u, v) of the MTM system (1.1) can be recovered from the solution of the Riemann–Hilbert problem 2 by using the following reconstruction formulas:

$$\left[-2iv'(x) + |u(x)|^2 v(x) + u(x)\right] e^{-\frac{i}{2}\int_x^{+\infty}(|u|^2+|v|^2)dy} = \lim_{|z|\to\infty} z\left[\widehat{M}(x;z)\right]_{21} \quad (4.9)$$

and

$$\overline{v}(x) e^{\frac{i}{2}\int_x^{+\infty}(|u|^2+|v|^2)dy} = \lim_{|z|\to\infty} z\left[\widehat{M}(x;z)\right]_{12}, \quad (4.10)$$

where the subscript denotes the element of the 2×2 matrix M.

Let us now outline the reconstruction procedure for (u, v) as a solution of the MTM system (1.1) in the inverse scattering transform. If the right-hand sides of (4.5) and (4.10) are controlled in the space $H^1(\mathbb{R}) \cap L^{2,1}(\mathbb{R})$, then $(\tilde{u}, \tilde{v}) \in H^1(\mathbb{R}) \cap L^{2,1}(\mathbb{R})$, where

$$\tilde{u}(x) = u(x) e^{\frac{i}{2}\int_x^{+\infty}(|u|^2+|v|^2)dy}, \quad \tilde{v}(x) = v(x) e^{-\frac{i}{2}\int_x^{+\infty}(|u|^2+|v|^2)dy}.$$

Since $|\tilde{u}(x)| = |u(x)|$ and $|\tilde{v}(x)| = |v(x)|$, the gauge factors can be immediately inverted, and since $H^1(\mathbb{R})$ is continuously embedded into $L^p(\mathbb{R})$ for any $p \geq 2$, we then have $(u, v) \in H^1(\mathbb{R}) \cap L^{2,1}(\mathbb{R})$. If the right-hand sides of (4.4) and (4.9) are also controlled in $H^1(\mathbb{R}) \cap L^{2,1}(\mathbb{R})$, then similar arguments give $(u', v') \in H^1(\mathbb{R}) \cap L^{2,1}(\mathbb{R})$, that is, $(u, v) \in H^2(\mathbb{R}) \cap H^{1,1}(\mathbb{R})$, in agreement with the function space used for direct scattering transform.

Remark 6 It follows from the limit (3.16) that $R(x; 0) = \widehat{R}(x; 0) = 0$ implying $M_+(x; 0) = M_-(x; 0)$ and $\widehat{M}_+(x; 0) = \widehat{M}_-(x; 0)$. More precisely, using (2.26), (2.27), (3.14), (3.15), and $\omega = z^{-1}$ we can derive

$$M(x; 0) = \begin{bmatrix} m_+^\infty(x) & 0 \\ 0 & n_+^\infty(x) \end{bmatrix}^{-1} \begin{bmatrix} 1 & \overline{v}(x) \\ u(x) & 1 + u(x)\overline{v}(x) \end{bmatrix} \begin{bmatrix} \widehat{m}_+^\infty(x) & 0 \\ 0 & \widehat{n}_+^\infty(x) \end{bmatrix}$$

and

$$\widehat{M}(x; 0) = \begin{bmatrix} \widehat{m}_+^\infty(x) & 0 \\ 0 & \widehat{n}_+^\infty(x) \end{bmatrix}^{-1} \begin{bmatrix} 1 & \overline{u}(x) \\ v(x) & 1 + \overline{u}(x)v(x) \end{bmatrix} \begin{bmatrix} m_+^\infty(x) & 0 \\ 0 & n_+^\infty(x) \end{bmatrix}.$$

In particular, the following holds:

$$[M(x; 0)]_{11} = \frac{\widehat{m}_+^\infty(x)}{m_+^\infty(x)} = e^{-\frac{i}{2}\int_x^{+\infty}(|u|^2+|v|^2)dy},$$

$$[\widehat{M}(x; 0)]_{11} = \frac{m_+^\infty(x)}{\widehat{m}_+^\infty(x)} = e^{\frac{i}{2}\int_x^{+\infty}(|u|^2+|v|^2)dy}.$$

In these formulas, we regain the same exponential factors as those in the reconstruction formulas (4.5) and (4.10). Hence, by substitution we obtain the following two decoupled reconstruction formulas:

$$u(x) = [M(x;0)]_{11} \overline{\lim_{|\omega|\to\infty} \omega[M(x;\omega)]_{12}},$$
$$v(x) = [\widehat{M}(x;0)]_{11} \overline{\lim_{|z|\to\infty} z[\widehat{M}(x;z)]_{12}}. \qquad (4.11)$$

Whereas Eqs. (4.4), (4.5), (4.9) and (4.10) are suitable for studying the inverse map of the scattering transformation in the sense of Theorem 2, the equivalent formulas (4.11) are useful in the analysis of the asymptotic behavior of $u(x)$ and $v(x)$ as $|x| \to \infty$.

4.3 Estimates on the Reflection Coefficients

In order to be able to solve the Riemann–Hilbert problems 1 and 2, we need to derive estimates on the reflection coefficients $r_\pm$ and $\widehat{r}_\pm$ defined by (4.2) and (4.7). We start with the Jost functions. In order to exclude ambiguity in notations, we write $m_\pm(x; z) \in H_z^1(\mathbb{R})$ for the same purpose as $m_\pm(x; \cdot) \in H^1(\mathbb{R})$.

Thanks to the Fourier theory reviewed in Proposition 1 in [29], the Volterra integral equations (2.11) and (2.20) with the oscillation factors $e^{\frac{i}{2}(\omega^{-1}-\omega)}$ and $e^{\frac{i}{2}(z-z^{-1})x}$ are estimated respectively in the limits $|\omega| \to \infty$ and $|z| \to \infty$, where $\omega := z^{-1}$, similarly to what was done in the proof of Lemma 3 in [29]. As a result, we obtain the following.

Lemma 8 Let $(u, v) \in H^{1,1}(\mathbb{R})$. Then for every $x \in \mathbb{R}^\pm$, we have

$$m_\pm(x; \omega^{-1}) - m_\pm^\infty(x)e_1 \in H_\omega^1(\mathbb{R}\setminus[-1, 1]),$$
$$n_\pm(x; \omega^{-1}) - n_\pm^\infty(x)e_2 \in H_\omega^1(\mathbb{R}\setminus[-1, 1]). \qquad (4.12)$$

and

$$\widehat{m}_\pm(x; z) - \widehat{m}_\pm^\infty(x)e_1 \in H_z^1(\mathbb{R}\setminus[-1, 1]),$$
$$\widehat{n}_\pm(x; z) - \widehat{n}_\pm^\infty(x)e_2 \in H_z^1(\mathbb{R}\setminus[-1, 1]). \qquad (4.13)$$

If in addition $(u, v) \in H^2(\mathbb{R})$, then

$$\omega\left[\frac{m_\pm(x;\omega^{-1})}{m_\pm^\infty(x)} - e_1\right]$$
$$- \left(\begin{matrix} -\int_{\pm\infty}^x \left[\overline{u}(u_x - \frac{i}{2}u|v|^2 - \frac{i}{2}v) - \frac{i}{2}u\overline{v}\right]dy \\ 2iu_x + u|v|^2 + v \end{matrix}\right) \in L_\omega^2(\mathbb{R}\setminus[-1, 1]), \quad (4.14a)$$

$$\omega \left[\frac{n_\pm(x;\omega^{-1})}{n_\pm^\infty(x)} - e_2 \right]$$

$$- \begin{pmatrix} \overline{u} \\ \int_{\pm\infty}^x \left[\overline{u}(u_x - \frac{i}{2}u|v|^2 - \frac{i}{2}v) - \frac{i}{2}u\overline{v} \right] dy \end{pmatrix} \in L_\omega^2(\mathbb{R}\setminus[-1,1]). \quad (4.14b)$$

and

$$z \left[\frac{\widehat{m}_\pm(x;z)}{\widehat{m}_\pm^\infty(x)} - e_1 \right]$$

$$- \begin{pmatrix} -\int_{\pm\infty}^x \left[\overline{v}(v_x + \frac{i}{2}|u|^2 v + \frac{i}{2}u) + \frac{i}{2}\overline{u}v \right] dy \\ -2iv_x + |u|^2 v + u \end{pmatrix} \in L_z^2(\mathbb{R}\setminus[-1,1]), \quad (4.15a)$$

$$z \left[\frac{\widehat{n}_\pm(x;z)}{\widehat{n}_\pm^\infty(x)} - e_2 \right]$$

$$- \begin{pmatrix} \overline{v} \\ \int_{\pm\infty}^x \left[\overline{v}(v_x + \frac{i}{2}|u|^2 v + \frac{i}{2}u) + \frac{i}{2}\overline{u}v \right] dy \end{pmatrix} \in L_z^2(\mathbb{R}\setminus[-1,1]). \quad (4.15b)$$

The following lemma transfers the estimates of Lemma 8 to the scattering coefficients a and $b_\pm$ by using the same analysis as in the proof of Lemma 4 in [29].

Lemma 9 *Let* $(u,v) \in H^{1,1}(\mathbb{R})$. *Then,*

$$a(\omega^{-1}) - a_0, \quad b_+(\omega^{-1}), \quad b_-(\omega^{-1}) \in H_\omega^1(\mathbb{R}\setminus[-1,1]), \quad (4.16)$$

and

$$a(z) - a_\infty, \quad b_+(z), \quad b_-(z) \in H_z^1(\mathbb{R}\setminus[-1,1]). \quad (4.17)$$

If in addition $(u,v) \in H^2(\mathbb{R})$, *then*

$$b_+(\omega^{-1}), \quad b_-(\omega^{-1}) \in L_\omega^{2,1}(\mathbb{R}\setminus[-1,1]), \quad (4.18)$$

and

$$b_+(z), \quad b_-(z) \in L_z^{2,1}(\mathbb{R}\setminus[-1,1]). \quad (4.19)$$

The following lemma transfers the estimates of Lemma 9 to the reflection coefficients $r_\pm$ and $\widehat{r}_\pm$. We give an elementary proof of this result since it is based on new computations compared to [29].

Lemma 10 *Assume* $(u,v) \in X_{(u,v)}$, *where* $X_{(u,v)}$ *is given by (1.5), and* a *satisfies Assumption 1. Then* $(r_+, r_-) \in X_{(r_+,r_-)}$, *where* $X_{(r_+,r_-)}$ *is given by (1.6).*

Proof Under the conditions of the lemma, it follows from Lemma 9 and from the definitions (4.2) and (4.7) that

$$r_\pm(\omega) \in \dot{H}^1_\omega(\mathbb{R} \setminus [-1, 1]) \cap \dot{L}^{2,1}_\omega(\mathbb{R} \setminus [-1, 1])$$

and

$$\hat{r}_\pm(\omega) \in \dot{H}^1_z(\mathbb{R} \setminus [-1, 1]) \cap \dot{L}^{2,1}_z(\mathbb{R} \setminus [-1, 1]).$$

It also follows from (4.2) and (4.7) that $r_\pm(\omega) = \hat{r}_\mp(\omega^{-1})$.

If $f(x) \in \dot{L}^{2,1}_x(1, \infty)$ and $\tilde{f}(y) := f(y^{-1})$, then $\tilde{f}(y) \in \dot{L}^{2,-2}_y(0, 1)$, which follows by the chain rule:

$$\int_1^\infty x^2 |f(x)|^2 dx = \int_0^1 y^{-4} |\tilde{f}(y)|^2 dy.$$

Since $\dot{L}^{2,1}(1, \infty)$ is continuously embedded into $\dot{L}^{2,-2}(1, \infty)$ and $\dot{L}^{2,-2}(0, 1)$ is continuously embedded into $\dot{L}^{2,1}(0, 1)$, we verify that $r_\pm(z) \in \dot{L}^{2,1}_z(\mathbb{R}) \cap \dot{L}^{2,-2}_z(\mathbb{R})$ and $\hat{r}_\pm(\omega) \in \dot{L}^{2,1}_\omega(\mathbb{R}) \cap \dot{L}^{2,-2}_\omega(\mathbb{R})$.

Finally, if $f(x) \in \dot{H}^1_x(1, \infty)$ and $\tilde{f}(y) := f(y^{-1})$, then $\tilde{f}(y) \in \dot{H}^{1,1}_y(0, 1)$, which follows by the chain rule $f'(x) = -x^{-2} \tilde{f}'(x^{-1})$ and

$$\int_1^\infty |f'(x)|^2 dx = \int_0^1 y^2 |\tilde{f}'(y)|^2 dy.$$

Combing all requirements together, we obtain the space $X_{(r_+, r_-)}$ both for (r_+, r_-) in z and for $(\hat{r}_+, \hat{r}_-)$ in ω, where $X_{(r_+, r_-)}$ is given by (1.6). □

Remark 7 It follows from the relations (3.7) and (3.11) that $r_+(\omega) = \omega r_-(\omega)$ and $\hat{r}_+(z) = z\hat{r}_-(z)$. Then, it follows from Lemma 10 and the chain rule that

if $r_+, \hat{r}_+ \in \dot{H}^1(\mathbb{R} \setminus [-1, 1]) \cap \dot{L}^{2,1}(\mathbb{R})$, then $r_-, \hat{r}_- \in \dot{H}^{1,1}(\mathbb{R} \setminus [-1, 1]) \cap \dot{L}^{2,2}(\mathbb{R})$

and

if $r_-, \hat{r}_- \in \dot{H}^{1,1}([-1, 1]) \cap \dot{L}^{2,-2}(\mathbb{R})$ then $r_+, \hat{r}_+ \in \dot{H}^1([-1, 1]) \cap \dot{L}^{2,-3}(\mathbb{R})$.

Therefore, we have $r_+, \hat{r}_+ \in \dot{H}^1(\mathbb{R}) \cap \dot{L}^{2,1}(\mathbb{R}) \cap \dot{L}^{2,-3}(\mathbb{R})$ and $r_-, \hat{r}_- \in \dot{H}^{1,1}(\mathbb{R}) \cap \dot{L}^{2,2}(\mathbb{R}) \cap \dot{L}^{2,-2}(\mathbb{R})$.

Remark 8 It may appear strange for the first glance that the direct and inverse scattering transforms for the MTM system (1.1) connect potentials $(u, v) \in X_{(u,v)}$ and reflection coefficients $(r_+, r_-) \in X_{(r_+, r_-)}$ in different spaces, whereas the Fourier transform provides an isomorphism in the space $H^1(\mathbb{R}) \cap L^{2,1}(\mathbb{R})$. However, the appearance of $X_{(u,v)}$ spaces for the potential (u, v) is not surprising due to the transformation of the linear operator L to the equivalent forms (2.7) and (2.17). The condition $(u, v) \in X_{(u,v)}$ ensures that $(Q_{1,2}, \hat{Q}_{1,2}) \in H^1(\mathbb{R}) \cap L^{2,1}(\mathbb{R})$, hence, the direct and inverse scattering transform for the MTM system (1.1) provides a transformation between $(Q_{1,2}, \hat{Q}_{1,2}) \in H^1(\mathbb{R}) \cap L^{2,1}(\mathbb{R})$ and $(r_+, r_-) \in X_{(r_+, r_-)}$, which is a natural transformation under the Fourier transform with oscillatory phase $e^{ix(\omega - \omega^{-1})}$. This allows us to avoid reproducing the Fourier analysis anew and to apply all the technical results from [29] without any changes, as these results generalize the classical results of Deift and Zhou [11, 37] obtained for the cubic NLS equation.

4.4 Solvability of the Riemann–Hilbert Problems

Let us define the reflection coefficient

$$r(\lambda) := \frac{\beta(\lambda)}{\alpha(\lambda)}, \quad \lambda \in \mathbb{R} \cup (i\mathbb{R}) \setminus \{0\}. \tag{4.20}$$

Recall the relations (3.7), (3.11), (4.2), and (4.7) which yield

$$\lambda^{-1} r(\lambda) = r_+(\omega) = \omega r_-(\omega), \quad \omega \in \mathbb{R} \setminus \{0\}. \tag{4.21}$$

and

$$\lambda r(\lambda) = \hat{r}_+(z) = z \hat{r}_-(z), \quad z \in \mathbb{R} \setminus \{0\}. \tag{4.22}$$

Also recall that $z = \lambda^2$ and $\omega = \lambda^{-2}$. By extending the proof of Propositions 2 and 3 in [29], we obtain the following.

Lemma 11 *If* $(r_+, r_-) \in X_{(r_+, r_-)}$, *then*

$$r(\lambda) \in L^{2,1}_\omega(\mathbb{R}) \cap L^\infty_\omega(\mathbb{R}), \quad r(\lambda) \in L^{2,1}_z(\mathbb{R}) \cap L^\infty_z(\mathbb{R}), \tag{4.23}$$

and

$$\lambda^{-1} r_+(\omega) \in L^\infty_\omega(\mathbb{R}), \quad \lambda \hat{r}_+(z) \in L^\infty_z(\mathbb{R}). \tag{4.24}$$

Proof Let us prove the embeddings in $L^2_z(\mathbb{R})$ space. The proof of the embeddings in $L^2_\omega(\mathbb{R})$ space is analogous. Relation (4.22) implies $|r(\lambda)|^2 = |\hat{r}_+(z)||\hat{r}_-(z)|$ and

$$r(\lambda) = \begin{cases} \lambda^{-1} \hat{r}_+(z), & |z| \geq 1, \\ \lambda \hat{r}_-(z), & |z| \leq 1. \end{cases}$$

Since $\hat{r}_+, \hat{r}_- \in L^{2,1}(\mathbb{R})$, Cauchy–Schwarz inequality implies $r(\lambda) \in L_z^{2,1}(\mathbb{R})$. Since $\hat{r}_+ \in H^1(\mathbb{R})$ by Remark 7, $r(\lambda) \in L_z^\infty(\mathbb{R} \setminus [-1, 1])$. In order to prove that $r(\lambda) \in L_z^\infty([-1, 1])$, we will show that $\lambda \hat{r}_-(z) \in L_z^\infty([-1, 1])$. This follows from the representation

$$z\hat{r}_-(z)^2 = \int_0^z \left[\hat{r}_-^2(z) + 2z\hat{r}_-(z)\hat{r}_-'(z)\right] dz$$

and the Cauchy–Schwarz inequality, since $\hat{r}_- \in \dot{H}^{1,1}(\mathbb{R}) \cap L^2(\mathbb{R})$. Similarly, $\lambda \hat{r}_+(z) \in L^\infty(\mathbb{R})$ since $\hat{r}_+ \in H^1(\mathbb{R}) \cap L^{2,1}(\mathbb{R})$. □

Remark 9 By using the relations (3.8), we obtain another constraint on $r(\lambda)$:

$$1 - |r(\lambda)|^2 = \frac{1}{|\alpha(\lambda)|^2} \geq c_0^2 > 0, \quad \lambda \in i\mathbb{R}, \tag{4.25}$$

where $c_0^{-1} := \sup_{\lambda \in i\mathbb{R}} |\alpha(\lambda)| < \infty$, which exists thanks to Lemma 6.

Under Assumption 1 as well as the constraints (4.23) and (4.25), the jump matrices in the Riemann–Hilbert problems 1 and 2 satisfy the same estimates as in Proposition 5 in [29]. Hence these Riemann–Hilbert problems can be solved and estimated with the same technique as in the proofs of Lemmas 7, 8, and 9 in [29]. The following summarizes this result.

Lemma 12 *Under Assumption 1, for every $r(\lambda) \in L_\omega^2(\mathbb{R}) \cap L_\omega^\infty(\mathbb{R})$ satisfying (4.25), there exists a unique solution of the Riemann–Hilbert problem 1 satisfying for every $x \in \mathbb{R}$:*

$$\|M_\pm(x;\omega) - I\|_{L_\omega^2} \leq C \|r(\lambda)\|_{L_\omega^2}, \tag{4.26}$$

where the positive constant C only depends on $\|r(\lambda)\|_{L_\omega^\infty}$. Similarly, under Assumption 1, for every $r(\lambda) \in L_z^2(\mathbb{R}) \cap L_z^\infty(\mathbb{R})$ satisfying (4.25), there exists a unique solution of the Riemann–Hilbert problem 2 satisfying for every $x \in \mathbb{R}$:

$$\|\widehat{M}_\pm(x;z) - I\|_{L_z^2} \leq \widehat{C} \|r(\lambda)\|_{L_z^2} \tag{4.27}$$

where the positive constant $\widehat{C}$ only depends on $\|r(\lambda)\|_{L_z^\infty}$.

The potentials u and v are recovered respectively from M and $\widehat{M}$ by means of the reconstruction formulas (4.5) and (4.10), whereas the derivatives of the potentials u' and v' are recovered from the reconstruction formulas (4.4) and (4.9). At the first order in terms of the scattering coefficient (see, e.g., [3]), we have to analyze the integrals like

$$\lim_{|\omega| \to \infty} \omega [M(x;\omega)]_{12} \sim \frac{i}{2\pi} \int_\mathbb{R} \overline{r_-(\omega)} e^{-\frac{i}{2}(\omega - \omega^{-1})x} d\omega \tag{4.28}$$

in the space $H_x^1(\mathbb{R}) \cap L_x^{2,1}(\mathbb{R})$. In order to control the remainder term of the representation (4.28) in $H_x^1(\mathbb{R}) \cap L_x^{2,1}(\mathbb{R})$, we need to generalize Proposition 7 in [29] for the case of the oscillatory factor

$$\Theta(s) = \frac{1}{2}\left(s - \frac{1}{s}\right).$$

The following lemma presents this generalization in the function space

$$X_0 := H^1(\mathbb{R}\setminus[-1,1]) \cap \dot{H}^{1,1}([-1,1]) \cap \dot{L}^{2,-1}([-1,1]).$$

The proof of this lemma is a non-trivial generalization of analysis of the Fourier integrals.

Lemma 13 *There is a positive constant C such that for all $x_0 \in \mathbb{R}_+$ and all $f \in X_0$, we have*

$$\sup_{x \in (x_0,\infty)} \|\langle x \rangle \mathcal{P}^\pm [f(\diamond)e^{\mp ix\Theta(\diamond)}]\|_{L^2(\mathbb{R})} \leq C \|f\|_{X_0} \tag{4.29}$$

where $\langle x \rangle := (1+x^2)^{1/2}$ and the Cauchy projection operators are explicitly given by

$$\mathcal{P}^\pm[f(\diamond)](z) := \lim_{\varepsilon \downarrow 0} \frac{1}{2\pi i} \int_\mathbb{R} \frac{f(s)}{s - (z \pm i\varepsilon)} ds, \quad z \in \mathbb{R}.$$

In addition, if $f \in X_0 \cap \dot{L}^{2,-1}(\mathbb{R})$, then

$$\sup_{x \in \mathbb{R}} \|\mathcal{P}^\pm[f(\diamond)e^{\mp ix\Theta(\diamond)}]\|_{L^\infty(\mathbb{R})} \leq C \left(\|f\|_{X_0} + \|f\|_{\dot{L}^{2,-1}(\mathbb{R})}\right). \tag{4.30}$$

Furthermore, if $f \in L^{2,1}(\mathbb{R}) \cap \dot{L}^{2,-1}(\mathbb{R})$, then

$$\sup_{x \in \mathbb{R}} \|\mathcal{P}^\pm[(\diamond-\diamond^{-1})f(\diamond)e^{\mp ix\Theta(\diamond)}]\|_{L^2(\mathbb{R})} \leq C \left(\|f\|_{L^{2,1}(\mathbb{R})} + \|f\|_{\dot{L}^{2,-1}(\mathbb{R})}\right). \tag{4.31}$$

Proof Consider the decomposition

$$f(s)e^{\mp ix\Theta(s)} = f(s)e^{\mp ix\Theta(s)}\chi_{\mathbb{R}_-}(s) + f(s)e^{\mp ix\Theta(s)}\chi_{\mathbb{R}_+}(s),$$

where χ_S is a characteristic function on the set $S \subset \mathbb{R}$. Thanks to the linearity of $\mathcal{P}^\pm$, it is sufficient to consider separately the functions f that vanish either on $\mathbb{R}_+$ or on $\mathbb{R}_-$. In the following we give an estimate for $\mathcal{P}^+[f(\diamond)e^{-ix\Theta(\diamond)}\chi_{\mathbb{R}_+}(\diamond)]$. The other case is handled analogously.

Fix $x > 0$ and consider the following decomposition:

$$f(s)e^{-ix\Theta(s)}\chi_{\mathbb{R}_+}(s) = h_I(x,s) + h_{II}(x,s), \tag{4.32}$$

with

$$h_I(x,s) = e^{-ix\Theta(s)}\frac{1}{2\pi}\int_{x/4}^{\infty} e^{ik(s-s^{-1})}\mathfrak{a}[f](k)dk$$

and

$$h_{II}(x,s) = e^{-i\frac{x}{4}(s-s^{-1})}\frac{1}{2\pi}\int_{-\infty}^{x/4} e^{i(k-\frac{x}{4})(s-s^{-1})}\mathfrak{a}[f](k)dk,$$

where

$$\mathfrak{a}[f](k) := \int_0^{\infty} e^{-ik(s-s^{-1})}\frac{1+s^2}{s^2}f(s)ds. \tag{4.33}$$

The following change of coordinates

$$y(s) = s - s^{-1}, \quad s(y) = \frac{y}{2} + \sqrt{1 + \frac{y^2}{4}},$$

$$s'(y) = \frac{1}{2} + \frac{y}{4}\left(\sqrt{1+\frac{y^2}{4}}\right)^{-1} = \frac{s(y)^2}{1+s(y)^2}$$

shows that $\mathfrak{a}[f](k) = \mathfrak{F}[\tilde{f}](k)$, where the function $\tilde{f}$ is given by

$$\tilde{f}(y) = f(s(y)), \quad y \in \mathbb{R}$$

and $\mathfrak{F}$ denotes the Fourier transform

$$\mathfrak{F}[\tilde{f}](k) = \int_{-\infty}^{\infty} e^{-iky}\tilde{f}(y)dy.$$

We obtain

$$\|\tilde{f}\|_{L^2(\mathbb{R})}^2 = \int_{\mathbb{R}} |f(s(y))|^2 dy = \int_0^{\infty} \frac{1+s^2}{s^2}|f(s)|^2 ds \leq \|f\|_{X_0}^2$$

and

$$\|\tilde{f}'\|_{L^2(\mathbb{R})}^2 = \int_{\mathbb{R}}\left(\frac{s(y)^2}{1+s(y)^2}\right)^2|f'(s(y))|^2 dy = \int_0^{\infty}\frac{s^2}{1+s^2}|f'(s)|^2 ds \leq \|f\|_{X_0}^2.$$

It follows that $\tilde{f} \in H^1(\mathbb{R})$ and thus by Fourier theory $\mathfrak{a}[f](k) \in L_k^{2,1}(\mathbb{R})$. Using the inverse Fourier transform

$$\mathfrak{F}^{-1}[g](y) = \frac{1}{2\pi} \int_{\mathbb{R}} e^{iyk} g(k) dk,$$

we find for $s > 0$:

$$f(s) = \tilde{f}(y(s)) = \mathfrak{F}^{-1}[\mathfrak{a}[f]](y(s)) = \frac{1}{2\pi} \int_{\mathbb{R}} e^{ik(s-s^{-1})} \mathfrak{a}[f](k) dk. \quad (4.34)$$

Addressing the decomposition (4.32), we obtain for the functions h_I thanks to $s'(y) < 1$:

$$\|h_I(x,\cdot)\|_{L^2(\mathbb{R}_+)}^2 \leq \left\| \frac{1}{2\pi} \int_{x/4}^{\infty} e^{iky} \mathfrak{a}[f](k) dk \right\|_{L_y^2(\mathbb{R})}$$

$$= \int_{x/4}^{\infty} |\mathfrak{a}[f](k)|^2 dk \leq \frac{C}{1+x^2} \|\mathfrak{a}[f]\|_{L^{2,1}(\mathbb{R})}^2. \quad (4.35)$$

On the other hand, the function $h_{II}(x,\cdot)$ is analytic in the domain $\{\mathrm{Im}(s) < 0\}$ and additionally for $s = -i\xi$ with $\xi \in \mathbb{R}_+$ we have

$$|h_{II}(x,s)| \leq C \|\mathfrak{a}[f]\|_{L^{2,1}(\mathbb{R})} e^{-\frac{x}{4}(\xi+\xi^{-1})}.$$

Therefore, $\|h_{II}(x,\cdot)\|_{L^2(i\mathbb{R}_-)}$ is decaying exponentially as $x \to \infty$. Now we have

$$\|\mathcal{P}^+[f(\diamond)e^{-ix\Theta(\diamond)}\chi_{\mathbb{R}_+}(\diamond)]\|_{L^2(\mathbb{R})}$$
$$\leq \|\mathcal{P}^+[h_I(x,\diamond)\chi_{\mathbb{R}_+}(\diamond)]\|_{L^2(\mathbb{R})} + \|\mathcal{P}^+[h_{II}(x,\diamond)\chi_{\mathbb{R}_+}(\diamond)]\|_{L^2(\mathbb{R})}$$

Since $\mathcal{P}^+$ is a bounded operator $L^2(\mathbb{R}_+) \to L^2(\mathbb{R})$ it follows by (4.35) that

$$\|\mathcal{P}^+[h_I(x,\diamond)\chi_{\mathbb{R}_+}(\diamond)]\|_{L^2(\mathbb{R})} \leq \|h_I(x,\cdot)\|_{L^2(\mathbb{R}_+)}^2 \leq C\langle x \rangle^{-1} \|f\|_{X_0}^2.$$

Using a suitable path of integration and the analyticity of h_{II} we find that

$$\mathcal{P}^+[h_{II}(x,\diamond)](z) = -\mathcal{P}_{i\mathbb{R}_-}[h_{II}(x,\diamond)](z),$$

where

$$\mathcal{P}_{i\mathbb{R}_-}[h](z) := \frac{1}{2\pi i} \int_{-\infty}^{0} \frac{h(is)}{is - z} ds, \quad z \in \mathbb{R},$$

for a function $h: i\mathbb{R}_- \to \mathbb{C}$. Since $\mathcal{P}_{i\mathbb{R}_-}$ is a bounded operator $L^2(i\mathbb{R}_-) \to L^2(\mathbb{R})$ (see, e.g., estimate (23.11) in [4]) and because $\|h_{II}(x,\cdot)\|_{L^2(i\mathbb{R}_-)}$ is decaying exponentially as $x \to \infty$, the proof of the estimate (4.29) is complete.

In order to prove the estimate (4.30), we first note that for $z \leq 0$

$$|\mathcal{P}^+[e^{-ix\Theta(\diamond)}f(\diamond)\chi_{\mathbb{R}_+}(\diamond)](z)|$$

$$\leq \int_0^\infty \frac{|f(s)|}{s}ds$$

$$\leq \left(\int_0^1 \frac{|f(s)|^2}{s^2}ds\right)^{1/2} + \left(\int_1^\infty \frac{1}{s^2}ds\right)^{1/2}\left(\int_1^\infty |f(s)|^2 ds\right)^{1/2}$$

$$\leq C\left(\|f\|_{X_0} + \|f\|_{\dot{L}^{2,-1}(\mathbb{R})}\right). \tag{4.36}$$

Thus it remains to estimate $|\mathcal{P}^+[e^{-ix\Theta(\diamond)}f(\diamond)\chi_{\mathbb{R}_+}(\diamond)](z)|$ for $z > 0$. First, we will derive a bound for the special case $x = 0$ and by (4.39) below we will see that the same bound holds for any $x \in \mathbb{R}$. Therefore, using (4.34) we decompose

$$f(s) = h_+(s) + h_-(s), \qquad h_\pm(s) := \pm\frac{1}{2\pi}\int_0^{\pm\infty} e^{ik(s-s^{-1})}\mathfrak{a}[f](k)dk,$$

where $h_\pm$ has an analytic extension within the domain $\{s \in \mathbb{C} : \text{Re}(s) > 0, \pm\text{Im}(s) > 0\}$ and for $\xi > 0$ we have

$$|h_\pm(\pm i\xi)| \leq C\|e^{-k(\xi+\xi^{-1})}\|_{L^2(\mathbb{R}_+)}\|\mathfrak{a}[f]\|_{L^2_k(\mathbb{R}_\pm)}$$

$$= \frac{C}{\sqrt{2}}\sqrt{\frac{\xi}{1+\xi^2}}\|\mathfrak{a}[f]\|_{L^2_k(\mathbb{R}_\pm)}. \tag{4.37}$$

Using a residue calculation we obtain for $z > 0$

$$\mathcal{P}^+[f(\diamond)\chi_{\mathbb{R}_+}(\diamond)](z) = \lim_{\varepsilon\downarrow 0}\frac{1}{2\pi i}\int_0^\infty \frac{h_+(s) + h_-(s)}{s-(z\pm i\varepsilon)}ds$$

$$= \mathcal{P}_{i\mathbb{R}_+}[h_+](z) - \mathcal{P}_{i\mathbb{R}_-}[h_-](z) + h_+(z).$$

Thanks to the bound (4.37), the summands $\mathcal{P}_{i\mathbb{R}_+}[h_+](z)$ and $\mathcal{P}_{i\mathbb{R}_-}[h_-](z)$ are estimated in the following way,

$$\sup_{z\in\mathbb{R}_+}|\mathcal{P}_{i\mathbb{R}_\pm}[h_\pm](z)| \leq \int_0^\infty \frac{|h_\pm(\pm i\xi)|}{\xi}d\xi$$

$$\leq C\int_0^\infty \frac{1}{\sqrt{\xi}\sqrt{1+\xi^2}}d\xi\,\|\mathfrak{a}[f]\|_{L^2_k(\mathbb{R}_\pm)}$$

$$\leq C\|\mathfrak{a}[f]\|_{L^2_k(\mathbb{R}_\pm)}.$$

In addition, for $z > 0$ we have $|h_+(z)| \leq \|\mathfrak{a}[f]\|_{L^1_k(\mathbb{R}_+)}$ so that the triangle inequality implies:

$$\sup_{z \in \mathbb{R}_+} |\mathcal{P}^+[f(\diamond)\chi_{\mathbb{R}_+}(\diamond)](z)| \leq C \left(\|\mathfrak{a}[f]\|_{L^1(\mathbb{R})} + \|\mathfrak{a}[f]\|_{L^2(\mathbb{R})} \right). \tag{4.38}$$

Now, let us reinsert the factor $e^{-ix\Theta(s)}$. From the definition of $\mathfrak{a}$ it follows that multiplication by $e^{-ix\Theta(s)}$ is equivalent of a shift of $\mathfrak{a}[f](k)$ in the k-variable,

$$\mathfrak{a}[e^{-ix\Theta(\diamond)} f(\diamond)](k) = \mathfrak{a}[f(\diamond)] \left(k + \frac{x}{2} \right). \tag{4.39}$$

Thus, the $L^1(\mathbb{R}) \cap L^2(\mathbb{R})$-norm with respect to k of $\mathfrak{a}[e^{-ix\Theta(\diamond)} f(\diamond)](k)$ does not depend on x. Therefore, (4.38) yields

$$\sup_{z \in \mathbb{R}_+} |\mathcal{P}^+[e^{-ix\Theta(\diamond)} f(\diamond)\chi_{\mathbb{R}_+}(\diamond)](z)| \leq C \|\mathfrak{a}[e^{-ix\Theta(\diamond)} f(\diamond)]\|_{L^1(\mathbb{R}) \cap L^2(\mathbb{R})}$$

$$= C \|\mathfrak{a}[f]\|_{L^1(\mathbb{R}) \cap L^2(\mathbb{R})} \leq C \|f\|_{X_0}, \tag{4.40}$$

which, together with (4.36), completes the proof of (4.30).

Finally, the bound (4.31) follows from $\|\mathcal{P}^\pm\|_{L^2 \to L^2} = 1$ and the fact that $(s - s^{-1}) f(s) \in L^2_s(\mathbb{R})$ if $f \in L^{2,1}(\mathbb{R}) \cap \dot{L}^{2,-1}(\mathbb{R})$. □

The first term in (4.28) is estimated with a similar change of coordinates $y := \omega - \omega^{-1}$ and further analysis in the proof of Lemma 13. However, it is controlled in $H^1_x(\mathbb{R}) \cap L^{2,1}_x(\mathbb{R})$ if the scattering coefficient r_- is defined in $X_{(r_+,r_-)}$ according to the bound

$$\left\| \int_\mathbb{R} \overline{r_-(\omega)} e^{-\frac{i}{2}(\omega - \omega^{-1})x} d\omega \right\|_{H^1_x(\mathbb{R}) \cap L^{2,1}_x(\mathbb{R})} \leq C \|r_-\|_{X_{(r_+,r_-)}}. \tag{4.41}$$

By using the estimate (4.41) and the estimates of Lemma 13, we can proceed similarly to Lemmas 10, 11, and 12 in [29]. The following lemma summarize the estimates on the potential (u, v) obtained from the reconstruction formulas (4.4)–(4.5) and (4.9)–(4.10).

Lemma 14 *Under Assumption 1, for every* $(r_+, r_-) \in X_{(r_+,r_-)}$ *and* $(\hat{r}_+, \hat{r}_-) \in X_{(r_+,r_-)}$, *the components* $(u, v) \in X_{(u,v)}$ *satisfy the bound*

$$\|u\|_{H^2 \cap H^{1,1}} + \|v\|_{H^2 \cap H^{1,1}}$$

$$\leq C \left(\|r_+\|_{X_{(r_+,r_-)}} + \|r_-\|_{X_{(r_+,r_-)}} + \|\hat{r}_+\|_{X_{(r_+,r_-)}} + \|\hat{r}_-\|_{X_{(r_+,r_-)}} \right), \tag{4.42}$$

where the positive constant C *depends on* $\|r_\pm\|_{X_{(r_+,r_-)}}$ *and* $\|\hat{r}_\pm\|_{X_{(r_+,r_-)}}$.

Lemma 10 proves the first assertion of Theorem 2. Lemma 14 proves the second assertion of Theorem 2 at $t = 0$. It remains to prove the second assertion of Theorem 2 for every $t \in \mathbb{R}$.

4.5 Time Evolution of the Spectral Data

Thanks to the well-posedness result of Theorem 1 and standard estimates in weighted L^2-based Sobolev spaces, there exists a global solution $(u, v) \in C(\mathbb{R}, X_{(u,v)})$ to the MTM system (1.1) for any initial data $(u, v)|_{t=0} = (u_0, v_0) \in X_{(u,v)}$. For this global solution, the normalized Jost functions (2.2) can be extended for every $t \in \mathbb{R}$:

$$\begin{cases} \varphi_\pm(t, x; \lambda) = \psi_1^{(\pm)}(t, x; \lambda) e^{-ix(\lambda^2 - \lambda^{-2})/4 - it(\lambda^2 + \lambda^{-2})/4}, \\ \phi_\pm(t, x; \lambda) = \psi_2^{(\pm)}(t, x; \lambda) e^{ix(\lambda^2 - \lambda^{-2})/4 + it(\lambda^2 + \lambda^{-2})/4}. \end{cases} \quad (4.43)$$

where $(\varphi_\pm, \phi_\pm)$ still satisfy the same boundary conditions (2.3). Introducing the scattering coefficients in the same way as in Sect. 3, we obtain the time evolution of the scattering coefficients:

$$\alpha(t, \lambda) = \alpha(0, \lambda), \quad \beta(t, \lambda) = \beta(0, \lambda) e^{-it(\lambda^2 + \lambda^{-2})/2}, \quad \lambda \in \mathbb{R} \cup (i\mathbb{R}) \setminus \{0\}. \quad (4.44)$$

Transferring the scattering coefficients to the reflection coefficients with the help of (3.7), (3.11), (4.2), and (4.7) yields the time evolution of the reflection coefficients:

$$r_\pm(t, \omega) = r_\pm(0, \omega) e^{-it(\omega + \omega^{-1})/2}, \quad \omega \in \mathbb{R} \setminus \{0\} \quad (4.45)$$

and

$$\hat{r}_\pm(t, z) = \hat{r}_\pm(0, z) e^{-it(z + z^{-1})/2}, \quad z \in \mathbb{R} \setminus \{0\}. \quad (4.46)$$

It is now clear that if $r_\pm$ and $\hat{r}_\pm$ are in $X_{(r_+, r_-)}$ at the initial time $t = 0$, then they remain in $X_{(r_+, r_-)}$ for every $t \in \mathbb{R}$. Thus, the recovery formulas of Lemma 14 for the global solution $(u, v) \in C(\mathbb{R}, X_{(u,v)})$ to the MTM system (1.1) hold for every $t \in \mathbb{R}$. This proves the second assertion of Theorem 2 for every $t \in \mathbb{R}$. Hence Theorem 2 is proven.

Remark 10 Adding the time dependence to the Riemann-Hilbert problem 1 we find the time-dependent jump relation $M_+(x, t; \omega) - M_-(x, t; \omega) = M_-(x, t; \omega) R(x, t; \omega)$, where

$$R(x, t; \omega) := \begin{bmatrix} r_+(\omega) \overline{r_-(\omega)} & \overline{r_-(\omega)} e^{-\frac{i}{2}(\omega - \omega^{-1})x + \frac{i}{2}(\omega + \omega^{-1})t} \\ r_+(\omega) e^{\frac{i}{2}(\omega - \omega^{-1})x - \frac{i}{2}(\omega + \omega^{-1})t} & 0 \end{bmatrix}.$$

The same phase function as in $R(x, t; \omega)$ appears in the inverse scattering theory for the sine-Gordon equation. A Riemann-Hilbert problem with such a phase function was studied in [7], where the long-time behavior of the sine-Gordon equation was analyzed.

Remark 11 In the context of the MTM system (1.1), it is more natural to address global solutions in weighted H^1 space such as $H^{1,1}(\mathbb{R})$ and drop the requirement $(u, v) \in H^2(\mathbb{R})$. The scattering coefficients $r_\pm$ and $\hat{r}_\pm$ are then defined in the space X_0. However, there are two obstacles to develop the inverse scattering transform for such a larger class of initial data. First, the asymptotic limits (2.14a) and (2.23a) are not justified, therefore, the recovery formulas (4.4) and (4.9) cannot be utilized. Second, without requirement $r_\pm, \hat{r}_\pm \in L^{2,1}(\mathbb{R})$, the time evolution (4.45)–(4.46) is not closed in X_0 since $r_-, \hat{r}_- \in L^{2,-2}(\mathbb{R})$ cannot be verified. In this sense, the space $X_{(u,v)}$ for (u, v) and $X_{(r_+,r_-)}$ for (r_+, r_-) and $(\hat{r}_+, \hat{r}_-)$ are optimal for the inverse scattering transform of the MTM system (1.1).

5 Conclusion

We gave functional-analytical details on how the direct and inverse scattering transforms can be applied to solve the initial-value problem for the MTM system in laboratory coordinates. We showed that initial data $(u_0, v_0) \in X_{(u,v)}$ admitting no eigenvalues or resonances defines uniquely the spectral data (r_+, r_-) in $X_{(r_+,r_-)}$. With the time evolution added, the spectral data (r_+, r_-) remain in the space $X_{(r_+,r_-)}$ and determine uniquely the solution (u, v) to the MTM system (1.1) in the space $X_{(u,v)}$.

We conclude the paper with a list of open questions.

The long-range scattering of solutions to the MTM system (1.1) for small initial data for which the assumption of no eigenvalues or resonances is justified can be considered based on the inverse scattering transform presented here. This will be the subject of the forthcoming work, where the long-range scattering results in [6] obtained by regular functional-analytical methods are to be improved.

The generalization of the inverse scattering transform in the case of eigenvalues is easy and can be performed similarly to what was done for the derivative NLS equation in [27]. However, it is not so easy to include resonances and other spectral singularities in the inverse scattering transform. In particular, the case of algebraic solitons [20] corresponds to the spectral singularities of the scattering coefficients due to slow decay of (u, v) and analysis of this singular case is an open question.

Finally, another interesting question is to consider the inverse scattering transform for the initial data decaying to constant (nonzero) boundary conditions. The MTM system (1.1) admits solitary waves over the nonzero background [2] and analysis of spectral and orbital stability of such solitary waves is at the infancy stage.

Acknowledgements A.S. gratefully acknowledges financial support from the project SFB-TRR 191 "Symplectic Structures in Geometry, Algebra and Dynamics" (Cologne University, Germany).

References

1. I.V. Barashenkov and B.S. Getmanov, "Multisoliton solutions in the scheme for unified description of integrable relativistic massive fields. Non-degenerate $sl(2, C)$ case", Commun. Math. Phys. **112** (1987) 423–446.
2. I.V. Barashenkov and B.S. Getmanov, "The unified approach to integrable relativistic equations: Soliton solutions over nonvanishing backgrounds. II", J. Math. Phys. **34** (1993), 3054–3072.
3. R. Beals and R. R. Coifman, "Scattering and inverse scattering for first order systems", Comm. Pure Appl. Math. **37** (1984), 39–90.
4. R. Beals, P. Deift, and C. Tomei, *Direct and Inverse Scattering on the Line*. American Mathematical Soc., 1988.
5. T. Candy, "Global existence for an L^2-critical nonlinear Dirac equation in one dimension", Adv. Diff. Eqs. **7–8** (2011), 643–666.
6. T. Candy and H. Lindblad, "Long range scattering for the cubic Dirac equation on $\mathbb{R}^{1+1}$", Diff. Integral Equat. **31** (2018), 507–518.
7. P.-J. Cheng, S. Venakides, and X. Zhou, "Long-time asymptotics for the pure radiation solution of the sine-Gordon equation", Comm. PDEs **24**, Nos. 7-8 (1999), 1195–1262.
8. A. Contreras and D.E. Pelinovsky, "Stability of multi-solitons in the cubic NLS equation", J. Hyperbolic Differ. Equ. **11** (2014), 329–353.
9. A. Contreras, D.E. Pelinovsky, and Y. Shimabukuro, "L^2 orbital stability of Dirac solitons in the massive Thirring model", Comm. in PDEs **41** (2016), 227–255.
10. S. Cuccagna and D.E. Pelinovsky, "The asymptotic stability of solitons in the cubic NLS equation on the line", Applicable Analysis, **93** (2014), 791–822.
11. P.A. Deift and X. Zhou, "Long-time asymptotics for solutions of the NLS equation with initial data in weighted Sobolev spaces", Comm. Pure Appl. Math. **56** (2003), 1029–1077.
12. D.J. Gross and A. Neveu, "Dynamical symmetry breaking in asymptotically free field theories", Phys. Rev. D **10** (1974), 3235–3253.
13. H. Huh, "Global strong solutions to the Thirring model in critical space", J. Math. Anal. Appl. **381** (2011), 513–520.
14. H. Huh, "Global solutions to Gross–Neveu equation", Lett. Math. Phys. **103** (2013), 927–931.
15. H. Huh and B. Moon, "Low regularity well-posedness for Gross–Neveu equations", Comm. Pure Appl. Anal. **14** (2015), 1903–1913.
16. R. Jenkins, J. Liu, P.A. Perry, and C. Sulem, "Global well-posedness for the derivative nonlinear Schrödinger equation", Comm. in PDEs **43** (2018), 1151–1195.
17. R. Jenkins, J. Liu, P.A. Perry, and C. Sulem, "Soliton resolution for the derivative nonlinear Schrödinger equation", Comm. Math. Phys. **363** (2018), 1003–1049.
18. D.J. Kaup and A.C. Newell, "On the Coleman correspondence and the solution of the Massive Thirring model", Lett. Nuovo Cimento **20** (1977), 325–331.
19. D. Kaup and A. Newell, "An exact solution for a derivative nonlinear Schrödinger equation", J. Math. Phys. **19** (1978), 789–801.
20. M. Klaus, D.E. Pelinovsky, and V.M. Rothos, "Evans function for Lax operators with algebraically decaying potentials", J. Nonlin. Sci. **16** (2006), 1–44.
21. E.A. Kuznetsov and A.V. Mikhailov, "On the complete integrability of the two-dimensional classical Thirring model", Theor. Math. Phys. **30** (1977), 193–200.
22. J. Liu, P.A. Perry, and C. Sulem, "Global existence for the derivative nonlinear Schrödinger equation by the method of inverse scattering", Comm. in PDEs **41** (2016), 1692–1760.

23. J. Liu, P.A. Perry, and C. Sulem, "Long-time behavior of solutions to the derivative nonlinear Schrödinger equation for soliton-free initial data", Ann. Inst. H. Poincaré C - Analyse non-linéaire **35** (2018), 217–265.
24. A.V. Mikhailov, "Integrability of the two-dimensional Thirring model", JETP Lett. **23** (1976), 320–323.
25. S. J. Orfanidis, "Soliton solutions of the massive Thirring model and the inverse scattering transform", Phys. Rev. D **14** (1976), 472–478.
26. D.E. Pelinovsky, "Survey on global existence in the nonlinear Dirac equations in one dimension", in *Harmonic Analysis and Nonlinear Partial Differential Equations* (Editors: T. Ozawa and M. Sugimoto) RIMS Kokyuroku Bessatsu **B26** (2011), 37–50.
27. D. E. Pelinovsky, A. Saalmann, and Y. Shimabukuro, "The derivative NLS equation: global existence with solitons", Dynamics of PDE **14** (2017), 271–294.
28. D. E. Pelinovsky and Y. Shimabukuro, "Orbital stability of Dirac solitons", Lett. Math. Phys. **104** (2014), 21–41.
29. D. E. Pelinovsky and Y. Shimabukuro, "Existence of global solutions to the derivative NLS equation with the inverse scattering transform method", Inter Math Res Notices **2018** (2018), 5663–5728.
30. A. Saalmann, "Asymptotic stability of N-solitons in the cubic NLS equation", J. Hyperbolic Differ. Equ. **14** (2017), 455–485.
31. S. Selberg and A. Tesfahun, "Low regularity well-posedness for some nonlinear Dirac equations in one space dimension", Diff. Integral Eqs. **23** (2010), 265–278.
32. M. Soler, "Classical, stable, nonlinear spinor field with positive rest energy", Phys. Rev. D. **1** (1970), 2766–2769.
33. W. Thirring, "A soluble relativistic field theory", Annals of Physics **3** (1958), 91–112.
34. J. Villarroel, "The DBAR problem and the Thirring model", Stud. Appl. Math. **84** (1991), 207–220.
35. Y. Zhang, "Global strong solution to a nonlinear Dirac type equation in one dimension, Nonlin. Anal.: Th. Meth. Appl. **80** (2013), 150–155.
36. Y. Zhang and Q. Zhao, "Global solution to nonlinear Dirac equation for Gross–Neveu model in 1 + 1 dimensions, Nonlin. Anal.: Th. Meth. Appl. **118** (2015), 82–96.
37. X. Zhou, "L^2-Sobolev space bijectivity of the scattering and inverse scattering transforms", Comm. Pure Appl. Math., **51** (1998), 697–731.
38. X. Zhou, "Inverse scattering transform for systems with rational spectral dependence", J. Diff. Eqs. **115** (1995), 277–303.